World Survey of Climatology Volume 7

CLIMATES OF THE SOVIET UNION

World Survey of Climatology

World Survey of Climatology Volume 7

Climates of the Soviet Union

by

Paul E. LYDOLPH

Department of Geography
University of Wisconsin-Milwaukee
Milwaukee, Wisc. (U.S.A.)

ELSEVIER SCIENTIFIC PUBLISHING COMPANY
Amsterdam-Oxford-New York 1977

ELSEVIER SCIENTIFIC PUBLISHING COMPANY
335 Jan van Galenstraat
P.O. Box 211, Amsterdam, The Netherlands

Distributors for the United States and Canada:

ELSEVIER/NORTH-HOLLAND, INC.
52 Vanderbilt Avenue
New York, N.Y. 10017

Library of Congress Cataloging in Publication Data

Lydolph, Paul E
Climates of the Soviet Union.

(World survey of climatology ; v. 7)
Bibliography: p.
Includes index.
1. Russia--Climate. I. Title. II. Series.
QC981.W67 Vol. 7 [QC989.R9] 551.6'08s
ISBN 0-444-41516-5 [551.6'9'47] 76-46298

With 263 illustrations and 138 tables

Printed in The Netherlands

World Survey of Climatology

Editor in Chief: H. E. LANDSBERG

Volume 1 General Climatology, 1
Editor: H. FLOHN

Volume 2 General Climatology, 2
Editor: H. FLOHN

Volume 3 General Climatology, 3
Editor: H. FLOHN

Volume 4 Climate of the Free Atmosphere
Editor: D. F. REX

Volume 5 Climates of Northern and Western Europe
Editor: C. C. WALLÉN

Volume 6 Climates of Central and Southern Europe
Editor: C. C. WALLÉN

Volume 7 Climates of the Soviet Union
by P. E. LYDOLPH

Volume 8 Climates of Northern and Eastern Asia
Editor: H. ARAKAWA

Volume 9 Climates of Southern and Western Asia
Editors: H. ARAKAWA and K. TAKAHASHI

Volume 10 Climates of Africa
Editor: J. F. GRIFFITHS

Volume 11 Climates of North America
Editors: R. A. BRYSON and F. K. HARE

Volume 12 Climates of Central and South America
Editor: W. SCHWERDTFEGER

Volume 13 Climates of Australia and New Zealand
Editor: J. GENTILLI

Volume 14 Climates of the Polar Regions
Editor: S. ORVIG

Volume 15 Climates of the Oceans
Editor: H. VAN LOON

Acknowledgements

The author wishes to express his gratitude to many Soviet colleagues in the fields of climatology and meteorology without whose cooperation this book would never have been possible. Foremost among these was F. F. Davitaya who not only supplied many materials himself but who also intervened on many occasions to get other colleagues and institutions to supply information. Other Soviet colleagues who were particularly helpful were the late B. L. Dzerdzeevskii, Kh. P. Pogosyan, A. A. Borisov, M. I. Budyko, and V. Maslov.

In the initial stages of the work Dr. Helmut Landsberg was extremely helpful in locating and supplying copies of rare sources of information. My graduate assistants in years past, Thomas Fedor and Hlib Hayuk, did a lot of spadework to dig out sources of information. Librarians in the office of interlibrary loan services at the University of Wisconsin-Milwaukee outdid themselves to procure materials wherever they existed, both in the libraries of the United States and those of the Soviet Union. Personnel of the Library of Congress and at the Atmospheric and Geophysical Sciences Libraries of the United States provided endless materials by means of interlibrary loan, as did a number of university libraries too numerous to mention.

For the tremendous amount of cartographic work, the author is indebted to the Cartographic Services Laboratory at the University of Wisconsin-Milwaukee, its director Don Temple, and his assistants, Donna Arndt, Denise Kuhn, Lynn Michalik and Rae Van Wybe. My good student, Kay Reynolds, aided in the final stages of proofreading and indexing.

In a number of ways the University of Wisconsin-Milwaukee provided some financial support for the effort, primarily through the Graduate Research Committee and the Department of Geography. For this I am grateful.

I am particularly appreciative of my wife, Mary, who typed the manuscript in all of its various drafts, labored over the compilation of tables, helped in editorial work and indexing, and generally kept me going through nine long years of research and writing.

Contents

Chapter 10. WIND

Chapter 11. CLIMATE DISTRIBUTION

Chapter 1

Introduction

The Soviet Union is a northern country. At its extremities it stretches latitudinally from approximately 35°N in Central Asia to almost 78°N at Cape Chelyuskin on the Arctic coast of Siberia. Arctic islands carry the territory even farther northward. Much of the country lies north of the 50th parallel, and therefore compares more to Canada than to the United States. The only other land mass on earth that resembles Eurasia climatically is North America, and the similarity between the two is not very close because of different arrangements of land and water, mountain systems, pressure patterns, and the like, which will be discussed in detail later.

The Soviet Union is also a very large country. It stretches almost half way around the earth at these high latitudes, from about 20°E to 170°W longitude. Much of its land lies great distances from the sea, and the long northern and eastern shores might as well lie away from the sea for they derive little benefit from it. Throughout a large part of the year the Arctic is frozen and climatically acts much as a snow-covered land surface. The Pacific during the winter has little influence because of the constancy of offshore winds from the Asiatic high. And the long southern margin of the country almost everywhere is rimmed by high mountains which effectively block potentially significant flows of warm moist air from the Indian Ocean and adjacent seas. Thus, high latitude and extreme continentality are the prime determinants of the climate over much of the Soviet Union (Fig.1-1).

Soviet climatological thought

In general, the country lacks heat. And where heat becomes more adequate, it lacks moisture. It is little wonder, then, that the Soviets have bent every effort to investigate ways of wringing every drop of moisture and every calorie of heat from their meager climatic storehouse to sustain the world's third most populous nation. Operating under such pressures of necessity, Soviet climatology has reached high levels of achievement. Without a doubt, Soviet climatologists are foremost in the world in many aspects of climatic studies. This is particularly true in studies of heat and water balance, which are so necessary to a total understanding of climate and to ways of deriving maximum use of limited resources of heat and moisture.

They have made many controlled experiments with all sorts of amelioration schemes, and in all such studies they have treated the effects of amelioration quantitatively as parts of

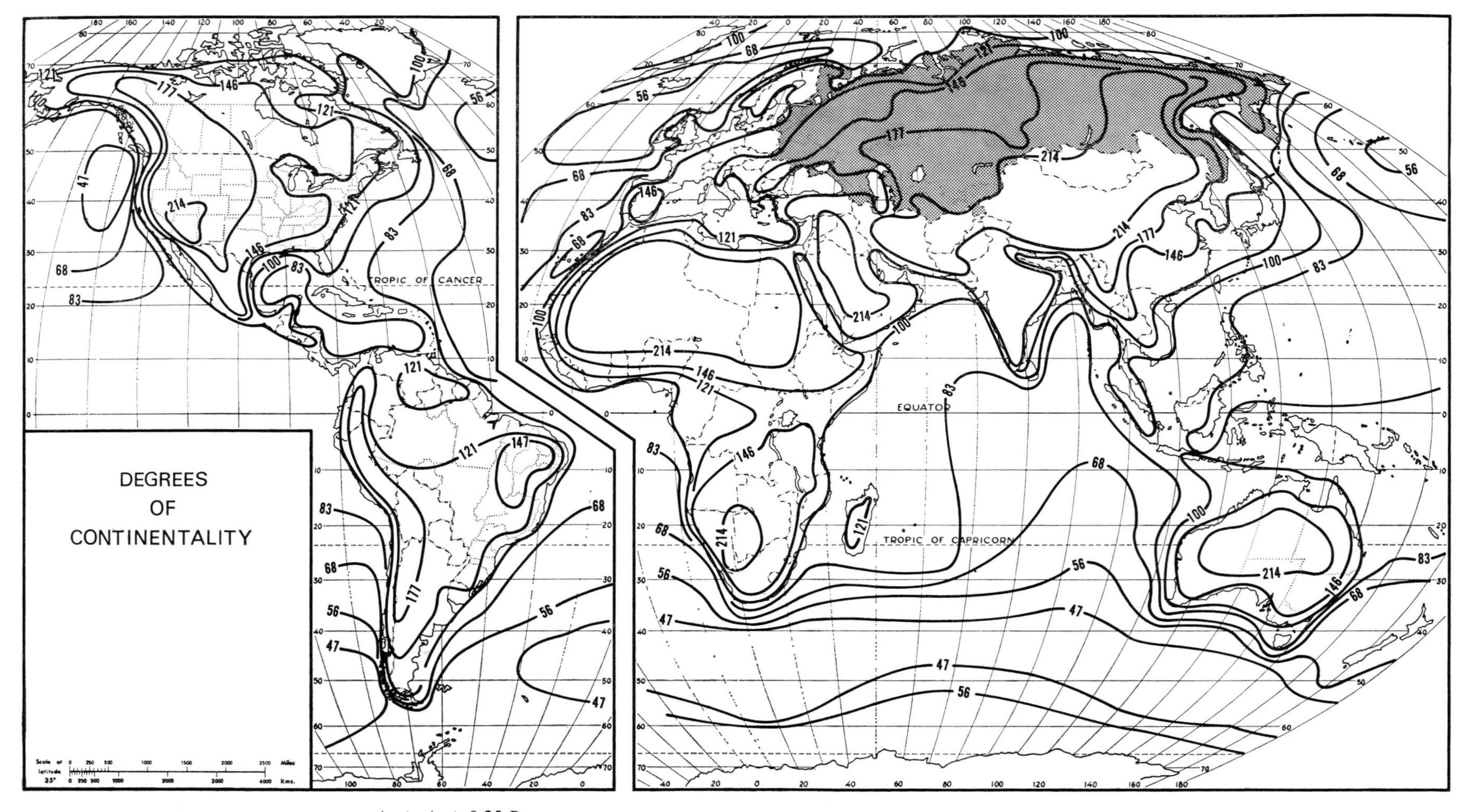

Fig.1-1. Degrees of continentality. $K = 100 \dfrac{Ag + As + 0.25\,Do}{0.36\varphi + 14}$, where K = continentality, Ag = annual amplitude of temperature, As = diurnal amplitude of temperature, Do = relative humidity deficit, = latitude. K ranges from 31 to > 214%. (After IVANOV, 1959.)

the heat and water balances. In this way they have gained more proper prospectives of the climatic effects of such things as irrigation, drainage, shelter belts, and snow retention measures. They have studied the circulation systems of the atmosphere, at various scales, and have related them to heat and water budgets over broad areas, again with the view in mind of completely understanding the causes of weather with the hope that some weather processes can be controlled. The same motive has prompted them in recent years to expend considerable effort on paleoclimatic studies, with the idea in mind that if they can identify the keys to climatic change in the past they will be able to specify the "pressure points" which might be within man's power to trigger successions of events that lead to desired changes of climate in the immediate future.

Working in large groups with almost unlimited resources for compiling information, the Soviets have produced some magnificent works in the forms of atlases and series of regional volumes on the climate of the U.S.S.R., as well as a great number of monographs on a wide variety of topics, that serve as a wealth of basic information for climatic studies of all kinds. Therefore, there is little lack of information on the climate of the Soviet Union.[1] The task of compiling a survey volume is rather one of sifting through a mass of specific materials and selecting and reducing them to a coherent account of the climate of the country. The present volume has drawn heavily on the many excellent Soviet materials available, and they are cited at length herein. (See also LYDOLPH, 1971.)

Content and organization of this volume

The climate of any area consists of two things; the individual types of weather that occur from day to day, and the statistical abstractions which present mean conditions and their deviations. An understanding of the first is necessary in order to appreciate the actual happenings and the reasons they occur. A presentation of the second is necessary to envision long-term means which largely determine the potentials of an area and relate to so many activities of man. Unfortunately, any regionalizations that can be made using these two bases generally do not coincide in their boundaries. In fact in the Soviet Union the two sets of boundaries seem to be more perpendicular than parallel to one another. Climatic zones based on resultant heat and moisture conditions tend to run east–west, while those based on genetic factors tend to run north–south, or at best northeast–southwest.

Weather processes are a continuum in kind which cut across zones of climate determined by variations in degree of the products of synoptic situations. Air masses, fronts, and storms transgress boundaries set by moisture and heat balances which are so obviously reflected in the natural phenomena of vegetation and soils; and conversely, zones fixed by heat and moisture, vegetation, and soil relations vary drastically internally in their makeup of day-to-day weather. Thus a variety of climatic results may relate genetically to a given flow pattern or sequence of weather occurrences, and zones based upon resultant balances of moisture and heat may be comprised of patchworks of weather events that vary greatly from one place to another within zones. Therefore, it is impossible in a

[1] Although there is a conspicuous absence of pressure- and wind-maps in all Soviet publications, data have been made available for compilation of such maps for this volume.

single discussion to cover adequately both the climate as it is related to causal factors and the climate as it affects other phenomena. Yet an understanding of both of these facets is requisite to a complete understanding of the climate of a given region. This book will, therefore, hit the climate from both angles, the causes of climate and the effects of climate. After an introduction to an understanding of country-wide distribution of primary climatic controls is presented, a major portion of the book (chapters 3–7) will be given over to a description, region by region, of types of weather that occur, the combinations of which make up the climate of each region. It is necessary to discuss weather processes within separate regions, since the country is so large that it is impossible to generalize weather processes across the entire territory simultaneously. Particularly in the peripheral regions of the country weather processes may be varied and somewhat unique to each region with only subtle connections from one region to another. In these regional discussions heavy emphasis will be placed upon synoptic situations as explanatory features. Some analyses of resultant climatic means and deviations and special phenomena that occur in each area will be discussed as they relate to local factors of landscape and atmospheric circulation. But the regional discussions will not attempt to give full presentations of the statistical abstractions of climate which result from successions of weather types. Rather, the regional climatic descriptions will dwell on integrated presentations of the various climatic factors as they occur in associations with one another. In many cases such presentation will be more qualitative than quantitative. Illustrations worked out by E. E. Fedorov and his colleagues showing frequencies of weather types will be utilized to give visual impressions of annual regimes of weather.[1] Precise statistical information will be left to the climatic tables in the Appendix at the end of the book which include much of the data recorded for stations whose locations are shown on the map in the Appendix.

The regional chapters, then, will dwell on atmospheric dynamics, their immediate results, and in-place characterizations of climate. The presentation of broad areal patterns of individual climatic elements will be left to the chapters which follow that contain discussions on a country-wide basis of all factors of the thermal, moisture, and wind aspects of climate (chapters 8–10). This presentation will rely heavily on maps and other graphic materials to present an integrated picture of individual components of climate over the entire territory of the Soviet Union.

This section of the book will be culminated in a brief presentation of a regional breakdown of the country into climatic types which are based on heat and water balances and are closely related to natural vegetation and soils and agricultural potentials of the land (Chapter 11). This can serve as a basis for evaluating the climate anywhere in the country with regard to possibilities of practical applications.

The applications of climate and associated climatic adaptations, ameloriation schemes, and climatic change will not be dealt with in this volume, since the primary purpose here is to present general climatic information that can be used as the basis for other studies which deal in depth with applied climatology and climatic change.

[1] For a concise explanation of Fedorov's weather types see LYDOLPH, 1959.

References and further reading

Alisov, B. P., 1956. *Klimat SSSR* (Climate of the USSR). Moscow University, Moscow, 127 pp.

Anonymous, 1958–1963. *Klimat SSSR* (Climate of the USSR). Gidrometeoizdat, Leningrad, 6 regional volumes.

Anonymous, 1960. *Atlas Sel'skogo Khozyaystva SSSR* (Atlas of agriculture of the U.S.S.R.). Moscow (pp. 9–48 are climatic maps).

Anonymous, 1964–1970. *Spravochnik po klimatu SSSR* (Climatological handbook of the USSR). Gidrometeoizdat, Leningrad, 34 regional volumes each divided into 5 separately bound parts.

Borisov, A. A., 1965. *Climates of the USSR*. Aldine, Chicago, Ill., 225 pp.

Borisov, A. A., 1970. *Klimatografiya Sovetskogo Soyuza* (Climatography of the Soviet Union). Leningrad University, Leningrad, 310 pp.

Budyko, M. I. (Editor), 1966. *Sovremennye Problemy Klimatologii* (Contemporary problems in climatology). Gidrometeoizdat, Leningrad, 450 pp. (Translated as *Modern Problems of Climatology*, by the Foreign Technology Division, Wright-Patterson Air Force Base, Ohio, 29 November 1967, 500 pp. AD 670 893.)

Budyko, M. I., 1971. *Klimat i Zhizn* (Climate and Life). Gidrometeoizdat, Leningrad, 471 pp. (Translated as a volume of the *International Geophysics Series*, Academic Press, New York, N.Y., 1974.)

Davitaya, F. F. (Editor), 1959–1960. *Klimaticheskiy Atlas SSSR* (Climatic Atlas of the USSR). Gidrometeoizdat, Leningrad, 2 vols. (never released).

Fedorov, Ye. K. (Editor), 1967. *Meteorologiya i Gidrologiya za 50 let Sovetskoy Vlasti* (Meteorology and hydrology during 50 years of Soviet regime). Gidrometeoizdat, Leningrad, 278 pp.

Gerasimov, I. P. (Editor), 1964. *Fiziko-Geograficheskiy Atlas Mira* (Physical-geographical atlas of the world.) Akad. Nauk SSSR, Moscow. (pp.20–44 are world maps of climate and pp.202–220 are USSR maps of climate; there are also about 15 pp. of text on climate; all legends and textual materials have been translated into English by Theodore Shabad and published in a double issue of *Soviet Geography: Review and Translation*, May–June, 1965).

Ivanov, N. N., 1959. Poyasa kontinental'nosti zemnogo shara (Zones of continentality on the earth) *Izv. Vses. Geogr. O-vo.*, 5: 410–423.

Konyukova, L. G., Orlova, V. V. and Shver, Ts. A., 1971. *Klimaticheskie kharakteristiki SSSR po mesyatsam* (Climatic characteristics of the USSR by months). Gidrometeoizdat, Leningrad, 144 pp. (This brief atlas was compiled from data collected for *Spravochnik po klimatu SSSR*, cited above: Anonymous, 1964–1970. The foreword describes the particulars of *Spravochnik po klimatu SSSR* and the six-volume *Klimat SSSR* series; Anonymous, 1958–1963. Data were collected from 4,000 meteorological stations and 11,000 posts over a period of 70–80 years.

Lydolph, P. E., 1959. Fedorov's Complex Method in Climatology. *Ann. Assoc. Am. Geogr.*, 49: 120–144.

Lydolph, P. E., 1971. Soviet Work and Writing in Climatology. *Sov. Geogr. Rev. Transl.*, XII: 637–666.

Voyeykov, A. I., 1948–1957. *Izbrannyye sochineniya* (Collected works). Vol. I, 1948, 751 pp.; Vol. II, 1949, 227 pp.; Vol. III, 1952, 502 pp.; Vol. IV, 1957, 359 pp. Nauka, Moscow.

Chapter 2

General Analysis of Climatic Controls

Soviet climatologists contend that the only primary controls of climate are solar radiation and the nature of the surface underlying the atmosphere. These two factors decide all the components of heat exchange which determine the magnitude of moisture exchanges and atmospheric circulation. Since atmospheric circulation is a product of insolation and the nature of the underlying surface, it is not itself a primary control. Perhaps this is true for the entire earth which can be treated as a closed system, but any part of the earth's surface will experience great gains or losses of heat and moisture through advective processes, and, in the middle latitudes at least, observed climatic elements, such as temperature, clouds, and precipitation, are more directly responsive to the motions of the air than to on-site radiational exchanges.

This is particularly true in the higher middle latitudes in winter when little insolation is received anywhere and snow cover reflects a high percentage of that which is received. During winter over much of the Soviet Union net radiation is negative. The air is losing heat to the ground and therefore the influences of the surface upon the air are minimal and do not differ significantly from one place to another. At the same time temperature and pressure gradients are at their maxima in winter so that the atmospheric circulation is most intense, with large meridional components that bring widely varying air masses into close juxtaposition. Summer weather is more the result of a combination of advective factors and on-site radiational exchanges, and therefore zonal characteristics of the earth's surface must be kept in mind when talking about summer climatic controls. But even in summer, day-to-day weather occurrences in the Soviet Union are probably more the result of atmospheric circulation than anything else.

Therefore, discussion of the causes of climate will center largely on atmospheric circulation and synoptic situations. The various aspects of radiational exchange will be treated in the latter part of this book as climatic elements along with temperature, precipitation, and the like, which will determine a climatic classification scheme that will be based on bio-climatic potentials. But we must not lose sight of the fact that relations between the atmosphere and the underlying surface flow on a two-way street. As air crosses the broad expanses of the U.S.S.R. it is obviously going to undergo modification as determined by the characteristics of the underlying surface. Therefore, some rudimentary information about the climatically active surface will have to be presented early along with circulation and air mass characteristics.

Atmospheric circulation over the U.S.S.R.

General atmospheric circulation over the Soviet Union is controlled by a relatively simple pattern of pressure centers which vary drastically from winter to summer because of the large size of the land mass, but the simplicity of the mean monthly pressure patterns belies a complexity of air masses and synoptic situations that, in the European part of the country at least, is probably unmatched anywhere else on earth.

In winter mean sea level isobars reveal the buildup of a huge high pressure system over the Asian land mass, which at its culmination is generally centered over the Mongolian Peoples Republic (Fig.2-1). Its precise position at any one time is greatly influenced by the mountain-rimmed plateaus and basins of northwestern Mongolia and the Sayan and Altay mountain systems of south-central Siberia which induce stagnation of air and enhance radiational cooling. The high is obviously primarily thermally induced. It is extremely shallow and frequently is protruded through by the higher mountain peaks in the region. Even at the 850-mbar level it no longer exists but is replaced by the southwestern side of an elongated trough that lies across the northeastern part of the country in a northwest–southeast orientation from Novaya Zemlya (Malye-Karmaeuly) to the Sea of Okhotsk (Fig.2-2). This trough intensifies aloft until at the 500-mbar level an enclosed low occupies a great part of eastern Siberia and the Far East (Figs.2-3, 2-4).

This upper trough represents a juncture of the Icelandic low in the northwest and the Aleutian low in the east which are at the height of their development at this time of year, and even at the surface pinch in on the Asiatic high from the two sides (Fig.2-1). Particularly the Icelandic low exerts a great influence on the winter weather of the Soviet Union, as it generally dominates the northern half of the country all the way from the western border eastward through the Ob Basin of western Siberia and even the northern part of the Central Siberian Uplands between the Yenisey and the Lena rivers. Thus, the so-called Siberian high does not dominate all of Siberia during the wintertime, and it generally is not centered in Siberia. A much better designation of this high would be "Asiatic high" or "Asiatic maximum," which is the term most often used by Soviet climatologists.

As shown in Fig.2-1, the Asiatic high has a broad concavity on its northwestern side where the isobars show a definite cyclonic curvature. The high consists mainly of a core area that is generally centered somewhere southwest of Lake Baykal and two elongated protrusions, one extending northeastward across the middle Lena River Basin into the deep mountain valleys of the Yana and Indigirka, and the other extending westward along the 50th parallel across northern Kazakhstan into southeast European Russia and eastern Ukraine. This western protrusion is Voyeykov's so-called "great axis" which in the winter divides a generally northeasterly flow of air along the northern Black Sea coast from the southwesterly flow of air across much of European Russia and thereby acts as a great divide much of the time between the type of weather that is taking place in the Black Sea area and the type which is taking place further north. When it is well formed, this great axis protrudes westward clear to the western coast of Europe and effectively blocks penetration of air and cyclonic systems from the Mediterranean region. It often acts as a wedge which divides routes of cyclonic storms and sends them either northeastward into the Ob Estuary or southeastward into the mountains of Central Asia.

The Asiatic high appears to be sustained by intrusions of anticyclones moving from the

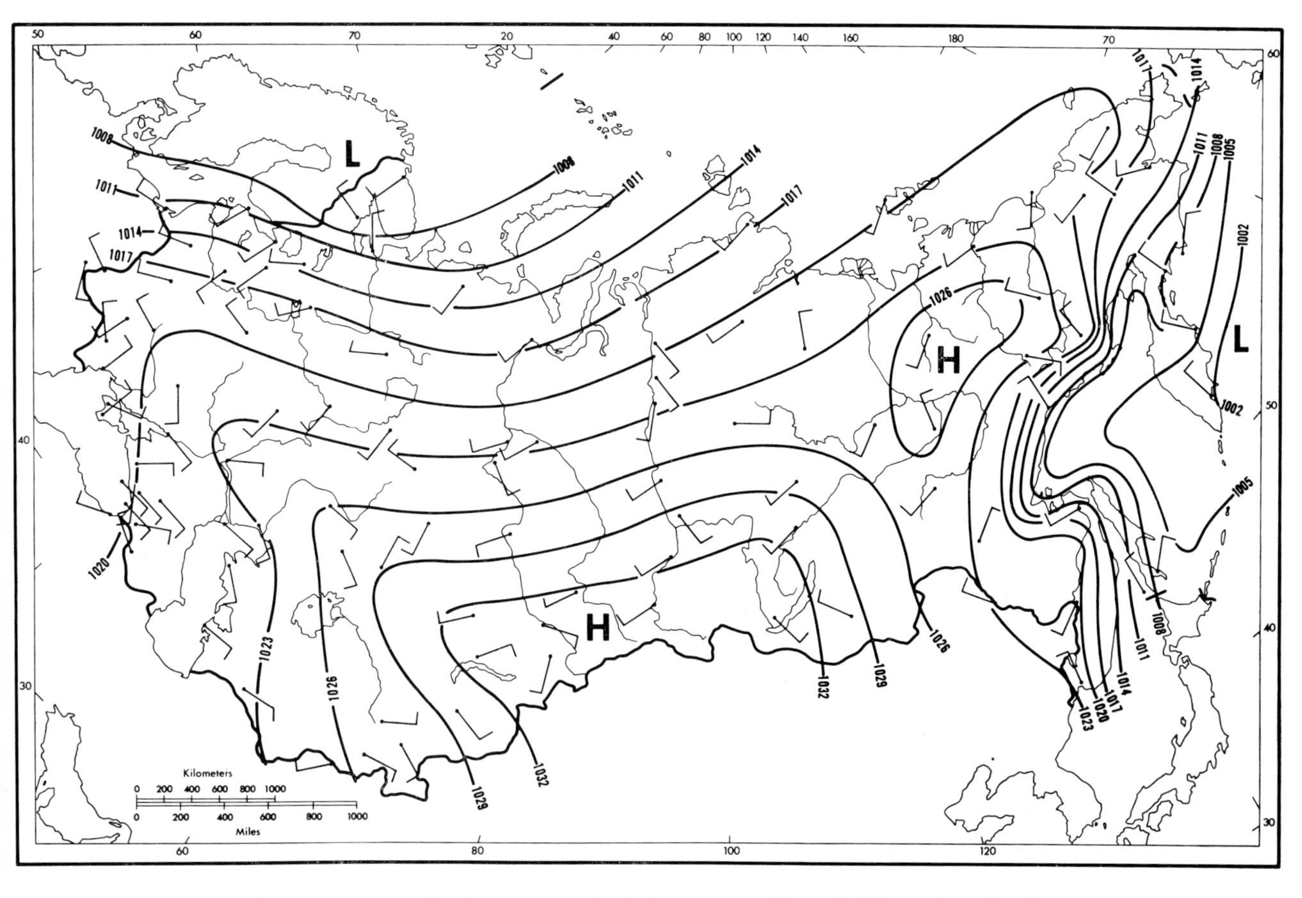

Fig.2-1. Mean sea level pressure and winds, January. (Compiled from the tables in the Appendix.)

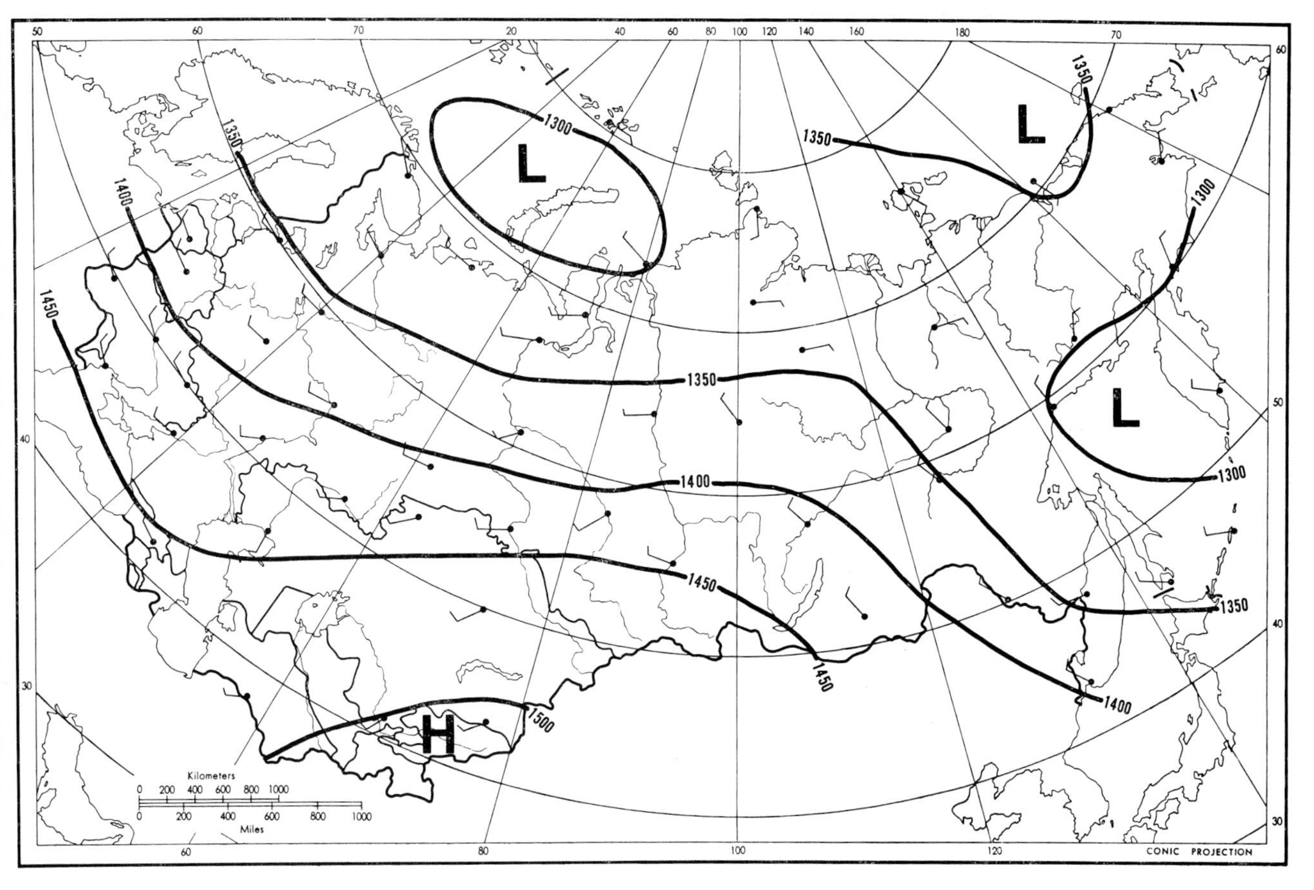

Fig.2-2. Mean heights (m) at 850 mbar, January. (Compiled from the tables in the Appendix.)

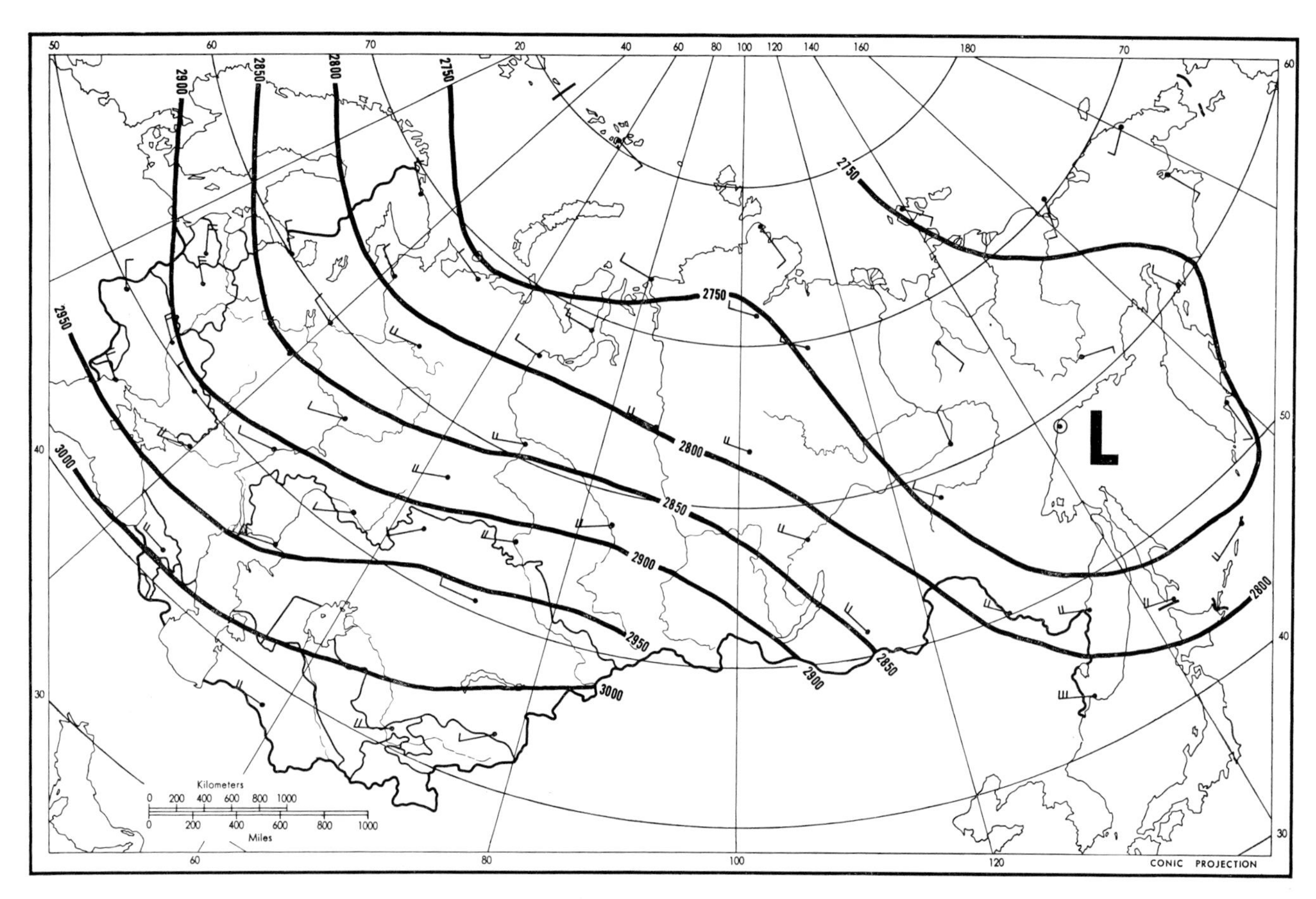

Fig.2-3. Mean heights (m) at 700 mbar, January. (Compiled from the tables in the Appendix.)

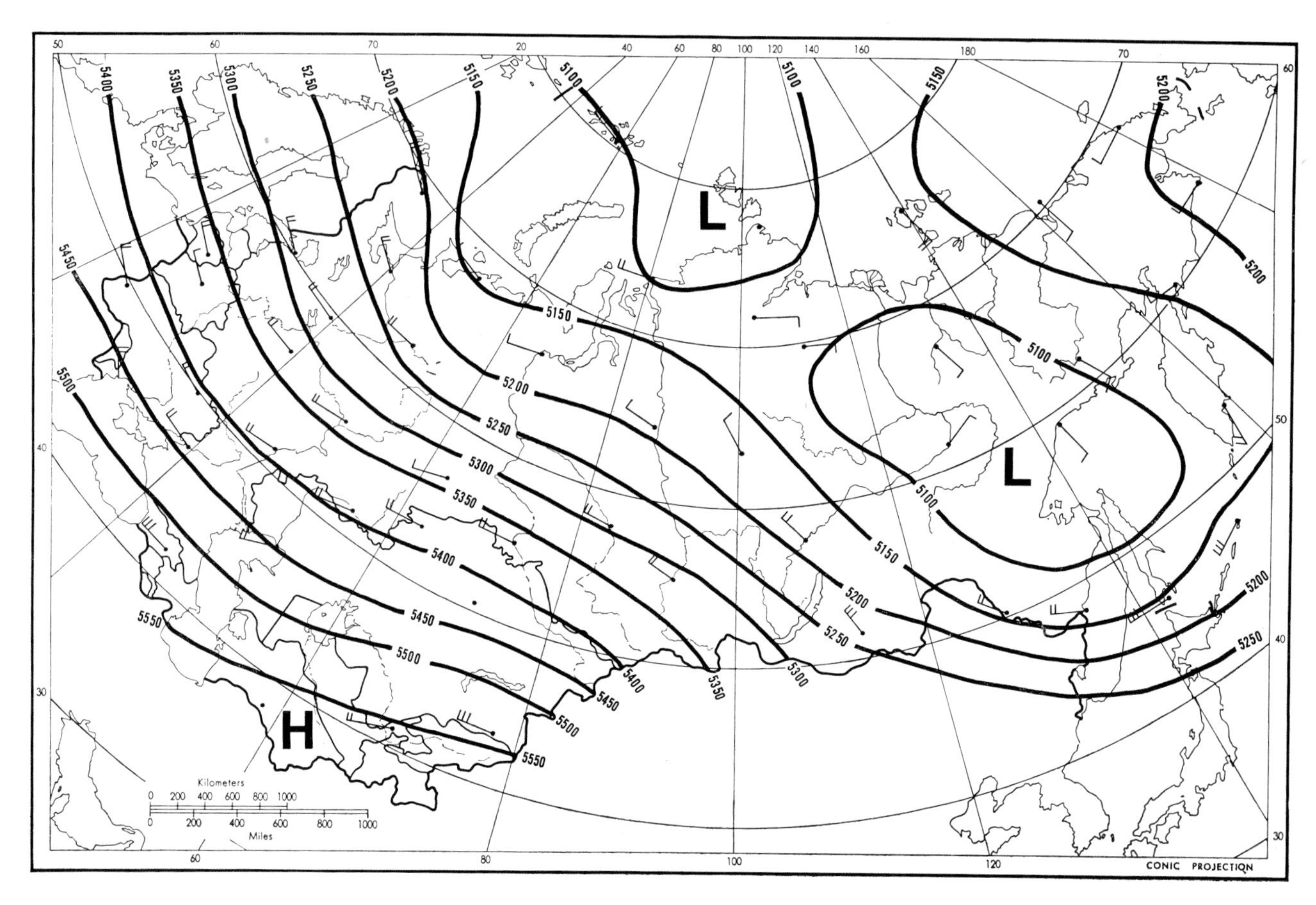

Fig.2-4. Mean heights (m) at 500 mbar, January. (Compiled from the tables in the Appendix.)

Arctic region southward through both European Russia and Siberia which eventually merge with the main body of the cold high and stagnate in the southern portions of the U.S.S.R. This imparts a pulsating nature to the Asiatic high which usually is stronger in one part than another, and at times in fact seems to be broken into different pieces which are interspersed by weak frontal systems and cyclonic storms. Therefore, the Asiatic high, though it is more consistent than most high pressure systems on earth, is not completely dominant during the winter. Even the central region of the high around Lake Baykal experiences some snowfall during the winter, meager though it may be. The European extension of the high consists not so much of air moving westward from Siberia as it does from the intrusions of Arctic anticyclones from the northwest and north which stagnate in southern European U.S.S.R.

Occasionally during winter blocking anticyclones develop over the Ob and Yenisey basins in the region that on the average lies in the northwest concavity of the Asiatic high. These have warm cores and extend upward to the least at 500-mbar level. A study was made of them during the period November–February, 1952–1969 by KALACHIKOVA (1973). During this period the area between 60° and 120°E longitude and 55° and 70°N latitude experienced 179 warm deep stationary anticyclones that occupied the area a total of 8.3% of the time, which is equivalent to the frequency of such high pressure cells in the northeastern Pacific during the cold season. The greatest number occurred during February.

The formation of these blocking highs seems to be tied in with the general circulation of the entire Northern Hemisphere during periods when upper air flow experiences great meridional movements. They are associated with the establishment of a ridge at the 500-mbar surface which extends northeastward from the lower Volga to the Taymyr Peninsula and on into the northeastern extremity of the country where it joins with a ridge that extends westward from North America across the Bering Straits into the Kolyma Valley (Fig.2-5). The ridge is flanked on both sides by deep troughs, one over Europe and one extending west-southwestward from the Sea of Okhotsk to Lake Balkhash, which convolute the high altitude planetary frontal zones and their associated jet streams all the way from the Arctic to 35°N. During such occurrences eastern Asia is under the influence of two jet streams, a subtropical one which flows eastward across northern China and Japan and an extratropical one which flows from northeast to southwest across central Siberia from the region of New Siberian Islands along the Arctic coast or from Yakutia to the southern Altay Mountains (Fig.2-5). Under such a flow arrangement independent nuclei of thermal anticyclones move across the Bering Straits into northeastern Asia.

During blocking situations all of eastern Asia is under high surface pressure created by two anticyclones, a warm deep one in northern Siberia and a cold shallow one (the usual Asiatic maximum) under the high level trough in southern Siberia and Mongolia. Such conditions bring on extraordinarily cold, partly cloudy weather over much of eastern Asia and stormy weather with abundant snowfall in the Far East basins.

February marks the height of the development of the winter circulation over the Soviet Union. A pattern similar to that on the January map prevails through much of November to March, with some decrease in distinctiveness at the two ends of the season. Also toward the beginning and end of the season the core of the Asiatic high appears to jump back and forth in location and frequently shows more than one central position. In spring, particularly, this is true, and each successive reformation of the cell shows the high to be weakening and the center to be shifting westward.

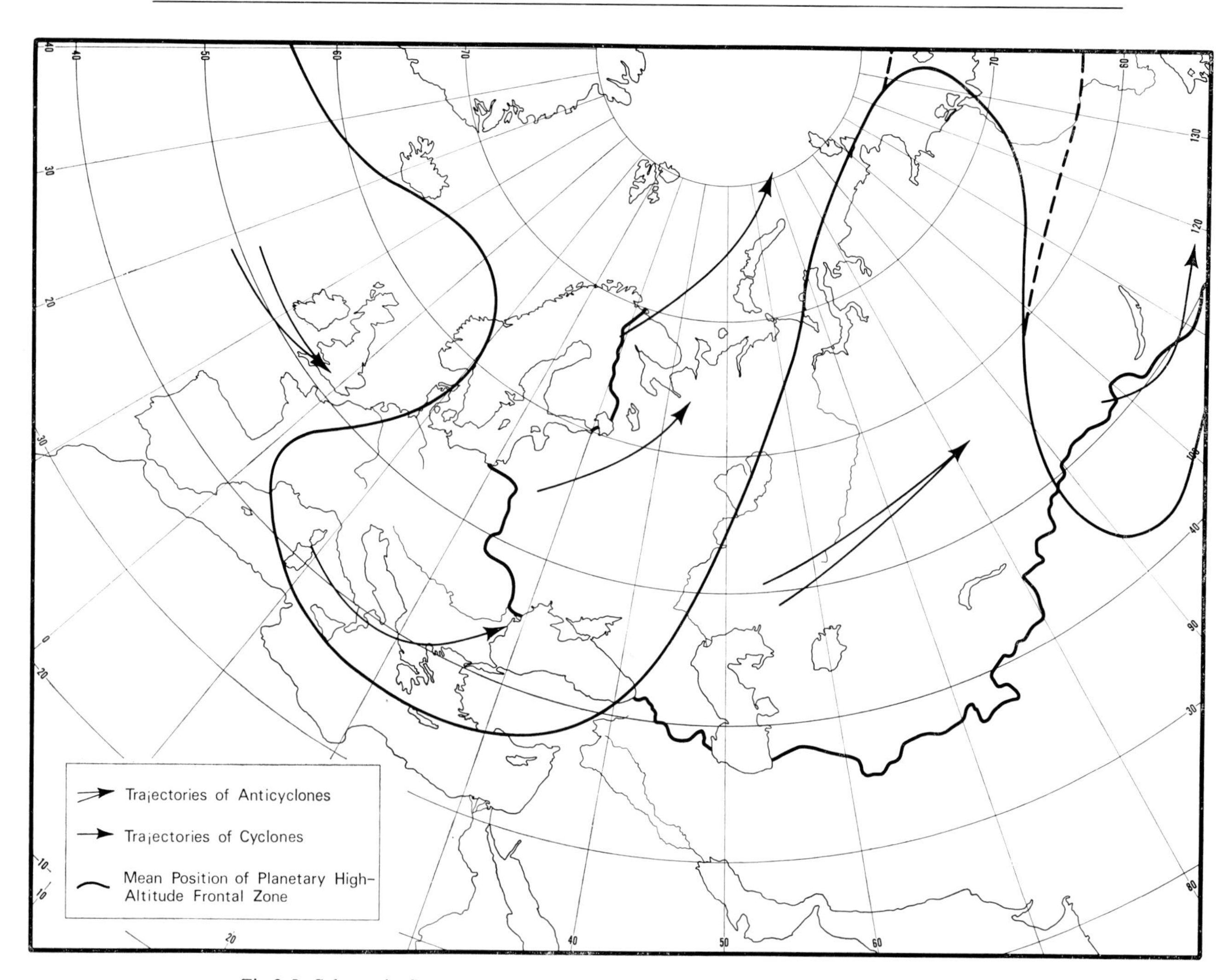

Fig.2-5. Schematic flow at the 500-mbar surface during a period of a blocking anticyclone. (After KALACHIKOVA, 1973.)

During the last week of February, large falling tendencies in surface atmospheric pressure set in over the southern Urals and spread for hundreds of miles in all directions eventually covering much of European U.S.S.R. to the southern borders of the country and to the Arctic and enveloping much of central Asia and western Siberia almost as far east as Lake Baykal (Fig.2-6). This falling tendency prevails through much of March and April albeit at a decreasing rate. During April, a major change takes place in the sea level air pressure pattern (Fig.2-7). The Asiatic high weakens and shifts its center westward to a position in northeastern Kazakhstan. The northeastern extension of the high in eastern Siberia and the Far East dissipates and is replaced by a low pressure extending inland from the Sea of Okhotsk in the east. Broad sections in the northern and western parts of the country show intermediate pressures with very little pressure gradients. It appears that what is taking place at this time of year is a shift from the dominance in the east of the Asiatic high centered over Mongolia to the dominance in the west of the Azores high protruding across Europe. During April, as the Asiatic high phases out and the Azores high strengthens and expands, for a short time an intermediate position is occupied by relatively higher pressure over northern Kazakhstan and the southern Urals region.

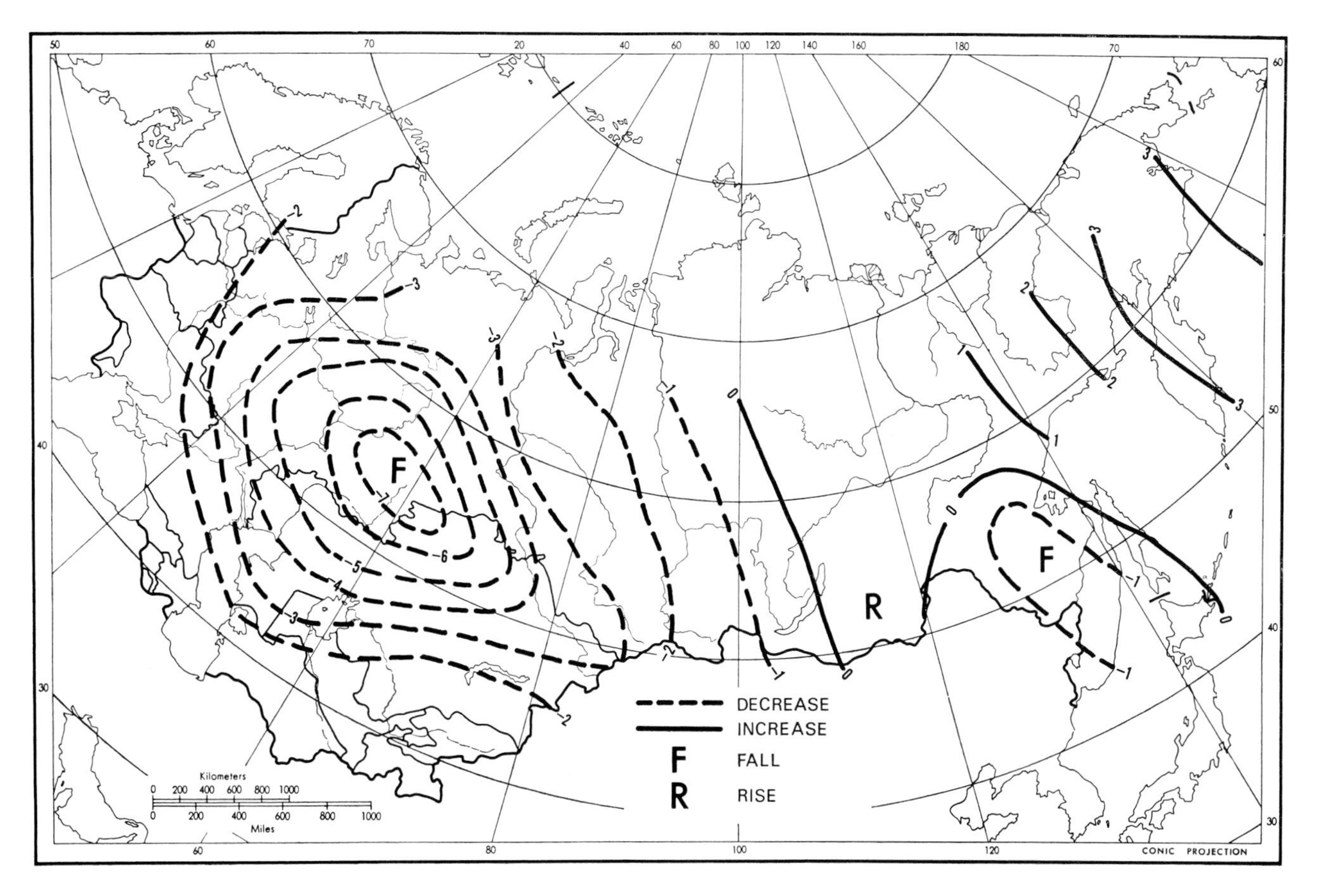

Fig.2-6. Five-day sea level pressure changes: February 25–March 1 to March 2–6. (After LAHEY et al., 1958.)

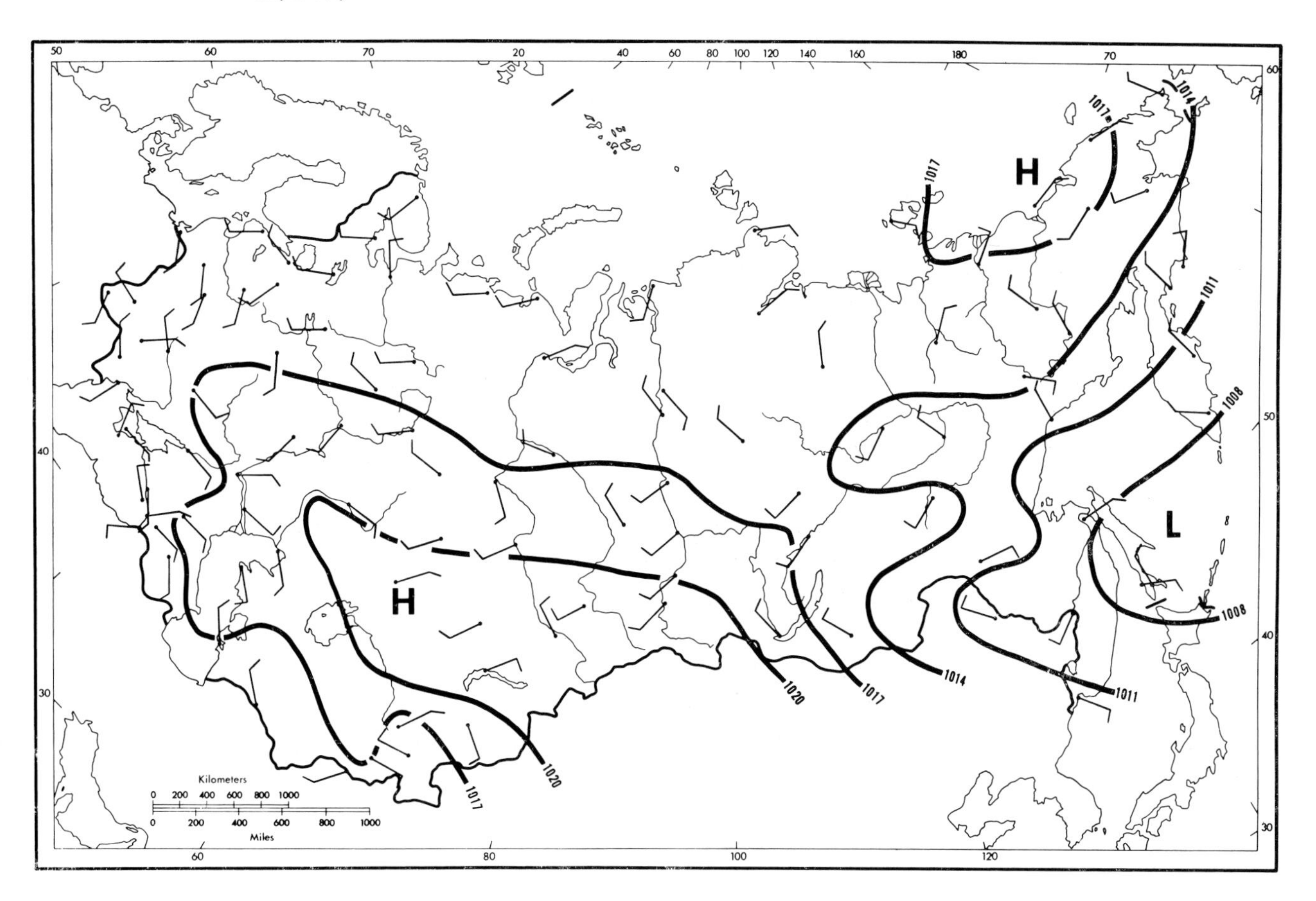

Fig.2-7. Mean sea level pressure and winds, April. (Compiled from data in tables of the Appendix at end of book.)

During mid-May large pressure decreases occur over almost all of Asiatic U.S.S.R., while rising tendencies begin in European U.S.S.R. This is the beginning of the end of the Asiatic high and the beginning of the dominance of the summertime Azores high in the west. During the latter third of May the Azores high spreads eastward clear into Central Asia. It remains fairly prominent in the middle of the continent until about June 10, after which thermal effects from extreme continentality bring on a summertime pattern of diffuse low pressure over much of Asia.

Throughout much of the summer a weak ridge extends eastward from the Azores high across central European U.S.S.R. as far east as central Kazakhstan (Fig.2-8). This ridge that appears on summer mean maps is maintained by the advances eastward of individual cells of high pressure out of the Azores anticyclone as well as by Arctic anticyclones that move from north to south across the plain and stagnate in the southeastern portion of European U.S.S.R. In European U.S.S.R., as the western nose of the wintertime Asiatic high is replaced by the eastern nose of the summertime Azores high, surface winds generally shift from southerly or southwesterly to westerly or northwesterly.

As the pressure pattern changes over the land during spring and early summer, so does it over the sea. The Icelandic and Aleutian lows weaken and no longer exert direct influence on the Eurasian land mass. The July map shows that the broad shallow low over Asiatic Russia is bordered on the north by a shallow high pressure system in the Arctic and on the east by a similar situation in the Pacific. Thus, the north and east in the Soviet Union

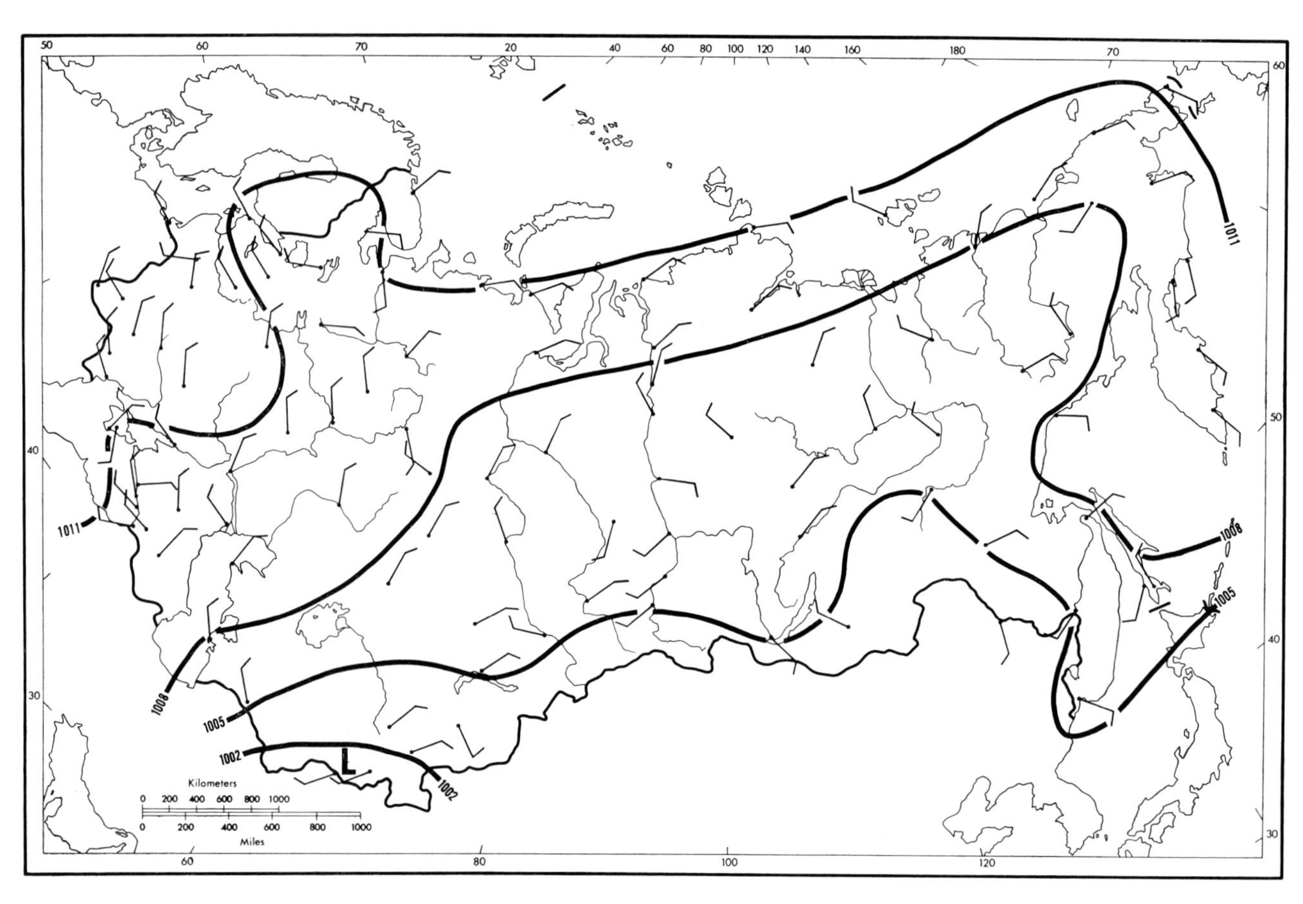

Fig.2-8. Mean sea level pressure and winds, July. (Compiled from the tables in the Appendix.)

experience reversals in pressure patterns between winter and summer, with high pressure over the land in winter and low pressure over the land in summer. This results in a true monsoonal reversal of resultant wind directions with the seasons, although, except along the immediate coasts, this general tendency is frequently obscured by air motions associated with moving highs and lows and the effects of local wind regimes.

In spite of the fact that the Arctic is frozen over right up to the coast during middle and late winter, the winter air over the Arctic ice remains considerably warmer than that over the snow-covered land mass to the south because of the transmittal of heat from the unfrozen Arctic water through the 3- or 4-m cover of ice into the atmosphere above. Therefore, during winter temperatures are warmer and air pressures are generally lower over the Arctic Ocean than over bordering Siberia, and winds generally circulate from the land to the sea. In spring, as the land heats up and air temperatures over the melting sea continue to hover around the freezing point, the pressure gradient is reversed and an atmospheric circulation sets up from sea to land. During the winter, then, there is generally an atmospheric surface pressure trough just off shore in the Arctic which is conducive to the positioning of the Arctic front just off the Siberian coast at this time of year, while during summer this is not so evident and segments of fronts tend to lie more over the heated interior.

Seasonal reversals of atmospheric flow occur all along the Pacific coast as well. Except for the western part of European U.S.S.R., surface air circulation during the winter tends to be from the interior of the country toward the peripheries. Therefore, the land mass is dominated by continental air. The lack of amelioration by the sea during the winter is one of the outstanding characteristics of the climate of the U.S.S.R., and largely accounts for the severity of the winter, particularly in the eastern regions.

The contour pattern at 700 mbar shifts from a January situation, which shows a weak trough over western European U.S.S.R. and a weak ridge over central Siberia, to a July situation which shows a generally zonal flow across the country as the cool surface high over the Arctic turns into a weak low at the 700-mbar level (Fig.2-9). The thermal effects on the circulation pattern accumulate upward until at 500 mbar a weak trough appears over the Ob Estuary and a closed low over the Kara Sea to the north (Fig. 2-10). These rather subtle changes in pressure patterns at the surface and aloft produce significant changes in the directions of movement of cyclones and anticyclones during the two seasons, which will be discussed later.

Mid-August marks the culmination of the summer circulation pattern, after which significant pressure changes begin to take place at the surface in European U.S.S.R. By mid-September a closed high has formed which is much elongated in an east–west direction and stretches all the way from the Carpathian Mountains to Lake Baykal. This is the beginning of the shift from the dominance of the Azores high from the west to the dominance of the Asiatic high from the east. Already by October a definite high pressure cell becomes centered east of Lake Balkhash (Fig.2-11). As the Asiatic high strengthens and expands, the center shifts eastward until in late December it occupies a position southwest of Lake Baykal. With some shifting of position between southwest and southeast of Lake Baykal, the Asiatic high remains quite consistent until mid-March.

In the Arctic autumn is considerably warmer than spring, due to the lag induced by the heat relationships of the changing states of water. The same is true to a certain extent over much of the land mass, since spring is delayed and shortened by the thawing period

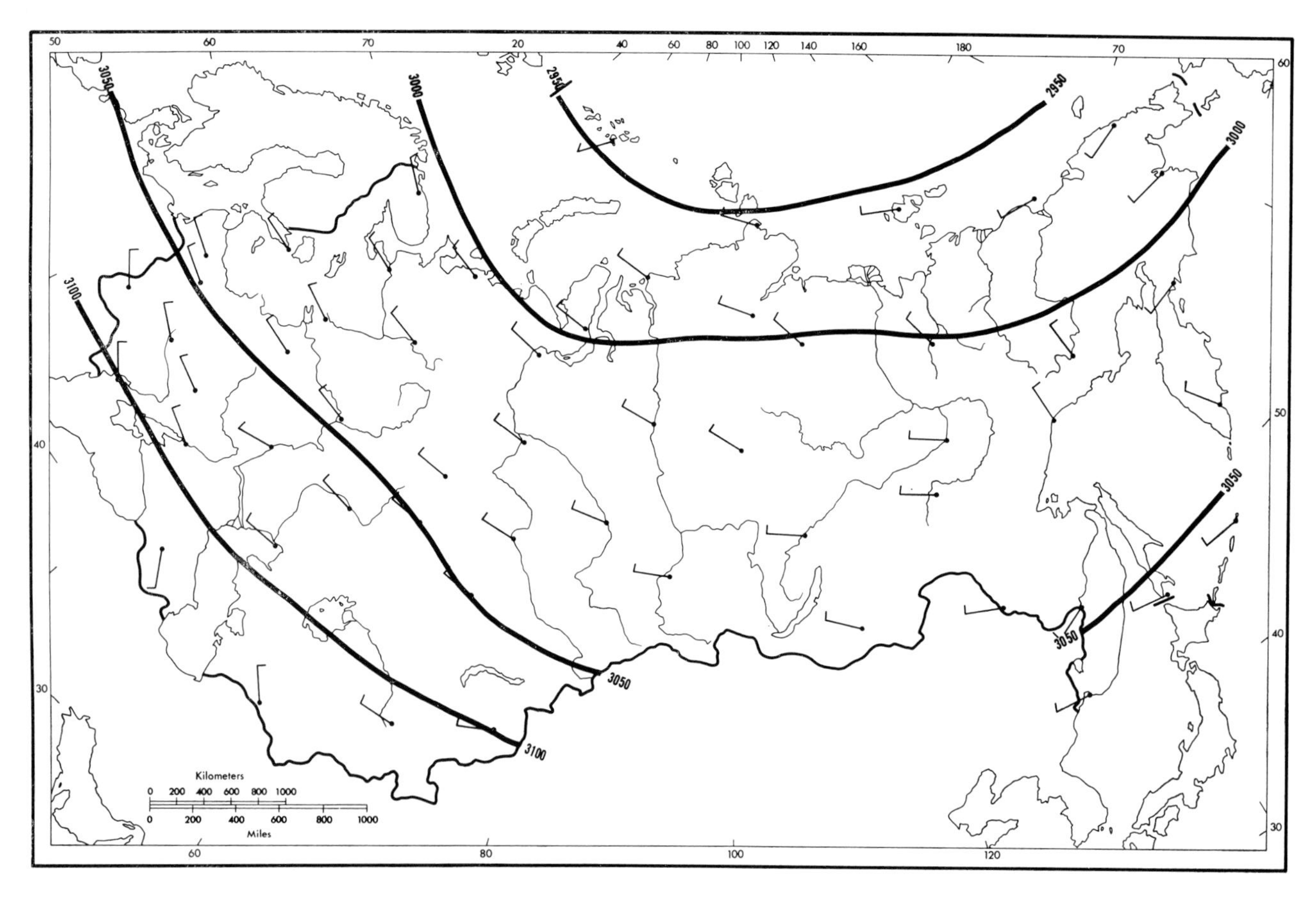

Fig.2-9. Mean heights (m) at 700 mbar, July. (Compiled from the tables in the Appendix.)

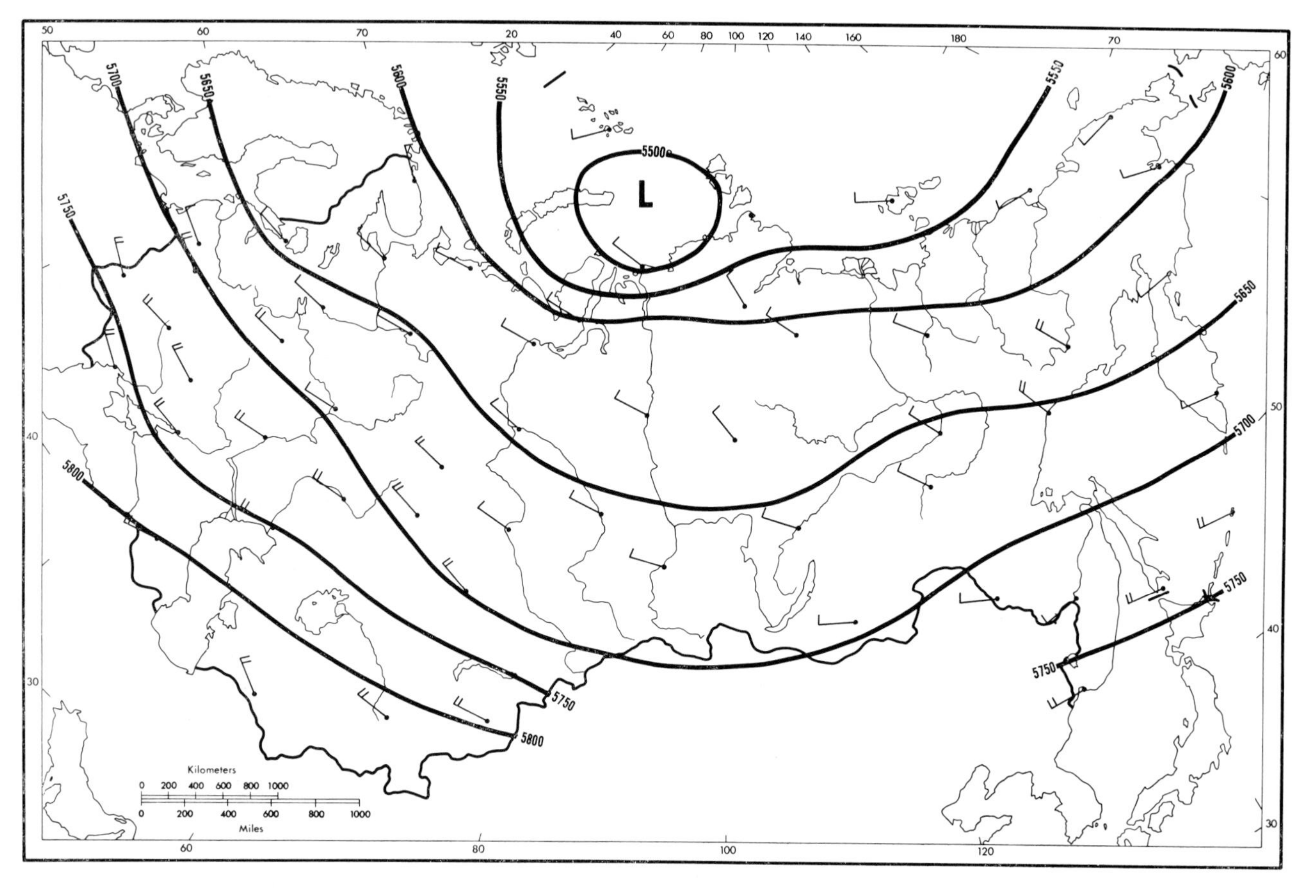

Fig.2-10. Mean heights (m) at 500 mbar, July. (Compiled from the tables in the Appendix.)

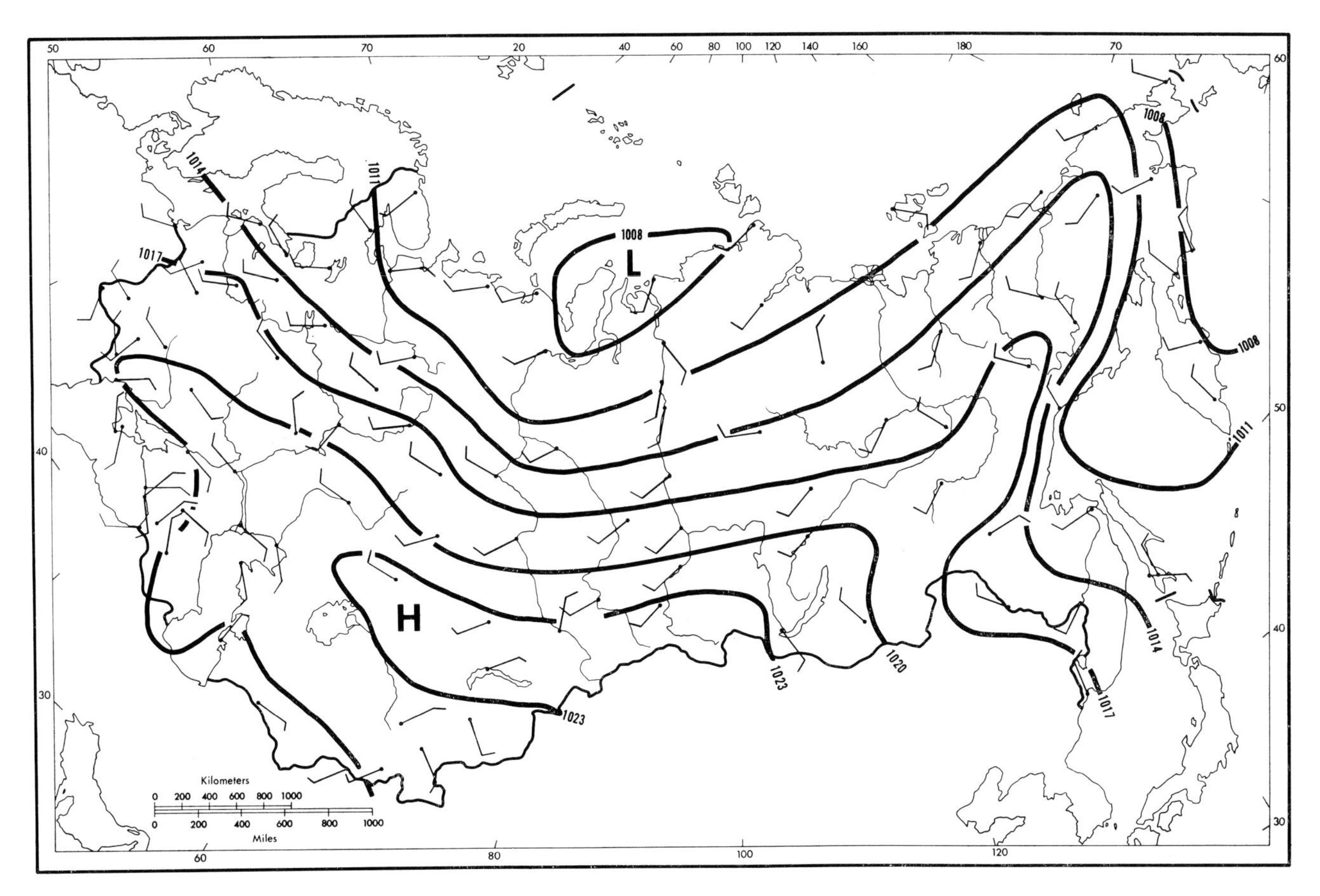

Fig.2-11. Mean sea level pressure and winds, October. (Compiled from the tables in the Appendix.)

required by the winter snow cover. These temperature differences in the atmosphere between spring and fall induce significant differences in the atmospheric circulation of the two transitional seasons. Generally, pressure zones and frontal positions are farther north in autumn than in spring.

Fronts, cyclones, and anticyclones

Winter

During winter the strong dominance of a high pressure cell over Eurasia causes the majority of fronts and cyclone tracks to be positioned along the peripheries of the land mass (Fig.2-12). The Polar front generally lies south of the U.S.S.R. with a western segment in the Mediterranean–Asia Minor–Middle East area and an eastern segment off the China coast across the Japanese Islands and extending northeastward into the Aleutian area. The western segment spawns many cyclonic storms which affect the Soviet Union. Those forming in the eastern Mediterranean typically move northeastward across the Black Sea–Caucasian area, through the Ukraine and lower Volga regions, eventually into western Siberia. Thus they pull the Polar front far northward into the country.

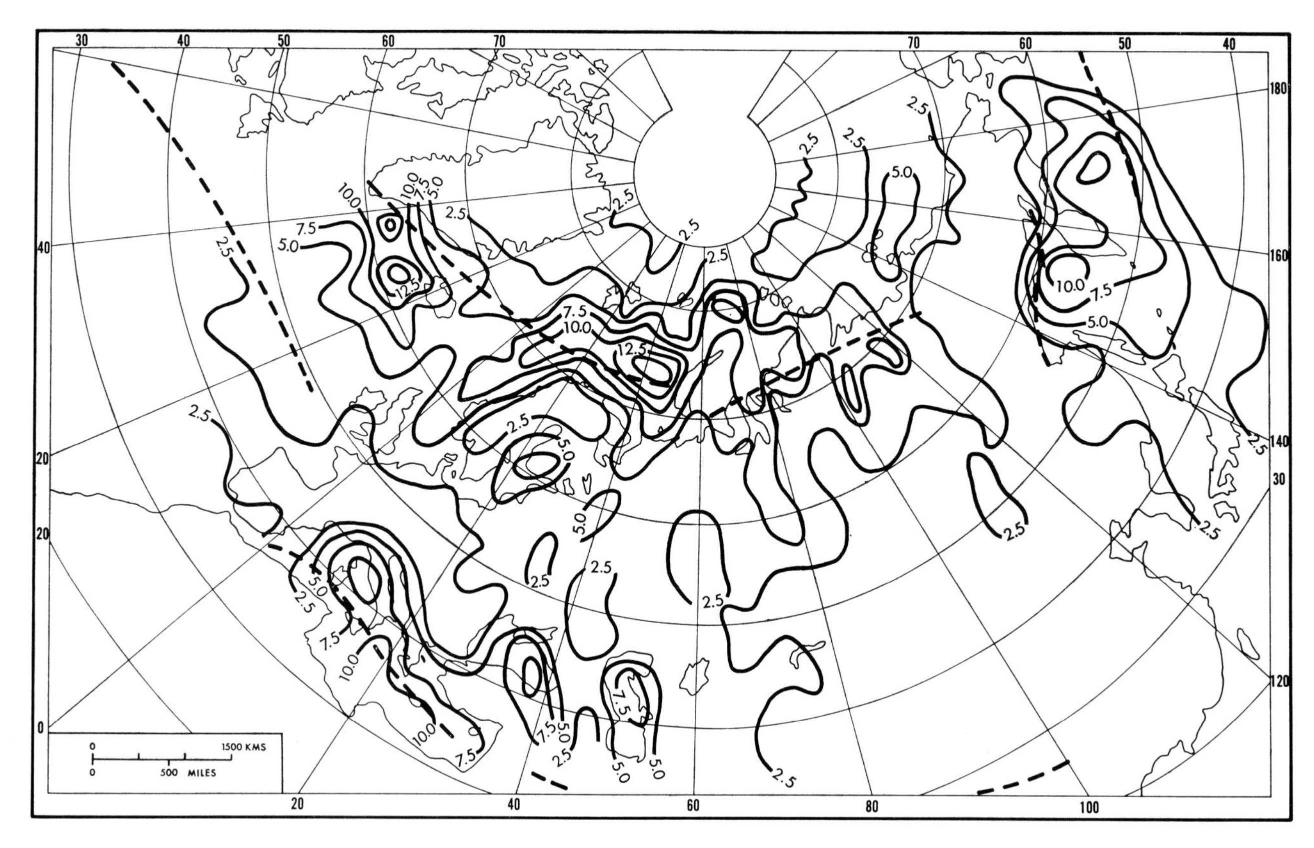

Fig.2-12. Percent of all days with cyclonic centers positioned within 5° grid sections, and average frontal positions, January. (Compiled from BUGAEV and DZHORDZHIO, 1940; BULINSKAYA, 1963; and BORISOV, 1970.)

Cyclones that form in the Middle East often find their way into Soviet Central Asia and cause the Polar front to swing northward into that area. But cyclones forming along the eastern segment of the Polar front in winter move northeastward across the sea and have little effect on the Soviet Union (Fig.2-13). The fronts which affect most of the Soviet Union at this time of year are segments of the Arctic front, which generally are positioned along the northern shores of the land mass but which sweep across the breadth of the country, particularly the European part, as far south as the Black Sea and Caspian Sea. The Arctic front on the average is broken into three segments, the Atlantic–European Arctic front which extends north of Scandinavia into the Barents Sea, the Asiatic Arctic front which is positioned near the coast of western and central Siberia along the Kara Sea and Laptev Sea, and the Okhotsk Arctic front along the northern shore of the Okhotsk Sea in the Far East. The average positions of these segments of the Arctic front mark the contrast during winter between the temperatures of the air over the cold snow-covered continent and the somewhat warmer air over the Arctic. This is particularly true over the Barents Sea, which is partially ice free in winter, but even in the Kara and Laptev seas the atmosphere gains some heat from the water beneath the ice. This is true also over the frozen Okhotsk Sea, and the front along its northern coast is strengthened by the topography of the land mass to the north which is quite mountainous. Air flow from the north piles up behind the coastal mountains and creates an extremely strong pressure gradient directed southward across the coast.

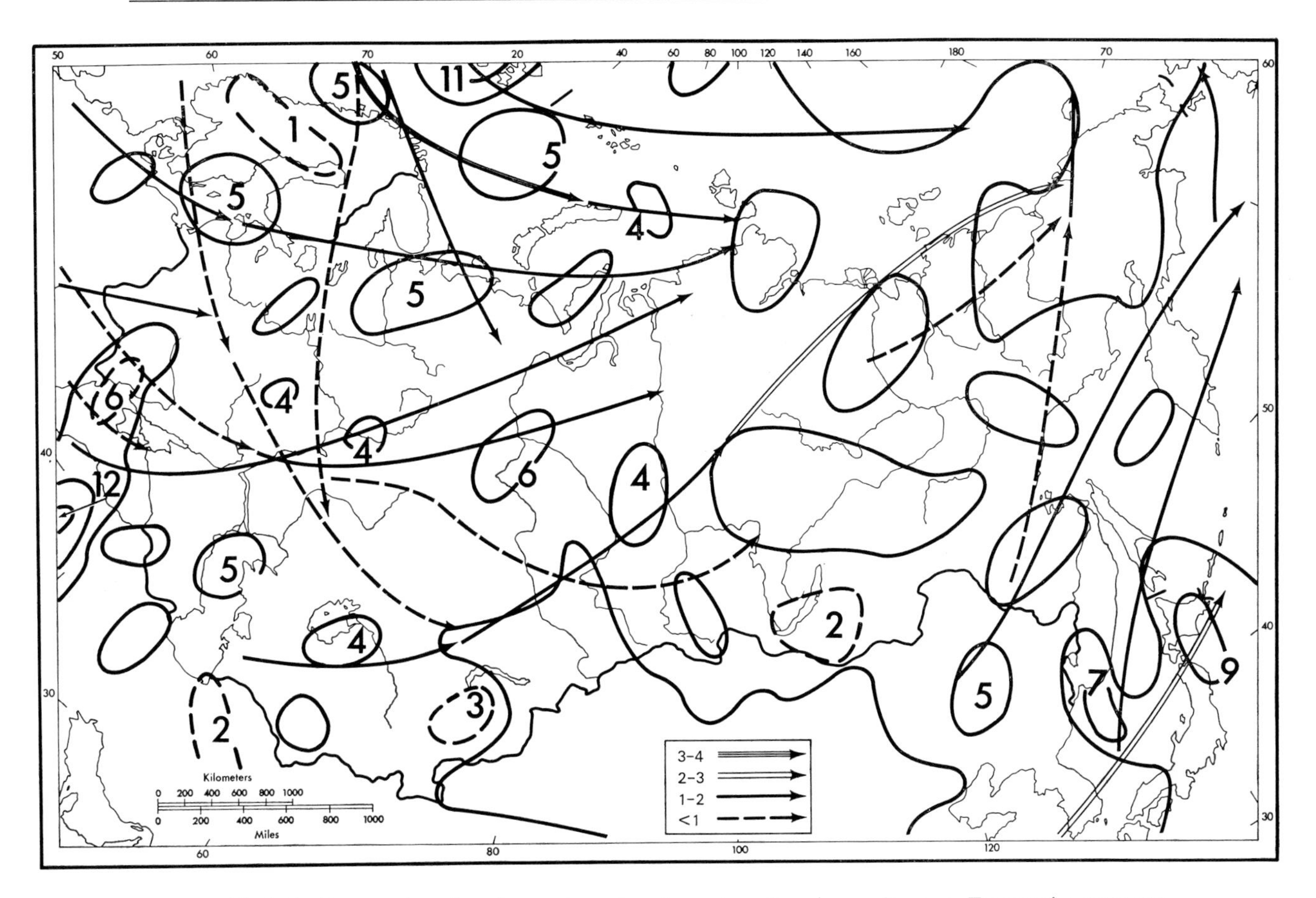

Fig.2-13. Frequencies of cyclogenesis and main routes of cyclones, January. Frequencies represent numbers of individual cases during 20 years within 5° grid sections. (After KLEIN, 1957, and BORISOV, 1970.)

Many of the cyclonic storms which affect the Soviet Union in winter have had their origins either in the Icelandic low region of the North Atlantic or in the Mediterranean and are shunted along the northern and southern borders of European U.S.S.R. by the western protrusion of the Asiatic high. In the Soviet area their frequencies are highest by far in the Barents Sea which acts as an area of cyclogenesis and rejuvenation of cyclones which move from the Icelandic low around the northern tip of Scandinavia eastward toward the Ob Estuary. Sometimes they continue on across the northern portion of the Taymyr Peninsula as far east as the delta of the Lena River. Minimum sea level pressures in this region vary from 960 mbar northwest of Scandinavia to 985 mbar east of the Ob Estuary.

Secondary frequency maxima occur in the south over the Black and Caspian seas which also act as significant areas of cyclogenesis during the winter. Minimum sea level pressures here vary from 1,000 to 1,005 mbar. The southern cyclones, many of which enter the region from the Mediterranean, generally travel either northeastward into western Siberia or eastward into the northern foothills of the mountains of Central Asia. Cyclogenesis takes place rather frequently in western Siberia at this time of year along fronts which trail southwestward from storms that have moved eastward from the Barents Sea or northeastward from the Black Sea and occluded in the Ob Estuary region.

The mid-Baltic is another area of significant cyclogenesis during the winter, and a secondary cell of significantly high cyclonic frequency is found over southern Finland which greatly affects the northwestern portions of the U.S.S.R. In the Far East high cyclonic frequency in the northern part of the Okhotsk Sea represents the western extremity of the Aleutian low at this time of year. Sea level pressures as low as 970 mbar have been observed here. Much of the cyclogenesis in this area takes place farther south over the Sea of Japan and the Japanese Islands and does not affect the Soviet Union except for the southern portion of Maritime Kray, Sakhalin Island, the Kuril Islands, and portions of Kamchatka Peninsula.

The Ob Estuary is the meeting place for cyclonic storms that have come in from the west along the Arctic coast and from the southwest from the Black Sea and Caspian Sea areas. It is therefore one of the stormiest regions in winter, and much of the West Siberian Lowland is marked by great interdiurnal changes of weather. To a lesser degree the same can be said about the northwestern part of eastern Siberia, at least as far east as the Lena Delta. The winter weather in these regions contrasts greatly with the constantly cold, clear, calm conditions that are experienced most of the winter farther south and east in the region of the Asiatic maximum.

The fate of the cyclonic storms that skirt along the foothills of the Central Asian mountains depends to a great extent on the conditions of the west wind jet at about 10 km height in the atmosphere associated with a major break in the tropopause whose geographical position seems to be closely tied to the high mountain region along the southern border of the U.S.S.R. and the seasonal shifting of the planetary circulation in this part of the world (Fig.2-14). During winter the tropopause break and its associated jet stream tend to be located on the southern slopes of the mountains in the Near and Middle East, while during the summer they are generally positioned on the northern slopes of the mountains. The fact that the tropopause is broken over the mountains, and a high speed westerly jet results, strengthens the role that the mountains play in constituting a barrier to meridional atmospheric flow. The mountains themselves are high enough and continuous enough to prevent much transfer of warm moist air from the southern fringes of Asia into the interior of the continent, but the jet stream above the mountain ranges carries the influence to even greater heights in the atmosphere. As the jet stream undergoes convolutions from day to day, cyclonic storms moving into the area from the Mediterranean tend to strengthen or weaken, depending on their positions with respect to the overlying jet. The effects are felt most in winter when cyclonic storms tend to frequent the area.

During winter anticyclones move into the U.S.S.R. primarily from the Arctic all along the Arctic coast into all parts of the country (Fig.2-15). They are particularly frequent in the northwest, where they describe a curved path concave to the east from the Kola Peninsula southeastward to the Aral Sea, and in the northeast, where they describe a slightly curved path concave to the west from the Laptev and East Siberian Seas southwestward toward the Amur Valley. These traveling highs generally stagnate in the southern parts of Siberia, Central Asia, and southeast European Russia and are a major factor in maintaining the Asiatic maximum during the winter.

The most frequent occurrences of high pressure centers lie in the south and east where traveling highs stagnate (Fig.2-16). This is particularly true southwest of Lake Baykal, the core of the Asiatic maximum, which has experienced sea level pressures as high as 1,075 mbar, and in its northeastern extension where the stagnation of cold air and extreme temperature inversions in deep mountain valleys frequently form separate cores within

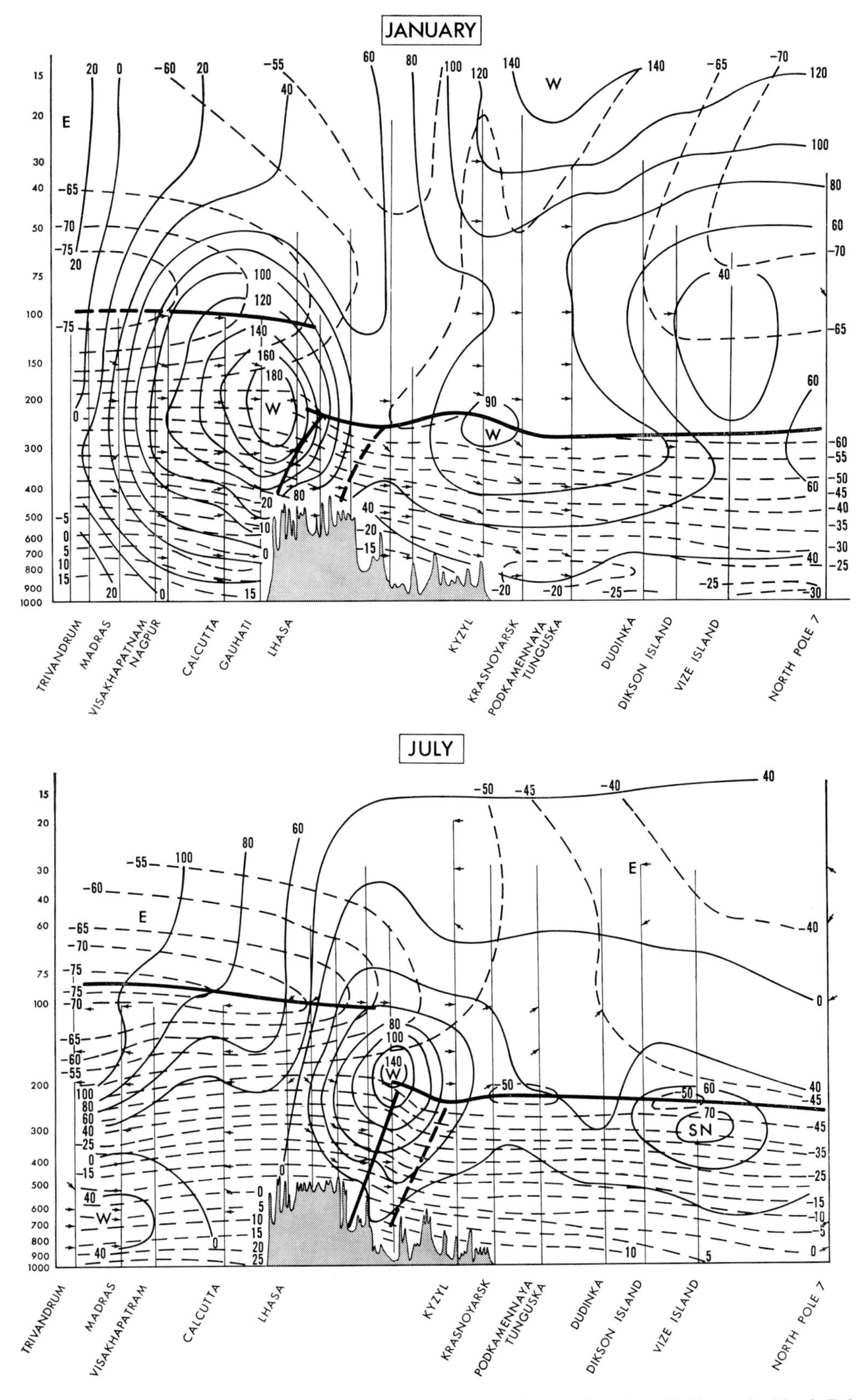

Fig.2-14 Mean vertical cross-sections of the atmosphere from Trivandrum, India to the North Pole, January and July. (After POGOSYAN and UGAROVA, 1959.)

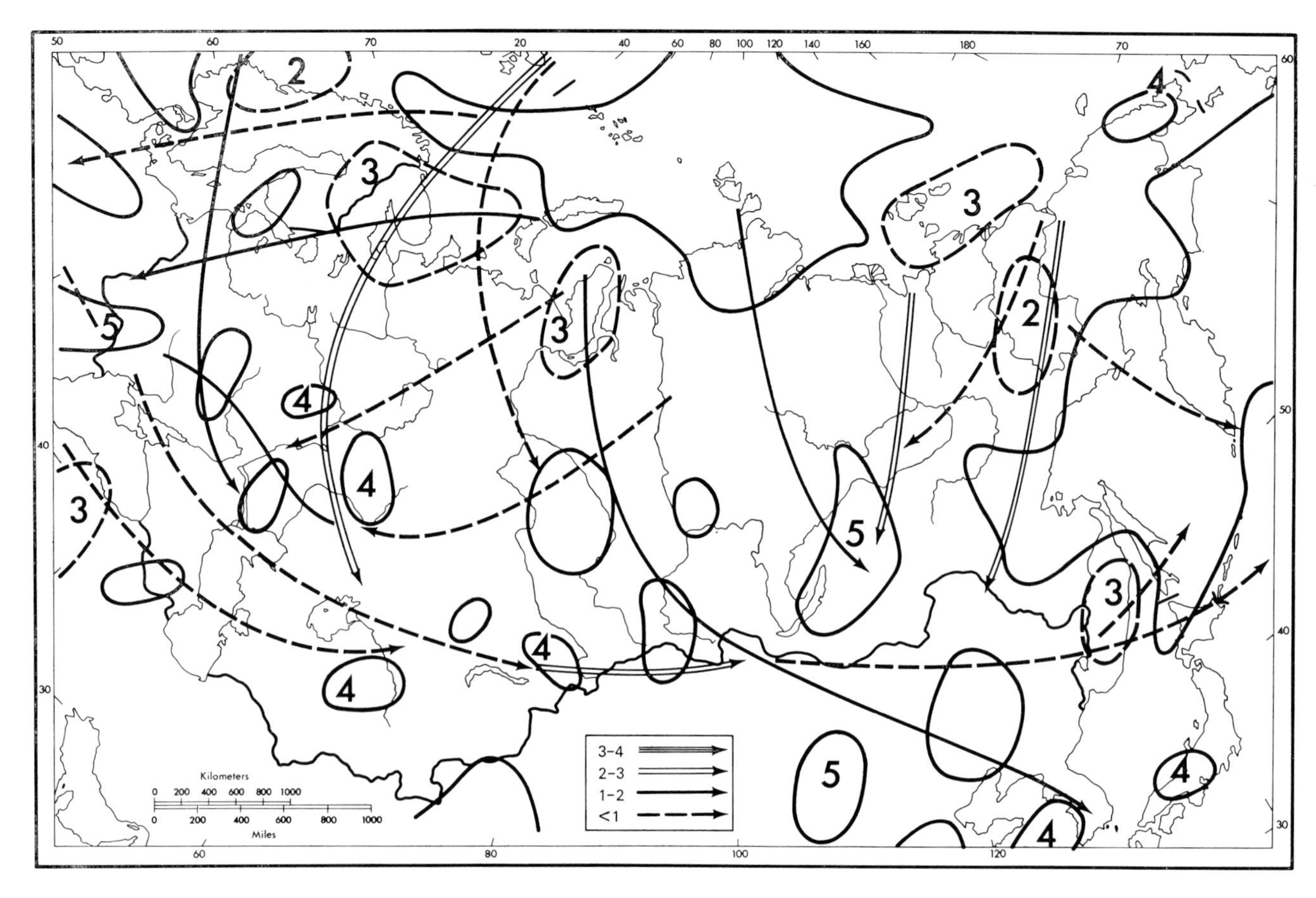

Fig.2-15. Frequencies of anticyclogenesis and main routes of anticyclones, January. Cases during 20 years within 5° grid sections. (After KLEIN, 1957, and BORISOV, 1970.)

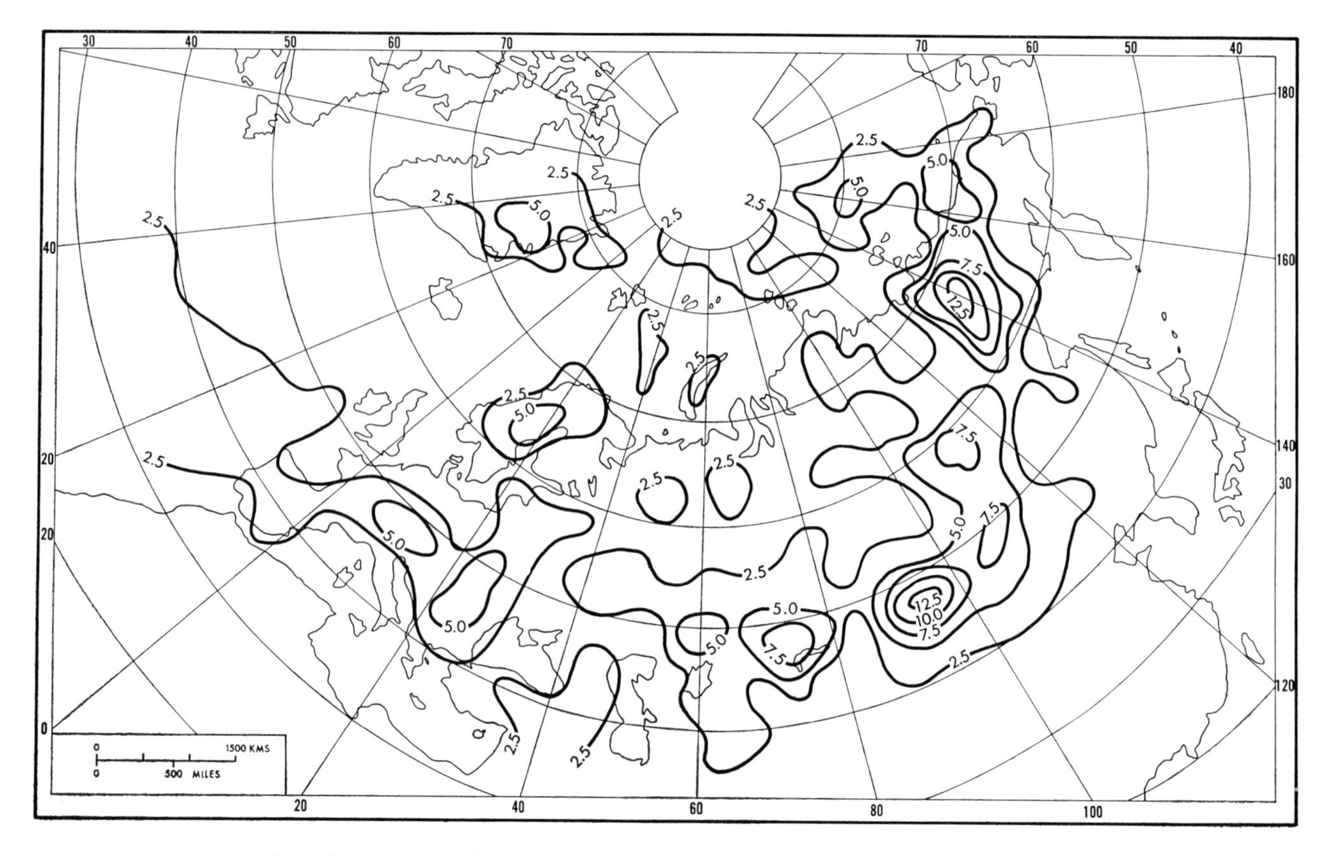

Fig.2-16. Percent of all days with anticyclonic centers positioned within 5° grid sections, January. (After BULINSKAYA, 1963.)

the Asiatic maximum during the winter. The westward extension of the Asiatic maximum, or "great axis," is apparent as far west as the middle Don River, and an eastward extension of the Azores high from the Atlantic across much of Europe is prevalent throughout most of the western portions of the U.S.S.R. A separate cell of high anticyclonic frequency is centered over the Carpathian Mountains, and in the northwest Scandinavia acts as a significant center for high pressure systems.

Summer

During spring the frequency of cyclonic storms in the Barents Sea decreases and the area of maximum frequency shifts eastward to the Ob Gulf and then southeastward in early summer into interior central Siberia south of the Taymyr Peninsula as the interior of the land mass warms rapidly under intense and prolonged insolation (Fig.2-17). Frequencies of cyclones over the Black and Caspian seas diminish significantly. In the Far East the Aleutian low weakens as a great increase in cyclonic frequency develops in the Amur Valley. In general, cyclones are more evenly distributed across the land mass during the summer than during the winter when they are more concentrated along the peripheries of the continent, and new centers of activity form over broad areas of central European Russia and western Siberia. During summer the greatest cyclonic frequency anywhere in the country is found over the upper and middle reaches of the Amur River where the

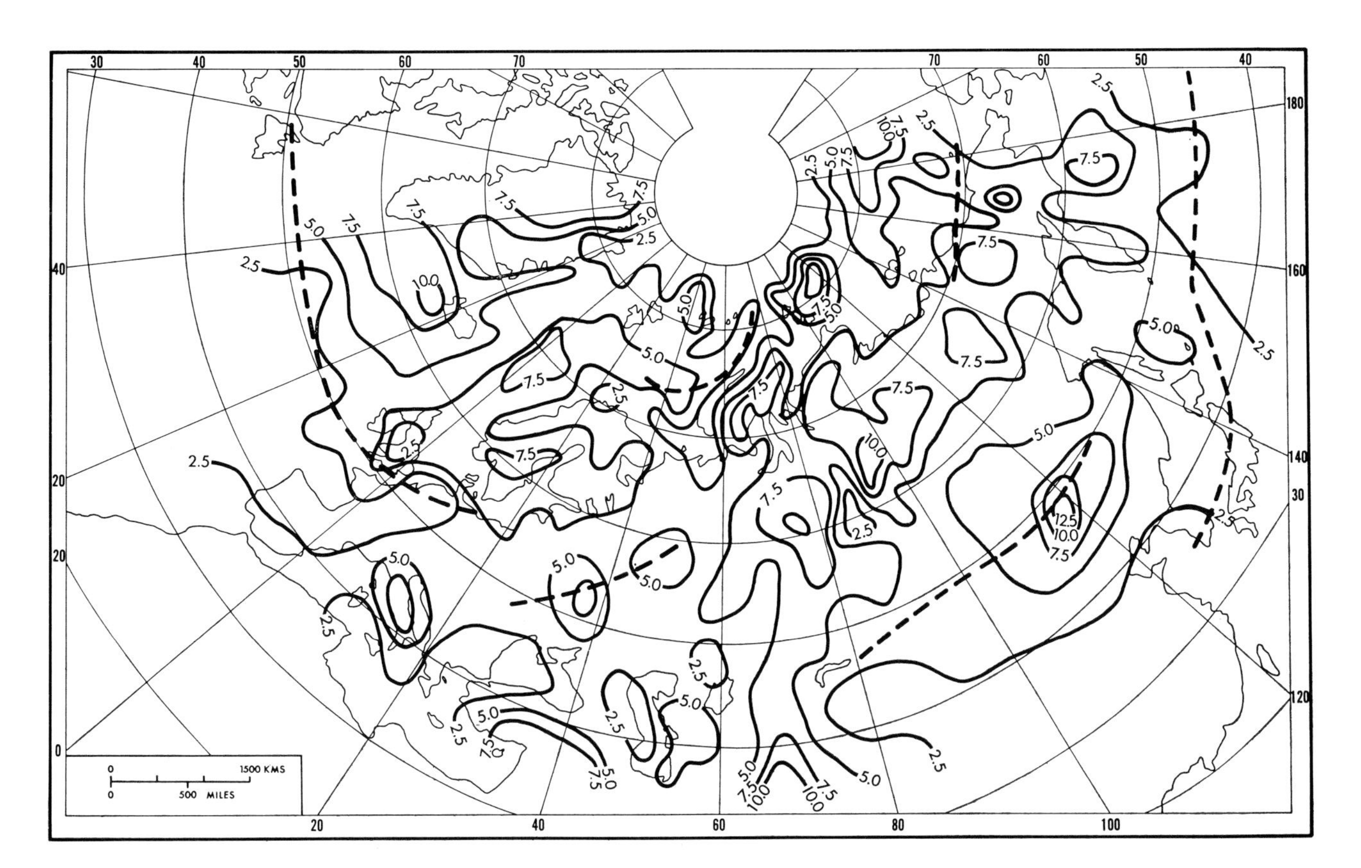

Fig.2-17. Percent of all days with cyclonic centers positioned within 5° grid sections and average frontal positions, July. (Compiled from Bugaev and Dzhordzhio, 1940, Bulinskaya, 1963, and Borisov, 1970.)

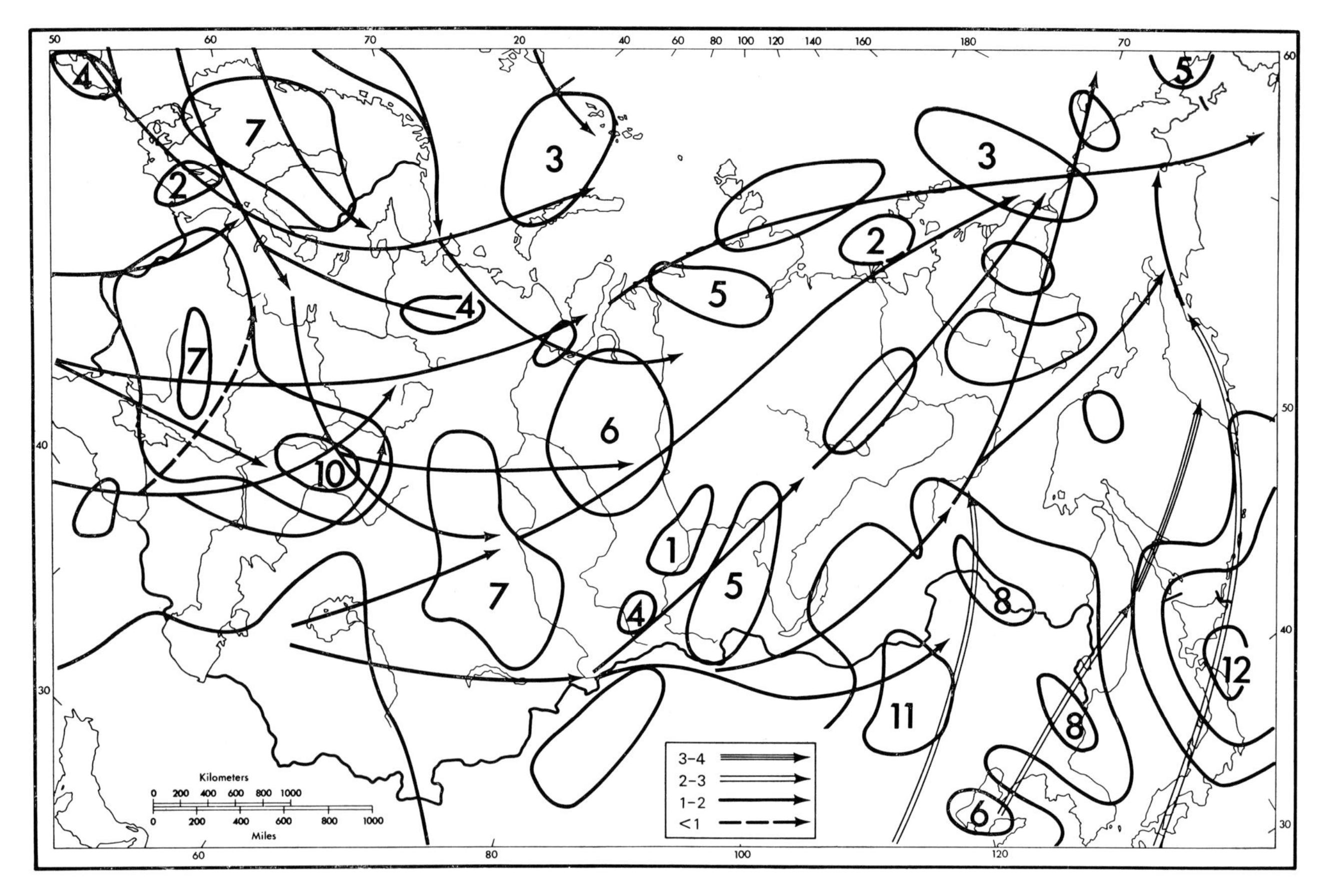

Fig.2-18. Frequencies of cyclogenesis and main routes of cyclones, July. Cases during 20 years within 5° grid sections. (After Klein, 1957, and Borisov, 1970.)

eastern segment of the Polar front becomes active over the land mass rather than over the ocean as is the case during winter.

During summer more areas within the interior of the U.S.S.R. become significant centers for cyclogenesis than is the case during winter when so many of the cyclones affecting the country enter from the outside. In general the most frequent cyclogenesis takes place in the areas which show the greatest frequency of cyclonic centers (Fig.2-18). During summer the area of high cyclonic incidence in the middle Amur Valley shows the greatest cyclogenesis activity of any part of the country, and in the west the broad region of increased cyclonic activity in central European U.S.S.R. is also marked by a high cyclogenesis region which stretches all the way from the western borders eastward to the southern Urals. Western Siberia also shows high cyclogenesis, as does north-central Kazakhstan. It appears that throughout much of the country, except for the northwest and the extreme southwest, there is as much or more cyclogenesis going on in summer as in winter. Cyclone frequency is greater in summer than in winter over much of the country except the northwest, the extreme south, and spots along the Arctic coast (Fig. 2-19). This is true of deep cyclones as well as of all cyclones. Bulinskaya's (1963) maps show that throughout much of the interior of the country, stretching all the way from the western border through eastern Siberia, summer has roughly twice as many cyclonic centers as winter which are enclosed by at least 2 isobars.

The areas of greatest cyclonic frequency in July mark the areas of most active fronts. The

Fig.2-19. Differences in cyclonic frequencies, January and July. (Compiled from Figs. 2-11 and 2-16.) Positive values denote a July maximum, negative values a January maximum.

greatest frontal activity by far during the summer is associated with two segments of the Polar front. An eastern segment on the average lies along a west-southwest–east-northeast zone through the Mongolian Peoples Republic, Manchuria, and adjacent parts of the Soviet Union. A western segment is oriented southwest–northeast across south-central European U.S.S.R. (Fig.2-17). Cyclones forming along these two segments of the Polar front generally move northeastward (Fig.2-18). Those forming in south-central European U.S.S.R. generally end up in the Ob Estuary, while those forming along the Mongolian front move either east-northeastward across the Sea of Okhotsk or northeastward across the entire length of the Soviet Far East into the Chukchi Peninsula where they may be rejuvenated along a segment of the Arctic front which lies along the northern coast at this time of the year.

The Arctic front does not affect the Soviet Union in summer nearly as much as it does in winter. In the west its position is usually well north of the Arctic coast, but occasionally it swings as far south as the central part of European U.S.S.R. Cyclonic storms are fairly active along it at this time of the year, and many of them move almost directly west–east near the Arctic coast into the Ob Estuary. But they are not as frequent or as intense as they are in winter.

Anticyclones move into the country during the summer primarily from the Arctic and also from the west along the southern fringes of the country (Fig.2-20). Both of these systems tend to stagnate in the southern portions of the country where they maintain the strength of the high pressure ridge that protrudes eastward from the Azores high through

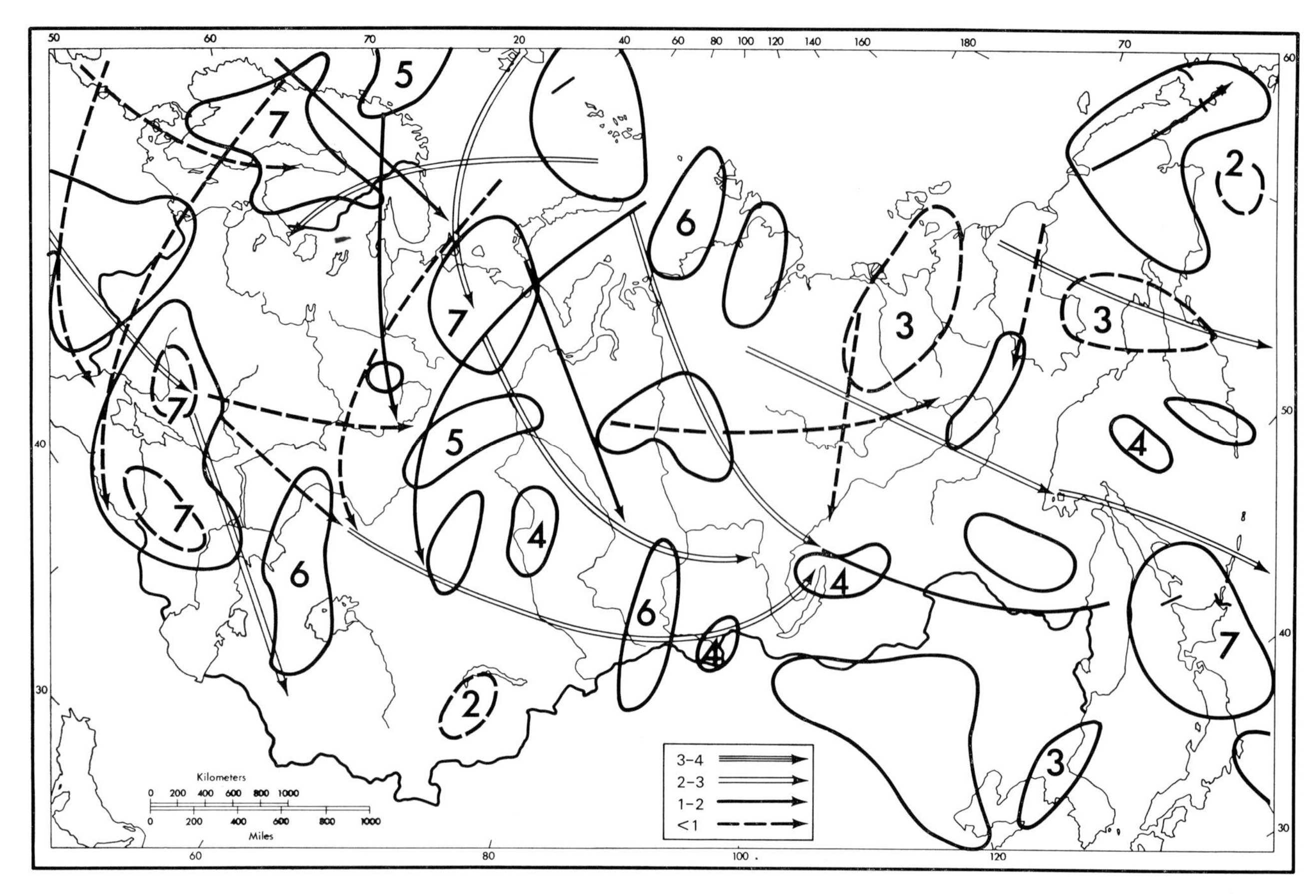

Fig.2-20. Frequencies of anticlogenesis and main routes of anticyclones, July. Cases during 20 years within 5° grid sections. (After KLEIN, 1957, and BORISOV, 1970.)

the Ukraine and Kazakhstan as far east as Lake Baykal. East of the Yenisey anticyclones from the Arctic proceed southeastward across the eastern extremity of the U.S.S.R. and off the Pacific coast to the sea. The regions of greatest anticyclogenesis within the Soviet Union during summer lie in the south around the Black and Caspian seas, in a strip across the mid-section of the country from the middle Urals to Lake Baykal, and along the Arctic coast of the Barents and Kara seas. There appears to be more anticyclogenesis throughout much of the country during July than during January. This is particularly true in the Black Sea–Caspian Sea area and in the Arctic.

The July map of anticyclone occurrence is much more nondescript than is the January one. No counterpart of the winter Asiatic maximum exists in July. Relatively high frequencies tend to occur in a discontinuous zone across the southern portions of the country all the way from the western border to Transbaykalia, and another zone of high frequency occurs along the Arctic coast generally over the water. But none of these frequencies is anywhere near as high as those in the area of the Asiatic maximum in winter, and most of the country during July experiences only scattered occurrences (Fig.2-21). Sea level air pressures reach their highest values in summer in the Central Asian and west Siberian areas where readings of 1,030–1,035 mbar have been observed. This compares to winter values as high as 1,075 mbar southwest of Lake Baykal and 1,070 mbar in areas around the Altay Mountains and the lower Ob Valley.

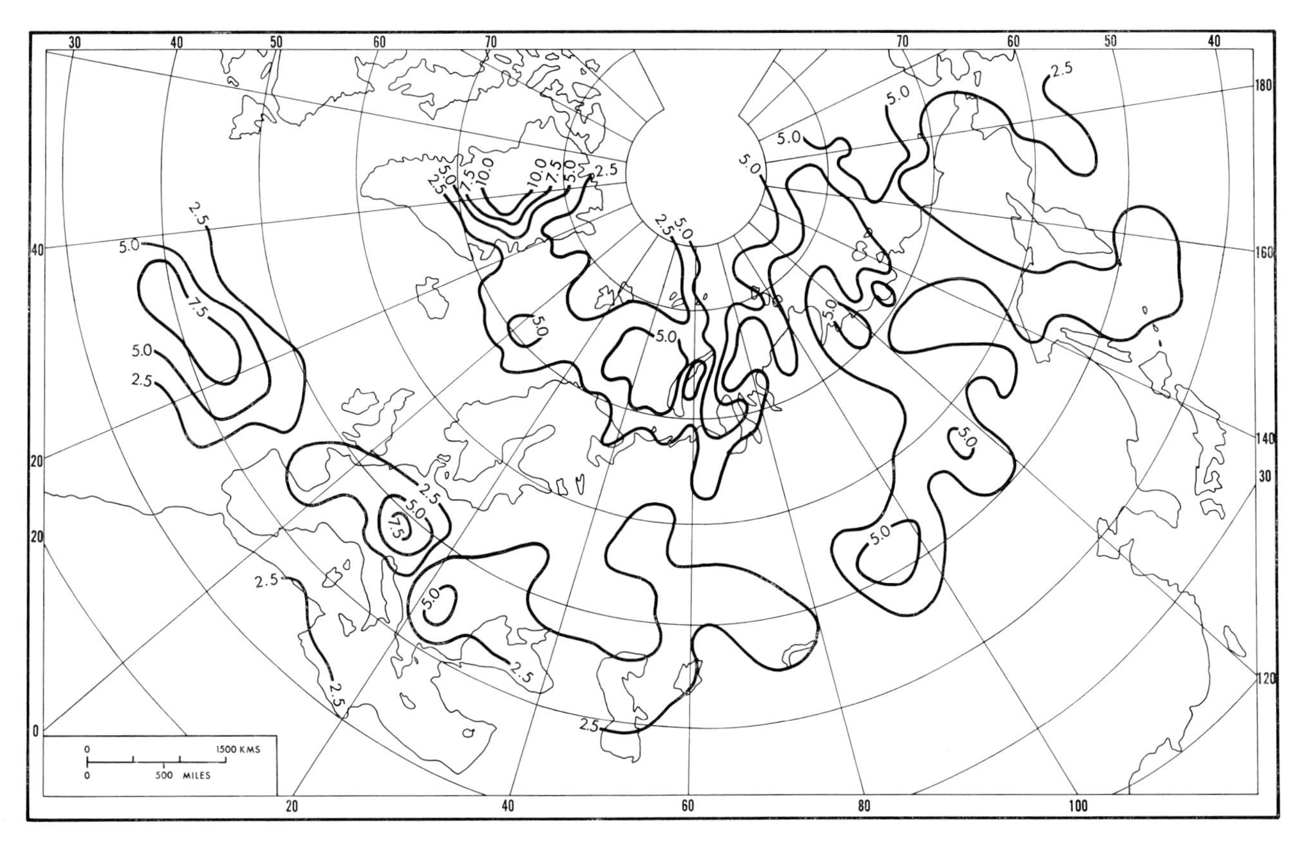

Fig.2-21. Percent of all days with anticyclonic centers positioned within 5° grid sections, July. (After BULINSKAYA, 1963.)

Air masses and regionalization

For the sake of convenience, a discussion of the synoptic climatology of the Soviet Union must be done on a regional basis, since the country is so large and since rather different combinations of things take place in different parts of it. True, it is difficult to single out any one portion of the country and treat it independently of all the rest, since atmospheric circulation in one part of the country generally has some effect on atmospheric circulations in other parts. However, certain peripheral areas particularly, such as the Far East, Central Asia, and the Caucasus, experience considerably different types of weather than does the bulk of the country across the broad plain of eastern Europe and western Siberia. Within the plain itself there are significant differences, but different conditions are not bounded by sharp lines. Climatic changes take place gradually across broad transition zones, and such phenomena as fronts and cyclonic storms transgress these zones from one region to another. Such a continuum is difficult to subdivide. Nevertheless, it is expedient to divide the plain into a number of parts so that each can be dealt with in detail without becoming overwhelmed by the breadth of the area under discussion.

The country will be broken into gross regions which roughly reflect the dynamics of the atmosphere. In part, these regions will be based on the paths frequented by cyclonic storms. Since many of these paths, particularly in Asiatic Russia, run from southwest to northeast, the boundaries of the regions will run somewhat in this direction also. However, the regions cannot be based entirely on the paths of cyclonic storms, since the same storm will bring one type of weather association to an area on one side of the storm and an entirely different type of weather association on the other side of the storm. This is true because cyclonic storms are convergence zones toward which air masses are moving from different directions from peripheral areas. A storm moving northeastward from the Black Sea to the Ob Estuary will pull in continental Arctic air from the northeast or maritime Arctic air from the north into the northwestern side of the storm in northern European Russia, while on the southeastern side of the storm a southerly or southwesterly flow of air will bring in continental temperate or other air masses from the south into western Siberia. Thus, the temperature, humidity, cloudiness, precipitation, and the entire weather type will be very different between northern European Russia and western Siberia when the two are being affected by different parts of the same cyclonic system. Therefore, a cyclonic path may be more of a dividing line than a unification line as far as regions based on weather types are concerned.

About the only way to arrive at a valid regionalization based on synoptic conditions is to draw boundaries between frequencies of occurrence of different types of air masses. BORISOV (1965) has conveniently worked out data on air mass frequencies and has divided the Soviet Union into regions on this basis. His work seems to incorporate much that has been done in the Soviet Union on air mass distributions and is the best summary available. Therefore, with some modifications, this system of regionalization will be used in the following chapters to discuss the occurrences of weather and climate across the country. Modifications are necessary in order to depict adequately areas such as the Transcaucasus, which in Borisov's regionalization are simply thrown into the "European south", which includes the Black Sea steppes, lower Volga, North Casp'an Lowland, and North Caucasian Foreland, as well as the Caucasus Mountains and the Transcaucasus, very disparate regions climatically and otherwise (Fig.2-22).

The air mass climatology of the Soviet Union, particularly of the European part, is quite complicated, because it is affected by so many air masses, many of which are not too distinguishable one from another. In addition, temperature and humidity conditions among air masses may reverse themselves seasonally in a given region. Since the country is so large and relatively compact, the dominant air mass in most places is of relatively local continental origin. Such air masses are made up of air which has invaded the country from various directions and stagnated over the interior until modification processes have destroyed much of the original characteristics. In most cases, the Soviets refer to this air as continental temperate air. In much of the rest of the world it would be known as continental Polar air, but this is really a misnomer, as the Soviets point out, since the air does not originate in the polar regions, and in the case of the Soviet Union this type of air does not move into the country from higher latitudes, as it does, for instance, in the United States. If one keeps in mind that "temperate" applies to mean conditions and allows for widely varying extremes, the use of the word to describe air masses of middle latitude origins is quite valid.

Although continental temperate air varies somewhat from one end of the Soviet Union to the other, it is generally cold and dry in winter and relatively warm and dry in summer. In winter it emanates from the stagnant core of the Asiatic high and spreads in all directions from the great axis which extends westward through the southern plain. In summer, it relates to the relatively warm stable air of the eastern projection of the Azores anticyclone or originates locally from a transformation of air masses moving in from the

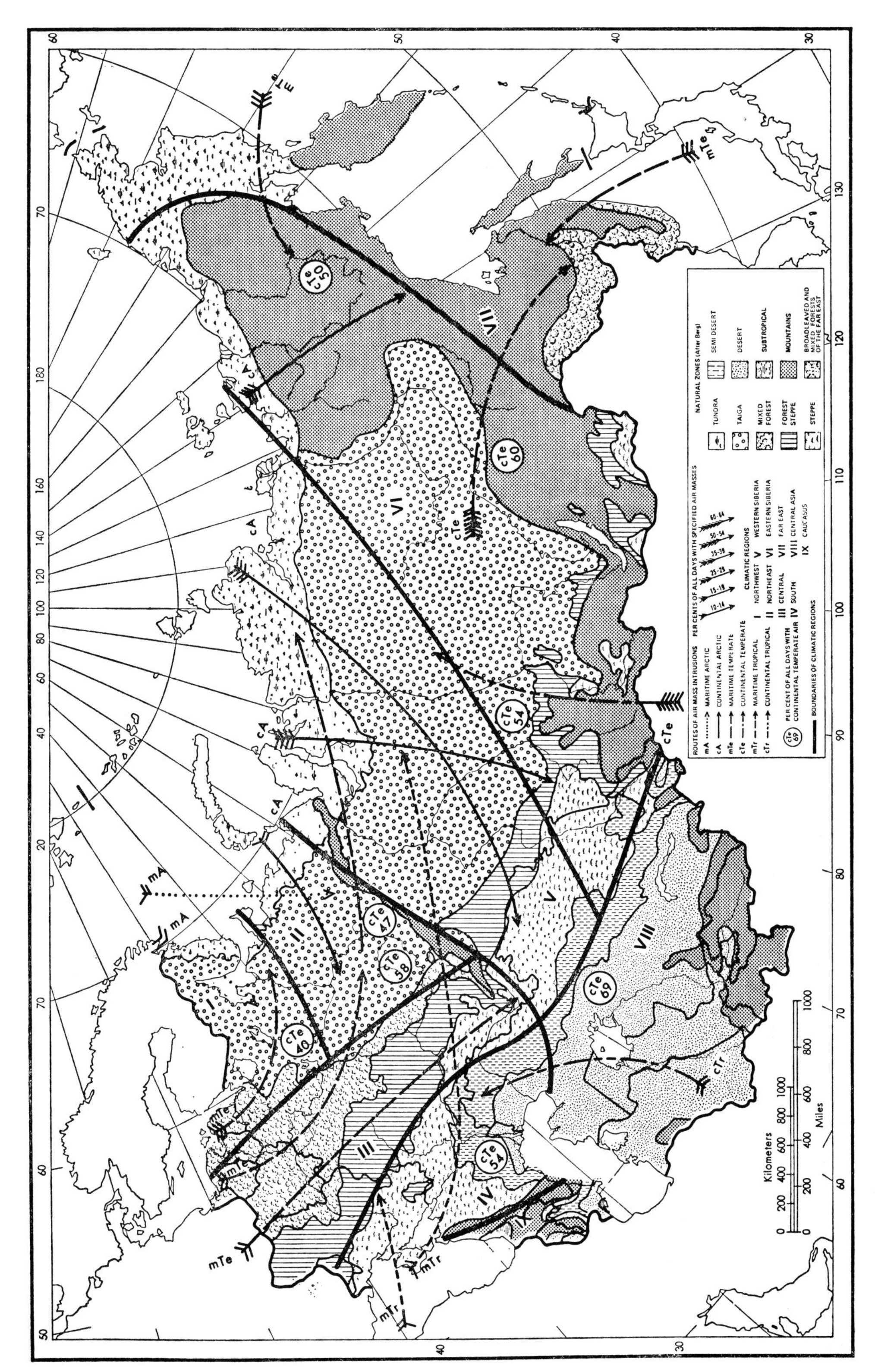

Fig.2-22. Air mass regions of the USSR. (Modified from BORISOV, 1970.)

peripheries of the country. Its frequency of occurrence as compared to all air masses varies from about 35% in the northwest along the Baltic to as much as 65% or more in continentally isolated regions of Kazakhstan and Central Asia. Since continental temperate air dominates all of the country most of the time, Borisov's regionalization is based on the second most dominant air mass in each region.

In the western part of the country maritime temperate air enters the Soviet Union from the Baltic Sea and eastern Europe. In the north it occupies the Baltic republics as much as 30% of the time, but its frequency decreases southward to less than 15% at Odessa on the Black Sea coast. In the western part of the country maritime temperate air is comparatively mild, warm in winter and cool in summer, with a rather high humidity content and a great deal of cloudiness. These conditions modify eastward across the Russian plain until the air becomes indistinguishable from continental temperate air somewhere between the 40th and 60th meridians east. However, under correct circulation conditions in the southern parts of cyclonic storms which are moving northeastward across central Siberia, maritime air from the Atlantic has been known to remain distinguishable as far east as Lake Baykal where in the wintertime it produces copious snowfall on the northwestern slopes of mountains in Irkutsk Oblast and Buryat A.S.S.R.

Maritime temperate air also enters the east coast of the country from the Pacific, almost exclusively in summer when the Pacific monsoon is flowing onto the land. Mountainous topography in the Soviet Far East quickly modifies the maritime air into continental temperate air as it penetrates the continent. But in middle and late summer the coastal regions of the Soviet Far East as well as Sakhalin Island and Kamchatka Peninsula are dominated by maritime temperate air with its associated cool temperatures, high humidity, and extremely cloudy and foggy conditions. The Kuril Islands, of course, are dominated by this air all year round.

Maritime Arctic air is limited primarily to the north European part of the country which is bordered by the open waters of the Barents Sea. This air enters the continent with a northwesterly flow, either in the rear of cyclonic storms or in well developed anticyclones that sometimes form over Scandinavia. It is most active during transitional seasons, when it is generally accompanied by fresh winds, much cloudiness, and snowfall. During the summer, it occurs further east along the Arctic coast as open water appears between the continent and the pack ice to the north. During this season it commonly occupies the tundra area from the Laptev Sea eastward to the Bering Straits and accumulates in the Kolyma and other river valleys along the northern side of the various mountain ranges in this area. At this time of year it is associated with instability showers. The frequency of maritime Arctic air ranges from about 23% of the entire year in northwest European Russia to less than 1% in eastern Siberia and the Soviet Far East.

Continental Arctic air reaches the northeastern part of European Russia and much of western Siberia in association with northeastern flows around the northern sides of cyclonic storms which are progressing northeastward across the central part of the country. This air also penetrates Yakutia and the Far East from the north. It is most frequent in winter when the Arctic seas are frozen right up to the continent. But it also occurs in summer when it flows off the Arctic ice pack without much modification from the patches of open water along the coast. Since the winds in summer along the Arctic coast are generally from the north much Arctic air penetrates the continent at this time of year. The frequency of occurrence of continental Arctic air varies from about 49% of the time in eastern Siberia to less than 5% in the southern Ukraine. It is generally cold, dry, and relatively stable for the season. In winter it is the coldest air to reach central European Russia. But in summer the heating from the land may cause instability in the lower layers, even in the tundra regions of the north, where frequent instability showers occur in association with this type of air. Continental Arctic air masses generally are shallow, varying from as much as 2,200 m in winter to only a few hundred meters in summer.

Continental tropical air forms over the Middle Asian and Kazakh regions during summer as well as over the southeastern part of European Russia. It also forms in small pockets in the eastern Transcaucasus, and farther west it reaches the Ukraine from Asia Minor, Arabia, North Africa, and the Balkans. During autumn it frequently flows into southern European U.S.S.R. from Central Asia. In modified form, continental tropical air also reaches the southwestern part of the Soviet Union in winter from North Africa, Arabia, and Asia Minor. In the southern part of the Soviet Far East continental tropical air is observed only in the summer when it moves northward in the warm sector of cyclonic storms along the Mongolian front. Its source region is Mongolia and north-central China. Only occasionally does it penetrate north of the Chinese–Soviet border. This continental tropical air in the Far East is much drier than are the air masses arriving in the west from Asia Minor. Except in the Turanian lowland, continental tropical air is rather rare in the Soviet Union, but in the Central Asian desert it occurs more than 25% of the time. In the southern Ukraine it occurs about 9% of the time, and there are rare occurrences even in the European north which account for about 1% of all air masses.

Maritime tropical air arrives in the U.S.S.R. at all seasons via the Mediterranean and Black Sea, but always in a modified form. Nevertheless, it is distinctive from other air masses in southern European U.S.S.R., being generally more humid and less stable. It is usually brought in from a southwesterly direction in the warm sectors of cyclonic storms which have formed along the east European branch of the Polar front. Occasionally in summer maritime tropical air from the Pacific may be identified along the coast of the Soviet Far East, but it is completely absent in this region in winter.

The general flow of the six types of air masses is shown in Fig.2-22. Table 2-I shows the frequencies of occurrences of these air masses at central points in the eight major regions outlined by Borisov. Boundaries of these regions are also shown on Fig.2-22, but in modified form. The expansive plain of European U.S.S.R. and western Siberia will be treated as a single gross region, because of its indivisible nature with respect to synoptic processes, and subregions will be used to depict separate sections of it which differ significantly one from another in air mass frequencies and weather types. The Caucasus will be treated as a separate region instead of as only a part of an extensive region which Borisov labeled simply "the South." One can hardly imagine that conditions south of the great Caucasus can be depicted by data recorded at Dnepropetrovsk in the south-central Ukraine, as Borisov has implied.

Boundaries between regions in Siberia and the Far East have been given even a more southwest–northeast orientation than Borisov did, to fit more closely the orientation of major routes of cyclonic storms in these areas, and the elongated region of the Far East has been subdivided into two parts to facilitate discussion of the disparate extremities of that region. Since the regions in Fig.2-22 are not identical with those for which Borisov computed air mass frequencies, such data cannot be shown for all regions in Fig.2-22.

The lines and arrows depicting air mass movements and regions on Fig.2-22 are underlain by a distribution of patterns that depicts the type of surface cover, which is very important in all considerations of heat and moisture exchange and general air mass modification. These are BERG's (1950) natural zones, which are based primarily on natural vegetation. When considering radiational exchanges and air mass modifications during the winter, one must keep in mind that much of the surface in the U.S.S.R. has some snow cover, and therefore the surface cover depicted by Berg's natural zones on Fig.2-22 should be supplemented with the snow cover map (Fig.9-41). A somewhat more detailed map of resultant climatic types which will relate to surface cover will be presented in the last chapter of the book after a consideration has been made of all regions and all separate climatic elements. But interactions between climate and the earth's surface operate in both directions, and therefore it is deemed necessary to introduce early some concept of the surface features of the Soviet Union as causative factors in weather phenomena.

The country will be divided, then, into five primary regions of a climate genesis nature for detailed discussion of synoptic and dynamic aspects of climate across the country. These discussions will be followed by descriptions of the distributions of separate weather elements across the entire country so that the reader can get an integrated picture of both the causes and the results. The five regions as outlined on Fig.2-22 are large and do not make allowances for microclimatic effects, or even the macroclimatic effects of mountains. Within each region containing areas of high relief, some attention will be given to strong local variations in climate. In preparation for that, a map of so-called "microclimatic correction regions" is presented in Fig.2-23. This has been devised by the well-known Soviet microclimatologist, I. A. Gol'tsberg.

TABLE 2-I

CHARACTERISTICS OF THE CLIMATIC REGIONS OF THE U.S.S.R.
(After BORISOV, 1965)

Regions (recording stations)	Frequency of different air-masses (%)						Frequency (%) of days with		Number of fronts per year			
	cTe	mTe	cA	mA	cTr	mTr	depressions	anti-cyclones	cold	warm	occluded	total
Northwestern (Petrozavodsk)	36.2	27.2	12.0	23.4	1.0	0.2	38	23	77	68	110	255
Northeastern (Syktyvkar)	46.7	13.7	19.0	19.0	1.6	0.1	17	41	76	62	81	219
Central (Moscow)	52.6	20.7	8.7	12.1	5.4	0.5	27	51	97	74	131	302
Southern (Dniepropetrovsk)	53.7	14.3	5.0	6.5	9.0	11.5	14	48	99	76	88	263
West Siberian (Tobol'sk)	52.4	3.8	33.3	5.6	4.9	0	29	49	170	149	181	500
East Siberia (Yakutsk)	51.0	0	48.6	0	0.4	0	19	42	90	61	110	261
Far East (Vladivostok)	46.3	25.8	24.0	0.1	0.9	2.9	26	33	65	33	86	184
Turano-Kazakhian (Samarkand)	63.6	0	8.8	0.1	27.5	0.1	27	40	96	53	37	186

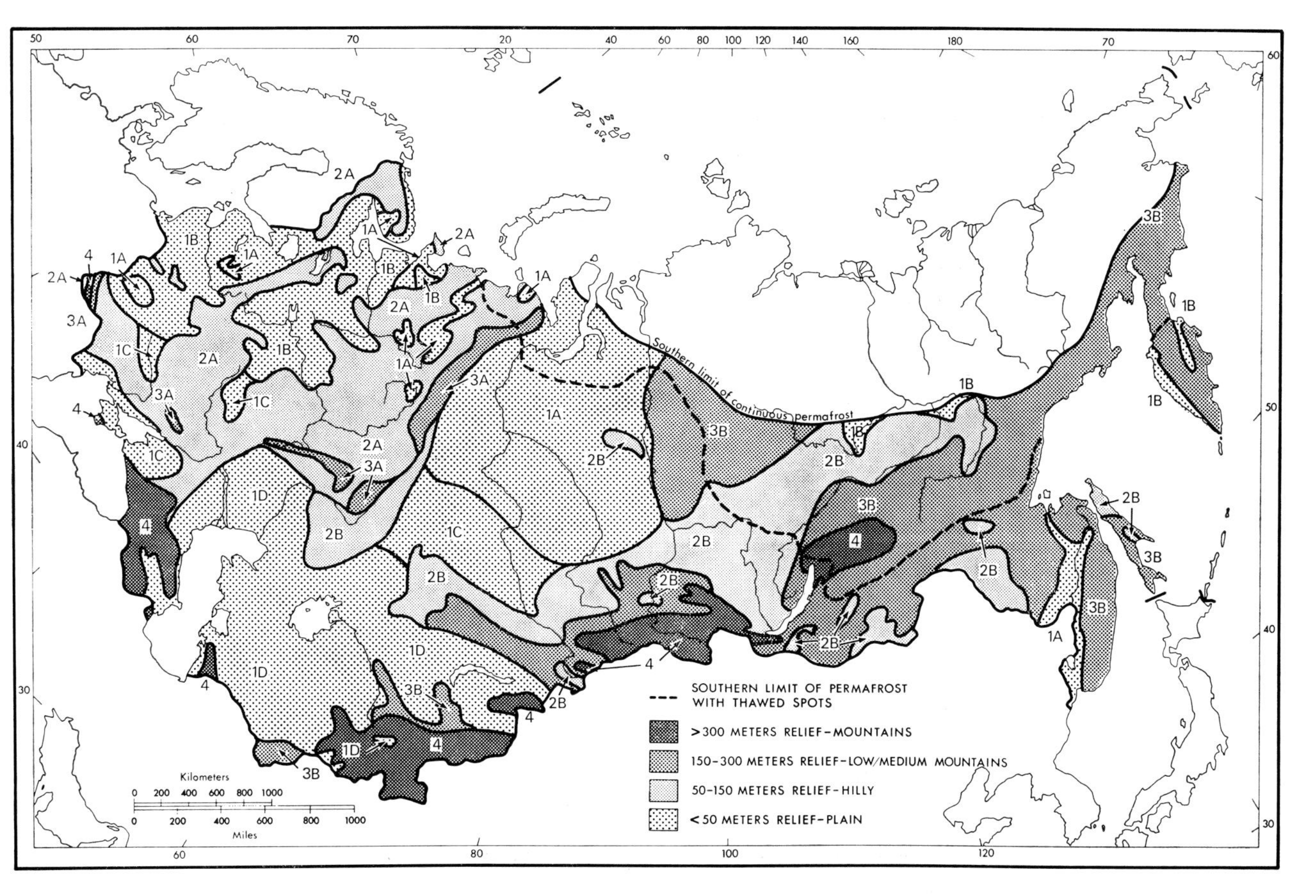

Fig.2-23. Microclimatic correction regions. Subregions: *1A*-waterlogged lowlands; *1B*-undulating plains with variegated soils; *1C*-plains with small soil differences (chernozems); *1D*-plains in deserts, semi-deserts, and dry steppes with gray and chestnut soils; *2A*-hilly relief of European U.S.S.R.; *2B*-hilly relief of Asiatic U.S.S.R.; *3A*-low mountain relief of European U.S.S.R.; *3B*-low and medium mountains; *4* mountains. (After GOL'TSBERG, 1967.)

References and further reading

ANONYMOUS, 1952. Normal Weather Charts for the Northern Hemisphere. *U.S. Weather Bur. Tech. Pap.*, 21: 74 pp.

ANONYMOUS, 1961. *Aeroklimaticheskiy Atlas Severnogo Polushariya* (Aeroclimatic Atlas of the Northern Hemisphere). I. *Pressure Field and Wind.* Gidrometeoizdat, Leningrad.

BERG, L. S., 1950. *Natural Regions of the USSR.* Macmillan, New York, N.Y., 436 pp.

BORISOV, A. A., 1965. *Climates of the USSR.* Aldine, Chicago, Ill., 255 pp.

BORISOV, A. A., 1970. *Klimatografiya Sovetskogo Soyuza* (Climatography of the Soviet Union). Leningrad University, Leningrad, 310 pp.

BUGAEV, V. A., 1936. O svyazi tsiklonicheskikh traektoriy s polozheniem arkticheskogo fronta zimoy (On the relation between cyclone trajectories and the position of the Arctic front in winter). *Meteorol. Gidrol.*, 8: 11–18.

BUGAEV, V. A. and DZHORDZHIO, V. A., 1940. Klassifikatsiya vozdushnykh mass SSSR (Classification of air masses in the USSR). *Meteorol. Gidrol.*, 12: 33–45.

BULINSKAYA, N. A., 1963. *Atlas Baricheskikh Kharakteristik Tsiklonov i Antitsiklonov* (Atlas of pressure characteristics of cyclones and anticyclones). Akad. Nauk SSSR, Moscow, 194 pp.

GOL'TSBERG, I. A. (Editor), 1967. *Mikroklimat SSSR.* Gidrometeoizdat, Leningrad. (Translated as *Microclimate of the USSR.* Israel Program for Scientific Translations, Jerusalem, 1969, No.5345, 236 pp.)

GUTERMAN, I. G. (Editor), 1963. *Aeroklimaticheskiy Atlas Kharakteristik vetra Severnogo Polushariya*

(Aeroclimatic atlas of wind characteristics in the Northern Hemisphere). Nauchno-issledovatel'sky Institut Aeroklimatologii, Moscow.

GUTERMAN, I. G. (Editor), 1963. *Klimaticheskiye Kharakteristiki vetra v Isobaricheskikh Poverkhnostyakh* (Climatic characteristics of winds at isobaric surfaces). Nauchno-issledovatel'sky Institut Aeroklimatologii, Moscow.

IL'INSKIY, O. K., 1959. K voprosy o severnoy vetvi zonal'nogo perenosa nad Aziey v zimniy period (Toward the question of the northern branch of zonal flow across Asia during winter). *Meteorol. Gidrol.*, 8: 13–14.

KALACHIKOVA, V. S., 1973. Blocking anticyclones over Siberia during the cold half of the year and the possibility of forecasting them. *Meteorol. Hydrol.*, 2: 61–67.

KAMINSKIY, A. A., 1932. *Klimat SSSR, Chast II. Davlenie vozdukha i veter v SSSR, vyp. 1 i 2. Davlenie vozdukha po mesyachnym srednim i napravlenie vetra v SSSR* (Atlas of average monthly surface air pressures and wind directions in the USSR). Gl. Geofiz. Obs., Leningrad, 44 plates.

KLEIN, W. H., 1957. Principal tracks and mean frequencies of cyclones and anticyclones in the Northern Hemisphere. *U.S. Weather Bur. Res. Pap.*, 40: 60 pp.

LAHEY, J. F., BRYSON, R. A. and WAHL, E. W., 1958. *Atlas of Five-Day Normal Sea-Level Pressure Charts for the Northern Hemisphere.* University of Wisconsin Press, Madison, Wisc.

LIR, E. S., 1936. Tipy sezonnykh tsirkulyatsiy atmosfery nad evraziey i atlantikoy (Seasonal types of atmospheric circulation over Eurasia and the Atlantic). *Meteorol. Gidrol.*, 1: 3–17; 3: 3–9; 4: 7–18; 5: 21–35; 6: 3–15; 7: 14–25.

MAKHOVER, Z. M., 1967. O prichinakh ustoychivogo polozheniya tsentra aziatskogo antitsiklona nad Mongoliey (On the causes for the stable position of the center of the Asiatic anticyclone over Mongolia). *Nauchnoissled. Inst. Aeroklimatol. Tr.*, 38: 1–131.

MAKSIMOV, I. V. and KARKLIN, V. P., 1969. Sezonnyye i mnogoletniye izmeneniya geograficheskogo polozheniya i intensivnosti sibirskogo maksimuma atmosfernogo davleniya (Seasonal and long-term changes in the geographical position and intensity of the Siberian maximum of atmospheric pressure). *Izv. Vses. Geogr. O-vo.*, 101: 320–330.

MAKSIMOV, I. V. and KARKLIN, V. P., 1970. Sezonnyye i mnogoletniye izmeneniya glubniny i geograficheskogo polozheniya aleutskogo minimuma atmosfernogo davleniya za period s 1899 po 1951 g. (Seasonal and long-term changes in the intensity and geographical position of the Aleutian Low from 1899 to 1951). *Izv. Vses. Geogr. O-vo.*, 102: 422–431.

POGOSYAN, KH. P., 1959. *Obshchaya Tsirkulyatsiya Atmosfery* (General circulation of the atmosphere). Gidrometeoizdat, Leningrad, 259 pp.

POGOSYAN, KH. P., 1960. *Struiynye Techeniya v Atmosfere* (Jet streams in the atmosphere). Gidrometeoizdat, Moscow, 182 pp.

POGOSYAN, KH. P. and UGAROVA, K. F., 1959. Vliyanie tsentralnoaziatskogo gornogo massiva na formirovanie struynye techeniy (The influence of the Central Asian mountain massive on jet streams). *Meteorol. Gidrol.*, 11: 16–25.

PUTNINS, P. and STEPANOVA, N. A., 1956. *Climate of the Eurasian Northlands.* Technical assistant to Chief of Naval Operations for Polar Projects, OP-03A3, Washington, D.C., 104 pp.

SHCHERBAKOVA, E. YA. and BRELINA, A. YU., 1956. Izmeneniya teplo- i vlago-soderzhaniya vozdushnykh mass v protsesse transformatsii v umerennykh shirotakh evrazii (Changes in heat and water content of air masses during transformation processes in the middle latitudes of Eurasia). *Gl. Geofiz. Obs. Tr.*, 62: 3–28.

VOYEYKOV, A. I., 1948. *Izbrannyye Sochineniya* (Collected Works). Nauka, Moscow, Vol. I, 751 pp.

WAHL, E. W. and LAHEY, J. F., 1969. *A 700 mb Atlas for the Northern Hemisphere.* University of Wisconsin Press, Madison, Wisc., 147 pp.

ZANINA, A. A., 1938. Materialy dlya klassifikatsii letnikh sezonov (Materials for the classification of summer seasons). *Gl. Geofiz. Obs. Tr.*, 16: 16–39.

Chapter 3

European U.S.S.R. and Western Siberia

Definition of the region

The broad plain of European U.S.S.R. and western Siberia (Fig.3-1) must be considered pretty much as a unit as far as the dynamics of weather are concerned, for the same circulation features generally affect the entire area. This is a region which is characterized by activity generated by the Arctic and Polar fronts, and it stands in stark opposition to eastern Siberia, which, during the winter at least, is dominated by the Asiatic high. During winter, in regard to both cyclonic activity and snow cover the west Siberian plain is more akin to the European plain of the U.S.S.R. than it is to the uplands of eastern Siberia.

Much of the time European U.S.S.R. and western Siberia lie under the eastern limb of an upper level atmospheric trough, so that the predominant route of cyclonic storms across the region proceeds from southwest to northeast (Figs.2-13, 2-18). Since the east–west position of the upper level trough shifts somewhat, its amplitude is somewhat obscured on mean maps (Figs.2-3, 2-4, 2-9, 2-10). It is much more evident on charts for individual days (Fig.3-2). When the upper level trough has large magnitude and short wavelength the movements of surface features commonly prescribe horseshoe-shaped paths southeastward from Scandinavia into central European Russia and then northeastward into western Siberia. Under such conditions segments of the Arctic front swing around the northern Urals as a hinge point in a counter-clockwise fashion (Fig.3-3). Farther south waves generated along the Polar front in the eastern Mediterranean and Asia Minor move northeastward into western Siberia to join cyclonic storms along the Arctic front in the region between the Ob Gulf and the Lena Delta.

The eastern boundary of this region is drawn as a southwest–northeast trending line which generally marks the northwestern edge of the Asiatic high in winter and thus the eastern limits of penetration of cyclonic storms from the southwest at this time of year. This boundary is not at all the traditional one that is usually drawn between western and eastern Siberia, which is usually fixed approximately by the topographic break between the very flat Ob Basin to the west of the Yenisey River and the more rugged rolling uplands to the east of the river. More precisely the boundary between west and east Siberia is generally considered to be the political boundary of the western border of Krasnoyarsk Kray. Compared to that boundary, the present climatic boundary which has been drawn includes in eastern Siberia the entire Altay mountain system and a significant portion of the plain to the west which is dominated by the western extension of the Asiatic high in

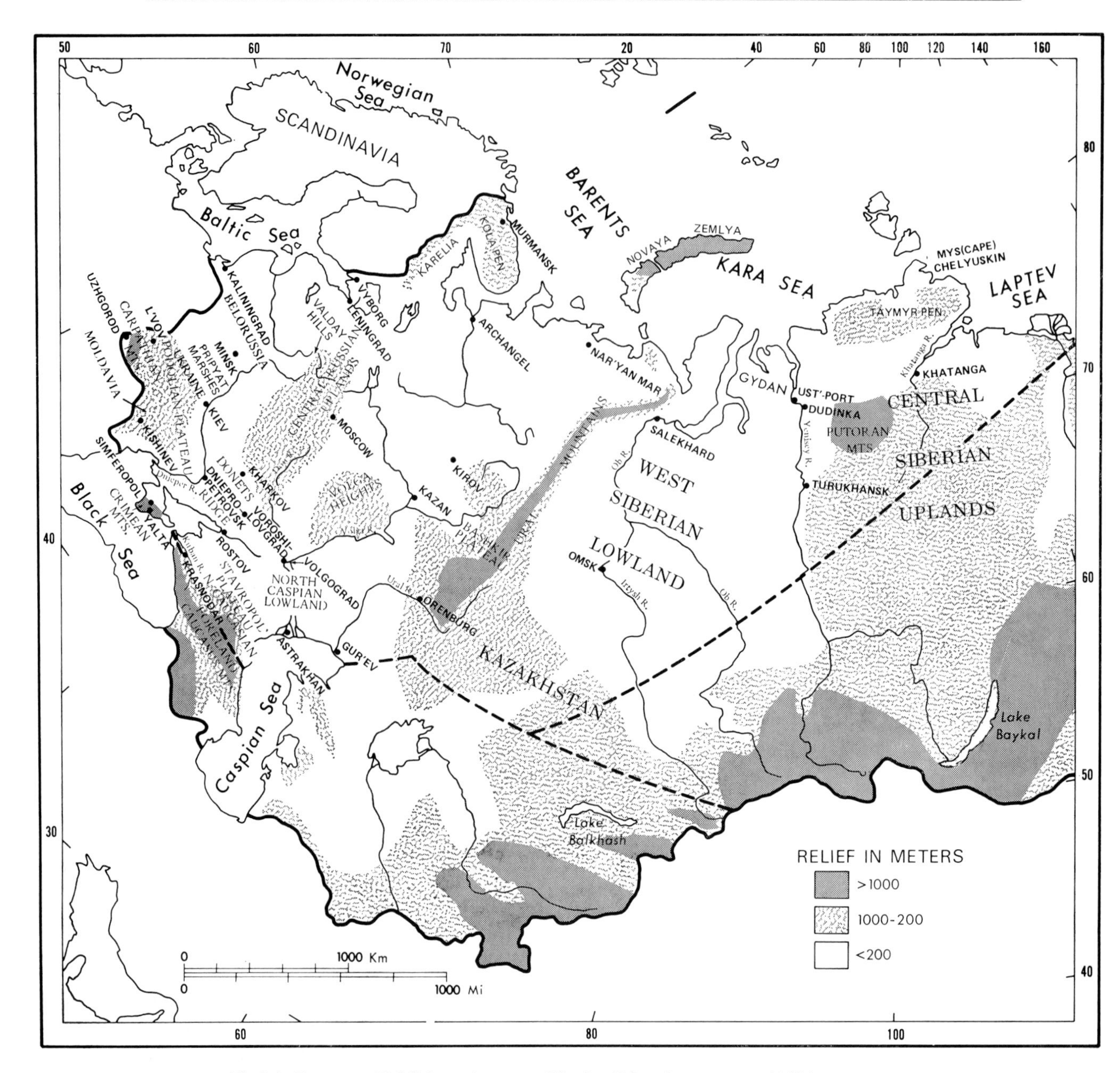

Fig.3-1. European U.S.S.R. and western Siberia. (After ANONYMOUS, 1969.)

winter, and places within western Siberia the northwestern quarter of the Central Siberian Uplands including the entire Taymyr Peninsula and the Arctic coastal lowlands as far east as the Lena Delta and beyond. This violation of the traditional concepts of western and eastern Siberia is justified climatically because during winter northwest of the line is an area which is decidedly the domain of the Arctic front with its attendant cyclonic storms, changeable weather, and considerable winds while southeast of the line lies the domain of the Asiatic high with its relatively calm, cold, clear, consistent weather. A pronounced trough in the upper atmosphere, that on the average becomes established somewhere between the 850- and 700-mbar levels and extends upward unto the atmosphere, is positioned such that in winter cold Arctic air is constantly fed into the area

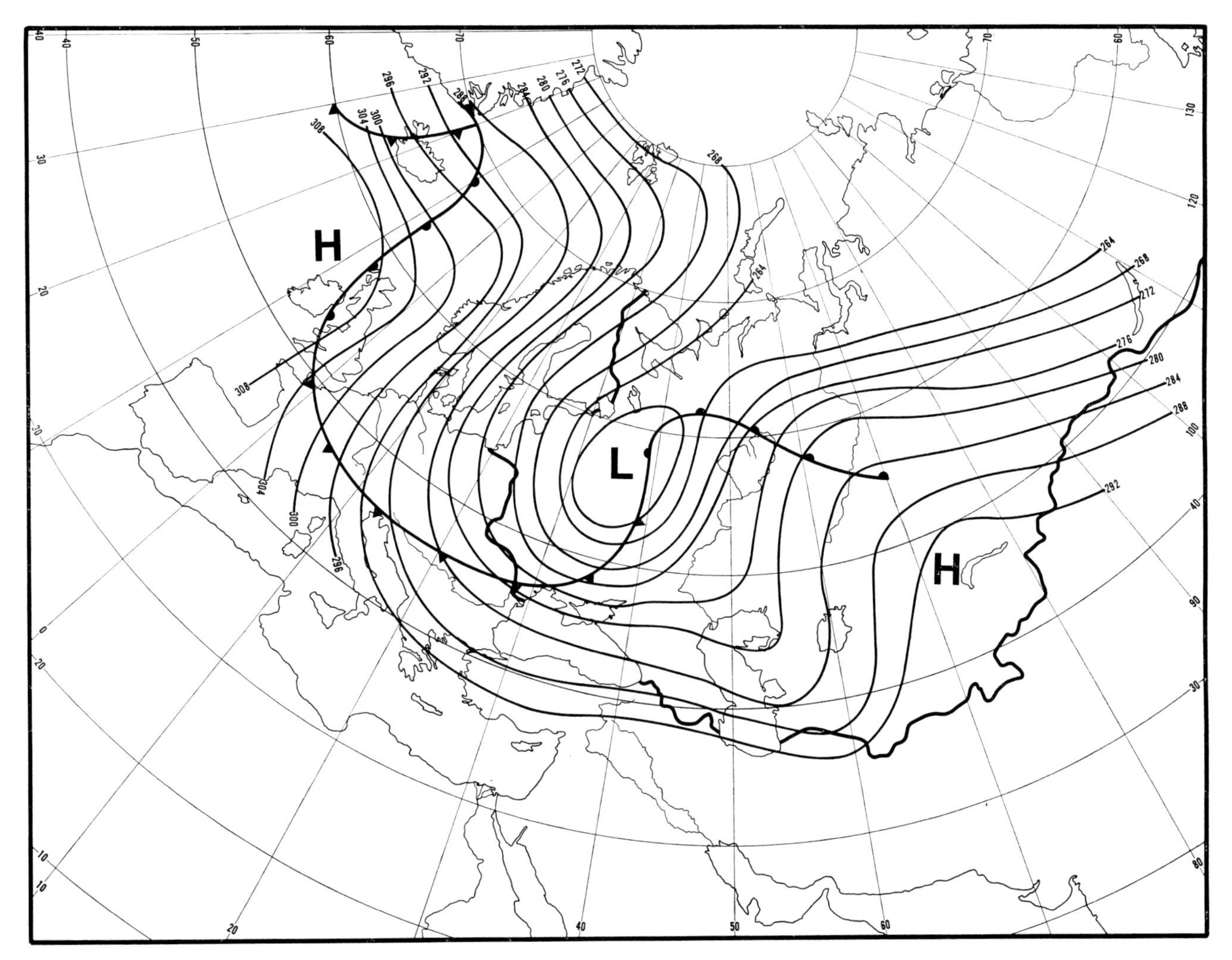

Fig.3-2. 700-mbar contours at 06h00, February 1, 1949, in tens of meters, and underlying surface fronts. (After BUGAEV et al., 1957.)

southeast of this boundary, but to the northwest air is usually brought into the region from the west or even the southwest (Figs.2-2, 2-3, 2-4). Contrasts between the two regions are not so great in summer.

Circulation features

Cyclogenesis and cyclone tracks

European U.S.S.R. and western Siberia lie under the profound influence of two high level planetary frontal zones with their associated jet streams. Cyclonic activity takes place in the lower troposphere along the Arctic and Polar fronts underlying these jet stream zones (Fig.3-4). Fig.3-5 well illustrates the two storm tracks, one in the north and one in the south, separated by a high pressure cell over much of European U.S.S.R. The two tracks are most distinct during a situation such as this when they are separated by a

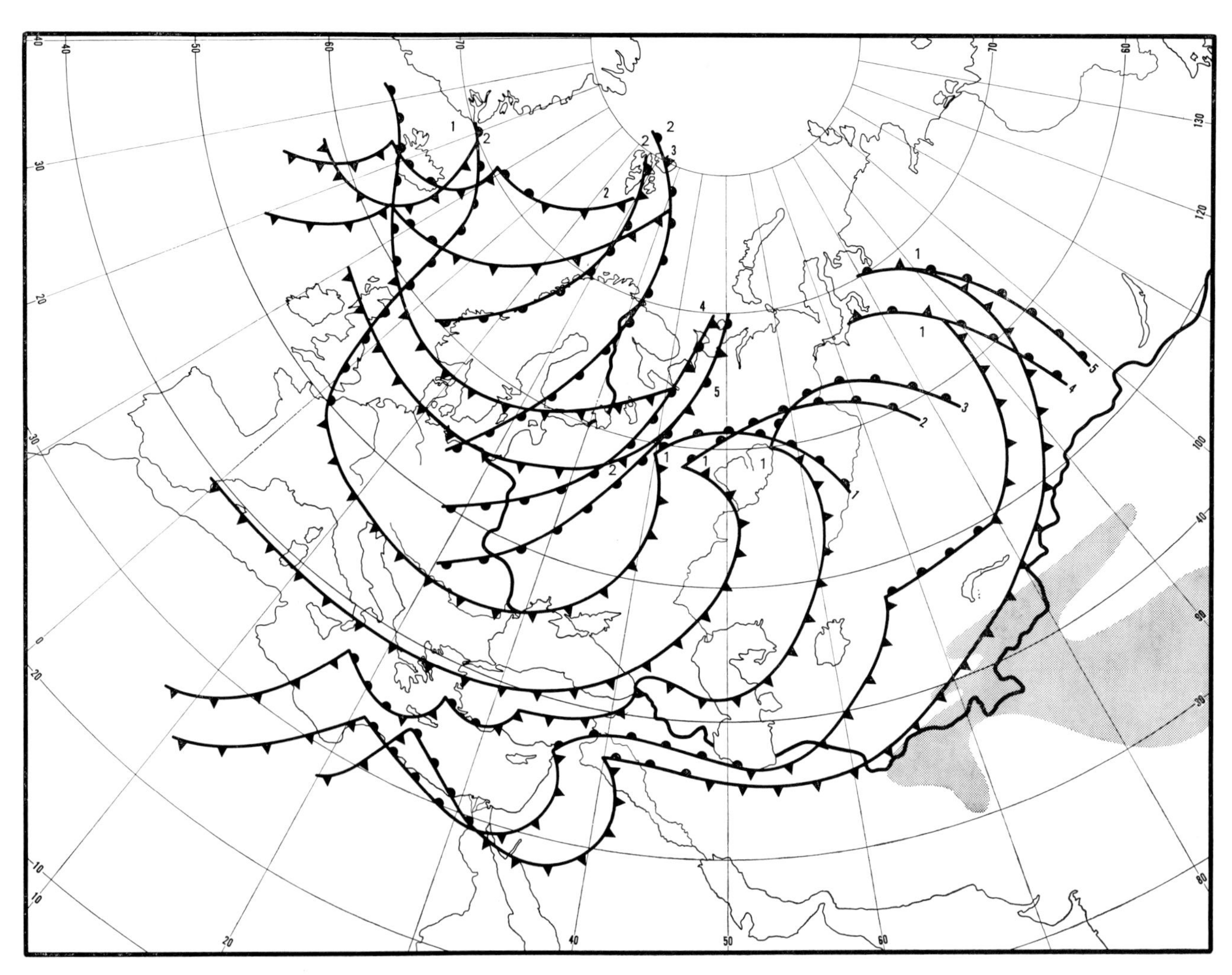

Fig.3-3. Successive positions of the Arctic Front during the period February 1–5, 1949. (After BUGAEV et al., 1957.)

blocking high and the upper flow is rather strongly zonal to the north and to the south of the high (Fig.3-5). Often the two frontal systems are quite indistinguishable on individual maps. Even though their average positions lie along the north and south peripheries of the U.S.S.R., and in fact during some seasons are off the edges of the U.S.S.R. entirely, they do swing back and forth across the Eurasian plain and come into close juxtaposition with one another, producing complicated frontal systems, cyclonic storms, air mass situations, and vertical atmospheric structures.

Movements of frontal systems and cyclonic storms are generally slow, as compared to, for instance, eastern North America, and frequently stagnate for a good part of a day or more in their progression across the Eurasian plain. Such slow movement gives ample opportunity for surface modification of air masses that have entered the region from the outside and thereby renders all air masses indistinguishable from their source regions and less distinctive from each other. As mentioned earlier, the two-front system and the variety of air masses entering the Eurasian land mass make for one of the most complicated air mass and frontal situations on earth in the plain of European U.S.S.R. and western Siberia. And although air masses are less contrasting across fronts than they are in some regions of the world, and thereby fronts are not so intensely developed, this only makes for greater complexity and greater difficulty in analysis.

Not only is there a great variety of air masses many of which do not differ perceptively one from another, but there is much in the way of secondary frontal and cyclonic formation and regeneration of cyclonic storms as they pass from the Polar frontal influence to the Arctic influence. Many of the cyclonic storms that affect the northwestern part of the country have regenerated along the Arctic front in the Barents Sea and are not new storms at all but old storms which have come in from the Icelandic low to the west and have pretty much run a course of life before they get to the Soviet Union. Often they are fully occluded by this time. However, they regenerate over the northern shore along the Barents Sea and continue along the Arctic coast into western Siberia as far east as the Taymyr Peninsula. Similarly, cyclonic storms which have their genesis in the Central Asian area east of the Caspian and move northeastward along the Polar front, often lose their steam about the time they get into the steppes of southwestern Siberia when they have run the course of the Polar front. However, these storms may continue northeastward and join up with storms along the Arctic front in the northern part of the region. Here they may be regenerated and may produce heavy snow falls in the northeastern corner of western Siberia and the uplands of central Siberia.

Most of the cyclonic storms tend to end up in the vicinity of the Ob Gulf no matter where they come from. One major track which is especially active during winter comes in from the west skirting the Arctic coast from an area of cyclogenesis in the Barents–White Sea area. Some of these originate even further west over the Norwegian Sea and move eastward across Scandinavia before reaching the U.S.S.R. A second major track comes up from the southwest where cyclogenesis over the Black Sea and north Caspian area feeds cyclonic storms into the southern part of western Siberia, where more cyclogenesis may take place. A significant number of storms form a complete horseshoe path southeastward from Scandinavia into central European Russia and then northeastward again into the Ob Basin. Still others start far southward in eastern Turkey and northern Iran, circumvent the southern end of the Caspian Sea and move northeastward across Central Asia and central Kazakhstan into western Siberia.

During summer the paths tend to be less concentrated across European Russia. Cyclonic storm tracks at this time of year may move across the European part of the U.S.S.R. at almost any latitude. But the paths definitely tend to converge eastward on the Lower Ob–Yenisey area. Thus the Arctic coast of the Soviet Union from the western border in the Kola Peninsula eastward to beyond the Lena Delta is a region of high cyclonic occurrence throughout the year. In the south the area around the Black Sea and Caspian Sea forms a region of secondary maximum although cyclonic frequencies here are only about 2/3 as high as those along the Arctic coast. Both regions contrast sharply with the broad interior plain which experiences significantly fewer cyclonic occurrences.

Synoptic situations and associated air masses

Fig.3-6 illustrates a synoptic situation similar to that in Fig.3-5, with a major cyclone moving eastward along the Arctic front in the north and a complicated series of waves forming in the complex topography of the eastern Mediterranean–Black Sea–Caspian

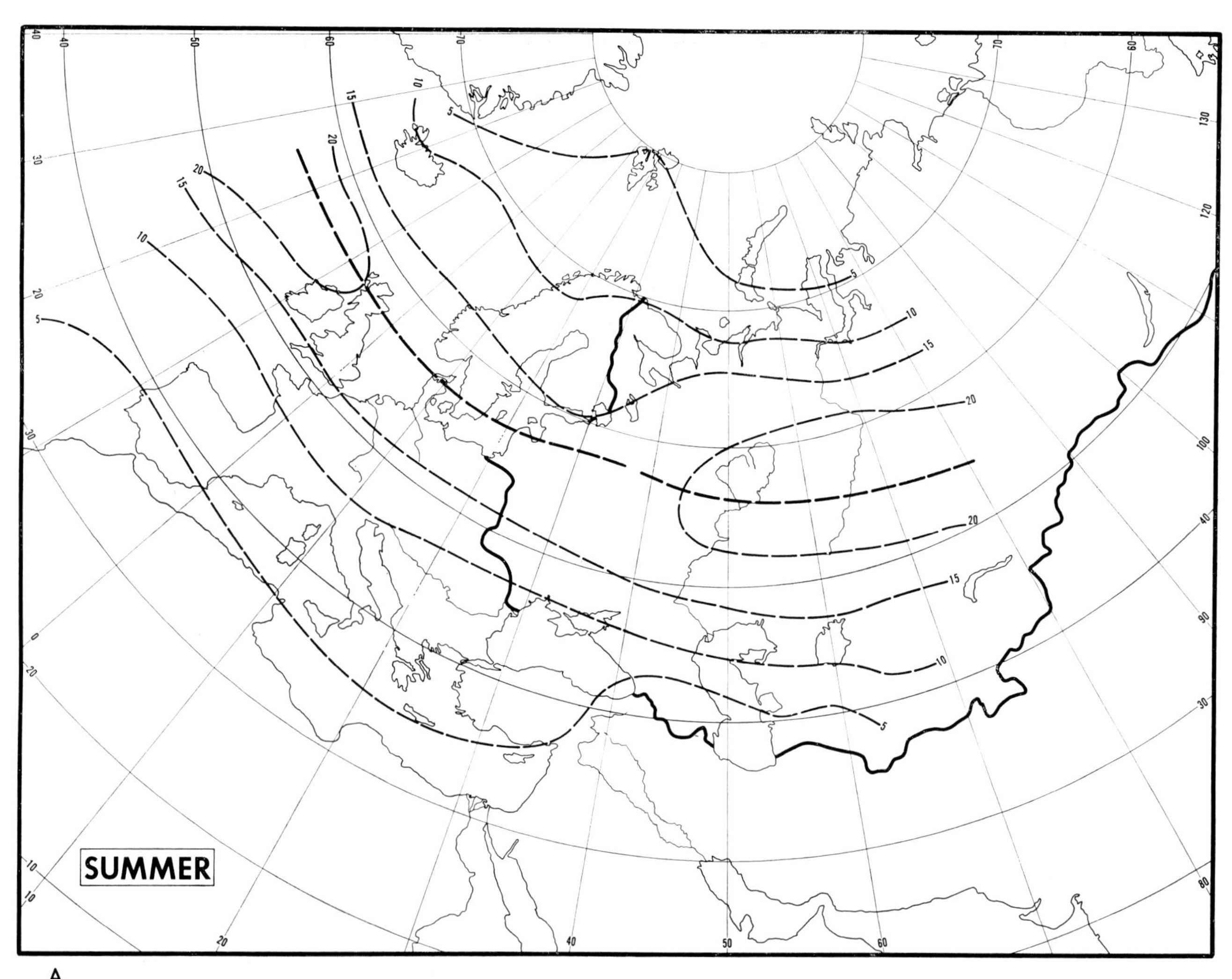

Fig.3-4. Frequency of position of the planetary high level frontal zone (%). (After PRIKHOT'KO et al., 1967.)

area along a complex dual system of an old segment of Arctic front that has moved down from the north and a segment of the Polar front which is moving up from the south. The high mountains, plateaus, plains, and intervening water bodies in this region are conducive to the generation of many waves along these frontal segments which generally move eastward and then northeastward into the Ob Basin sometimes generating into full-fledged cyclonic storms and frequently regenerating in the Ob Gulf region as they merge with cyclonic storms that have moved in along the Arctic coast from the west. Under such conditions the central part of the European plain is occupied by so-called continental temperate air that has moved in from Siberia and northern Kazakhstan in an easterly trajectory along the northern periphery of the southern frontal zone. Some of this air is carried northeastward along the leading edge of the northern cyclone into the lower Ob region, so that continental temperate air dominates almost all of western Siberia. Maritime Arctic air is brought in on the western side of this northern cyclone from the north-northwest across Scandinavia into the Murmansk–Karelia–White Sea area where it merges with the continental temperate air to the south. In the far south the

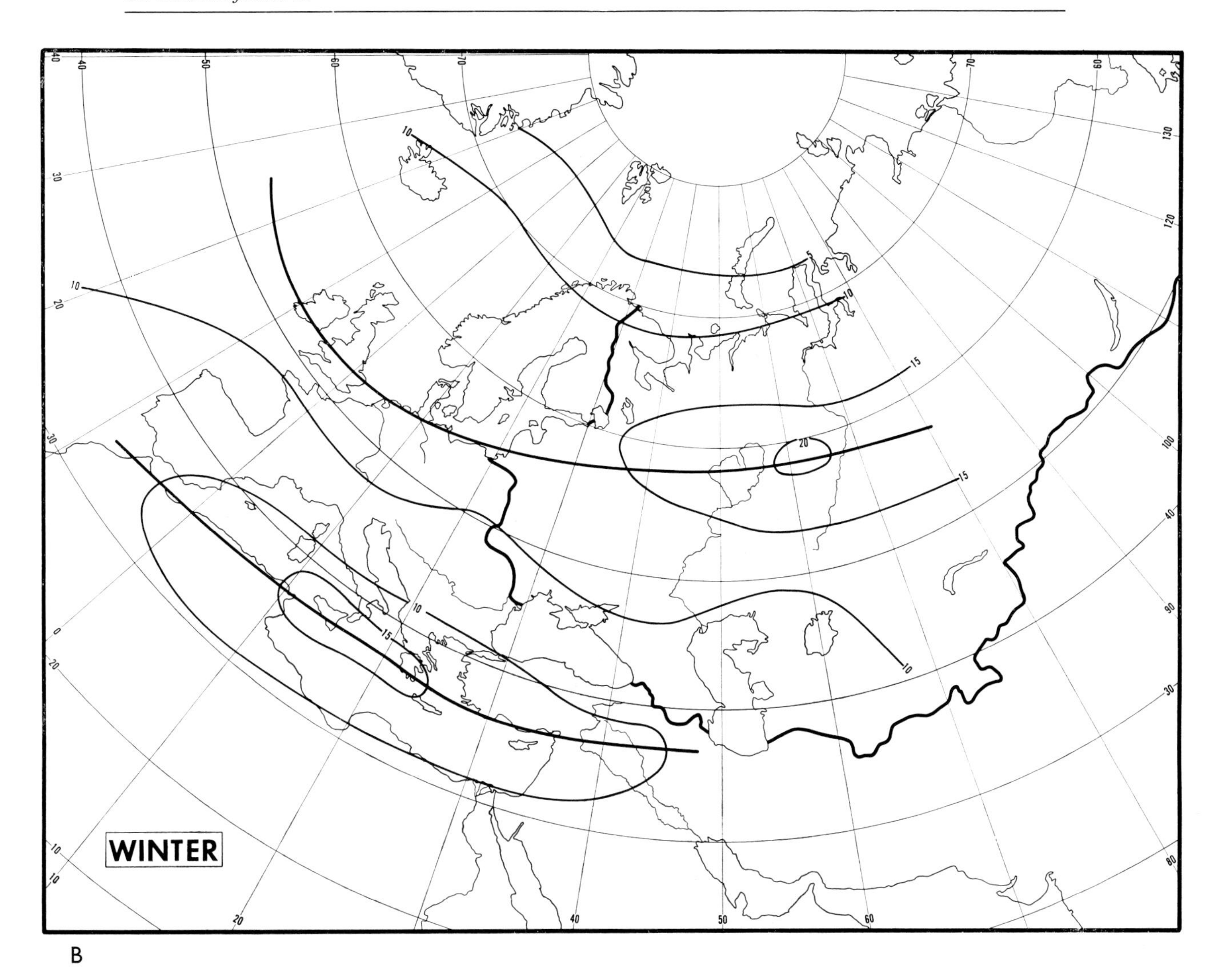

B

Black Sea–Caucasian–Caspian–Central Asian region is experiencing air flows from the south that bring in a modified type of subtropical air which becomes indistinguishable with the continental temperate air as it moves northward.

Fig.3-7 represents a frequently occurring winter situation. The Asiatic high is occupying Kazakhstan, and storms are following the northern track across European Russia into western Siberia. There is a broad strong flow of maritime temperate air from the Atlantic and North Sea regions across Europe into the northern 2/3 of European U.S.S.R. and much of western Siberia and well into eastern Siberia. Of course this air modifies rapidly as it moves eastward. The southern plain is dominated by continental temperate air which has come around the southwestern periphery of the Asiatic high.

Fig.3-8 shows almost an opposite situation. The blocking high in western U.S.S.R. essentially eliminates maritime temperate air from the country. A very well developed low pressure center north of the Caspian Sea is the dominant circulation, and this feeds in a strong flow of continental Arctic air from northeast to southwest clear into the Black Sea area. Much of western Siberia and European U.S.S.R. at this time is occupied by

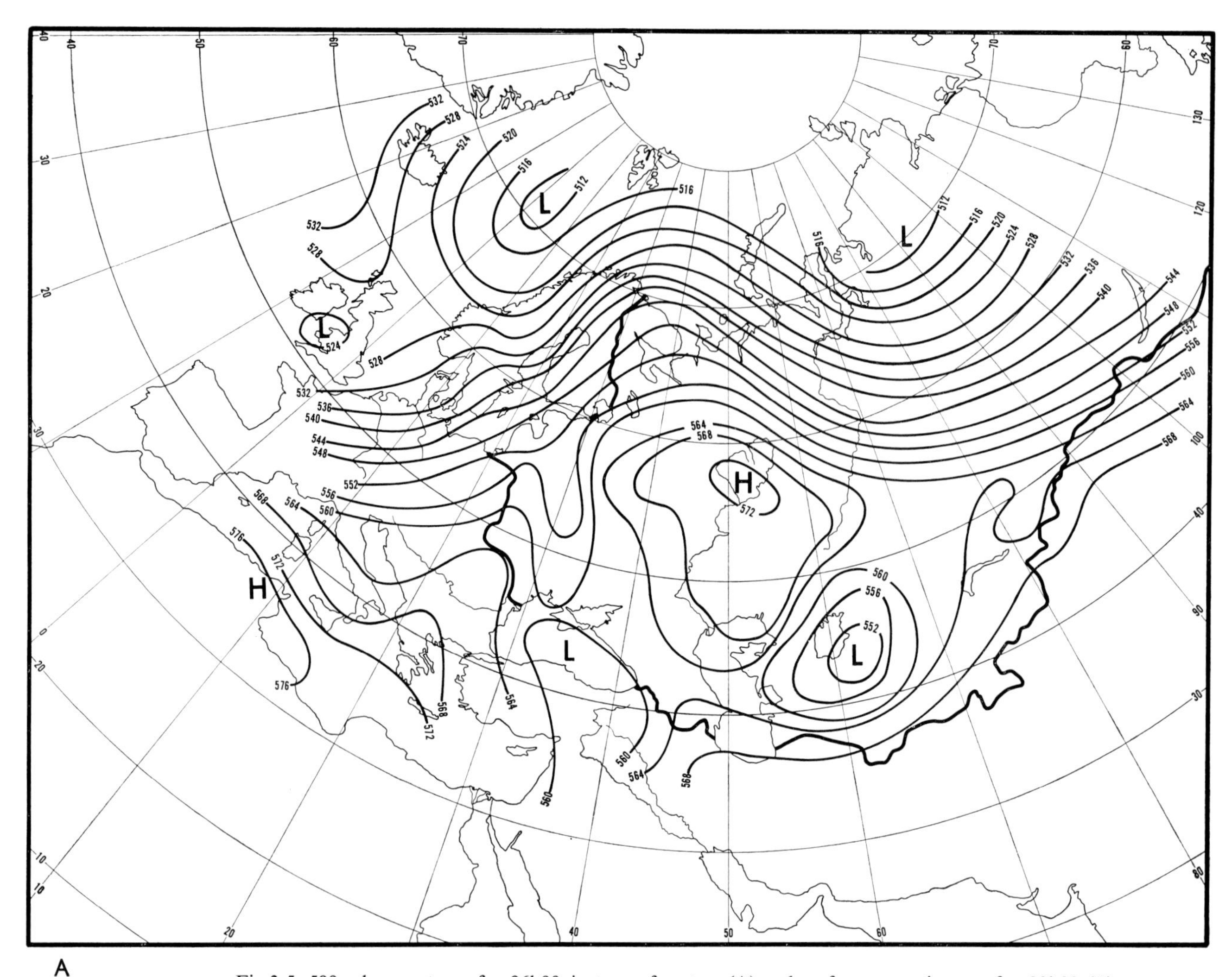

A

Fig.3-5. 500 mbar contours for 06h00, in tens of meters (A) and surface synoptic map for 03h00 (B), 6 April 1953. (After BUGAEV et al., 1957.)

continental Arctic air. The cold low north of the Caspian intensifies aloft as does the warm blocking high over the Baltic. This is well illustrated at the 500-mbar level. This type of circulation brings the coldest weather to much of European U.S.S.R. during the winter.

A similar situation is shown in Fig.3-9 which illustrates conditions that bring late spring frost to the southern plains. In this case the air flow behind the cyclonic storm is almost directly north–south and the air intruding behind the front is of a maritime Arctic character. During this time of year, and particularly later in summer, there is little distinction of Arctic air masses into marine or continental types. In this particular map much of European U.S.S.R. is occupied by maritime Arctic air and much of western Siberia by continental temperate air of local origin which is moving northward ahead of the cold frontal system.

Perhaps more typical of an early summer situation is Fig.3-10 which shows drought conditions in central European U.S.S.R. under the influence of a rather stationary high pressure cell, with frontal activity to the north and south, especially in western Siberia.

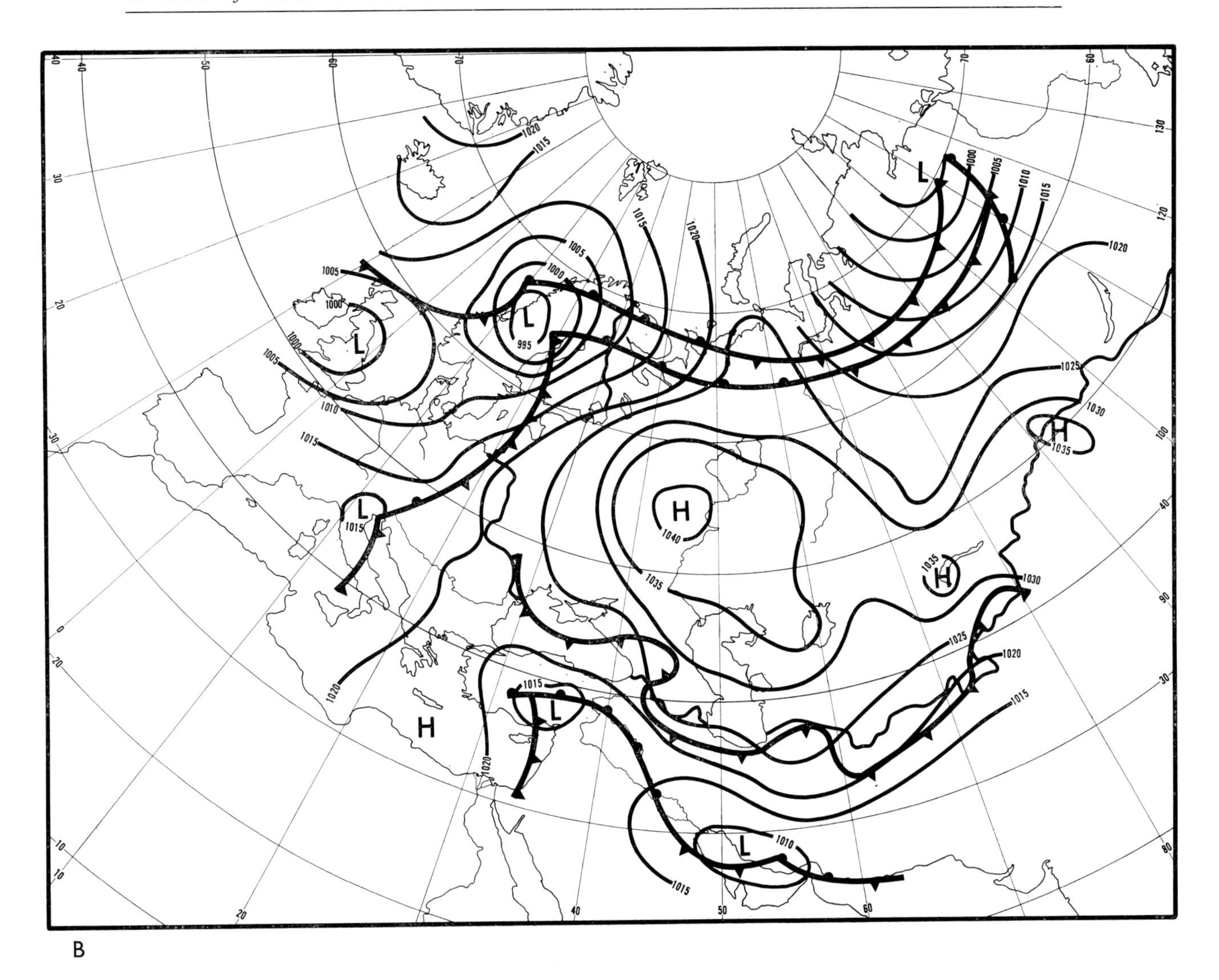

B

Particularly there is strong frontal activity in the interior south of the Taymyr Peninsula which at this time of year is one of the regions that experiences the greatest frequency of cyclonic storms. Maritime Arctic and maritime temperate air are being fed into the high pressure system in European U.S.S.R. and the same mixture occupies much of western Siberia as well. Rapid heating at the surface is taking place as the air moves eastward, and this drops the relative humidity and makes for rather dry conditions under the high pressure cell in European U.S.S.R.

Fig.3-11 depicts probably the most typical summer situation, with the eastern end of the Azores high nosing into the southern part of European U.S.S.R. and cyclonic activity taking place to the northeast leading into the Ob Gulf. Maritime temperate air is being fed into the southern part of the plain along the northern periphery of the Azores high, and Arctic air is entering the northern part of the European plain as well as the western part of the West Siberian Lowland. The eastern part of western Siberia is experiencing a southerly flow of continental temperate air.

Fig.3-12 illustrates the synoptic situation which during summer can bring extremely hot

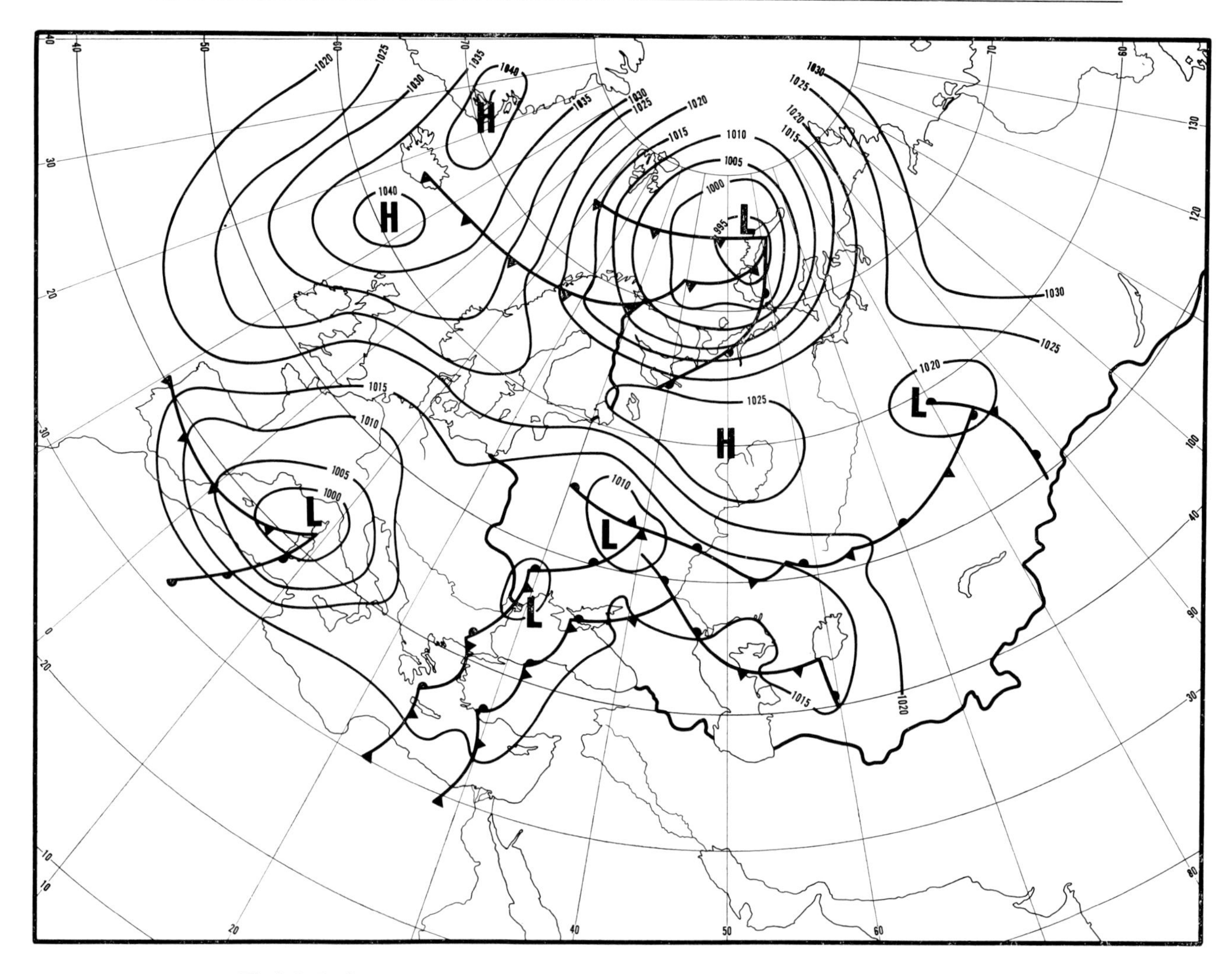

Fig.3-6. Surface synoptic map for January 15, 1960. (After POKROVSKAYA, 1969.)

weather to the southern plain. Air which circulates slowly southward from the Arctic coast warms rapidly and experiences a rapid fall in relative humidity as the initially low absolute humidity is rapidly outstripped by the capacity of the air to hold moisture. Clear dry weather occurs throughout much of European U.S.S.R. and surface temperatures rise rapidly under the influence of strong insolation in contact with a relatively dry surface. Arctic air rapidly transforms into continental tropical air. This is a typical situation for the occurrence of the so-called "sukhovey", a hot dry type of weather that is very desiccating to plants. Although this phenomenon occurs most frequently in Central Asia, it is more damaging in the southern part of the European plain where agriculture is being carried on throughout much of the region without the benefit of irrigation (Fig.9-58).

During autumn the Icelandic low reestablishes itself, and cyclonic activity increases in the Barents Sea area. Fig.3-13 illustrates an extremely well developed cyclonic storm centered over the Barents Sea which has undergone a period of development that has caused a series of fronts to radiate south and eastward following each other across the

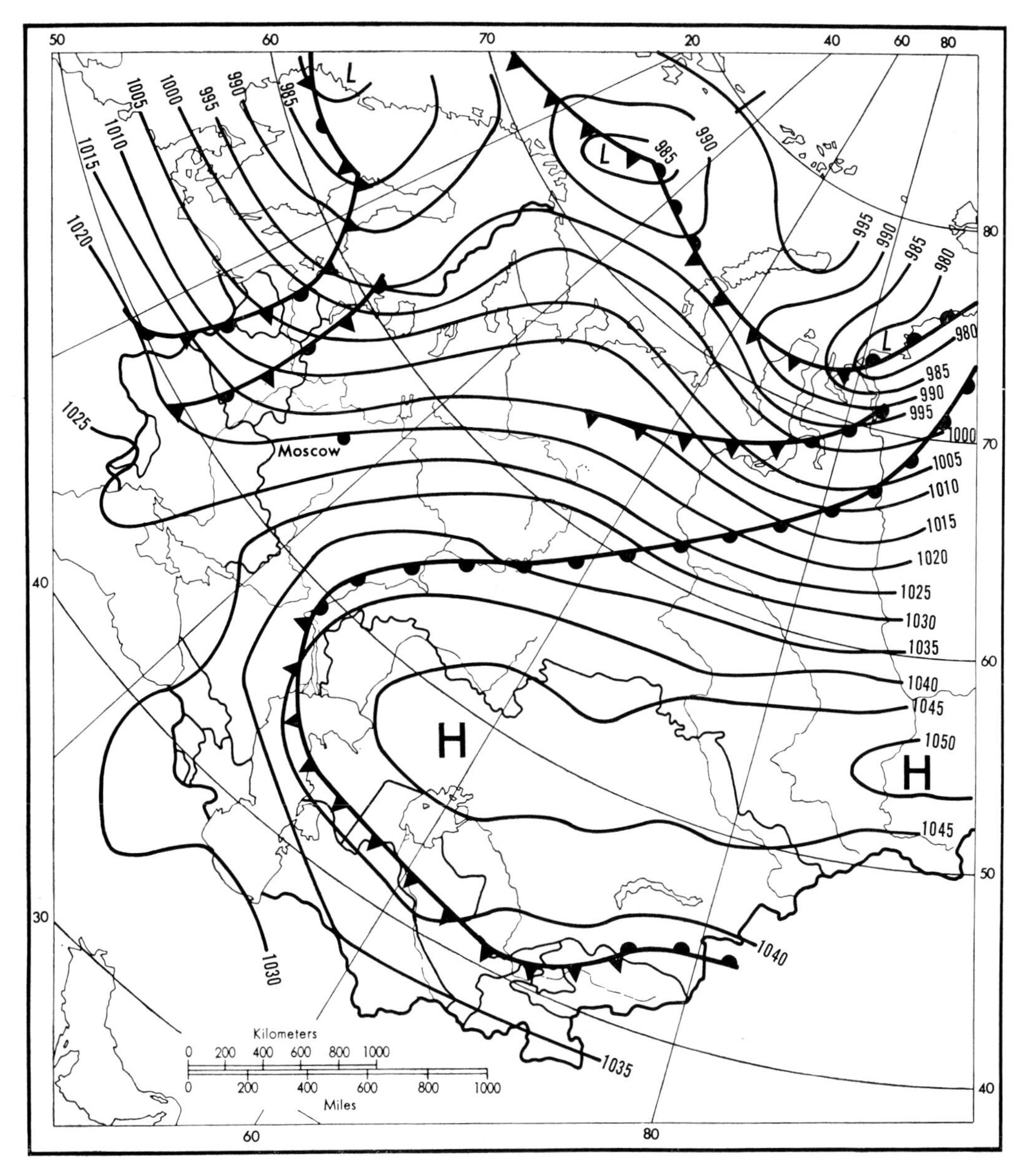

Fig.3-7. Surface synoptic map for January 31, 1948. (After ALISOV, 1956.)

plain. Secondary waves are forming along the front in the north European plain. Under these conditions a strong flow of maritime Arctic and maritime temperate air dominates the entire European U.S.S.R. and western Siberia. Weak high pressure cells in the south are composed of maritime temperate air that has been modified into continental temperate. Relatively cold air with rising pressures enters the European part of the U.S.S.R. in the rear of cyclonic storms that are moving eastward across the plain. Most of this air enters the continent from the northwest, the north, or the northeast. In most cases it enters as air streams which eventually stagnate somewhere in the central or southern parts of the plain to build up pressure and form high pressure cells. But occasionally a well formed high pressure cell may move en masse onto the continent and settle itself in the southern plain. Fig.3-14 illustrates a high pressure cell of Arctic air moving south-

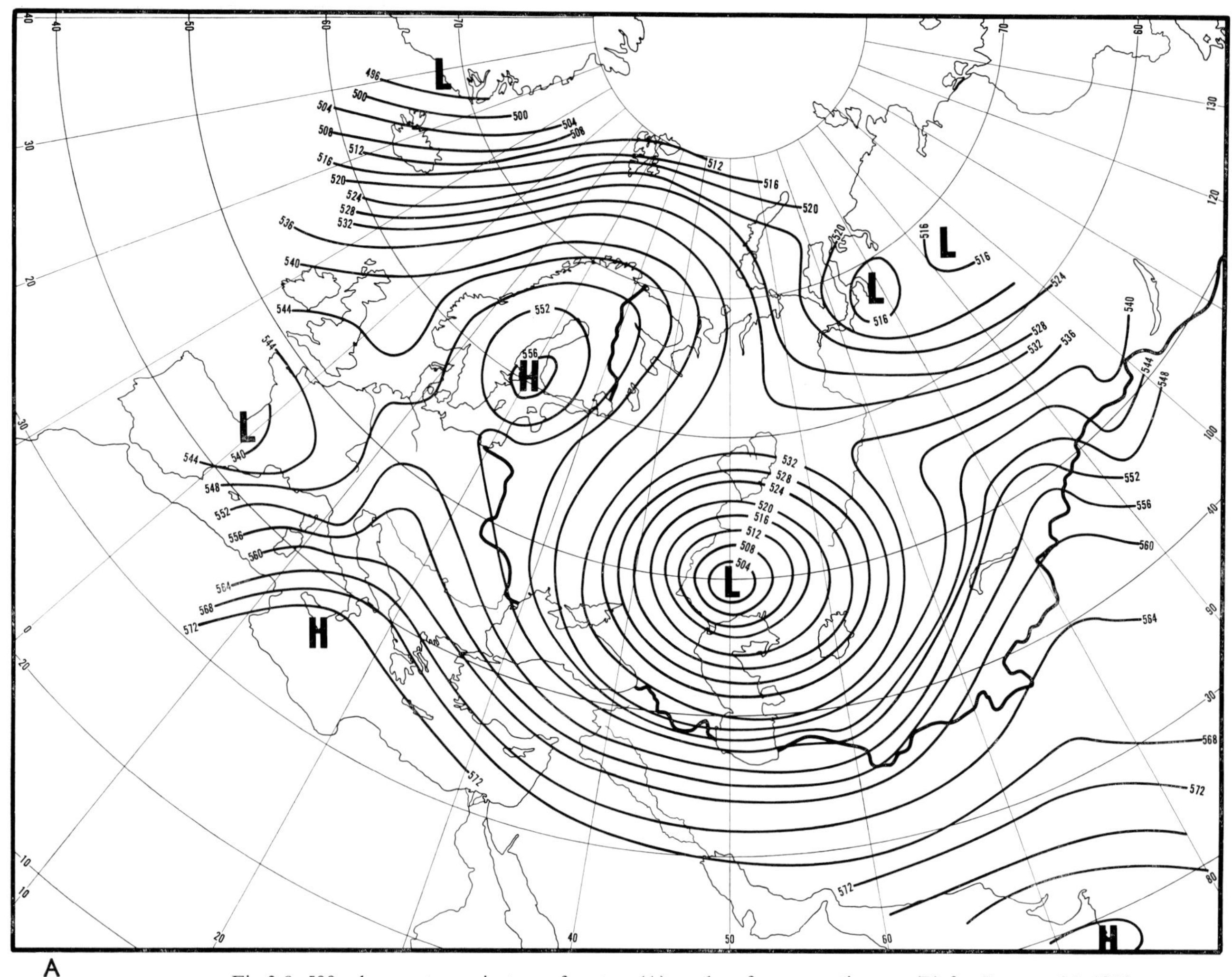

Fig.3-8. 500-mbar contours, in tens of meters (A), and surface synoptic map (B) for January 25, 1951. (After BUGAEV et al., 1957).

westward from the Kara Sea across western Siberia and the central Urals into central European U.S.S.R. This is most indicative of a winter situation when Arctic high pressure cells have a tendency to move southwestward into European U.S.S.R. (Fig.2-15). Although these are not preferred routes of travel, they do occur occasionally during winter. In summer this type of movement is essentially absent (Fig.2-20).

Anticyclonic cells tend to move from northwest to southeast across the plain at all times of the year. But cold air more commonly streams onto the land mass in narrow flows from the northwest, north, or northeast around the peripheries of well established cyclones or anticyclones which are moving in an easterly direction across the plain. Fig.3-15 illustrates such streams of cold air intrusions during summer around various pressure cells. Routes *2*, *3*, *4*, and *5* enter the continent between the southwestern periphery of cyclonic storms located in the northeastern part of European U.S.S.R. and the eastern nose of the Azores high located in the south, while the various routes depicted by number *1* circulate around a high pressure cell that is stagnated in the Urals region.

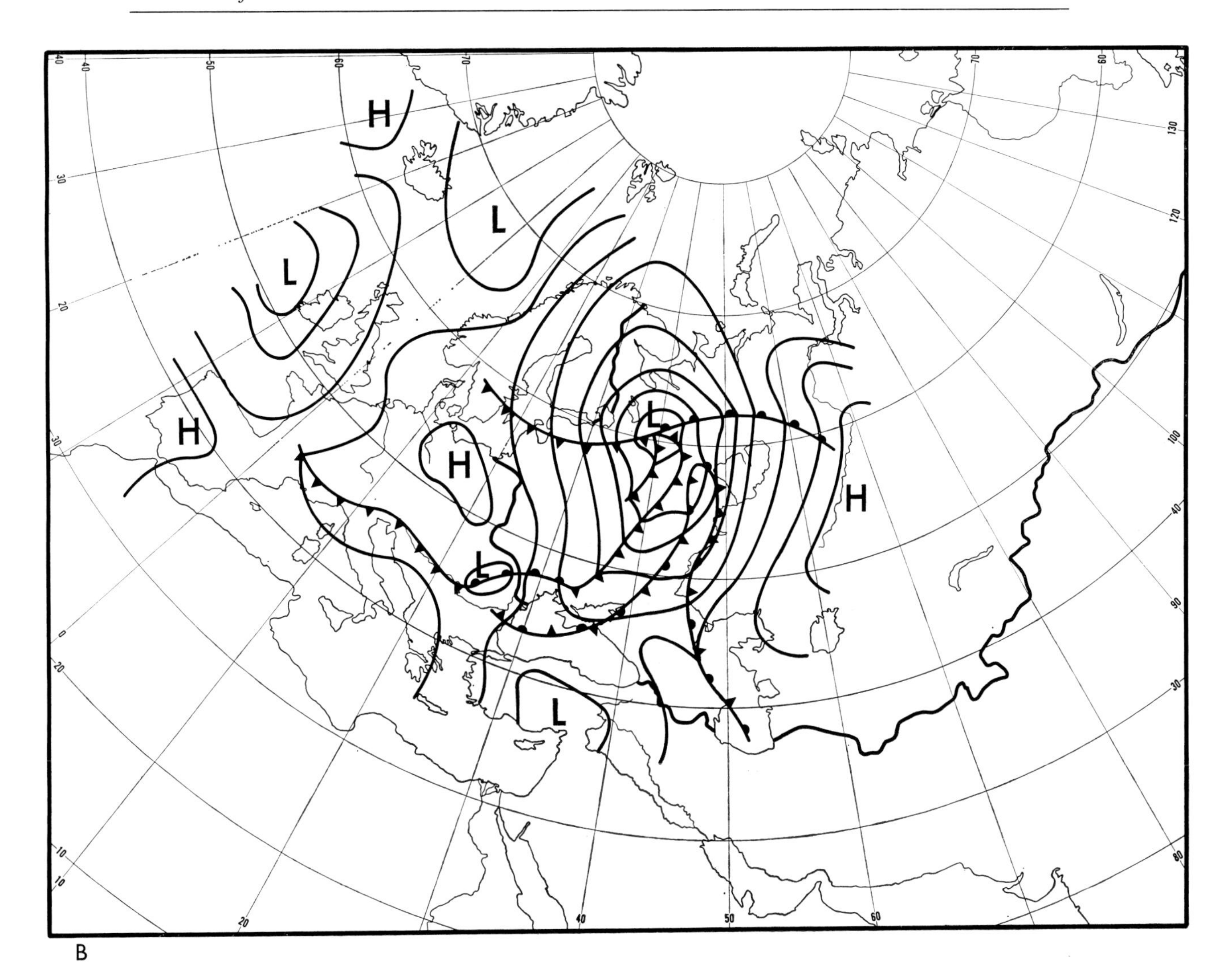

B

Air flows

Thus it is seen that a large storm system or a large anticyclone somewhere over the area can induce the flow of different types of air masses into different parts of the plain. For instance, a large storm system centered in the Moscow region will bring in a strong flow of maritime temperate air from the Baltic or maritime Arctic air from the Barents and White seas into the northwestern part of the plain which will produce widespread cloudiness of the stratus type with perhaps prolonged drizzly weather, cool temperatures, and high humidities. The same storm will induce a flow of continental temperate or even maritime tropical air out of the southwest across the Black Sea into the eastern Ukraine and lower Volga area that will bring on a characteristic association of weather phenomena which, depending upon the season, is much different than that in the northwest.

In an attempt to depict sequences of weather associations which are typical for the Eurasian plain, Soviet meteorologists and climatologists have developed elaborate air flow studies. These became quite popular in the 1930's and have continued sporadically ever since. Probably the best known classifications of air flows are those by LIR (1936),

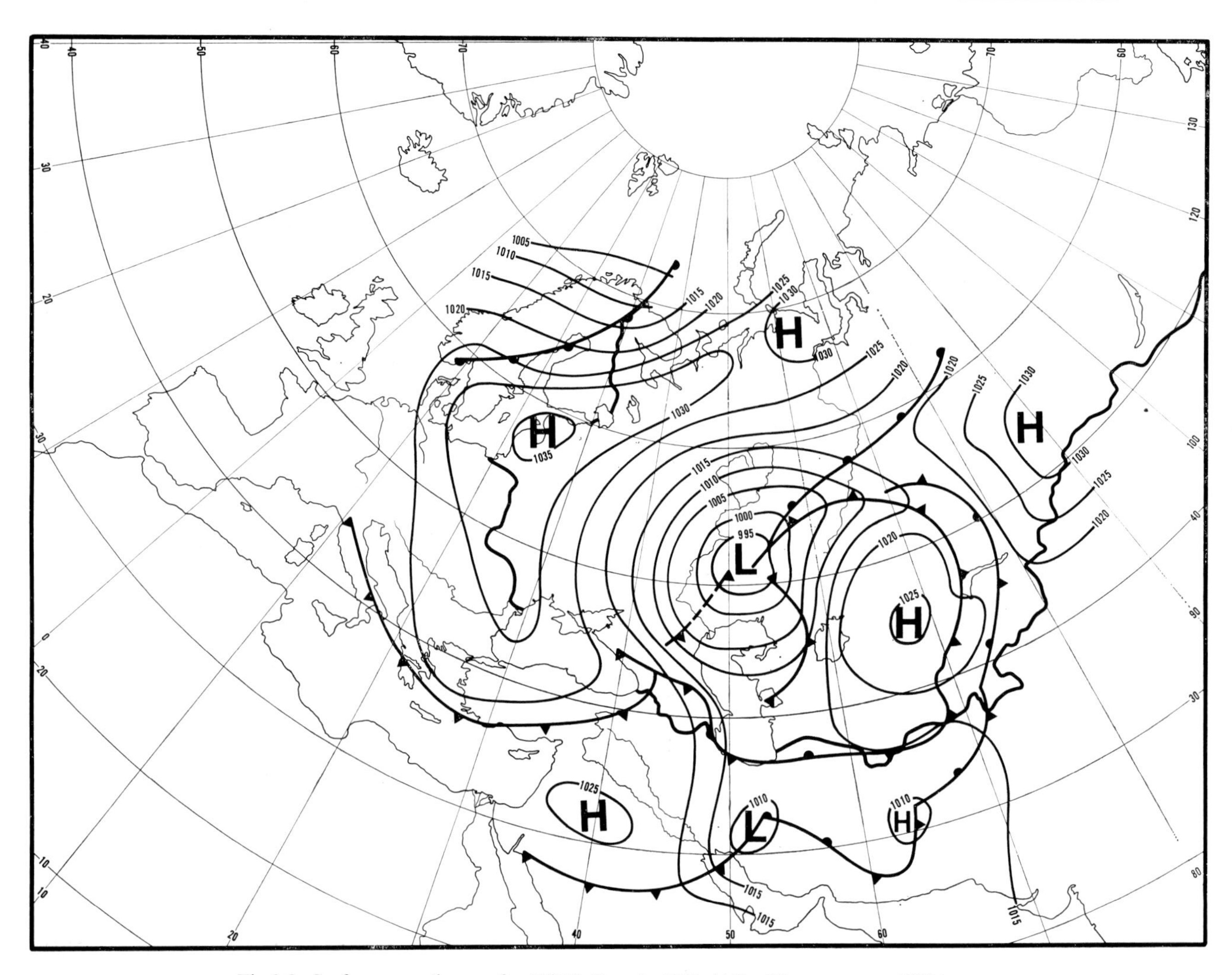

Fig.3-9. Surface synoptic map for 07h00, June 1, 1935. (After NEZDYUROVA, 1939.)

POGOSYAN (1936), YAROSLAVTSEV (1936), and VANGENGEYM (1946). Some aspects of these have been summarized in BLUMENSTOCK and PROZOROWSKI (1956).

Lir's works were probably the most exhaustive. Using synoptic maps for 26 years, she devised a scheme which recognized 17 types of flow plus certain variants. Thus, it is quite a complicated system, and it is unfortunate that in working out her flow situations she used a method of averaging the field of mean flow across areas without respect to curvature. Therefore, it is misleading to use her curved arrows to judge cyclonic or anti-cyclonic processes, and it is impossible to reconstruct the synoptic situations from her flow maps.

The most simple scheme is that by VANGENGEYM (1946), which divides flows simply into westerly, easterly, and meridional. This is the scheme which seems to have caught on most among more recent Soviet authors, and it is the basis for the discussion of the circulation factor of climate in LEBEDEV (1958).

Vangengeym contended, with a great deal of truth, that the weather that occurs at any given time in any given part of the east European plain is controlled to a great extent by

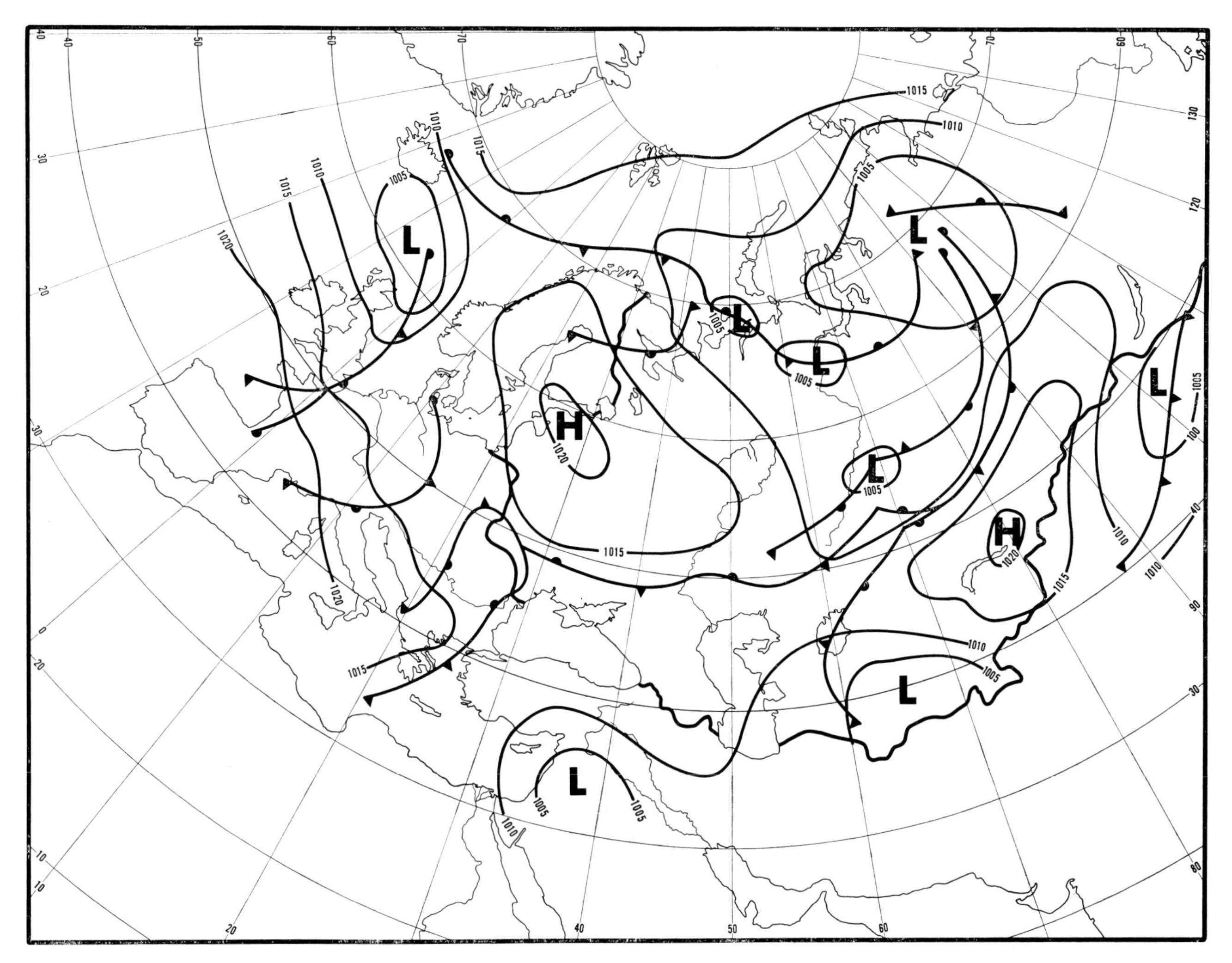

Fig.3-10. Surface synoptic map for June 24, 1946. (After ALISOV, 1956.)

whether the air flow has a predominantly western, eastern, or meridional component. Western flow generally brings in more humid maritime types of air which lead to greater cloudiness and precipitation and moderate temperatures. Easterly air flow on the other hand generally brings in continental temperate or continental Arctic air which is dry and cold. Of course, the entire latitudinal expanse of the plain does not experience the same type of air flow simultaneously. Easterly air flow, say around a high pressure cell in the southern Urals, might bring cold dry weather into the middle Volga region, while the continuation of the anticyclonic circulation northward and then northeastward again might activate cyclonic storms along the Arctic front in the northern fringes of the plain and bring above average cloudiness and precipitation to that region.

The effect that air flow has on precipitation is determined by the type of air mass that is being advected by the flow, and this varies by region and by time of year. Fig.3-16 shows that with westerly circulation in June a large central part of the plain receives well above average precipitation while certain peripheral areas, particularly the extreme south, receive below average precipitation. In August the same flow gives a somewhat different pattern of precipitation deviations.

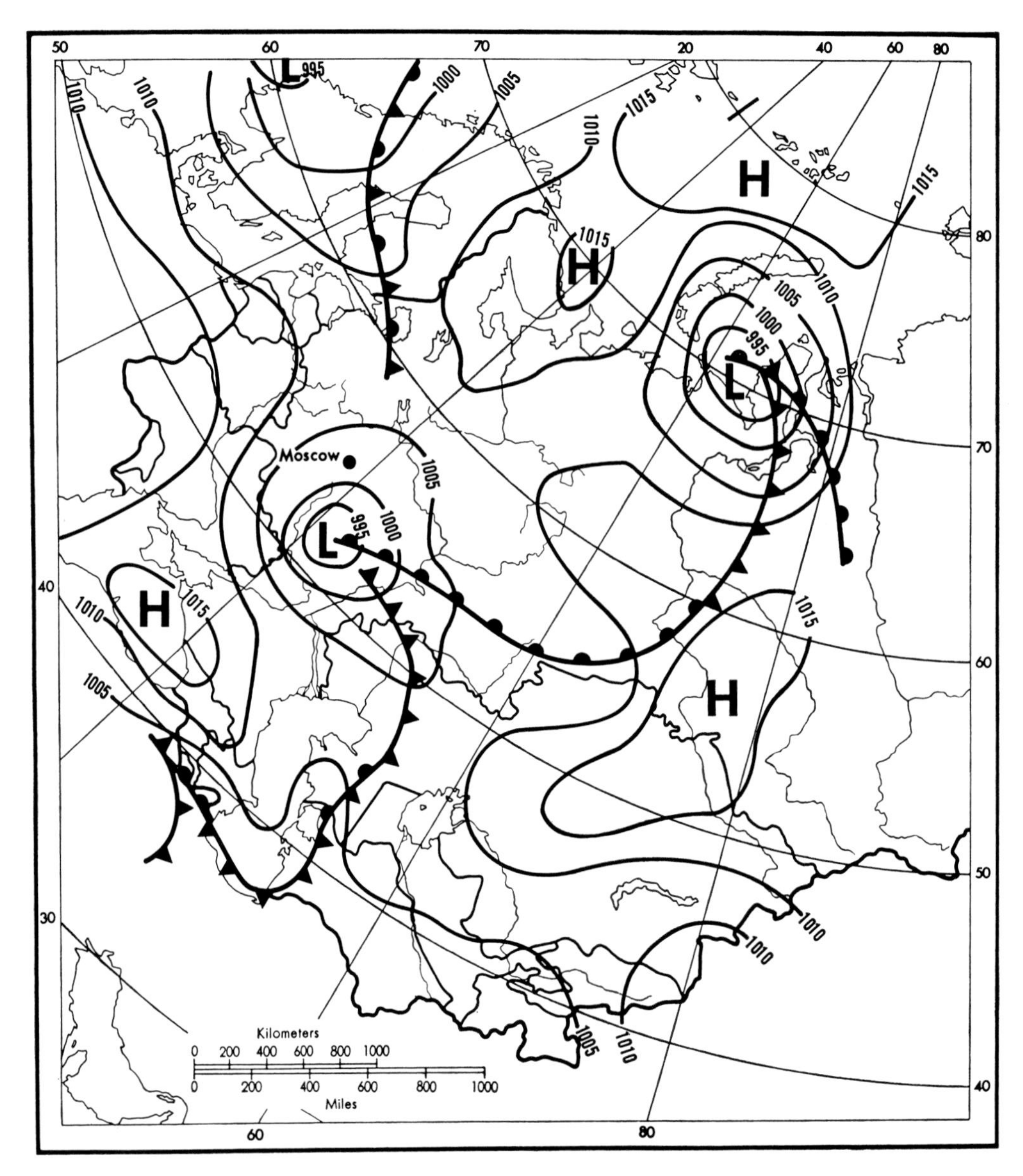

Fig.3-11. Surface synoptic map for July 15, 1953. (After ALISOV, 1956.)

Easterly flow gives a completely different distribution, with generally very dry conditions in the south and above normal precipitation in the north. The flow of air from the east and the south tends to move northwestward across the plain, so that the western portion of the northern plain is affected more by the warm dry air from the Caspian region than is the northeastern sector of the plain. This is illustrated by the precipitation distribution on the August map with easterly flow.

There seems to be stronger correlation between air flow and precipitation in the southern part of the plain where precipitation generally is at a deficit and moisture supply is critically balanced with various components of atmospheric dynamics. A change in air flow or air mass type in this region will generally show a quicker response in precipitation and other processes than farther north where things are not so critically balanced. Data at Kiev show a fairly strong positive correlation between precipitation and either westerly

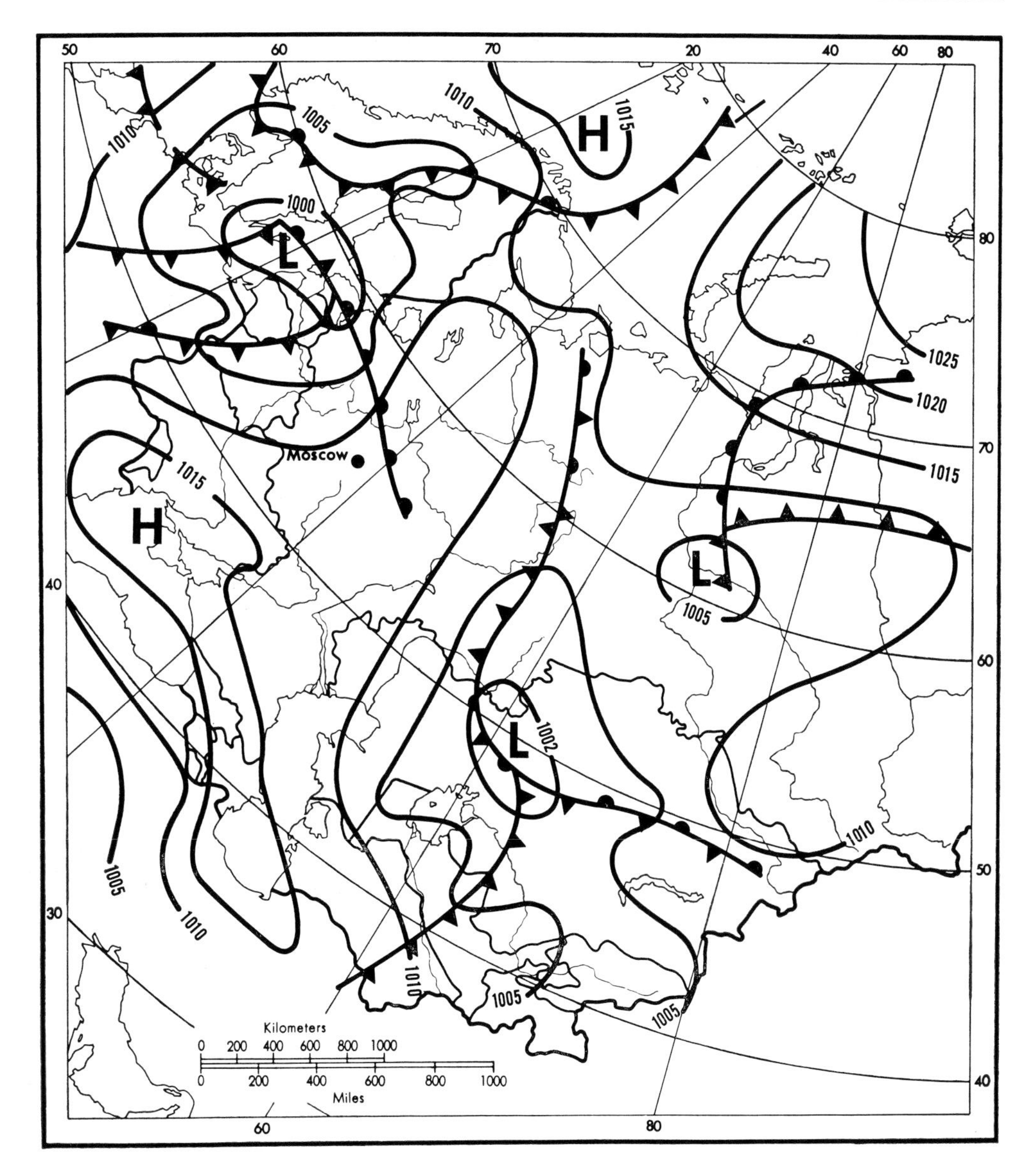

Fig.3-12. Surface synoptic map for August 15, 1954. (After ALISOV 1956.)

or meridional circulation and a very strong inverse relation between precipitation and easterly circulation. Moscow shows weaker correlations than Kiev, but they are still quite discernible. At Leningrad correlations are only slight or lacking.

By the same token the direction of air flow has a great influence on temperature departures from normal (Fig.3-17). These are accentuated in the northern part of the plain where during winter westerly flow may produce temperatures 15°–20°C above normal while northerly flow drops temperatures 25°–30°C below normal. The coldest flows of air onto the east European plain during the winter are flows of continental Arctic air from the northeast which circulate counter-clockwise around the northern sides of cyclonic storms that are moving eastward along the Arctic front. Under such conditions temperatures at Moscow can drop as low as −42°C.

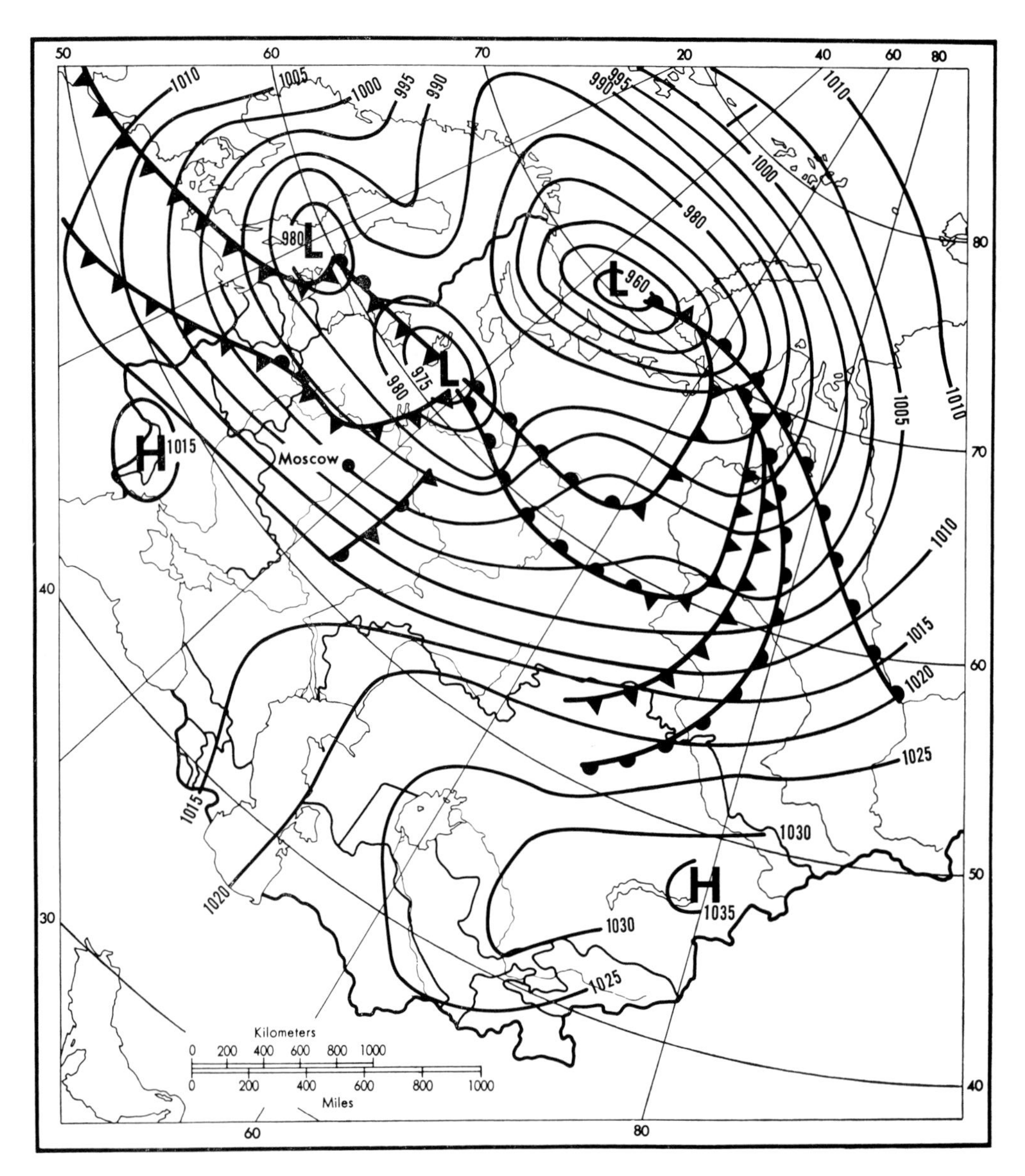

Fig.3-13. Surface synoptic map for October 15, 1955. (After POKROVSKAYA and BYCHKOVA, 1967.)

Table 3-I shows the frequency of occurrence of Vangengeym's three types of flow. During the course of the entire year westerly flow is somewhat dominant, with 40% of all occur-

TABLE 3-I

MEAN ANNUAL NUMBER OF DAYS WITH FORMS OF CIRCULATION
(After LEBEDEV, 1958, p.46)

Flow	Months: I	II	III	IV	V	VI	VII	VIII	IX	X	XI	XII	Year
West	12	10	11	10	9	12	12	15	15	14	13	12	145
East	12	11	12	11	11	7	8	9	8	10	11	13	123
Meridional	7	7	8	9	11	11	11	7	7	7	6	6	97

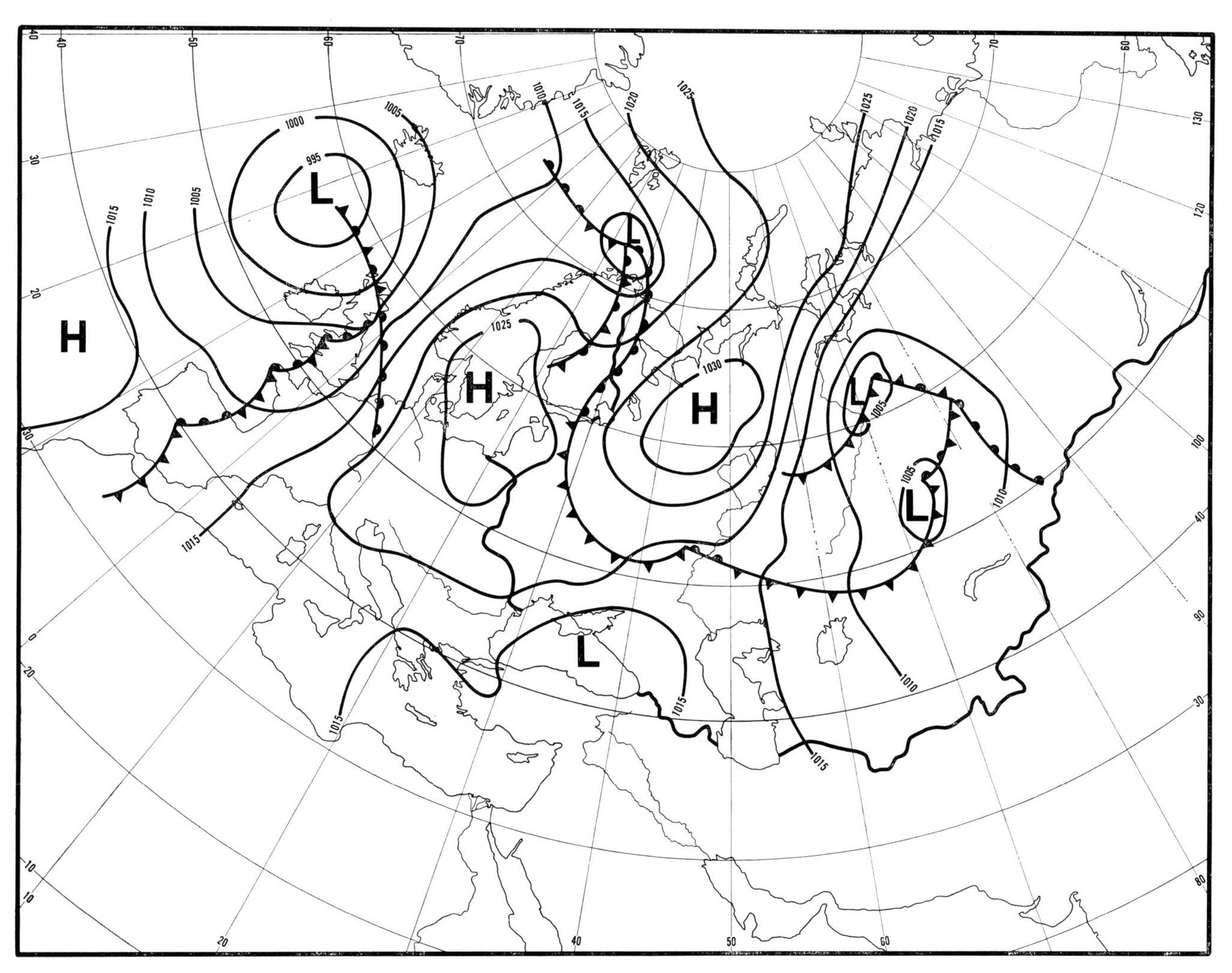

Fig.3-14. Surface synoptic map for September 5, 1949. (After POKROVSKAYA, 1969.)

rences. Easterly flow occurs 34% of the time, and meridional flow 26%. Westerly flow is most dominant toward the end of summer when it constitutes 50% of the flow, and at a minimum in the second half of winter and spring when it constitutes only about 33%. Easterly flow is most dominant during the winter months when it occurs about 40% of the time and is least apparent in the first half of summer and the beginning of autumn when it occurs about 26% of the time. Meridional flow is most in evidence during the end of spring and the first half of summer when it occurs more than 1/3 of the time, but it drops off rapidly in late summer and reaches a minimum in November and December. Pogosyan dealt primarily with upper air flows and divided the flow at the 500-mbar level into four distinct types (POGOSYAN and SAVCHENKOVA, 1950). These are much simpler than surface flows and tend to generalize those classification schemes based on surface flows. Therefore, they have been quite useful. Type 1 is primarily zonal flow with only a slight tendency toward wave formation over the Eurasian plain (Fig.3-18). This type of flow is conducive to rapid west–east movement of maritime temperate air from the Atlantic across the Eurasian plain. It is the most prevalent of the flow types, occurring approximately 45% of the time during the year.

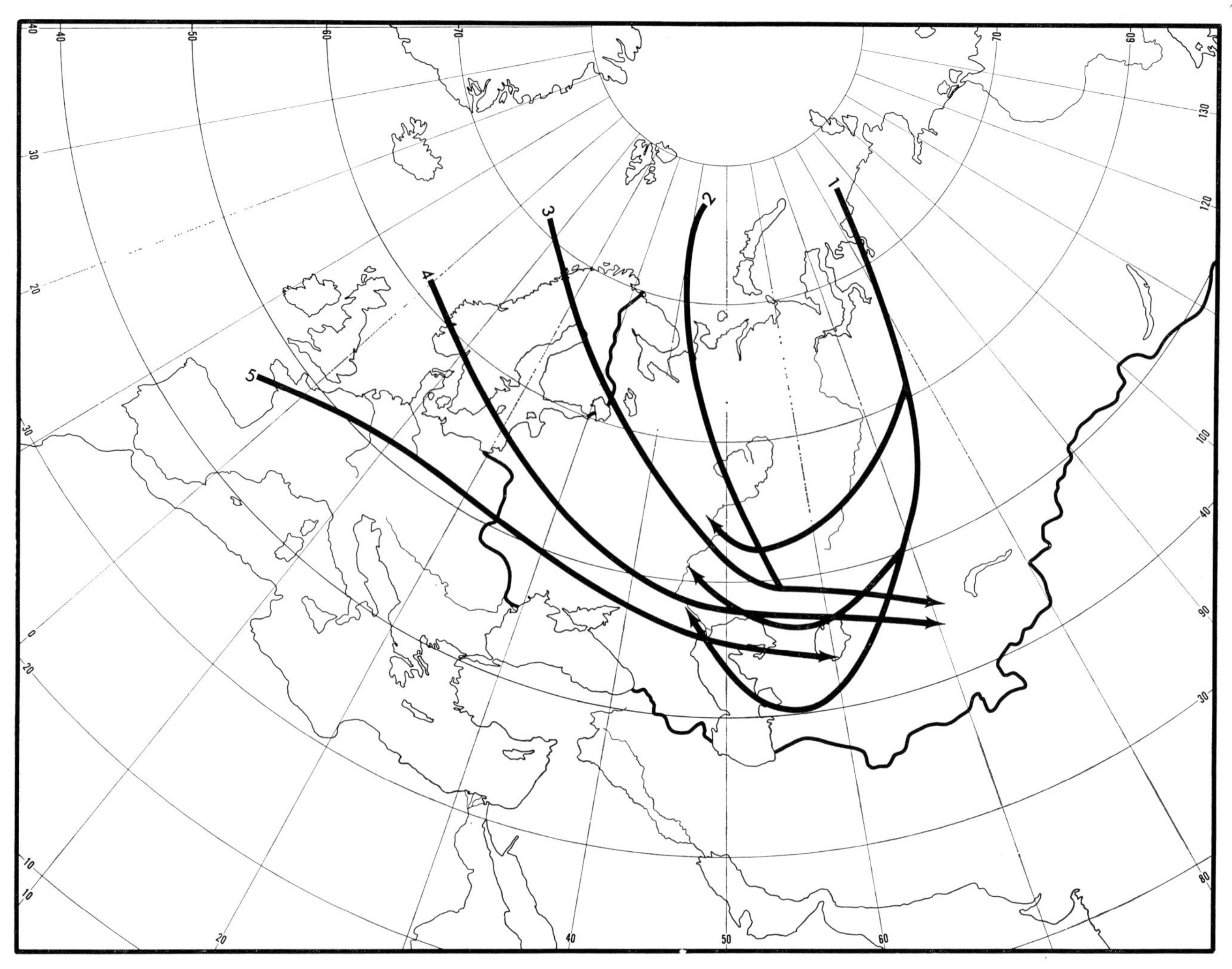

Fig.3-15. Trajectories of cold air intrusions into European USSR during summer. (After ZVEREV, 1957.)

Flow type II is illustrated in Fig.3-19. Here a well developed trough extends from the Ob Gulf south-southwestward to the Black Sea. Maritime Arctic air is brought into the Eurasian plain across Scandinavia from the northwest along the western edge of the trough and is modified into continental temperate air in the central part of European U.S.S.R. From there it is carried northeastward into western Siberia. This is primarily a winter circulation.

The third type of flow is illustrated in Fig.3-20. This is just about the opposite of flow type II. A warm blocking high develops over the east European plain which causes air to flow from south to north along its western periphery and from north to south along its eastern side. In the west continental tropical and maritime tropical types of air masses are carried far northward, while along the eastern periphery of the high in western Siberia air moves from the Arctic into the land mass. This is primarily a late summer–early fall phenomenon and may bring on an occurrence of what is known as "old women's summer", which is comparable to the so-called "Indian summer" of eastern North America during late autumn.

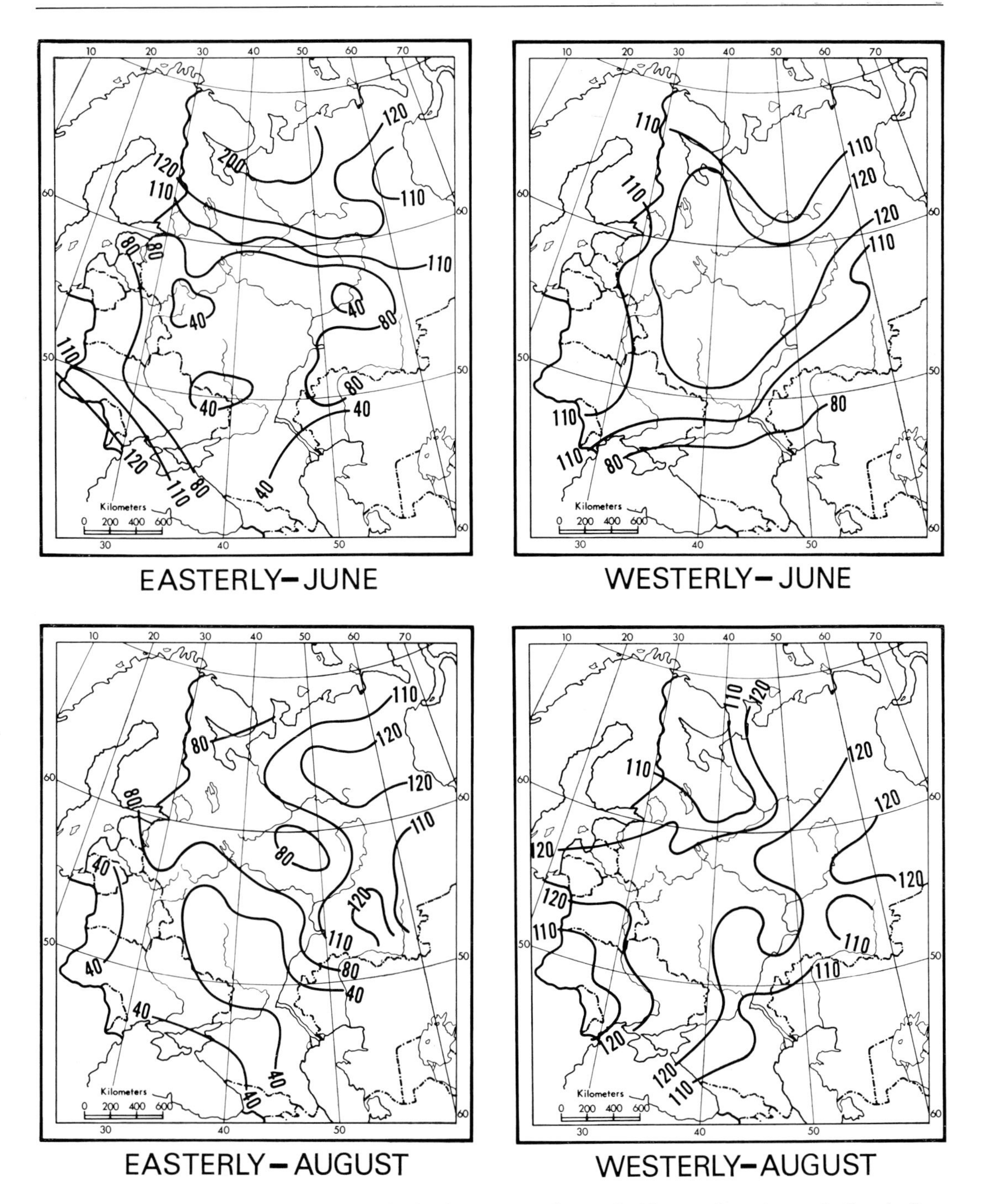

Fig.3-16. Distributions of precipitation in percents of normal with westerly and easterly flow in June and August. (After LEBEDEV, 1958.)

Type IV flow is the main easterly type of flow (Fig.3-21). A blocking high stagnates over Scandinavia with its elongated axis paralleling the Arctic coast. The center of the continent is dominated by low pressure which pushes far westward into European U.S.S.R. Air circulates from northeast to southwest through the mid-section of the continent between the two pressure cells and brings very cold continental Arctic air into central

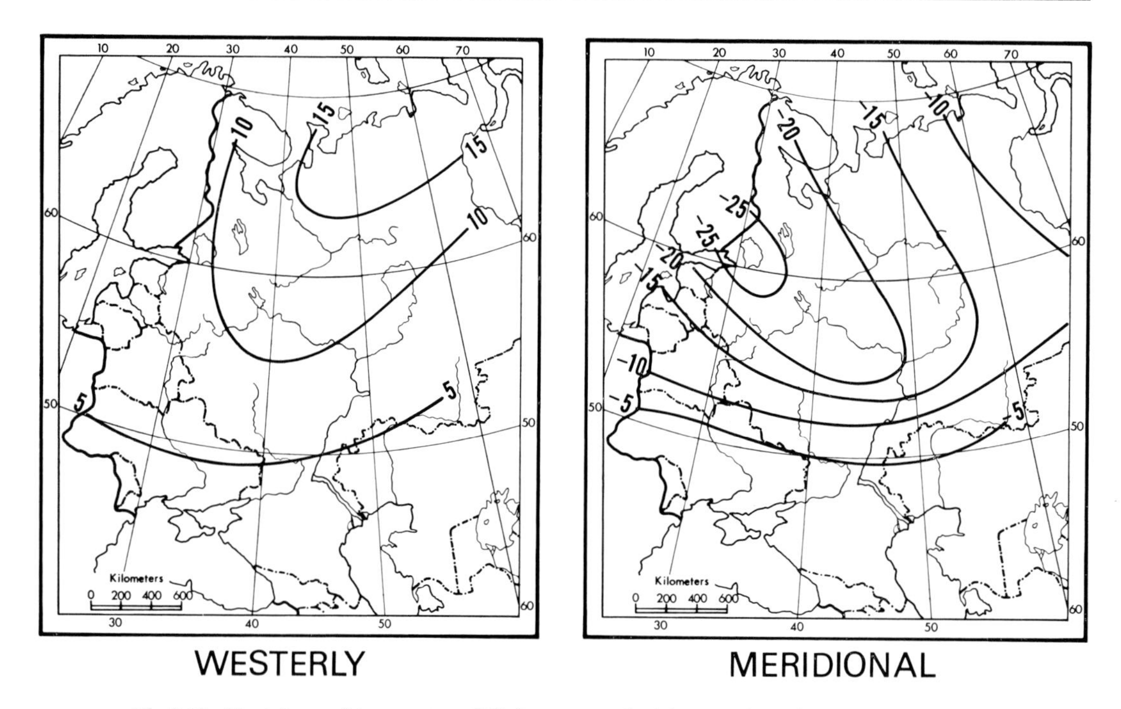

Fig.3-17. Deviations of temperature (°C) from normal with westerly and meridional flow during winter. (After LEBEDEV, 1958.)

European Russia. This is primarily a winter type flow. When it does occur occasionally in summer it may bring very hot dry clear weather to central European U.S.S.R. The Arctic air coming in from the northeast originally has very low absolute humidity. Intense warming in contact with the surface under the influences of strong radiation through clear dry air causes surface air temperatures to rise rapidly over the land mass as the flow proceeds southwestward and the relative humidity drops to very low values.

As can be seen from Table 3-II, the zonal flow from the west occurs 45% of the time and the flow around an upper level trough over the east European plain occurs 31% of the time. These are by far the two most dominant flows, and hence the air masses associated

TABLE 3-II

ANNUAL FREQUENCIES OF CIRCULATION TYPES ACCORDING TO POGOSYAN AND SAVCHENKOVA, IN PERCENTS OF TOTAL TIME
(After BLUMENSTOCK and PROZOROWSKI, 1956)

	Annual Percent Occurrence
Type I	45
Type II	31
Type III	12
Type IV	12

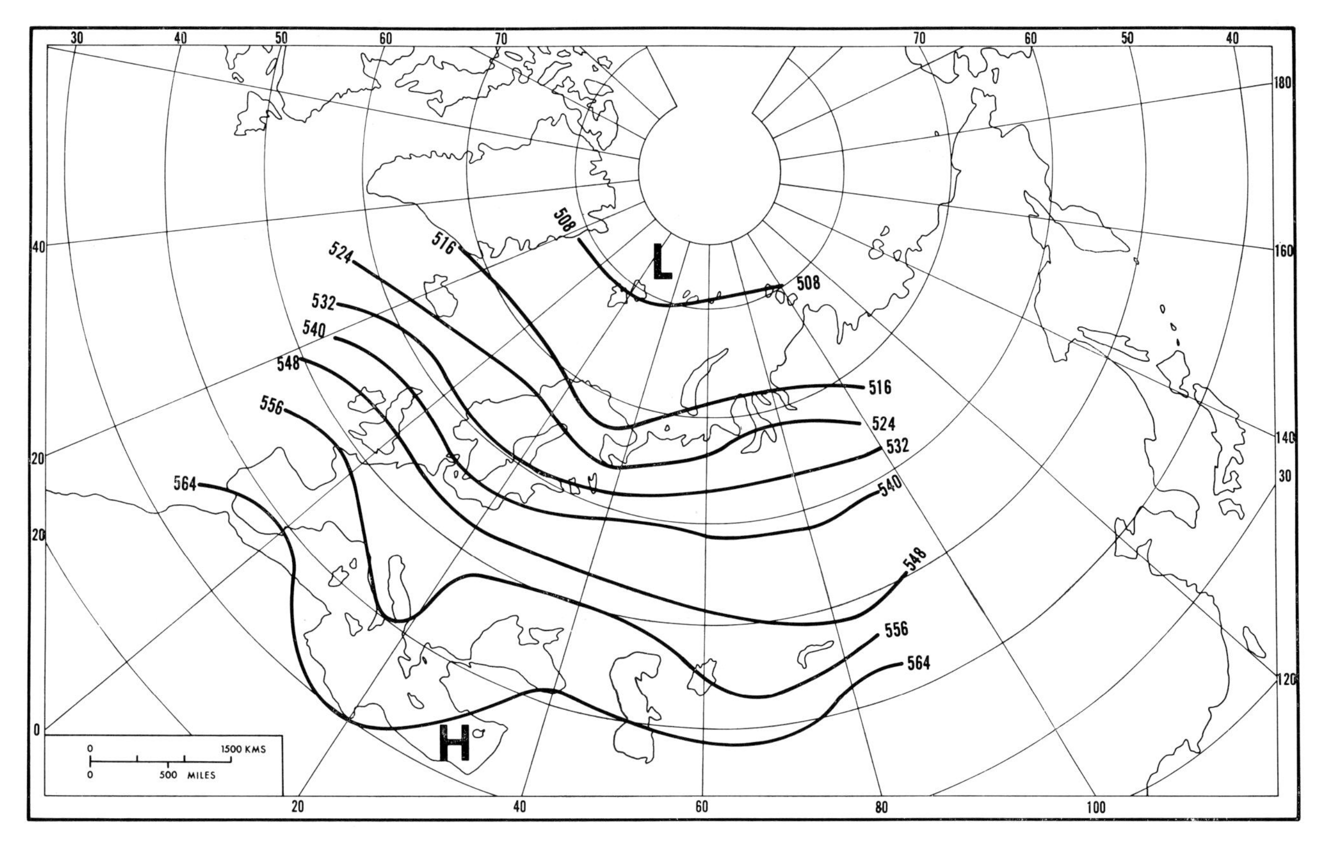

Fig.3-18. Atmospheric circulation type I. Mean 500-mbar contours for April 18–23, 1944, in tens of meters. (After POGOSYAN and SAVCHENKOVA, 1950.)

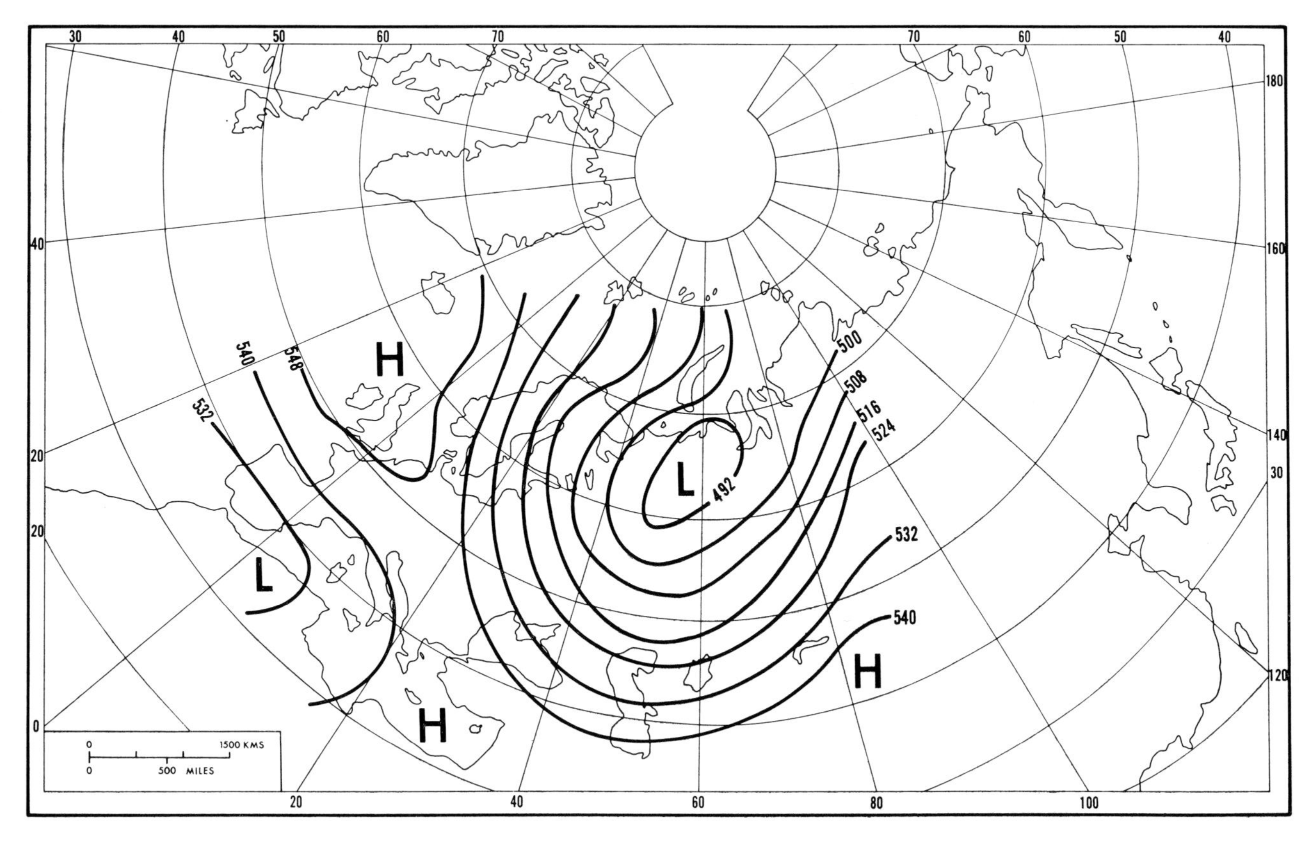

Fig.3-19. Atmospheric circulation type II. Mean 500-mbar contours for February 16–23, 1938, in tens of meters. (After POGOSYAN and SAVCHENKOVA, 1950.)

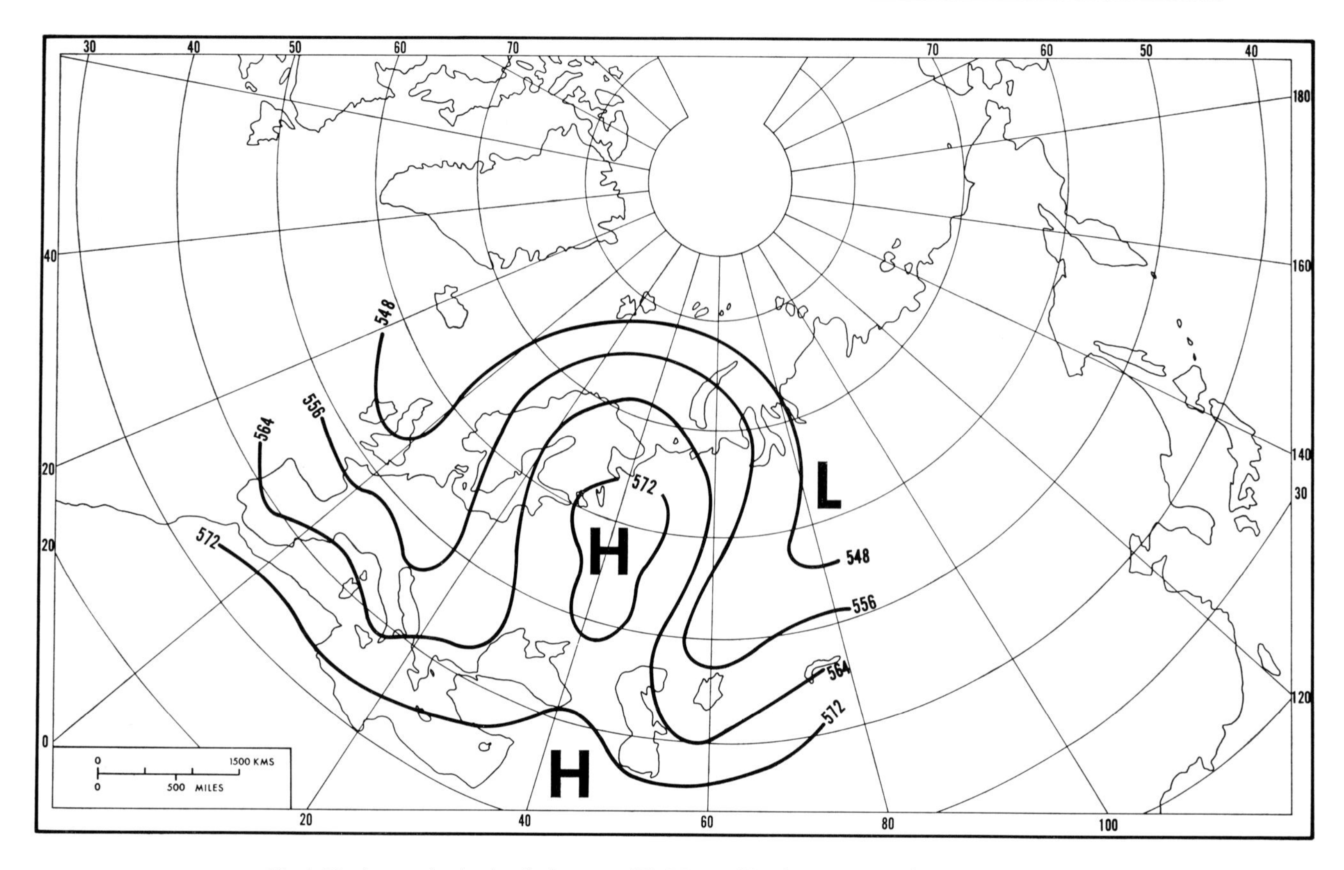

Fig.3-20. Atmospheric circulation type III. Mean 500-mbar contours for August 28–September 3, 1938, in tens of meters. (After POGOSYAN and SAVCHENKOVA, 1950.)

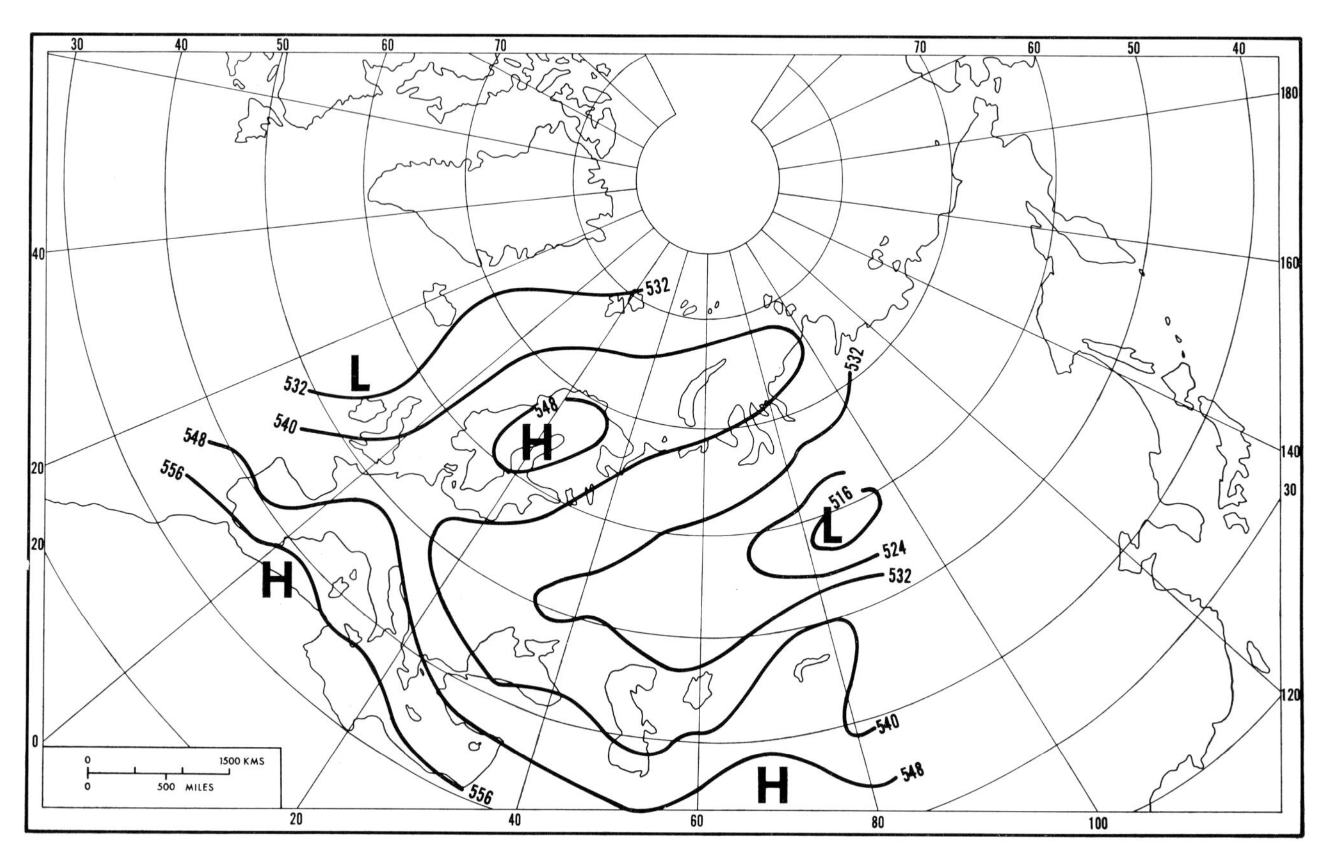

Fig.3-21. Atmospheric circulation type IV. Mean 500-mbar contours for December 13–18, 1938, in tens of meters. (After POGOSYAN and SAVCHENKOVA, 1950.)

with them are the most dominant types of air masses in the region. This would include the maritime temperate air from the Atlantic and maritime Arctic air from the Norwegian Sea. Of course, as these air masses cross the continent they modify greatly and eventually all become continental temperate in character. The type III air flow with the blocking high over the east European plain occurs only 12% of the time, as does the easterly flow associated with the blocking high over Scandinavia. Thus, the occurrence of such air masses as continental tropical or maritime tropical from the southwest and continental Arctic from the northeast occur much less frequently than do the air masses previously mentioned.

Air masses

The European part of the Soviet Union is affected by six different air masses in varying degrees. These were listed with their frequencies at regional centers for the entire year in Table 2-I. Their frequencies of occurrence, of course, vary greatly across the plain and from one time of year to another.

As can be seen in Fig.3-22, during January continental temperate air is the most frequently occurring air mass throughout the entire east European plain. It is twice as dominant in the southeastern part of the plain as in the northwest. In the north Caucasus and adjacent North Caspian Lowland more than 24 days during January are dominated by continental temperate air. This area is isolated by distance from other types of air masses and is under the influence of the western end of the Asiatic high which is the source for much of the continental temperate air. The northwest part of the plain is exposed to intrusions of maritime temperate air from the west, and along the Baltic coast this air mass occurs just as frequently in winter as does continental temperate air. However, the occurrence of maritime temperate air decreases rapidly east-southeastward. Maritime Arctic air enters from the Barents and Norwegian seas and proceeds south-southeastward across the plain where it gradually modifies into continental temperate air. Its occurrence in January averages 6 days in the north and only 1 in the south. Continental Arctic air enters from Siberia and the frozen Arctic east of Novaya Zemlya and proceeds southwestward across the European plain. Its occurrence in January also amounts to approximately 6 days in the north and 1 day in the south.

Continental Arctic air is the coldest air mass during winter. It is usually extremely stable, often with a subsidence inversion at a height of 1–2 km above the ground. The surface mixing ratio is usually less than 1 g/kg. Maritime Arctic air is slightly more moist, with surface mixing ratios averaging about 1.2 g/kg. It is usually stable to a height of about 1 km, since it has moved across land that is colder than its source region in the Arctic, but it is usually conditionally unstable aloft and often gives rise to snow flurries. It usually does not display a subsidence inversion. Continental temperate air is usually more dry and stable than maritime Arctic air, but not appreciably colder. It is coldest when it is mixed with continental Arctic air further north. Maritime temperate air is the warm air mass for winter. In the western and southern parts of the plain it may produce frequent thawing even in midwinter. It is stable air with an average mixing ratio of 1.5–2 g/kg at the surface. It is generally associated with overcast skies of low lying stratoform clouds.

May is the month of greatest meridional flow and hence the month of greatest occurrence

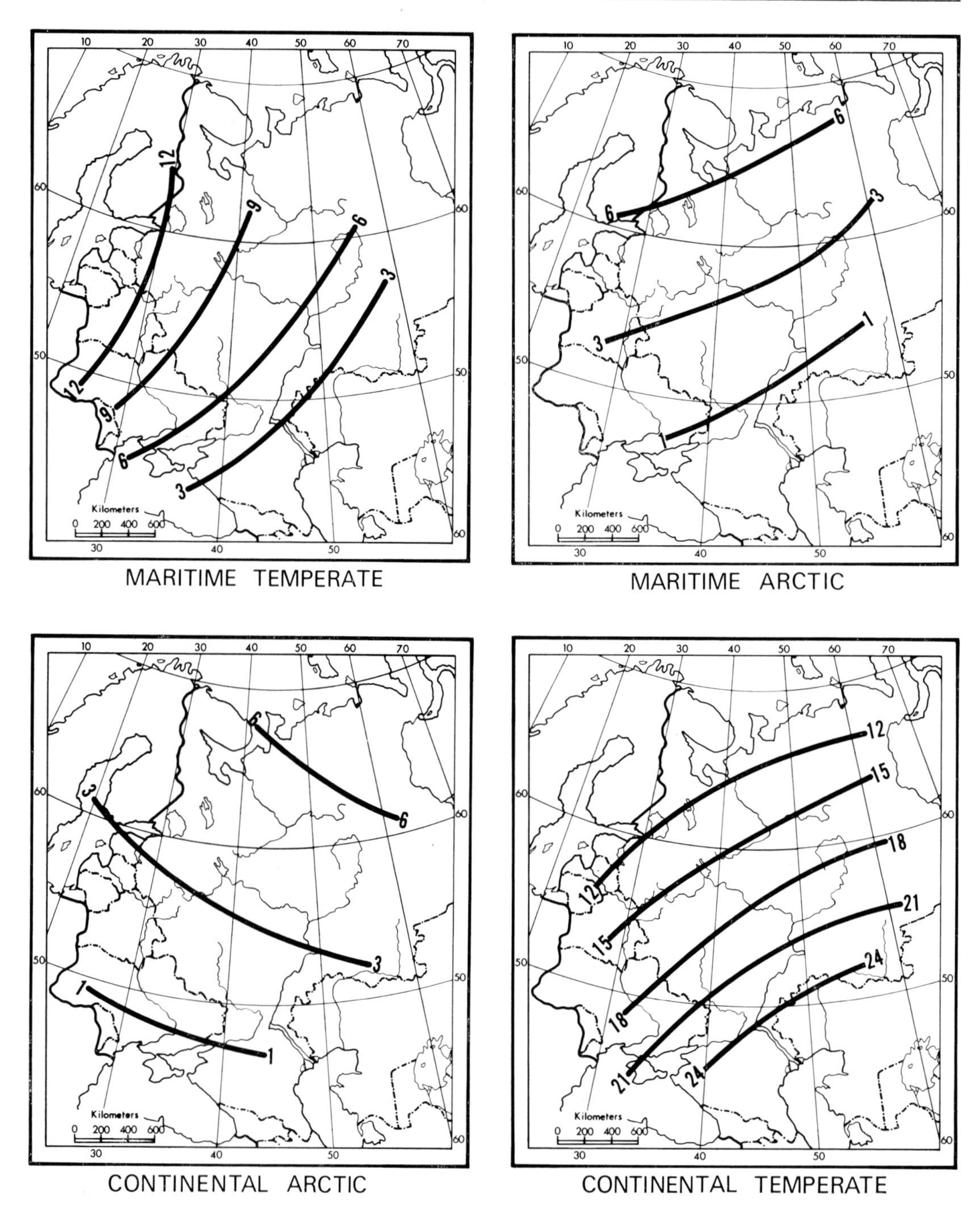

Fig.3-22. Frequencies of air masses during January, in days. (After FEDEROV and BARANOV, 1949.)

of Arctic air in European U.S.S.R. Continental Arctic air occurs about 18 days during May in the northeastern part of the European plain and 3 days in the far south. Apparently the intensification of cyclonic activity along the Polar front in the lower Volga–southern Urals area in late spring induces the more active intrusion southward of continental Arctic air from the north Urals area along the western peripheries of cyclonic storms. This air intrudes clear into the Black Sea steppes in the south. The frequency of maritime Arctic air remains at about 6 days in the northwest and 1 in the southeast. Continental temperate air still occurs about 24 days in the southeast, but in the north its occurrence is reduced to only about 6 days. Maritime temperate air during May occupies the western part of the plain 6 days. Its frequency diminishes rapidly east-southeastward until at Moscow it occupies the area only a little more than 3 days.

During summer the cold air source for European U.S.S.R. is the Arctic sea surface and immediately adjacent land along northernmost Eurasia. Arctic air masses during this time of year are not distinguishable into continental and maritime. Cool air enters the continent from the north Atlantic as maritime temperate air. Warm air develops in the central portions of the east European plain and western Siberia, and a hotter source exists around the northern end of the Caspian Sea and eastward into Kazakhstan and Central Asia. In addition, Asia Minor may act as a continental tropical air source during the summer, and the Balkans and adjacent areas of the Mediterranean may act as a source for maritime tropical air. There is such rapid modification of moving air masses across the warm dry surface of the southern plains in summer that no matter what the origins of the air masses they are quickly modified into a rather hot dry continental temperate type of air. Therefore, one has to specify as to the freshness of maritime temperate or Arctic air during the summer when characterizing their properties.

Fig.3-23 shows the frequency distribution of the most dominant air masses in the European Plain in summer. Again, over much of the plain, continental temperate air is dominant, although its distribution is considerably different than it is during January. The maximum frequency of occurrence lies in the central portions of the plain and frequencies diminish both southeastward as continental tropical air takes over and northwestward as maritime temperate and Arctic air become dominant. Arctic air masses are considerably more frequent in summer than in winter as the relatively warm land induces air flows from north to south across the plain. Continental tropical air occurs more than 1/3 of the time in the North Caspian and North Caucasian areas and can be traced as far northward as the southern parts of Archangel Oblast. In the west maritime temperate air occupies the region with about the same frequency as it did during the winter. Its frequency in the eastern part of the plain is somewhat less in summer than in winter because the modification process takes place more rapidly during summer.

During summer maritime temperate air generally enters the east European plain with a surface mixing ratio of about 12 g/kg. This is the moistest air during the summer on the east European plain except for occasional intrusions of maritime tropical air from the southwest. The air is usually conditionally unstable, and rapid heating at the surface may produce considerable cumulus activity with occasional thundershowers. As warming takes place over the land the moisture is mixed upward, and surface mixing ratios drop to about 8 g/kg by the time the air has been modified into continental temperate air. A mean temperature lapse rate of about 8°C/km becomes established to a height of about 5 km, and the air remains conditionally unstable.

Arctic air masses in summer enter the northern coast with surface mixing ratios of about 7 g/kg. They are usually conditionally unstable and may produce short showers as they move inland and are heated by the earth's surface. As the air becomes modified into continental temperate, surface mixing ratios drop to about 5 g/kg and the air becomes stable.

Continental tropical air moves north-northwestward from the Caspian and north Caucasian area. It is quite dry but has a steep lapse rate, commonly about 9°C/km, so

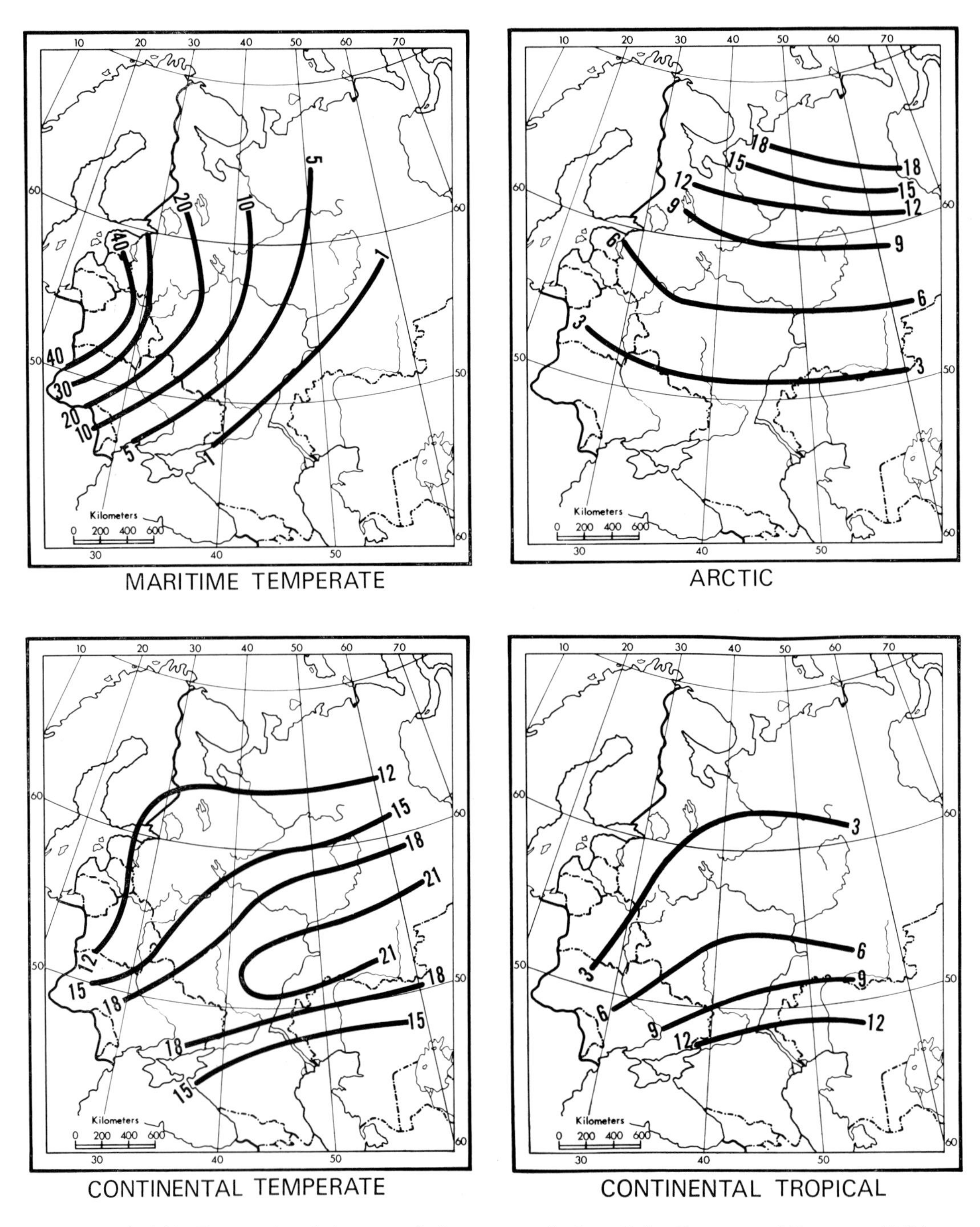

Fig.3-23. Frequencies of air masses during summer, in days. (After FEDEROV and BARANOV, 1949.)

that the air is conditionally unstable and often produces convective clouds and even some thunderstorms despite the low mixing ratios and high condensation levels.

During September continental Arctic air from the northeast decreases to 9 days in the northeast and 1 day in Moscow. Maritime Arctic air from the north reaches its greatest frequency with 15 days in the White Sea area and 3 days in the far south. Thus, there are considerable differences in air flows across the plain between May and September. The major air flow in September seems to be straight north–south from the Barents Sea to the

Black Sea, while in May the flow is from northeast to southwest. During September the flow of maritime temperate air from the west is somewhat stronger than during May. It occupies more than 9 days along the western border. This reduces to 0 in the Volga Delta. The dominant air mass over much of the plain in September is still continental temperate.

Moisture flux

The net results of complex flows and complex frontal and pressure situations in European U.S.S.R. were revealed to a certain extent by a study on moisture content and moisture flux that was made during the 1950's over European U.S.S.R. by DROZDOV and GRIGOR'EVA (1963). A full discussion of the study is contained in Chapter 9 in this book. It is quite obvious from Figs.9-5 and 9-6 that on the average throughout the year total moisture flux across European U.S.S.R. is almost exactly zonal from west to east. Hence, the primary moisture source for the region is the Atlantic Ocean. However, the moisture flux varies considerably from year to year and from one period to another within a year, depending upon the predominant flow pattern, and it fluctuates in direction much more at lower altitudes than at higher altitudes. About half the moisture flux takes place in the first 1,500 m of the air where the moisture content is higher, and it is just this boundary layer of air that undergoes the most fluctuation.

During winter there is a significant meridional component of air flow from south to north throughout the entire plain all the way from the Black Sea to the Arctic coast. Obviously in winter the Mediterranean and adjacent water bodies are significant sources of moisture for European U.S.S.R. During summer this is not so much the case, although the Black Sea might supply some moisture to cyclonic storms that are regenerated in the area along the Polar front. More significant in summer probably is the influx of moisture from the Arctic in the northern part of the plain particularly from the Barents Sea.

In addition in summer a considerable amount of moisture is contributed to the air by evaporation from the underlying land. The moisture evaporated from the land is not transferred upward instantaneously and therefore the vertical lapse rate of moisture in summer is generally much greater than it is in winter. This leads to much greater atmospheric instability in summer than in winter and enhances the instability role of surface heating during the summer.

It has been calculated that precipitation derived from evaporation from the land amounts to about 10–20% of total precipitation during the year at the western edge of the U.S.S.R. and 30–40% in the southeastern part of European U.S.S.R. The largest amount of precipitation stimulated by evaporation occurs in June, but percentagewise the role of evaporation is greatest in May when over the entire territory of European U.S.S.R. evaporation is responsible for 65% of the precipitation that falls. During the course of the year local evaporation increases precipitation by approximately 28% over European U.S.S.R.

Thus evaporation plays a very important role in precipitation even though only small amounts of locally derived moisture are actually precipitated. The addition of moisture at the surface increases the conditional instability of the air and thereby stimulates a disproportionate increase in precipitation. This ratio is much higher during the warm months of the year. It has been calculated that during the period April–August the

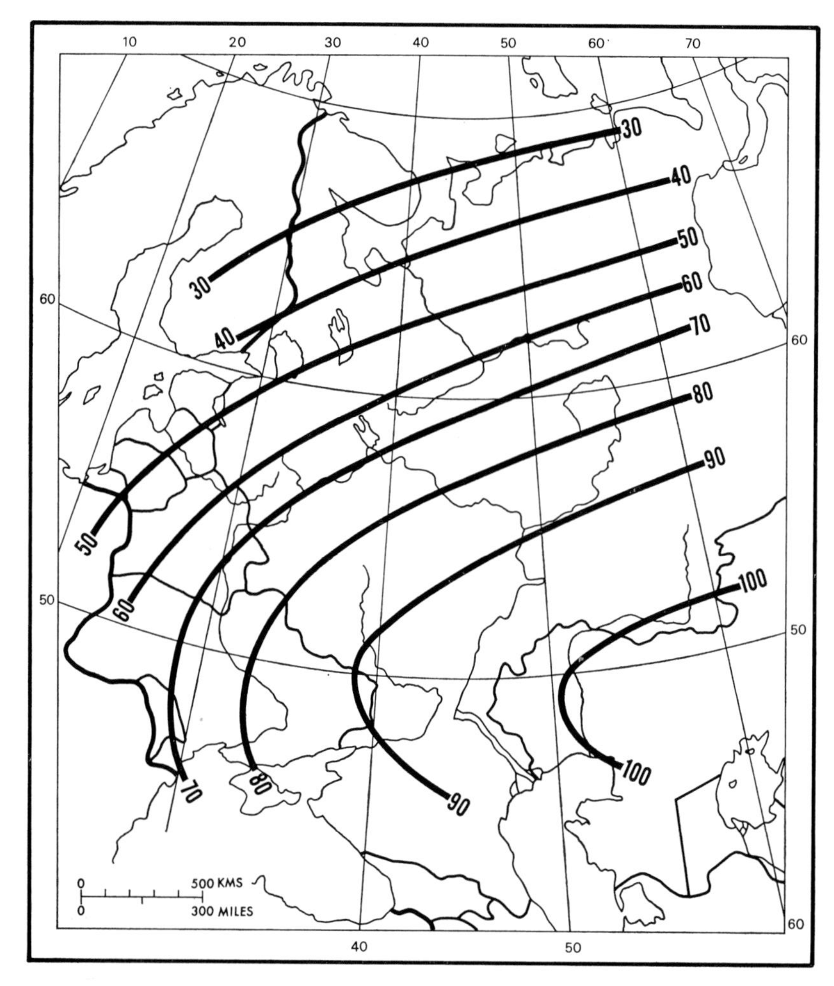

Fig.3-24. Percentage of total precipitation stimulated by evaporation from the surface, April–August. (After ZHAKOV, 1963.)

percentage of the precipitation that is stimulated by evaporation from the land amounts to 100% in the North Caspian Lowland and as much as 30% along the Arctic coast (Fig.3-24). These figures reduce to 50% in the North Caspian Lowland and 10% along the Arctic coast during the period September–October, and during winter the role of evaporation is essentially nil.

Subregions

The territory of the European part of the U.S.S.R. and western Siberia which has been defined here according to general circulation features is so extensive that a description of its weather and climate necessitates a breakdown into subregions. Obviously, an area that stretches latitudinally from 43° to 78°N, and more in island groups in the Arctic, is going to experience tremendous differences in temperature as well as in other elements of climate. Also, the east–west expanse of the plain experiences different degrees of continentality which induce wide differences in temperature and moisture conditions. The relief of the plain, too, although not great, is enough to induce significant influences on temperature and precipitation, as well as on some other elements of climate, such as radiation and surface air movement. Such features as the Central Russian Uplands, the Volga Heights, the Podolian Plateau, etc. rise a hundred meters or more above adjacent lowlands and cause differences in precipitation on the order of 10% between their prevailingly windward and leeward sides. The Ural Mountains, of course, which at their highest rise to almost 1,900 m, greatly influence the climate, and will be discussed briefly in a separate section. The same is true of the Carpathian and Crimean Mountains on the southern fringes of the plain.

During winter temperature is more affected by atmospheric circulation, relief, and nearness to surrounding water bodies than it is to latitude since at this time of year the influx of radiation is small and more than 60% of it is reflected by snow cover. Therefore, during winter isotherms in the western part of the Soviet Union run more meridionally than latitudinally (Fig.8-29). Orenburg at a latitude of 54°45'N has a colder January than does Archangel at latitude 64°32'N. Astrakhan at the mouth of the Volga at latitude 46°21'N is colder in January than Vyborg on the Finnish border at 60°10'N. The same contrast holds for number of days with thawing during the winter. Along the western border thaws are experienced about 50 days between the first of December and the last of February. This diminishes to about 5 days along a line from Archangel to Orenburg and to zero along the lower Ob. During December–February periods of thawing completely dissipate the snow cover on about 15 days along the western border and about 1 day through the central region around Moscow. But in the northeastern part of European U.S.S.R. and western Siberia the snow cover is never completely dissipated during this period.

During summer the temperature gradient is somewhat the opposite, with the increased continentality of the eastern regions generally causing higher temperatures than in the western part of the country which is dominated more by maritime temperate air from the Atlantic. There are also east–west differences in precipitation and moisture conditions, although these are not as great as the north–south differences.

Therefore, it is necessary to divide the plain in both north–south and east–west directions. But the boundaries devised for any subdivision of the plain are going to be rather arbitrary and indistinct since generally there is not enough topographic variety to cause abrupt changes in climate. The Urals might be an exception to this, but not a striking one. Many climatic subdivisions of the Eurasian plain have been devised, some of them simple, some of them very complex. The one used here has striven for simplicity and yet a differentiation of distinctions wherever they exist. The low broken Urals, although they do not form a very distinguishable climatic boundary between northeast European U.S.S.R. and western Siberia, do affect climatic statistics within themselves, because of elevation and exposure factors, and therefore they will be treated under a separate heading although they have not been accorded subregional status on the map in Fig.2-22.

The south

Nature of the region

The southern part of the plain in European U.S.S.R. is relatively far removed from its main source of moisture, the Atlantic Ocean, and suffers from some degree of moisture deficit much of the time. The southern two-thirds of the region is made up of the so-called Black Sea steppes along the northern coast of the Black and Azov seas, the North Caucasian Foreland, the North Caspian Lowland, and the Crimean Peninsula, most of which, except for the Crimean Mountains along the southeastern third of the Crimean Peninsula, have the primary aspect of flat, low lying, dryish plains. The droughty nature increases toward the east and south. Odessa in the west receives 389 mm of precipitation per year, and Gur'ev in the east receives only 164. The most humid part of this southern plain is the so-called Kuban District along the Kuban River in the western part of the North Caucasian Foreland, where the city of Krasnodar receives 610 mm of precipitation per year, an amount sufficient for many crops without irrigation.

The northern third of the region is a more rolling plain, becoming locally rather hilly, particularly around the western sides of major streams such as the Dnieper and the Volga. Precipitation generally is somewhat greater than in the south. Khar'kov averages 519 mm per year. Nevertheless this part of the plain also is consistently hampered by moisture deficits and is occasionally subjected to severe drought. Khar'kov suffers from some degree of drought more than 20% of all years and the frequency increases to 60% at Volgograd. Moderate to severe droughts occur more than 10% of the time at Khar'kov and more than 50% of the time in the North Caspian Lowland (Fig.9-56).

Air masses

The entire region is dominated by locally derived continental temperate air masses which form in stagnating high pressure cells that have generally entered the region during the summer from the west as the eastern extension of the Azores high or from the north as an intrusion of Arctic air which rapidly modifies on its way southward. In winter much of the air originates from the western extension of the Asiatic high. Voyeykov's famous "great axis" of high pressure generally extends from east to west across the northern half of the region in winter which causes the air to flow from northeast to southwest over much of the area during that time of year. Hence, during winter the area receives little benefit from the Black and Caspian seas except on occasions when southerly cyclonic storms move into the area. The northern portions of these seas freeze over during middle and late winter which further reduces their influences on the adjacent northern shores. At Dniepropetrovsk in the western part of the region on the average throughout the year continental temperate air occurs about 54% of the time. Maritime temperate air on the other hand occurs only 14% of the time. Continental Arctic air occurs 5% of the time, maritime Arctic air 6.5% of the time, continental tropical 9% of the time, and maritime tropical air 11.5% of the time (Table 2-I). As one progresses eastward through the region toward the North Caspian Lowland the influences of continental temperate and continental tropical air increase and the influences of maritime temperate and maritime tropical air decrease.

None of these air masses are very fresh when they get into the southern plain and they are not strikingly distinguishable one from another. During winter maritime temperate air usually is pulled into the region around the southwestern periphery of a cyclonic storm that is well developed in the northern part of the plain. In summer maritime temperate air generally enters around the northeastern periphery of the Azores high. By the time the air has reached the southern plain it has been modified considerably and brings only a modest amount of moisture to the area.

The Arctic air masses from the north also are greatly modified but they do bring the coldest temperatures to the region during the winter. During summer the dry clear air from the Arctic may heat so much across the plain that by the time it reaches the southern part of the plain it is just as hot as the air already in the region. In fact, the famous sukhovey phenomenon typically occurs after an intrusion of Arctic air from the north which builds up a stagnant high pressure cell in the south and heats up intensely under strong insolation through the very clear dry air.

Continental tropical air enters the region from the southeast from Central Asia around the northern end of the Caspian Sea. This does not occur as frequently as at first thought, and much of the hot air in the North Caspian Lowland in summer is locally derived. However, when a relatively strong southeasterly flow brings in desert air it creates some of the hottest temperatures experienced during the summer.

Maritime tropical air is brought into the southern plain, particularly along the northern Black Sea coast, in the warm sectors of cyclonic storms that move from southwest to northeast across the area. This is the moistest air that enters the region, but it enters rather infrequently and it is not as moist or unstable as typical maritime tropical air in other parts of the world. Much of this air is derived over the eastern Mediterranean and has not had a long trajectory over a warm water surface. The proximity of Africa and Asia Minor limits its development.

Cloudiness and precipitation

The southern plain during the year averages less cloudiness than the plain farther north. The amount of sky cover averages about 6.3–6.7 tenths through the year. It is much cloudier in winter than in summer. During January much of the Black Sea coast registers more than 8 tenths sky cover, whereas during July it averages 3.9–4.9. The winter maximum of cloudiness relates to the frequency of cyclonic storms at that time of year, which are associated with widespread stratus clouds. During summer much of the cloud is cumuloform due to convection.

Precipitation on the plain varies from about 380 mm per year in the southwest to more than 500 mm along the northern border to about 160 mm in the extreme east. Everywhere the precipitation is insufficient to satisfy potential evaporation which ranges from about 600 to 900 mm per year. Throughout most of the plain the maximum precipitation falls in early summer. June is the month of maximum in most cases. As one proceeds northward the maximum lags into July. There is also a July maximum around the northern end of the Caspian. On the southeast coast of the Crimean Peninsula precipitation is at a maximum during winter when the southern coast is frequented by cyclonic storms. However, there is a secondary June and July maximum caused by mild thundershowers. Yalta, which has been exalted as the Soviet area of Mediterranean climate, is not really Mediterranean in type. Although it has a January–February maximum, it has precipitation every month of the year and a secondary maximum in June and July. Its total annual precipitation is 560 mm, which is as much as anything on the plain to the north.

The region receives a moderate number of thunderstorms. Much of the plain experiences from 23 to 27 days with thunderstorms per year, although the frequency diminishes to about 11 in Gur'ev along the north Caspian coast. Everywhere the maximum frequency occurs in summer, generally in June or July. Occasionally, hail accompanies the storms. Sometimes thunderstorms are organized in meridionally-oriented squall lines that move eastward in the Rostov–Krasnodar area. It appears that this phenomenon occurs about once per year southeast of the Sea of Azov and about five times per year in the Donets area farther north. Thunderstorms are a rare occurrence in winter. Simferopol on the northern plain of the Crimean Peninsula has experienced only three winter thunderstorms during its period of record, which began in 1891.

Snowstorms occur about 5 days during the winter along the southern coast and from 20 to 30 days along the northern border of the region. It falls very rarely along the immediate Black Sea coast, and then it is generally in the form of light flurries associated with the intrusion of cold air from the northeast. Snow cover throughout the region is generally thin during the winter, and there is an extensive region in the central Black Sea steppes including the Crimean Peninsula which has no snow cover at all at least 50% of the years. Frequent thawing and refreezing in this southern region may produce ice crusts on the ground which may be added to by pre-warm frontal ice storms associated with cyclonic storms advancing from the south. The Donets Ridge north of the eastern end of the Sea of Azov and the Stavropol' plateau in the north Caucasian plain are areas of unusually high occurrence of glaze ice. In both regions about 30 days each winter experience such conditions.

Wind

One outstanding feature of the climate of this southern region is the relatively high wind speeds which sweep across these treeless plains (Fig. 10-3). The winds whip up the fine dry soil in the area and cause frequent dust storms that are particularly severe in the north Caucasian plain, the lower Don region, and the Black Sea steppes including the northern plain of the Crimean Peninsula (Fig.9-59). Some of the most severe soil erosion takes place in these regions during periods in winter when the snow cover is thin or absent and strong easterly or southeasterly winds circulate around the western end of the Asiatic high. The dust generally circulates northwestward through the Ukraine into the western part of the Soviet Union and sometimes can be identified as far north as Scandinavia.

In the northern part of this southern region dust is frequently found interbedded with layers of snow in late winter, signifying the alternation of dust storms and snow storms throughout the winter. A strong wind with blowing snow is known as "buran" throughout the Soviet Union. A dust storm is known as "chernaya burya", or "black buran". Over thousands of years the dust has been deposited throughout much of the northwestern Ukraine where it has accumulated to thicknesses of as much as 30 m or more in the form of loess soil. The Soviets have planted many shelter belts in this southern region to combat the actions of the wind.

Occasionally winds reach hurricane force, and very rarely, perhaps once in five years, a tornado occurs somewhere on the plain. Accounts are very scattered in the literature and there are no statistics on numbers or characteristics of tornadoes, but a few photographs of funnel clouds verify their existence.

Foehn winds sometimes descend the slopes of the Crimean Mountains on either the southeast or the northwest depending upon the pressure pattern. They occur along the southeastern slopes when a strong high builds up over the plain north of the Black Sea and a strong cyclone forms in the eastern Mediterranean. They occur on the northern slopes either with an anticyclone centered over central Kazakhstan with southeasterly winds around its western nose across the Black Sea or with a low pressure centered over the central Mediterranean and southerly winds circulating around the eastern end of the low across the Black Sea. In the northern Crimea foehns occur on an average of 69 days per year, 42 cases of which are associated with Mediterranean cyclones and the other 27 of which are associated with high pressure centers in Kazakhstan. On either side of the mountains foehns are most numerous from March to June when the atmosphere circulation is most vigorous in this region.

The most characteristic feature associated with foehns in the Crimea is a sharp drop in relative humidity. Temperature rises may also take place but generally they are not spectacular. However, occasionally the temperature rise is significant. Along the southern coast temperatures have increased as much as 16°C with a foehn and relative humidity has dropped as much as 70%. When foehns are descending the southern slopes the south coast is frequently 20°C warmer than the northern steppes of the Crimea. Wind speeds at heads of canyons are generally 15–20 m/sec, but gusts may reach 40 m/sec. Strong winds from an easterly direction may bring foehn winds up to 40 m/sec in Sevastopol' on the western coast of Crimea. Occasionally in winter cold air descends the south slopes from the northern plains and, although it warms adiabatically, it does not reach temperatures as high as those already along the coast. Under these conditions a bora wind is experienced.

Weather types

The variations in weather across the southern plain are represented by Figs.3-25–3-28 which show the frequencies of weather types at the stations of Kishinev in central Moldavia in the west, at Voroshilovgrad in the eastern Ukraine, at Gur'ev on the northern coast of the Caspian, and at Yalta on the southeastern shore of the Crimean Peninsula. These graphs, first devised by E. E. Federov, will be used extensively throughout the regional discussions. The graphs can be augmented with precise data from the tables in the Appendix at the end of the book.

From Fig.3-25 it can be seen that at Kishinev summers are relatively long and warm, and winters are relatively mild. Cloudiness is considerable and is greatest during the winter. Precipitation is moderate and reaches a maximum in June. Mean temperatures range from approximately −8°C in January to +20°C in July, and an extreme minimum temperature of about −30°C has been observed in February and an extreme maximum temperature of approximately 35°C has been observed in July. Farther east at Voroshilovgrad the summers are hotter, the winters are slightly colder, and the precipitation is somewhat reduced. At Gur'ev the summers are much hotter, although no longer, and the winters are somewhat colder. Precipitation is much reduced, as is cloudiness throughout the year.

Yalta on the Crimean coast shows a pattern of weather types that is much different from the other three. It has a long warm summer with a short period of humid tropical weather, and it has very mild winters. A considerable amount of cloudiness exists during the

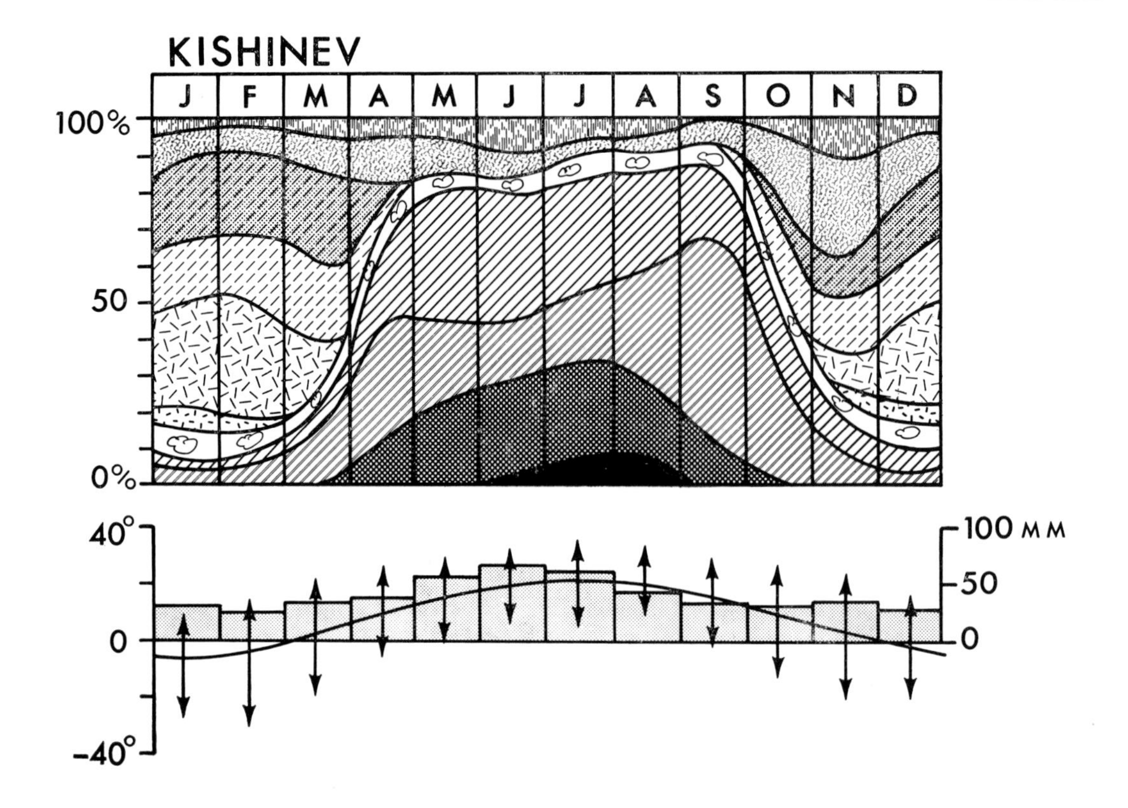

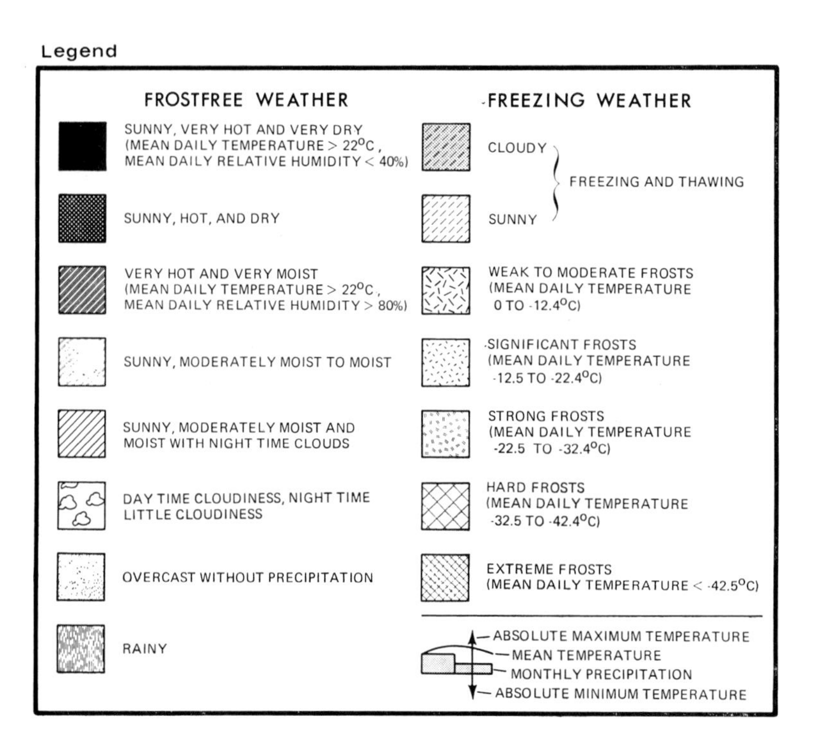

Fig.3-25. Frequencies of weather types, Kishinev. (After GERASIMOV, 1964.)

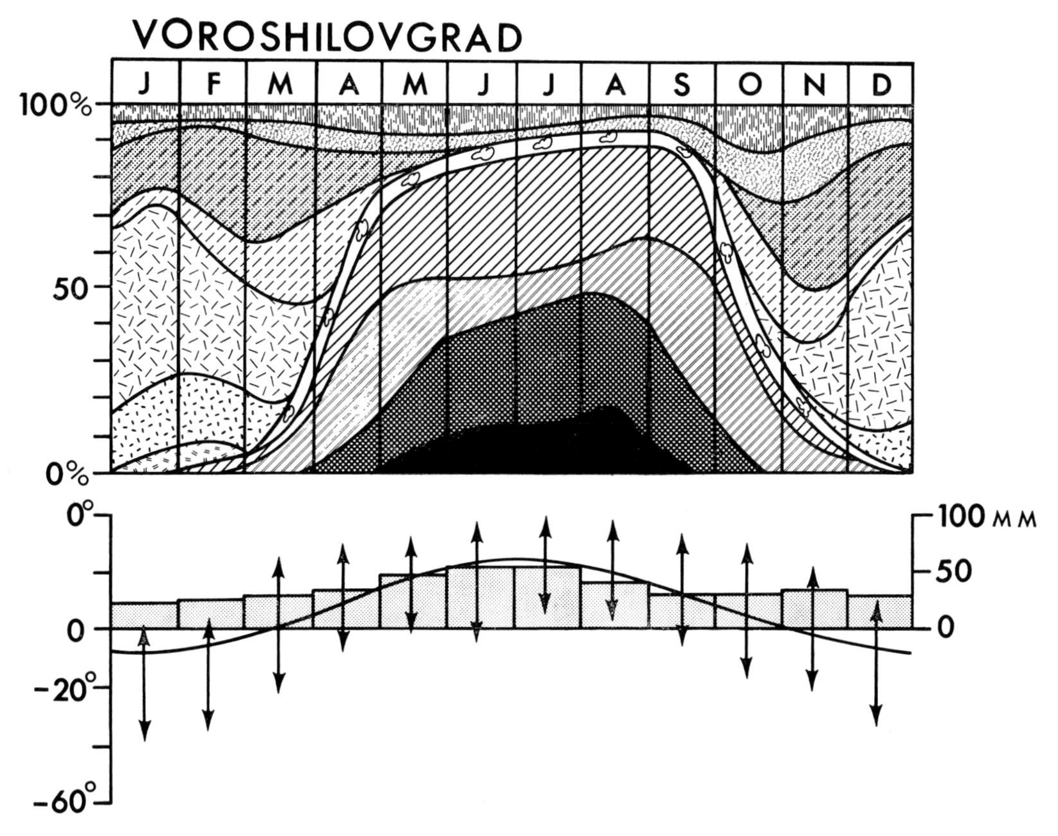

Fig.3-26. Frequencies of weather types, Voroshilovgrad. (After GERASIMOV, 1964.) For legend Fig.3-25.

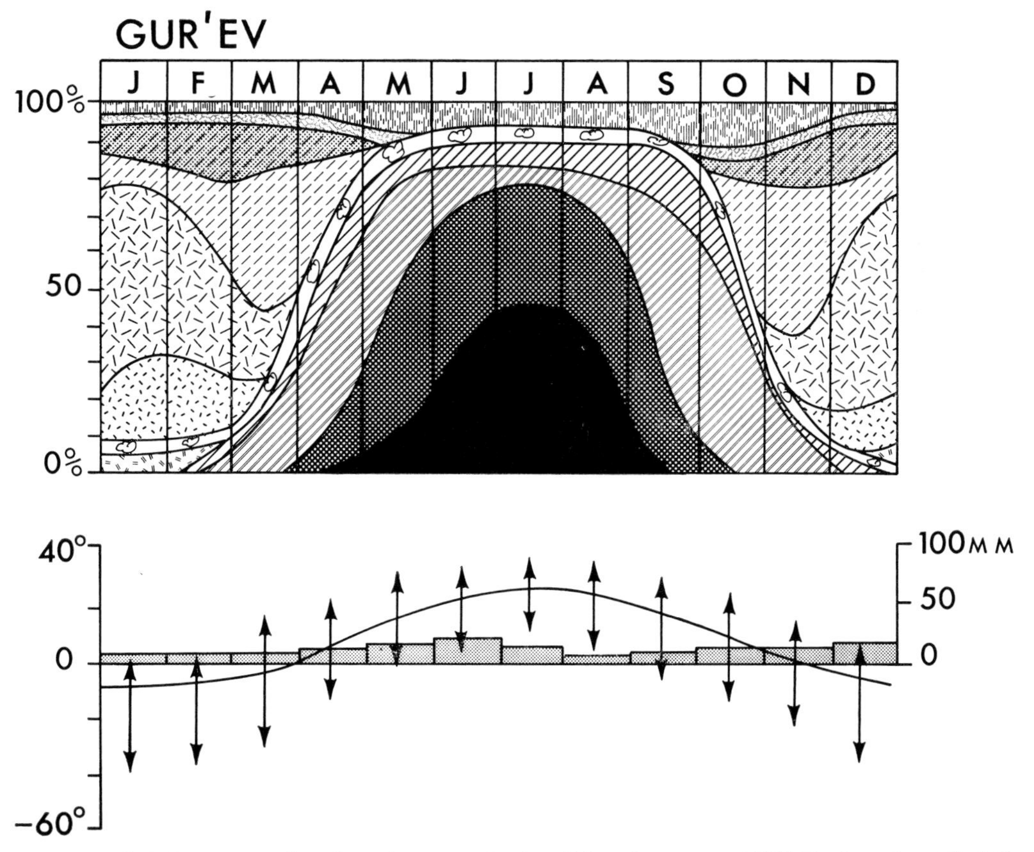

Fig.3-27. Frequencies of weather types, Gur'ev. (After GERASIMOV, 1964.) For legend see Fig.3-25.

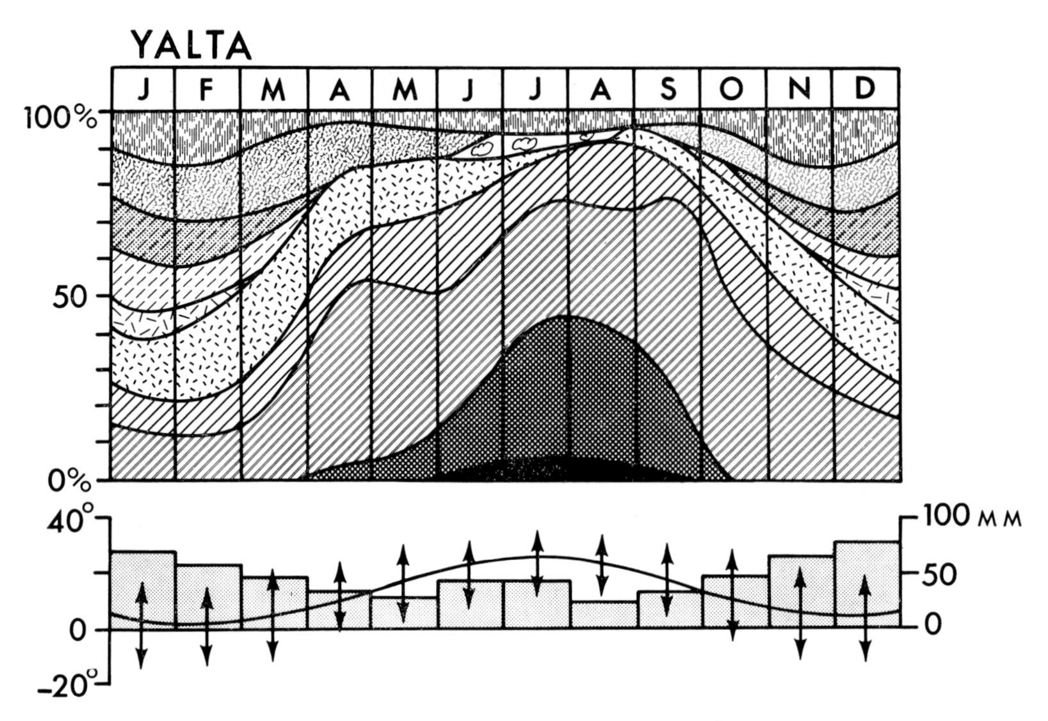

Fig.3-28. Frequencies of weather types, Yalta. (After GERASIMOV, 1964.) For legend see Fig.3-25.

wintertime. Precipitation is significant throughout the year but reaches a maximum in December, with a secondary maximum in June and July. Temperature ranges are relatively small.

The central region

The central zone of the European plain of the U.S.S.R. is a meeting place for air that comes in from the west, the north, and the south. Like the southern plain it also is dominated by continental temperate air (Table 2-I). But it is very strongly influenced by maritime temperate air, particularly in its western portions, and Arctic air masses enter the region much more frequently than they do farther south. Continental tropical air is still significant during the summer, but maritime tropical air is almost absent, reaching the region only occasionally during very strong southwesterly flows of air in warm sectors of cyclonic storms.

Air mass contrasts in the region are relatively weak, and frontal systems are indistinct and very complicated. Most of the moisture that comes to the region comes from the west or northwest, so that there is a gradual decrease in precipitation and soil moisture east-southeastward across the plain. The northwestern part of the area is heavily forested with a mixture of broadleaf deciduous trees and coniferous evergreens, but these give way gradually to a mixture of grasses and scattered trees in the southeastern half of the region, the area of the so-called "wooded steppe". In the far southeastern portion of the region around the southern Urals the area is open grasslands with no trees. Moscow sits in the north-central portion of this region and marks just about the northwestern limit of drought conditions. Northwest of Moscow the effects of drought are hardly ever felt.

Temperatures contrast greatly from west to east across the region also. This is particularly true in winter when the maritime temperate air from the Atlantic greatly meliorates temperatures in the western part of the region but does not influence them so much in the east. Also the greater cloud cover in the maritime air in the west greatly reduces radiational heat losses at night thereby holding minimum temperatures much higher than farther east. The temperature gradient across this region during winter is directed west–east and is independent of latitude. During the three winter months L'vov averages from $-1°$ to $-4°C$, while in Moscow temperatures average $-8°$ to $-10°C$, and in Orenburg they average $-12°$ to $-15°C$. Absolute minima temperatures during winter are $-33°C$ at L'vov, $-42°C$ at Moscow, and $-44°C$ at Orenburg.

During summer temperatures are generally higher in the eastern part of the region because of the greater continentality and drier land surface. In the west where there is much marsh and swamp as much as 80% of the annual insolation may be used for evaporation, and only 20% is available for air temperature increases. This typifies great portions of Belorussia and adjacent northwestern Ukraine in the Pripyat Marshes, portions of the Baltic states, and the Valday Hills area northwest of Moscow. In these regions summers are quite cool. Minsk averages only 17.6°C in July. Its absolute maximum temperature is 35°C. Moscow is a little warmer with a July average of 17.8°C and an absolute maximum of 37°C. This, in spite of the fact that Moscow is somewhat north of Minsk. Farther east at about the same latitude Kazan' has a July average of 20.0°C and an absolute maximum of 39°C. In the southeast Orenburg averages 22.0°C in July and has an absolute maximum of 41°C.

Precipitation in the region varies from an annual total of 655 mm at L'vov in the southwest to 575 mm at Moscow in the center to 358 mm at Orenburg in the southeast. In the west it exceeds potential evaporation, but eastward across the plain moisture deficits set in. At Moscow potential evaporation is about the same as precipitation or even a little more; at Orenburg it is about twice as much.

Almost everywhere in the region maximum precipitation comes in July, although it lags into August along the northern border. This is not the best rainfall regime for agriculture, since spring and early summer may experience some deficits of moisture while the harvest season may be hampered by wetness. In general fall is significantly wetter than spring, but during the fall period, usually in September or early October, the singularity "old women's summer" usually occurs throughout much of the plain. This brings stable clear dry conditions for several days on end as a high pressure cell builds up and stagnates in the southern plain. Warm air is fed into the mid-section of the plain in the rear of the high and also from aloft through the subsidence process.

About 3/4 of the annual precipitation falls during the summer half-year. This is primarily of the showery type, associated with weak fronts and sometimes with thunder and lightning. Kaliningrad on the Baltic experiences 17 thunderstorms during the months April–September. L'vov in the southwest experiences 26 during the same period. Moscow experiences 23, and Orenburg 22. Occasionally, particularly in spring, intense squall lines develop which move across the region toward the southeast. Heavy hail may fall from severe thunderstorms.

The winter half-year receives only about 1/4 the annual precipitation. During winter the precipitation is more of a prolonged continuous type with rather light intensity associated with cyclonic storms and overcast skies. The annual regime of number of days with precipitation is almost exactly opposite to that of the annual regime of amount of precipitation. Normally the winter months experience from 1/2 to 2/3 of the days with precipitation while summer months experience no more than 1/3 to 1/2. For instance, at Moscow May has only 12 days with precipitation while December has 19. This seasonal difference in rainy days is reflected in cloud cover, which averages about 8.4 tenths during January and about 7.2 tenths during July. Hardly ever are there any thunderstorms during winter, although once in a while they do occur. At Moscow one winter out of every three may experience thunder. This increases to one in every 2 years along the western border of the region and decreases to 1 in every 5 years along the Volga.

Much of the winter precipitation comes as snow, particularly in the eastern parts of the region. Much of the region receives snowfall on 25–30 days during the winter, although along the western border higher winter temperatures in the maritime air causes much of the precipitation to fall as rain. In this area only about 10 days during the winter experience snow. Snow depths are moderate. They range from less than 20 cm in the west to about 80 cm in the Bashkir Plateau on the western side of the southern Urals. Around Moscow snow depth averages about 50 cm during its period of deepest cover (Fig.9-41). Snow lies on the ground on the average of 105 days per year at Kiev, 146 days at Moscow, and 146 days at Orenburg. In the northwestern part of the region Kaliningrad on the Baltic coast has a snow cover only 7 days per year on the average, with a maximum depth of only 16 cm.

Wind speeds in the forest zone are generally lower than they are in the steppe zone to the south. However, on occasion a strong cyclonic storm may bring in winds of gale force, and very rarely tornadoes occur. In the vicinity of Moscow one occurred in June 1904, another one in September 1945, a rather weak one in August 1951, and a more severe one in August 1956 (Kolobkov, 1960, pp.55–80.)

Carpathians

In the southwest corner of the central region lies the mid portion of the Carpathian Mountains which in the Soviet Union rise to slightly above 2,000 m. The climate, of course, is altered by the elevation as well as by obstruction to surface air flow and the initiation of local winds. The Soviet portion of the Hungarian plain southwest of the mountains has milder winters because the mountains partially block cold northeastern air from the western protrusion of the Asiatic high. Uzhgorod in Transcarpathian Ukraine averages −2.9°C in January and has experienced an absolute minimum temperature of −28°C, whereas L'vov north of the mountains averages −3.8°C in January and has experienced an absolute minimum temperature of −33°C. During summer the surface air in Uzhgorod is significantly warmer and more humid than in L'vov.

Foehn winds occur from 10 to 15 days per year, most frequently during the winter half-year. They can occur on both sides of the mountains with a high pressure centered over the mountains, but the strongest foehns generally occur on the northern slopes in the warm sector of intense cyclonic storms which cross the region either from the Atlantic or from the Mediterranean during winter. The foehns generally produce rather intense inversions on the lower mountain slopes during winter. Warming on the northern slope during a foehn flow from the southwest is due both to adiabatic heating from descending air and advection of warm air from the Balkan Peninsula.

A typical sequence of weather with the passage of a cyclonic storm in winter is first warm frontal rain, followed by freezing rain as the cold air begins to move into the rear of the storm, and finally snow sometimes reaching blizzard proportions on the mountain slopes. During summer the region is frequented by thundershowers, often in association with weak fronts.

Annual precipitation varies from about 700 mm on the Transcarpathian plain around Uzhgorod to as much as 1,500 mm on the wetter crests and southern slopes of the mountains. Thus, the mountains are heavily forested, and the forests go right up to the top since the mountains are not so high that temperatures preclude tree growth. Although there is a definite summer maximum of precipitation in the Carpathians winter is by no means dry. Midwinter floods are rather common occurrences due to heavy snowfall in the mountains or rainfall in the lowlands.

Weather types

The variations of weather types across the central region are illustrated by Figs.3-29–3-32. Winter temperatures at Kiev and Minsk are very similar, whereas Moscow shows the influence of greater continentality with colder temperatures. During summer Moscow and Kiev have relatively warm conditions, while Minsk in the west is more tempered by the maritime air from the Atlantic. Cloud cover is relatively abundant at all three stations, with spring and fall being the cloudiest seasons. There is a midsummer maximum of precipitation at all three stations.

Orenburg in the southeastern part of the region shows much more continentality than the other three stations. It has long hot summers and very cold winters. Precipitation is much

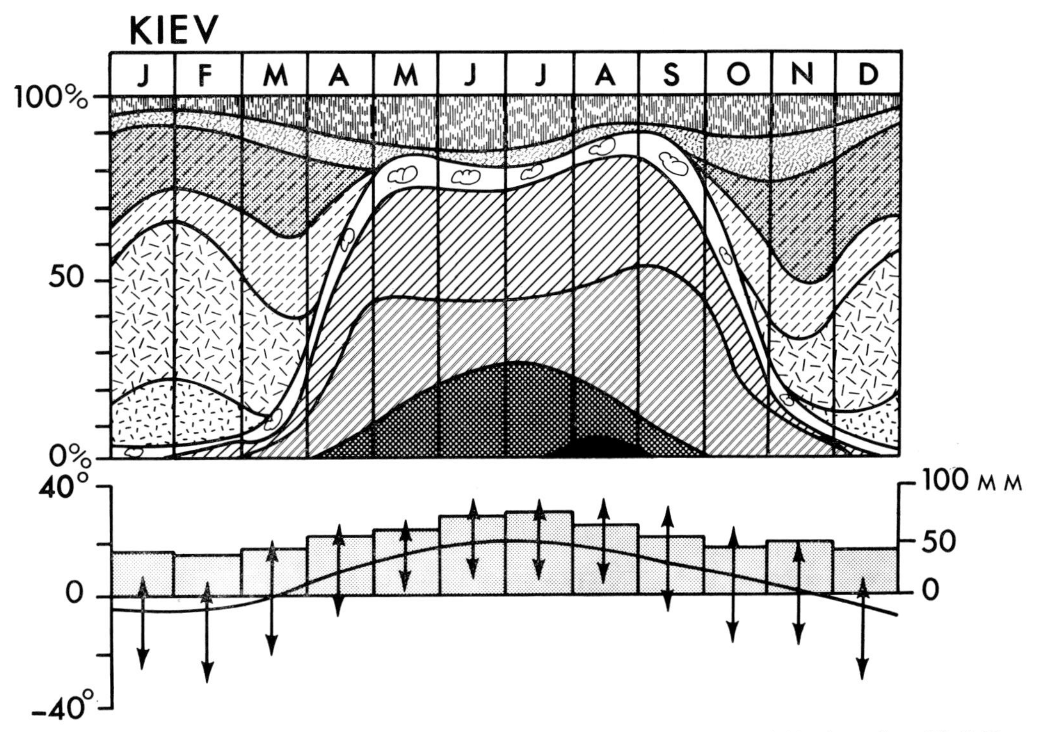

Fig.3-29. Frequencies of weather types, Kiev. (After GERASIMOV, 1964.) For legend see Fig.3-25.

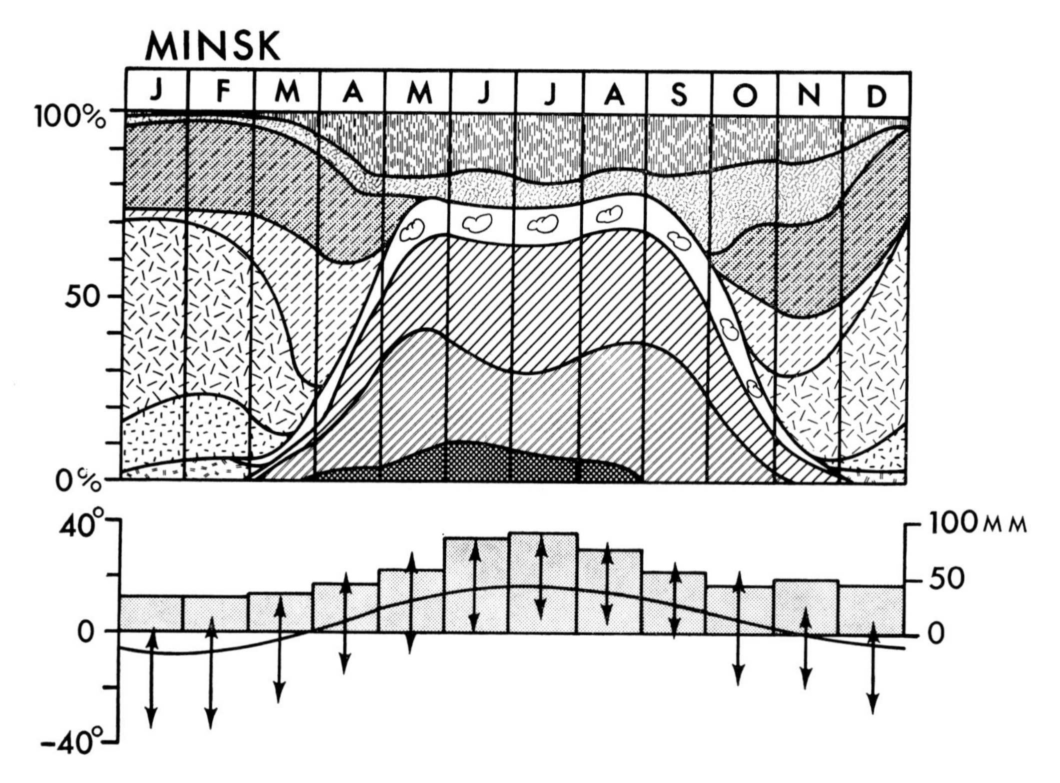

Fig.3-30. Frequencies of weather types, Minsk. (After GERASIMOV, 1964.) For legend see Fig.3-25.

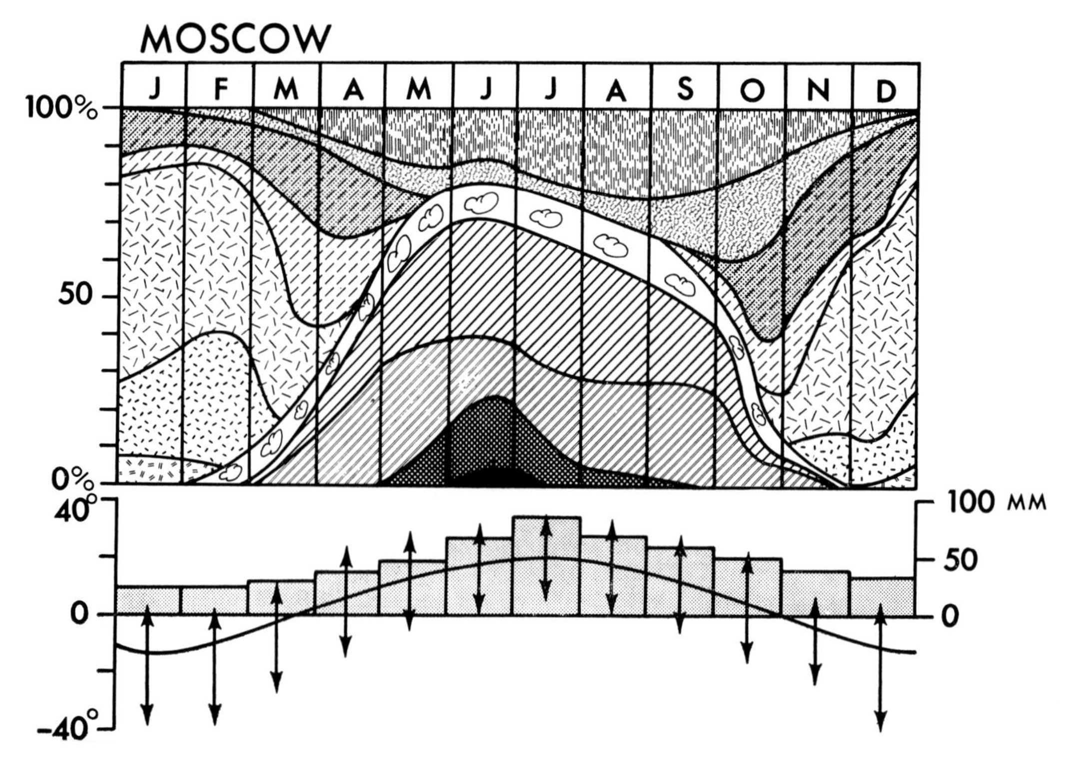

Fig.3-31. Frequencies of weather types, Moscow. (After GERASIMOV, 1964.) For legend see Fig.3-25.

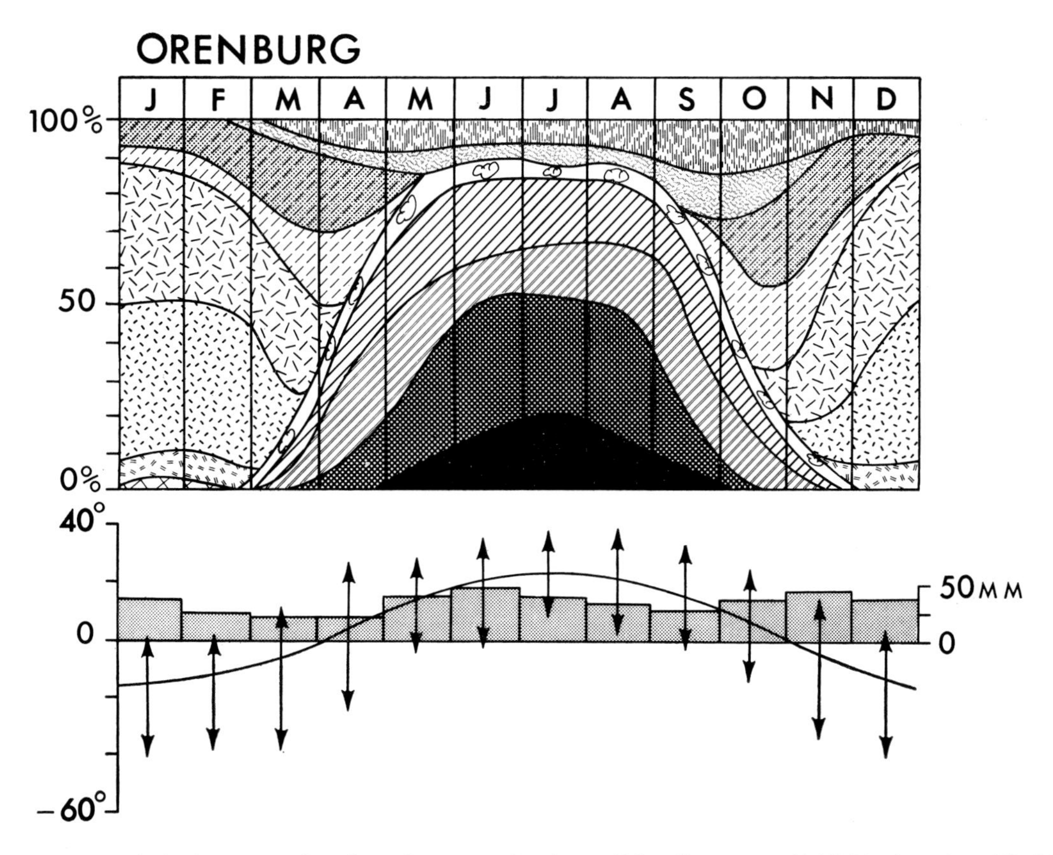

Fig.3-32. Frequencies of weather types, Orenburg. (After GERASIMOV, 1964.) For legend see Fig.3-25.

reduced throughout the year and shows the influence of early summer thundershowers and winter cyclonic storms which concentrate on the track around the southern end of the Urals.

The European northwest

The northwestern part of European U.S.S.R. is the only part of the entire country which experiences a dominance of maritime air. Maritime temperate air from the Atlantic occurs 27.2% of the time and maritime Arctic air from the Barents and Norwegian seas occupies the area 23.4% of the time (Table 2-I). For its latitude this is the most temperate part of the country, experiencing the least extremes. Adding to the tempering effects of the marine air are large amounts of cloudiness which are present year round but are greatest in winter. Summers are cool and moist and winters are relatively mild and moist. At Leningrad July averages 17.7°C and January —7.5°C. Farther north at Murmansk July averages only 12.8°C and January averages —9.9°C. Winters at Murmansk are anomalously warm for the high latitude because of the influence of the Gulf Stream which comes all the way around Scandinavia and keeps the Barents Sea ice free during the winter.

Precipitation is only modest in this northern region, but since the temperatures are cool it is quite sufficient. Most of the area is forested with coniferous trees. The northern half of the Kola Peninsula is tundra. Many lakes and marshes and swamps abound in this glacial country. Thus, in summer the land is usually wet, and most of the insolation during the long summer days is used to evaporate moisture rather than to warm the air.

Much of the precipitation comes as prolonged light continuous precipitation in conjunction with cyclonic storms that frequent the north coastal area all year. During summer there are convective showers set off primarily by the surface heating of maritime Arctic air as it moves inland from the Barents Sea and southward across the land. August is typically the month of maximum precipitation and normally has two to four times as much as the driest month which generally falls between January and March. But, like the central region, winter has significantly more days with precipitation than does summer.

Fall is usually considerably warmer than spring, particularly along the coast where spring temperatures are held down by the water which is warming up much less rapidly than the land and fall temperatures are held up by the water which is cooling off less rapidly than the land. Along the north coast sea fogs are very common in middle and late summer. Farther inland fog is more common in winter because it is primarily of the radiation type. Figs.3-33 and 3-34 for Leningrad and Murmansk typify the weather conditions in this region.

The European northeast

The European northeast differs from the European northwest primarily in its increased degree of continentality. Maritime temperate air masses occur only half as frequently as they do in the northwest and maritime Arctic air masses also occur less frequently. Con-

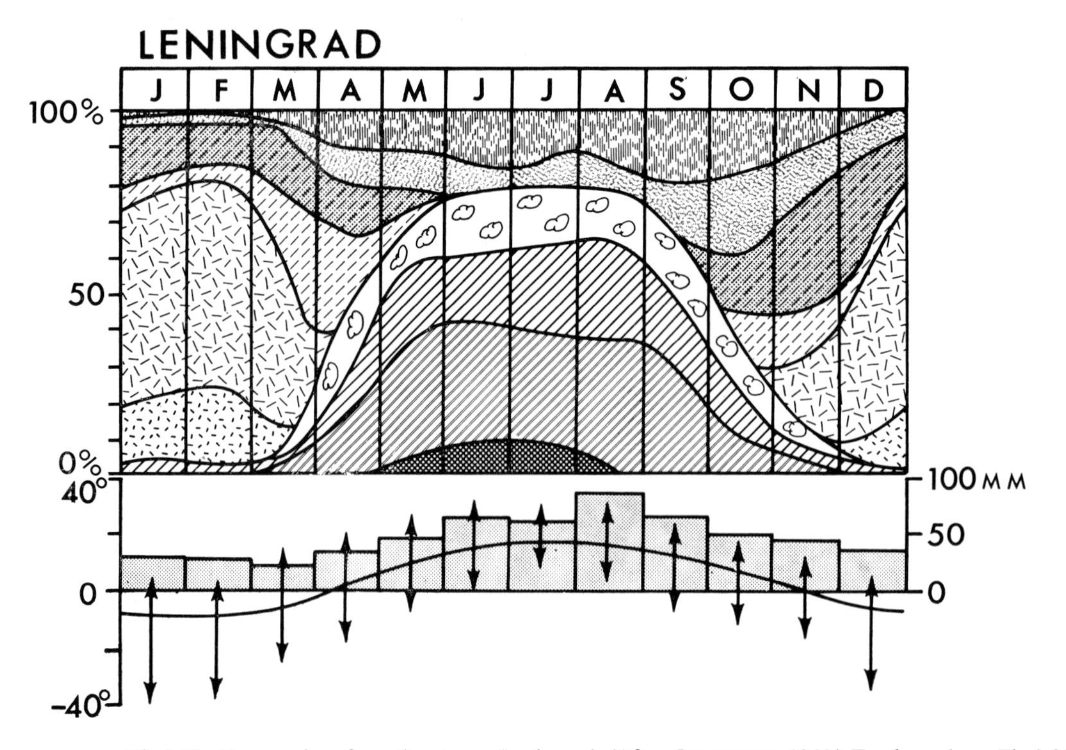

Fig.3-33. Frequencies of weather types, Leningrad. (After GERASIMOV, 1964.) For legend see Fig.3-25.

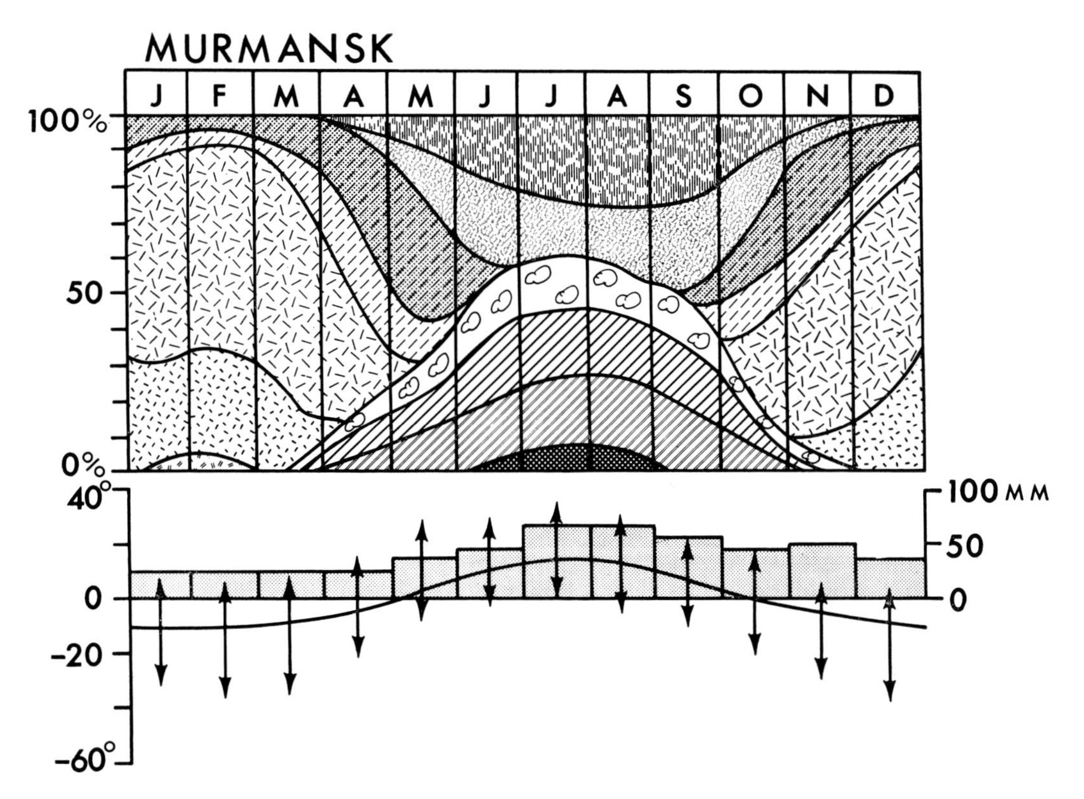

Fig.3-34. Frequencies of weather types, Murmansk. (After GERASIMOV, 1964.) For legend see Fig.3-25.

tinental Arctic air is just as frequent as maritime Arctic air in the northeast, and continental temperate air occupies the region almost 50% of the time.

Weather types are quite similar between east and west, but produce greater extremes in temperatures in the eastern part of the plain. In the north of the region Nar'yan-Mar has experienced temperatures as low as −51°C. Frosts are experienced all summer long. The average temperature in July is only 12°C. In the southern part of the region Kirov has experienced temperatures as low as −45°C. July averages 18°C. Ordinarily the frost free period here is 122 days, but can be much shorter, and freezing temperatures have been observed in June and September.

A significant difference between the western and eastern parts of the plain is the amount of thawing during winter. In the west between 5 and 50 days experience thawing, while in the east only 0 to 5 days do.

Precipitation totals here are very similar to those in the western part of the northern plain, and again the area is primarily forested with coniferous forests, except for a northern strip of tundra. Snow storms occur 30–40 days during the winter throughout much of this region but as the Arctic coast is approached their frequency increases rapidly to more than 100 days in some places. Everywhere snow storms occur with predominantly southerly winds. This is true throughout the European plain north of about the 50th parallel. Much of the region accumulates snow to a depth of 70 cm or more during winter, and the western slopes of the middle Urals accumulate it to 90 cm in some places, which is the thickest snow cover at low elevations in the Soviet Union.

The Urals

Although the Ural Mountains have not been accorded the status of a separate region, they obviously cause the climate to change locally with elevation and induce enough perturbations in the west–east air flow across the Eurasian plain to produce significant differences in weather on the two sides of the range. Although they are relatively low and broken, there are many places where they rise to 1,500–2,000 m, and such elevations have very significant effects on radiation, temperature, precipitation, and atmospheric circulation.

Unfortunately, there are few data at higher elevations in the Urals. Those that do exist indicate that during winter the topography of the Urals is significant in stagnating cold air, so that there are great temperature differences between enclosed lowland basins and upper slopes. Strong temperature inversions are the rule. In summer on the other hand there is usually a rather steep lapse rate along the mountain slopes, averaging approximately 8°C/km. This is partially due to greater turbulence caused by the roughness of the topography as compared to the plains on either side and partially due to heat losses through greater evaporation of precipitation and greater cloud cover than in the plains.

One of the outstanding differences between the Urals and the plains on either side is the amount of cloudiness. At the observation station of Taganay in the south-central Urals at an elevation of 1,102 m overcast skies occur more than 60% of the time every month of the year. October shows the greatest frequency when overcast skies occur about 75% of the time. During the average year 230 days record fog at Taganay. This high fog frequency must relate to stratus clouds that have been formed by orographic uplift and are lying on the ground at this elevation. Snow storms occur 97 days per year at Taganay, and maximum snow cover averages about 90 cm.

The Urals do not form a climatic boundary between the east European plain and west Siberian plain but rather sharpen up the gradients of climatic change which are taking place from west to east across the plain. There seems to be some rain shadow effect of the Urals on the eastern side, although this may relate more to favorite routes of cyclonic storms than to the mountains themselves. But precipitation on the western side of the West Siberian Lowland next to the mountains is often about 300 mm per year less than on the windward western side of the mountains and also is significantly less than the eastern part of the West Siberian Lowland. This difference between precipitation in the western part and the eastern part of the West Siberian Lowland undoubtedly relates to favored paths of cyclonic storms which come up from the southwest and swing northeastward away from the Urals into the lower Ob' and Yenisey valleys.

The Urals stretch over 20° of latitude and therefore show great differences in climate from north to south. The northern Urals, southward to about 65°N latitude, are a single sharp range averaging between 1,300 and 1,900 m elevation. They rise abruptly from the very flat plains on either side and form a rugged rocky ridge which is a wild and treeless wind-swept tundra land. South of 65°N the mountains lower and divide into a number of discontinuous ranges as tree growth softens the slopes and greatly reduces wind speeds. This is the region of mixed forests with coniferous evergreen trees generally on higher elevations and deciduous trees in the lower valleys. South of about 58°N the trees begin to thin out as the climate becomes drier and all but the upper slopes are taken over by steppe grasses. The very southern part of the Urals has the aspect of semi desert.

Western Siberia

East of the Urals the climate of western Siberia represents the culmination of weather processes that have been described in the northwestern and northeastern regions of European U.S.S.R. Storms that move eastward along the Arctic coast terminate in this area, as do many of the storms which move up from the southern plains. On the other hand, the region is more influenced by the Siberian high in winter than is the European U.S.S.R. and it is more subject to intrusions of air masses from the Arctic at all times of the year. The influences of maritime temperate air are small, as are the influences of maritime Arctic air (Table 2-I). Continental types of air predominate almost all the time. Continental temperate air which has been over the continent long enough to lose its original characteristics occupies the region more than half the time and continental Arctic air occupies it another third. Frequently during summer continental tropical air from Kazakhstan and Central Asia enters the region from the south.

Of course this is a very large area, and combinations of air masses and weather types vary greatly from one end to the other. The region as defined includes not only most of the West Siberian Lowland but also a large northwestern portion of the Central Siberian Uplands eastward to the delta of the Lena River and beyond. Thus, the region stretches half way across the continent along the Arctic coast, as well as latitudinally from about 48°N in northern Kazakhstan to about 78°N at Cape Chelyuskin at the northern tip of the Taymyr Peninsula. The rough topography of the included portion of the Central Siberian Uplands contrasts greatly with the flat topography of the West Siberian Lowland and induces the stagnation of cold pools of air in low lying areas during winter with the attendant formation of extreme temperature inversions.

But the weather in this large region is unified by the movement of cyclonic storms throughout the area, which impart rapid and intense fluctuations in weather that stand in sharp contrast to the more stable conditions to the southeast. In fact, the northern half of the West Siberian Lowland is known to be one of the two areas on earth that record the greatest day-to-day-variations in weather phenomena, particularly temperature. There are great meridional movements of air masses across this region. Cold Arctic air masses alternate with southerly air masses from Kazakhstan or the southern plain of European U.S.S.R. As fronts sweep through the area these air masses interchange very rapidly and rival the frequency and magnitude of changes that take place in interior North America east of the Rocky Mountains.

The fluctuations in weather are particularly intense at the beginning of winter, in November and December, and again in spring, particularly in May. Temperatures rise very rapidly in late spring away from the coast as intense solar radiation warms the interior land mass. The warming effect is delayed by the melting of snow and ground and the evaporation of water during early spring so that there is no gradual rise in air temperature commensurable to the rise in insolation. Rather, after the snow is gone, temperatures rise very rapidly in late spring as maximum insolation is approached. This is typical of high latitudes where the elevation of the sun above the horizon changes so rapidly in spring and fall that the transitional seasons are almost non-existent. Fall is in more evidence than spring because it is not affected so much by changing states of water.

The tundra zone

The northern fringe of the territory which is deeply indented along the Arctic coast by the long broad estuaries of streams such as the Ob', the Yenisey, and the Khatanga is too cool and windy in summer for tree growth and hence is covered with tundra vegetation. This zone extends southward to Salekhard in the west and Khatanga farther east. Weather types along the immediate coast are characterized by the diagram for Mys Chelyuskin which shows an almost non-existent summer and long severe winters (Fig. 3-35).

The entire Arctic coast experiences a monsoonal reversal of winds by seasons because of the great temperature contrasts between land and sea. During summer surface temperatures in the open water along the coast are held consistently near 0°C by melting ice flows while the land warms strongly under constant sunlight which continues for about 70 days without sunset during mid summer. Winds blow inland from the sea with average speeds during summer of 6–7 m per second, which is significantly greater than even in the steppe regions in the south. These carry the cool humid influence of the sea onto the land and produce much cloudiness and fog. North of the Arctic Circle the region receives only 28% of its possible solar radiation. The coastal area during summer receives about 240 h of sunshine. As cloudiness reduces somewhat inland, the hours of sunshine increase to about 300 in the southern parts of the tundra, which is the greatest number of sunshine hours during the summer of any part of western Siberia. Fog occurs about 80 days per year right along the coast, and more than half of it occurs during the three

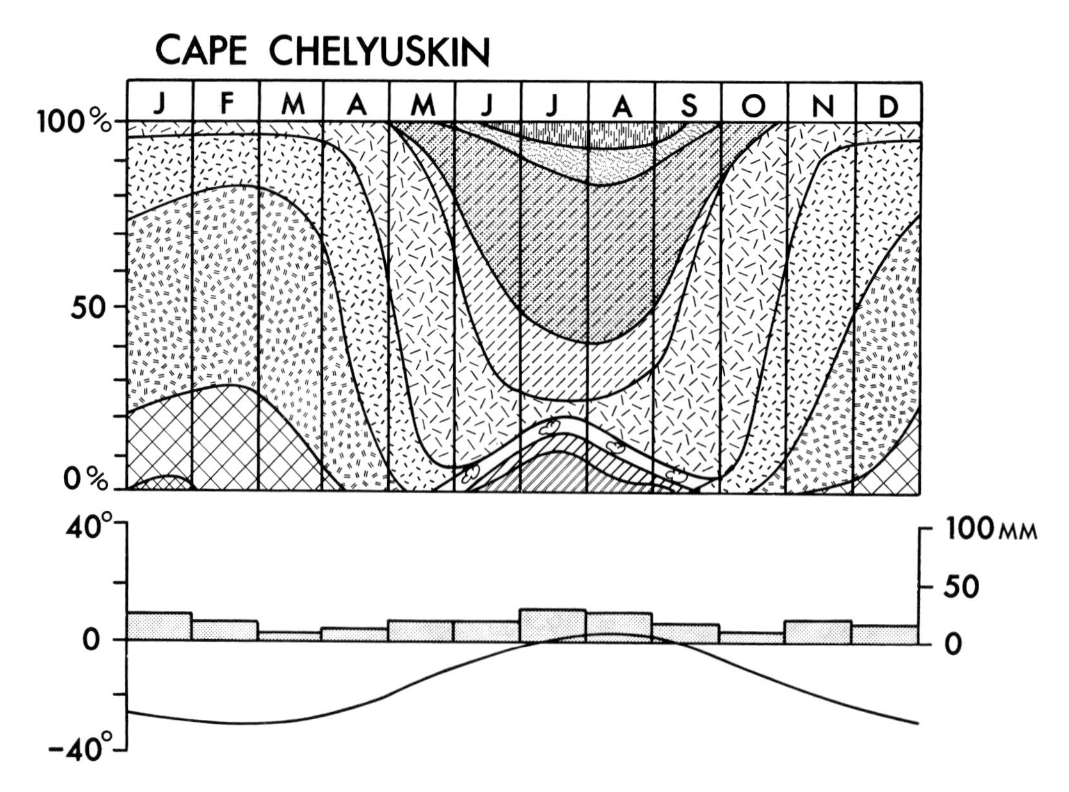

Fig.3-35. Frequencies of weather types, Cape Chelyuskin. (After GERASIMOV, 1964.) For legend see Fig.3-25.

summer months. The fog diminishes rapidly southward and changes to a winter maximum as the fog type changes from advection along the coast to radiation inland. The cloudiest time of year is autumn when the water is most open along the coast and air masses blowing southward from the Arctic toward the land are already becoming quite cold. October experiences up to 25 days with low overcast clouds.

Clouds, fog, and misty rain accompanying intrusions of cold air from the sea hold summer temperatures right around freezing in the immediate coastal area. At Mys Chelyuskin July temperatures average 0.8°C. Temperatures have fallen as low as −6°C during July. Farther inland, Khatanga averages 11.8°C in July and has experienced a minimum of only −1°C. However, temperatures can rise sharply with the advection of southerly air into the region around the western sides of anticyclones during the summer. Under such conditions even along the coast maximum temperatures may rise to 20°C or more. Mys Chelyuskin has experienced a temperature of 24°C in July. At Khatanga the absolute maximum temperature is 34°C. The influence of such high temperatures on tundra vegetation is generally more fatal than favorable since the transpiration of moisture is retarded by the low temperature of the soil, and as a consequence the leaves become desiccated and the phenomenon of "opal" (scorching) occurs.

During winter the winds blow northward from the very cold land toward the Arctic which is being warmed somewhat by the conduction of heat from the unfrozen water beneath the ice. Wind speeds throughout much of the region average 7–9 m/sec, with maxima going beyond 30 m/sec in the forest tundra in the southern part of the zone and more than 40 m per second along the coast. Such winds produce extremely low wind chill in spite of the fact that the actual temperatures are not as low as they are further inland (Fig.8-39). This region experiences one of the lowest wind chill factors during winter anywhere on earth. It has been reported that at Dudinka along the southern border of the tundra zone in winter inhabitants do not leave their houses for 2 or 3 days on end during gales for fear of getting lost and freezing to death. Meteorological observers during storms rope themselves together so as not to lose their way, and they sometimes take more than an hour to cover a distance of 30–40 m (BORISOV, 1965, p.141).

Winds sweep across the treeless tundra and drift the thin snow cover into the lowlands and leave much of the upland areas relatively bare during winter. This allows the soil to freeze to great depths which perpetuates permafrost conditions throughout the entire region. The driven snow becomes so compacted that it is common for man and animals to walk across it without sinking in.

Added to the severity of the winter is the dreariness of the skies which are cloud-covered much of the time. During January much of the tundra zone averages at least 8 tenths sky cover day in and day out. In the estuaries of the Ob and Yenisey more than 210 days per year experience overcast.

Precipitation is frequent in the tundra but of light intensity, because of the low capacity of the consistently cold air to hold moisture. Throughout much of the region 160–170 days of the year experience some precipitation. The annual total amounts to about 250 mm along the coast and 350 mm further inland. The precipitation maximum usually comes in late summer when frequent instability showers occur in the cool Arctic air that is being warmed as it moves southward over the land surface. However, this surface convective activity seldom reaches thunderstorm proportions. Throughout much of the region thunderstorms occur only 2 to 5 times per year, and along the coast the frequency reduces to less than 1 per year.

About 45% of the annual precipitation in the tundra zone falls as snow. The number of days with snow storms amounts to about 140 per year in the western part of the zone and 100 in the eastern part. Snow cover forms during the first decade of October and lasts until the first decade in June. In the eastern part of the tundra it may even form towards the end of September. Because of the small amount of precipitation during the winter the snow cover is not deep. Along the coast it amounts to about 20–30 cm, and in the forest tundra to the south it may reach 50–55 cm.

The forest zone

The broad central part of the west Siberian region is covered by coniferous forests, which occupy about 70% of the ground surface. Here winter temperatures are colder than they are near the coast, winds are less severe, clouds are somewhat reduced, and precipitation generally is somewhat increased, with a much more pronounced summer maximum. Summer temperatures are much warmer than in the tundra zone due to the more complete dominance of continental air masses. Summer showers are frequent, often triggered by the passing of weak fronts. The precipitation maximum generally falls in August and the minimum in late winter or spring.

From 1/4 to 1/3 of the annual precipitation falls as snow, and a snow cover forms early in winter. Snow accumulation is thicker than it is either to the north or to the south and reaches a maximum in the middle Yenisey Valley of about 90 cm which rivals the western slope of the middle Urals for the deepest snow cover in the U.S.S.R. at low elevations during winter. It is generally thickest on the western slope of the Putoran Mountains between Dudinka and Turukhansk, which receives the greatest amount of precipitation during the winter six months anywhere in western Siberia. This occurs generally with upslope motion along cold fronts or fronts that are occluding in the area. In extreme winters the snow cover here may reach depths of more than 2 m.

Under such snow covers there is little freezing of the soil, and there are extensive patches of ground that are not underlain by permafrost. For example, Turukhansk, with a mean annual surface air temperature of −8°C, has no permafrost, whereas much farther south Ulan-Ude in the Transbaykal area, with an annual average surface air temperature of only −2.2°C, does have permafrost. In the Transbaykal area the snow cover is very thin and often the ground is swept bare by the winds (BORISOV, 1965, p.152).

However, much of the forest zone is underlain by permafrost, commonly to depths of 200 m or more. The frozen ground is used as a refrigerator at shipping points such as Ust-Port where shafts are sunk into the ground to depths of about 18 m where they connect with side tunnels and chambers to store and preserve products.

The diagram for Turukhansk typifies much of the forest zone (Fig.3-36). This is a short relatively warm summer and a long very cold winter, with considerable cloudiness throughout the year. Precipitation is modest and reaches a maximum in August.

The steppe zone

The southern part of the west Siberian region is a natural grassland, much of which has been plowed up and put under cultivation. This steppe land forms a zone 300–400 km in

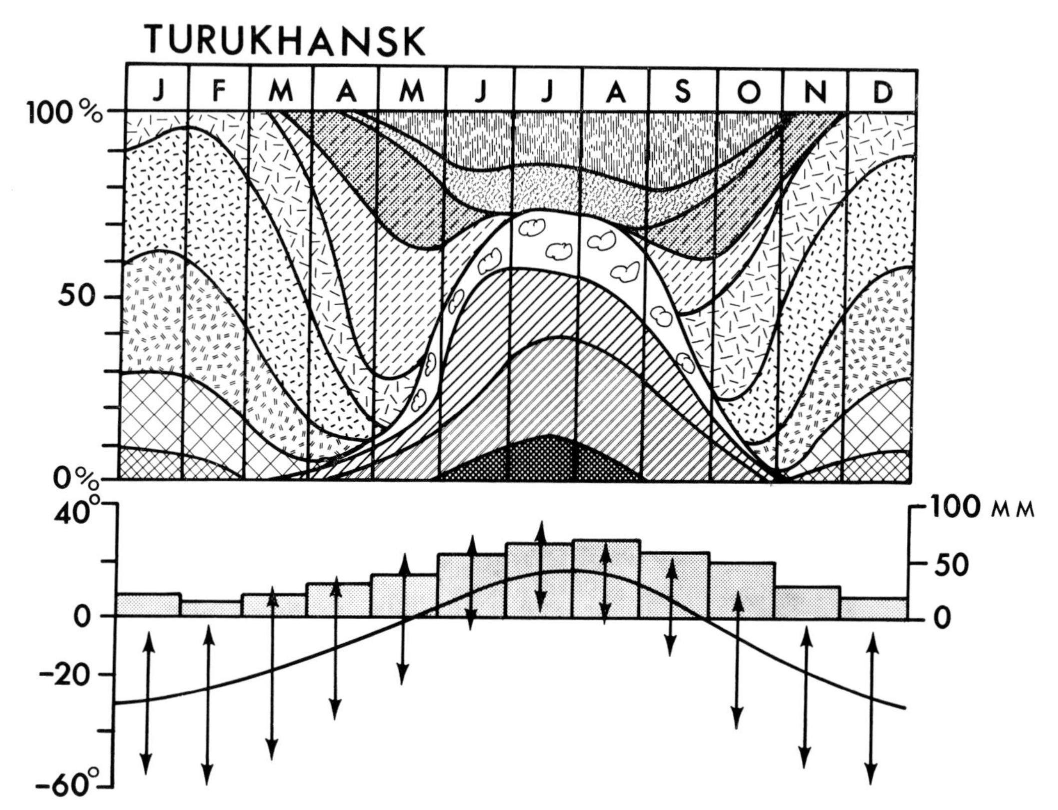

Fig.3-36. Frequencies of weather types, Turukhansk. (After GERASIMOV, 1964.) For legend see Fig.3-25.

width straddling the 52nd parallel. To the north of the steppe lies a narrow zone of forest steppe, and to the south in north central Kazakhstan begins the semidesert. The region is sprinkled with salt lakes interspersed with meadow lands in the north and dry takyrs, or salt flats, in the south. The region is generally under the influence of anticyclonic activity which induces a high frequency of dry clear weather and extreme continentality.

Annual precipitation ranges from about 400 mm in the north to 200 mm in the south. Potential evaporation generally is at least twice that much. Therefore, the area suffers from a continuous threat of drought. The precipitation that does fall is concentrated more in summer than it is further north, and particularly in early summer, June and July. Therefore the reduced moisture resources are somewhat better distributed through the summer for agricultural purposes. However, the zone is frequented by drought and sukhovei, particularly during spring just when the plants are undergoing their most rapid growth. The spring droughts are prefaced by meager snowfalls in late winter and early spring, the blowing away of snow by strong winds, the decrease of the snow cover by vaporization during clear weather in late winter and early spring, and the deep freezing of the soil which increases the surface runoff during spring. Dust storms are common to the area due to the dry, fine soil conditions and the relatively high average wind speeds, which are somewhat greater than they are in the forests to the north.

The predominance of anticyclonic weather makes for very cold and severe winters with average temperatures around −19°C and minimum temperatures as low as −50°C. Winter precipitation is meager, and the snow cover generally does not reach more than 20–30 cm. Strong winds drift the snow so that it lies very unevenly on the ground. The thinness of the snow cover generally rules out winter grain cultures, and it is also significant with respect to the moistening of the soil during spring. Much of the spring rain and melt water is lost to runoff because of the ground being frozen to depths of 1.5–2 m. Much of the water that does not run off in spring is evaporated as temperatures rise sharply. The thawing period generally lasts for no more than 10 days.

The summer maximum of precipitation generally occurs in the form of brief sharp showers. Thunderstorms throughout much of the area occur with a frequency of about 20 per year, most of which occur in midsummer. For instance, Omsk receives on the average 7 thunderstorms during July. It has received as many as 15 in July. Summers in the steppes are generally hot and dry with frequent strong winds and a great number of clear days. Temperatures in July average 19°–20°C with maxima up to 40°C. However, with strong Arctic outbreaks during summer the lower areas of the relief may experience frost on the ground. This is particularly true in the hilly relief of the Kazakh Folded Country in northeastern Kazakhstan. Only the southeastern part of this steppe region is absolutely free of summer frosts during the entire period of record. On the average the frost free period throughout the region ranges from 115 to 120 days. Since the crops are primarily cereals which are somewhat frost resistant, the growing season is somewhat longer than that.

The diagram for Omsk is typical of the steppe region (Fig.3-37). It illustrates the long cold winters and relatively warm summers with an asymmetrical cloud regime, autumn being considerably cloudier than spring. The hottest period of the year generally falls during May–July when skies are clearest. The precipitation maximum comes in early summer and the minimum in late winter and spring.

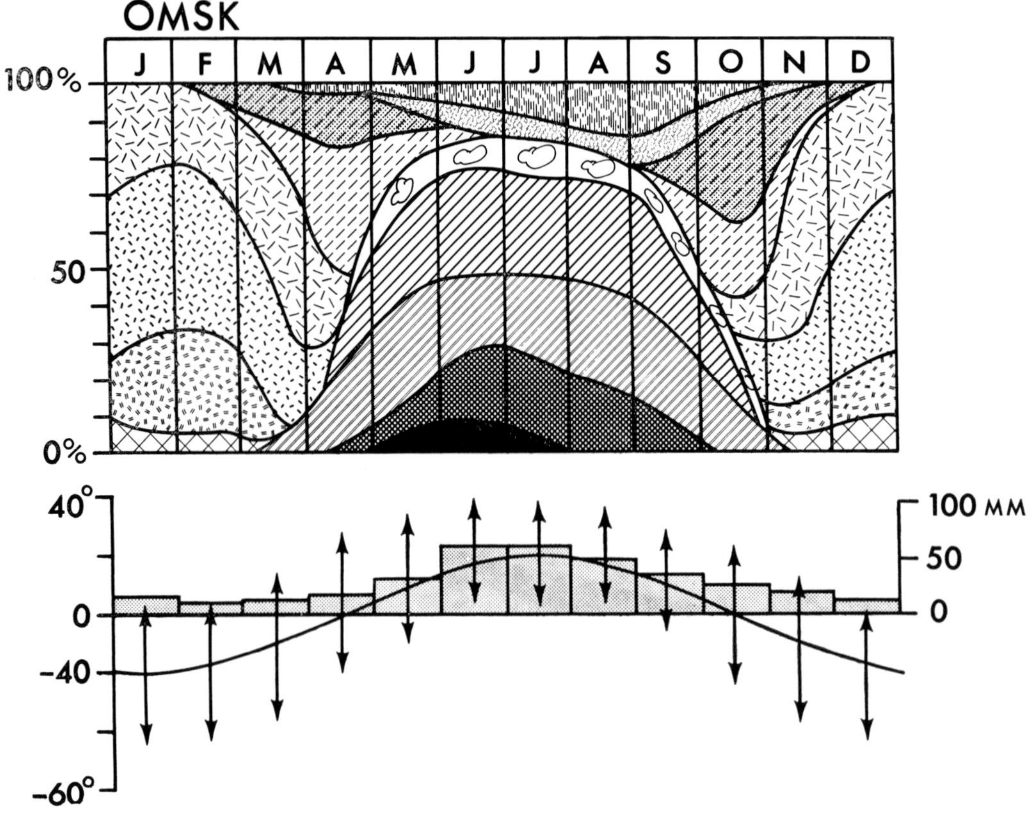

Fig.3-37. Frequencies of weather types, Omsk. (After GERASIMOV, 1964.) For legend see Fig.3-25.

References and further reading

ALIȘOV, B. P., 1956. *Klimat SSSR* (Climate of the USSR). Moscow University, Moscow, 127 pp.

ANAPOL'SKAYA, L. E., 1958. Rezhim skorostey vetra zapadnoy Sibiri i Kazakhstana (Wind speed regime in western Siberia and Kazakhstan). *Ge. Geofiz. Obs. Tr.*, 85: 81–101.

ANDRIANOV, M. S., 1938. Tsirkulyatsionnye protsessy atmosfery v periody veseney-letney vegetatsii neurozhay nykh godov za minuvshee 45-letie na evropeyskoy territorii soyuza (Circulation processes in the atmosphere during spring-summer vegetative periods during crop failure years within the past 45 years in European USSR). *Meteorol. Gidrol.*, 4: 39–53.

ANONYMOUS, 1953. *Mikroklimaticheskie i Klimaticheskie Issledovanie v Prikaspiskoy Nizmennosti* (Microclimatic and climatic investigations in the Caspian lowland). Akad. Nauk SSSR, Inst. Geografiya, 167 pp.

ANONYMOUS, 1969. *Atlas SSSR* (Atlas of the USSR). Gl. Upr. Geod. Kartogr. Sov. Ministr. SSSR, Moscow, 2nd ed., 199 pp.

ARKHANGEL'SKIY, V. I., 1938. Sinoptocheskie usloviya obrazovaniya radiats-ionnykh tumanov na territorii evropeyskoy chasti SSSR (Synoptic conditions generating radiation fog in European USSR). *Meteorol. Gidrol.*, 2: 52–68.

ASKNAZIY, A. I., 1938. Eshche k voprosii o letney tsirkulyatsii atmosfery na evropeyskoy territorii SSSR (Still more on the question of the summer circulation of the atmosphere over European USSR). *Meteorol. Gridrol.*, 3: 59–72.

BASHTAN, N. S., 1970. Tsirkulyatsionnye faktory klimata severozapada evropeyskoy chasti soyuza (Circulation factors of climate in northwest European USSR). In: O. A. DROZDOV, *Formirovaniye klimata*. Uch. Zapiski No.342, Ser. geogr. nauk vyp. 20, Leningradskogo Universiteta, pp.53–67.

BAYBAKOVA, E. M., DZERDZEEVSKIY, B. L., FEL'DMAN, YA. I., CHUBUKOV, L. A. and SHVAREVA, YU. N., 1958. Struktura klimata v pogodakh ravnin aziatskoy chasti SSSR i ee svyaz' s obshchey tsirkulyatsiey atmosfery (Structure of climate in weather types on the plain of the Asiatic part of the USSR and its connection with the general atmospheric circulation). In: G. D. RIKHTER (Editor), *Voprosy Fizicheskoy Geografii*. Akad. Nauk, Moscow, pp.7–46.

BLUMENSTOCK, D. I. and PROZOROWSKI, O. PH., 1956. *The Synoptic Climatology of the Moscow Basin.* Air Force Cambridge Research Center, 123 pp., with tables and diagrams. Mimeographed.

BORISOV, A. A., 1956. O kol'ebonyakh klimata kryma za istoricheskoye vremya (On the fluctuations of climate in the Crimea during historical times). *Izv. Vses. Geogr. O-vo.*, 88: 532–541.

BORISOV, A. A., 1965. *Climates of the USSR.* Aldine, Chicago, Ill., 255 pp.

BORISOV, A. A., 1972. Klimaticheskie osobennosti Kaliningradskoy oblasti (Climatic peculiarities of Kaliningrad Oblast). *Izv. Vses. Geogr. O-vo.*, 104: 102–108.

BUCHINSKIY, I. E., 1963. *Klimat Ukrainy v Proshlom, Nastoyashchem i Budushchem* (Climate of the Ukraine in the past, present, and future). Gossel'khozizdat Ukrainskoy SSR, Kiev, 308 pp.

BUCHINSKIY, I. E., 1970. *Zasukhi, Sukhovei, Pyl'nye Buri na Ukraine i Borba s Nimi* (Drought, sukhovei, and dust storms in the Ukraine and the struggle with them). Urozhay, Kiev, 236 pp.

BUGAEV, V. A., 1936. O svyazi tsiklonicheskikh traektoriy s polozheniem arkticheskogo fronta zimoy (On the correspondence of cyclone trajectories and the position of the Arctic front in winter). *Meteorol. Gidrol.*, 8: 11–18.

BUGAEV, V. A. and DZHORDZHIO, V. A., 1940. Klassifikatsiya vozdushnykh mass SSSR (Classification of air masses in the USSR). *Meteorol. Gidrol.*, 12: 34–45.

BUGAEV, V. A., DZHORDZHIO, V. A., KOZIK, E. M., PETROSYANTS, M. A., PSHENICHNYY, A. YA., ROMANOV, N. N. and CHERNYSHEVA, O. N., 1957. *Synopticheskie Processy Sredney Azii* (Synoptic processes in Central Asia). Izd. Akad. Nauk Uzb. SSR, 477 pp.

BUT, I. V. and YAKOVLEV, V. V., 1936. Grozy v azovo-chernomorskom krae (Thunderstorms in the Azov-Black Sea territory). *Meteorol. Gidrol.*, 8: 3–10.

CHEKHOVICH, S. N., 1936. Sinopticheskie usloviya meteley na dorogakh ETS pri yuzhnykh tsiklonicheskikh vtorzheniyakh (Synoptic conditions conducive to snowstorms on the roads of European USSR during intrusions of southern cyclones). *Meteorol. Gidrol.*, 9: 10–32.

CHERNOVA, V. F., 1968. Nyryayushchie tsiklony i struynye techeniya (Cyclone intrusions and the jet stream). *Tr. Gl. Upr. Gidrometeorol. Sluzhby*, 22: 70–75.

DAVITAYA, F. F. (Editor), 1955. *Agroklimaticheskie i Vodnye Resursy Rayonov Osvoeniya Tselinnykh i Zalezhnykh Zemel'* (Agroclimatic and water resources in the region of virgin and idle lands). Gidrometeoizdat, Leningrad, 463 pp.

DROGAYTSEV, D. A., 1957. Formirovanie anomalii osadkov na Ukraine (The formation of precipitation anomalies in the Ukraine). *Akad. Nauk SSSR, Izv. Ser. Geogr.*, 3: 15–22.

DROZDOV, M. P., 1936. Usloviya letney transformatsii vozdushnykh mass v rayone moskvy (Basic transformations of air masses during summer in Moscow region). *Meteorol. Gidrol.*, 5: 36–40.

DROZDOV, O. A. and GRIGOR'EVA, A. S., 1963. Vlagooborat v atmosfere. Gidrometeoizdat, Leningrad. (Translated as *The Hydrologic Cycle in the Atmosphere*, Isr. Progr. Sci. Transl., Jerusalem, 282 pp.)

DZERDZEEVSKII, B. L. (Editor), 1957. *Sukhovei ikh Proiskhozhdenie i Bor'ba s Nimi*. Akad. Nauk, Moscow. (Translated as *Sukhoveis and Drought Control*. Isr. Progr. Sci. Transl., Jerusalem, 1963, 366 pp.)

EGURKO, V. B., 1956. Preemstvennost sinopticheskikh protsessov privodyashchikh k rezkim ponizheniyam temperatury po evropeyskoy territorii SSSR (Successions of synoptic processes which bring about rapid falls in temperature in European USSR). *Gl. Geofiz. Obs. Tr.*, 65 (127): 27–40.

FEDEROV, E. E. and BARANOV, A. I., 1949. Klimat ravniny evropeyskoy chasti SSSR v pogodakh (Climate of the plains of the European part of the USSR in weather types). *Tr. Inst. Geogr.*, XLIV: 412 pp.

GALAKHOV, N. N., 1959. *Izuchenie Struktury Klimaticheskikh Sesonov Goda* (A study of the structure of the climatic seasons of the year). Akad. Nauk, Moscow, 182 pp.

GAVRILENKO, N. M., 1955. Sinopticheskie usloviya prodolzhitel'nykh osadkov v Kievskoy oblasti (Synoptic situations causing prolonged precipitation in Kiev region). *Meteorol. Gidrol.*, 4: 31–34.

GERASIMOV, I. P. (Editor), 1964. *Fiziko-Geograficheskiy Atlas Mira* (Physical-geographical atlas of the world). Akad. Nauk, Moscow, 298 pp.

GOL'MAN, S., 1939. Sinopticheskie usloviya proryva yuzhnykh tsiklonov na territoriyu Kazakhstana i zapadnoy Sibiri v letnee polugodie (Synoptic conditions during invasions of southern cyclones into Kazakhstan and western Siberia during the summer half-year). *Meteorol. Gidrol.*, 1: 36–46.

GOL'TSBERG, I. A. and DROZDOV, O. A. (Editors), 1956. *Klimaticheskie Resursy Tsentral'nykh Oblasty Evropeyskoy Chasti SSSR i Ispol'zovanie ikh v Sel'skokhozyaystvennom Proizvodstve* (Climatic resources of the central regions of the European part of the USSR and their utilization in agriculture). Gidrometeoizdat, Leningrad, 312 pp.

GRIGOR'EVA, A. S., 1959. Perenos vodyanogo para nad evropeyskoy territoriey SSSR v razlichnye mesyatsy (The flux of water vapor over European USSR, by months). *Gl. Geofiz. Obs. Tr.*, 89: 3–19.

GRIN, A. M., 1965. *Dinamika Vodnogo Balansa Tsentral'nochernozemnogo Rayona* (Dynamics of the water balance of the Central Black Earth region). Nauka, Moscow, 145 pp.

GUK, M. I., POLOVKO, I. K. and PRIKHOT'KO, G. F., 1958. *Klimat Ukrainskoy RSR* (Climate of the Ukraine). Radyans'ka shkola, Kiev, 70 pp.

KAUSHILA, K. A., 1963. Regime of Global Radiation in Lithuanian SSR. *All-Union Sci. Meteorol. Conf. Trans.*, 6: 470–477 (Leningrad, translated.)

KOLOBKOV, N. V., 1937. Uragan v Moskve 28 Maya 1937 g. (Hurricane winds in Moscow on 28 May 1937). *Meteorol. Gidrol.*, 8: 89–93.

KOLOBKOV, N. V., 1960. *Klimat Moskvy i podmoskov'ya* (Climate of Moscow and the Moscow region). Moskovskiy rabochiy, Moscow, 105 pp.

KOLOBKOV, N. V., 1961. O vliyanim Kuybyshevskogo vodokhranilishcha na meteorologicheskie usloviya i klimat v priberezhnoy zone (On the influence of the Kuybyshev reservoir on meteorological conditions and climate in the shore zone). *Izv. Vses. Geogr. O-vo*, 93: 511–514.

KONSTANTINOVA, A. P. and GOYSY, N. I., 1966. *Atlas Sostavlyayushchikh Teplovogo i Vodnogo Balansa Ukrainy* (Atlas of the components of the heat and water balances of the Ukraine). Gidrometeoizdat, Leningrad, 170 pp.

KORNILOV, B. A. and MUKHINA, L. I., 1969. *Usloviya Melioratsii v Taezhmoy Zone Zapadnoy Sibiri* (Conditions for meliorating the taiga zone of western Siberia). *Izv. Akad. Nauk SSSR, Ser. Geogr.*, 4: 61–68.

KOTLYAKOV, V. M., 1968. *Snezhnyy Pokrov Zemli i Ledniki* (Snow cover on the earth and glaciers). Gidrometeoizdat, Leningrad, pp.102–110.

KURGANSKAYA, V. M., 1936. Sinopticheskie usloviya meteley na dorogakh zapadnoy Sibiri i Kazakhstana pri yuzhnykh tsiklonicheskikh vtorzheniyakh (Synoptic conditions accompanying snowstorms on the roads of western Siberia and Kazakhstan during intrusions of southern cyclones). *Meteorol. Gidrol.*, 1: 18–40.

KURSANOVA-ERV'E, I. A., 1939. Sinopticheskie usloviya fenov v krymu (Synoptic conditions during foehns in Crimea). *Meteorol. Gidrol.*, 9: 34–51.

LEBEDEV, A. N., 1958. *Klimat SSSR, Vypusk 1, Evropeyskaya Territoriya SSSR* (Climate of the USSR, volume 1, the European territory of the USSR). Gidrometeoizdat, Leningrad, 367 pp.

LEBEDEVA, O. N., 1938. Proiskhozhdenie i kharakter ottepeley evropeyskoy chasti SSSR (The occurrence and character of thaws in the European part of the USSR). *Gl. Geofiz. Obs. Tr.*, 16: 41–87.

LEYST, E. E., 1904. Moskovskiy uragan (Moscow hurricane). *Ezhemes. Byull. Gl. Geofiz. Obs.*

LIR, E. S., 1936. Tipy sezonnykh tsirkulyatsiy atmosfery nad evraizey i atlantikoy (Seasonal types of atmospheric circulation over Eurasia and the Atlantic). *Meteorol. Gidrol.*, 1: 3–17; 3: 3–9; 4: 7–18; 5: 21–35; 6: 3–15; 7: 14–25.

LIR, E. S., 1940. Asnovnye cherty sezonnykh tsirkulyatsii vozdukha na yugovostoke evropeyskoy territorii SSSR (Basic distinctions of seasonal circulations of air over southeast European USSR). *Meteorol. Gidrol.*, 5 and 6: 24–37.

MALIK, S. A., 1940. Shkvaly v Rostovskoy oblasti i Krasnodarskom Krae (Squalls in Rostov Oblast and Krasnodar Kray). *Meteorol. Gidrol.*, 9: 109–114.

MAMONTOV, N. V., 1966. O fevral'sko–martovskoy anomalii v raspredelanii temperatury na yuge zapadnoy sibiri (The February–March anomaly in the distribution of temperature in the southern part of western Siberia). *Vopr. Prikl. Klimatol. Zapadn. Sib.*, 42(2): 60–67.

MIKHEL'SON, V. A., 1904. Uragan v Moskve 16(29) iyunya 1904 g. (Hurricane in Moscow on 16 (29 new calendar) June, 1904). *Ezhemes. Byull. Gl. Geofiz. Obs.*,

MUCHNIK, V. M., 1938. Vliyanie chernogo morye na pogody pribrezhnoy polosy Odessy v letnee vremya (The influence of the Black Sea on the weather of the shore zone of Odessa during the summer). *Meteorol. Gidrol.*, 8: 68–72.

NEZDYUROVA, T., 1939. Sinopticheskie usloviya vesennikh zamorozkov na territorii kalininskoy oblasti (Synoptic conditions during spring frosts in Kalinin Oblast). *Meteorol. Gidrol.*, 4: 42–56.

NIKOLAEV, S. D., 1972. Izmeneniya klimata Chernogo morya v golotsene po izotopnokislorodnym dannym (Changes in the Holocene climate of the Black Sea based on isotopic oxygen data). *Vestn. Mosk. Univ., Geogr.*, 6: 84–88.

NIKOLAEV, V. A. and RAYNER, YU. L., 1956. Peski chernykh zemel' zimoy (Black earth sands in winter). *Izv. Akad. Nauk SSSR, Ser. Geogr.*, 1: 86–88.

NUTTONSON, M. Y., 1947. *Ecological Crop Geography of the Ukraine and the Ukrainian Agro-Climatic Analogues in North America.* American Institute of Crop Ecology, Washington, D.C., 24 pp.

ORLOVA, V. V., 1955. Vlagooborot vegetatsionnogo periods zasushlivogo goda v zapadnoy Sibiri v soyazi s proiskhozhdeniem osadkov (Moisture exchanges during the vegetation period in dry years in western Siberia in relation to the origins of precipitation). *Gl. Geofiz. Obs. Tr.*, 50(112): 21–38.

ORLOVA, V. V., 1956. Osobennosti vlago-oborota zapadnoy Sibiri vo vlazhnye i zasushlivye mesyatsy (Particular moisture exchanges in western Siberia during moist and dry months). *Gl. Geofiz. Obs. Tr.*, 62(124): 52–61.

ORLOVA, V. V., 1962. *Klimat SSSR, Vypusk 4, Zapadnaya Sibir* (Climate of the USSR, volume 4, western Siberia). Gidrometeoizdat, Leningrad, 360 pp., with maps.

PCHELKO, I. G., 1946. Ob odnom sluchae shtorma nad tsentral'nymi rayonami evropeyskoy territorii soyuza (On one case of storm winds over the central regions of European USSR). *Meteorol. Gidrol.*, 1: 39–47.

PIVOVAROVA, Z. I., 1960. Osnovnye kharakteristiki radiatsionnogo rezhima evropeyskoy territorii SSSR. *Gl. Geofiz. Obs. Tr.*, 115: 77–94. (Translated by I. A. DONEHOO and published as *The Chief Characteristics of the Radiation Regime of the European Territory of the USSR.* U.S. Weather Bureau, Washington, D.C., 1961, 20 pp.)

POGOSYAN, KH. P., 1936. Inversiya szhatiya v usloviyakh severo-zapada ETS (Subsidence inversions in northwestern European USSR). *Meteorol. Gidrol.*, 10: 41–51.

POGOSYAN, KH. P. and SAVCHENKOVA, E. I., 1950. O chislovom vyrazheniy vida atmosfernoy tsirkulyatsiy (The quantitative representation of atmospheric circulation types). *Meteorol. Gidrol.*, 3: 5–13.

POKROVSKAYA, T. V., 1969. *Klimat Goroda Gor'kogo* (Climate of the city of Gorky). Gidrometeoizdat, Leningrad, 224 pp.

POKROVSKAYA, T. V. and BYCHKOVA, A. T., 1967. *Klimat Leningrada i ego Okrestnostey* (Climate of Leningrad and its environs). Gidrometeoizdat, Leningrad, 199 pp.

POKROVSKAYA, N. D. and POLTARAUS, B. V., 1972. Vazhneyshie tipy pogody v orlovskoy oblasti na fone osobennostey klimata (Principal weather types in Orel Oblast against the background of climatic characteristics). *Vestn. Mosk. Univ., Ser. Geogr.*, 4: 81–88.

PONOMARENKO, S. I., 1968. K voprosy o vozniknovenii ochensilnogo livnya (On the question of the origin of very intense rainfall). *Gl. Upr. Gidrometeorol. Sluzhby, Tr.*, 32: 34–43.

PRIKHOT'KO, G. F., TKACHENKO, A. V. and BABICHENKO, V. N., 1967. *Klimat Ukrainy* (Climate of the Ukraine). Gidrometeoizdat, Leningrad, 413 pp.

SAPOZHNIKOVA, S. A., 1964. *Agroklimaticheskiy Atlas Ukrainskoy SSR* (Agroclimatic atlas of the Ukraine). Urozhay, Kiev, 82 pp.

SAVINA, S. S., 1963. *Gidrometeorologicheskii Pokazatel Zasukhi i ego Raspredelenie na Territorii Evropeyskoy Chasti SSSR* (Hydrometeorological indices of drought and their distribution on the European part of the USSR). Akad. Nauk, Moscow, 103 pp.

SEMENOV, V. G., 1960. *Vliianie Atlanticheskogo Okeana na Rezhim Temperatury i Osadkov na Evro-*

peyskoy Territorii SSSR (The influence of the Atlantic Ocean on temperature and precipitation regimes in European USSR). Gidrometeoizdat, Moscow, 148 pp.

SEMENTSOVA, A. R., 1958. Vostochno-evropeyskaya vetv' polyanogo fronta kak faktor klimata (The east European branch of the polar front as a climatic factor). *Izv. Vses. Geogr. O-vo.*, 90: 543–545.

SHCHERBAKOVA, E. YA. and BRELINA, A. YU., 1956. Izmenenie teplo- i vlago-soderzhaniya vozduzhnykh mass v protsesse transformatsii v umerennykh shirotakh evrazii (Changes in heat and moisture content of air masses during processes of transformation in temperate latitudes in Eurasia). *Gl. Geofiz. Tr.*, 62(124): 3–28.

SHPAK, I. S. and BULAVSKAYA, T. N., 1967. Raspredelenie snezhnogo pokrova i vysotnye gradienty osadkov i snegozapasov v rayone zakarpatskoy stokovoy stantsii (The distribution of snow cover and elevation gradients of precipitation and snow resources in the Transcarpathian experimental watershed). *Ukr. Nauchno-Issled. Gidrometeorol. Inst. Tr.*, 66: 59–69.

SLABKOVICH, G. I. (Editor), 1968. *Klimaticheskiy Atlas Ukrainskoy SSR* (Climatic Atlas of the Ukraine). Gidrometeoizdat, Leningrad, 232 pp.

SMOLYAKOZ, P. SH., 1947. *Klimat Tatarii* (Climate of the Tatar ASSR). Tatgosizdat, Kazan, 108 pp.

SOROKINA, L. P., 1971. Raspredelenie i nekotorye statisticheskie kharakteristiki vnutrisutochnoy izmenchivosti temperatury vozdukha na territorii zapadno-sibirskoy ravniny v zimny period (The distribution and some statistical characteristics of the interdiurnal variability of winter air temperatures in the west Siberian plain). *Dokl. Inst. Geogr. Sib. Dal'nego Vostoka*, 31: 29–38.

TEMNIKOVA, N. S., 1938. Tipizatsiya zamorozkov evropeyskoy territorii soyuza, ikh dlitel'nost' i povtoryaemost' (Frosts in European USSR, their durations and frequencies). *Meteorol. Gridrol.*, 6: 26–44.

TEMNIKOVA, N. S., 1959. *Klimat Severnogo Kavkaza i Prelezhashchikh Stepey* (The climate of the northern Caucasus and adjacent steppes). Gidrometeoizdat, Leningrad, 368 pp.

VANGENGEYM, G. YA., 1946. O kolebanii atmosfernoy tsirkulyatsii nad severnym polyshariem (On the fluctuations of atmospheric circulation in the Northern Hemisphere). *Izv. Akad. Nauk SSSR, Ser. Geogr.*, 5.

YAROSLAVTSEV, I. M., 1936. Tipy-gomologi osnovnykh vozdushnykh mass nad tsentral'nymi rayonami ETS (Typology of the basic air masses in the central regions of European USSR). *Meteorol. Gidrol.*, 7: 26–42.

YASHINA, A. V., 1972. Osnovnye printsipy tipizatsii zim tsentral'no evropeyskoy lesostepi po faktoram formirovaniya snezhnogo pokrova (Basic principles for the typology of winters in the central European wooded steppe based on the factors of snow cover formation). *Izv. Akad. Nauk SSSR, Ser. Geogr.*, 5: 92–101.

ZAMORSKIY, A. D., 1939. Sluchay okrashennogo snegopada v Rostov-na-Donu (Occurrences of extreme snowfall in Rostov-on-Don). *Meteorol. Gidrol.*, 6: 46–50.

ZATS, V. I., LUKYANENKO, O. YA. and YATSEVITCH, G. V., 1966. *Gidrometeorologicheskii Rezhim Yuzhnogo Berega Kryma* (The hydrometeorological regime on the south shore of Crimea). Gidrometeoizdat, Leningrad, 120 pp.

ZAVARINA, M. V. and MICHEL, V. M., 1968. Klimaticheskie kharakteristiki oblakov v zone nedostatochnogo uvlazhneniya evropeyskoy territorii SSSR (Climatic characteristics of clouds in the zone of moisture deficit in European USSR). *Gl. Geofiz. Obs. Tr.*, 219: 110 pp.

ZHAKOV, S. I., 1960. Osadki iz atlanticheskogo umerennogo vozdukha na evropeyskoy territorii SSSR v teplyy period (Precipitation from Atlantic temperate air over the European territory of the USSR during the warm period). *Izv. Vses. Geogr. O-vo.*, 92: 463–465.

ZHAKOV, S. I., 1963. Zavisimost' osadkov evropeyskoy territorii SSSR ot ispareniya s poverkhnosti kontinenta (Precipitation on the European territory of the USSR due to evaporation from the surface of the continent). *Izv. Akad. Nauk SSSR, Ser. Geogr.*, 2: 75–79.

ZVEREV, A. S., 1957. *Sinopticheskaya Meteorologiya* (Synoptic meteorology). Gidrometeoizdat, Leningrad, 559 pp.

Chapter 4

Eastern Siberia

Definition of the region

Eastern Siberia is characterized by the highest degree of continentality on earth. As the region is defined here it is primarily the Soviet portion of the domain of the Asiatic high in winter (Fig.4-1). Thus, the region's climatic boundaries vary considerably from what is usually considered to be eastern Siberia in general geographical terms. This is particularly true in the northwest where the boundary is marked by a long line trending northeastward from north central Kazakhstan across the middle Yenisey and lower Lena Valley to the Yana–Indigirka interfluve along the Arctic coast. This oblique line approximates the average position of the northwestern periphery of the Asiatic high during January (Fig.2-1), and thereby separates the consistently cold, clear, calm weather of eastern Siberia from the more stormy, changeable weather to the west. On the southeast the climatic region of eastern Siberia is distinguished from the climatic region of the Far East by a boundary which also runs obliquely across the map generally along the drainage divide between the Arctic and the Pacific. This line separates the cold, clear, calm weather of eastern Siberia from the more changeable, windy weather along the coast which experiences monsoonal reversals of atmospheric flow throughout much of its extent.

The area so defined is primarily mountainous, with the most rugged topography along the southeast. The mountains are interspersed here and there with broad lowland basins, and the northwestern part of the region is characterized by rolling uplands lying generally between 500 and 1,000 m. The nature of the topography has great influence on the specific locations of atmospheric pressure cells and, of course, exerts local influences on all climatic elements.

Pressure systems and weather

Winter

During winter most of the region is dominated by the Asiatic high, while the regions to the west are frequented by cyclonic storms and fronts, and to the east the Pacific fringes are strongly influenced by the western portions of the Aleutian low. Fig.4-2 is very characteristic of the synoptic situation at this time of year. Typically the main core of the

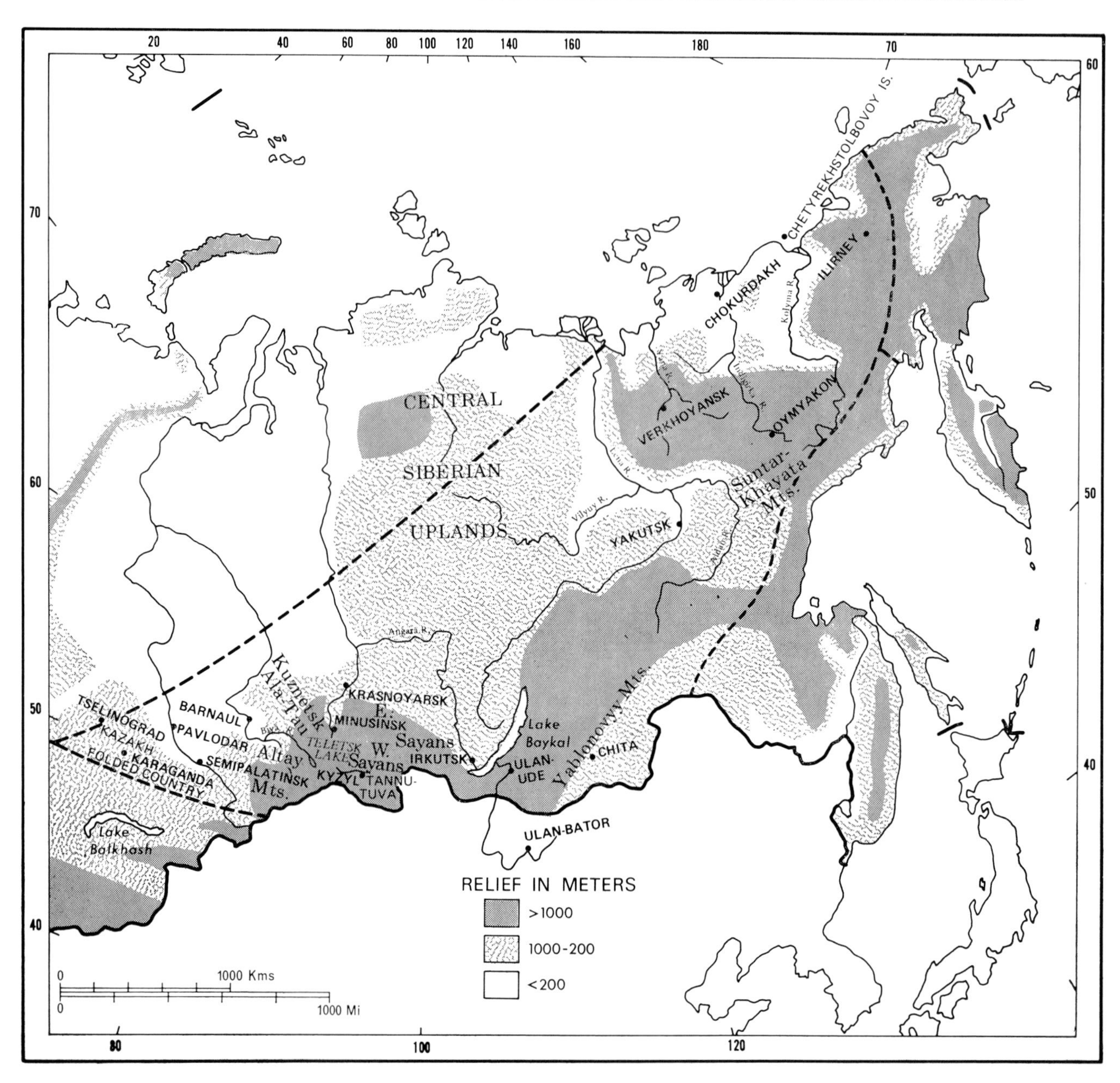

Fig.4-1. Eastern Siberia: boundaries, topography, and climatological stations. (After ANONYMOUS, 1969.)

Asiatic high is located somewhere near Lake Baykal, and there is a northeastern extension which frequently culminates in a secondary node somewhere in the Yana–Indirgirka Kolyma region of the extreme northeast. A westward extension from the Lake Baykal core lies across northern Kazakhstan and separates the cyclonic activity of European U.S.S.R. and western Siberia in the north from that coming in from the Mediterranean and Caucasian track into southern parts of Central Asia.

As has been pointed out in Chapter 2, the surface Asiatic high in winter does not extend very high into the atmosphere. On the average, even the 850-mbar chart shows a pronounced trough over the eastern part of the country. The western limb of this trough covers much of eastern Siberia and directs a nearly constant stream of cold Arctic air

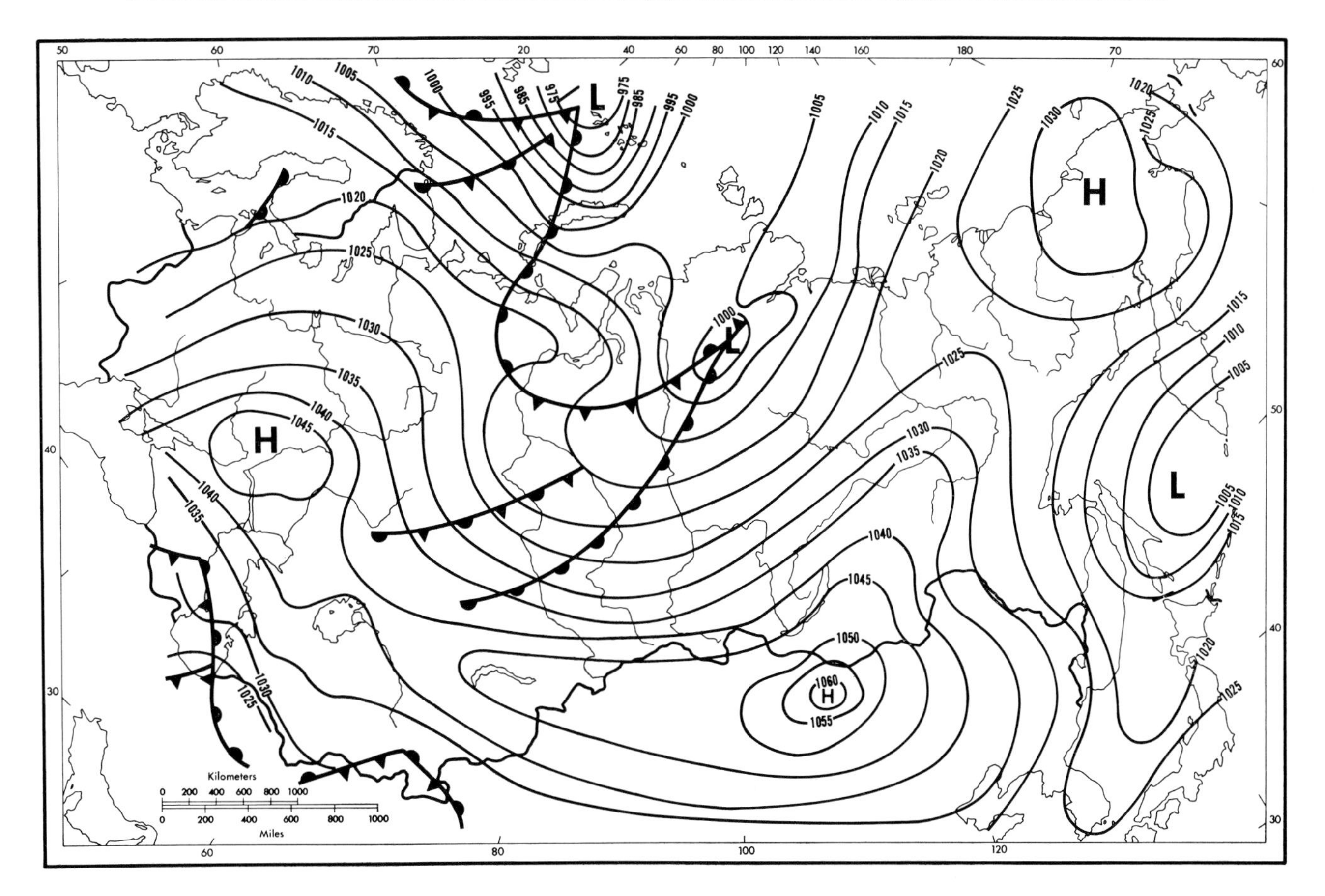

Fig.4-2. Surface synoptic conditions at 21h00, January 3, 1957. (After ZVEREV, 1957.)

onto the continent during the winter which becomes even colder in contact with the frozen, snow-covered land surface. Thus, although the surface winds through much of eastern Siberia during the winter are prevailingly from the southwest, the upper winds, generally beginning no more than 1–2 km above the surface, are very consistently from the north or northwest. This constant feeding of cold Arctic air onto the continent continually strengthens and regenerates the surface Asiatic high. Therefore, although the surface high is augmented to a great extent by extreme surface cooling, it also owes much of its strength and constancy to this steady infeeding of fresh Arctic air.

Just what the surface air pressures are in given parts of the area depends to a great extent upon the topography, as do other weather elements such as temperatures and winds. Generally, the lowest temperatures and calmest winds are found in the protected basins between mountain ranges. Extreme temperature inversions are characteristic of most of the lowlands in eastern Siberia throughout much of the winter. Thus, there is typically an increase in temperature with height through the first 500–1,000 m above the surface of basin floors.

Topographic peculiarities usually position the main core of the Asiatic high about midway between Lake Baykal and Lake Balkhash in the basins of Tannu-Tuva and the adjacent Great Lakes Basin of northwestern Mongolia. There is an unusually great tendency for this area to act as a center for anticyclonic formation much of the year (MAKHOVER, 1967). It appears that western Mongolia acts as a center for anticyclones no less than 91% of the time in January and as much as 56% of the time even in July. These are the two extreme values, and values for the other months are distributed symmetrically around them.

The basin floors in this area lie at elevations of 600–800 m above sea level, and they are surrounded by mountains which rise to 2,000–3,500 m (Fig.4-3). During winter cold air settles into these intermontane basins and lodges there while sea level air pressure equivalents have been known to rise as high as 1,075 mbar. Fronts and other weather systems traversing the area lift off the ground and ride over the cold air domes in the basins (Fig.4-4). Under these conditions there are frequently two inversions, one at the surface induced by surface cooling and one at the top of the cold air dome produced by warm air in the warm sector of the overriding cyclonic storm.

An analysis has been made of radiosonde data taken in Ulan-Bator during the winter of 1958, which was a rather normal winter. Although this is farther east in Mongolia and does not represent such enclosed topographic conditions, it is the only radiosonde

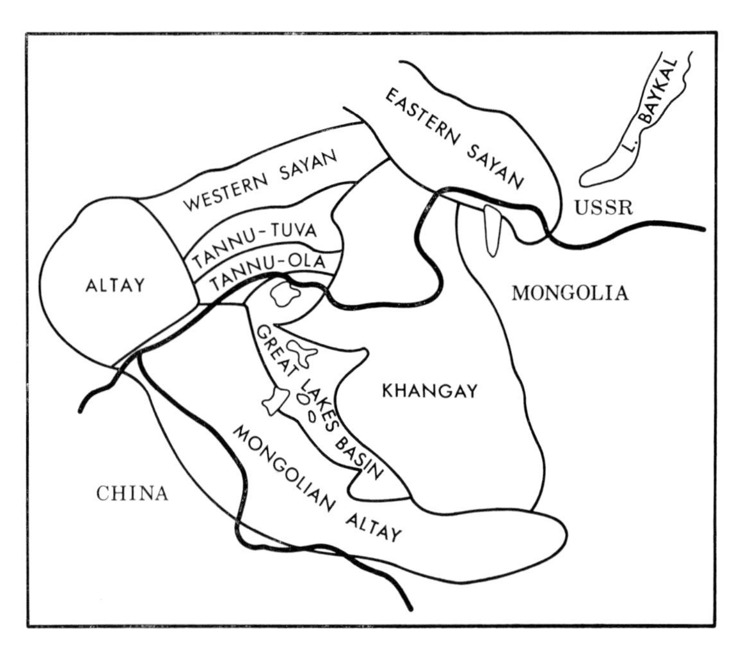

Fig.4-3. Topographic features influencing the core of the Asiatic high. (After MAKHOVER, 1967.)

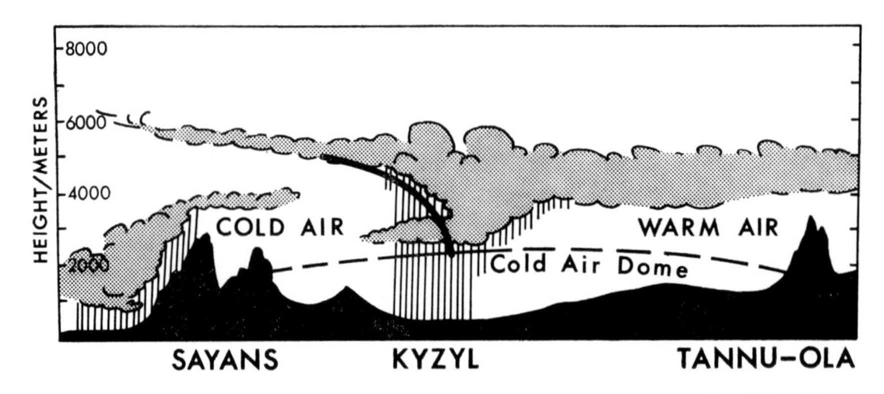

Fig.4-4. Upper cold front riding across cold air dome in Tannu-Tuva Basin. (After ZVEREV, 1957.)

station in the area. It was found that during January 1958 a surface inversion was evident 94% of the time during morning hours and 87% of the time during evening hours. Average thickness of the inversion was 960 m in the morning and 720 m in the evening. During morning hours the average temperature increase through the inversion was 14.2°C, giving an average temperature lapse rate through the inversion of −1.48°C per 100 m. In the evening the temperature difference on the average was 10.7°C and the lapse rate was −1.49°C/100 m. The inversion was typically made up of two layers, a lower one with a rapid increase in temperature with height and a higher one with a less rapid temperature increase. The lower layer typically had its base on the surface of the ground and had a thickness of 220–250 m. The average lapse rate throughout this thickness was −4.05°C per 100 m in the morning and −3.45°C per 100 m in the evening. During the morning hours 73% of the observations showed a temperature increase through this lower layer of more than 10°C. A maximum temperature change of 21.4°C was obtained on one of the days. During evening hours 63% of the observations showed temperature increases of more than 10°C.

It can be assumed that in the western basins of Mongolia conditions are even more extreme. NIKOL'SKIY (1957) took 307 radiosonde readings at Kyzyl in 1957 and found that the inversion thickness averaged about 200 m during winter with an average temperature increase through the inversion of 10.6°–13.5°C.

There does not seem to be very good areal correspondence between frequency and intensity of surface inversions and magnitude of sea level air pressure. The inversions often seem to be more pronounced farther northeast in the mountain basins east of the Lena River, whereas sea level air pressures are generally greatest southwest of Lake Baykal. In a study during 1940–1941, FLOHN (1947) found that at Yakutsk the average inversion during winter had a mean temperature difference of 17.9°C between bottom and top, with extremes ranging from 4° to 27°C. The average height of the maximum temperature in the inversion was 1,720 m, and this ranged from 520 m up to 4,210 m. Flohn assumed that the inversions were even more intense over Verkhoyansk. The maps that he drew with limited amounts of data show frequencies of inversions of at least 95% over this northeastern part of the country during winter. Unfortunately he did not have data for northwestern Mongolia, so this area on his maps was left blank.

It appears that in these early studies that were carried out by various people during the 1930's and early 1940's there was a tacit assumption that the center of the so-called Siberian high was generally in the Yakutsk area and that inversion characteristics and other accompanying features were arranged in a concentric pattern around Yakutsk. This misconception was due to lack of data. However, a later study cited by BERLYAND (1966) shows inversion characteristics distributed similarly to those found by Flohn (Fig.4-5). The thickest, most intense inversions accompanied by the lightest surface winds seem to be in the northeastern part of the country, primarily east of the Lena River.

Perhaps the greater intensity and thickness of temperature inversions in the northeast can be explained by additional amounts of subsidence which take place here due to the fact that this narrowing portion of the continent acts as a zone of surface divergence during the winter with surface winds blowing both to the north toward the Arctic and to the south toward the Sea of Okhotsk. Thus, inversions here are combinations of subsidence from above and cooling from below, while farther southwestward they are primarily due to radiational cooling at the surface alone.

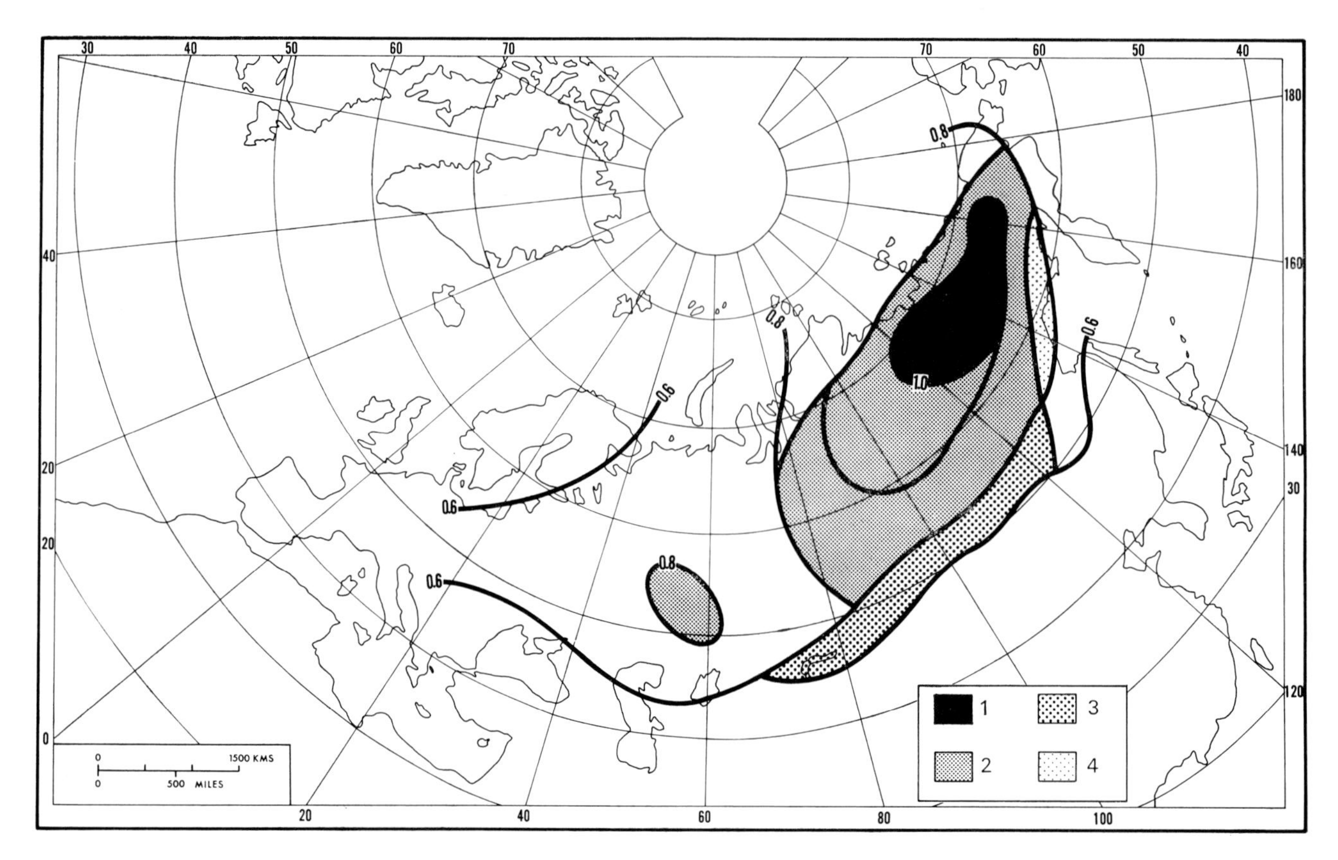

Fig.4-5. Distribution of temperature inversions and surface wind speeds, January: *1* = Average thickness of inversion >1 km, inversion intensity >10°C, average surface wind speed <2 m/sec; *2* = inversion thickness >0.8 km, inversion intensity >8°C, surface wind speed <4 m/sec; *3* = inversion thickness >0.6 km, inversion intensity >8°C, surface wind speed <2 m/sec; *4* = inversion thickness <0.8 km, inversion intensity <8°C, surface wind speed >4 m/sec. Isolines show height in kilometers of maximum temperature in inversion. (After BERLYAND, 1966.)

The typical sounding for Yakutsk shows a rapid increase in temperature with height through the first 1 or 2 km of the air and then a near isothermal layer through the next 2 km or so (Fig.4-6). The lower portion of the inversion with the rapid increase of temperature with height is the result of the combined effects of adiabatic warming from above and radiational cooling at the surface. The isothermal layer above this represents a thickening of the inversion by the continued subsidence of upper air. This subsiding action causes some temperature increase throughout an exceedingly thick layer; the cold surface temperatures in this particular instance are not repeated again until one reaches a height of approximately 7 km in the atmosphere.

A separate node of high pressure generally exists over the northeastern extremity of the country within the northeastern extension of the main Asiatic high which emanates from the core southwest of Lake Baykal. During January pressures in this high cell in the northeast average 1,032 mbar, while southwest of Lake Baykal they average 1,045 mbar. However, the eastern node of the Asiatic high shows about the same number of anticyclonic centers during January as does the central core southwest of Lake Baykal (Fig. 2-16).

In between these two cells of high anticyclonic frequency lies a corridor of lower frequen-

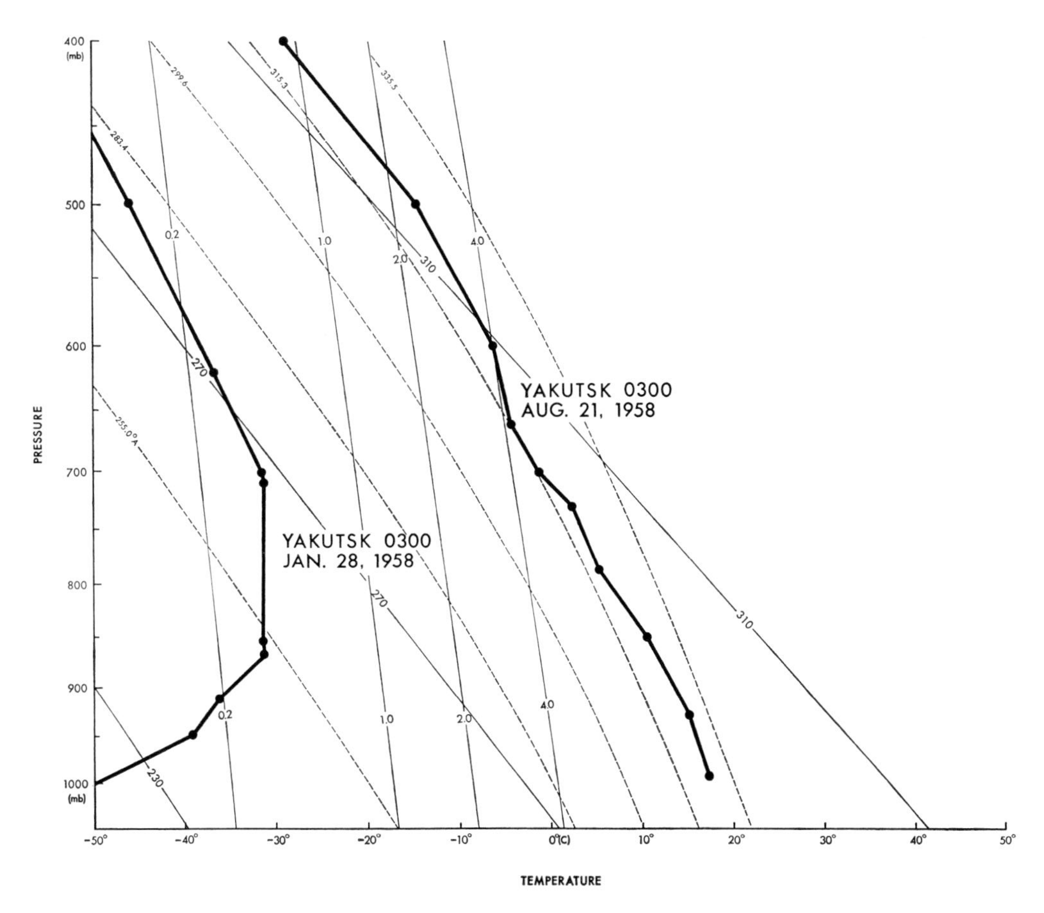

Fig.4-6. Typical vertical temperature distributions at Yakutsk. (After POGOSYAN, 1962.)

cies amounting to only about 1/4 as high values as on either side. This corridor acts as a favorite route of travel for cyclonic storms during winter which may penetrate southeastward out of the Kara Sea across the continent to the northwestern coast of the Sea of Okhotsk where they commonly join with cyclonic storms over the Sea of Okhotsk to be regenerated in the western end of the Aleutian low.

This preferred route of cyclonic storms in midwinter follows the western limb of the upper air trough from northwest to southeast and then eastward into the Sea of Okhotsk (Figs.2-3, 2-4). Locally the storms are influenced by topography. Most cyclones appear to skirt the Putoran Massif on the northeast and then follow lower elevations between high lands to the south and to the northeast. They cross the Lena–Aldan–Vilyuy lowland and then have only the narrow Dzhugdzhur Mountains to cross as they approach the coast of the Sea of Okhotsk. During the incursions of families of cyclonic waves across Yakutia the Asiatic high is fragmented into a number of lesser centers (Fig.4-7). The influences of these cyclonic storms is evidenced by significantly higher amounts of cloudiness during January in this corridor than on either side (Fig.9-11). The corridor stands out even more significantly for percentage of days with 8 to 10 tenths of sky cover. An elongated area here has this much sky cover more than 60% of the time, while to the northeast it drops down to 40% and to the south to less than 20% along the Amur. There is a very steep gradient southward.

The corridor also experiences somewhat higher winter precipitation. January precipita-

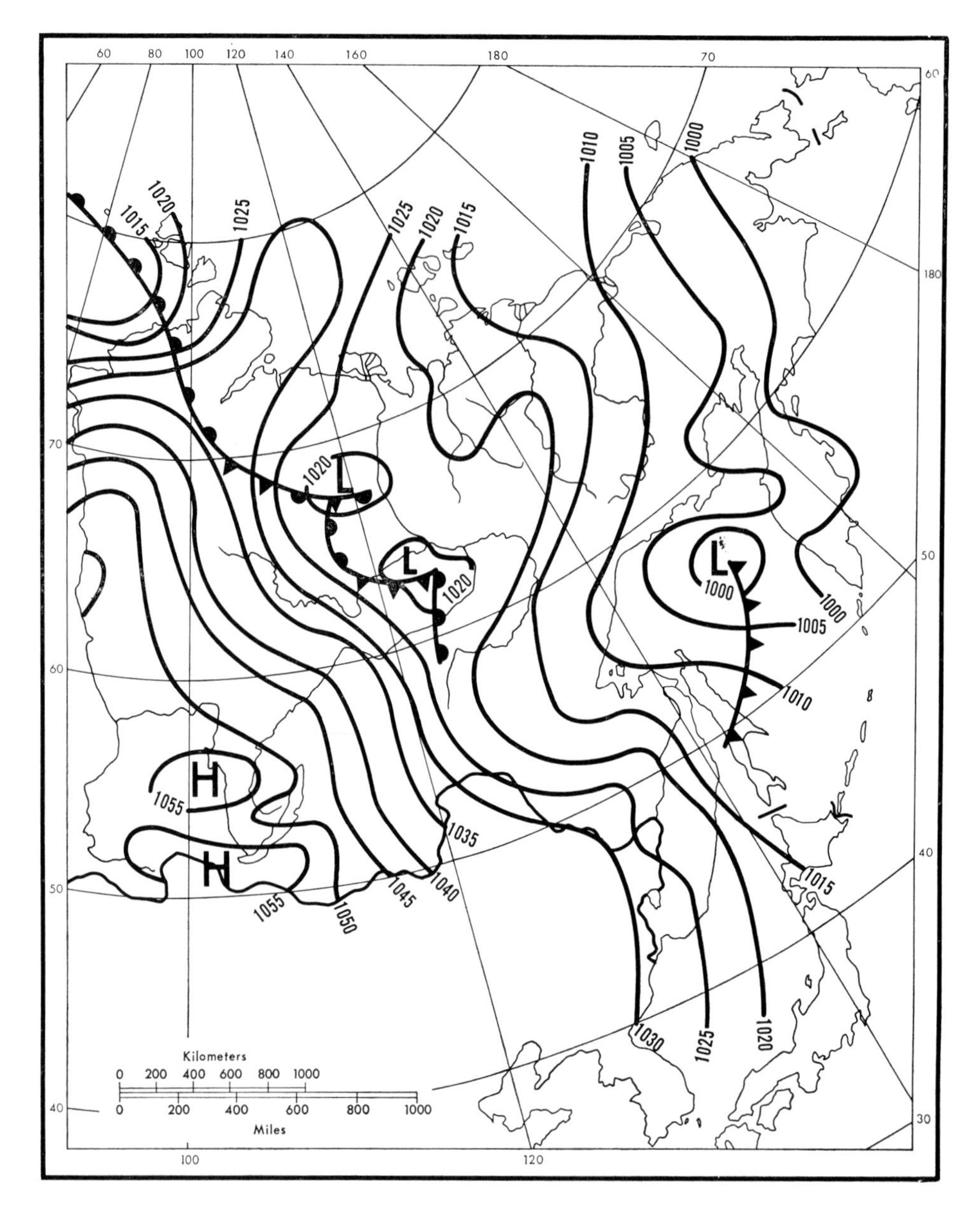

Fig.4-7. Surface synoptic map for the evening of February 14, 1953. (After ALISOV, 1956.)

tion amounts to between 10 and 20 mm here, whereas on either side it drops below 10 mm. Along the Lena River downstream from Yakutsk there is even less than 5 mm of precipitation during January. There is also less than 5 mm in some localized spots along the Amur.

The greater part of eastern Siberia is quite dry in winter, particularly during the second half of winter. In Yakutia and Transbaykal a whole month or more might pass during winter without any precipitation at all. Transbaykal is significantly drier in January than is Prebaykal. The January precipitation picks up again almost everywhere in the Far East.

Transbaykal has about the thinnest snow cover in the country outside of the very southerly regions of European U.S.S.R. and Central Asia. East of Lake Baykal the Amur region generally has no more than 20 cm of snow on the ground during the ten days of greatest depth. This compares to 50–60 cm just to the north in the cyclonic corridor. The snow cover diminishes again northeastward to a minimum of about 30 cm over the upper Yana Valley. Thus, the cyclonic corridor represents an eastward extension of heavier snow cover from the area of maximum snow cover around the middle Yenisey Valley where more than 90 cm are accumulated during the height of the winter. There is particularly little snow in the enclosed basins of Transbaykal. There is also little snow in the tundra zone of the north, and it lies very unevenly on the ground because of strong winds along the Arctic coast. The light snow cover over much of the region leads to deep permafrost. In Yakutia permafrost may extend to a depth of more than 200 m, and along the shore of the Taymyr Peninsula it may extend down to 500 m.

There is frequent ice crystal fog during winter. Usually the vapor pressure in the relatively warm air of the inversion is higher than that in the cold surface air at the base of the inversion and therefore a downward movement of moisture is usually taking place even though the surface air might be saturated. In the cold temperatures that prevail, much of the moisture flux takes place as a slow settling of ice crystals from the inversion top to the inversion base. This may cause a barely audible rustling sound on clear cold nights when it is most perceptible. The Yakuts refer to this as the "whisper of the stars".

Occasionally during winter cyclones move across the central latitudes of Siberia from west to east. Such occurrences bring abnormally warm weather to the region. Very infrequently there is an anomalous east-to-west movement of a cyclonic storm along the Arctic coast. This is brought on by a warm upper level anticyclone over the Arctic and a cold upper level cyclone over eastern Siberia. Under such conditions cyclones may form along the eastern sector of the Arctic front which is lying right along the north coast and move from the Chukotsk Sea westward sometimes as far as the Taymyr Peninsula. Whatever the type of cyclonic activity in eastern Siberia during the winter it always brings about a warming trend, not only through advection of warmer air into the region, but also through a vertical mixing downward of much warmer air from the inversion above the surface.

Spring

Spring brings on an intensification of zonal circulation which causes the most rapid progression of cyclones from west to east through the middle latitudes of Siberia at this time of year. These storms bring relatively strong winds to the region. Temperature rises intensively, particularly from March to April. The snow cover disappears quickly, sometimes actually evaporating under the effects of strong sunshine in the dry clear air, even with temperatures significantly below freezing.

Throughout most of eastern Siberia spring and early summer are very dry. Relative humidity is at a minimum during April and May, and precipitation is very light. Particularly bad spring droughts generally occur in the southern part of Transbaykal, the Minusinsk and Tuva basins and in the basins of eastern Yakutia. There are extreme temperature contrasts between day and night, particularly in the interior regions. Strong freezes occur throughout spring generally with intrusions of cold air in deep cyclones. In the northern parts of eastern Siberia even in May there are frequent days without any thawing at all. As is typical of continental areas in high latitudes, spring is very short in eastern Siberia. Its beginning is delayed until the snow and soil surface melt and the topsoil begins to dry out. Then air temperatures rise very rapidly under clear skies and almost 24-h daylight as the June solstice is approached.

Summer

During summer the west-to-east circulation weakens, and cyclonic activity strengthens as a diffuse low pressure system engulfs much of the interior parts of Asia at the surface. The upper level trough which is so pronounced over eastern Siberia during winter changes to a mild ridge, which causes movements of surface systems to proceed mainly from west-southwest to east-northeast throughout all but the very northeasterly part of the region. In Transbaykal precipitation falls primarily from cyclones which form along the Mongolian sector of the Polar front at this time of year and move east-northeastward. Farther north in Yakutia cyclones form along the Arctic front (Fig.4-8). Sometimes during summer there is a high level warm anticyclone over Yakutia and a high level depression over Mongolia. Under such conditions particularly strong cyclonic activity develops over the southern half of eastern Siberia. At these times cyclones along the Mongolian front may stagnate or even move slowly westward and bring intense rainfall to the southern part of Transbaykal and Yakutia.

Anticyclones generally move into eastern Siberia during the summer from the west. Masses of air may also be brought in from the north and northwest in the rears of cyclones. This air is quickly transformed into continental temperate air over the land surface, which at this time of year is kept quite wet by thawing ground. During daytime hours there is a considerable development of cumulus clouds and shower precipitation from such air. The combination of increased cyclonic activity and increased convection during summer causes eastern Siberia to be somewhat cloudier than it is during winter and causes a summer maximum of precipitation which over most of the region occurs during July or August.

Although summer is short in eastern Siberia it is generally warm and sometimes even hot. Over much of the territory maximum temperatures reach 35°C, and in the southern part of Transbaykal they reach 40°C. Nevertheless, night temperatures are generally quite low, and even in July there may be significant frosts in many lowland areas. The diurnal range of temperature is therefore quite great. The growing season is longest and most favorable for crops in the steppes west of the Altay Mountains in Barnaul, Pavlodar, and Tselinograd regions and surrounding areas. It is also quite favorable in the Krasnoyarsk, Minusinsk, and Tuva basins and in the southern basins of Transbaykal. Over much of eastern Siberia in summer fog is fairly frequent, except along portions of the Arctic coast and some of the major river valleys of Prebaykal.

Autumn

In autumn the upper circulation returns to a zonal flow over eastern Siberia, although meridional intrusions of cold air from the north become more prevalent as the season progresses toward winter. The west-to-east movement in fall is slower than in spring. During fall cyclones sometimes move out of Kazakhstan northeastward into eastern Siberia where they often regenerate along the Arctic front over the Central Siberian Up-

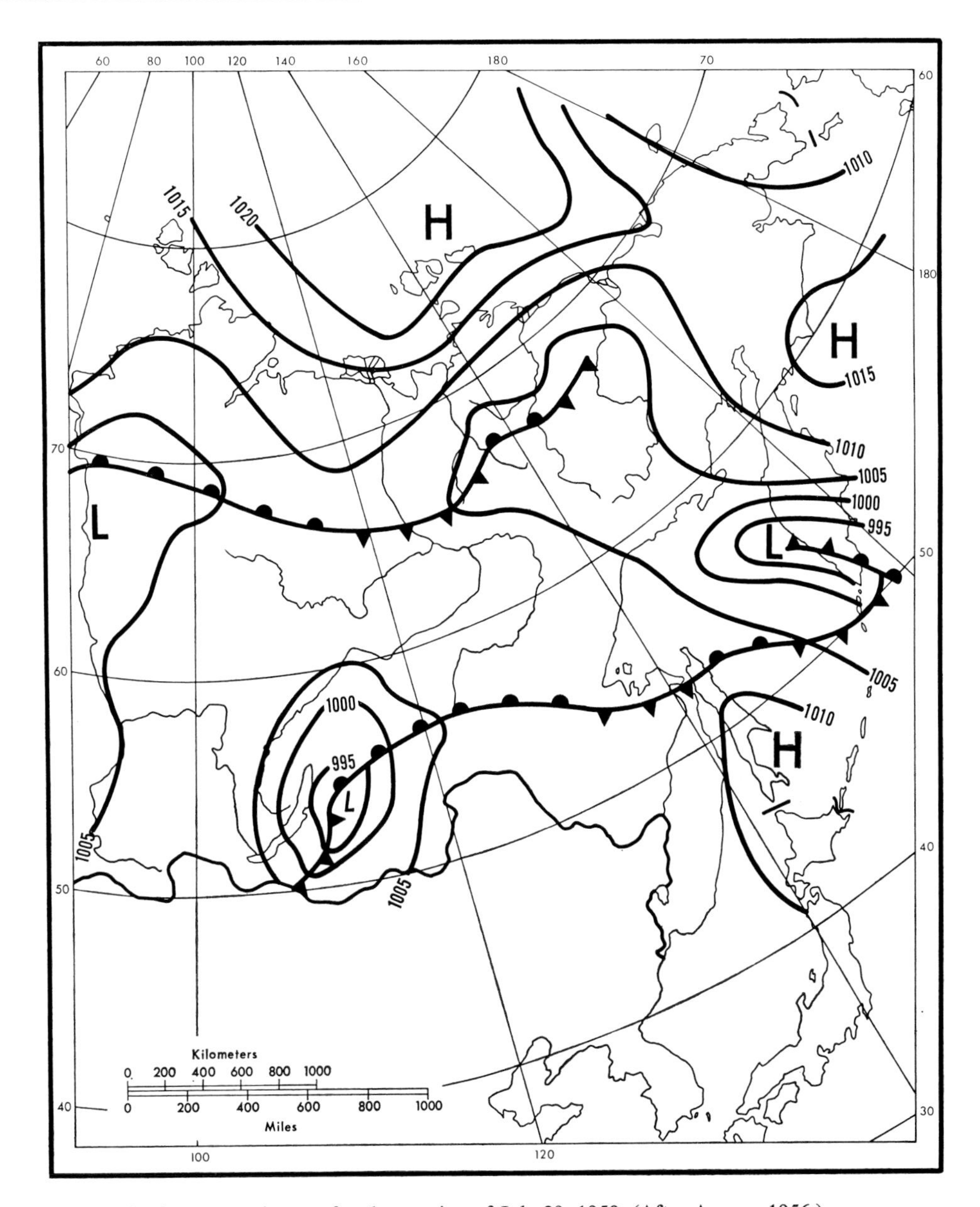

Fig.4-8. Surface synoptic map for the evening of July 29, 1950. (After ALISOV, 1956.)

lands. The warm sectors of such storms bring warm air into eastern Siberia, but after the storms pass deep intrusions of Arctic air follow which may penetrate to the southern border of the country. Similar intrusions of cold Arctic air may be brought in by cyclonic storms which move from northwest to southeast from the Kara Sea to Lake Baykal. Such processes bring on widespread cooling across eastern Siberia. With such occurrences the beginnings of the Asiatic high become discernible as early as the first of October. The winter circulation definitely takes over about the middle of November when the Asiatic high becomes well established.

Regional description

Although the unity of the surface atmospheric pressure pattern, particularly in winter, makes eastern Siberia one of the most homogeneous climatic regions in the U.S.S.R., the great areal extent of the region and diversity of topography do make for considerable climatic differences from one part of the region to another. The entire southern fringe of the region is occupied by rugged topography; high mountains with steep slopes alternate with intervening basins at low elevations, and here the greatest climatic diversity exists.

In extreme southwest, west of the Altay Mountains, the plain, which generally lies at elevations between 100 and 200 m above sea level, is semiarid and contains a great number of salt lakes. The driest part of this steppe region lies along the Irtysh River where precipitation processes are impeded by the Kazakh Folded Country to the south. Semipalatinsk has an average annual precipitation of only 264 mm. A fairly pronounced summer maximum is centered in July (Fig.4-9). Much of the summer precipitation is of a heavy showery type, a large portion of which runs off and does little good. These showers are produced by convection that has been triggered by weak fronts. The condensation level of the surface air is generally too high for surface heating alone to produce showers. A slight secondary maximum of precipitation occurs at Semipalatinsk during late fall and early winter when cyclonic activity is at its height. The region is particularly dry from January through April when the western extension of the Asiatic high holds sway. Winters are generally cold and clear and the snow cover is thin. January tempera-

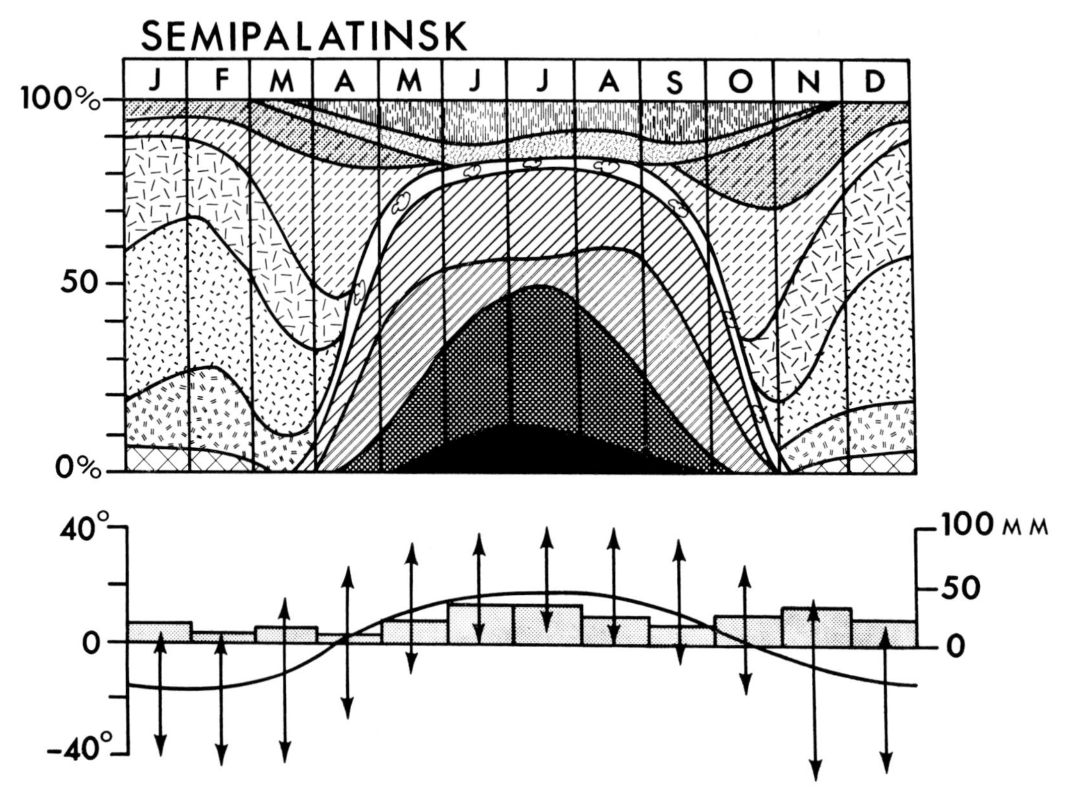

Fig.4-9. Frequencies of weather types, Semipalatinsk. (After GERASIMOV, 1964.) For legend Fig.3-25.

tures at Semipalatinsk average −16.2°C. The maximum depth of snow cover is only 30 cm. Southwest at Karaganda it is only 16 cm.

Precipitation increases eastward toward the Altay Mountains. Barnaul in the Kulunda steppe along the upper Ob' River averages 464 mm per year. The summer maximum is more pronounced here, with July receiving 79 mm. Snow cover reaches a maximum of 67 cm in Barnaul. On the southwestern slopes of the Altay Mountains it reaches as much as 100 cm.

Spring is short in this southwestern steppe where there is little snow to be melted and the ground is relatively dry. Strong insolation causes temperatures to rise rapidly in April and May. Summers are quite warm for the latitude. Semipalatinsk averages 22.1°C in July and has experienced temperatures of 42°C. This contrasts with an absolute minimum at Semipalatinsk of −49°C, which gives the area an absolute range of 91°C.

The Altay Mountains, of course, have variations of climate that are considerably different from the plain to the west. During winter their slopes between the elevations of 200 and 1,000 m above sea level are much warmer than the lowlands because of temperature inversions and, in places, foehns. Foehns in this region are most frequently associated with strong southerly winds preceding cold fronts which descend north-facing slopes and cause rapid rises in temperature and decreases in relative humidity. The foehns blowing into the southern end of Teletsk Lake on the Biya River are particularly well known. This area averages 129 foehns per year, with high frequencies all months except June, July and August. The maximum is in October, and a secondary maximum is in April. Higher in the mountains above the influences of inversions and foehns winters are severe. Temperatures may drop below −50°C in enclosed basins. Summer temperatures, of course, are greatly affected by elevation. At 1,000 m elevation July temperatures average only 14°–16°C. Night frosts are common even in July in enclosed depressions. The perpetual snow line lies at approximately 2,000 m in the northeast, 2,400 m in the southwest, and, 3,000 m in the southeast where snowfall is meager.

Most of the moisture-bearing winds that come to the Altay region come from the southwest so that the southwestern slopes of the mountains are considerably wetter than the northeast. Also some moisture may penetrate the mountains from the northwest where individual mountain ranges fan out to expose interior lowlands to airflow from that direction. The most isolated areas are in the southeastern plateaus region of the Altay. Here annual average precipitation may be no more than 200–300 mm, in contrast to 1,500 mm in the western Altay.

East of the Altay Mountains the Minusinsk and Tuva basins on the upper Yenisey River are outstanding for their dryness and extremes of temperature. Temperatures average below −20°C in January and as much as 20°C in July. Absolute temperatures at Minusinsk range from +39°C to −53°C. In the center of the Minusinsk Basin average annual precipitation is no more than 250 mm, and particularly little falls in winter. No more than 12% of the annual precipitation is received from November through March. Winter weather is very stable. Winds are generally light and blizzards are practically unknown. Snow cover is relatively insignificant. During winter the ground freezes to great depths. Permafrost exists and contains lenses of ice in the ground. The great dryness of the area is emphasized by foehns which generally flow down the eastern slopes of the Kuznetsk Ala-Tau along the western fringe of the basin.

The Tuva Basin to the south has even more extreme conditions than the Minusinsk Basin.

It lies at elevations from 527 to 638 m. Cold air stagnates in the basin during the winter, and the basin acts as one of the centers for the Asiatic high. Fronts which approach the basin from the west pass over the mountains and then ride across the cold air dome in the basin and do not descend onto the basin floor (Fig.4-4). About the only evidence of frontal passage in winter is a period of increased cloudiness and slight temperature rise. During such a frontal passage temperatures during the day may rise to −20°C. This is known as a Tuva thaw! The absolute minimum temperature in Kyzyl is −58°C. Winter precipitation is meager and the snow cover is very thin. The soil freezes to great depths. Because of the general lack of snow cover livestock usually graze throughout the winter. But during years when snowfall is heavier than usual livestock may starve to death. Winter grazing is enhanced by foehns which generally enter the basin from the south ahead of passing cold fronts. They are common throughout the winter but reach their greatest frequency in May. Their warming effects are due to a combination of adiabatic heating with descent, destruction of the inversion, and advection of warmer air into the region.

In spring temperatures in the Minusinsk and Tuva basins rise very quickly during March and April. At this time of year frequent foehns occur which often fill the air with dust and fine sand. Relative humidity during these occasions drops to 15 or 20%. The blowing sand damages leaves on trees and other vegetation and is particularly damaging to vegetables. July is the only month that completely escapes frosts. And during July and August about 60% of the yearly precipitation falls, generally in the form of heavy showers.

Further north the Yenisey Valley receives somewhat more precipitation and it is a little more evenly distributed through the year, although there is still a pronounced summer maximum, with August having the maximum amount (Fig.4-10).

Farther east, Lake Baykal has a profound influence on the climate in the immediate vicinity. Due to its great depth and heat storage the lake surface generally does not freeze over until sometime in January, and therefore its influence during early winter is very great. Even after ice forms over the surface a great deal of heat is derived from the underlying water. Isotherms during winter encircle the lake and show a difference of about 14°C between the surface air over the lake and the surface air behind the mountains surrounding the lake (Fig.4-11). This strong temperature difference distorts the general pressure pattern over the region so that the isobars also tend to encircle the lake (Fig.4-12).

The strong pressure gradients directed from the cold land to the relatively warmer lake in winter cause many bora-type winds to descend the short steep river valleys leading down into the lake all around its circumference. The best developed of these winds is the so-called "sarma" which blows from the northwest down the Sarma River Valley on the west coast of Lake Baykal. It blows most frequently in October to December when the unfrozen lake contrasts most sharply with cold air crossing the Primorskiy Range on the west side of the lake. On the average the sarma blows 113 days per year with wind speeds from 15 to 40 m/sec. After issuing onto the lake surface from the mouth of the Sarma River the winds generally die down within 20–30 km off the shore. When they occur in summer they bring a great deal of dustiness onto the lake. The winds reach their strongest velocities when the local baric gradients are strengthened by the general circulation over the area. This occurs when a high pressure cell lies just to the west of the lake

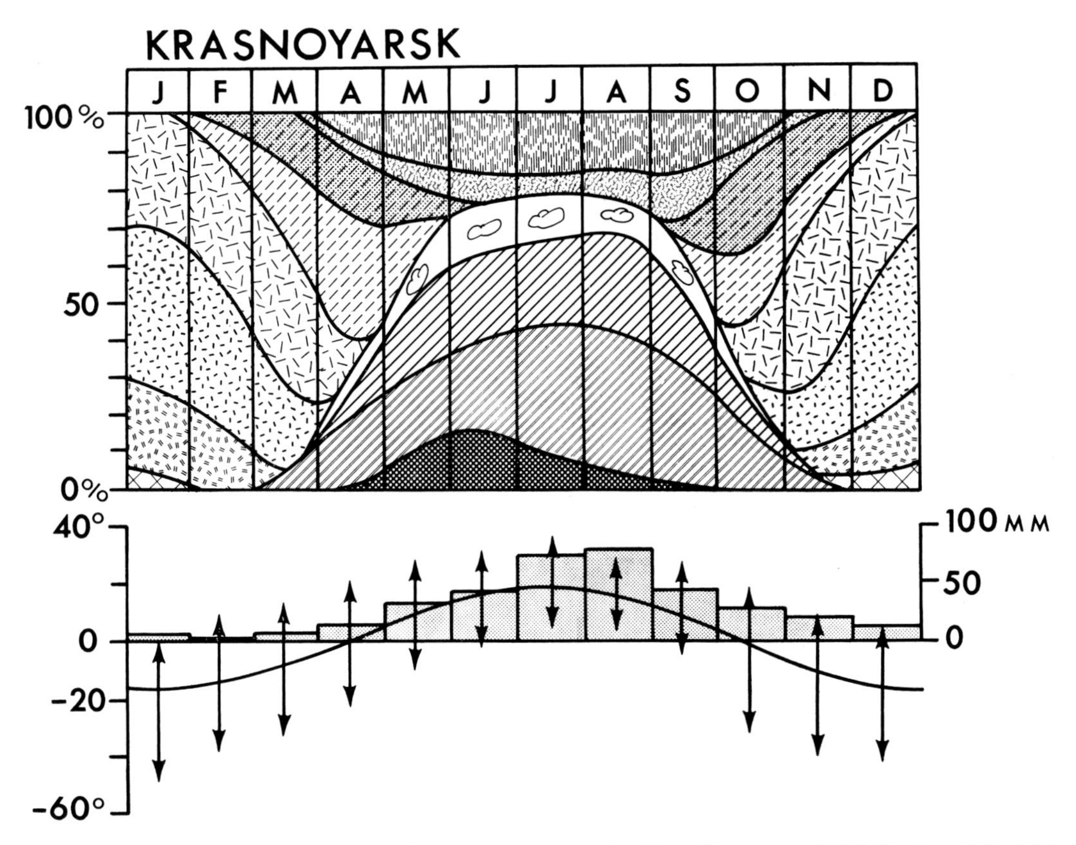

Fig.4-10. Frequencies of weather types, Krasnoyarsk. (After GERASIMOV, 1964.) For legend Fig.3-25.

and a low pressure occupies the Transbaykal or Amur districts. Under these conditions pressure gradients are steepened in the Lake Baykal area and isobars run essentially from north to south (Fig.4-13).

Strong winds in other valleys around the lake are given local names, but they are not so well developed or so well known as is the sarma. Not all valleys have well developed winds, however. At Irkutsk near the head of the Angara River wind speeds during December and January average less than 1 m/sec in 60 to 65% of the time. Strong winds blowing out of river mouths onto the lake cause great hazards to shipping, particularly in the narrow strait between the western shore and Ol'khon Island at the mouth of the Sarma River.

Since the surface of Lake Baykal is relatively warm during winter the cold temperature inversion so characteristic of eastern Siberia does not occur over the lake. Along the southern shore the slopes of the Khamar-Daban Range show an average temperature decrease with height of 0.3°C for every 100 m. On occasion, however, when air temperatures behind the Baykal Mountain Chain reach −50°C or lower and air pressure builds up to extreme values, cold air fills up the entire Lake Baykal depression and temperatures at the surface of the lake may then fall to −40°C. Steam fogs occur during such incursions of cold air over the lake, and are especially intensive in the southwest at the head of the Angara River where the water seldom freezes.

Spring warming is greatly impeded by the cool influence of the lake which becomes ice free only in the middle of May. Ice remains longest on the eastern shore where it is

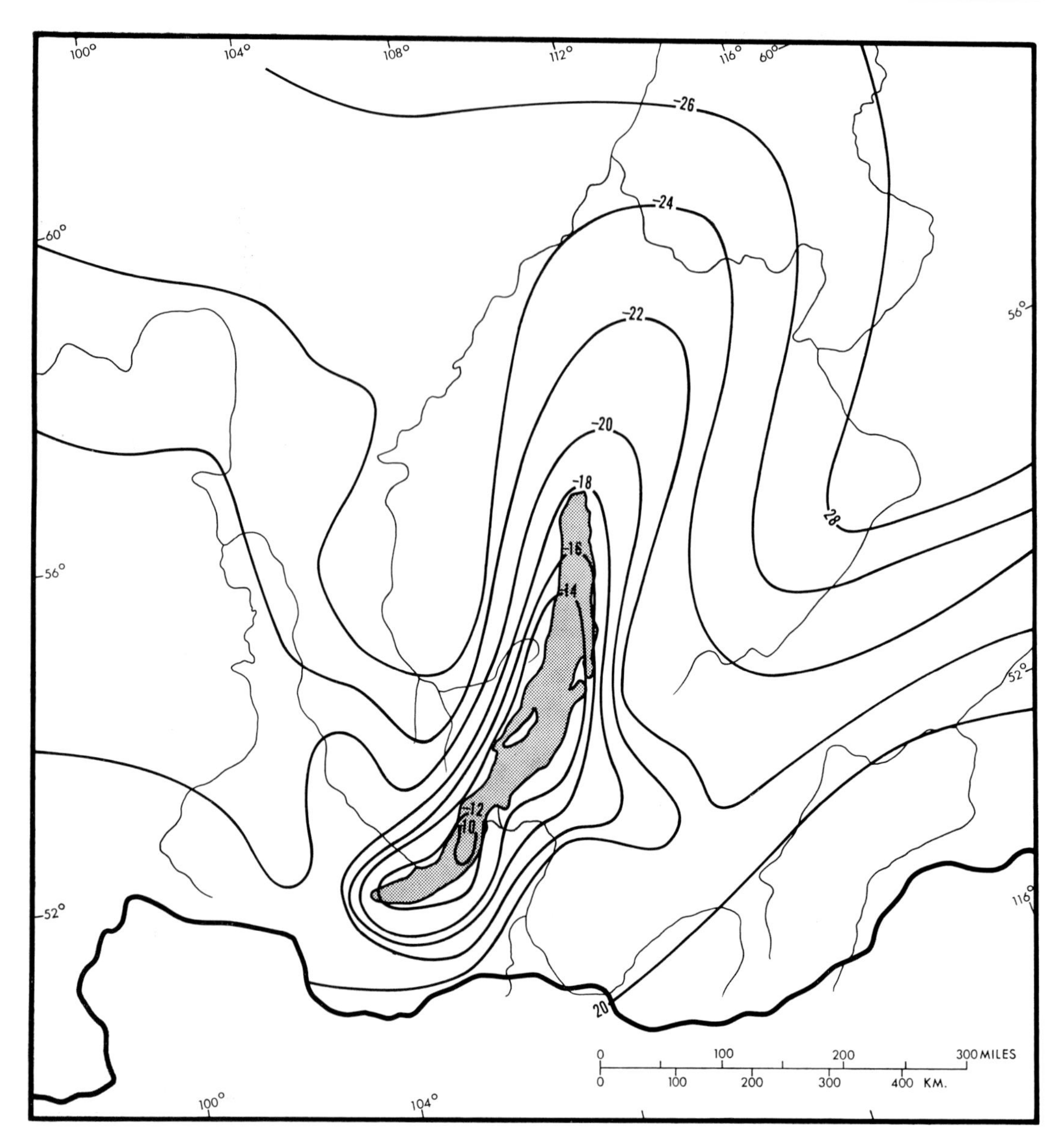

Fig.4-11. Isotherms around Lake Baykal in December. (After ALISOV, 1956.)

drifted by the winds after breaking up. Spring is considerably colder than fall in the area. Advection fogs are common in the second half of spring when warm air from the south moves over the cold lake surface.

During summer temperature and pressure gradients around the lake are not as strong as they are in winter, but the lake is still a very significant influence on these two patterns (Figs.4-14, 4-15). During July the lake surface averages from 4° to 6°C cooler than the surrounding land. Temperatures are highest over the lake in August when they average only 12°–14°C. Since the surface of the lake is colder than the air during summer, vertical temperature gradients are small over the lake. Isothermal conditions often occur to considerable heights. This prevents much vertical exchange of air over the lake. Fogs form frequently, particularly in the northern part of the lake. The temperature of the

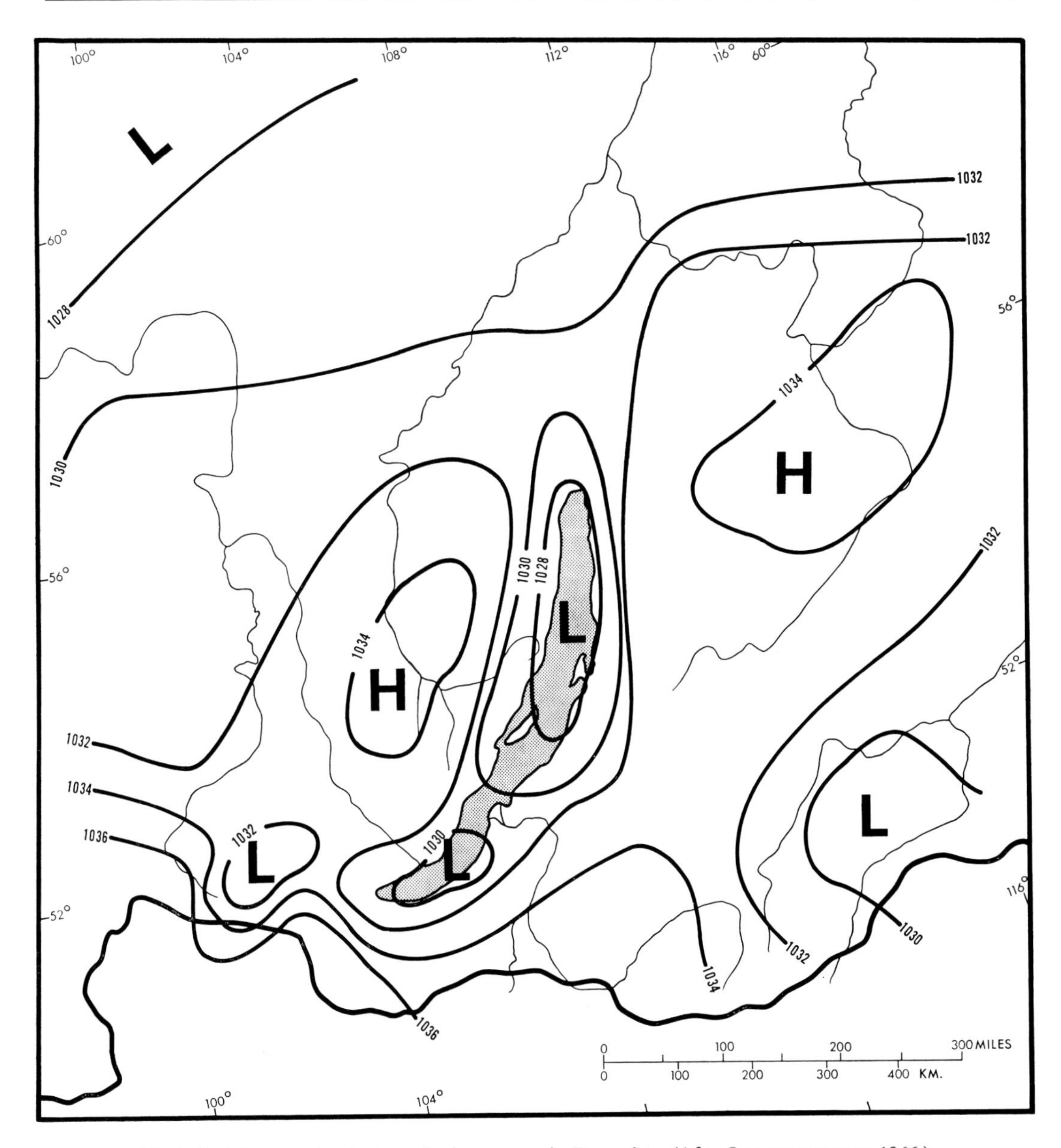

Fig.4-12. Mean sea level atmospheric pressure in December. (After LADEYSHCHIKOV, 1966.)

lake water is somewhat warmer in the south where the Selenga River brings in fresh warm water from the surrounding land. This is the primary inflow of water into the lake. The Baykal area experiences its greatest precipitation in July and August when cyclonic activity is at its peak. The region is affected by both west Siberian cyclones which move eastward into the Baykal region and cyclones forming along the Mongolian front south-east of the lake. With strong development of cyclonic storms along the Mongolian front, foehns blow during summer down the slopes along the southern coast of the lake and down the northern slopes of the Eastern Sayan Mountains. Sea breezes develop around the lake during summer during daytime hours. At night the cool waters of the lake do not contrast sufficiently with the land to produce significant land breezes. Sometimes during the night there are relatively strong cold winds blowing off the land onto the lake,

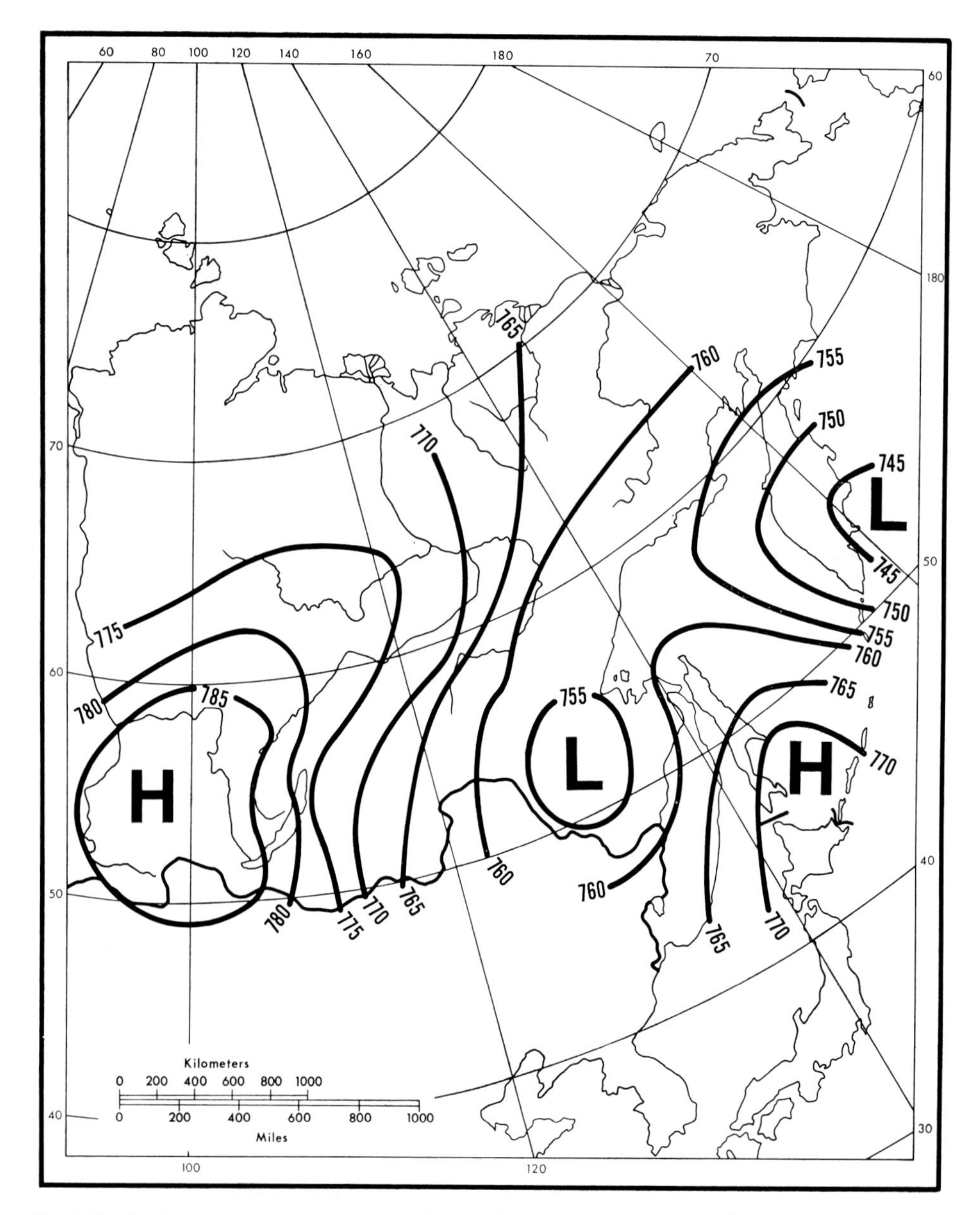

Fig.4-13. Typical surface synoptic conditions with strong development of "sarma". Isobars are in millimeters of mercury. (After SHCHERBAKOVA, 1961.)

but these winds are of a mountain-valley type, with the lake playing only a secondary role.

Some attempt has been made to assess the importance of the lake on precipitation in the surrounding region. Evaporation from the lake's surface is highest in November and December when the lake is still unfrozen and there is a great contrast between the relatively warm lake surface and the cold air masses moving across it. At this time of year there may be significant influences on local precipitation, but precipitation at this time of the year is very low throughout the area, and during summer when most of the precipitation falls the lake effect seems to be negligible.

East of Lake Baykal in the southern half of Transbaykalia the mountain ranges run from

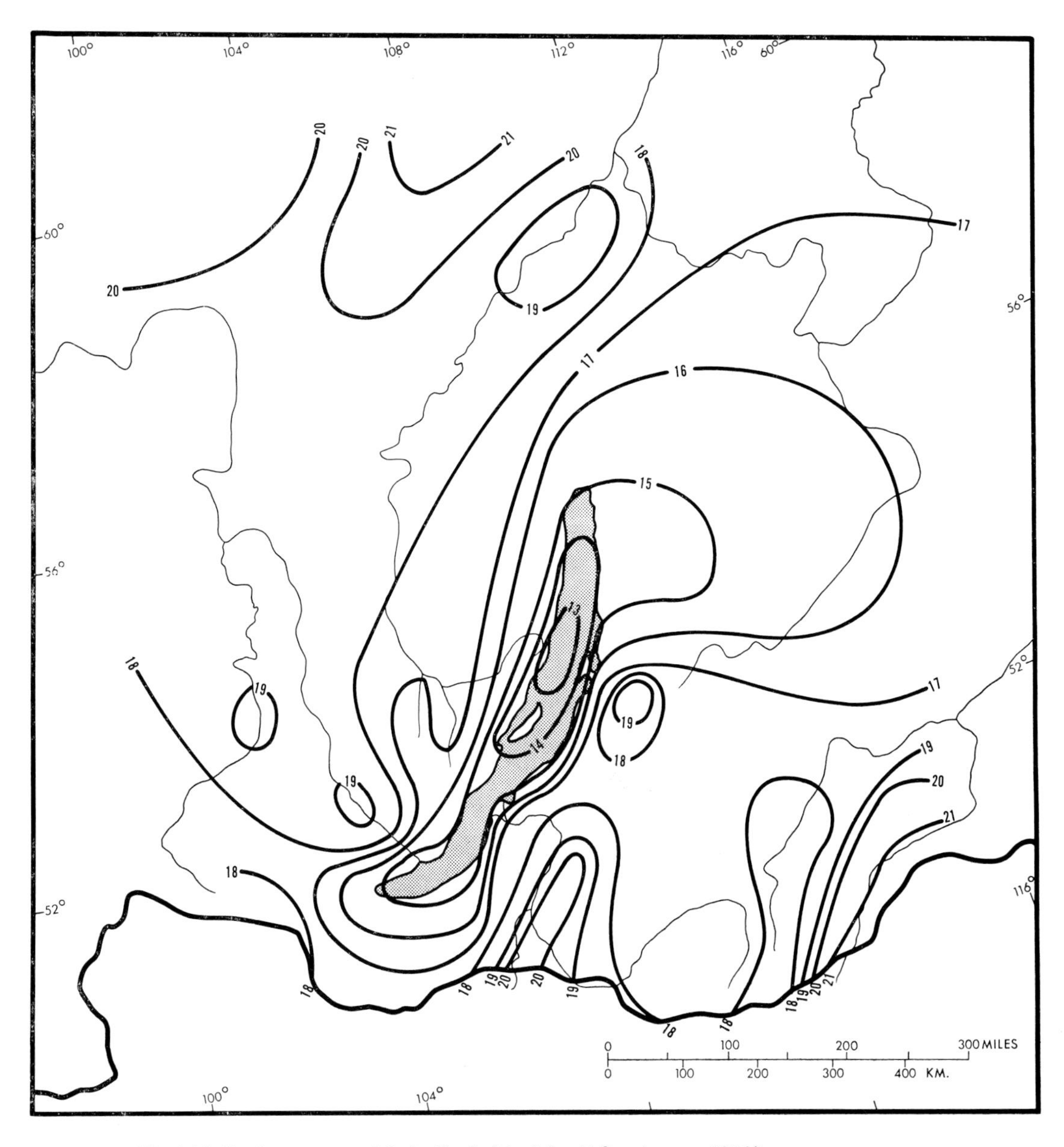

Fig.4-14. Isotherms around Lake Baykal in July. (After ALISOV, 1956.)

southwest to northeast at elevations between 1,200 and 2,500 m. These merge in the north with the Vitim, Patom, and Aldan plateaus, between which are mountains that rise in places to almost 3,000 m. The main range in the south is the Yablonovyy which separates the basin of Ulan-ude on the west from that of Chita on the east. These two basins lie generally between 500 and 700 m above sea level. Their climate is reminiscent of the Minusinsk and Tuva basins farther west. Semiarid conditions have led to the development of grasslands and chernozem soils. Throughout this region forests and steppe grasses alternate according to topography and exposure. Forests penetrate on northern slopes into extreme southern regions of Transbaykalia, and steppe grasses penetrate through the river floodplains up to 60° latitude.

The climate of the southern basins is well represented by the diagram for Chita (Fig. 4-16).

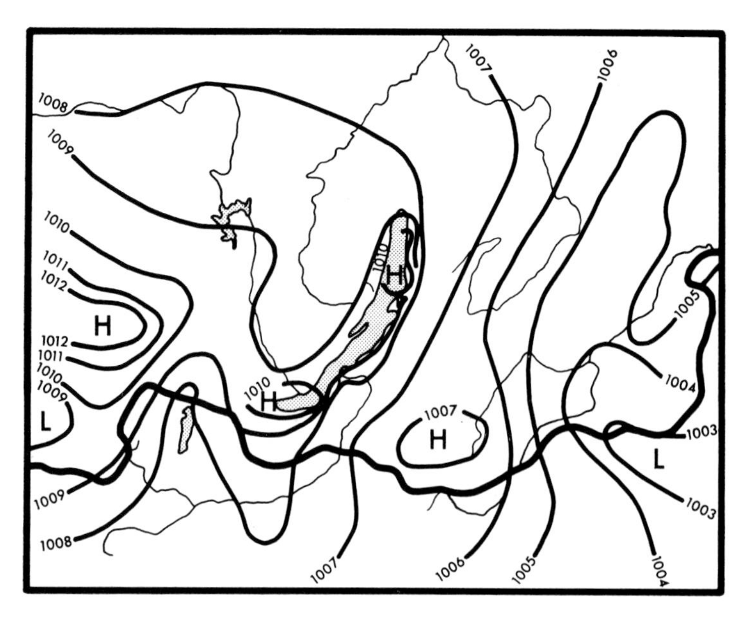

Fig.4-15. Mean sea level atmospheric pressure in July. (After LADEYSHCHIKOV, 1966.)

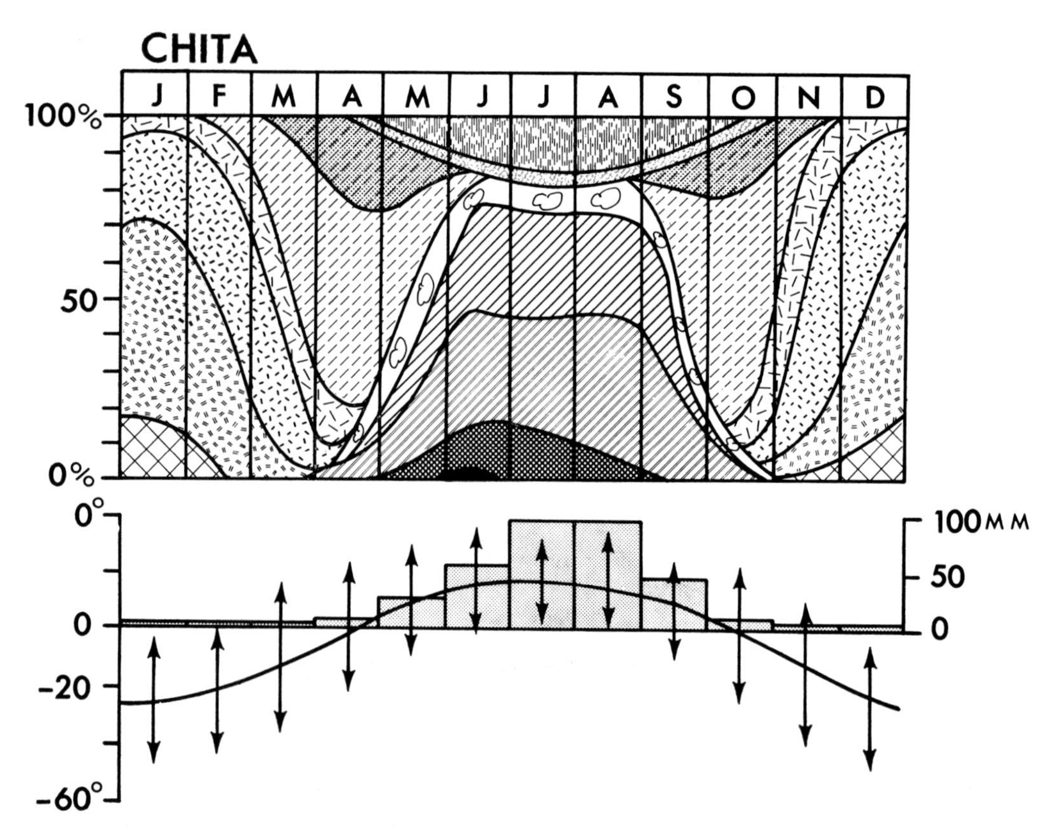

Fig.4-16. Frequencies of weather types, Chita. (After GERASIMOV, 1964.) For legend see Fig.3-25.

During the winter the basin is occupied by the eastern extension of the core of the Asiatic high with its attendant clear sunny skies, extreme cold, and lack of precipitation. Snow cover in this region is exceedingly thin, and the soil freezes to several meters depth. Extreme dryness continues through April, after which precipitation begins to increase as cyclonic activity develops along the Mongolian front. But atmospheric relative humidity remains low. During May it averages no more than 30% at 13h00. This is as dry as the air in Central Asia. Dust storms occur frequently during the dry first half of summer. Such conditions contrast greatly with the warm rainy weather of the second half of summer. July and August are the rainiest months at Chita and receive 183 mm of the total annual amount of 343 mm. Much of this late summer rainfall comes in downpours associated with cold fronts. Occasionally they can be caused by local heating of unstable air. Most of the showers come in the afternoon and evening. During one August almost 1/3 of the normal yearly precipitation fell during one 24-h period. These heavy showers are generally accompanied by thunderstorms. On the average July has 8 thunderstorms, and June and August each have 5. The rainy time of year is also by far the cloudiest time.

Extreme temperature conditions characterize the southern basins. Chita averages 18.5°C in July and −26.8°C in January. Its maximum and minimum temperatures have reached +41°C and −51°C. Diurnal ranges of temperature are also great, averaging over 20°C in late spring and early summer and 13° to 14°C in midwinter. The frost-free period in these southern basins is only about 100 days. Lowland areas may experience night frosts in June even in the most southerly regions. Because of this valley bottoms with the best soils are generally used only for hay, while grain is planted on slopes of poorer soils where it is more secure from summer frosts.

In the plateaus to the north there is much more winter cloudiness, and the precipitation is somewhat more evenly distributed through the year, although there are still pronounced late summer maxima (Fig.4-17). Winters are somewhat colder than the lowlands farther

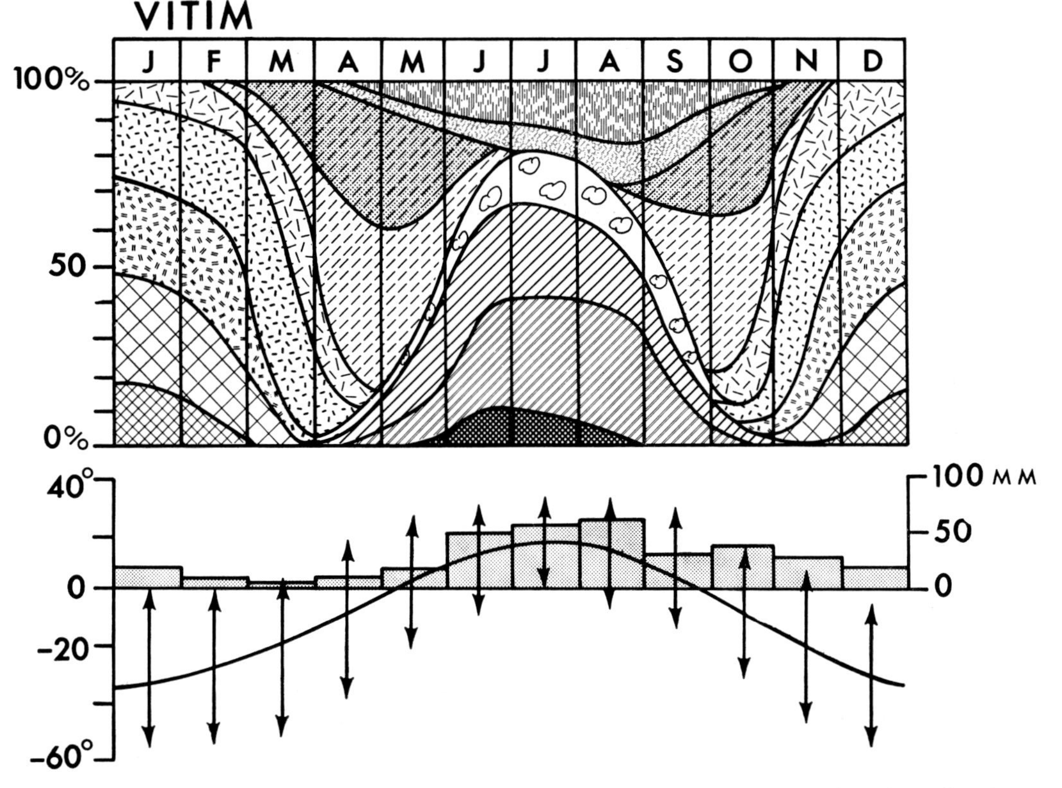

Fig.4-17. Frequencies of weather types, Vitim. (After GERASIMOV, 1964.) For legend see Fig.3-25.

south and summers are cooler and shorter. On the Vitim Plateau at elevations from 1,000 to 1,300 m the average temperature in July is around 13° –15°C. On the Aldan Plateau it is around 15°–17°C.

In the central part of Yakutia the Lena–Vilyuy Lowland lies at elevations of 100–500 m above sea level. Winters are severe with little snow and summers are moderately warm for the latitude and relatively dry. Much of the region is vegetated by forest steppe and is underlain by permafrost. At Yakutsk the annual precipitation is only 213 mm. The maximum falls in July and is produced by cyclonic storms along the Arctic front. October is the cloudiest month. Throughout the winter the region generally has southwest to westerly winds with frequent calms. Wind speeds of less than 1 m/sec occupy 65 to 75% of the time in winter. The average length of the frost-free period at Yakutsk is 95 days, which is almost as long as it is at Chita in the south. In early summer at this latitude the daylight period is almost 24 h. Therefore, in spite of its far northern location the region around Yakutsk is known for its Arctic agriculture.

East of the Lena River lie a series of mountain ranges which generally run from north to south whose intervening valleys open up onto broad lowlands along the Arctic coast. The mountain crests generally lie at elevations between 2,000 and 3,000 m, but occasionally they rise to a little more than 3,000 m. At these high latitudes such elevations are well above the tree line so that much of the territory is occupied by tundra vegetation. The area is noted for its extremely low winter temperatures and extreme annual temperature ranges in the valleys between the mountain ranges.

Verkhoyansk on the upper Yana River has often been called the cold pole of the world. Although colder temperatures have been recorded elsewhere on earth, nowhere have such cold temperatures been recorded at such low elevations. Verkhoyansk, at an elevation of only 137 m above sea level has experienced a temperature as low as −68°C. It averages −48.9°C in January. The coldness in winter due to its extreme continental location is added to by its site in a narrow valley between mountain ranges where cold air stagnates and extreme temperature inversions form. On the slopes of the Verkhoyansk Range at an elevation of 1,350 m above sea level the average January temperature is 21° higher than it is in Verkhoyansk. The air stagnation in the basin makes for very low wind speeds. Winds average less than 1 m/sec from November through March. A speed of 10 m/sec is very rare and during January does not exist at all.

The same general and local conditions that make for extreme cold in winter make for extreme warmth in summer for such a high latitude. Temperatures average 15.3°C in July and have reached 35°C. Thus, Verkhoyansk has an average annual temperature range of 64.2°C and an absolute range of 103°C. The diurnal temperature range is high in spring. April averages 24.8°C. But in December, with extreme surface temperature inversions and complete lack of sunlight, the daily range is no more than 4°C.

In the upper Indigirka Valley, Oymyakon boasts a little lower absolute minimum temperature and a little greater absolute temperature range than Verkhoyansk, but it sits at an elevation of 740 m above sea level. The January average temperature is −50.1°C and a temperature of −71°C has been observed. July averages 14.5°C and has reached 33°C. The increased elevation at Oymyakon produces longer winters and shorter summers than at Verkhoyansk (Figs.4-18, 4-19).

Throughout this eastern extension of eastern Siberia the precipitation is very light. Verkhoyansk averages 155 mm per year and Oymyakon 193. The maximum comes in July,

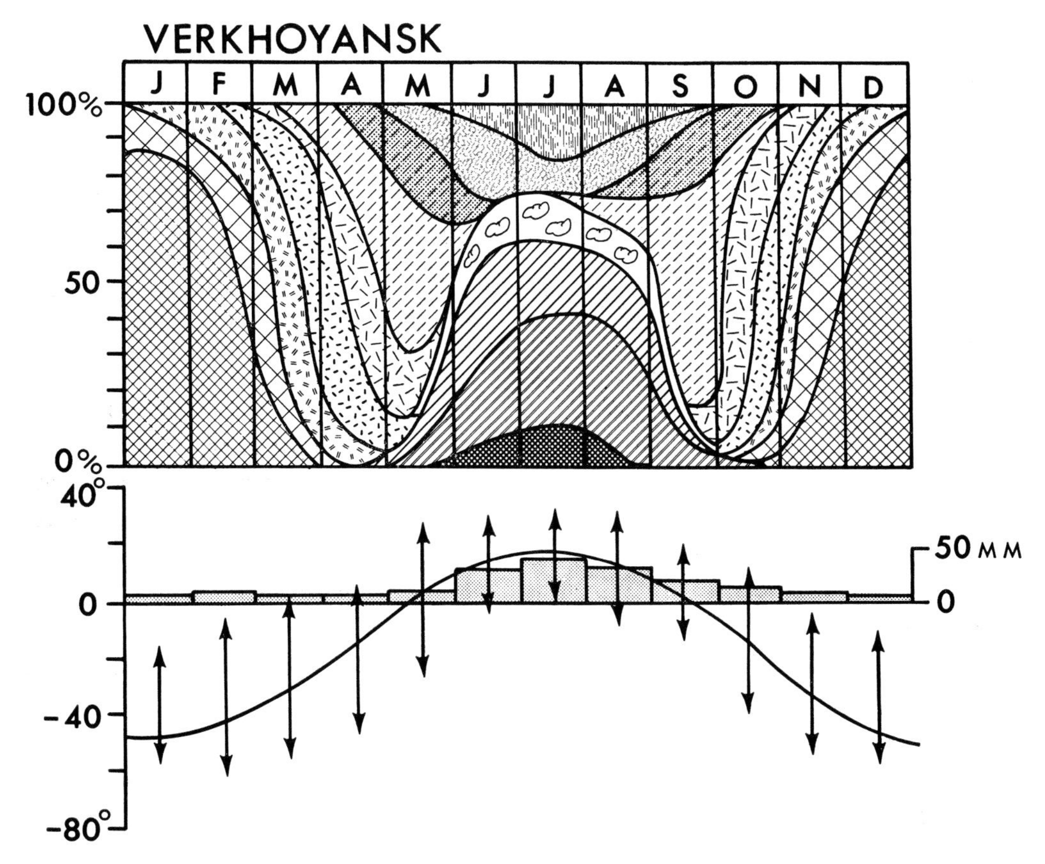

Fig.4-18. Frequencies of weather types, Verkhoyansk. (After GERASIMOV, 1964.) For legend see Fig.3-25.

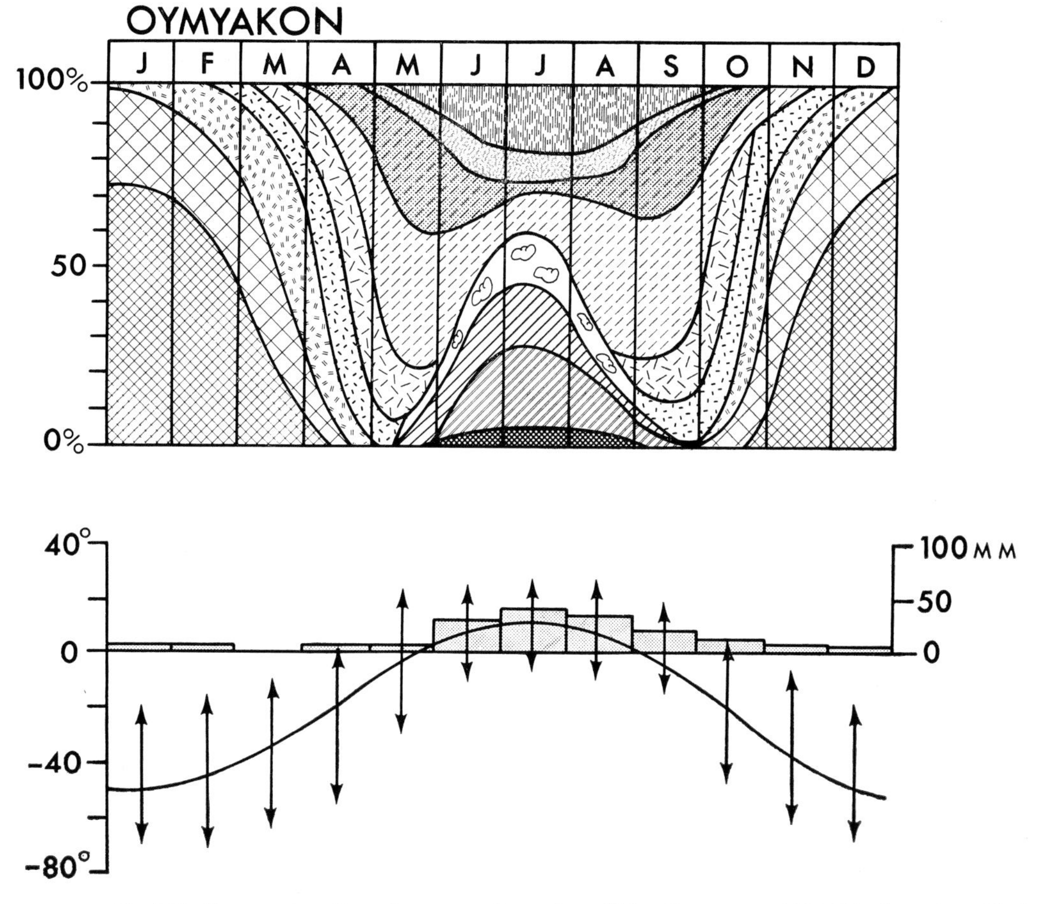

Fig.4-19. Frequencies of weather types, Oymyakon. (After GERASIMOV, 1964.) For legend see Fig.3-25.

with June or August being close behind. Winter and spring are the driest times of the year.

The low Arctic coastal plain to the north experiences considerably warmer winters, cooler summers, higher wind speeds, and greater cloudiness than do the interior basins. Chokurdakh about 150 km inland on the lower Indigirka River averages −36.2°C in January and 10.2°C in July. Wind speeds average 3–4 m/sec in winter and 6 m/sec in summer. There are 147 days per year with overcast skies, the maximum falling during June–October. 56 days per year experience fog, with the summer half-year having twice as many as the winter half-year. Precipitation continues low in this area because of the constantly cool atmospheric temperatures.

Off shore, Chetyrekhstolbovoy Island shows even warmer winters and cooler summers. January averages −29.5°C and July +2.5°C. The Arctic ice pack begins to melt back from the coast in July near the mouth of the Lena River and the melting spreads eastward and westward until in August and September most of the north coast is clear of ice to a distance of approximately 300 km. But in October the ice freezes back to the shore line. Since summer sea surface temperatures hover near freezing, the surface air temperatures on the island groups rarely rise much above freezing. The diagram for Tiksi well illustrates the long winters and short cool cloudy summers along the coast (Fig.4-20).

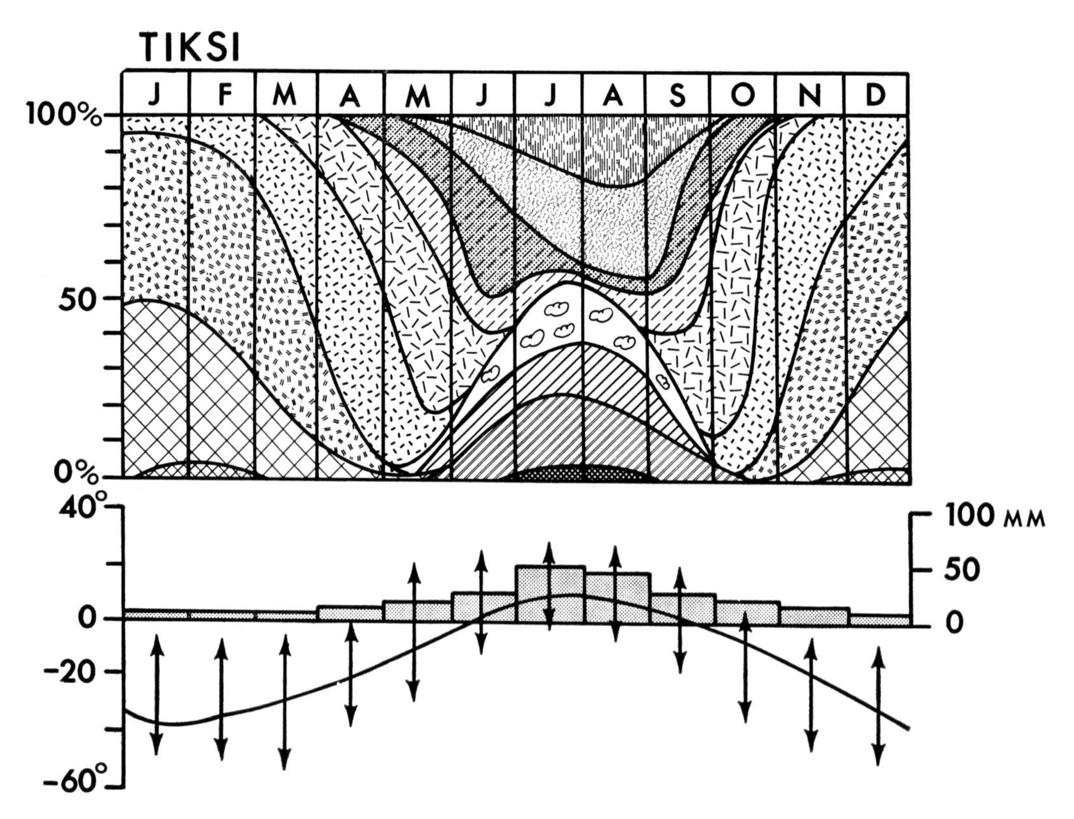

Fig.4-20. Frequencies of weather types, Tiksi. (After GERASIMOV, 1964.) For legend see Fig.3-25.

References and further reading

ALISOV, B. P., 1956. *Klimat SSSR* (Climate of the USSR). Moscow University, Moscow, 127 pp.

ANAPOL'SKAYA, L. E. and TYUKTIK, V. V., 1969. Raschetnyye skorosti vetra v gorakh Altaya i Sayan (Calculation of wind speeds in the Altay and Sayan mountains). *Gl. Geofiz. Obs. Tr.*, 246: 29.

ANONYMOUS, 1969. *Atlas SSSR* (Atlas of the USSR). Gl. Upr. Geod. Kartogr. Sov. Ministr. SSSR, Moscow, 2nd ed., 199 pp.

BERLYAND, M. YE., 1966. The climatological aspects of investigation of atmospheric contamination with industrial wastes. In: *Sovremennye Problemy Klimatologii.* Gidrometeoizdat, Leningrad. (Translated by Foreign Technology Division, Wright-Patterson Air Force Base, Ohio, as *Modern Problems of Climatology*, 1967, pp.300–315.)

BORISOV, A. A., 1965. *Climates of the USSR.* Aldine, Chicago, Ill., 255 pp.

FILIPPOV, A. KH. and KHUTORYANSKAYA, D. F., 1971. Statistical characteristics of thunderstorms in Yakutia. *Dokl. Inst. Geogr. Sib. Dal'nego Vostoka*, 31: 39–46.

FLOHN, H., 1947. Zum Klima der freien Atmosphere über Sibirien, II. Die regionale winterliche Inversion (The climate of the free atmosphere over Siberia, Part II. The regional characteristics of the winter inversion). *Meteorol. Rundsch.*, 1: 75–79.

GERASIMOV, I. P. (Editor), 1964. *Fiziko-Geograficheskiy Atlas Mira* (Physical-geographical atlas of the world). Akad. Nauk, Moscow, 298 pp.

KORNIENKO, V. I., 1969. Nekotorye dannye po vlagooborotu zabaykal'ya (Some data pertaining to the hydrological cycle in Transbaykalia). *Gl. Geofiz. Obs. Tr.*, 245: 114–120.

KORNIENKO, V. I., 1969. Sootnosheniye srednego i turbulentnogo potokov vlagi v atmosfere nad zabaykal'em (Interrelationship between mean and turbulent fluxes of moisture in the atmosphere over the Transbaykal region). *Gl. Geofiz. Obs. Tr.*, 245: 103–113.

KORNIENKO, V. I., 1969. Vliyaniye ispareniya s ozera baykal na osadki okruzhayushchikh rayonov (The influence of evaporation from Lake Baykal on precipitation in surrounding regions). *Gl. Geofiz. Obs. Tr.*, 247: 122–136.

LADEYSHCHIKOV, N. P. (Editor), 1966. Klimat ozera Baykal i pribaykal'ya (The climate of Lake Baykal and Pribaykalia). *Akad. Nauk SSSR, Sib. Otd., Tr. Limnol. Inst.*, 10(30): 174 pp.

MAKHOVER, Z. M., 1967. O prichinakh ustoychivogo polozheniya tsentra Aziatskogo antitsiklona nad Mongoliey (On the causes for the stable position of the center of the Asiatic Anticyclone over Mongolia). *Tr., Nauchnoizzled. Inst. Aeroklimatol.*, 38: 54–59.

MAKSIMOV, I. V. and KARKLIN, V. P., 1969. Sezonnyye i mnogoletniye izmeneniya geograficheskogo polozheniya i intensivnosti sibirskogo maksimuma atmosfernogo davleniya (Seasonal and long range changes in the geographical position and intensity of the Siberian maximum of atmospheric pressure). *Izv. Vses. Geogr. O-vo.*, 101: 320–330.

NEVYAZHSKIY, I. I. and BIDZHIEV, P. A., 1960. Eolovye formy rel'efa tsentral'noy yakutii (Eolian forms of relief in central Yakutia). *Akad. Nauk SSSR, Izv., Ser. Geogr.*, 3: 90–94.

NIKOL'SKIY, K. N., 1957. Osobennosti zimnego temperaturnogo rezhima Tuvinskoy avtonomnoy oblasti (The particulars of the winter temperature regime in Tuva A.O.). *Meteorol. Gidrol.*, 12:

OSOKIN, I. M., 1961. Nekotorye dannye o kratkovremennoy intensivnosti livney v g. Chita (Some incidences of brief intensive downpours in Chita). *Izv. Vses. Geogr. O-vo.*, 1961: 336–337.

POGOSYAN, KH. P., 1962. *Uchebnyy Sinopticheskiy Atlas* (Student synoptic atlas). Gidrometeoizdat, Leningrad.

PUTNINS, P. and STEPANOVA, N. A., 1956. *Climate of the Eurasian Northlands.* Technical Assistant to Chief of Naval Operations for Polar Projects, OP-03A3, Washington, D.C., 104 pp.

RUSANOV, V. I., 1961. Raspledelenie srednego godovogo kolichestva osadkov v tsentral'nom altae (Distribution of average annual precipitation in the central Altay). *Izv. Vses. Geogr. O-vo.*, 1961: 507–511.

SHASHKO, D. I., 1961. *Klimaticheskie Usloviya Zemledeliya Tsentral'noy Yakutii* (Climatic conditions for agriculture in central Yakutia). Akad. Nauk, Moscow, 264 pp.

SHCHERBAKOVA, E. YA., 1961. *Klimat SSSR, 5. Vostochnaya Sibir'* (Climate of the USSR, 5. Eastern Siberia). Gidrometeoizdat, Leningrad, 300 pp.

VITVITSKIY, G. N., 1972. O prirode letnikh osadkov Vostochnoy Sibiri (On the nature of summer precipitation in Eastern Siberia). *Akad. Nauk, Izv., Ser. Geogr.*, 1: 93–99.

ZVEREV, A. S., 1957. *Sinopticheskaya Meteorologiya* (Synoptic meteorology). Gidrometeoizdat, Leningrad, 559 pp.

Chapter 5

Far East

Introduction

The Soviet Far East as defined in Fig.5-1 is an elongated, irregular region of mountainous maritime fringes, peninsulas, and islands which stretches all the way from about 42°N to more than 70°N latitude and from 120°E to 170°W longitude. This is an area of much variability within itself, but it is set off fairly distinctively from eastern Siberia on the west by a marked difference in the degree of continentality. Although for a coastal area the climate still shows much continentality, particularly in winter, it contrasts sharply with the region just to the west which shows the highest continentality on earth.

It is a region of steep gradients in climate characteristics, from the ultimate of continentality in the Yana and Indigirka valleys of the interior, across the narrow mountain divides and down the steep seaward slopes to the nearly encircled seas bordering the northwestern Pacific. Wind velocities in winter along these mountainous coasts attest to the steepness of the temperature and air pressure gradients that exist across this transitional fringe between continent and ocean (Fig.5-2). As shown in Fig.8-39, this region in winter experiences some of the most extreme chill factors in the U.S.S.R. These are accounted for primarily by the high velocities of the winds, which stand in stark contrast to the utter calm of the interior basins which lie under the icy grip of the northeastern extension of the Asiatic high.

During summer the southern half of this territory, at least northward to the 60th parallel along the northern coast of the Sea of Okhotsk, is influenced profoundly by a monsoonal circulation which brings in maritime air from the Pacific and produces a summer climate in the region that is strikingly different from that farther inland.

The region here defined climatically as the Soviet Far East is considerably different from designations often referred to in other contexts as the Soviet Far East. The region has been defined in a more limited way than usual, more closely hugging the Pacific coast. The boundary drawn here between the Far East and eastern Siberia generally lies along the drainage divide between the Arctic and the Pacific, but this is not entirely true, particularly in the headwaters of the Amur River east of Lake Baykal. Here the boundary has been drawn in a very oblique southwest–northeast manner in order to approximate the progression of weather systems from southwest to northeast across the area and the inland penetration of Pacific marine air during summer.

Throughout this far-flung region the topography almost everywhere is mountainous. This coupled with the raggedness of the coastline, the many inlets, peninsulas, and

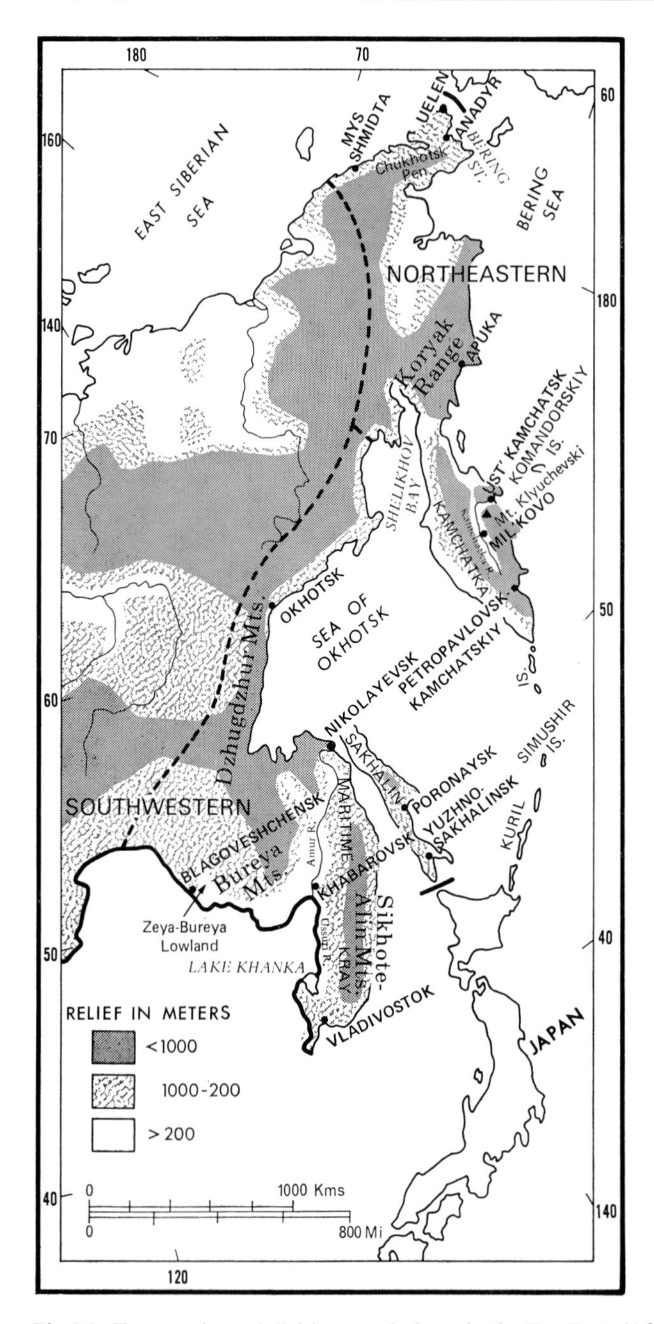

Fig.5-1. Topography, subdivisions, and places in the Far East. (After ZANINA, 1958, and ANONYMOUS, 1969.)

islands, makes for great variations in climatic elements from one place to another and renders almost impossible a generalization of the sequences of weather that move across the region. Maps of air pressure patterns and cyclonic and anticyclonic circulations derived for low elevations are largely useless because these levels in fact do not exist over the land. Therefore, we are often dealing with fictitious constructs which intersect mountain slopes and do not represent the air flow in intervening valleys. In addition, paucity of scientific investigation in the area leaves many things yet undiscovered. Nevertheless, an attempt will be made to generalize about the air flow across this region before the area is broken into subregions for detailed climatic description.

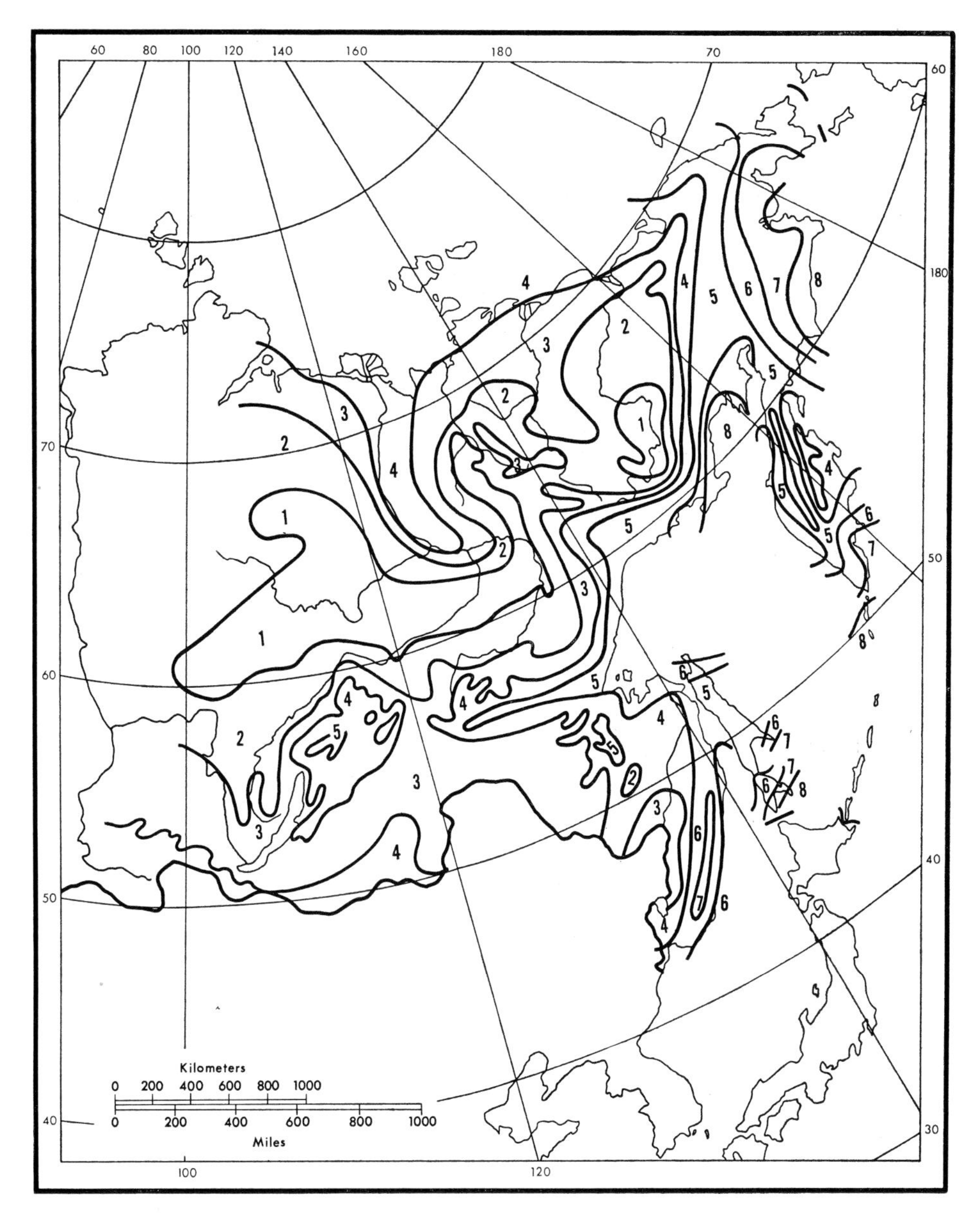

Fig.5-2. Wind speeds. For explanation of speed categories, see Fig.10-3. (After ANAPOL'SKAYA and TYUKTIK, 1969.)

Atmospheric circulation and precipitation processes

As has been shown in Figs.5-3 and 5-4, the air pressure pattern over the Soviet Far East undergoes drastic reversals in character between summer and winter. During winter the northeastern extension of the Asiatic high settles in over the cold land mass and forms an extremely strong pressure gradient down to the Bering and Okhotsk seas which are at this time of year dominated by the western end of a well developed Aleutian low. The peninsula of Kamchatka acts somewhat as a secondary center for high pressure between the sea on either side.

The surface high over the land mass is primarily thermally induced. The air over both the Arctic to the north and the Bering and Okhotsk seas to the south is considerably warmer than over the land in spite of the fact that these seas are frozen over during the winter. Considerable heat escapes from the unfrozen water underneath the ice to the air above to keep surface air temperatures on the average at least 20°–30°C higher than those over the land. Thus, surface air drainage tends to move out from the northeastern extension of the Soviet Union both toward the Arctic to the north and toward the seas to the south. This produces horizontal divergence in the lower layers of the troposphere over the center of the land mass which induces subsidence and further strengthens the northeastern core of the Asiatic high in winter.

The strong baric gradient between the surface high over the land and the surface low over the sea to the south and east produces consistently strong northerly winds across this coast during the winter which causes the winter climate in this area to be much more severe than expected along a sea coast at this latitude. Very little marine influence is felt at this time of year except in some of the more exposed peninsulas and island groups. And

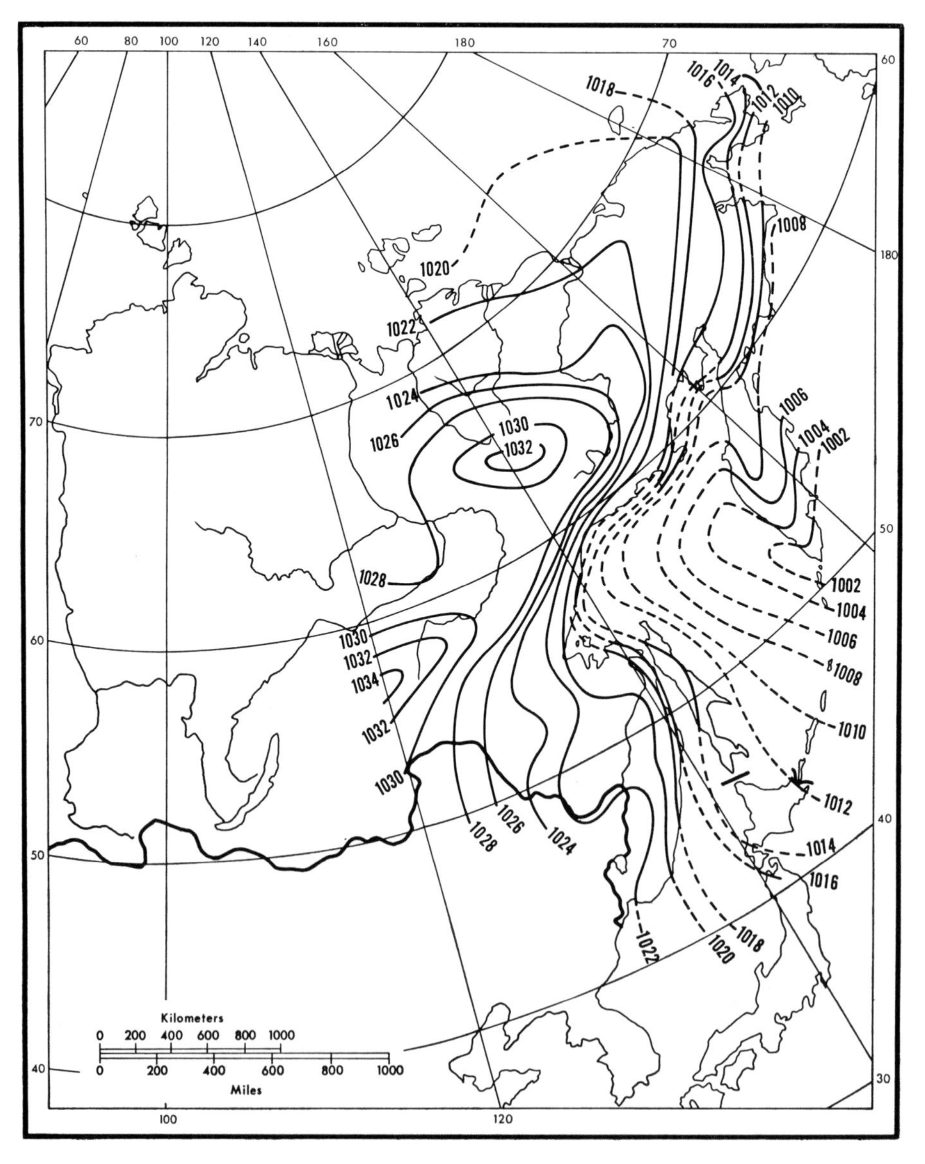

Fig.5-3. Mean atmospheric pressure (mbar) at sea level, January. (After ZANINA, 1958.)

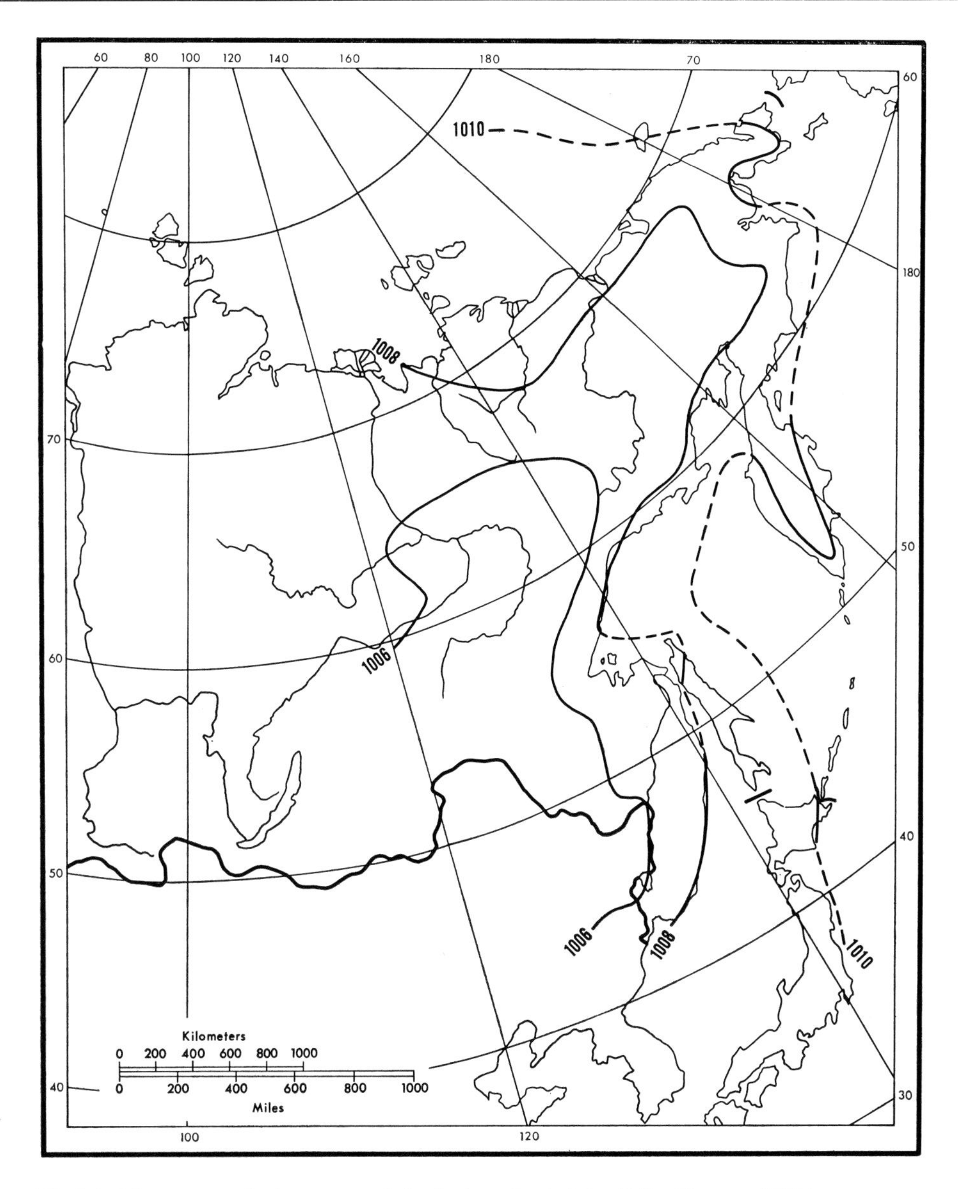

Fig.5-4. Mean atmospheric pressure (mbar) at sea level, July. (After ZANINA, 1958.)

even in the most marine exposed areas winters are still more severe than normal for the latitude.

Although the surface high over the continent is shallow, usually extending no more than 1 or 2 km into the troposphere, northeastern U.S.S.R. east of the Lena River experiences a consistent north-to-south flow across the entire region all the way from the central Arctic southward to the Bering and Okhotsk seas and beyond. This is assured because the area is generally overlain by the western limb of an upper air trough in winter that represents the western edge of the Aleutian low which, unlike the surface high over the continent, is dynamically originated and extends much higher into the atmosphere.

During summer the land heats up relative to the sea and a shallow low forms over the land. Atmospheric temperature and pressure gradients are much weaker than they were during winter so that the summer circulation is less well defined and less consistent over the entire region. For a while during spring a localized thermally-induced shallow high pressure system forms over the Sea of Okhotsk because of the contrast between the very cold surface of the sea, the northwestern $^{3}/_{4}$ of which is frozen over during the winter, and the land mass to the north and west which is experiencing very rapid warming in spring and early summer. Ice flows linger in the Sea of Okhotsk well into August, particularly along the west coast where they are carried southward by the counter-clockwise currents in the Sea of Okhotsk as far as the northeast coast of Sakhalin.

During summer the subtropical high in the western Pacific expands and moves northward to occupy much of the oceanic area of the northwest Pacific as the Aleutian low shrinks and gradually fills. Pressures in the subtropical high are augmented by Arctic air masses that move southward across the eastern extension of the U.S.S.R. to join and become part of the western end of the Pacific high. Thus, the Sea of Okhotsk is engulfed by atmospheric high pressure during summer as well as in spring, but the difference between the general nature of the high in summer and its localized nature in spring causes significantly different climatic effects in the southern part of the Soviet Far East. During spring, the marine air circulates around the localized cold high over the Sea of Okhotsk and comes into Maritime Kray and adjacent regions as relatively cold dry air which does not provide much precipitation at this time of year. But in middle and late summer the marine air circulating into Maritime Kray has traversed a long distance over water along the southern periphery of the Pacific high where it has picked up a great deal of warmth and moisture which provides the source for much precipitation in the southern part of the Soviet Far East during the latter half of summer and early fall.

Thus, the monsoonal influx of marine air into the southern part of the Soviet Far East is much different in spring and early summer than in late summer and early fall. And this has great climatic consequences. Spring and early summer in farming regions of Amur District and the Ussuri–Khanka Lowland can be absolutely droughty, while the harvest season in late summer and fall can be disastrously wet and cloudy. Thus, the precipitation and cloud regimes during the summer in the better farming districts of the Soviet Far East are just about exactly wrong; moisture deficits when the plants are experiencing their greatest growth and moisture excesses when it is time to harvest.

Little of the influx of moisture during the summer monsoon in Maritime Kray and Amur Oblast comes into these regions directly from the ocean to the southeast. Southeasterly flows are generally very shallow, no more than 0.5 to 2 km in depth, and usually do not penetrate beyond the southern and southeastern slopes of the Sikhote-Alin Mountains in Maritime Kray, except in the Ussuri–Khanka Lowland which lies west of the Sikhote-Alin and is open to the sea on the south. But generally the influx of monsoonal moisture takes place from the southwesterly quadrant in the warm sectors of cyclonic storms which are active along the Mongolian segment of the Polar front at this time of year. The subtropical maritime air from the Pacific has circulated around the western end of the Pacific high into central and northern China where it curves into a southwesterly flow that moves northeastward back to sea across the southern part of the Soviet Far East. Even as low as the 850 mbar level, the southeasterly flow in Maritime Kray virtually disappears (Fig.5-5). The total moisture flux across the Soviet Far East is always from the west at all seasons of the year (Fig.5-6).

Thus, the explanation of late summer precipitation in the southern part of the Soviet Far East is not as simple as it first appears. The southwesterly flow of moisture into the region from Pacific subtropical air across north China may be joined by maritime tropical air from as far away as the Indian Ocean, so that some of the summer precipitation in the Soviet Far East may be derived from the Indian Ocean as well as from the Pacific. However, as can be seen in Table 5-I, the bulk of the moisture precipitated in the southern part of the Soviet Far East during summer has its origins in the western Pacific in the area of the Philippines and Indochina. A significant amount of the precipitation at this time of year is derived from cyclones coming in from the west that carry moisture all the way from Central Asia, Kazakhstan, Siberia, and the Barents and Kara seas.

During summer in the Amur Valley a convergence line often is formed in the lower portions of the troposphere between the southerly and southwesterly flows of subtropical air which have circulated into the continent from the Pacific across northern China and more westerly flows of air just to the north which have come across Siberia. Thus, a rather active section of the Polar front normally lies in this area which sometimes stretches all the way from northern Mongolia to the Pacific coast and beyond. This is often termed the Mongolian branch of the Polar front. Cyclones form along this front and move eastward and northeastward typically across the lower Amur Basin, Sakhalin Island, and into the Okhotsk Sea. Occasionally they may be traced even farther east across Kamchatka. Typically a series of waves form on the front and follow one another, becoming partially occluded as they proceed eastward (Fig.5-7).

The warm fronts of these cyclonic storms are generally underlain by maritime temperate air on their northeastern sides and overrun by maritime subtropical air on their south-

TABLE 5-I

FREQUENCIES OF ARRIVAL OF VARIOUS AIR MASSES TO THE SOVIET FAR EAST DURING DAYS WITH PRECIPITATION (%)
(After DROZDOV and GRIGOR'EVA, 1965)

Air masses arrival from	Summer (1951, 1954, 1956, 1958)			Autumn (1951, 1954, 1956–1958)			Winter (1956–1959)			Spring (1951, 1954, 1956–1959)		
	850 mbar	700 mbar	500 mbar	850 mbar	700 mbar	500 mbar	850 mbar	700 mbar	500 mbar	850 mbar	700 mbar	500 mbar
The North Pacific	4	1	1	4	4	0	7	3	2	0	0	0
The West Pacific (subtropical)	0	0	0	5	3	0	1	1	0	6	2	0
The region of the Philippines	25	29	18	–	–	–	0	0	0	–	–	–
Southern China	0	0	0	15	9	12	10	6	6	13	8	6
Indochina	24	14	11	6	8	7	0	0	0	13	0	0
India and Burma	0	0	0	–	–	–	2	2	0	–	–	–
Southern Central Asia	14	18	25	12	12	12	18	16	18	8	8	15
Kazakhstan and Siberia	19	19	28	36	49	47	41	47	48	34	48	54
The Barents and Kara seas	11	11	12	20	18	15	18	17	17	14	27	23
The eastern Arctic	3	9	6	2	4	5	2	10	9	20	9	1

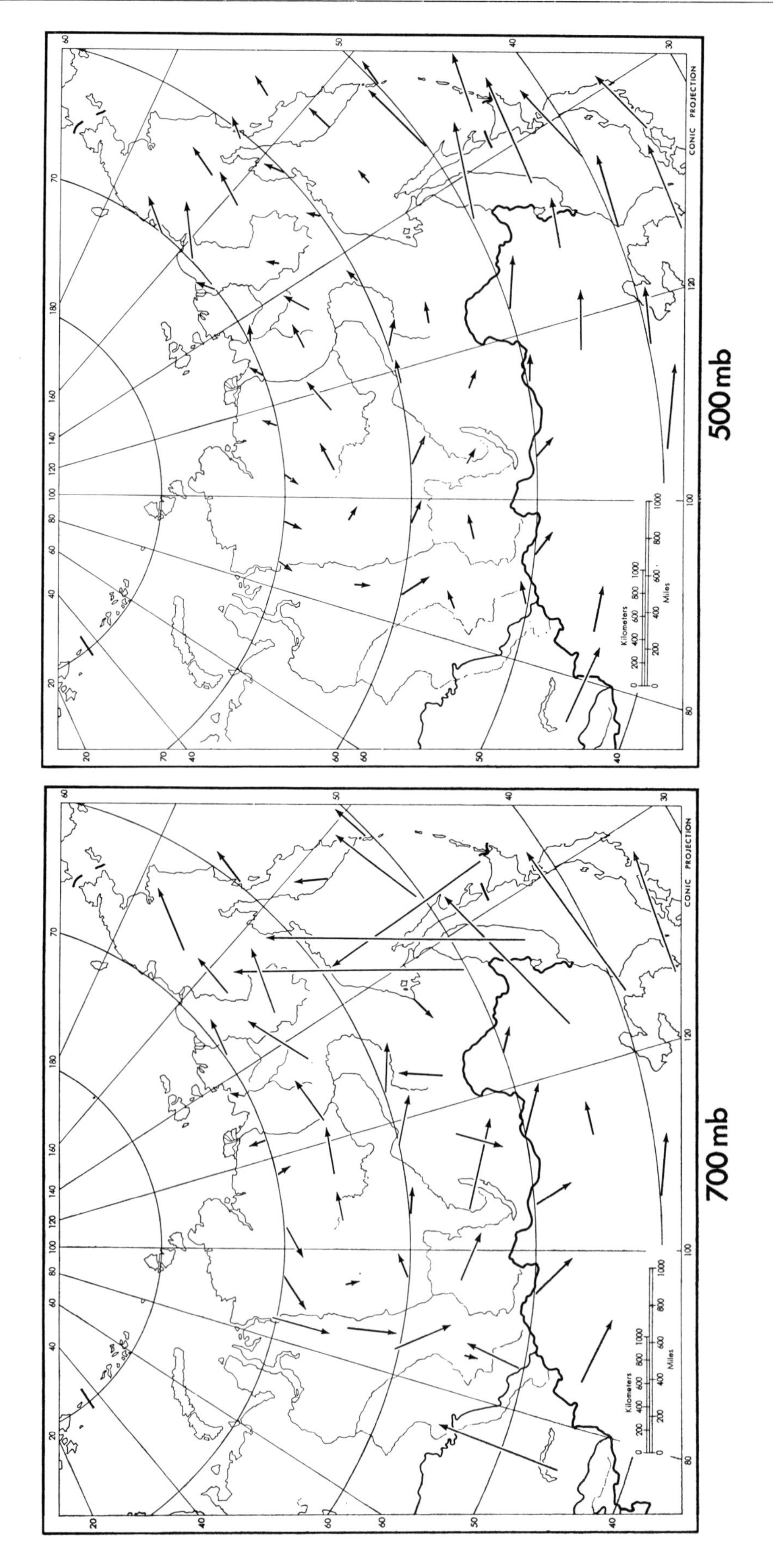
500mb
700mb
CONIC PROJECTION
Kilometers
Miles

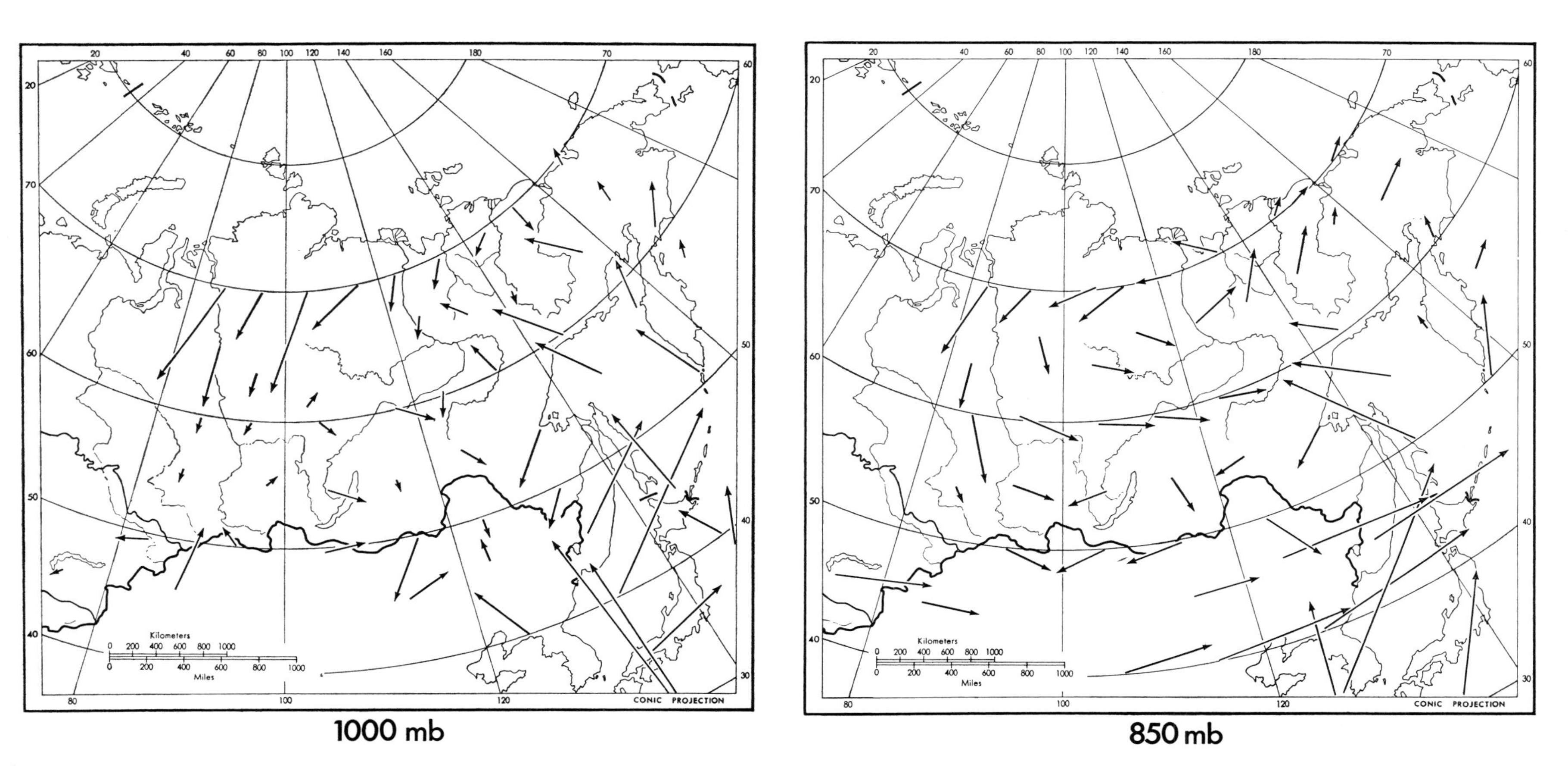

Fig.5-5. Average moisture flux in July, 1961. Lengths of arrows are proportional to magnitudes of flux. (After SOROCHAN, 1966.)

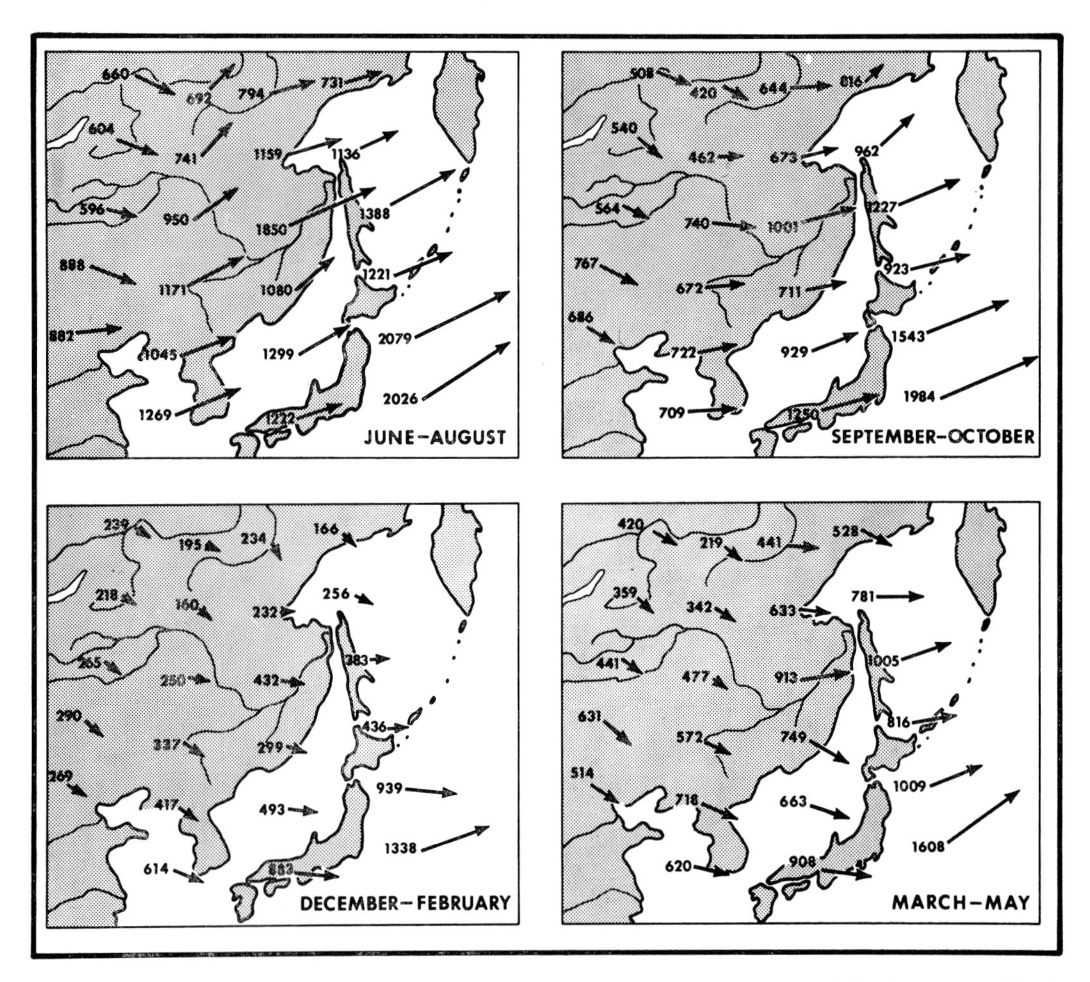

Fig.5-6. Total moisture flux across the Soviet Far East by seasons during 1951, 1954, and 1956–1958, in g/cm per second. (After DROZDOV and GRIGOR'EVA, 1965.)

western sides which provides the moisture for the precipitation. However, orographic uplift of the maritime temperate air underneath the fronts itself may cause precipitation on the windward eastern slopes of the Sikhote-Alin Mountains, the mountains of Sakhalin, and the Dzhugdzhur and Suntar-Khayata Mountains along the northwest and northern coasts of the Okhotsk Sea. This orographic precipitation in the maritime temperate air is generally of a prolonged drizzly nature and does not equal in amounts the precipitation derived from the humid subtropical air overriding the fronts, which frequently occurs in heavy thundershowers.

A penetration of maritime Pacific air onto the coast from the southeast is thus facilitated by the counter-clockwise circulations of cyclonic storms along the Polar front which pull in air from the southeast ahead of warm fronts. Frequently the most prolonged rain is associated with southeasterly flow at the surface, but most of the precipitation is being derived from the air overriding the front, which is coming from the southwest. And of course, the storm itself is proceeding from southwest to northeast. Frequently cyclonic storms coming in from the southwest are fully occluded by the time they reach the Soviet Far East, so that the southwesterly flow is completely off the ground in the Soviet Union and is not detectable as a surface flow. All this leads to the rather erroneous impression

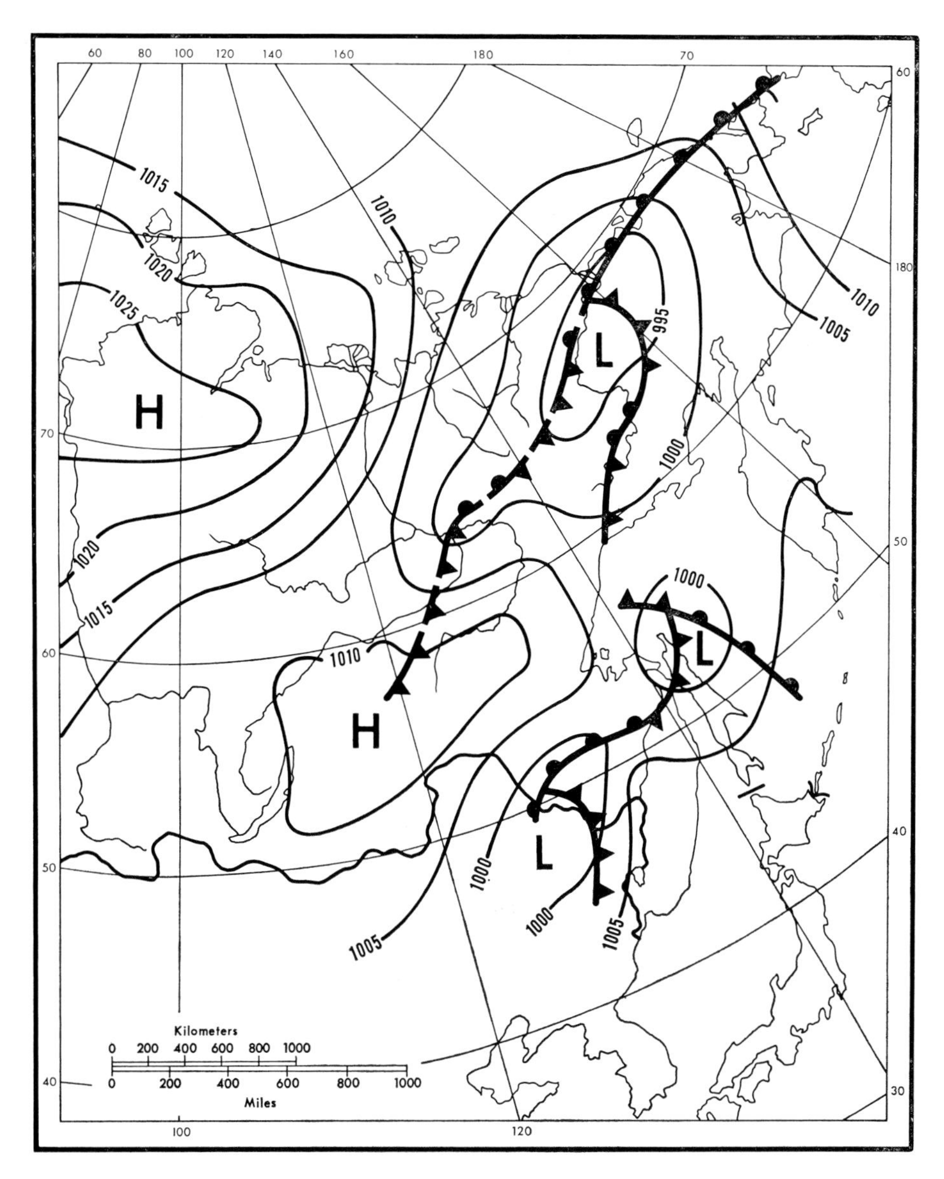

Fig.5-7. Cyclonic activity along the Polar and Arctic fronts in the Soviet Far East during summer. Synoptic map for the evening of August 9, 1953. (After ALISOV, 1956.)

that most of the moisture-bearing winds during summer in the Soviet Far East are coming from the southeast directly off the Pacific.

The second half of summer and early fall is the only time when the Pacific monsoon is of any great significance to the Soviet Far East. At all other times of the year the precipitation arriving in the Far East is due mainly to the entry of cyclones of western origin. During the second half of summer as much as 50% of the precipitation is derived from air masses of subtropical and equatorial origin in the Pacific, but in other seasons this is less than 10%, and then only in coastal areas.

In January the moisture flux is much reduced over the Soviet Far East and it is generally directed from land to sea (Fig.5-8). Near the surface over the mainland in January there is hardly any net flux at all. Magnitudes increase slowly upward, but they are small everywhere. The 850-mbar level in January shows a divergence line in the far northeastern

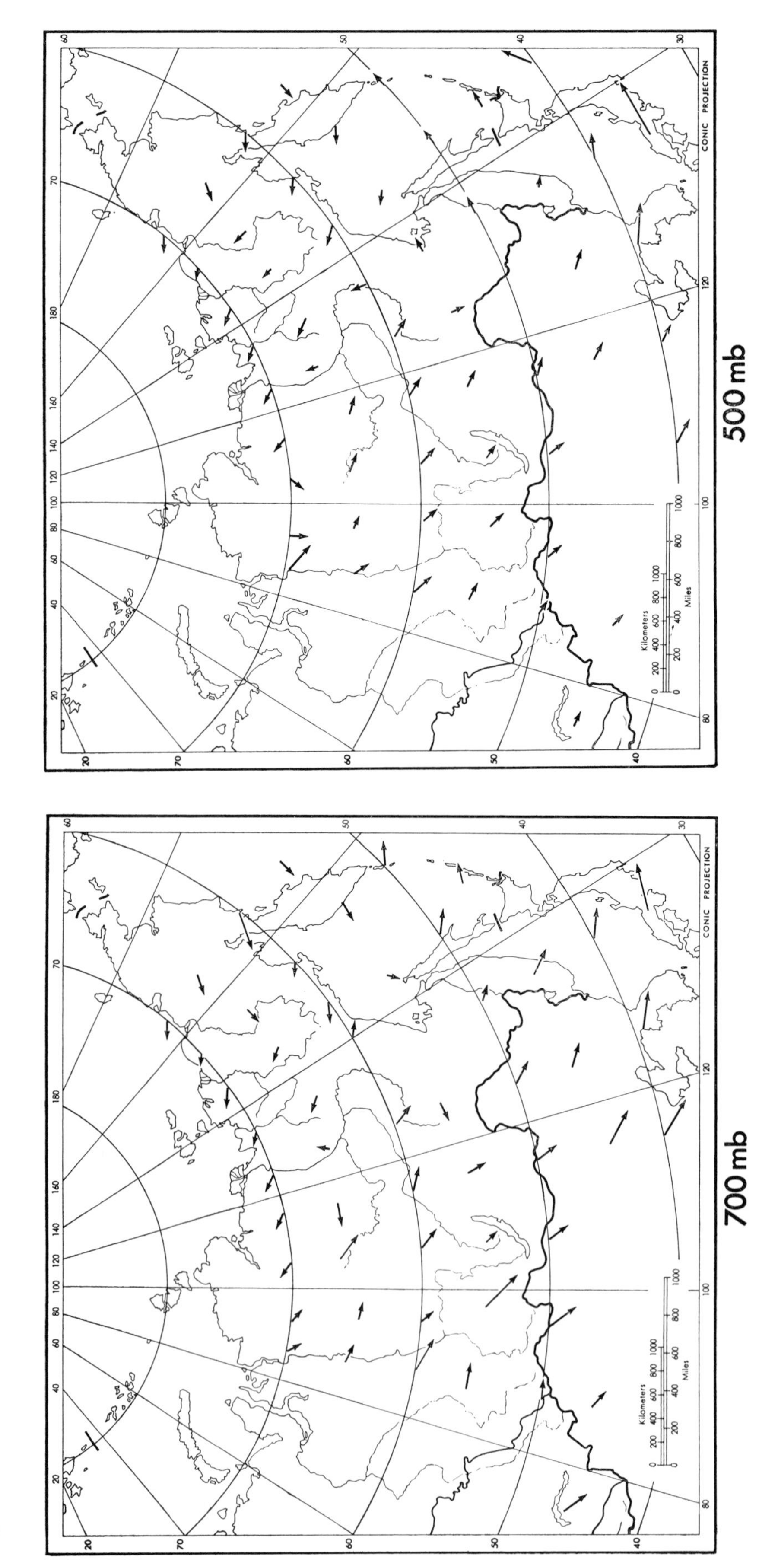

500 mb
700 mb
CONIC PROJECTION
Kilometers
Miles

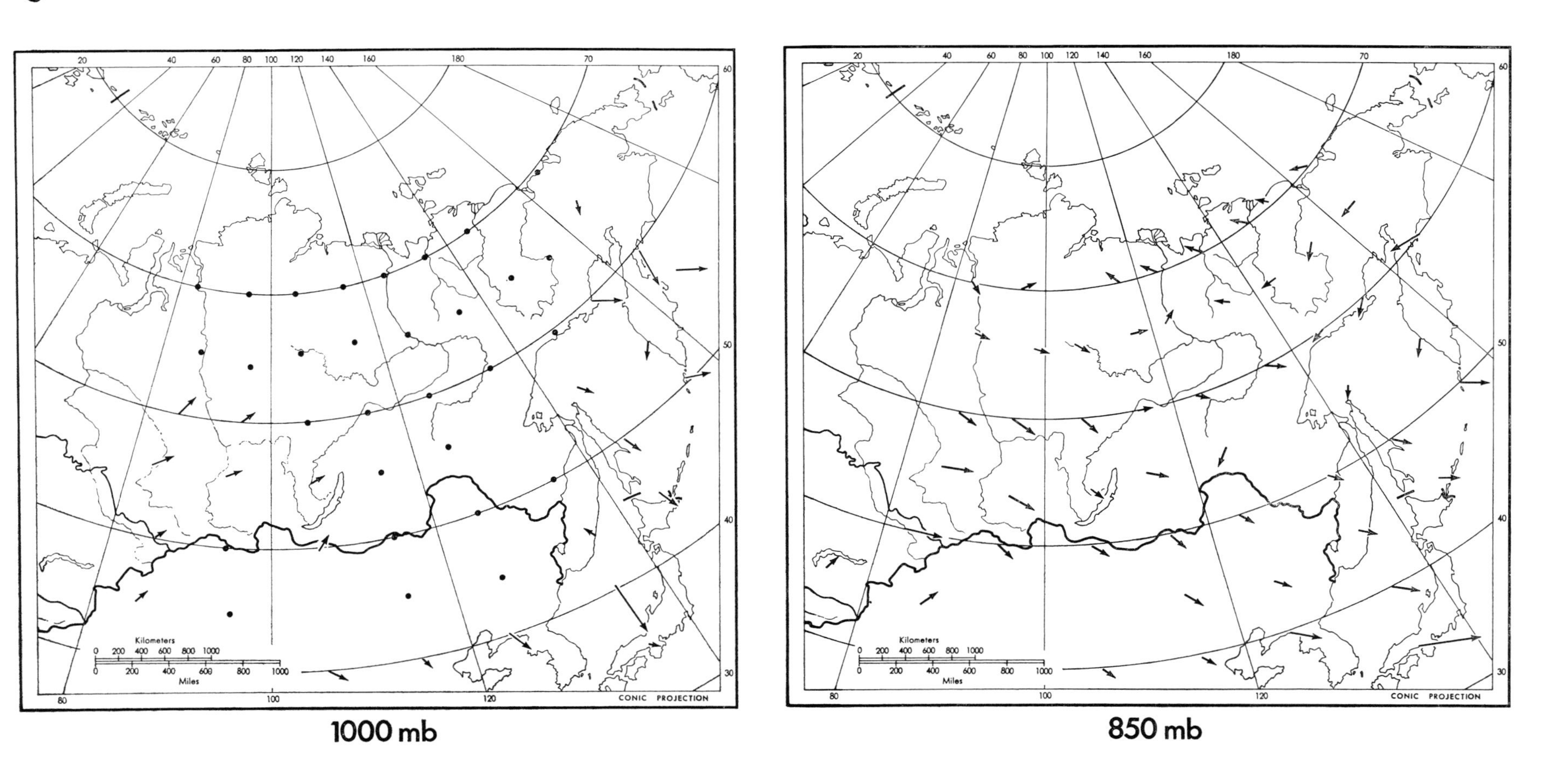

Fig.5-8. Average moisture flux in January, 1961. Lengths of arrows are proportional to magnitudes of flux. (After SOROCHAN, 1966.)

extension of the continent at about the Arctic Circle with moisture moving from land to sea along the Arctic coast and also from land to sea along the northern shore of the Sea of Okhotsk. The air over the land at this time of the year is considerably colder than that over adjacent water bodies even if they are frozen over and therefore there is a thermally induced divergence outward to sea in all directions. This undoubtedly strengthens the northeastern extension of the Asiatic high during this time of year by inducing subsidence in the lower troposphere.

During summer a segment of the Arctic front generally hangs in the central portions of the northeastern extension of the U.S.S.R. where a convergence zone is created between maritime flows of air onto the continent from both the south and the north (Fig.5-7). Cyclonic storms do form along this front, but they are not nearly as frequent as cyclonic storms along the Polar front to the south nor do they produce the amounts of precipitation that the southerly storms do. Nevertheless, occasionally humid subtropical air may be drawn far north into the warm sectors of these northern cyclones, even spilling over the mountains north of the Sea of Okhotsk and descending into the river valleys of the Kolyma and adjacent streams facing the Arctic coast. During such conditions the surface air becomes excessively warm and humid for this region, and heavy thundershowers may occur.

As one proceeds eastward from Maritime Kray across Sakhalin Island and the Sea of Okhotsk to the Kamchatka Peninsula and beyond, the influences of the cyclonic storms along the Polar front in summer become weaker and weaker, and the winter time formation of the western portion of the Aleutian low becomes a predominant factor in the precipitation processes. The period of maximum precipitation lags steadily from middle and late summer into fall and early winter as one proceeds from Maritime Kray to Kamchatka.

Petropavlovsk on the southeast coast of Kamchatka receives 1,335 mm of precipitation per year with the maximum occurring in November. A secondary maximum occurs in March. It appears that the transitional seasons are rainier in this area than either summer or winter. Precipitation is at a minimum in early summer when the western end of the Pacific high is engulfing the area, and it also diminishes in midwinter when a localized high tends to form over the interior of Kamchatka causing surface air to blow outward toward the sea in all directions.

Precipitation picks up somewhat in late summer on the southeast coast of Kamchatka, Sakhalin Island and portions of Maritime Kray due to the backwash of typhoons which proceed northeastward from the Japanese Islands toward the Bering Sea. Hardly ever do these storms come close enough to any portion of the Soviet Far East to cause wind damage, but their widespread clouds and precipitation may significantly alter precipitation regimes on exposed southeastern fringes of coastal protrusions. Typhoons coming up from the South China Sea frequently cause regeneration of cyclonic storms moving northeastward along the Polar front. In such cases the forward motion of the cyclonic storm is impeded and prolonged rainfall results. Through July–September 70–90% of total monthly rainfall may occur in 5–6 day periods with such occurrences. Very infrequently a typhoon may even wander into the Sea of Okhotsk to die.

During winter the east Asian Polar front is usually well off the coast, even east of the Japanese Islands, and the Arctic front is stabilized along the northern shore of the Sea of Okhotsk. The continent is normally occupied by high pressure (Fig.5-9). However,

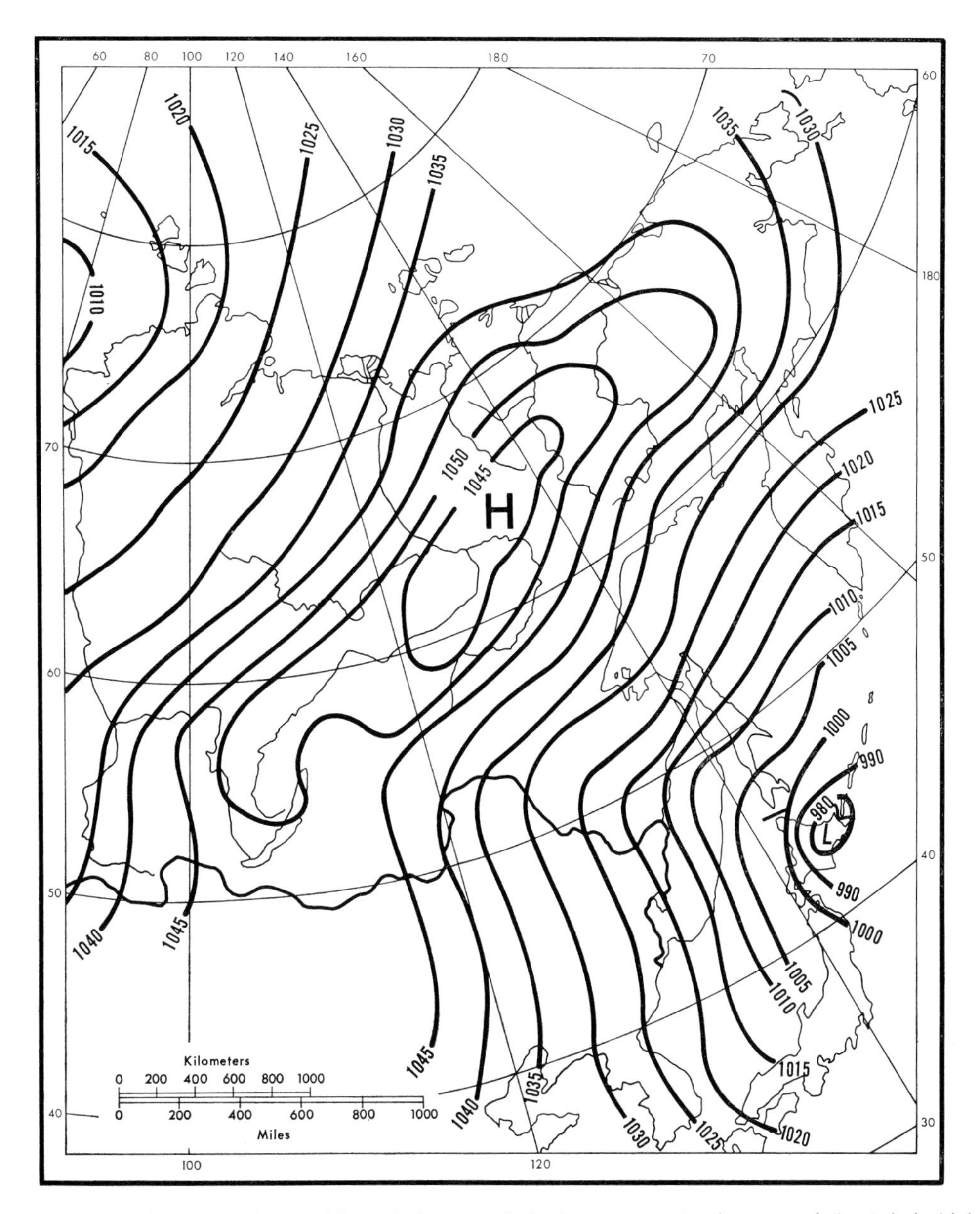

Fig.5-9. Typical synoptic conditions during a period of maximum development of the Asiatic high in the Soviet Far East. Evening map for January 13, 1953. (After ALISOV, 1956.)

even in midwinter the Asiatic high may be penetrated by portions of either of these fronts along which cyclonic storms may form and proceed eastward. A favorite path for cyclones is east-northeastward from Lake Baykal across the northern portion of the Sea of Okhotsk and north central Kamchatka into the Aleutian low over the Bering Sea (Fig.5-10). Therefore, on the average there is more cloudiness during winter through this intermediate territory than there is further south in the Amur Valley (Fig.8-11).

Old occluded cyclones sometimes move northward from the Polar front lying southeast of Japan and frequently regenerate along the Arctic front on the north coast of the Sea of Okhotsk. They commonly occur in series of four or five, during which time they bring much warm marine air in their forward edges which produces wet snowfalls and thawing along the coast and over peninsulas and islands. In some areas thawing can be quite

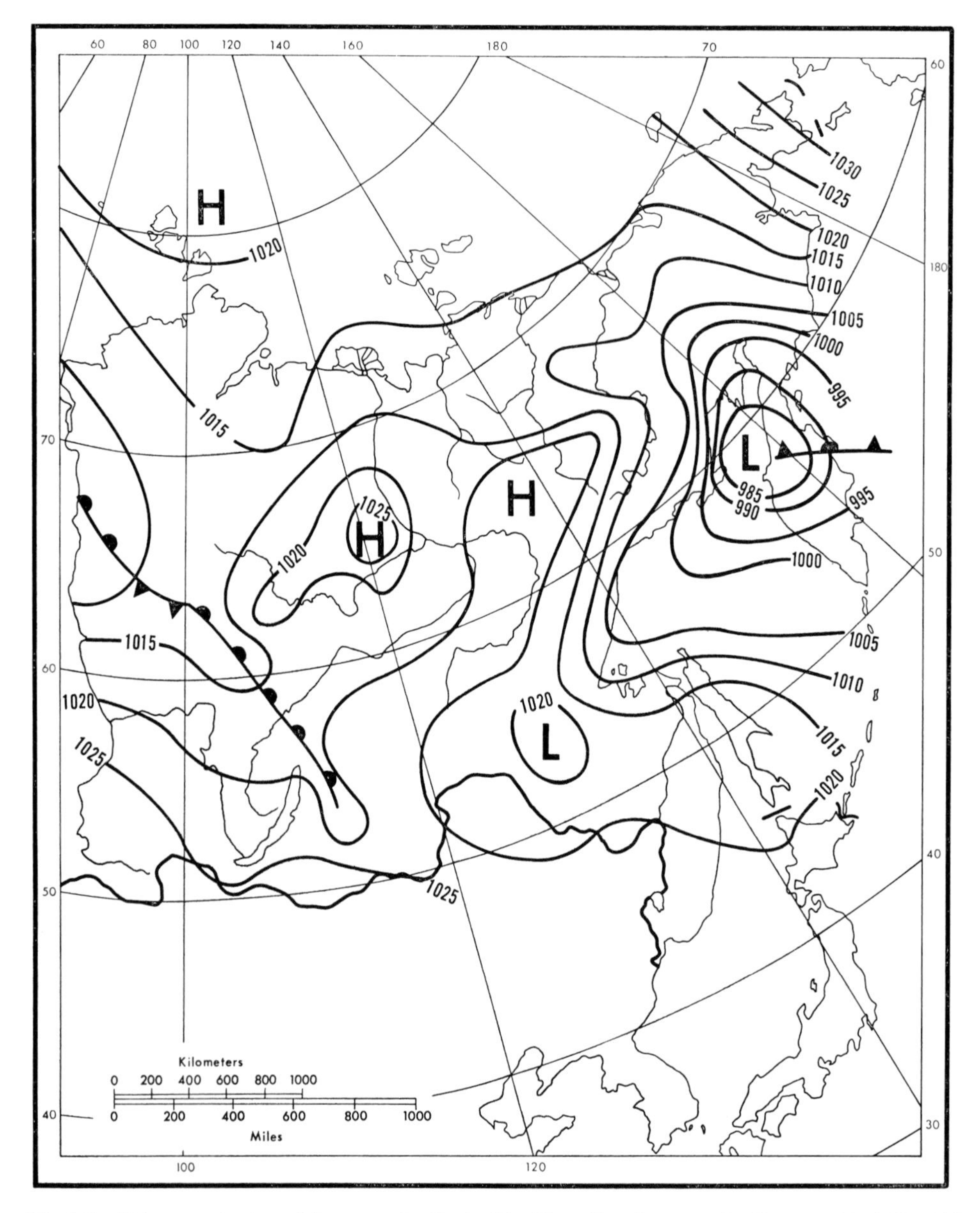

Fig.5-10. Winter cyclonic activity over the Soviet Far East. Evening map for January 12, 1952. (After ALISOV, 1956.)

intensive during winter. Cyclones moving into the Sea of Ohkotsk may bring such conditions to Sakhalin and to the lower Amur Valley. Those moving into the Bering Sea may bring such conditions to the east coast of Kamchatka, the Komandorskiy Islands, and the shore of the Chukhotsk Peninsula. Occasionally this warm marine air may even penetrate across the Koryak and Gydan ranges into the basin of the Kolyma River to produce strong snowfall in this region. Such intrusions of maritime temperate air may last for 7 or 8 days and bring relatively warm, moist, and low cloud conditions with a great deal of snowfall to much of the Far East. Practically always these occurrences are associated with strong winds and blizzards.

The Asiatic high begins to break up in March, and by the end of March there is already development of cyclonic activity in the basin of the Amur. Most rapid transition takes place during April and May when there is a great diminishing of air pressure. Spring throughout much of the area is characterized by deep cyclonic storms which bring strong winds and snowstorms.

In such mountainous topography along a ragged sea coast it is only logical that there should be many local wind regimes. Sea breezes are well developed during summer along much of the coast, and mountain–valley regimes are particularly well developed in the southern part of the territory, especially toward the end of spring after the snow cover has gone in the lower elevations and there is an intense heating differential between the rapidly warming lowlands and the still snow-covered upper slopes. During winter in valleys that are oriented in the direction of the prevailing winds there are rather continuous bora along the coasts in many places, particularly along the steep mountainous coast along the northwestern margins of Shelikhov Bay, the northeastward extension of the Sea of Okhotsk.

Foehn winds are important in many of the mountain areas. Some of the most remarkable are those which result from air which occasionally in winter blows from the Bering Sea northwestward across the mountainous eastern extension of the land mass to descend into the Kolyma Valley along the Arctic coast where relatively warm winds may afford welcome if brief respites from the icy grip of winter in this region.

Climatic regions

In this far-flung area it is only natural to expect the climate to vary greatly from one part to another. However, it is difficult to divide the area into discrete regions because within any one locality weather types are broken up by mountain ranges and proximity to water. Everywhere there is a great contrast between the immediate coastal area and the interior.

The Sea of Okhotsk has a profound influence on the climate of all the areas around it, especially in summer when it acts as a cold source because of its floating ice packs. This is particularly true in the western part of the sea where drifting ice that moves westward with currents in the northern part of the sea lodge in embayments around Shantar Island where ice may linger close in shore for 10 or 11 months during the year (Fig.5-11). This maritime cooling is also very effective on the northern end of Sakhalin Island and southward along the east coast where cold water and ice flows are carried southward in the Sakhalin current. To a lesser extent the same is true in the thin southward flowing current that finds its way through the Tatar Strait and hugs the coast of Maritime Kray in the Sea of Japan. Even along Kamchatka in summer the west coast is usually 2°–3°C colder than the east coast in spite of the fact that the east coast of Kamchatka is washed consistently by the cold Kamchatka current in the Pacific while the west coast of Kamchatka has indecisive currents flowing along it.

One important consequence of the summer coolness of the Okhotsk Sea is the extreme amount of fogginess, particularly in early summer, along most of the coastal areas. To a certain extent this is true also along the coast of Maritime Kray in the Sea of Japan and the coast of the far northeast along the Bering Sea, but generally the Okhotsk Sea sides of land masses are foggier because of the extreme coolness of the Okhotsk Sea. The fog varies greatly over short distances depending upon exposure to the drifts of the local

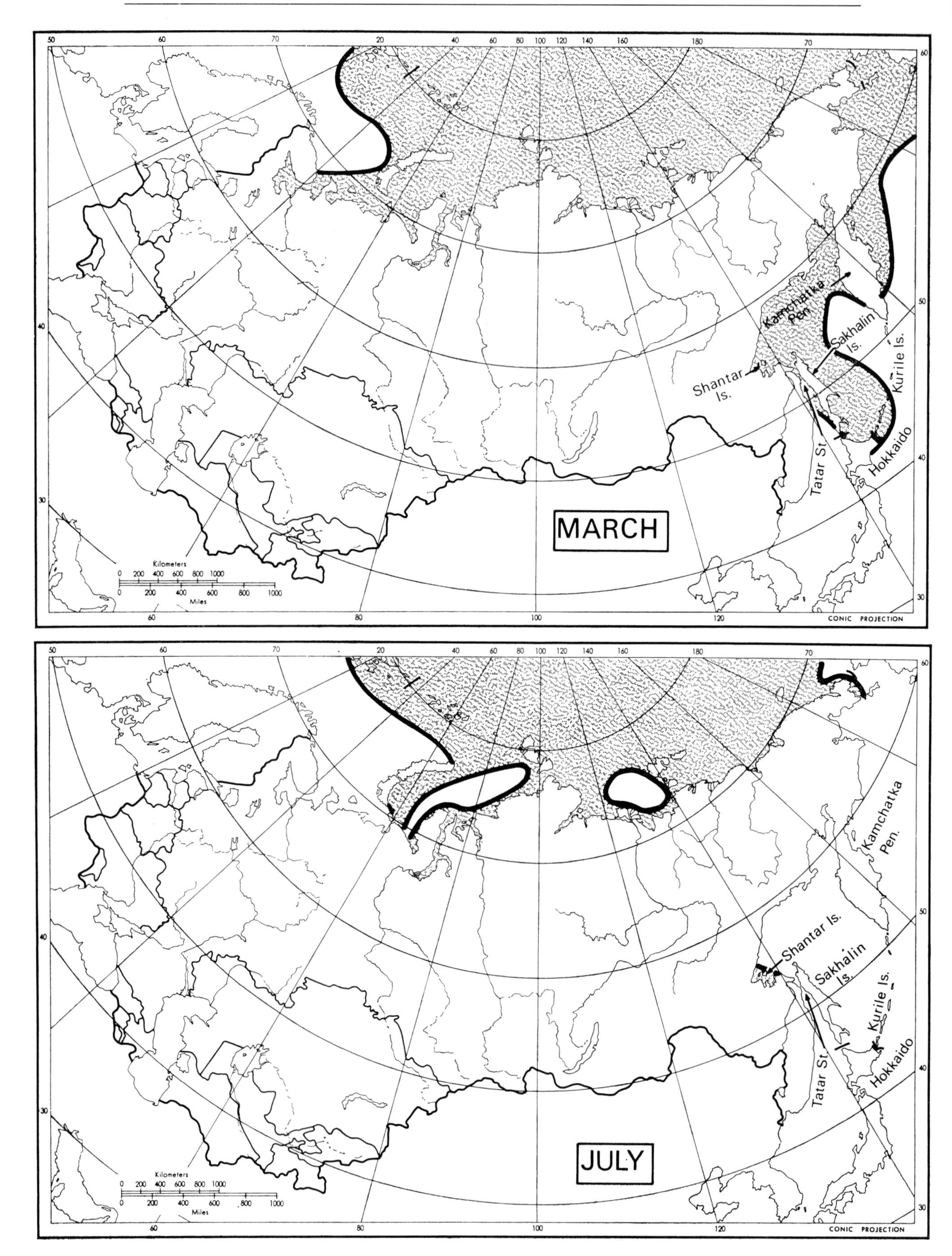

Fig.5-11. Sea ice conditions, March and July. (After BUDYKO, 1963.)

winds and their temperature contrasts with the underlying water, since all this summer fog is advection fog.

Since the southwestern part of the Soviet Far East has seasonal regimes of weather that are tied in closely with monsoon circulations to the south, and the northeastern part of the Soviet Far East has weather that is more directly related to seasonal developments and decay of the Aleutian low, the area will be divided into two parts for the convenience of discussion rather arbitrarily along a north–south line running from the north coast of Shelikhov Bay just to the west of Kamchatka and between Kamchatka and the Kuril Island chain. But it must be realized that within each of these gross areas there are widely varying weather types.

The southwestern region

Throughout this region the monsoonal air flow regime has profound effect. Precipitation is concentrated in the warm part of the year as a result of the influx of humid subtropical air from the southwest in conjunction with the procession of cyclonic storms northeastward along the Mongolian sector of the Polar front. Convective activity augments this frontal precipitation in interior lowlands, and orographic effects are added on the windward slopes of mountain ranges.

The most continental part of this region is the Zeya–Bureya Lowland to the north of the middle Amur Valley which is almost completely encircled by low mountain ranges that reach their highest elevation of slightly more than 2,000 m in the Bureya Range on the east. Blagoveshchensk on its southern margins shows a typically continental distribution of precipitation and temperature (Fig.5-12). There is a pronounced summer maximum of precipitation centered on July and very dry winters. The snow cover is meager and often blows away from much of the surface so that the ground freezes to great depths. Since there is little melt water in spring, the rivers typically do not flood in spring but later in summer as the monsoon rains and flash thundershowers occasionally hit with great intensity. The average annual precipitation is 534 mm, which generally supports a natural vegetation of mixed forests, but sections of the lowland can be quite droughty particularly in spring and early summer, and steppe grasses are frequently interspersed with groves of trees. On the mountain slopes surrounding the lowlands the forests become denser and change from a mixture of broadleaf and coniferous trees to more pure stands of coniferous with elevation.

In spite of the monsoon rains, a great deal of the time in summer is taken up by relatively clear dry weather and as much as 10% of the time in early summer can be quite droughty. On the other hand about 5% of the time during summer is occupied by stiflingly warm muggy weather with the intrusion of tropical air in the warm sectors of cyclonic storms. Relative humidities throughout the summer average approximately 75% and can reach 90 to 100% with temperatures that rise well above 30°C. The maximum temperature experienced in Blagoveshchensk has been 41°C. Typically 10 to 12 days per month during midsummer experience thunderstorms.

Winters in the middle Amur region are extremely cold, clear, and stable. This region has the least amount of winter cloudiness in the entire country. Blagoveshchensk averages only 3.9 tenths of sky cover during January. The temperature during January averages

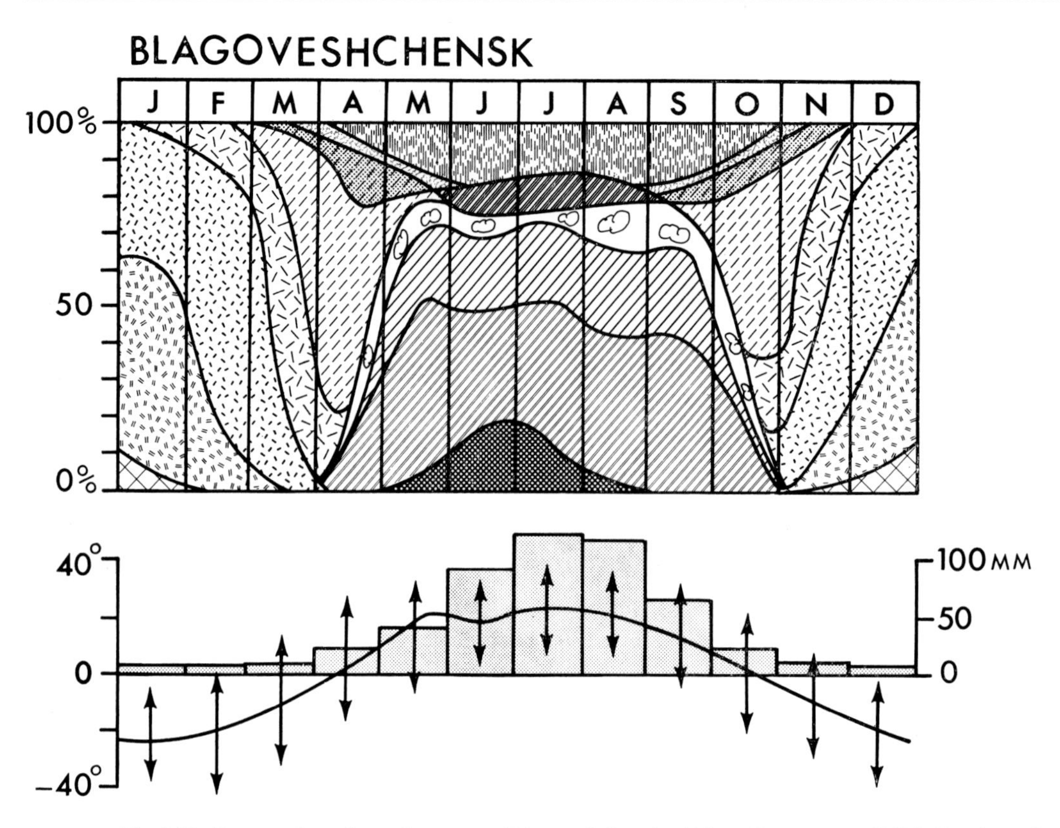

Fig.5-12. Frequencies of weather types, Blagoveshchensk. (After GERASIMOV, 1964.) For legend see Fig.3-25.

—24°C and has dropped as low as —45°C. A frost-free period during the year of only 144 days gives this area a rather short growing season.

East of the Bureya Mountains lies an elongated lowland that stretches north-northeast–south-southwest. Northeast of Khabarovsk it is known as the lower Amur Valley and southwest of Khabarovsk as the Ussuri–Khanka Lowland. Except for the far northern end near the mouth of the Amur, this lowland is everywhere bordered on the east by the Sikhote-Alin Mountains which rise at their greatest elevations to a little over 2,000 m. This is a fault block mountain range which presents a steep escarpment toward the Sea of Japan on the east and effectively blocks the lowland to the west from maximum influences from the sea. Winds in the lowland are primarily bidirectional, either from the north-northeast or the south-southwest. Marine air thus enters the lowland from either end and its effects are least felt in the central portions. This lowland is not quite as continental in character as is the Zeya–Bureya Lowland to the west. Nevertheless, it experiences more severe winters than anything in similar latitudes in the European part of the country. Even in the far south at Vladivostok January temperatures average no more than —15°C, and temperatures as low as —31°C have occurred. The port is closed by ice 110 days per year. Farther north in the center of the plain Khabarovsk averages —22.7°C in January and has experienced temperatures as low as —43°C. Still farther north at the mouth of the Amur Nikolayevsk averages —25.8°C in January and has experienced —47°C. This, even on the sea coast!

During July Khabarovsk averages 21°C and has experienced 40°C. It is not quite so warm in Vladivostok on the sea and maximum temperatures are delayed over what they are in the center of the plain. August is the warmest month with 20°C, and the highest temperature ever recorded at Vladivostok is 36°C.

The contrast in the degree of continentality between Khabarovsk and Vladivostok is most clearly borne out in the asymmetry of summer conditions of clouds and humidity along the sea coast as opposed to the more even distribution in the interior (Figs.5-13, 5-14). Along the sea coast early summer experiences much low cloudiness and fog; later the weather clears up perceptibly although more precipitation falls. As the precipitation diminishes in September autumn becomes the best time of year. Skies become clear and dry, and the leaves in the forests turn brilliant colors reminiscent of the Indian summer in eastern North America.

Vladivostok experiences 81 days per year of fog with greatest frequencies by far in May, June, and July when 13, 16, and 20 days respectively are observed. These early summer fogs are generally advected into the bay surrounding Vladivostok around the corner from the east in very shallow layers. The upper mountain terraces in the city often lie above the fog in the sunshine. The shallow fogs are most frequent during early morning hours just before sunrise and generally dissipate by 09h00 or 10h00. Farther inland, at Khabarovsk, fog occurs only 18 days per year and is distributed fairly evenly throughout the year. Winter fogs in inland basins are generally due to night time radiation and usually dissipate shortly after sunrise. Along the coast, fog generally occurs most frequently in

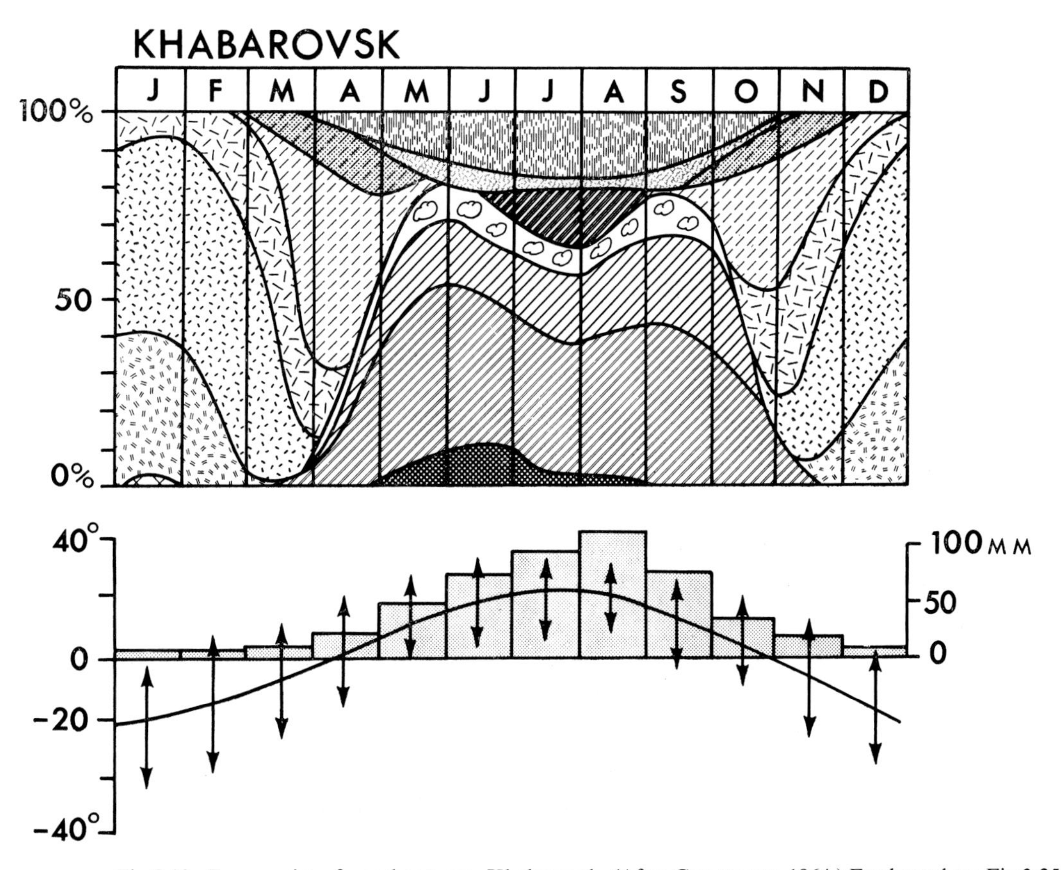

Fig.5-13. Frequencies of weather types, Khabarovsk. (After GERASIMOV, 1964.) For legend see Fig.3-25.

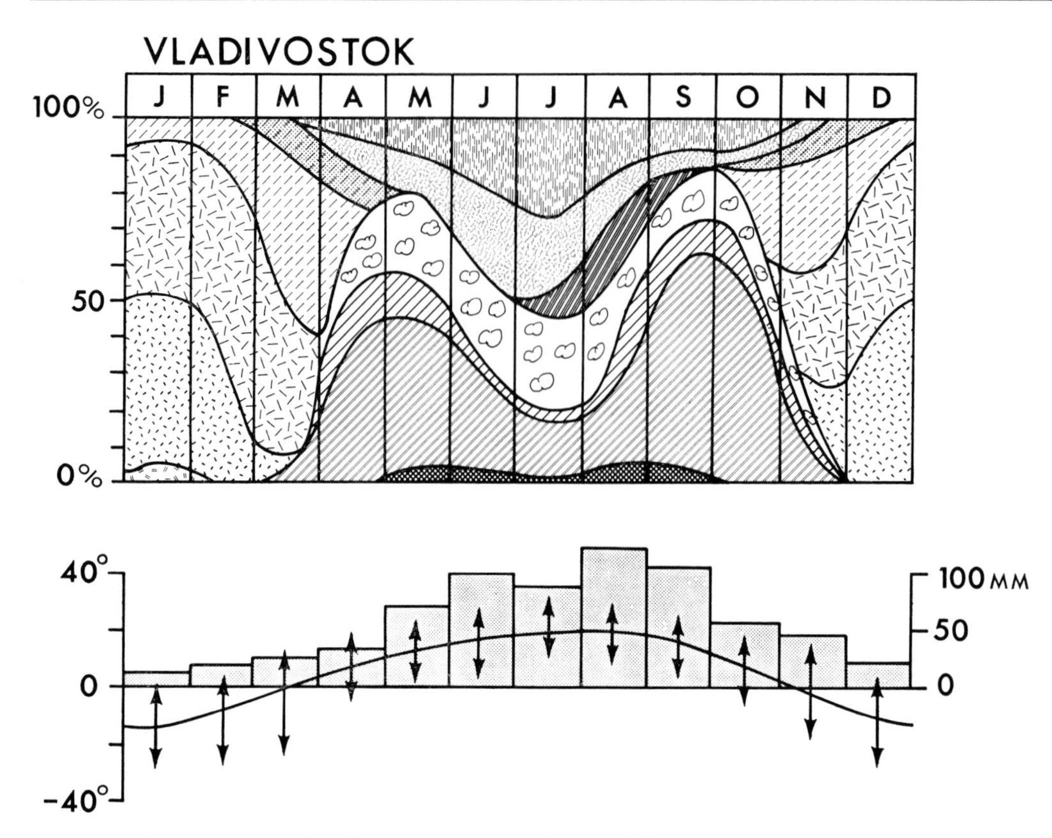

Fig.5-14. Frequencies of weather types, Vladivostok. (After GERASIMOV, 1964.) For legend see Fig.3-25.

embayments that are open to the south or southeast. On other exposures fog frequencies can be quite low. Nikolayevsk at the mouth of the Amur, which is sheltered to a great extent by Sakhalin Island, experiences only 25 days of fog per year.

As one moves eastward in the Soviet Far East the precipitation regime changes from one which is very concentrated in midsummer in the Zeya–Bureya Lowland to one which is more evenly distributed through the year and which has a maximum occurring later and later in the year. In contrast to Blagoveshchensk which receives its maximum precipitation in July and receives 76% of its annual precipitation during the months June–September, Vladivostok receives its maximum precipitation in August and receives 64% of its total annual precipitation during the four maximum months June–September. In general the total annual precipitation increases eastward also. Blagoveshchensk receives 534 mm, and Vladivostok receives 721 mm. However, this varies considerably in each area according to local conditions. The driest part of Maritime Kray is in the western portion of the Khanka Lowland.

Although wind speeds vary greatly over short distances, they generally increase eastward as the coast is approached. In Blagoveshchensk mean monthly wind speeds vary from 3 to 5.8 m/sec. In Vladivostok they vary from 6.3 to 8.1 m/sec. There is a rather pronounced early winter maximum of wind speeds in Vladivostok which are predominantly from the north. During the summer half-year winds are predominantly from the southeast. During winter Vladivostok has storm winds on the average from 9 to 11 days per month. These occur generally when cold clear air blows in from the north. The Vladivostok region, like the Blagoveshchensk region, is one of the sunniest parts of the Soviet Union during midwinter. The coldest winter weather in Vladivostok is associated with north winds and clear skies. It is just the opposite in summer when southerly winds and cloudy skies bring the coldest temperatures.

To the east of the Amur–Ussuri Lowland lie the Sikhote-Alin Mountains which form an unbroken ridge with peak elevations going slightly above 2,000 m. The orographic effects of the mountains greatly increase the precipitation during the summer monsoon, and, of course, reduce temperatures and greatly affect cloud and fog conditions. In general annual precipitation increases by about 20% for every 100 m increase in height in the Sikhote-Alin. This is particularly true on the southeastern slopes of the mountains which are affected most by monsoonal inflows of air from the Pacific. On days when weather is fine over the Japan Sea to the east and over the Ussuri Lowland to the west, mountain slopes above 350 m elevation are often continuously covered with clouds and mist. The wettest spots in the mountains receive more than 1,000 mm of precipitation per year. Much more snow falls during winter in the Sikhote-Alin than on either side, so that the snow cover becomes much deeper. The Sikhote-Alin show the greatest wind speed gradients anywhere in the U.S.S.R. across their crest from west to east. On the coast east of the mountains winds range up to 60–70 m/sec.

The Sikhote-Alin represent a very profound climatic divide between the lowland to the west and the sea coast to the east. During winter they block much of the cold interior air that finds its way southward through the Ussuri Valley and thereby allow winter temperatures on the coast to average about 10°C above those in the Ussuri Lowland. Typical January averages along much of the coast of Maritime Kray lie between −10° and −15°C. Vladivostok at the southern end of the coast is the coldest of the entire coastal region, probably because of the enclosed nature of the embayment in which it sits and its exposure to winds from the north which have come down the lowland west of the Sikhote-Alin Mountains.

In summer the coastal area of Maritime Kray does not benefit from the heating by radiation which is taking place in the lowland to the west of the mountains. A cold ocean current flows southward from the Tatar Strait and hugs the coast of Maritime Kray. Temperatures during July average 14° to 19°C along the coast whereas they average more than 20°C in the Ussuri Lowland.

Across the Tatar Strait lies Sakhalin Island, which with its double range of mountains continues the progression eastward of north–south oriented intervening highlands and lowlands. The whole island reflects its marine environment more than any part of the mainland does, and the east coast reflects it more than the west coast. In the interior of the island between the mountain ranges the climate is more continental than on either coast. The highest peaks of the mountains lie between 1,300 and 1,400 m above sea level. The southern extension of the island consists of the single western range of mountains, and the northern third of the island is rather low in elevation almost everywhere.

For its latitude and its marine location Sakhalin has a very severe climate. In winter it is greatly affected by strong cold winds coming off the continent which can drop temperatures even in the southern extremities of the island to −39°C. All except the southwestern tip of the island is engulfed in sea ice in late winter and early spring, and the northern half of the island is surrounded from December through May. Thus, throughout the long winter the island is really an extension of the mainland as far as climate is concerned. Generally in winter the west coast has a somewhat more severe climate than the east coast. However, the ice extends farther southward on the east coast of the island where the cold Okhotsk current flows southward throughout the year. On the west side of the island the moderately warm Tsusimski current flows northward much of the time to at least the latitude of Aleksandrovsk. Ice generally does not break up along the northern coast of Sakhalin until the end of June, and it drifts southward along the east coast even after that, sometimes as late as August.

High winds are most frequent in winter and autumn in association with depressions which are moving across the island from west to east. The wind regime of the island is essentially unaffected by typhoons, although they may produce widespread cloudiness and precipitation in late summer and autumn. On exposed capes on Sakhalin winds may reach speeds of 50–60 m/sec. A major track of cyclonic storms which have come across southern Siberia and the lower Amur Valley crosses Sakhalin and the southern half of the Sea of Okhotsk during winter and continues on across Kamchatka to the Aleutian low. These storms may bring heavy snowfalls to portions of Sakhalin, particularly the northern part, and snow cover in places may reach as much as 70 cm. Snow lies on the ground in most places on Sakhalin for at least 200 days during the year. In most places along the coasts the snow disappears in late April or early May. In the center of the island it melts 10–15 days later. Permafrost is widespread in the northern half of the island.

Spring is very cold and raw on much of Sakhalin because of the ice surrounding it. The first half of summer remains unpleasant with raw cold sea winds and frequent fogs. As summer progresses the sea temperatures and the southern part of Sakhalin warm up considerably but the northern half of the island remains chilly throughout the summer. At Yuzhno-Sakhalinsk on the southern end of the island August, the warmest month, averages 17.1°C. At the northern tip of the island this reduces to about 10°C.

Sea fogs are most common in late spring and early summer and are more prevalent along the east coast than the west. In summer the east is colder than the west coast. Whereas in winter Sakhalin receives cold winds from the land, in summer it receives them from the sea. Summer everywhere on Sakhalin is characterized by humid weather with relative humidities often above 85%. Sunshine is rare, thick clouds on the average cover 8/10 of the sky, fogs are frequent, and drizzle is common. The cool windy weather throughout the summer in the north is conducive to tundra type vegetation, while further south much of the island is heavily forested with coniferous trees.

The best weather on Sakhalin probably exists in the central part of the island which is partially protected from either the winter monsoon on the west or the summer monsoon on the east. However, this is a very continental part of the island and experiences temperature extremes. Maximum temperatures have reached 32°C and minimum temperatures −48°C. During winter a localized high tends to form over the middle portion of the island which causes winds locally to blow outward in all directions. This localized winter monsoon extends only to 500–800 m in height.

Weather along the southern coast of the central valley of Sakhalin is represented by the diagram for Poronaysk (Fig.5-15). Here the asymmetry of summer is well displayed. Early summer is cool with only moderate precipitation while late summer is warmer and has more precipitation. This is typical of marine locations. Maximum temperatures occur in August and maximum precipitation in September. Thus, the lag in maximum precipitation continues eastward from the mainland.

Eastward across the Okhotsk Sea the Kuril Island chain is a line of volcanic peaks sticking

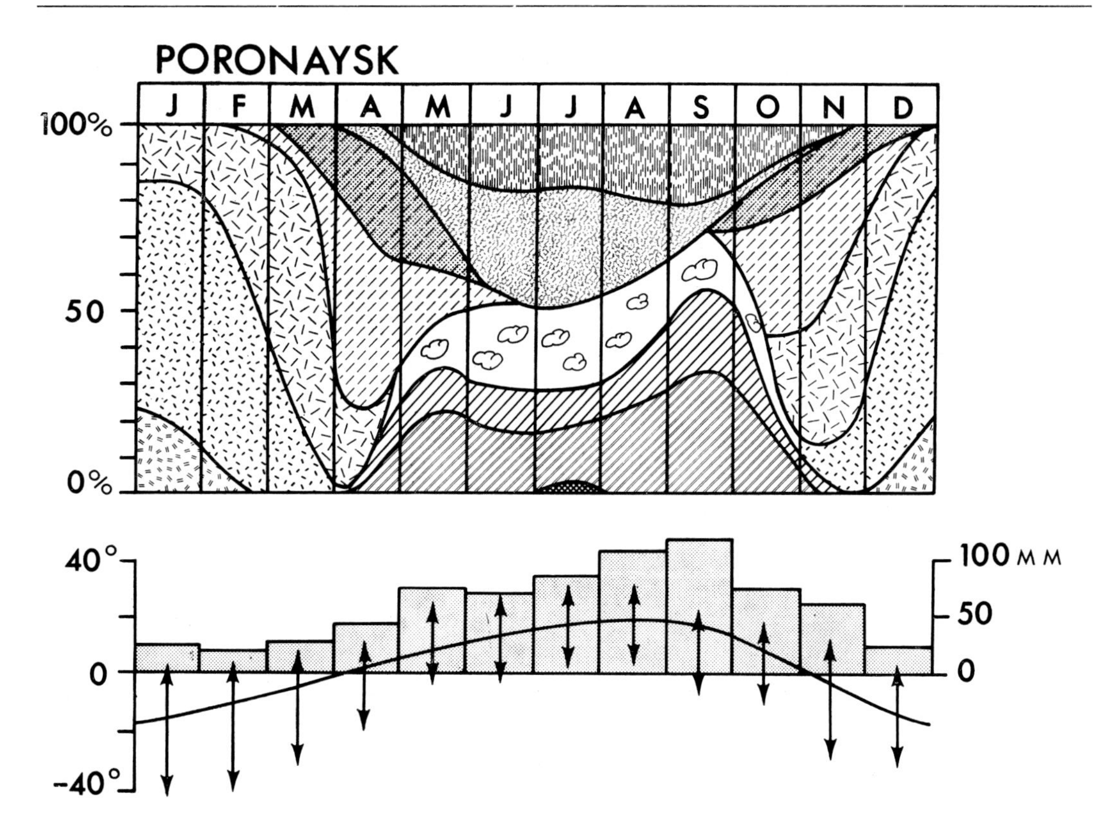

Fig.5-15. Frequencies of weather types, Poronaysk. (After GERASIMOV, 1964.) For legend see Fig.3-25.

out of the water from an underwater ridge. Everywhere their elevations are under 2,000 m except on one island in the northern end of the chain which rises to 2,339 m. They are engulfed on both the east and the west by a cold ocean current which comes down from the north along the east side of Kamchatka in the Bering Sea. This is generally known as the Kamchatka current on the east coast of Kamchatka and as the Kuril current along the Kuril Island chain. Sea ice generally engulfs the southern end of the chain next to Hokkaido in March when it reaches its most southerly extent after being driven southward by the Okhotsk current along the east coast of Sakhalin. However, the rest of the year it is absent, and north of here the islands are ice free year round. Surface water temperatures in the Kuril current average 0°–2°C in February and 8°–13°C in August.

Thus, the islands are completely dominated by marine air and have a cold marine climate which does not reach as low temperatures as Sakhalin to the west but never gets very warm at any time of the year. Simushir in the middle of the chain averages −5.2°C in February, the coldest month of the year, and 10.6°C in August, the warmest month. August temperatures range from 9°C in the north to 17°C in the south. The lower slopes of most of the islands are forested with light stands of birch, cedars, and other species. The volcanic peaks are often barren sands and ash or they are covered by a marine tundra type of vegetation.

Relative humidities in the Kuril Islands are constantly high, averaging 80% in January and 96% in August. The islands are swept by clouds and fog much of the year. Fog occurs on 20 to 26 days per month during June, July, and August. Mean precipitation at Simushir during the year is 1,461 mm, with all months being quite wet. September shows the highest total with 198 mm.

Winter in the Kuril Islands is characterized by exceedingly strong winds, great cloud cover, and heavy snowfalls. Summer also has strong winds, very high air humidities, and frequent rains. Autumn is distinguished by the strengthening of cyclonic activity in the western end of the Aleutian depression. Thus, the situation in fall is somewhat the opposite of that on the mainland where autumn generally sees a diminishing of cyclonic activity and precipitation.

The northwest and north coasts of the Sea of Okhotsk are still influenced by the summer monsoon from the south, but this is its northern limit, and rugged mountains immediately back from the coast in most cases block penetration by this air into the land mass. The Dzhugdzhur Mountains along the northwest coast rise to approximately 2,000 m above sea level and have their steepest slope facing the sea so that there is little watershed on the sea side of the divide. Therefore the monsoonal climate is limited to a very narrow strip of land in this area. Great temperature contrasts exist across this range during both seasons. The currents in the water in the northern part of the Okhotsk Sea flow westward so that floating ice lodges along this northwest shore later in summer than in any other part of the sea. Ice is usually in shore here until late June and in some years considerably later than that. Water temperatures in late summer and fall never get above 12°C in this area, and neither do surface air temperatures. At Okhotsk, August is the warmest month with a mean temperature of 12.9°C. However, the temperature has risen to 32°C with foehn winds coming down from the mountains. Predominantly southerly and southeasterly winds in summer bring sea fogs into the coast. The lower slopes of the mountains are forested by pines along much of this coast.

During winter the Okhotsk seaboard is dominated by continuously strong northerly winds blowing out of the northeasterly extension of the Asiatic high. Often they are above gale force, and they do not abate significantly from October through March. Although temperatures are not as cold as they are further inland, the wind chill factor along this coast is one of the worst in the entire country. January temperatures at Okhotsk average −24.5°C, and a minimum of −45°C has been observed. At the same latitude in the western part of the country Leningrad averages −7.6°C in January. Winters along this northwestern coast are quite clear and rather snowless. Okhotsk records only 6 mm of precipitation in February. Much of the yearly total of 378 mm falls between May and November. Springs are raw, cloudy, and foggy. The best time of year is autumn when sea temperatures have reached their maximum and air temperatures rise accordingly, fogs diminish, and the sky becomes clear (Fig.5-16).

It is interesting to note the contrasts in weather types between Okhotsk on the sea coast and Oymyakon about 300 miles to the north and inland over the mountains at an elevation of 740 mm (Fig.4-19). Oymyakon, of course, has a much colder winter and a much shorter, cooler summer. But there is also a difference in shape of the summer season. At Oymyakon the weather types are oriented rather symmetrically around mid summer, whereas at Okhotsk there is the skewness toward late summer and fall that is so characteristic of marine locations. August is the warmest time of the year and the rainiest, and September is the time of greatest sunshine. In contrast, spring is cool, cloudy, and foggy.

The northeastern region

Northeast of Shelikhov Bay the remote northeastern corner of the Soviet Union is a

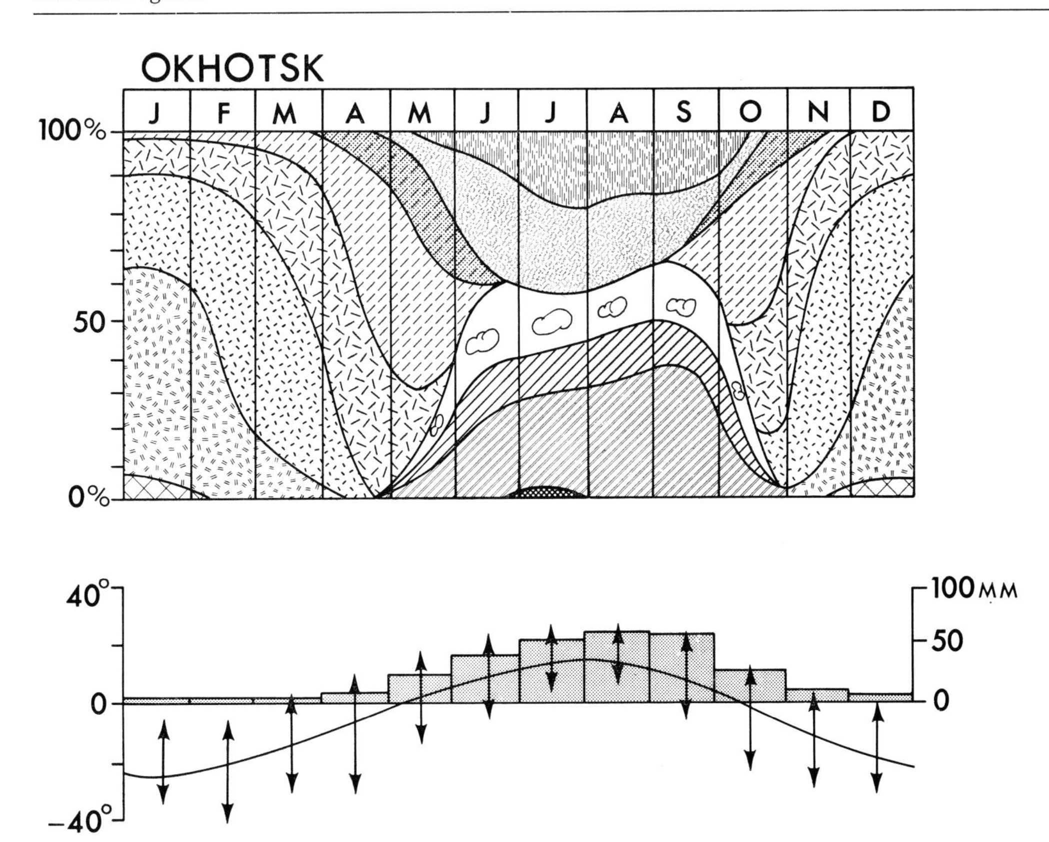

Fig.5-16. Frequencies of weather types, Okhotsk. (After GERASIMOV, 1964.) For legend see Fig.3-25.

jumble of mountains and intervening lowlands which together produce a very uneven coastline with deep embayments and rocky peninsulas and islands. Most of the mountains are well under 2,000 m in elevation, but even that is enough at these high latitudes to put their summits well above the tree line. The highest elevation is 2,562 m in the Koryak Range which joins the peninsula of Kamchatka to the mainland. The coastal lowlands too are generally devoid of trees because of the cold cloudy summers. Only some of the interior hills which get enough moisture and warmth during the summer support growth of alder, willow, aspen, and birch. The rest of the area is primarily windswept tundra. The area is completely surrounded by sea ice from December through April, and ice does not leave the Bering Straits until late July or early August. The Arctic coast is relatively free of ice during the months of August and September.

The southeastern coasts of the area during winter are influenced primarily by the western end of the Aleutian low, and the interior and northern coasts are influenced by the northeastern extension of the Asiatic high. During winter winds generally blow from the interior toward the coasts to the north and the south. In summer they usually blow inland from the coasts and carry cold damp air onto the peripheries of the continent which produce a great amount of overcast skies and low temperatures in the summer time. Many places experience killing frosts all through summer. From May to September moist easterly winds bring overcast, windy, foggy weather to the coast, while in the interior the weather may be sunny and dry with comparatively calm winds. Autumn becomes even more overcast, rainy, and windy as the Aleutian low begins to develop. This is in contrast to the sunny dry weather of autumn farther to the southwest.

Annual precipitation over much of this northeastern region totals 200 to 250 mm. However, along certain coastal areas precipitation may amount to as much as 450 mm. In the interior July is usually the month of maximum precipitation, and in the coastal areas it is generally August. However, winters are not dry, particularly along the southeastern coast where the development of the Aleutian low produces considerable amounts of prolonged cyclonic precipitation in winter. When the Aleutian low is well developed winters are generally more snowy along the coasts than inland, but during years when the Aleutian low is not so well developed there is generally more snow inland than on the coasts. Winter blizzards are common and sometimes continue for one to two weeks on end and reduce visibility to only a few meters. Average snow depth accumulates to 50–60 cm and may reach 100 cm in drifts produced by storm winds. On the mountains snow may persist into July or even August and some years it hardly melts before new snow falls in the autumn.

Some of the strongest pressure gradients and winds are observed between the Kolyma Range and the sea coast along the northwestern shore of Shelikhov Bay. Here true bora conditions may exist in winter much of the time in the short valleys that drain down the seaward slopes of the Kolyma Mountains.

Temperatures on the Bering sea coast at Apuka and Anadyr average about 10°C in July. North of the Bering Straits Uelen averages only 5.4°C. On the Arctic coast Mys Shmidta averages 3.6°C in July, and in the interior Ilirney averages 10.4°C. All these stations average far below zero in winter. The minimum temperature recorded at Ilirney is −58°C.

Anadyr', on an embayment of the east coast, is fairly typical of the region (Fig.5-17). Winters are long and cold, and summers are cool and rather cloudy. The most distinctive feature of the seasonality of the region probably is the contrast between spring and fall. Spring tends to be cold and relatively sunny, while fall is much warmer and cloudier. Maximum precipitation comes in August and maximum temperatures in July. However, temperatures are arranged very asymmetrically around July. September averages 3.9°C., while May averages −3.5°C. August is almost as warm as July, while June is 5.7°C colder.

The large peninsula of Kamchatka has somewhat of a climate regime of its own. It is greatly affected by the seasonal development and decline of the Aleutian low to the east, but superimposed on the general atmospheric circulation is a local monsoon over Kamchatka itself. The peninsula consists primarily of two mountain ranges with an interior valley between. The western range is the longest and most continuous and reaches an elevation of 1085 m in the west-central part of the peninsula. The eastern range is shorter and more discontinuous and consists primarily of individual volcanic peaks many of which are active. The highest is Mount Klyuchevski which reaches an elevation of 4,750 m. The volcanic cones in the eastern range in many places come relatively close to the sea coast, while along the western coast there is a fairly broad continuous coastal plain. In winter cold air develops over the interior valley and the peninsula tends to produce a southward dip of high pressure from the main high over the mainland to the north (Fig.5-3). Winds tend to blow outward from the center of Kamchatka toward both coasts during winter. During summer the interior is warmer than the coasts and winds tend to blow inland from both the east and the west. Thus, a convergence line forms over the central portion of the peninsula during much of summer.

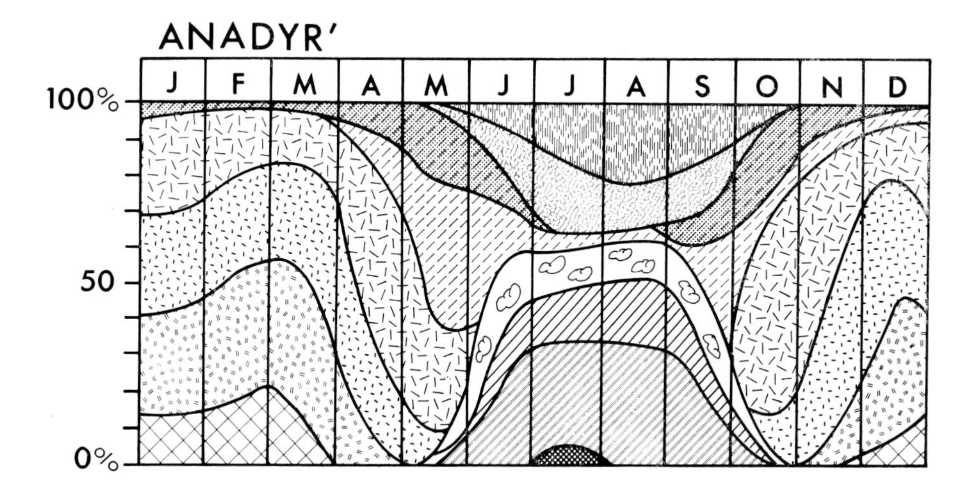

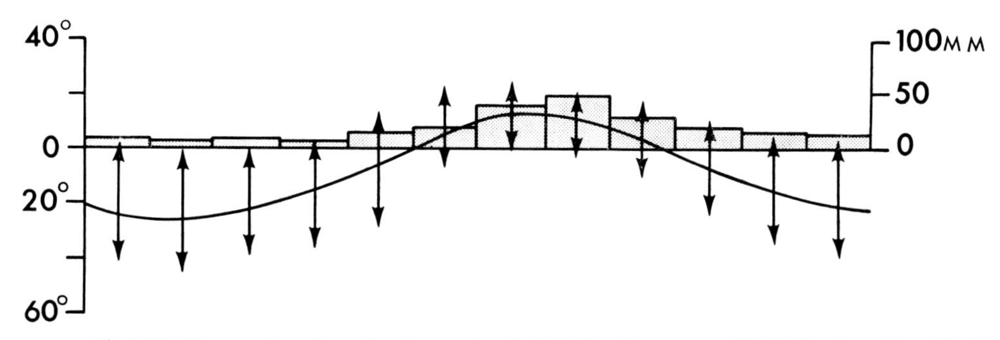

Fig.5-17. Frequencies of weather types, Anadyr'. (After GERASIMOV, 1964.) For legend see Fig.3-25.

The great influence of the western end of the Aleutian low on Kamchatka is reflected in the distribution of precipitation amounts and regimes on the peninsula. The southeastern coast receives well over 1,000 mm per year, while the northwestern coast receives less than 500 mm. Petropavlovsk–Kamchatskiy receives 1,335 mm of precipitation per year, most of which falls during autumn and winter. November has the maximum amount with 182 mm, and June has the minimum with 58 mm. The late fall maximum on the southeastern coast is probably due to a combination of the early part of the development of the Aleutian low in winter and the relatively high sea surface temperatures which have not cooled off too much from summer. Later in the winter cyclonic activity probably increases, but the decreasing sea and air temperatures cause a reduction in precipitation. Earlier in the fall cyclonic activity is not intense enough to make maximum use of the warmer sea surface temperatures.

Sea ice starts to extend down both the east and west coasts of Kamchatka in December. It extends farther southward along the west coast early in the winter, but as the season proceeds the ice in the Bering Sea extends farther south than that in the Okhotsk Sea. By January ice along the east coast of Kamchatka has extended more than half way down the coast, and by March it has reached nearly to Petropavlovsk–Kamchatskiy. In April it begins to recede northward again and by June it has disappeared from both coasts of Kamchatka. The presence of the sea ice and the lingering coldness of the water along the northeastern part of the coast causes this part of the coast to receive considerably less precipitation than the southeast part of the coast, although it is still quite moist in the northeast. Ust'Kamchatsk, in the central portion of the east coast averages 670 mm per year, as opposed to the 1,335 mm at Petropavlovsk in the southeast. And the regime shifts somewhat from south to north. Ust'Kamchatsk shows a maximum of precipitation in January. The minimum is still in June, as it is farther south.

Precipitation is least in the central valley of the Kamchatka River where it averages as little as 400 mm per year. And the maximum in the interior generally falls in late summer, as opposed to late fall or winter on the east coast. The interior valley obviously is sheltered to a great extent from the influences of cyclonic storms in the western part of the Aleutian low during winter and is more affected by convective activity and thundershowers in summer.

Kamchatka is known for its blizzards. These can occur either during the intrusion of warm air from the southeast associated with cyclonic storms and heavy snowfall or with cold air and clear skies coming off the continent from the north. High winds are encountered in either case which whip up the snow from the ground and cause low visibilities. Highest wind speeds are found in winter and spring and they are generally higher on the east coast of Kamchatka than on the west. On the east coast during winter wind speeds exceed 20 m/sec 3–5% of the time, while on the west coast they exceed 20 m/sec only about 1% of the time. They reach their greatest speeds at the southern tip of the peninsula on Cape Lopatka, which has a probability of a maximum wind of 56 m/sec once in 20 years.

Snowfall is greatest on the southeastern coast. During warm winters the depth of snow cover at Petropavlovsk reaches 130–200 cm. In exceedingly snowy winters it may reach as much as 3 m. Because of the great depth of snow in the southern half of Kamchatka the soil freezes only slightly, generally no more than 10 cm deep. The snow cover is much less over the interior of the peninsula, where in sheltered places horses are grazed throughout the winter. The heat from volcanic activity in some localities greatly reduces the depth of snow cover and raises the snow line on the mountain slopes. Generally the snow line on Kamchatka lies at about 1,600 m above sea level.

Both the east and west coasts of Kamchatka are rather cold for the latitude. This is particularly true in summer when the entire eastern coast and the southern half of the western coast are paralleled by cold ocean currents moving southward. Since ice lingers longer in the Sea of Okhotsk than it does in the Pacific, the west coast of Kamchatka is about 2°C colder during summer than the east coast is. However, it is less foggy and cloudy than the east coast. On the southeast coast Petropavlovsk averages 13.5°C in August. Generally, the frost-free period is from 4 to 4½ months, but frost can occur anytime during the summer. On the west coast, August temperatures average about 11°C and the frost-free period generally is no more than 3 months. Frosts can occur here anytime during summer also.

The interior of the peninsula shows a great deal more continentality than does either coast. During July Mil'kovo averages about 15°C, and temperatures as high as 30°C are sometimes reached. In January the central valley averages about −18°C. This compares with a minimum of −8.7°C at Petropavlovsk in February.

All things considered, the climate of the interior part of Kamchatka is generally preferred over that of the coastal areas because of its lower humidities, greater sunnyness, and warmer summers. The coastal areas are notedly damp, particularly along the southeast. In July and August relative humidities at Petropavlovsk average more than 80%. Fog occurs 10–11 days per month from June to August. Sky cover averages between 7 and 8 tenths from May through September.

Spring is generally the best time of year throughout Kamchatka, in spite of its relatively cold temperatures. At this time of year the air is drier and there is more sunshine.

Figs.5-18 and 5-19 represent the extremes of weather types on Kamchatka. Petropavlovsk on the east coast shows a typical oceanic regime influenced by the western end of the Aleutian low. Springs are cold and autumns are relatively warm. Maximum precipitation comes in November. Winters are not nearly as cold as they are in the interior at Mil'kovo. Summer temperatures at Mil'kovo are much more symmetrically arranged around midsummer, and the maximum precipitation falls in August.

East of Kamchatka lie the Komandorskiye Islands, which reach an elevation of 751 m on the larger island, Bering Island. The climate is completely influenced by the sea. Winters average about −2° to −4°C and summers are cold, very damp, and nearly always overcast. There is much fog. Relative humidity averages from 83 to 93% every month of the year. Maximum temperatures occur in August when they average 10.6°C. There are many days with precipitation, but amounts are only moderate. Bering Island averages 516 mm per year, and there is a very even distribution through the year. October shows the maximum amount with 72 mm, and February has the minimum with 23 mm. Because of the consistently low temperatures, the islands are completely tundra.

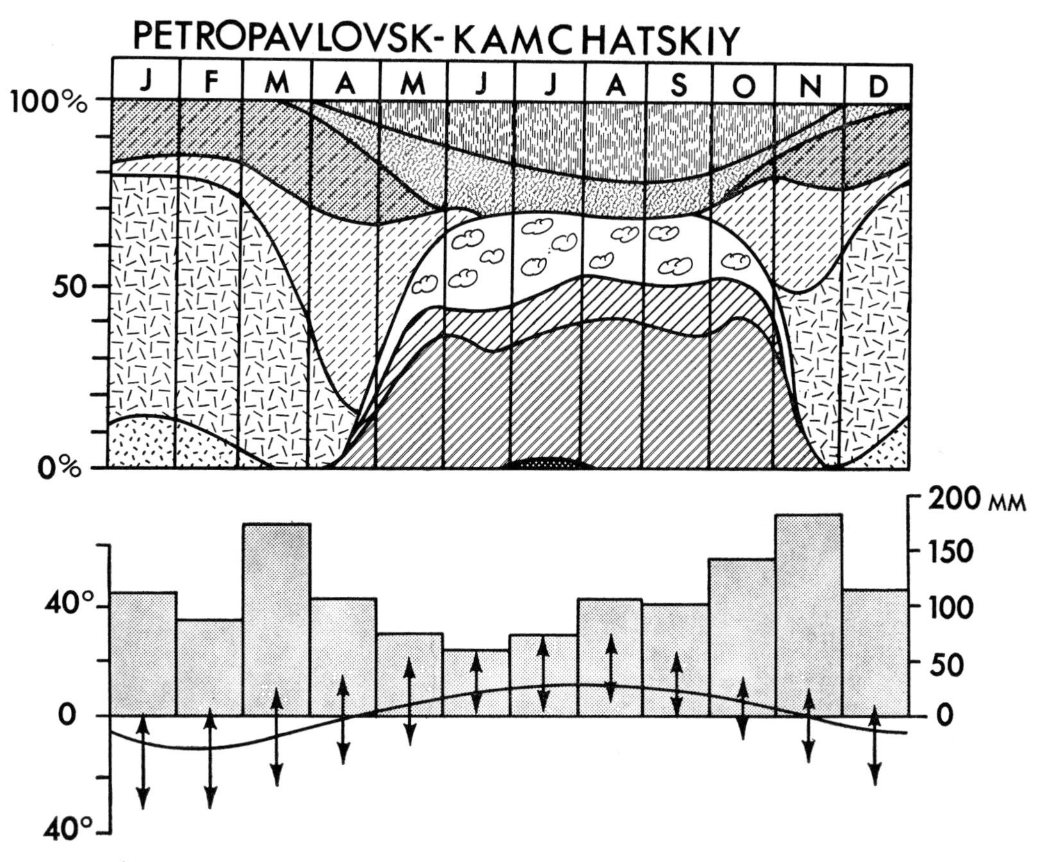

Fig.5-18. Frequencies of weather types, Petropavlovsk-Kamchatskiy. (After GERASIMOV, 1964.) For legend see Fig.3-25.

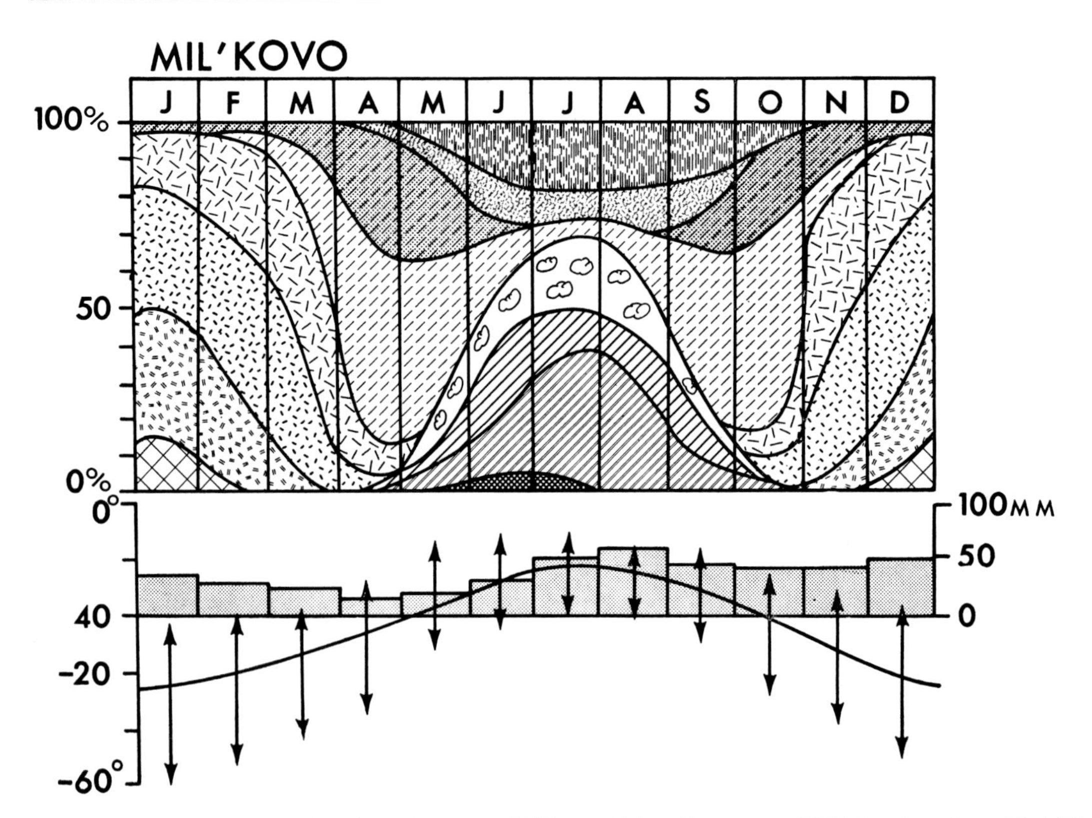

Fig.5-19. Frequencies of weather types, Mil'kovo. (After GERASIMOV, 1964.) For legend see Fig.3-25.

References and further reading

ALISOV, B. P., 1956. *Klimat SSSR* (Climate of the USSR). Moscow Univ., Moscow, 127 pp.

ANAPOL'SKAYA, L. E. and TYUKTIK, V. V., 1969. Vetrovoy rezhim gornykh rayonov vostochnoy sibiri, dal'nego vostoka, p-ova Kamchatka i o. Sakhalin (The wind regime in the mountainous regions of eastern Siberia, the Far East, the peninsula of Kamchatka, and the island of Sakhalin) *Gl. Geofiz. Obs. Tr.*, 246: 12–24.

ANONYMOUS, 1969. *Atlas SSSR* (Atlas of the USSR). Gl. Upr. Geod. Kartogr. Sov. Ministr. SSSR. Moscow, 2nd ed., 199 pp.

BORISOV, A. A., 1965. *Climates of the USSR*. Aldine, Chicago, 225 pp.

BUDYKO, M. I. (Editor), 1963. *Atlas Teplovogo Balansa Zemnogo Shara* (Atlas of heat balance of the earth). Gl. Geofiz. Obs., Moscow, 69 pp.

DROZDOV, O. A. and GRIGOR'EVA, A. S., 1963. *Vlagooborot v Atmosfere*. Gidrometeoizdat, Leningrad. (Transl. as *The Hydrologic Cycle in the Atmosphere*, Isr. Progr. Sci. Transl., Jerusalem, 1965, 282 pp.)

GERASIMOV, I. P. (Editor), 1964. *Fiziko-Geograficheskiy Atlas Mira* (Physical geographical atlas of the world). Akad. Nauk, Moscow, 298 pp.

GORSHKOV, V. E., 1958. O zimney tsiklonicheskoy deyatel'nosti nad rayonam kontinental'nogo Dal'nego Vostoka (On winter cyclonic activity over the continental regions of the Far East). *Izv. Akad. Nauk SSSR, Ser. Geogr.*, 3: 68–71.

LYAKHOV, M. E., 1961. O mussonnosti klimata Kamchatki (On the monsoon nature of the climate of Kamchatka). *Izv. Akad. Nauk SSSR, Ser. Geogr.*, 3: 47–49.

MAKSIMOV, I. V. and KARKLIN, V. P., 1970. Sezonnyye i mnogoletniye izmeneniya glubniny i geograficheskogo polozheniya aleutskogo minimuma atmosfernogo davleniya za period s 1899 po 1951g. (Seasonal and long-term variations in the intensity and geographical position of the Aleutian minimum of atmospheric pressure from 1899 to 1951). *Izv. Vses. Geogr. O-vo.*, 102: 422–431.

SOROCHAN, O. G., 1957. K voprosu o prirode letnikh osadkov mussonnoy oblasti vostochnoy Azii

(Toward the question of the nature of summer precipitation in the monsoon regions of eastern Asia). *Gl. Geofiz. Obs., Tr.*, 72: 92–109.

SOROCHAN, O. G., 1966. Mussony Azii (The monsoons of Asia). In: M. I. BUDYKO (Editor), *Sovremennye Problemy Klimatologii.* Gidrometeoizdat, Leningrad. (Translated as *Modern Problems of Climatology*, Foreign Technol. Div., Wright-Patterson Air Force Base, Ohio, FTD-HT-23-1338-67, 29 November 1967, pp. 215–235.)

ZANINA, A. A., 1958. *Klimat SSSR, Vypusk 6, Dalnevostochnye Rayony, Kamchatka, i Sakhalin* (Climate of the USSR, Volume 6, Far Eastern regions, Kamchatka, and Sakhalin). Gidrometeoizdat, Leningrad, 167 pp.

ZVEREV, A. S., 1957. *Sinopticheskaya Meteorologiya* (Synoptic meteorology). Gidrometeoizdat, Leningrad, 559 pp.

Chapter 6

Central Asia

Definition of the region

The area here defined as Soviet Central Asia is one that is fairly homogeneous and self contained climatically and is considerably more extensive than the region that is generally designated as Central Asia by Soviet geographers (Fig.2-22). The climatic region extends northward into central Kazakhstan where the northern boundary coincides approximately with the great axis of high pressure which in the winter extends westward from the core area of the Asiatic high (Fig.6-1). On the average during winter winds diverge northward and southward from this high pressure ridge so that Central Asia to the south experiences primarily northerly winds, while Siberia to the north experiences primarily southerly winds (Fig.6-2). Of course, at any time during winter Siberian air can invade Central Asia which is wide open to the north, but on the average the ridge of high pressure serves as something of a dividing line between air masses to the north and air masses to the south.

In summer, there is no such line of divergence of prevailing wind directions, but the northern boundary of Central Asia as defined here approximates the southern limit of effective cyclonic activity during the summer time which brings, in conjunction with increased convective activity, a maximum of precipitation to western Siberia at this time of the year (Fig.6-3). Thus, the boundary as drawn here separates the region to the south with predominantly spring and winter precipitation from the region to the north with predominantly summer and fall precipitation (Fig. 9-17).

The high mountains bordering Soviet Central Asia everywhere on the south effectively block the intrusion of air masses from territories lying south of the U.S.S.R. Thus, there is a natural climatic boundary here. The northwesterly extensions of the mountain masses which lie within the Soviet Union are part of the region itself and display a variety of climate which is much different from that of the desert plains to the north and west. Individual ranges which jut out from the main masses distort the surface air flow so that many local circulations are set up that are difficult to correlate with the general pressure pattern overlying the region. But gross features of the surface flow can be discerned from the forms of the drifting sand deposits which occupy large areas of the desert floor and penetrate up the river flood plains between the mountain spurs (Fig. 6-4).

On the west, the Caspian Sea forms a convenient boundary for Soviet Central Asia, although it is certainly true that air moves freely across at least the northern half of the sea and dry climate extends westward across the sea. However, the Caucasian Highlands

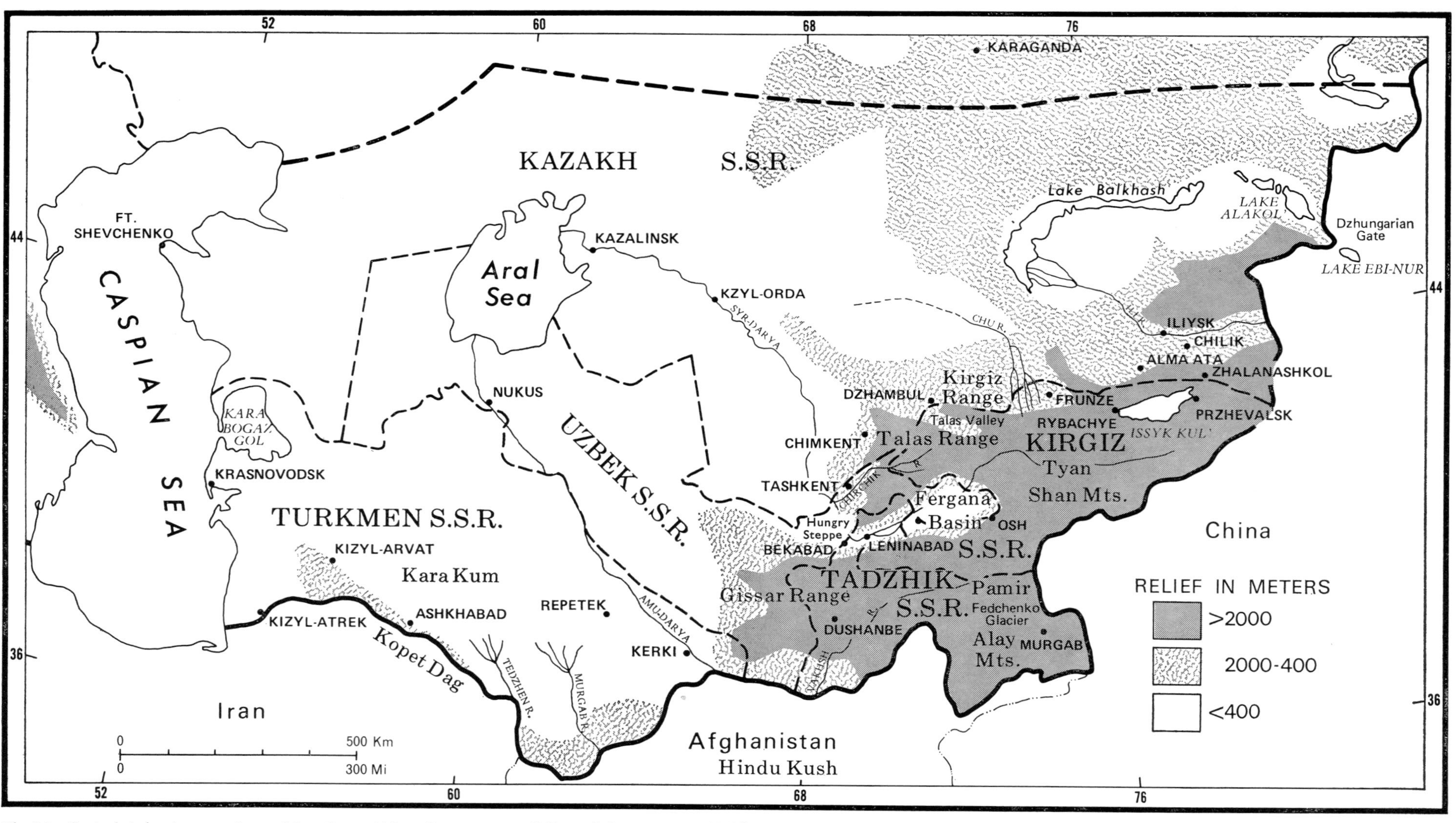

Fig.6-1. Central Asia: topography and locations. (After CHELPANOVA, 1963, and ANONYMOUS, 1969.)

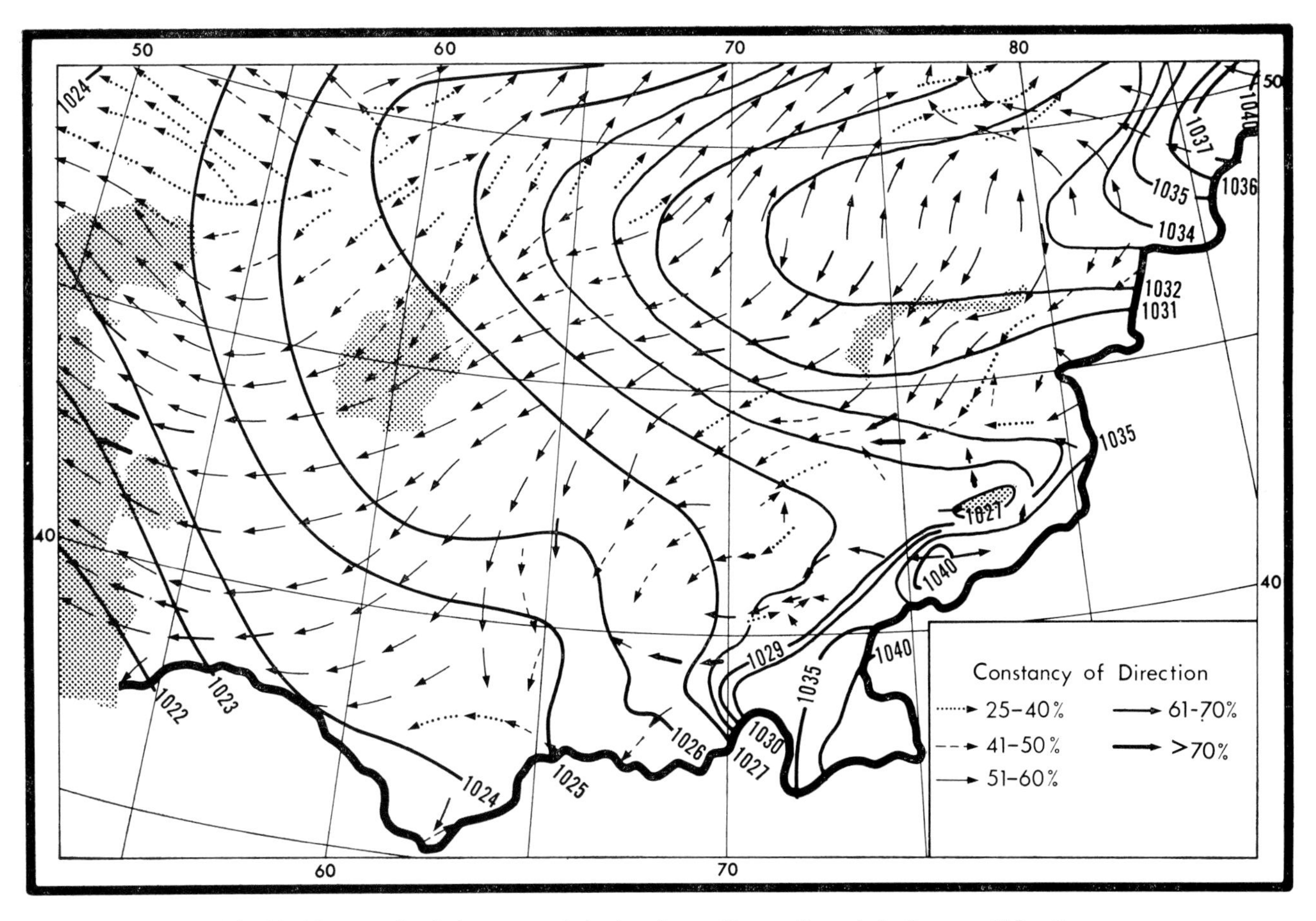

Fig.6-2. Mean sea level air pressure (mbar) and prevailing surface winds, January. (After CHELPANOVA, 1963.)

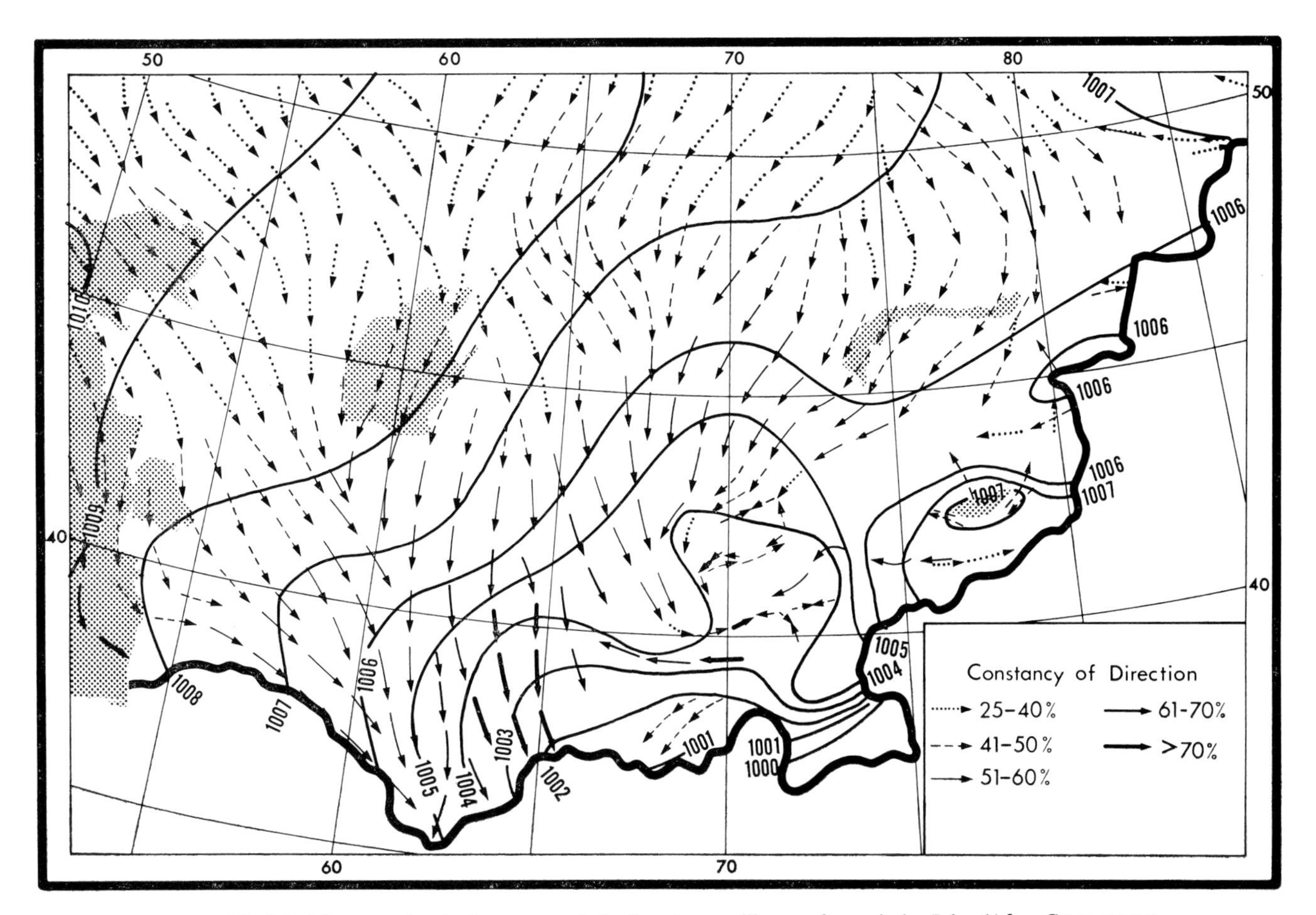

Fig.6-3. Mean sea level air pressure (mbar) and prevailing surface winds, July. (After CHELPANOVA, 1963.)

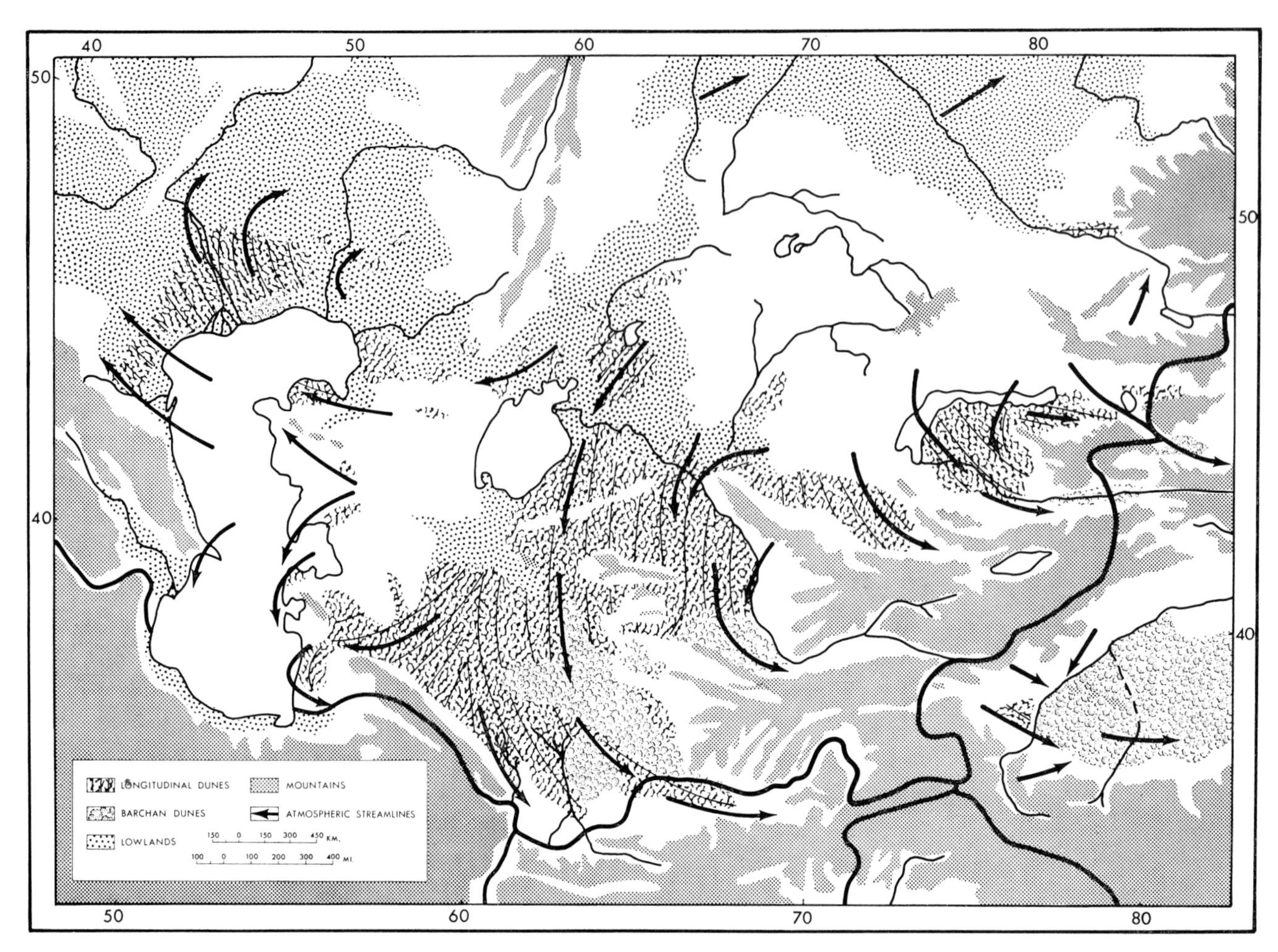

Fig.6-4. Surface air streamlines as revealed by sand dunes. (After FEDOROVICH, 1948.)

generally preclude movements of cyclonic storms into Central Asia from a westerly or northwesterly direction. And therefore most of the cyclonic activity that influences Central Asia enters the region from somewhere south of the Caspian. Thus, as far as weather types are concerned the Caspian Sea does function as a reasonable western boundary of Soviet Central Asia.

General atmospheric flow and synoptic processes

The dominant features of the sea level pressure pattern over Central Asia which are revealed by the mean maps for January and July in Figs.6-2 and 6-3 are the western periphery of the core area of the Asiatic high in winter and the southeastern periphery of the Azores high in summer. These pressure patterns induce a northeasterly flow of air across Central Asia in the winter and a northerly and northwesterly flow of air in summer which moves down the pressure gradient from the north Caspian and Central Kazakh area toward a thermal depression that tends to form at this time of year over the very southerly portion of the region, particularly in the headwater area of the Amu-Darya in the plateau country of southwestern Tadzhikistan. However, such mean maps obscure a considerable variety of surface air flow and pressure patterns at all seasons in the area, most of which do not extend very high into the atmosphere. Most cyclones and anticyclones in this area do not extend upward beyond 5 km and about half of them do not reach above 2 km. There is some tendency for them to extend higher in summer and fall than in winter and spring, but in general Central Asia is not an area of deeply developed pressure cells. In a study made during 1942–1951, it was found that some cold highs in winter and thermal depressions in summer do not even reach up to the 900-mbar level (BUGAEV et al., 1957, p.437).

Aloft the flow is strongly zonal at all seasons. This is shown on both the 700-mbar and 500-mbar maps (Figs.2-3, 2-4, 2-9, 2-10). The average flow is strictly zonal in summer, while in winter there is a tendency for a shallow upper level trough to develop from north to south across the east European plain extending into the north Caspian area. This places Central Asia during winter under the eastern limb of an upper level trough which tends to steer lower tropospheric systems from southwest to northeast. Frequently the trough intensifies in a much more meridionally elongated pattern which brings in strong southwesterly and southerly flow from Iraq, Iran, and Afghanistan. On these occasions cyclonic storms tend to form along the Iranian front in the Middle East and move northeastward into Soviet Central Asia where they occlude along the northwestern foothills of the Pamir-Alay and Tyan Shans.

The high mountain massif surrounding the Tibetan Plateau forces a split in the zonal westerlies of the upper troposphere which causes some of the air to go southeastward over northern India and some of it to go northeastward over northern China. This disturbance in the zonal flow induces patterns of divergence and convergence through Central and Eastern Asia which vary according to level (below, at, and above mountain tops) and with seasons as the general air flow west of the mountain massif varies from southwesterly in winter to westerly in summer. Some tentative maps of divergence and convergence patterns over Asia have been included in the paper by PETROSYANTS, 1971. But study of the area is still inconclusive with regard to the relation between upper air divergence and convergence and lower tropospheric weather phenomena.

The disturbance in the upper trospheric air flow produced by the Central Asian mountain massif also contributes to the positioning of the so-called "high level planetary frontal zone" which is signified by a break in the tropopause and its accompanying jet stream. Particularly in winter, this upper tropospheric feature is generally positioned somewhere over the Tyan Shans and provides a favorable environment for the strong development of the Asian sector of the Polar front at this time of year with its wave activity and cyclonic storms (Fig.6-5).

Bugaev has defined 12 types of surface air flow in Central Asia. Their percentage frequencies of occurrence during summer and winter halves of the year are shown in Table 6-I, where types "southwest periphery of anticyclone" and "southeast periphery of anticyclone" have been combined. The main types are shown diagrammatically in Fig. 6-6 where widths of arrows are proportional to relative frequencies. The most frequently occurring synoptic situation throughout the year is the southwest periphery of an anticyclone, which surprisingly enough occurs even a little more frequently in summer than in winter, although during summer it is outnumbered slightly by westerly intrusions. Westerly, northwesterly, and northerly intrusions occur most frequently in summer when the flow aloft is most zonal in character and there is no well developed ridge of high

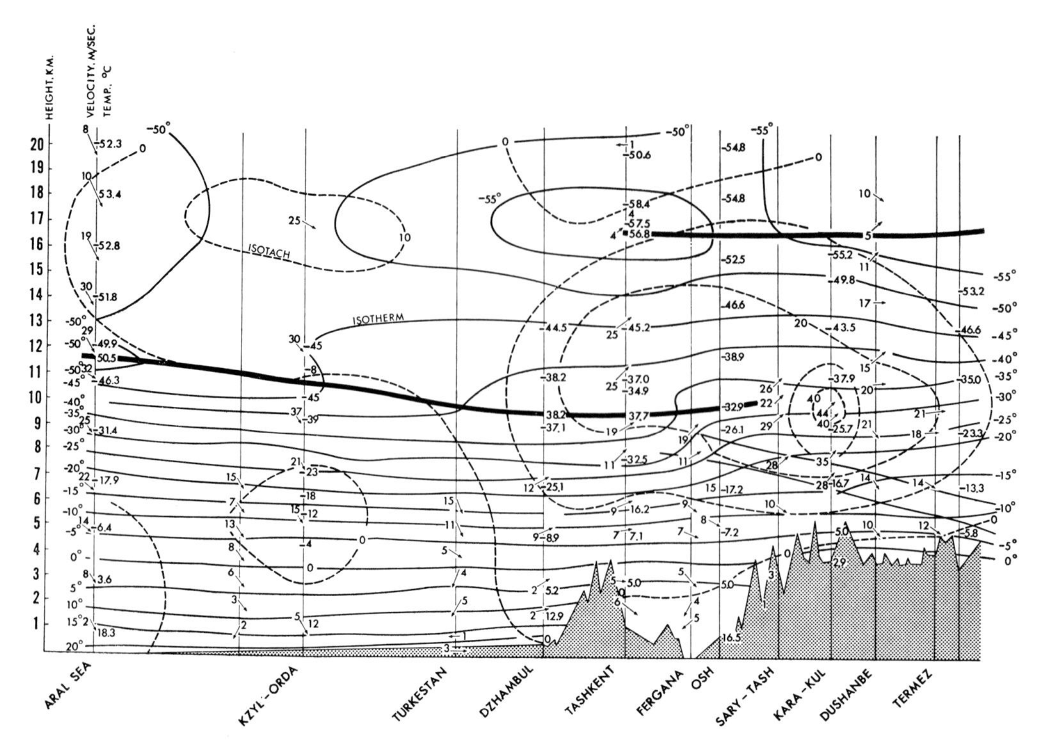

Fig.6-5. Vertical cross-section of the high level planetary frontal zone in Central Asia. (After BUGAEV et al., 1957.)

TABLE 6-I

PERCENT OF DAYS WITH VARIOUS TYPES OF CIRCULATION OVER CENTRAL ASIA DURING THE COLD AND WARM HALF-YEARS
(After BUGAEV et al., 1957)

	November–April	May–October
Cyclonic:		
South Caspian	10.8	4.3
Murgab	8.4	0.7
Upper Amu-Darya	4.1	1.6
Broad-scale intrusion of warm air	1.6	0.6
Cold intrusion:		
Northwest	15.9	22.6
North	8.4	11.1
Wave activity	6.1	1.7
Stationary cyclone in northern Central Asia	2.4	0.6
Periphery of anticyclone	29.6	24.5
Western intrusion	12.7	24.8
Thermal depression		7.5
Total	100	100

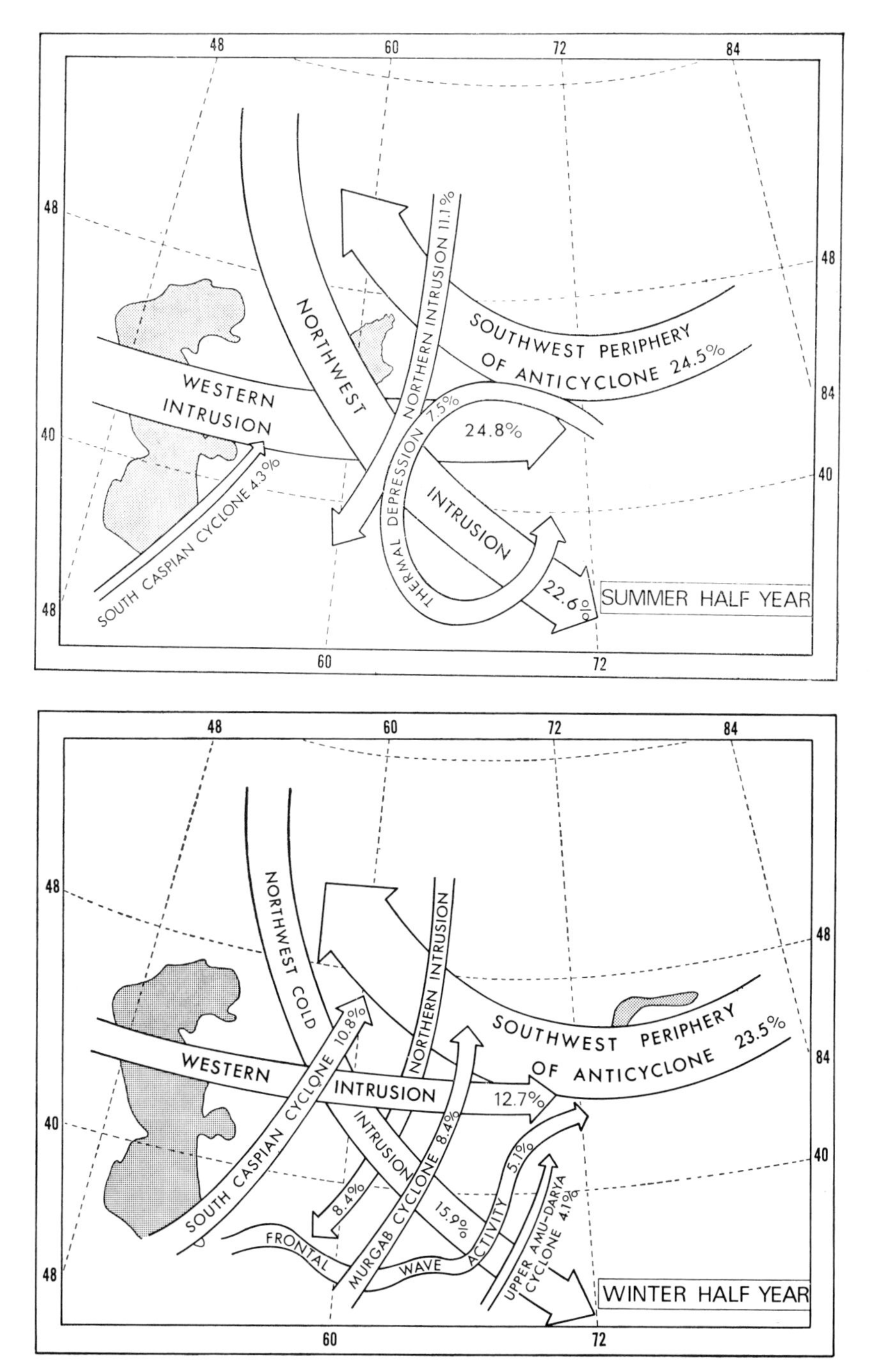

Fig.6-6. Percentage frequencies of types of air flow in Central Asia. (After BUGAEV et al., 1957.)

pressure across Kazakhstan to impede deep southerly penetrations of Arctic and Siberian air. Also in summer the strong heating of the desert surface in Central Asia creates generally favorable conditions for easy intrusions of air from the outside, although distinguishable formations of a thermal depression in Central Asia are relatively infrequent and generally limited to the very southeastern part of the region. Cyclonic storms, of course, are much more frequent in winter than in summer, but even in winter they occupy only 23.3% of the time. So even in winter the weather tends to be rather stable in Central Asia. During summer only a few of the south Caspian cyclones invade the area. None of the Murgab and upper Amu-Darya cyclones are formed at this time of the year. Such a variety of flows of course bring in a great variety of air masses. However, these air masses modify quickly as they cross the dry surface of the desert plains, and they merge with locally derived continental temperate or continental tropical air. The area is most open to the north and west. Air coming in from these directions has generally originated in the Arctic and traveled long distances over land before reaching Central Asia. The absolute humidity of such air is low to begin with and as the air heats in its contact with the surface the relative humidity becomes exceedingly low. Little of the westerly air comes directly from the Atlantic; Central Asia seems to benefit from the Atlantic moisture source about the least of any region in the U.S.S.R. Meridional flows of air are characterized by north-to-south flow at all seasons; southerly flow being largely blocked by the high mountains. The northerly flow is particularly prominent during the transitional seasons between the more stable conditions of winter dominated by the Asiatic high and summer dominated by thermal depressions and the eastern extension of the Azores high.

Most of the meridional flow is instigated by cyclonic circulations or frontal movements of one sort or another. Therefore, there typically occurs a sequence of events with one type of flow following closely upon the heels of another. This is particularly true in spring at the height of cyclonic activity in the southern part of the region. Typically dry tropical air is drawn in across the mountains from the Middle East ahead of cyclonic systems and cold intrusions from the north and northwest are drawn into the rear of the storms.

Generally cyclones proceed rather slowly across Central Asia so that flows persist for the good part of a day or even several days on end. The cold intrusions following cyclonic storms may initiate the beginning of a prolonged stable period of weather during which local anticyclones form over Central Asia, even in the summer time. The mountainous topography along the southern fringe greatly affects the conformation of fronts and induces many local flows of air which are highly significant parts of the climate of specific regions within the area. These will be taken up individually a little later.

Cyclonic storms

On the average about 32 cyclonic storms affect Central Asia each year. Most of these originate southwest, west, or south of the region and travel across it in a northeasterly direction toward the lower Yenisey region in central Siberia. A significant number are generated locally over Central Asia, and some of the storms that enter the region die within the region. The maximum frequency of cyclonic storms occurs in January and March. A slight dip occurs in February when the Asiatic high is at its maximum intensity. The lowest frequency occurs in August, and there is a slight secondary maximum in October.

There are significant distributional differences of cyclonic storms within the Central Asian region. Total cyclone occurrences for the entire year show a maximum zone extending from the southeastern part of the Caspian Sea northeastward across the mid-section of Central Asia (Fig.6-7). The maximum frequency in January lies in an east–west zone across the southern periphery of the region east of the Caspian Sea. As the

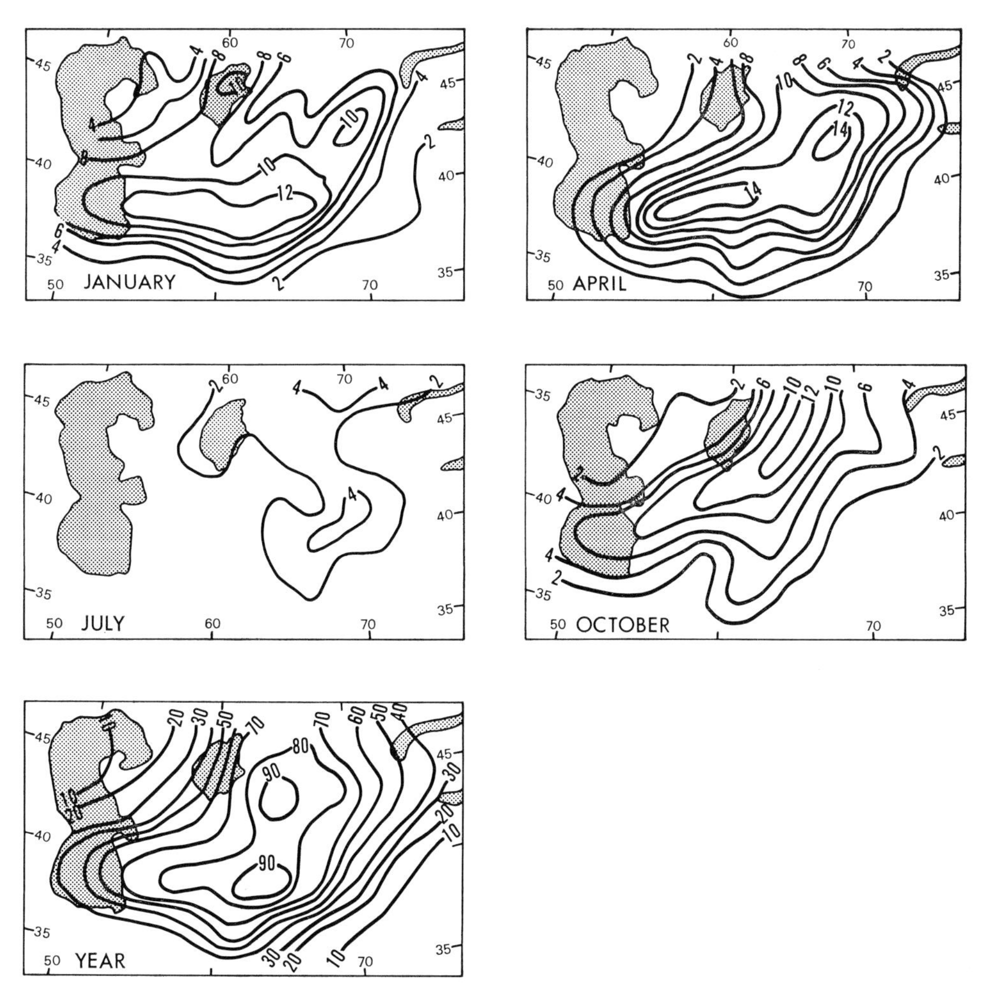

Fig.6-7. Cyclonic frequencies in Central Asia (in days). (After BUGAEV et al., 1957.)

season progresses this zone intensifies and expands northeastward so that by April a maximum frequency per unit area is reached. During that month central Turkmenia experiences cyclonic activity more often than every other day. At the same time, however, cyclone frequencies have diminished perceptibly in the northwest part of the region. The areal distribution in April contrasts sharply with that in October, which is the month of secondary maximum for the entire area. It can be seen that the autumn cyclonic storms represent an intensification of systems that affect the northern portion of the area during summer when the southern periphery is experiencing an almost complete lack of cyclones. While storms are intensifying in the north central portion of the region in October, the mountains to the southeast are experiencing their lowest frequency for the entire year.

The southern part of Central Asia is affected by essentially three types of cyclones, the so-called south Caspian, Murgab, and upper Amu-Darya. The south Caspian cyclones are the most frequent type in winter and the only type that penetrates the southern part of Soviet Central Asia in summer (Fig.6-6). They occur most frequently in January. The cyclones originate in the eastern part of the Mediterranean Sea and find their way into the Caspian where over a period of one or two days they regenerate along southwesterly extensions of cold fronts and then make their way northeastward toward the lower basin of the Yenisey. Ahead of these storms in the mountains of Central Asia are experienced intense developments of foehns and other strong local winds. These generally bring mid-winter thaws and may bring early break-ups of river ice. They may cause melting of the snow and refreezing of the melt water into a hard crust of ice on top of the snow. Freezing rain may also occur ahead of warm fronts. This is generally followed by snowfall and strong winds with drifting snow with passages of cold fronts which are often great hazards to railroads and other lines of transportation and communication. They tend to bring the most intensive snowfalls that are experienced in Central Asia. When there is no snow cover on the ground these cyclones develop into dust storms and sand storms even during winter months. This is all followed by deep intrusions of cold air which rapidly drop temperatures as cold fronts pass.

The Murgab cyclones come out of Iraq and Iran and travel down the basins of the Murgab and Tedzhen rivers. They are characterized by tropical air to great heights, rapidly rising surface temperatures, and the disappearance of snow cover. The northerly advance of the warm air in these cyclones frequently reaches as far north as the Aral Sea or Lake Balkhash. Precipitation along the cold front often is quite intense. Within 12 to 36 h after entering the Soviet Union these cyclones generally occlude in the Kirgiz Mountains or the Lake Balkhash region.

The upper Amu-Darya cyclones are found in the southern part of the Tadzhik Republic. They have the most easterly trajectory, having originated in the southeastern part of Iran or the western parts of Pakistan. They cross the high mountain regions of the Pamir Alay in Afghanistan where they carry tropical air to heights of 2–3 km. This brings on prolonged precipitation in the high mountains while at the same time the western part of Central Asia experiences clear weather and falling temperatures.

A typical sequence of events with an advancing cyclone from the southwest is illustrated by Figs.6-8–6-10. The whole sequence took place during 14–16 December, 1949 when a broadly based upper level trough dominated the region between the eastern end of the Mediterranean and the Pamir-Alay and Tyan Shan massifs (Fig.6-8). Initially a wave developed on the Polar front in southern Iran (Fig.6-9A). This intensified and moved northeastward with the upper level flow until at 15h00 on December 15 it had reached the Gissar Range in the Central Asian mountains where it began to occlude. By 03h00 on December 16 it had completed much of its occlusion process as a new cyclonic storm followed close on its heels from Afghanistan. During the period of its life span, December 14–16, the surface air in different parts of Central Asia followed rather intricate paths depending upon the topography but in general the trajectories described counterclockwise semicircles (Fig.6-10A). The air at the 700-mbar level generally moved from southwest to northeast but was guided to a considerable extent by individual mountain ranges which stuck above this level (Fig.6-10B). Higher up at the 500-mbar level the mountains did not obstruct the flow perceptibly as it described a smooth curve around the southeastern edge of the upper level trough (Fig.6-10C).

The higher ranges of the Pamir-Alay and Tyan Shan Mountains rise to more than 6,000 m and form formidable barriers to air flow. They therefore have profound influences on frontal configurations which generally separate flows of air that are no more than 2 or 3 km thick. This is particularly true of cold air masses which move in from the north or northwest and lodge against the foothills of the northwestern slopes of the mountains. In general the mountain ranges are oriented east–west and encompass

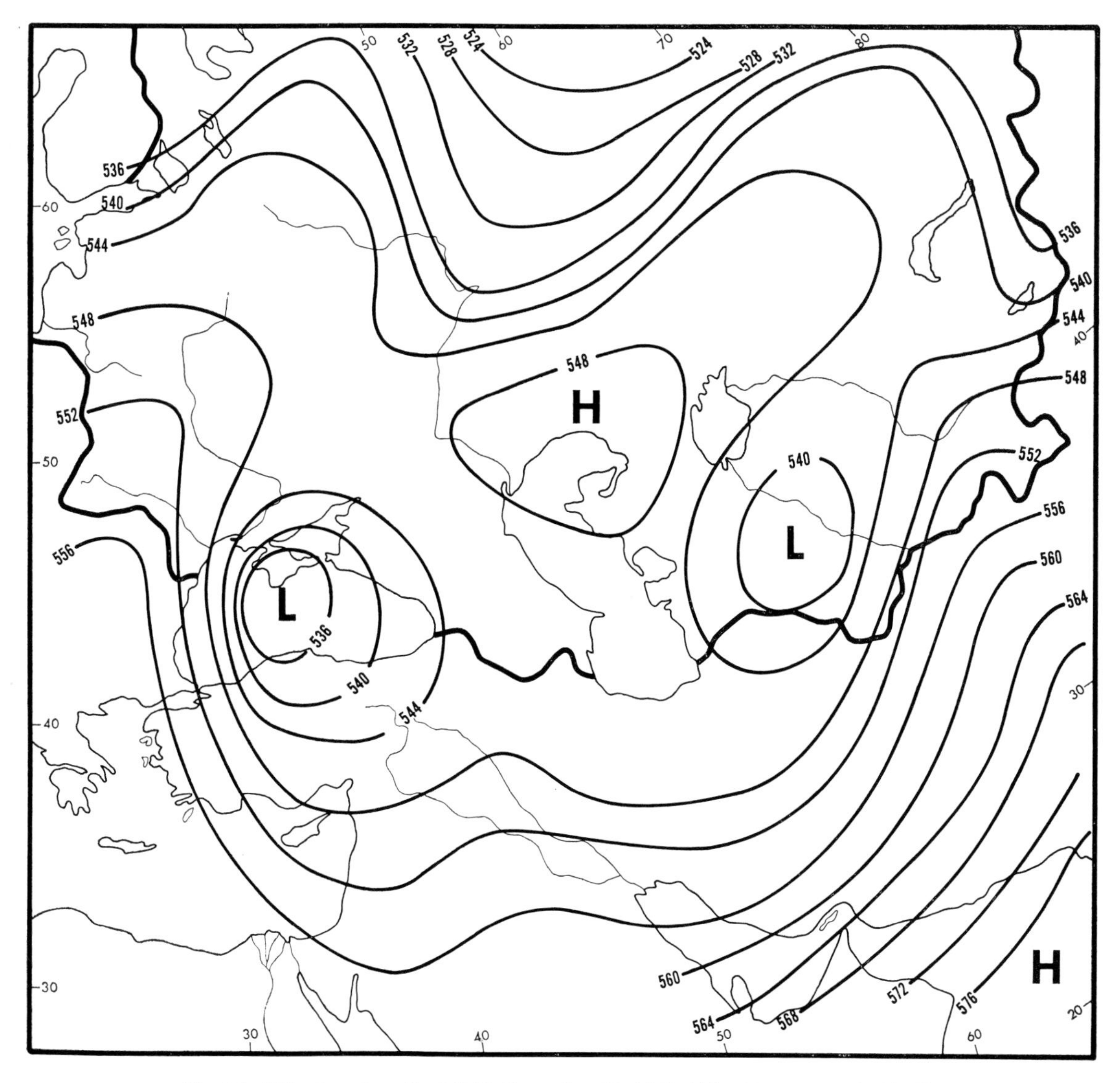

Fig.6-8. 500-mbar chart for 06h00, December 15, 1949. (After BUGAEV et al., 1957.)

rather broad basins which open toward the west. Chief among these are the upper Amu-Darya Basin in northern Afghanistan and the southern portions of the Uzbek and Tadzhik republics and the Fergana Basin farther north that lies along the upper Syr-Darya in the eastern extension of the Uzbek and surrounding republics. Frequently when an extensive frontal system lies along the outer reaches of these mountain ranges waves develop along the front whose positions are strongly influenced by orographic features, and as these waves progress along the front they tend to become occluded and this process also is strongly influenced by orography.

Fig.6-11 is a typical example of a wave formation in the Murgab and Tedzhen river basins along a front that lies parallel to the Kopet Dag and Hindu Kush mountain ranges. A strong push of warm air from Afghanistan and Iran is interacting with colder air to the north of the mountains. This sort of situation is conducive to the development of the so-called Murgab cyclone.

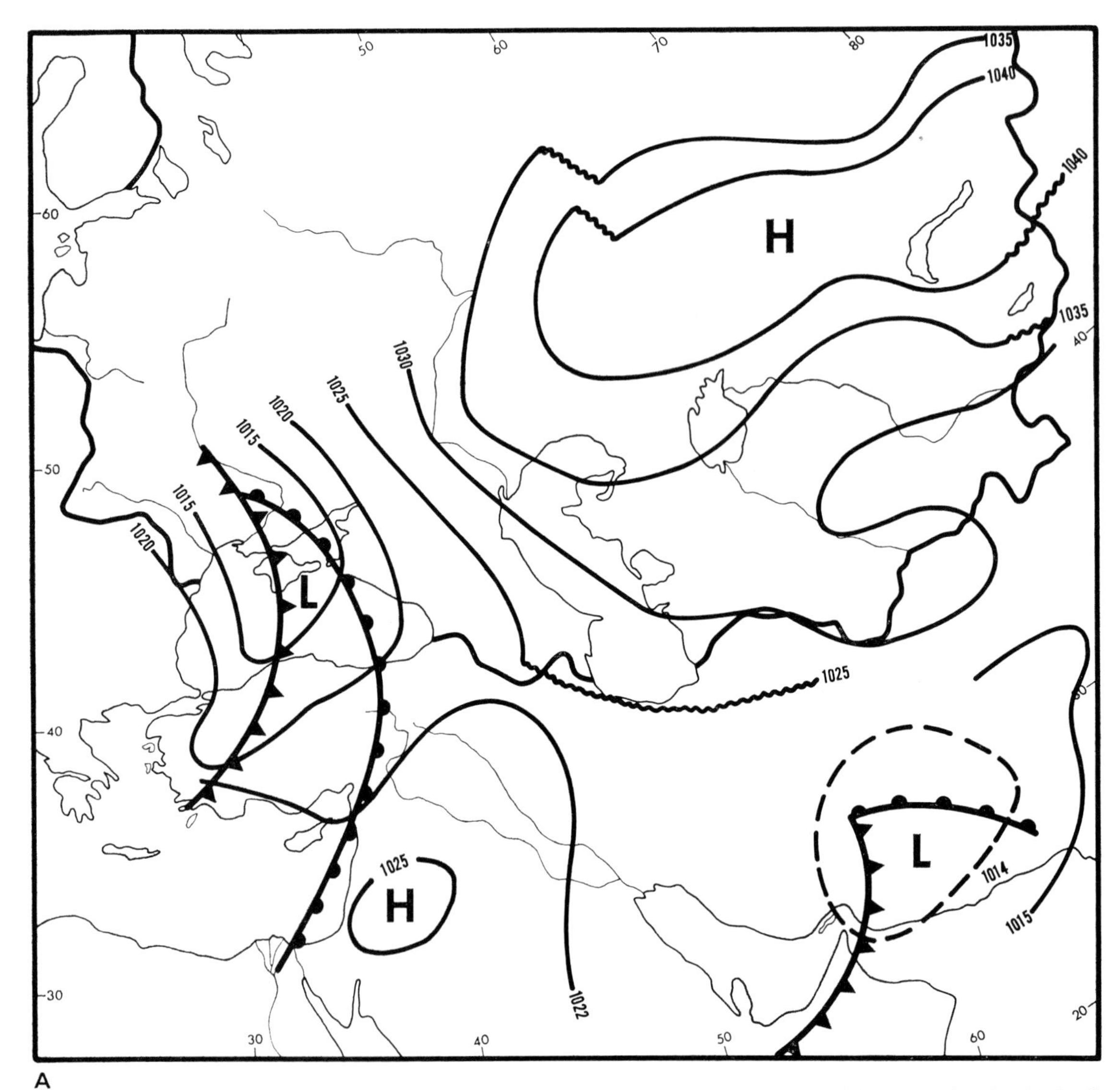

Fig.6-9. Surface weather maps: A. for 15h00, December 14, 1949; B. 03h00, December 15, 1949; C 15h00, December 15, 1949; D. 03h00, December 16, 1949. (After BUGAEV et al., 1957.)

Fig.6-12 is a typical case of wave formation along the upper Amu-Darya along a meridionally oriented front that has moved in from the west. Fig.6-13 shows a sequence of events that typically takes place when there is a strong push of tropical Iranian air northward into the upper Amu-Darya region. The progression northward of the front is obstructed greatly by the Gissar and associated ranges while segments of the front advance rapidly into the upper Amu-Darya regions to the south of the range and into the Fergana Basin to the north of the range. A somewhat similar situation is illustrated by Fig.6-14 except here the push of air is from the west behind a meridionally oriented cold front. As the front advances eastward it quickly sweeps across the Fergana Valley and the upper Amu-Darya Basin but again is retarded along the Gissar Range and the ranges north of the Fergana Valley. Such a situation generally leads to occlusion processes the details of which are shown in Fig.6-15 where an occlusion takes place across the Gissar Range. Similar processes may take place farther north although the mountains diminish

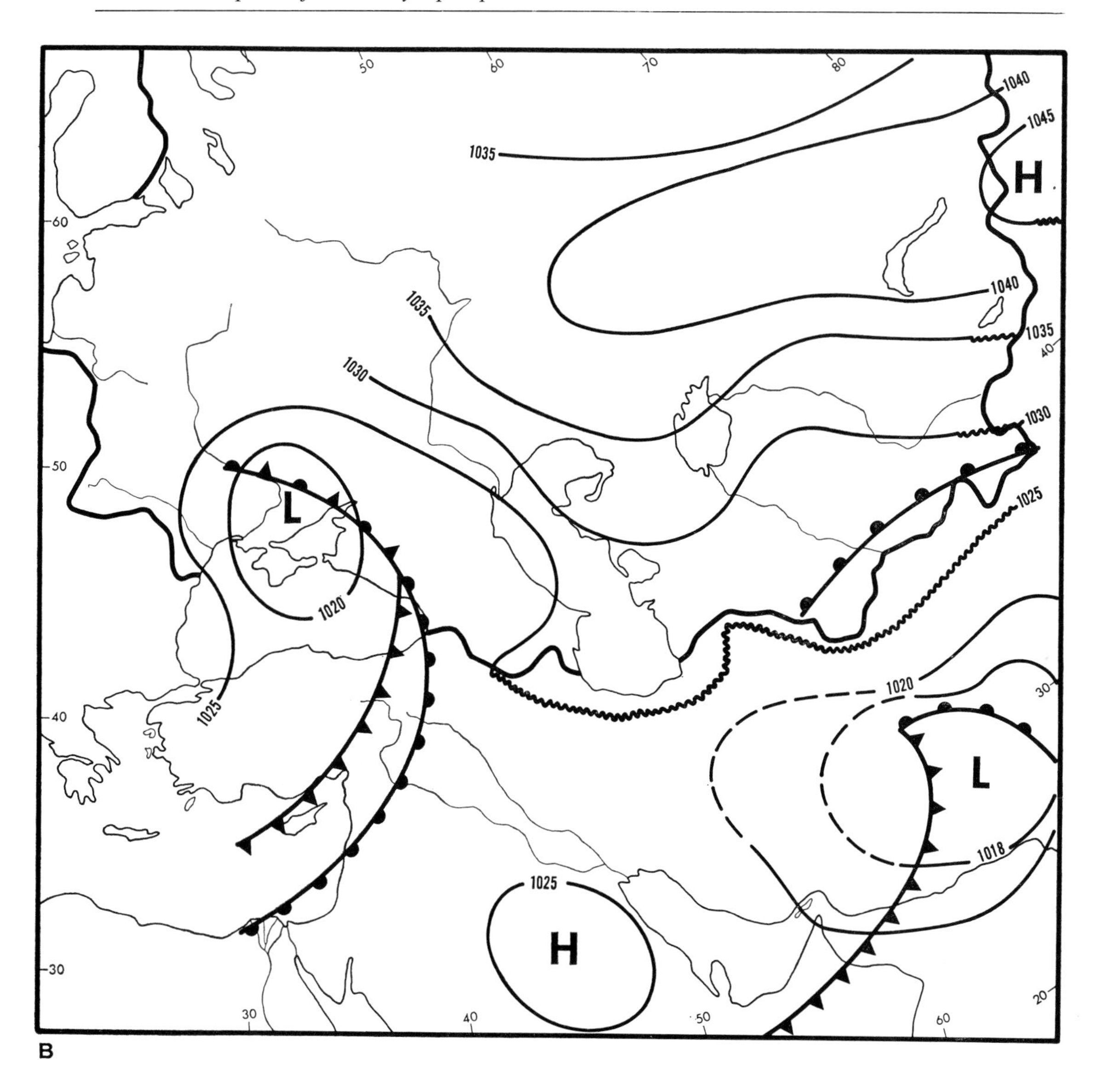

in height and intervening valleys are not so well pronounced. But even in northern Kazakhstan the so-called Folded Country with elevations generally between 1 and 2 km may exert important influences which bring on wave formation along cold fronts as they cross the region.

Local winds

The high values of radiational components induced by the dry atmosphere and desert surface of Central Asia bring about a great deal of turbulence in the lower troposphere and are conducive to surface air circulation which generally is stronger than those in more humid areas. The proximity to the desert plains of high mountains and large water bodies accentuates thermal and dynamic differences and tend to increase wind speeds and produce characteristic winds of local origin which are known by particular

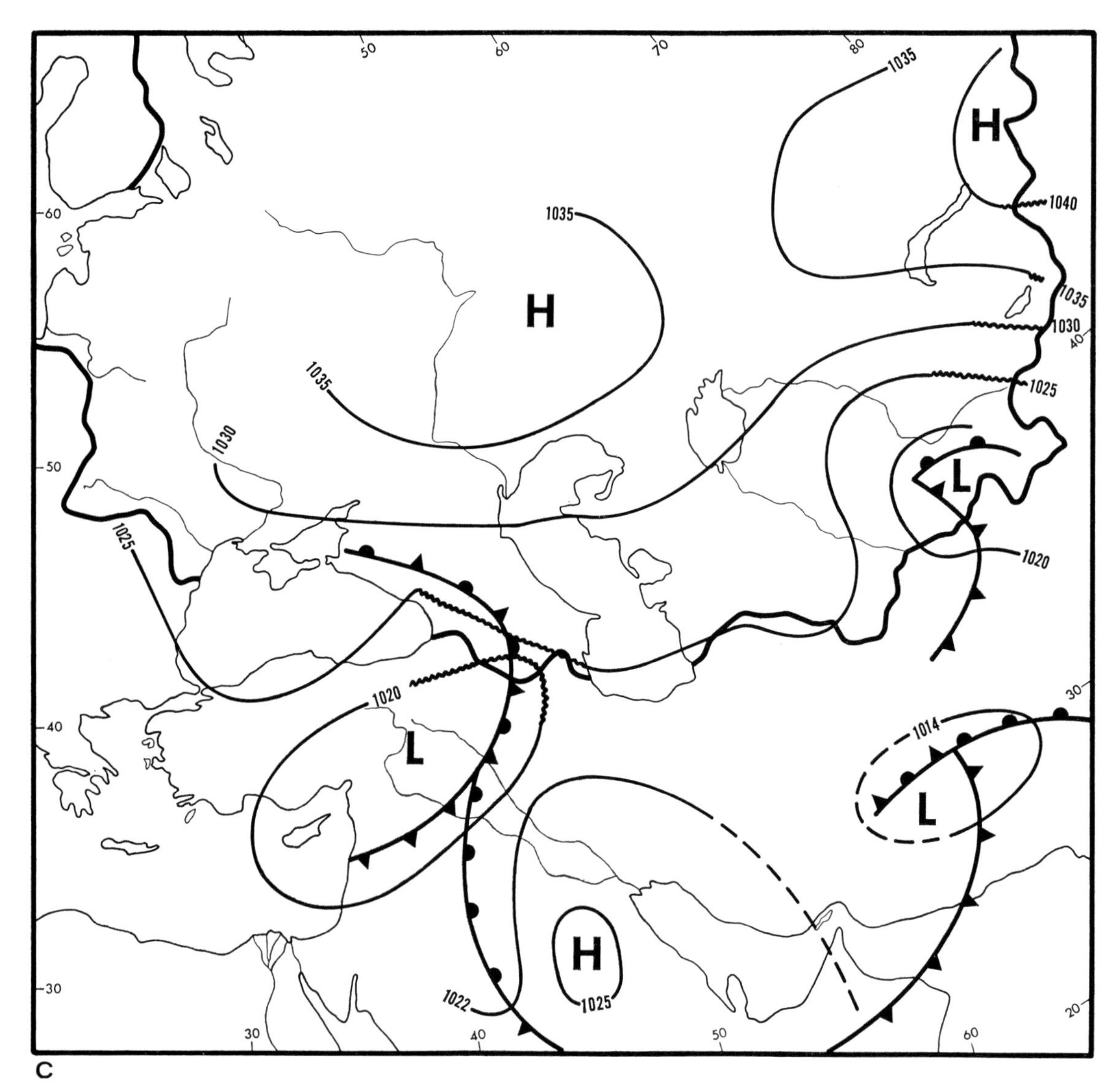

Fig.6-9. C, D (legend on p. 162)

names. Around the three large water bodies, the Caspian Sea, the Aral Sea, and Lake Balkhash, as well as around some of the smaller lakes, such as Issyk Kul', lake breezes develop, particularly in summer, which affect the immediate shores.

There is a definite seasonal reversal of winds over the Caspian, from southeast in winter to northwest in summer (Figs.6-2, 6-3). This is not primarily a function of land–water differences, but is due to changes in the primary controls of circulation during the two seasons, the easterly winds in winter being a product of the southwestern periphery of the Asiatic high and the northwesterly winds in summer a product of the eastern periphery of the Azores high. Nevertheless, a vertical cellular circulation tends to be set up across the east coast of the Caspian during much of the time. At Fort Shevchenko on the western promontory of the Mangyshlak Peninsula during January winds from the surface up to about 2 km in the air are generally from the southeast while those above 2 km are from the west. This is true from October through April. From June through August

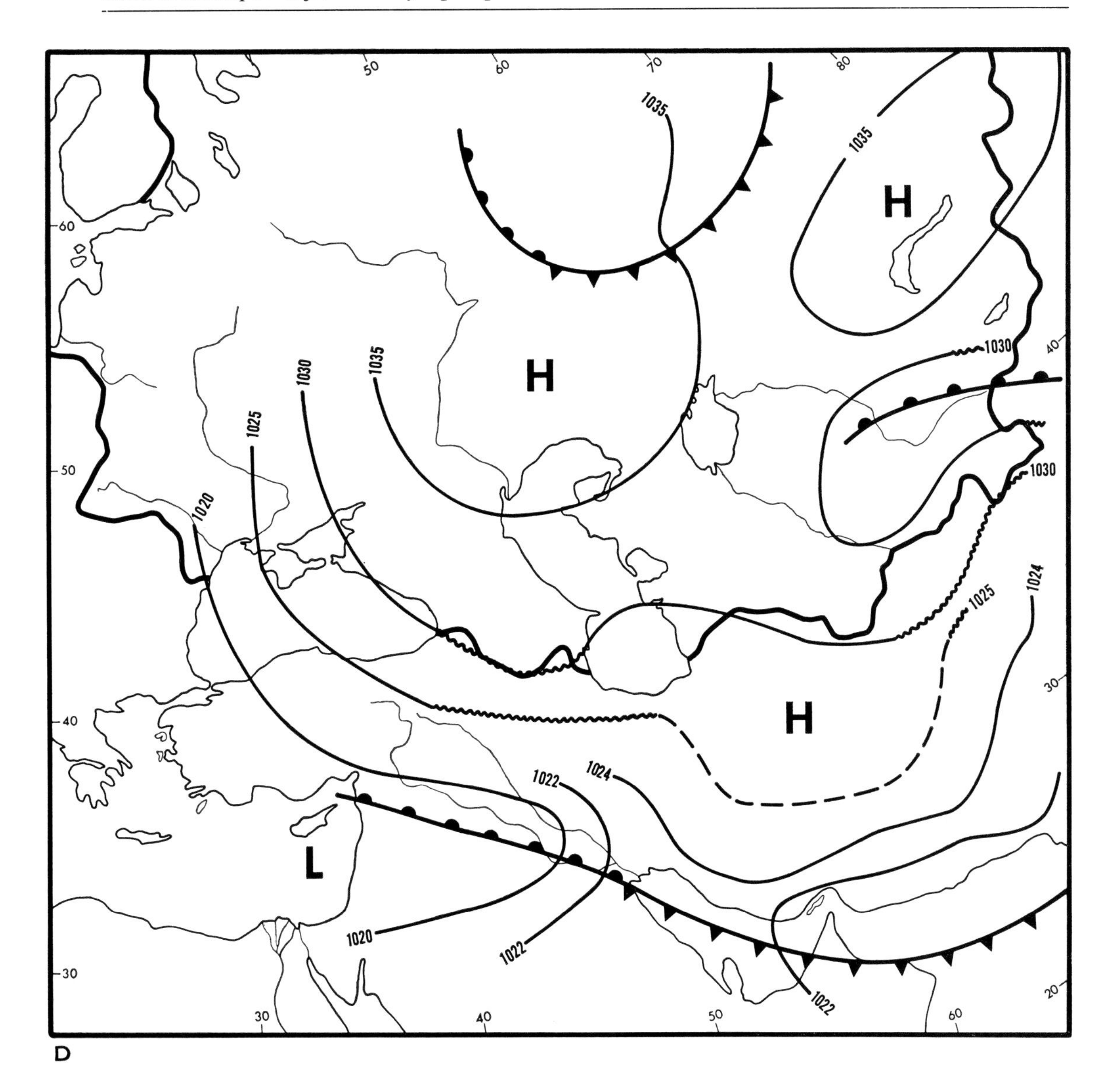

D

surface winds up to about 1 km are generally from the north or northwest and winds above 1 km are from the southeast. In August the surface wind flow from the north is very shallow, often being no more than 100 m in depth. (CHELPANOVA, 1963, pp.246–247.) The greatest frequency of storm winds recorded anywhere on the plains occurs at Fort Shevchenko where on the average 84 days per year experience wind speeds of more than 15 m/sec. This high frequency of storm winds exceeds even that at Novorossiysk on the Black Sea coast which is known for its bora winds.

Many of the stronger winds of the desert lowlands are associated with vigorous intrusions of colder air from the west and northwest behind rapidly moving cold fronts that generally bring little precipitation to Central Asia but often cause hazardous dust storms. Drifting sand is a potent threat to man-made constructions wherever they have been built adjacent to the open desert, and the fine dust rises into the air often to 3 km or higher where it remains for days as it drifts generally southward eventually to lodge

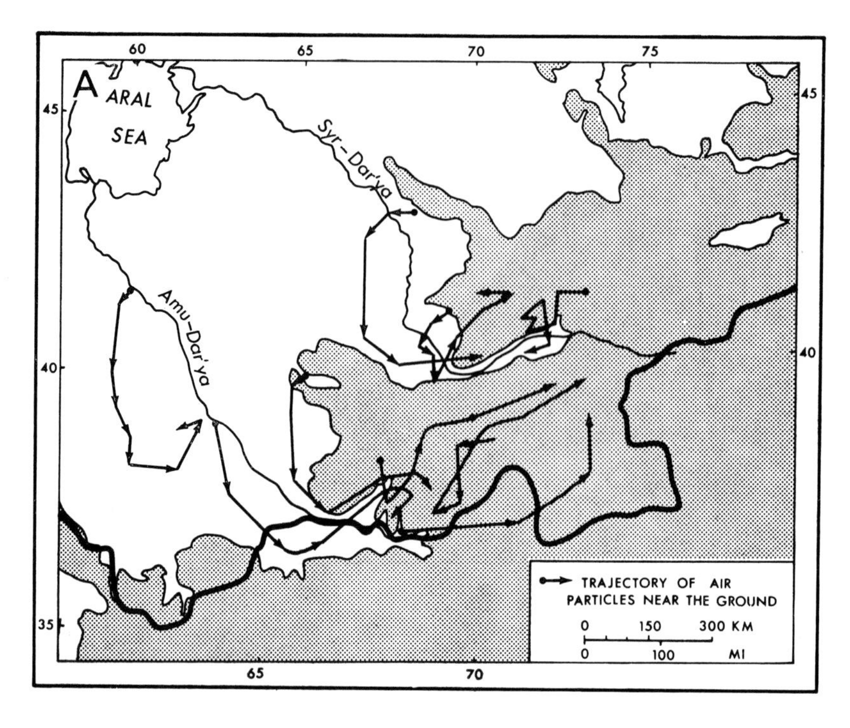

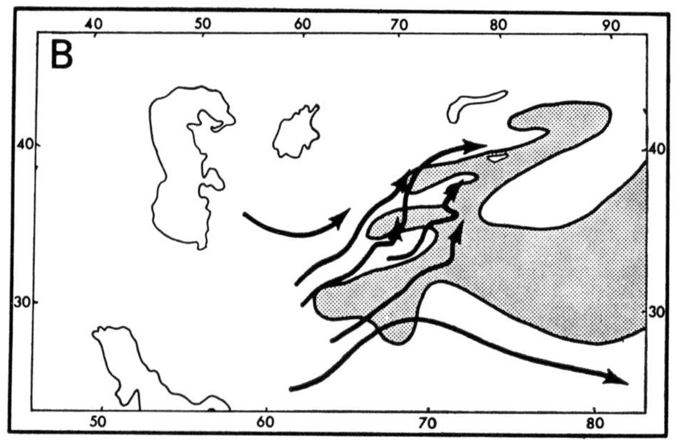

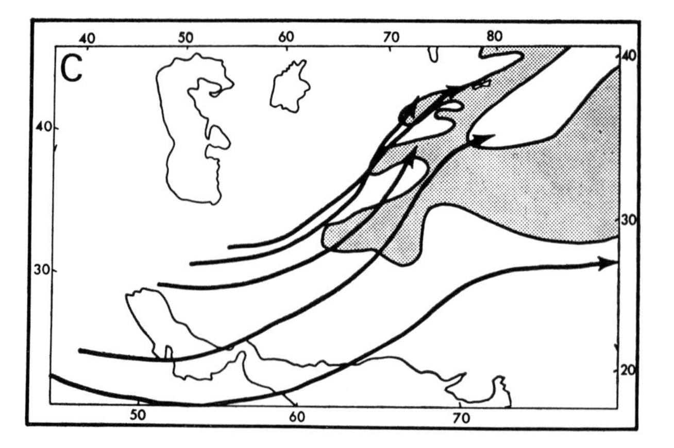

Fig.6-10. Air trajectories, December 14–16, 1949; A. at the surface; B. 700 mbar; C. 500 mbar. Shaded areas are mountains. (After BUGAEV et al., 1957.)

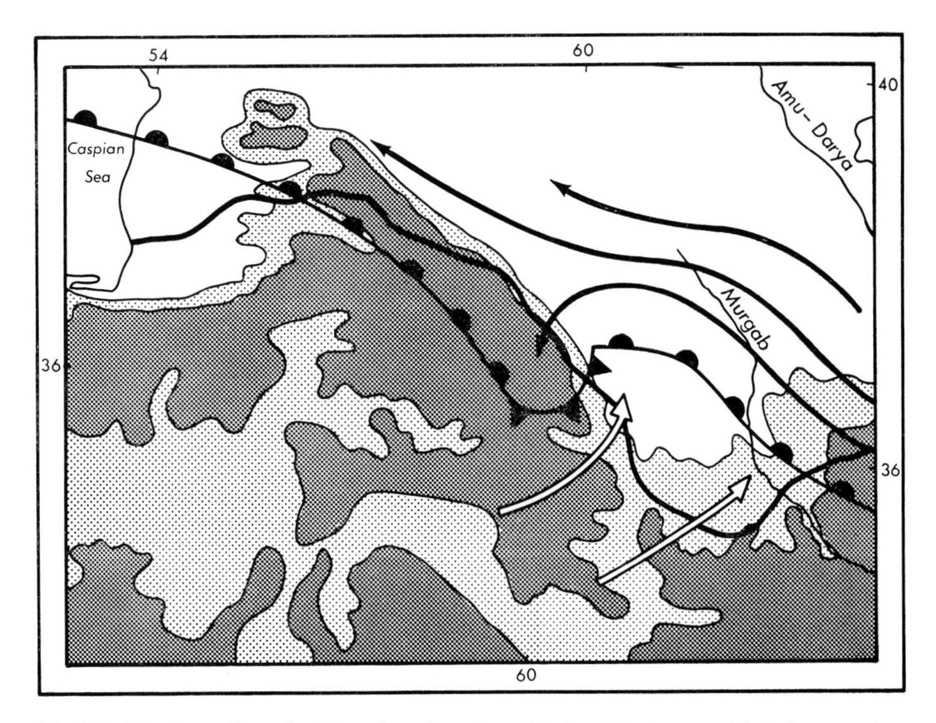

Fig.6-11. The formation of a Murgab cyclone brought about by topographic influences along a warm front advancing northward from Iran and Afghanistan. Map for 13h00, December 26, 1937. White space = plains; light stippling = foothills; heavy stippling = mountains. (After YAROSLAVTSEV, 1939a).

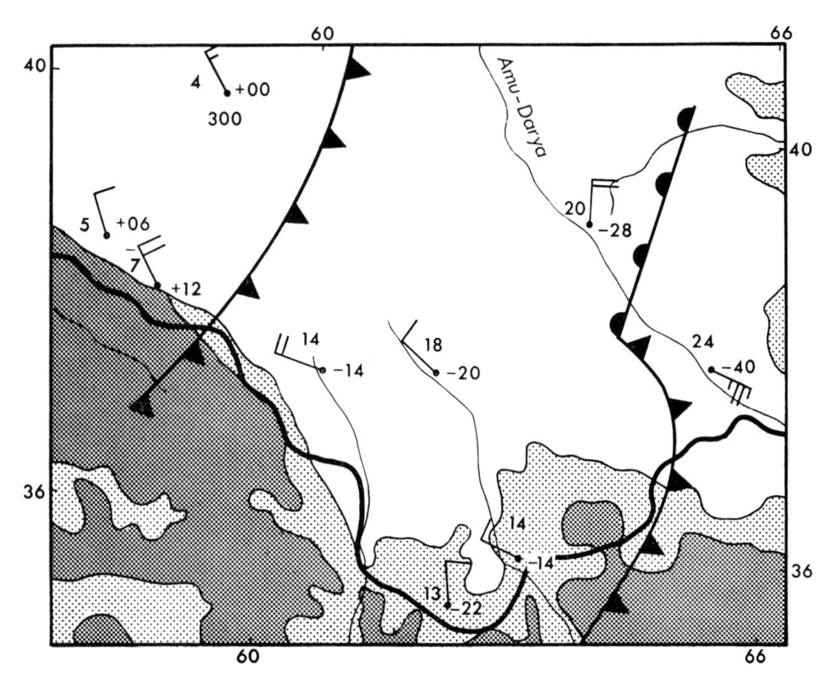

Fig.6-12. The formation of an upper Amu-Darya cyclone along a meridionally oriented cold front. Map for 13h00, March 15, 1938. Further explanation in Fig.6-11. (After YAROSLAVTSEV, 1939a.)

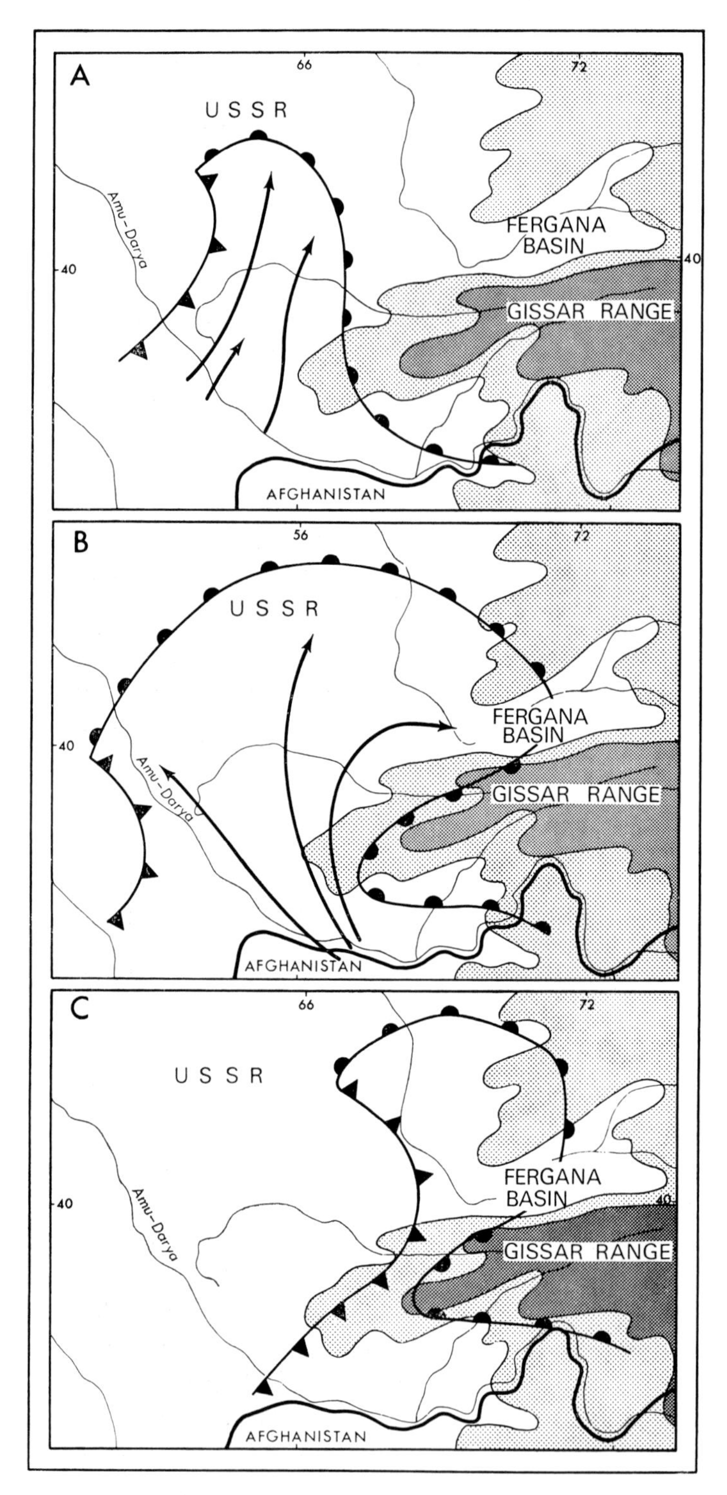

Fig.6-13. Topographic influences on a warm front advancing northward from Afghanistan. Maps for: A. 13h00, January 8, 1938; B. 19h00, January 9, 1938; C. 01h00, January 10, 1938. Further explanation in Fig.6-11. (After YAROSLAVTSEV, 1939a.)

against the foothills of the southern mountains. The atmosphere of Central Asia typically has a hazy, whitish hue because of the great amount of dust in the air. The deposits of dust along the northern foothills of the southern mountains have developed a thick deposit of fertile, friable soil, known as loess, which induced riverine settlements based on irrigation agriculture as early as 4,000 years ago and today sustains the important irrigated cotton agriculture of the area.

A characteristic post cold frontal wind is the Afghanets, a strong dry wind generally blowing from a westerly direction which produces huge dust storms and affects the southeastern part of Central Asia, particularly the upper reaches of the Amu-Darya after

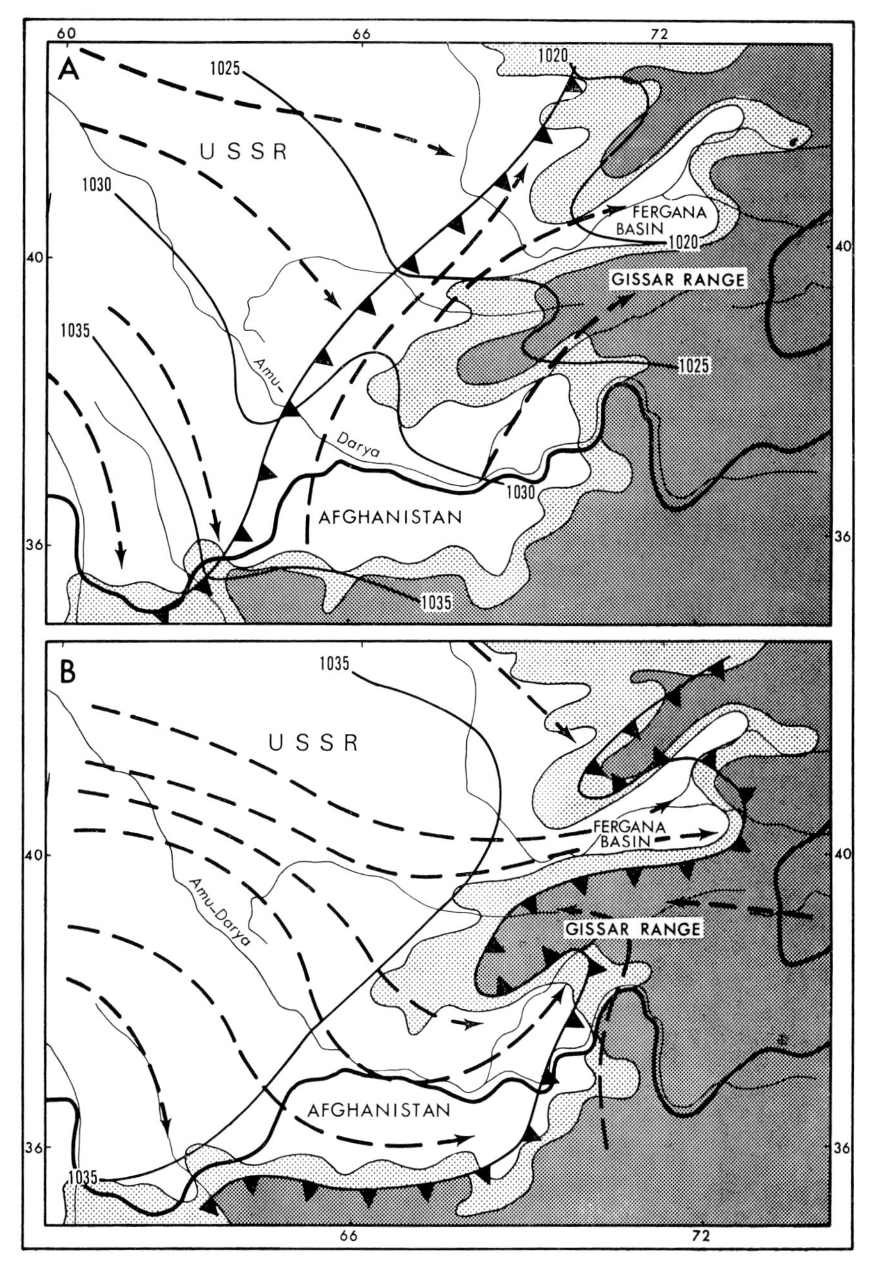

Fig.6-14. Topographic influences on an eastward advancing cold front. Maps for: A. 07h00, November 7, 1937; B. 13h00, November 7, 1937. Further explanation in Fig.6-11. (After YAROSLAVTSEV, 1939a.)

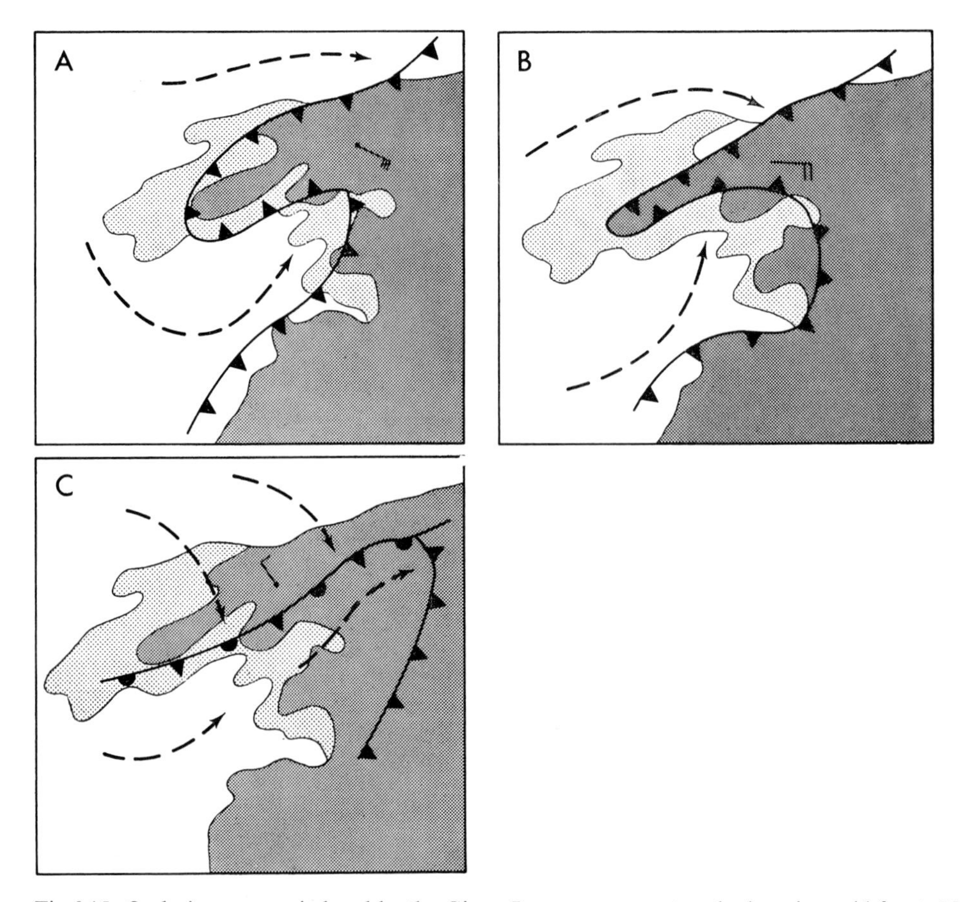

Fig.6-15. Occlusion process induced by the Gissar Range on an eastward advancing cold front. Maps for: A. 13h00, December 4, 1937; B. 19h00, December 4, 1937; C. 01h00, December 5, 1937. Further explanation in Fig.6-11. (After YAROSLAVTSEV, 1939a.)

crossing the Kara Kum. At times these winds penetrate up the Amu-Darya and its head water stream, the Vakhsh River, sometimes even into the uplands of the Tyan Shans. They also effect the Syr-Darya area, particularly the Hungry Steppe and the Fergana Basin. Their vertical extent may occasionally reach up to 3,500 m. Dust has been carried to that height and has been known to take 10 days to settle after the winds died down. In extreme cases the winds may last for 5 days. In September, 1930 strong winds reached upwards to 6 km, where the winds were strong anyway, and the turbulence was a great threat to aircraft. Electrical phenomena developed in the air due to static electricity that was built up by friction on dust particles. Such winds are known by many local names, such as the chang or the kara-buran. In all cases they are associated with intensive cold outbreaks from the north or northwest. Their effects are most felt in summer when they combine with the natural effects of the hot desert air to produce unusual turbulence and extreme dust conditions.

Dust storms tend to be most frequent and most severe in the southern part of the Turkmen Republic. Repetek averages 69 days with dust storms per year and has recorded as many as 105. Near the mouth of the Amu-Darya, Nukus averages 34 days per year with a maximum of 48. Tashkent on the other hand averages only 4 days per year with a maximum of 11, and Dushanbe in the southwestern part of the Tadzhik Republic averages 6 per year with a maximum of 12. The dry upper portions of the Pamir-Alay

Ranges experience continuously strong winds and frequent dust storms. Here dust storms can be expected to occur between 15 and 20 days per year on the average and as much as 40 days per year occasionally. (BALASHOVA et al., 1960, 215.)
The winds associated with frontal systems are intensified as the systems approach the mountains and the air flow becomes constricted between the fronts and the mountain slopes. In addition the mountains provide high altitude surfaces against which radiational interaction can take place to produce temperatures that are in great contrast to those existing at similar altitudes in the free atmosphere over adjacent plains. This sets up so-called compensatory winds, mountain and valley regimes, and so forth which frequently override the general circulation of the area and become the predominating winds. Some of these show a diurnal cycle, while others blow in one direction much of the time. On the higher mountain peaks where snow fields and glaciers exist year round, katabatic winds generally descend the valley heads all of the time.
Foehn winds are widespread throughout the mountains of Central Asia. Depending upon the general circulation over the region they can occur on almost any slope. Generally the strongest winds are experienced on northerly slopes in association with southerly and southeasterly winds ahead of cyclonic storms. But their aridifying effects due to rapidly rising temperatures and rapidly falling relative humidities may be felt more strongly on southern slopes. On occasions when the mountain massif is occupied by high pressure foehns may descend all mountain slopes in a generally subsiding fashion. Their effects then usually are not as intense as they are when a steady circulation from one direction forces the air to rise and precipitate moisture on windward slopes, thus releasing latent heat, and then warming adiabatically as they descend leeward slopes.
A different type of foehn may occur along the lower mountain slopes during winter when anticyclonic weather dominates the mountain region and strong temperature inversions are associated with weak pressure gradients and weak winds at intermediate elevations. Mountain breezes tend to develop during the evening and night hours that carry down warmer weather from the inversion layer to the underlying foothills.
Since most foehns occur ahead of cyclonic storms they are most frequent during winter. They generally occur from a southerly direction well ahead of frontal systems. During winter they bring thawing of the snow and even sublimation directly from solid to vapor. There is a saying in the Tyan Shans that "two days of foehns are worth two weeks of sunshine". This signifies the favorable effects that foehns have on life in general during the winter. Foehns in the Tyan Shans keep cattle pastures open year round. There are such favorable conditions for foehns on the northern slopes of the Kirgiz Range that at a height of about 3 km the snow cover during winter is very little and in some years lies only in patches, while at the same time on the southern slopes of the same range the snowcover may be 2 m or more thick at the same elevation. Relative humidities during a foehn may drop from 70–90% down to 5–20%. Temperatures may rise in January up to 20°C. In Turkmenia winter foehns have raised the temperature 25°C. Foehn winds at Ashkhabad on January 28, 1961 had speeds of 28 m/sec and temperatures of 21.5°C.
During summer foehns are very weak. Even if the air is descending and heating adiabatically, the descending air generally is no warmer than the air in the foothills which are receiving great amounts of radiation at this time of the year. Only in rare cases during summer when a cold front comes into close proximity with the mountains and the descent of air is very rapid is one able to discern a foehn condition clearly.

In several locations foehn winds reach unusual velocities and associated aridifying effects due to constrictions in air flow through mountain passes or around mountain edges. These outstanding winds generally are known by local names. One such wind is the "harmsil". This is a dry hot wind which generally blows from the east or southeast, often in the warm sector of a cyclonic storm moving in from the southwest. The air is constricted between the cold front on the west and the mountain slopes on the east and thus increases its speed as the cyclonic storm approaches the mountains. The southeasterly flow takes air away from western mountain slopes and causes descending air motion in the free atmosphere to compensate for the loss of air along the foothills. High wind speeds, low relative humidities, high temperatures, and great amounts of dust are characteristic. Sometimes even the absolute humidity drops. Harmsil winds in Tadzhikistan have raised temperatures to 47.8°C and dropped relative humidities to 8% with wind speeds of 6 m/sec. Under such conditions trees drop their leaves and fruit.

The harmsil is really a very strong foehn which occurs day and night. It occurs primarily in the southern mountain system, but it can also occur in the northern part of the Kirgiz Republic. The winds are very desiccating to vegetation. They may lower yields of cotton by 20 to 50% after blowing only a few hours. Also hazardous to vegetation is the rapid drop of temperature after the cold front passes which is a strong contrast that follows so rapidly on the heels of the very hot dry air ahead of the front. These winds can occur anytime of year but are generally most prevalent in summer and their effects are most damaging at that time of year.

Another foehn-type wind which is well known throughout Central Asia is the so-called "ursatevskiy". This is an east to southeast wind that blows with great force through the narrow western throat of the Fergana Basin onto the flat plain of the Hungry Steppe to the west. It takes its name from the railroad junction of Ursat'evskaya which lies in the southeastern portion of the Hungry Steppe usually directly in line with the winds issuing from the Fergana Basin. The winds occur almost always along the leading edge of the northeastern quadrant of a south Caspian or Murgab cyclone which is approaching the area from southwest and along the southwestern periphery of the Asiatic high which lies to the northeast. This general pressure pattern sets up the easterly flow of air which collects in the Fergana Basin and then funnels westward through the narrow neck at the western end of the basin. The foehn-like nature of the wind may be augmented by local winds which develop within the Fergana Basin due to differences in heating between the basin floor and mountain slopes.

Observations made at Ursat'evskaya over a 7-year period recorded 189 cases of such winds when speeds at the surface reached at least 12 m/sec. (CHANYSHEVA, 1966, pp. 89–99.) These 189 cases covered a total of 483 days for an average of 69 days per year with strong winds. Generally such conditions last the major part of a day or two or three days, but one such condition during the 7-year period lasted for 7 consecutive days. Since the winds are associated with strong north–south pressure gradients generated by the juxtaposition of the Siberian high and cyclonic storms, they are primarily winter and spring phenomena. During these 7 years of study January had the greatest frequency, with 29 occurrences in 7 years. July, August, and September each had only 1 or 2 cases during this 7-year period. During the 7-year period February had 6 cases with speeds of more than 30 m/sec, January 4 such cases, December 4 such cases, March 3 cases, and April 1 case. The rest of the months never experienced such high wind speeds. Thus, during the 7 years of study there were 18 periods when winds reached velocities of more than 30 m/sec. 43 cases recorded wind speeds between 21 and 29 m/sec, 75 cases recorded speeds between 16 and 20 m/sec, and 53 cases recorded between 12 and 15 m/sec. The strongest winds generally blow during the first half of the day with maximum speeds reached around 07h00. There generally is a conspicuous weakening of winds as evening approaches.

Maximum speeds of the ursatevskiy winds usually do not occur at Ursat'evskaya, but further east at Bekabad in the constricted Leninabad Pass at the western end of the Fergana Basin. They generally decrease in all directions from this spot, but decline most rapidly eastward. They seldom extend as far east in the Fergana Basin as the city of Fergana, although they may reach speeds of 7–10 m/sec at heights of 50 m above the ground over Fergana. Generally highest wind speeds are not found anywhere at ground level but somewhere above the ground. And generally these maximum wind speeds in the free atmosphere are observed during March. The lower limits of the high wind speeds generally start somewhere between ground level and 200 m above the ground and extend up to 1,500 m or so above the ground. The height of the upper limit of the winds varies considerably and has been observed as high as 6,400 m above the ground. The greatest wind speed recorded was 60 m/sec at a height of 1,360 m above ground over Bekabad. The next highest wind speed was 52 m/sec at only 240 m above ground over Ursat'evskaya. Although on occasion the ursatevskiy wind has been characterized as a cold katabatic airstream descending from the Fergana Basin to the Hungry Steppe, in fact there is usually a considerable rise of temperature accompanying the winds. Sometimes it is not very significant, but on occasion it has amounted to as much as 19°C with a rise of 12.5°C in a 24-h period. The city of Tashkent seems to experience the highest rises of temperature during such occasions in spite of the fact that Tashkent does not lie in the path of these winds. However, foehn winds of a less spectacular nature blow southwestward down the Chirchik River into Tashkent whenever the pressure pattern is such that the ursatevskiy winds are blowing in the south. Ursatevskiy-type winds blow from northeast to southwest down the Chirchik Valley and cause the trees to lean southwestward along the river. Such winds may blow from 1 to 4 days and cause serious sandstorms in the Chirchik Valley. On February 27, 1940 such a wind raised the temperature to 24.4°C and dropped the relative humidity to 19%. Similar conditions are common in the Talas Valley farther northeast between the Kirgiz Range on the north and the Talas Range on the south. When a strong easterly flow is set up air moves down the Talas Valley to the west and brings rapidly rising temperatures to the pass through which the railroad runs between Chimkent and Dzhambul.

In spite of the fact that the most spectacular winds in Leninabad Pass blow westward out of the Fergana Basin toward the Hungry Steppe, it is more common throughout the year for this pass to experience winds blowing from west to east. This is particularly true during the summer half year from March through October when fairly strong west winds generally blow during daytime hours along a line from Leninabad to Kokand. This is something of a valley breeze blowing upslope from the Hungry Steppe. It is known as the "kokandets". The prevailing nature of westerly winds in this area is attested by the many windbreaks of poplar trees which all lean toward the east and which all have very few branches on their western sides.

During winter also strong westerly winds may blow into the Fergana Basin after an occurrence of ursatevskiy winds when the cyclonic storm has moved on northeastward and a strong outbreak of cold Siberian air is pulled into its rearward portion. The typical sequence of events during winter with the passage of a cyclonic storm from southwest to northeast along the western edge of the Fergana Basin is initially strong easterly winds associated with high temperatures and low relative humidities abruptly followed by less strong but still gusty cold winds that blow from west to east into the basin.

Northeastward along the northern foothills of the Tyan Shans strong winds of foehn character blow wherever deep narrow river valleys lead down the northern slopes to the plains below. Such winds are particularly strong and frequent at the mouths of canyons of some of the headwater streams of the Chu and Ili rivers. They are particularly well known in the Kurdayskom Pass between Alma Ata and Frunze where the winds are known as the "kastek" and in the lower course of the Chilik River at the town of Chilik south of the junction of the Chilik River with the Ili. During winter the Chilik experiences storm winds every 3 or 4 days. During the summer these diminish to only 4 or 5 days per month. During the course of a year Chilik experiences 79 days with storms. Farther downstream the city of Iliysk at the point where the Ili River cuts through some low mountains experiences storm winds 32 days per year.

By far the most significant and most well known winds in Central Asia are those that blow from either direction through the Dzhungarian Gate on the Soviet–Chinese border east of Lake Balkhash between Lake Alakol' on the northwest and Lake Ebi-Nur on the southeast in China. The Dzhungarian Gate is only 10 km wide at its narrowest near Lake Zhalanashkol. Zhalanashkol has strong winds 100 days per year with maximum velocities of 70 m/sec. January and December have strong winds about 18 days per month. After temperatures have been 20°–30°C below zero thaws occur and clouds dissipate.

The southeasterly winds are the best known since they create intense, broad-scale foehn conditions on the Soviet side of the border as they approach Lake Alakol'. These are known as the "ibe" because they blow out of the Lake Ebi-Nur region. The local inhabitants in the Alakol'skoy Valley call these storm winds the "evgey". The winds that blow from the northwest across Lake Alakol' into the Dzhungarian Gate are known as the "saykan".

During the late 1950's and early 1960's when the railroad was being laid through the pass to the Chinese border weather records were kept that show that the ibe frequently reached speeds of up to 60 m/sec and on one occasion the anemometer at Surazhskogo registered a wind of 80 m/sec. Wind blows there as if it were blowing though a wind tunnel. The meterological station had to be built to withstand such high winds, and the observation tower had to be connected to the station with a cable so that it wouldn't blow down. The speed of the winds is such that not only snow and sand but also light gravel is lifted to considerable heights in the air. Visibility is reduced to less than 5 m. (CHELPANOVA,1963, p.262.)

Cold bora-type winds are characteristic of either end of the Issyk-Kul' Basin in the eastern Tyan Shans. This oval shaped mountain basin elongated east–west parallel to the mountain ranges lies at an elevation of 1,609 m. It is completely surrounded by mountains which on the north and south rise to more than 4,500 m. Eastern and western ends of the basin rim are a little lower, and when a strong push of cold air enters the mountain region from the northwest the air eventually builds up and spills over the passes at either end of the basin. The spillover usually takes place first at the western end of the basin where the cold westerly winds that descend onto Lake Issyk-Kul' are known as the "boom" after the Boomsk Gorge in the upper Chu Valley or "ulan" after the Ulan River. A little later as the cold air reaches the eastern mountains it spills over the Santash Pass at the eastern end of the basin and descends onto Lake Issyk-Kul' as a cold easterly to northeasterly wind known as the "santash". The occurrence of such winds is most frequent in spring and winter when cold outbreaks of air are most prevalent.

During a two-year study of santash winds they were observed 60 times per year at Przheval'sk with 46.6% of the occurrences happening in spring and 26.5% in winter. Only 12% occurred in summer. (PODREZOV, 1965, pp.36–37.) These frequencies included all winds above 10 m/sec. Such winds generally last from 1 to 12 h but they may last up to 24 h or even 2 or 3 days. The winds contribute to the droughty nature of the Issyk-Kul' Basin which is cut off from most moisture sources anyway. The strong winds are generally limited in height. On the average they reach no more than 750 m into the air, although on occasion they may go up to 3 km. In all events the maximum velocity is found within 500 m above the surface. The maximum velocity probability of the santash is 25 m/sec every year and 36 m/sec every 20 years. Somewhat greater velocities are achieved by the ulan winds at Rybachye in the western end of the basin where a maximum of 33 m/sec can be expected every year and of 45 m/sec every 20 years.

Moisture conditions

Moisture deficit is the key climatic characteristic of Soviet Central Asia. This is the driest part of the Soviet Union. Annual precipitation totals vary across the plain from about 200 mm along the northern fringes of the region to less than 30 mm in the Hungry Steppe southwest of Tashkent (Fig.9-14). And yet, as Fig.9-3 shows, the atmospheric moisture content in Central Asia during summer is some of the highest in the country in spite of the fact that the surface is very dry and precipitation rarely falls at this time of the year. The high value of the moisture content can be explained by the generally high temperatures and high moisture holding capacity of the air. Of course relative humidities are rather low and evaporation is great from water surfaces such as the Caspian and Aral Seas and from irrigated farm land.

The distribution of moisture content shown in Fig.9-3 would indicate that the southern part of the Caspian Sea and the Aral Sea have significant effects on the moisture content of the atmosphere. The Caspian during the course of the year evaporates 389.6 km^3 of water. About half of this moisture is carried southward toward the central Asian mountains and constitutes about 5% of the total advection of moisture in this direction. This locally derived moisture increases the vapor pressure at the ground by about 0.2 to 0.3 mm and increases relative humidity at the ground by 2 to 3%. Thus the role played by the Caspian in the moderation of the surface air in Central Asia is quite small, but nevertheless it exerts an appreciable influence on precipitation. It has been estimated that the combined influences of additional water vapor and conditional instability in the lower layers of the atmosphere produced by local evaporation increase precipitation on the average by about 40 mm on the northern slopes of the Central Asian mountains, by about 10–20 mm in the interior of the mountains, and by 8–11 mm on the plains of Central Asia.

The evaporation from the Caspian lowers the air temperature at the surface, which is also beneficial to precipitation formation. This probably increases the role of evaporation by a factor of about 1.5. And within a restricted belt the total moisture advected toward the mountains of Central Asia is about four times as much as that computed for the whole of Central Asia and raises the atmospheric moisture content in this belt by 20%. Therefore the presence of a basin like the Caspian Sea is not negligible with respect to precipitation formation.

On the other hand, the portion of the yearly precipitation throughout Central Asia that is actually derived from local evaporation is only about 4%, and even in spring it does not get above 8%. (DROZDOV and GRIGOR'EVA, 1965, p.222.) Throughout much of the area March is the month of maximum precipitation, and therefore spring is the time of year when there is the maximum amount of moisture available in the soil to be evaporated.

The large inland bodies of water, the Caspian, Aral Sea, and Lake Balkhash seem to have little direct influence on precipitation totals in their immediate vicinities. Their shorelines are just as dry as areas farther inland. The areas adjacent to all three water bodies generally receive no more than 100 mm of precipitation per year.

Precipitation increases in the southern mountains where southwestern slopes that are exposed to winter cyclonic storms may receive 1,000 mm or more. In a few spots as much as 1,600 mm is received. The wettest area in Central Asia appears to be the southwestern exposure of the Gissar Range. Generally the wettest zones in the mountains lie between 1,500 and 2,500 m above sea level. Above 2,500 m precipitation decreases again as the mountains rise above most of the winter storms. The eastern portion of the high Pamirs probably receives no more precipitation than does the driest spot in the desert. It is a barren windswept area, treeless and snowless.

Precipitation regime

Much of the precipitation that falls in the southern part of Soviet Central Asia is associated with cyclonic storms which are most frequent in spring and winter. Throughout the southern plains March is usually the rainiest month. It is also usually the month with the greatest number of rainy days, although December and January are about as important in this respect as March is. As one moves northward there is generally a lag in the month of maximum precipitation to April. This is true of much of the plains immediately bordering the northern range of the Tyan Shans. In the foothill areas themselves precipitation maxima may be delayed into May. Much of the area around Lake Balkhash has a May maximum.

Along the northern border of Soviet Central Asia, as defined here, the month of maximum precipitation changes rather abruptly from April or May to June or July as the primary controls switch from the winter cyclonic storms of the south to the summer cyclonic storms of the northern steppes. The summer rainfall of the northern steppes is also augmented by convective activity in air which is not so dry as farther south.

The surface air over the heated desert in summer is subjected to a great deal of convective mixing but the condensation level is so high that the convective currents die out before it is reached. Generally a thermal low in Central Asia extends no higher than 2 km into the air above which the eastern extension of the Azores high produces a subsidence inversion. Thus, much of the southern desert is relatively rainless from June through Sep-

tember. Occasionally thunderstorms do occur during the summer but they bring little rainfall. Often the rain falling from such storms evaporates before it hits the ground.
In the mountains the precipitation regime is frequently reversed from what it is on the adjacent plains. Cumulus cloud build up along the mountain slopes in summer due to differential heating between the mountain slopes and the free atmosphere adjacent to them brings summer showers to areas which in winter generally lie above most of the stratus clouds associated with cyclonic storms. The elevation at which the switch-over takes place is around 2,000–2,500 m. For instance, in the Tadzhik Republic, Dushanbe at an elevation of 824 m receives the maximum number of days with rain in March and April. Obi-Garm at an elevation of 1,807 m on the south slope of the Gissar Range experiences the maximum number of rainy days in April. Fedchenko Glacier at an elevation of 4,169 m experiences the most rainy days in June, and Murgab in the central Pamirs at an elevation of 3,640 m receives its maximum number of rainy days in June and July.

Snow and ice

On the plains snowfall is infrequent and snow accumulation is very light or even absent. Krasnovodsk on the eastern shore of the Caspian experiences only 6 days of snowfall per year. Zeagly in the central Kara Kum on the average has 10 days per year. Ashkhabad has 15 days, and Tashkent has 24. The number of snowy days and depths of snow on the ground, of course, increase northward. Throughout the desert plains southwest of the Syr-Darya, the northern Aral Sea, and northern Caspian maximum snow cover does not exceed 10 cm and is generally much less. In many places it is absent altogether. Krasnovodsk never experiences a measurable snowcover. Kizyl-Arvat farther east on the Transcaspian railroad has a brief snowcover on only 10% of all years. Farther east Kerki on the middle Amu-Darya has a snowcover only 2% of all years. Tashkent experiences a snow cover 35% of all years. Various parts of the Fergana Basin experience snow cover from 25 to 40% of the years. The northern fringe of the Central Asia region as defined here accumulates snow to maximum depths of 20–30 cm. However, it lies very unevenly on the ground since the winds are consistently strong.
In the mountains a great deal of precipitation falls as snow. Obi-Garm at an elevation of 1,807 m receives snowfall on 53 days per year and rainfall on 70 days, but during the winter months of December–February days with snow far outnumber those with rain. In January snow falls on 14 days and rain only 2 days. Naryn at an elevation of 2,049 m in the central Tyan Shans experiences snowfall 49 days per year and rainfall 60 days. In December and January it experiences 8 days of snowfall per month and no days with rain whatsoever. It experiences some snowfall every month except July and August. The station of Tyan Shan at an elevation of 3,672 m in the central Tyan Shans experiences snowfall on 136 days per year and rain on 35. From October through February it receives no rain at all but experiences 7 to 12 days per month of snowfall. It experiences even more frequent snowfall in summer when orographic effects and cumulus activity set off frequent snow showers. In May no less than 18 days experience snowfall and only 2 days experience rain. In June 17 days experience snowfall and 6 days experience rain. Only in July and August do rainy days exceed snowy days, but even then snow falls on 10 to 11 days per month. On Fedchenko Glacier at an elevation of 4,169 m 198 days per year experience snowfall and only 6 days experience rain. From December through March

every month experiences 22 to 23 days with snowfall. August and September experience 6 and 7 days, respectively. In the eastern Pamirs Murgab at an elevation of 3,640 m experiences snowfall each month of the year but snowy days add up to only 35 days per year because of the extreme dryness of the air in this region.

Intermediate slopes that are exposed to the winter cyclonic storms in the south accumulate snow to considerable depths. On exposed slopes and mountain passes at elevations of 2,000–4,000 m snow frequently accumulates to 2–3 m. As a whole the mountains accumulate snow to depths of approximately 50–100 cm. The snow pack at these elevations generally completely melts during the summer and provides much of the water for the streams that flow northwestward into the desert to provide irrigation water during the summer time. It has been estimated that the annual snow pack provides 65% of the annual discharge of these streams, while glaciers supply about 25% and rain supplies about 10% (BORISOV, 1965, p. 214). If this winter precipitation in the high mountains fell in the form of rain it of course would run off immediately and be lost for irrigation purposes if it were not impounded in giant reservoirs. As it is, the climate of the high mountains with its heavy snow pack which reaches its greatest rate of melting just when it is most needed on the dry lowlands forms a complementary and integral part of the climate and utilization of the adjacent desert plains.

The perpetual snow line in the Central Asian mountains is very high for the latitude because of the dryness of the region and relative warmth as compared to mountain masses such as the Alps. Generally the peripheral ranges of the mountains have snow lines from 3,000 to 3,800 m on northern slopes and 4,000 to 4,300 m on southern slopes, but in the drier interior ranges of the massifs the snow line is higher. In the eastern Pamirs it is over 5,000 m. This is comparable to tropical mountains. In some instances where southern slopes receive much more snowfall than northern slopes the perpetual snow line is lower on the southern slopes. For instance, Peter I Range generally has a snow line at about 3,800–4,100 m on the southern slope and one at approximately 4,900 m on the northern slope.

The higher peaks of the Pamir Alay and Tyan Shans are occupied by mountain glaciers. Fedchenko Glacier is known around the world as perhaps the most extensive mountain glacier on earth. It has a total length of 71.2 km, a width that varies from 1,700 to 3,100 m, and a surface area of about 900 km^2. The ice reaches a thickness of 500 m. At these high elevations little precipitation falls, but it has been determined that a great deal of condensation takes place directly from the air onto these glacial surfaces, which significantly replenishes the ice.

Ice storms and ice-encrusted ground present serious problems in Central Asia. Such conditions generally exist from 1 to 11 days per year throughout the northern half of Central Asia and occasionally in the southern half as well. Much of the ice-encrusted surface is caused by melting and refreezing of the snow cover brought on by alternations between warm foehn conditions and cold intrusions of Siberian air accompanying the passage of cyclonic storms. Usually there is a quick succession from warm conditions ahead of storms to cold conditions behind them. If there is a significant snow cover on the ground it may incorporate a number of different ice crusts within it because of the alternation of freezing and thawing. Such conditions not only provide hazards to transportation but make grazing impossible and may bring about widespread starvation of livestock which generally depend upon open grazing all during the winter in the Central Asian region.

Thunderstorms and hail

Thunderstorms are infrequent throughout much of Central Asia. During summer the upper air is too stable and the lower air is too dry most of the time for thunderstorms to develop, and during the winter the cyclonic storms generally are associated with stratoform clouds. Most of the thunderstorms that do occur take place during April–June in the southern desert, May–August in the northern desert, and May–July in the mountains. Most of them bring only little rainfall. During a year's time the number of thunderstorms ranges from 6 at Kazalinsk along the lower Syr-Darya to 20 at Iliysk on the middle Ili River in the northern desert, from 23 at Tyan Shan observatory to 32 at Susamyr in the northern and central Tyan Shans, from 5 at Krasnovodsk on the eastern shore of the Caspian and at Repetek southwest of Chardzhou to 20 at Ashkhabad in southern Turkmenia and 20 at Osh in the Fergana Basin which represent the southern desert zone, and from 2 at Fedchenko Glacier to 26 at Dushanbe mountain station in the Pamir Alay region (CHELPANOVA, 1963, pp. 442–443).
Hail comes infrequently to Central Asia and when it does it is usually in association with the thunderstorms that arrive early in the year. The desert plains generally experience 1 occurrence of hail during a period of 2 to 3 years. This usually occurs during February–May. In the foothills hail occurs from 1 to 5 times per year generally from March to August. Generally hail is not a great hazard in Central Asia.

Clouds and fog

Central Asia is known for its cirrus. This is particularly true in the desert plains.Cloud cover increases somewhat in the foothill areas and particularly in the high mountains where orographic effects and additional convective activity are conducive to frequent cloud and fog formation. On the plains most of the clouds are of a stratiform type associated with cyclonic storms which are by far most frequent in late winter and early spring. Tashkent shows a maximum sky cover of 7.4 tenths in February and March. This diminishes to only 1.5 tenths at 07h00 and 1.1 tenths at 13h00 in August. Throughout most of the desert plains maximum of cloudiness occurs during the morning hours and minimum cloudiness during evening hours. Overcast skies occur from 55 to 60% over much of the desert plains during January and from less than 5 to 20% or more during July. The clearest region during July is the area of the thermal depression in the very southeastern part of the plains along the Tedzhen, Murgab, and upper Amu-Darya valleys. During winter the clearest area is through the mid-section of the desert extending in a broad zone northeastward from Ashkhabad to Kzyl-Orda along the lower Syr-Darya.
In the northern foothills Alma-Ata shows a somewhat different annual and diurnal regime of cloudiness. Maximum cloudiness at 07h00 is reached in March with 7 tenths sky cover, and the minimum is in September with 3.8. Thus the annual range of cloudiness is somewhat less in the foothills than in the desert plains. And maximum cloudiness during the day generally occurs later in the foothills than in the plains. In March Alma-Ata shows a cloudiness of 7.1 tenths at 13h00. The afternoon cloudiness increases through spring and early summer to a maximum in May of 7.6 tenths sky cover. The minimum sky cover at 13h00 occurs in September with 5 tenths coverage. Maximum sky

cover for the entire record at Alma-Ata occurs at 19h00 during June when 7.6 tenths of the sky is covered on the average. Obviously, Alma-Ata experiences considerably more convective activity than does Tashkent with more cumuloform clouds that shifts its maximum cloudiness later in the season and later in the day than that of Tashkent. However, the number of days with overcast skies shows a regime very similar to that of Tashkent, with a maximum of 13 days in March and a minimum of 3 in August. Frequency of overcast conditions are about the same during the winter at the two stations but during the summer Tashkent is considerably clearer than is Alma-Ata. (See the climatic tables of the Appendix.)

In the Fergana Basin the city of Fergana shows a cloud regime somewhat midway between that of Tashkent and Alma-Ata. Generally it is cloudier than Tashkent throughout the year. It is cloudier than Alma-Ata during the winter but less cloudy during the summer. Maximum cloudiness occurs at 07h00 in February when the sky is 7.9 tenths covered. Minimum cloudiness occurs at 13h00 in August when 1.7 tenths of the sky is covered. Generally there is little difference in sky cover between 07h00 and 13h00. During January half the days are overcast, while July experiences only 1 overcast day during the entire month.

The upper Vakhsh Valley, as exemplified by Dushanbe, shows a cloud regime very similar to that of the Fergana Valley. It has a little more cloudiness during winter and significantly less cloudiness during summer than Fergana does. Cloudiness reaches a maximum of 8 tenths sky cover at 07h00 in March and a minimum of 1 tenth sky cover at 19h00 in August. During January half the days are overcast while in August only two-tenths of 1 day on the average is overcast.

The east coast of the Caspian has considerably more summer cloudiness than does the area further inland. Krasnovodsk has an average of 2 tenths sky cover at 19h00 hours in September. Minimum overcast is reached in August when on the average one day is overcast. On the other hand, Krasnovodsk has significantly less winter cloudiness than Dushanbe or Fergana. December through March average 10–11 days per month with overcast skies. Northward throughout the plains cloudiness tends to increase during the summer as more cyclonic storms and greater activity take place. Karaganda just to the north of the region experiences 3 days in August with overcast skies. It is somewhat cloudier throughout the year; 109 days experience overcast skies, as compared to 74 at Krasnovodsk, 90 at Dushanbe, and 88 at Fergana.

As was the case with precipitation, cloudiness in the high mountains shows a rather opposite regime from that on the plains below. Tyan Shan at an elevation of 3,672 m shows a maximum number of cloudy days in May and a minimum number of clear days in May and June. Much of the cloudiness here is of the cumulus type due to orographic uplift and convective activity induced by differential heating between the mountain slopes and the free atmosphere over the valley below.

Fog does not occur frequently throughout most of Central Asia because of the generally low relative humidities and rather high average wind speeds. Dew also is a rather infrequent occurrence over most of the desert plains during summer because of the same reasons. Most of the desert plains experience from 15 to 25 days per year with fog. Some of the exposed promontories along the east coast of the Caspian receive as many as 35 days, and the Fergana Basin generally receives 25 to 30 days. The foothill areas tend to receive somewhat more. Alma-Ata observatory receives 43 days per year. Very little fog is received in some of the higher basins and mountain valleys. Rybachye in the western end of the Issyk-Kul' Basin experiences only 2 days of fog per year, and Murgab in the Pamir Alay Mountains in southeastern Tadzhikistan experiences only 3 days per year.

Throughout most of the region winter is the foggiest time of the year. This extends into spring in the northern plains, but in the southern deserts almost without exception December and January show the highest fog frequencies. Very little fog is experienced during the summer half year except in some of the high mountains where convective activity produces cumulus clouds whose bases lie on the surfaces of the upper slopes. For instance, Tyan Shan observatory experiences two days of fog per month from April through September, which accounts for 11 of its 15 days of fog per year.

Thermal conditions

With its relatively low latitude and relatively sunny skies with respect to the rest of the Soviet Union, the Central Asian area receives the greatest amount of insolation during the year of any region of the country. Total direct and scattered radiation varies from about 160 kcal./cm^2 in the southeastern part of the region to about 120 kcal./cm^2 in the north. Albedos of the desert surface are relatively high, and therefore much of the area loses from 35 to 40 kcal./cm^2 to reflection as soon as the radiation hits the ground. The effective outgoing radiation is also quite high because of the dry surface and dry atmosphere, generally amounting to between 45 and 55 kcal./cm^2 per year. Thus, the radiation balance is not so outstandingly high, but it is still the highest within the Soviet Union. It ranges from approximately 45 kcal./cm^2 in the south to about 27 or 28 kcal./cm^2 in the north (Fig.8-17).

Thus, compared to the rest of the Soviet Union, Soviet Central Asia is rather well endowed with heat resources. However, many of these resources are not too well utilized. Over the dry desert surface much of the absorbed heat is transferred upward as sensible heat into the atmosphere and causes hot surface air temperatures during the summer. Average temperatures during July range from 32°C or more in southern Turkmenistan to 25°C along the northern margins of Central Asia. Absolute maximum temperatures reach more than 50°C in the central Kyzyl-Kum southeast of the Aral Sea. Throughout the entire lowland region they are greater than 40°C. Daytime soil surface temperatures are even hotter. They have been recorded as high as 70°C. Even in January sand surface temperatures have reached more than 47°C. The midday heat generally does not penetrate far into the soil. At a depth of 10 cm temperatures generally are 20°–30°C cooler than they are at sand surfaces during the heat of the day. At night, of course, great radiational cooling takes place from the land surface through the dry clear air, which yields high diurnal ranges of temperature. For instance, at Ashkhabad the mean diurnal range of temperature during October is 16.5°C, the greatest for the year. During that month temperatures may rise to 40°C and fall to −5°C. During July in Ashkhabad temperatures have risen to 47°C and fallen to 12°C.

The components of the heat budget may be greatly altered by irrigation which lowers surface temperatures, reduces the diurnal ranges of temperature, and makes for greater use of heat resources available for plant growth and other forms of man's activities.

The albedo of cotton fields is around 20% as compared to 35% for sandy deserts, and the effective outgoing radiation is reduced by 10–15% because of reduced surface temperature and increased humidity of the surface air. The result is that the radiation balance over a field is frequently twice as great as that over a sandy desert. Surface air temperatures average about 3°C cooler over irrigated areas than over unirrigated areas during summer, and a small temperature inversion is generally observed over irrigated plots throughout the day and night. The loss of heat to evaporation over irrigated plots leads to even lower nighttime temperatures than over the desert. Relative humidities in the surface air often are 20% higher over irrigated plots than over dry desert surfaces.

Over irrigated plots more heat typically is absorbed than is received from insolation, since heat is advected in from surrounding desert areas. The vegetation decreases the wind and evaporation, but transpiration increases, and the vapor pressure of the surface atmosphere increases by 4–5 mm. Irrigation greatly reduces the harmful effects of the sukhovey by reducing air temperatures, increasing humidities, and reducing wind speeds. Complete sukhoveys generally are not observed over irrigated plots. Over desert surface during summer they prevail from 50 to 60% of the time. Some degree of sukhovey occurs over the open desert from 70 to 100% of the time during July.

The hottest days in Central Asia are generally accompanied by a strengthening of the wind, which it has been determined is related to excessive heat energy absorbed by the dust in the air. If the dustiness considerably exceeds the average, the emission of heat from the dust to the air may exceed by several times the emission of heat directly from the land surface to the air.

During the transitional seasons the reductions of temperature are small over irrigated plots, and so the influence of irrigation on the length of the frost-free period is negligible. On the plains the average length of the frost-free period varies from about 225 days in the far southern portion of the desert to about 150 days in the north (Fig.8-45). Successful cotton growing generally requires an average daily temperature above 14°C. Such conditions last for 6 to 7 months in the southern part of the desert. This is quite favorable for cotton growing. However, northeast of the Syr-Darya the growing season generally is too short for cotton.

During winter temperatures can get quite cold in Central Asia because of open exposure to the north. Average temperatures during January range from +3°C in the extreme south to −15°C in the north. Annual minimum temperatures average about −16°C in the south and −30°C or colder in the north (Fig.8-38). Absolute minimum temperatures have reached −30° in the south and −40° and −50°C in the north (Fig.8-36). Alma-Ata along the northern foothills of the Tyan Shans has experienced a temperature as low as −48°C. Even in the sheltered Fergana Basin temperatures have dropped to −28°C at the city of Fergana. And in the extreme south Dushanbe has experienced temperatures of −28°C.

The larger water bodies afford considerable temperature amelioration during the winter. Absolute minimum temperature at Fort Shevchenko on the western promontory of the Mangyshlak Peninsula on the eastern side of the northern part of the Caspian Sea is only −26°C. Over the central Aral Sea temperatures during January average −5°C, which is 2° warmer than the southern coast of the sea and 6° warmer than the northern coast.

Radiation and temperature vary greatly with altitude in the Central Asian mountains.

On the average direct radiation increases by as much as 30% during the first 1,500 m of ascent because of the thinning and purification of the air. Generally during summer there are no cloud formations below this elevation. At greater heights insolation would depend upon cloud conditions. Outgoing radiation is also increased in the mountains because of high albedos, averaging about 30% during summer. During winter when much of the mountain surfaces are covered by snow albedos may rise to as much as 90%. The radiation balance reaches about 45 kcal./cm^2, which is about the same as on the very southern portions of the plain. The radiation balance is greatest in the foothills up to about 1,200 m in elevation. Above that clouds and precipitation tend to lower the radiation balance. Above 2,500 m the radiation balance rises again, but it remains less than it was in the foothills. The daily ranges of temperature are great in the high mountains because of the thin air and high values of incoming radiation during the day and outgoing radiation during the night. Also, there are great differences in temperature between sun and shade during the day in the high mountains. According to mountain climbers in Central Asia the side of one's face turned toward the sun becomes dark and dry like parchment while the side of the face in the shade almost freezes.

During winter typically the foothill areas are warmer than the lower plains because of strong temperature inversions which persist to elevations of about 1,500 m. Often cold outbreaks of air are no thicker than 500 m when they reach the southern mountains, so that the foothill areas are above these coldest air masses. Also, during winter foehn winds are common throughout much of the region which greatly alter temperature conditions over short periods of time. The temperature at any particular time may be related more to vertical movements of air than to radiational factors. In some mountain basins winters are unusually mild because of local conditions. Issyk Kul'Basin is outstanding in this respect because during most winters the temperature of the surface water in the lake does not fall below 2.75°C and ice forms only along the shore and in enclosed embayments. By contrast on the northern plains Lake Balkhash freezes over from November to mid April, and the Aral Sea freezes over and is closed to shipping about 5 months of the year. The lower portion of the Syr-Darya freezes from December to late March and the Amu-Darya below Nukus is frozen from 2 to 2.5 months during the year. Along the upper portions of the Amu-Darya ice forms only along the banks.

Changing weather with altitude in the Alma-Ata region of the northern Tyan Shans is illustrated by the graphs in Fig.6-16. The reduced severity of winter temperatures in the intermediate altitudes is illustrated nicely by Medeo at an elevation of 1,529 m, which shows considerably fewer days of significant frost than does Alma-Ata below at an elevation of 848 m. Not until one rises to Verkhniy Gorel'nik at an elevation of 2,254 m does one run into winter severity equal to that of Alma-Ata. However, at that elevation cloud conditions have become considerably different from what they were below at Alma-Ata. This is accentuated still higher up at Myn-Dzhilki. Summer temperatures, of course, are highest at lower elevations and decrease steadily upward, as does the length of the summer period.

One of the hottest parts of the desert is illustrated by Repetek (Fig.6-17). However, the growing season is longest around Kizyl-Atrek (Fig.6-18). This is in the southwestern corner of the region in the small basin of the Atrek River which is protected on the northeast from cold air by the northwestern extension of the Kopet-Dag. Here the growing season is almost year round and during midsummer some days would classify

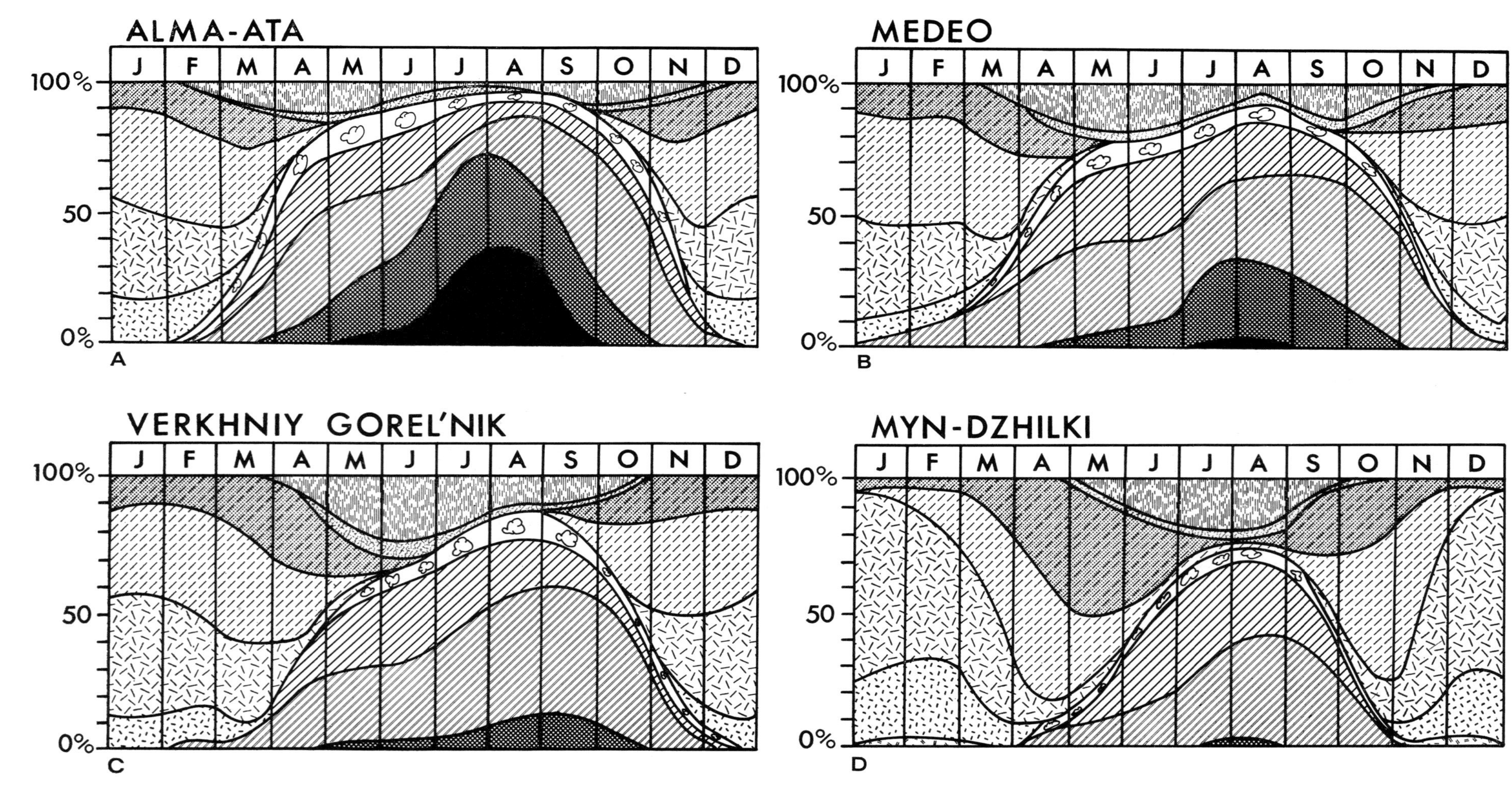

Fig.6-16. Frequencies of weather types: A. Alma-Ata, 848 m; B. Medeo, 1529 m; C. Verkhniy Gorel'nik, 2,254 m; D. Myn-Dzhilki, 3,036 m. (After GERASIMOV, 1964.) For legend see Fig.3-25.

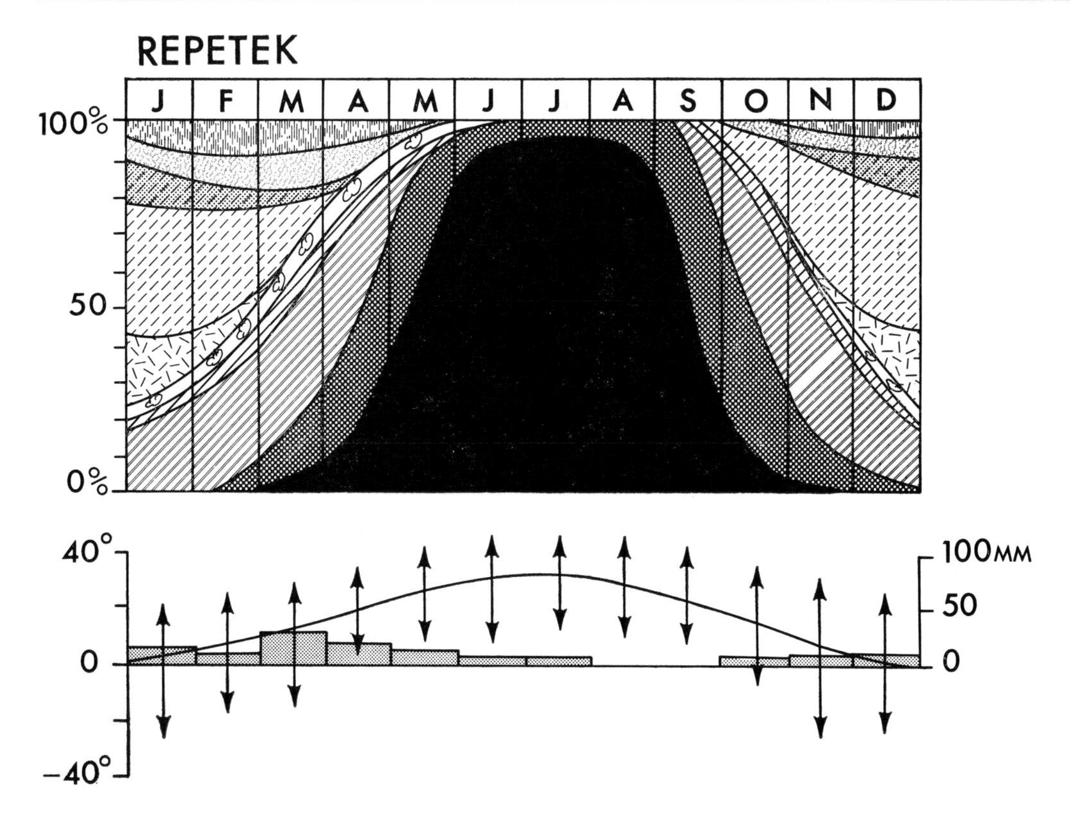

Fig.6-17. Frequencies of weather types, Repetek. (After GERASIMOV, 1964.) For legend see Fig.3-25.

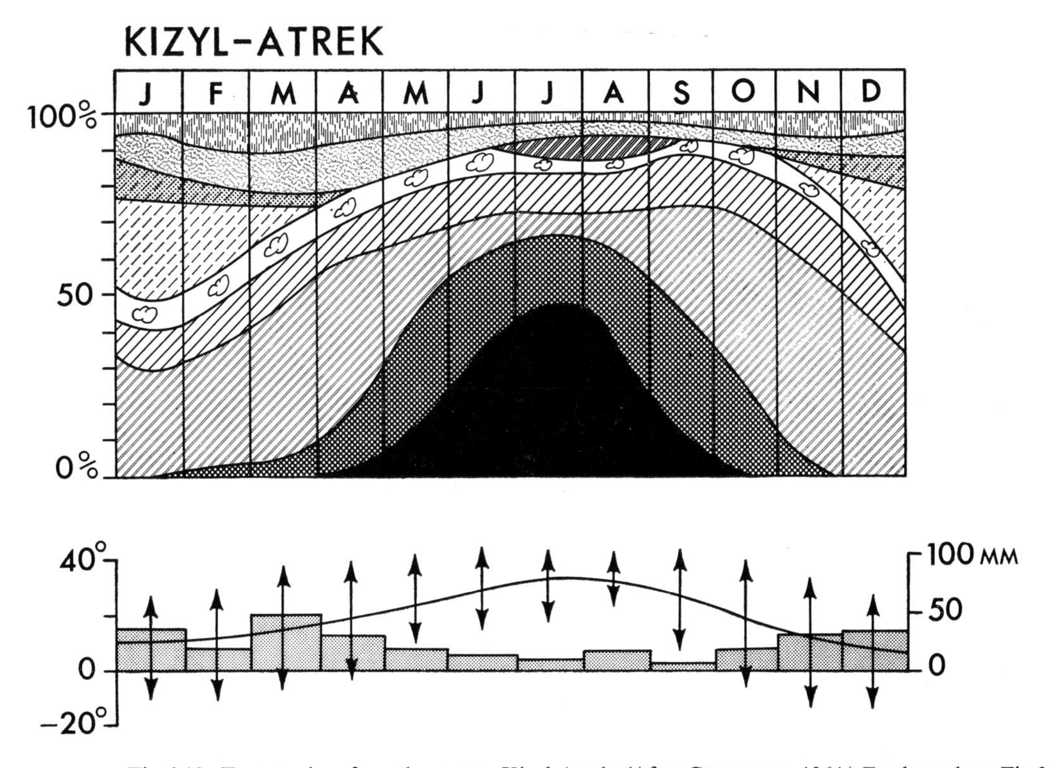

Fig.6-18. Frequencies of weather types, Kizyl-Atrek. (After GERASIMOV, 1964.) For legend see Fig.3-25.

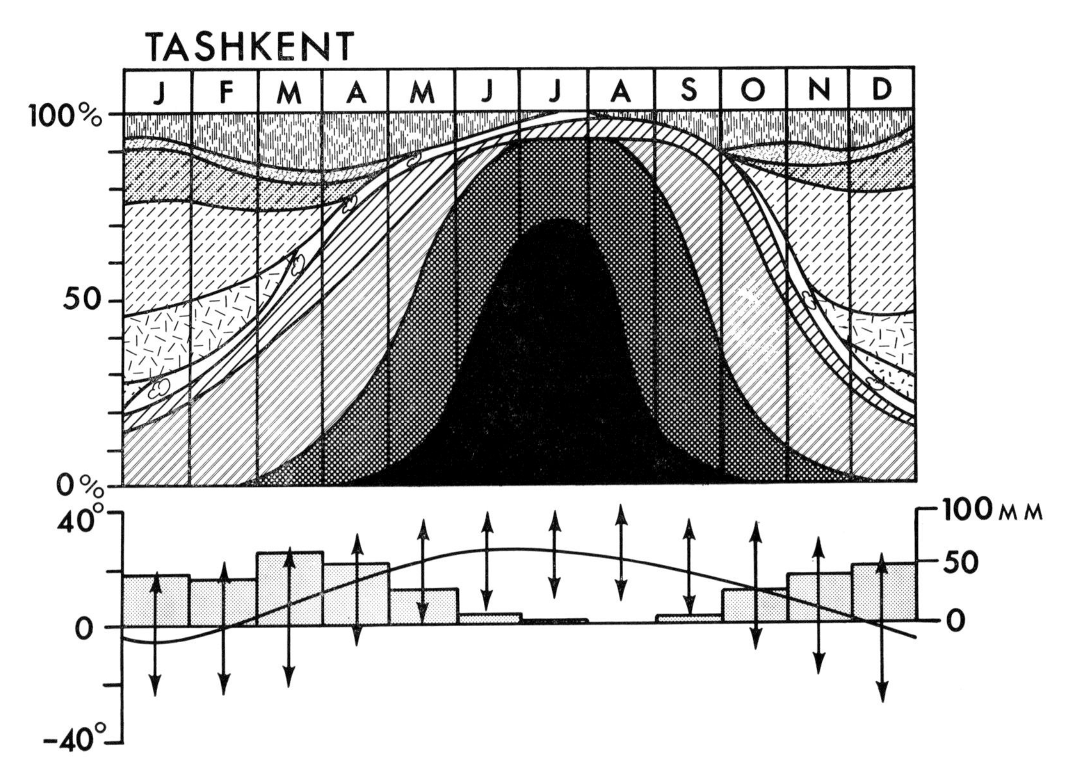

Fig.6-19. Frequencies of weather types, Tashkent. (After GERASIMOV, 1964.) For legend see Fig.3-25

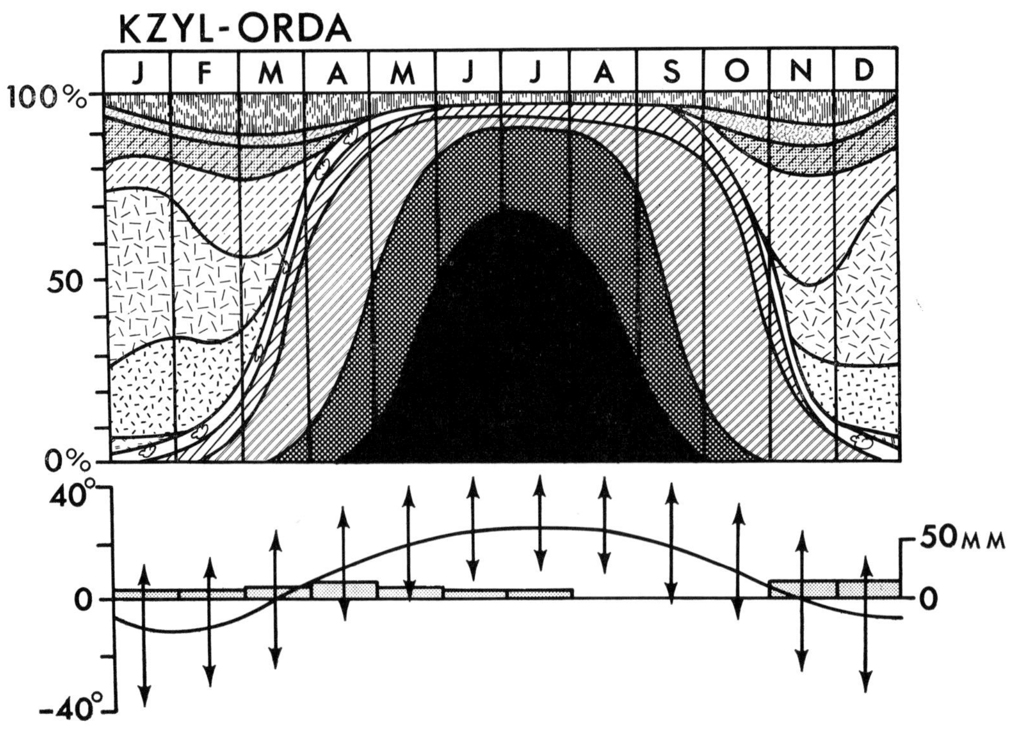

Fig.6-20. Frequencies of weather types, Kzyl-Orda. (After GERASIMOV, 1964.) For legend see Fig.3-25.

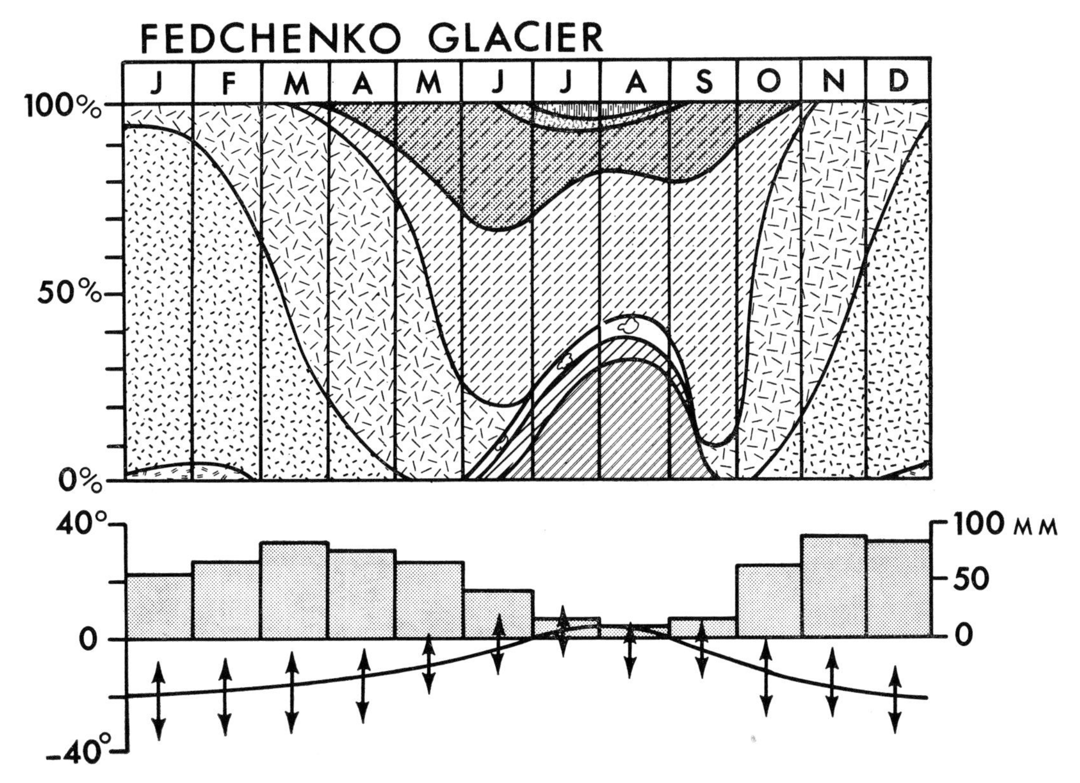

Fig.6-21. Frequencies of weather types, Fedchenko Glacier. (After GERASIMOV, 1964). For legend see Fig.3-25.

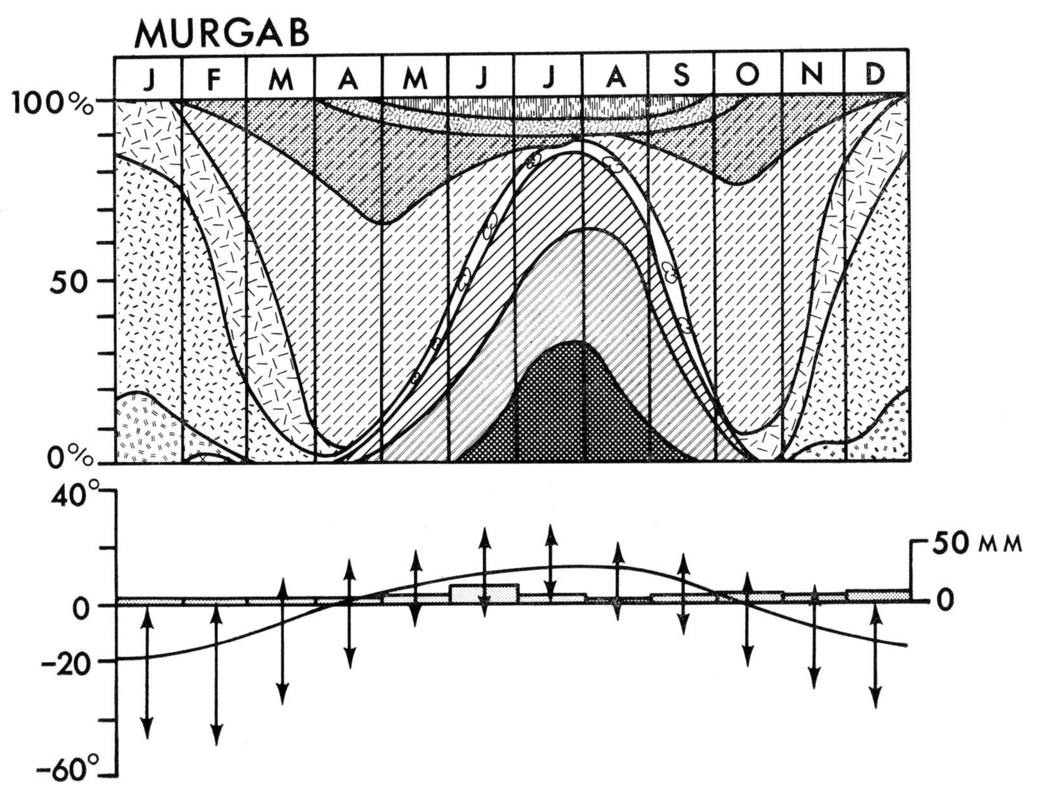

Fig.6-22. Frequencies of weather types, Murgab. (After GERASIMOV, 1964.) For legend see Fig.3-25.

as humid tropical weather.

Tashkent is typical of much of the southern desert. Here the summers are relatively hot and relatively long, and winters are somewhat cool. The precipitation regime shows a strong concentration during winter and spring with an absolute maximum falling in March (Fig.6-19). Farther northwest Kzyl-Orda reflects the rapidly increasing severity of the winters and the reduction of precipitation as one gets farther away from the Central Asian mountains and the frequented routes of winter cyclonic storms (Fig.6-20). In the high mountains Fedchenko Glacier represents the ultimate in cold prolonged winters and near absence of summers (Fig.6-21). Farther southeast Murgab shows a greater seasonality, with colder temperatures during the winter and warmer temperatures during the summer, because of its drier nature and ice-free surface (Fig.6-22). There is a great contrast in precipitation amounts between Glacier and Murgab. Fedchenko Glacier catches rather heavy snowfall because it is somewhat open toward the west, while Murgab on the eastern slopes of the Pamirs is almost completely isolated from moisture-bearing winds from the west.

References and further reading

ALISOV, B. P., 1956. *Klimat SSSR* (Climate of the USSR). Moscow Univ., Moscow, 127 pp.

ANONYMOUS, 1954. *Atlas Tipichnykh Sinopticheskikh Protsessov nad Sredney Aziey* (Atlas of typical synoptic processes over Central Asia). Inst. Mat. Mekh. V. I. Romanovsk., Akad. Nauk SSR, Tashkent.

ANONYMOUS, 1969. *Atlas SSSR* (Atlas of the USSR). Gl. Upr. Geod. Kartogr. Sov. Ministr. SSSR, Moscow, 2nd ed., 199 pp.

AYZENSHTAT, B. A., BALASHEVA, E. N. and ZHITOMIRSKAYA, O. M., 1958. *Klimaticheskoe Opisanie Golodnoy Stepi* (Climatic treatise about the Hungry Steppe). Gidrometeoizdat, Leningrad, 74 pp.

AYZENSHTAT, B. A., 1967. Investigation of the thermal balance of Central Asia. In: M. I. BUDYKO (Editor), *Modern Problems of Climatology*. Transl. Div., Wright-Patterson Air Force Base, Ohio, pp.96–136.

BABUSHKIN, L. N., 1960. *Agroklimaticheskoe Rayonirovanie Khlopkovoi Zony Sredney Azii* (Agroclimatic regions of the cotton zone of Central Asia). Gidrometeoizdat, Leningrad, 133 pp.

BALASHOVA, E. N., ZHITOMIRSKAYA, O. M. and SEMENOVA, O. A., 1960. *Klimaticheskoe Opisanie Respublik Sredney Azii* (Climatic treatise on the republics of Central Asia). Gidrometeoizdat, Leningrad, 241 pp.

BALASHOVA, E. N., ZHITOMIRSKAYA, O. M., KARAUL'SHCHIKOVA, N. N. and SABININA, I. G., 1963. *Klimaticheskoe Opisanie Zeravshanskogo Rayona* (A climatic study of Zeravshan region). Gidrometeoizdat. Leningrad, 120 pp.

BAYDAL, M. KH., 1964. *Dolgosrochnye Prognozy Pogody i Kolebaniy Klimata Kazakhstana* (Long range forecasting of weather and climatic fluctuations in Kazakhstan). Gidrometeoizdat, Leningrad, 446 pp.

BORISOV, A. A., 1965. *Climates of the USSR*. Aldine, Chicago, Ill., 125 pp.

BOROVIKOV, A. M., GRUDZINSKIY, M. E. and KHRGIAN, A. KH., 1958. O meteorologicheskikh usloviyakh vysokogornogo Tyan'-Shanya (On meteorological conditions of the high Tyan Shans). *Tsentr. Aerolog. Obs. Tr.*, 21: 175–199.

BUGAEV, V. A., 1946. *Klimat Sredney Azii i Kazakhstana* (Climate of Central Asia and Kazakhstan). Izd. Akad. Nauk Uzbek. SSR, Tashkent, 23 pp.

BUGAEV, V. A., DZHORDZHIO, V. A., KOZIK, E. M., PETROSYANTS, M. A., PSHENICHNYY, A. YA., ROMANOV, N. N. and CHERNYSHEVA, O. N., 1957. *Sinopticheskie Protsessy Sredney Azii* (Synoptic processes in Central Asia). Izd. Akad. Nauk Uzbek. SSR, Tashkent, 477 pp.

CHANYSHEVA, S. G., 1966. *Mestnye Vetry Srednei Azii* (Local winds of Central Asia). Gidrometeoizdat, Leningrad, 120 pp.

CHELPANOVA, O. M., 1963. *Klimat SSSR, vypusk 3, Srednyaya Aziya* (Climate of the USSR, volume 3, Central Asia). Gidrometeoizdat, Leningrad, 447 pp.

DMITRIEV, A. A., 1963. Artificial control of the climate of large and small areas (based on experience in the coastal area of Issyk-Kul' Lake). *Izv. Akad. Nauk. Ser. Geogr.*, 1: 45–49. JPRS 19145, 1963.

DROZDOV, O. A. and GRIGOR'EVA, A. S., 1963. *Vlagooborot v Atmosfere*. Gidrometeoizdat, Leningrad. (Transl. as *The Hydrologic Cycle in the Atmosphere*, Isr. Progr. Sci. Transl., Jerusalem, 1965, 282 pp.)

FEDOROVICH, B. A., 1948. Rel'ef peskov Azii kak otobrazhenie protsessov tsirkulyatsii atmosfery (Sand dunes in Asia as indicators of atmospheric circulation). *Probl. Fiz. Geogr.*, 13: 91–109.

GERASIMOV, I. P., 1964. *Fiziko-Geograficheskiy Atlas Mira* (Physical Geographical Atlas of the World). Akad. Nauk, Moscow, 298 pp.

KOZIK, E. M. and IL'INOVOY, E. S. (Editors), 1968. Voprosy regional'noy sinoptiki i klimatologii Srednei Azii (Problems of regional synoptics and climatology of Central Asia). *Tr. Sredneaziat. Nauchno-issled. Gidrometeorol. Inst.*, 33(48): 108 pp.

KRENKE, A. N., 1973. Usloviya sushchestvovaniya sovremennogo oledeneniya Sredney Azii (Climatic basis for contemporary glaciation in Central Asia). *Izv. Akad. Nauk. Ser. Geogr.*, 1: 20–34.

MOROZOVA, M. I., PETROSYANTS, M. A. and CHERNYSHEVA, O. N., 1959. Osobennosti vozdushnykh techeniy nad Pamirom i zapadnym Tyan'-Shanem (Particular air flows over the Pamirs and western Tyan Shans). *Meteorol. Gidrol.*, 9: 3–12.

PETROSYANTS, M. A., 1971. Effect of orography on the general circulation of the atmosphere. In: B. L. DZERDZEEVSKII and KH. P. POGOSYAN (Editors), *General Circulation of the Atmosphere*. Isr. Progr. Sci. Transl., Jerusalem, pp.235–268.

PODREZOV, O. A., 1965. Veter Issyk-Kul'skoy kotloviny san-tash (San-Tash winds of the Issyk-Kul' Basin). *Meteorol. Gidrol.*, 6: 35–37.

POGOSYAN, KH. P. and UGAROVA, K. F., 1959. Vliyanie tsentral'noaziatskogo gornogo massiva na formirovanie struynye techeniy (Influence of the Central Asian mountain massive on the formation of jet streams). *Meteorol. Gidrol.*, 11: 16–25.

PONOMARENKO, P. N., 1970. O vertikal'nom raspredelenii osadkov v Kirgizii (On the vertical distribution of precipitation in Kirgizia). *Meteorol. Gidrol.*, 7: 86–90.

UTESHEV, A. S. (Editor), 1959. *Klimat Kazakhstane* (The climate of Kazakhstan). Gidrometeoizdat, Leningrad, 368 pp.

YABLOKOV, A. A., 1968. O snezhnom pokrove zapadnogo Tyan'-Shanya (On the snow cover in the western Tyan Shans). *Izv. Vses. Geogr. O-vo.*, 100: 158–161.

YAROSLAVTSEV, I. M., 1939a. Nekotorye osobennosti sinopticheskogo analiza v sredney Azii (Some particular synoptic analysis in Central Asia). *Meteorol. Gidrol.*, 6: 26–33.

YAROSLAVTSEV, I. M., 1939b. Materialy k tipizatsii sinopticheskikh protsessov nad Sredney Aziey (Materials for typifying synoptic processes over Central Asia). *Meteorol. Gidrol.*, 10–11: 46–56.

ZVEREV, A. S., 1957. Sinopticheskaya Meteorologiya (Synoptic meteorology). Gidrometeoizdat, Leningrad, 559 pp.

Chapter 7

The Caucasus

Topography

The climate of the Transcaucasian Lowlands is exceptional in the Soviet Union because of its subtropical nature. This is due primarily to the mountainous topography and the proximity of the Black Sea in the west (Fig.7-1). The most climatically significant mountains are the Great Caucasus Range which stretch unbroken from the northeast coast of the Black Sea east-southeastward to the Caspian. They lower at both ends but are continuously high throughout much of their length where passes generally lie above 2,500 m. Throughout much of the range peaks reach 4,000 m and more. The highest elevation of all is 5,642 m on Mount Elbrus in the west-central portion of the range. The range acts both as a significant barrier to horizontal air movements and as a high mountain mass which provides for vertical zonations of climate on its slopes. Permanent snow fields and glaciers exist on the higher peaks, while mild climatic conditions prevail in the lower valleys, particularly on the southern side of the mountain mass. The range acts as a most effective screen during winter when cold air masses to the north generally are no more than 1,500–2,000 m thick. During summer warm air masses on either side of the range generally extend up to 6,000 m or more and tend to spill over the passes through the range.

South of the Great Caucasus lies a synclinal valley which stretches all the way between the Black and Caspian seas. This is divided into western and eastern lowlands by the transverse Surami Range which crosses the west-central portion of the valley in a south-southwest–north-northeast orientation. Although elevations in the Surami Range generally are no higher than 1,000 m, the range forms a significant divide between the marine influences of the Black Sea to the west, with their associated warm moist conditions in the Colchis Lowland of Georgia, and more continental conditions to the east, with the associated dry hot climate of the Kura Lowland in Azerbaydzhan.

The southwestern end of the Surami Range merges with the Lesser Caucasus, which, like the Great Caucasus, are oriented in a northwest–southeast direction and form the southwesterly rim of the synclinal lowland. Their elevations are generally no more than 2,000 m, but locally peaks range from 2,500 to 2,900 m, and in southwestern Azerbaydzhan to as much as 3,724 m.

On their southwestern side the Lesser Caucasus merge with the Armenian Plateau which extends westward and southward across the international boundaries into Turkey and Iran. This volcanic plateau generally lies at elevations between 1,500 and 3,000 m. Thus

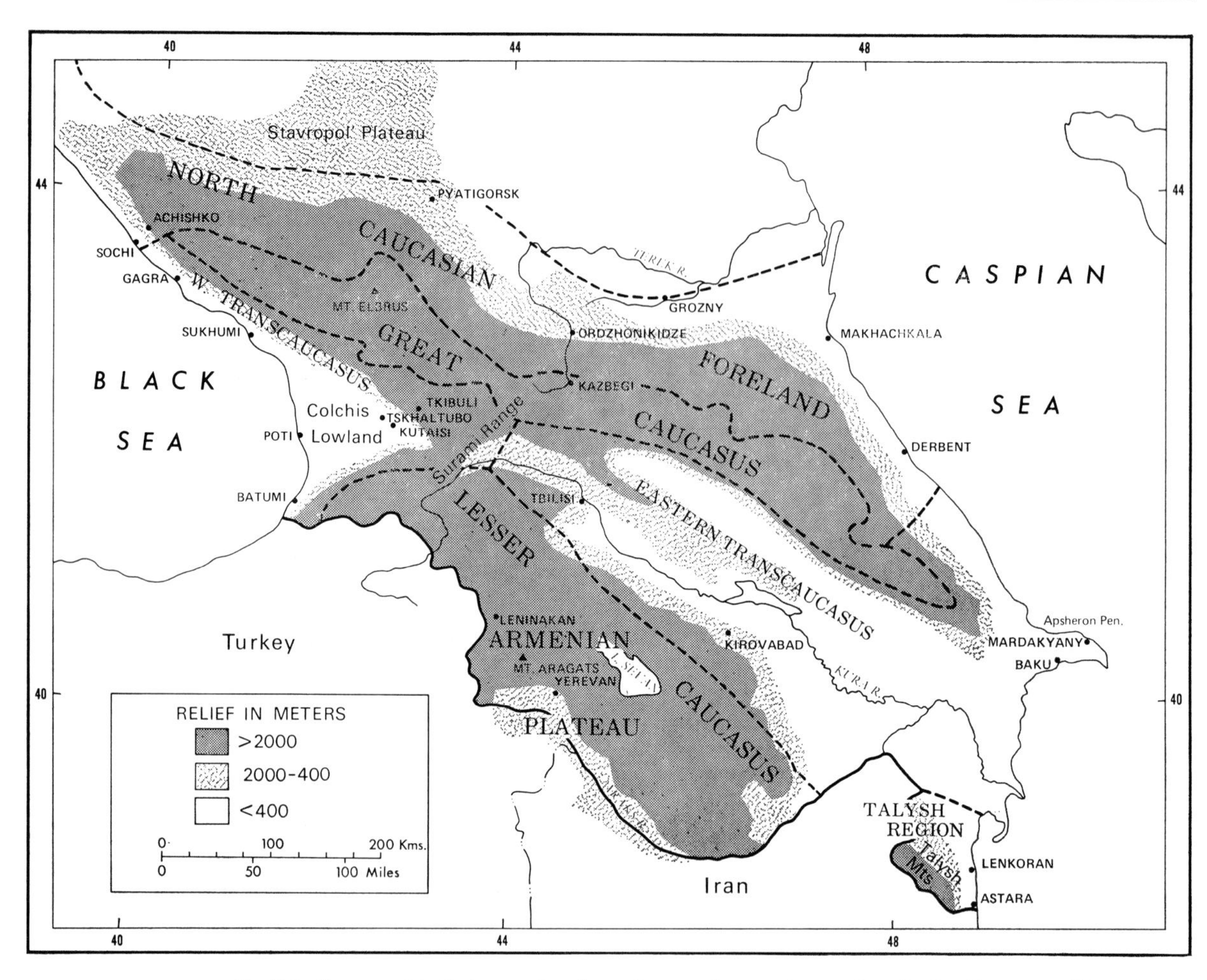

Fig.7-1. Topographic features, climatic regions, and reporting stations in the Caucasus. (After ANONYMOUS, 1969; ZANINA, 1961.)

it is at elevations similar to the Lesser Caucasus and thereby masks their structure as a distinctive mountain range. Locally the plateau rises to more than 3,500 m; at its highest elevation, on the volcanic cone of Mount Aragats, it reaches an elevation of 4,090 m. In the center of the plateau the large Lake Sevan lies at an elevation of 1,900 m. Lowest elevations are found along the international border where the Araks Valley lies between 800 and 900 m.

The Caucasus region thus consists of the diverse topographic subdivisions of the Great Caucasus in the north, the Colchis Lowland in the west, the Kura Lowland in the east, the Surami Range in the center, and the Lesser Caucasus and the Armenian Plateau in the south. Each of these areas exhibits a high degree of variability in elevation and exposure to various flows of air and great diversity in climatic characteristics. In the rougher areas climate varies greatly locally. Although the crest of the Great Caucasus forms somewhat of an air mass boundary between north and south, the climate on the northern slopes of the range is influenced in many ways by the range itself, and therefore the northern foothills and plain immediately adjacent to them will be included in this discussion of the Caucasus. The climate of this so-called "North Caucasian Foreland is much different from that of the Transcaucasus and varies greatly between its eastern and western portions. This east–west division is significantly influenced by a broad upwarp in its central portions known as the Stavropol' Plateau which rises to elevations of more than 800 m.

Atmospheric circulation

The Caucasus are dominated by the zone of subtropical high pressure in summer and the western extension of the Asiatic high in winter. On these major circulation features are superimposed profound local influences of relief and the Black and Caspian seas. Since the upper air flow is generally from the west, influences from the Black Sea are much greater than those from the Caspian, which are limited primarily to shoreline areas. The interior of the Transcaucasus, particularly the Armenian Plateau, which is well isolated from marine influences at either end, induces a local high pressure system in winter and a low pressure system in summer due to thermal effects. Mean sea level pressures for January show a cold anticyclone centered over eastern Turkey with a steep pressure gradient northeastward toward the Kura Lowland and northwestward toward the Colchis Lowland (Fig.7-2). July isobars show a heat low centered over almost exactly the same region (Fig.7-3). The surface form of the Armenian Plateau, which is generally higher on its edges than in its center, enhances its formation of continental air masses.

Both westerly and easterly flow occur across the Caucasus region, depending primarily upon the synoptic situation over the plains to the north. Westerly flow occurs usually around the southern periphery of a surface low pressure and under the southern end of an upper trough (Fig.3-7). Easterly flow is associated with a surface high pressure that is centered in the middle Volga region and spreads over much of the European plain (Fig.3-4). This type of flow is essentially absent during the summer.

When westerly flow occurs a cold front usually strings southwestward out of the low pressure center on the northern plain and wraps itself around the northwestern end of the Great Caucasus. Cold air then intrudes both to the north of the Great Caucasus and into the Colchis Lowland to the south. Often fronts become stationary in this position and the progression of the synoptic situation is delayed in the Caucasian region in comparison to the areas to the north and to the south. Cold air piles up at the surface in the Colchis Lowland and creates a wedge for warmer air from the Black Sea and Asia Minor to ride up and form clouds and prolonged precipitation.

A similar situation may occur with easterly flow around the southeastern end of the Great Caucasus where cold air from the east intrudes both to the north of the range and into the Kura Lowland to the south. At times during winter when an intense high forms north of the Great Caucasus cold air intrudes both from the east and from the west into the Transcaucasus and a cold front wraps around the Great Caucasus from both ends nearly meeting in the center across the Surami Range (Fig.7-4). Such a condition is known locally as an orographic occlusion. At such times the Great Caucasus appear as a warm island almost completely surrounded by colder air at low elevations.

Cyclonic storms along the Asia Minor segment of the Polar front affect the region all year round, but are much more frequent and more intense during winter when local cyclogenesis is augmented by movements into the area of cyclones from the Medi-

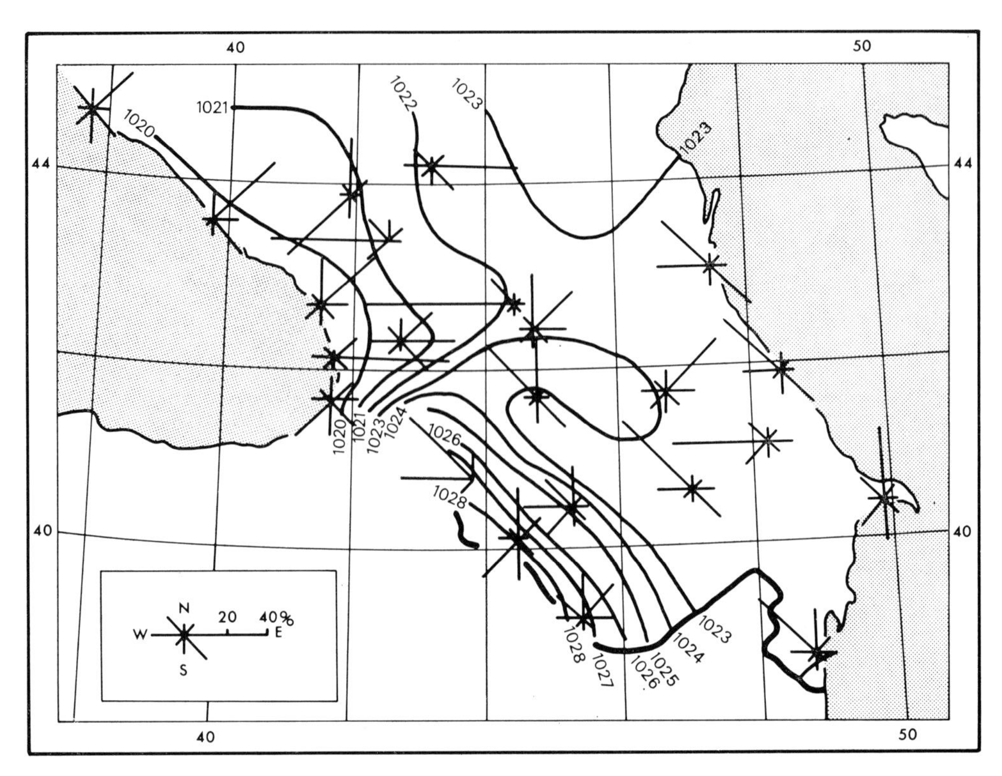

Fig.7-2. Sea level pressures (mbar) and wind roses (in frequency percentages), January. (After ZANINA, 1961.)

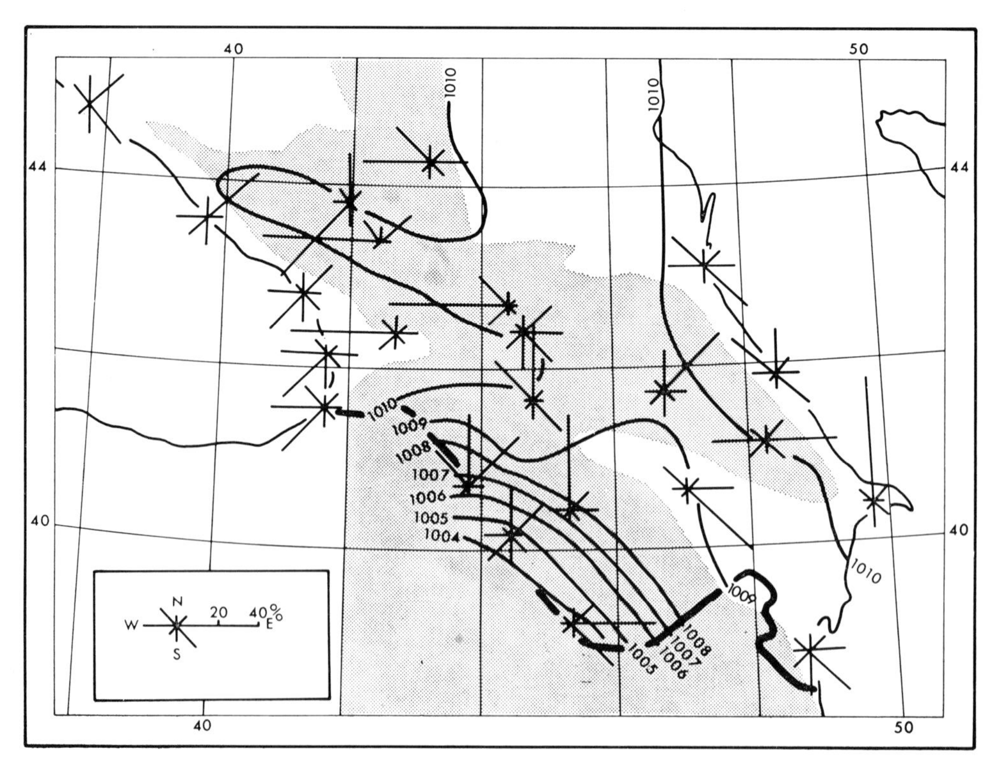

Fig.7-3. Sea level pressures (mbar) and wind roses (in frequency percentages), July. (After ZANINA, 1961.)

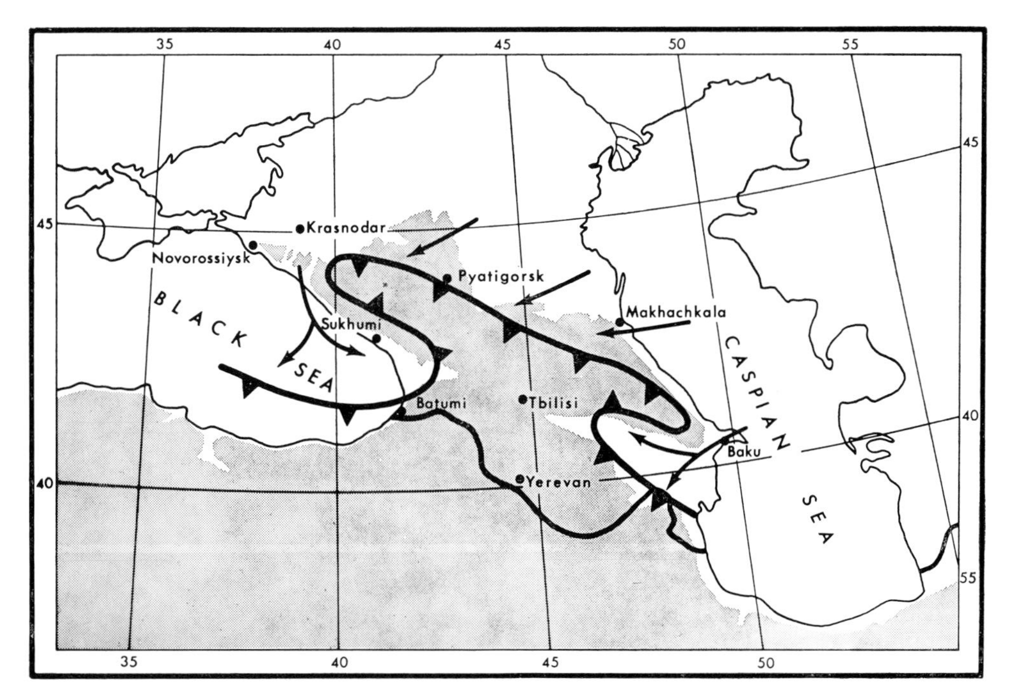

Fig.7-4. Orographic occlusion. (After ZVEREV, 1957.)

terranean track across Asia Minor and the Black Sea (Fig.7-5). During winter the cyclonic storms are generally fully occluded by the time they reach the Caucasus, so that at ground level maritime air from the west is meeting the locally derived continental air over the central Transcaucasus. But aloft a southwesterly flow of tropical air out of Asia Minor and the eastern Mediterranean provides much of the moisture for the heavy precipitation that falls in the western part of the Transcaucasus. During summer open waves are characteristic along the Polar front in this region. Their weak frontal surfaces may trigger convective showers in the warm relatively unstable air masses of the western Transcaucasus.

Local circulations and associated weather

Because of the high relief and the proximity of the Black and Caspian seas, local regimes of circulation abound in the Caucasus. Foehns are a common occurrence in many parts of the area and bora exist in the northwestern and southeastern extremities of the Great Caucasus, as well as in some of the more dissected edges of the Armenian Plateau. Diurnal breeze regimes are set up in many places along the coasts at either end of the Caucasus, and mountain–valley regimes are prevalent in many areas particularly in summer when the general circulation over the area is weak. Because of topographic features, winds in certain areas are almost unidirectional or bidirectional (Figs.7-2, 7-3). This is particularly true in some of the mountain passes and along the Caspian shoreline of the northeastern slopes of the Great Caucasus.

Foehns are most common in the western Transcaucasus as they descend the western slopes of the Surami Range. Although they can occur at any time of the year, they

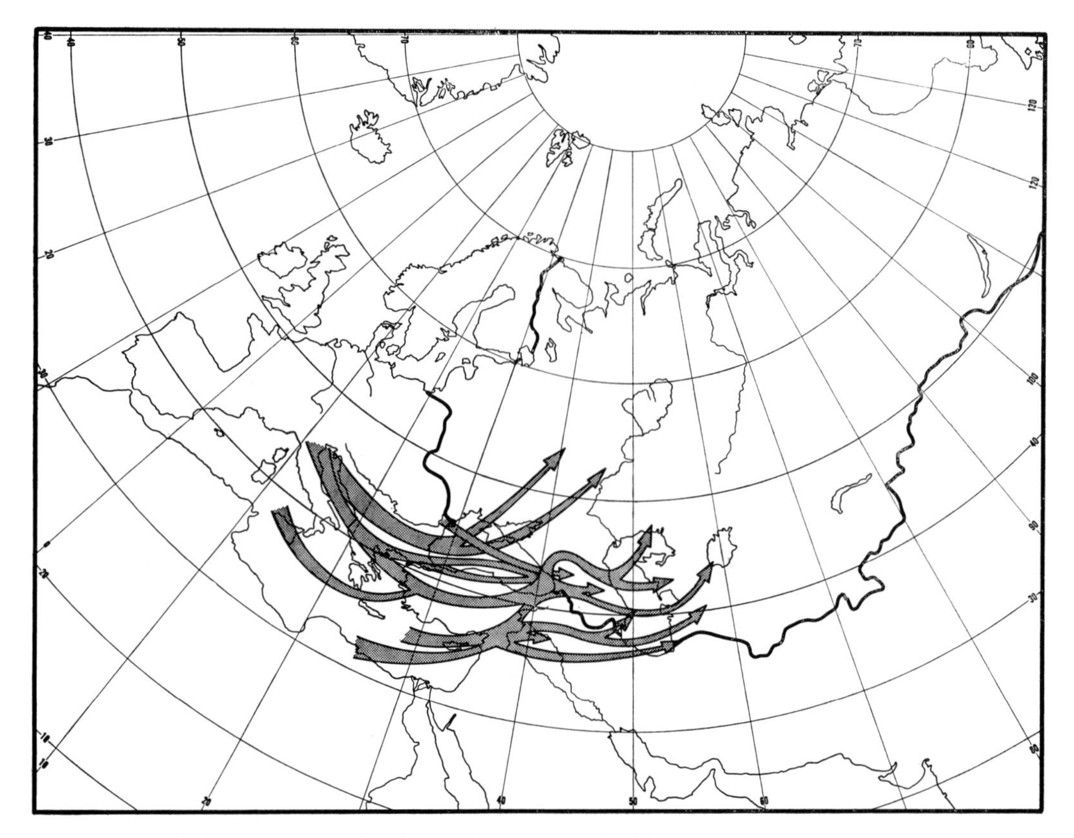

Fig.7-5. Cyclone trajectories in winter. (After ZANINA, 1961.)

are particularly frequent in winter when the general atmospheric circulation over the region is more conducive to an easterly flow. This occurs usually around the southern periphery of a high pressure cell centered somewhere over the middle Volga Valley. Or it may be associated with a strong westward push of the western extension of the Asiatic high. Foehn effects are most noticeable around Kutaisi at the northeastern edge of the Colchis Lowland where more than 100 days per year experience foehns. During the height of their occurrence in winter in this area they can occupy as much as 50% of the time. In Kutaisi they occur almost exclusively from the east or east-southeast. The easterly flow is usually shallow, no more than 1–2 km deep, above which westerlies prevail. Surface winds have maximum speed probabilities of 36 m/sec every year and 55 m/sec once in 20 years. During one extreme case in December the temperature in Kutaisi rose to 17.5°C and the relative humidity fell to 8%. This same flow continued southwest to Batumi and caused a temperature of 38°C. During winter foehns may cause evaporation of snow. During summer in Kutaisi they appear as very hot and desiccating winds which after several hours may cause vegetation to drop leaves. They usually last 2–10 h, but occasionally they may last up to a week, and after such duration much vegetation withers and dies. They thus bring on sukhovey conditions.

Foehns are less frequent in the eastern Transcaucasus than in the west, but they do occur fairly frequently in certain parts of the Kura Valley along the southwestern slopes of the Great Caucasus and also along the northeastern and eastern slopes of the Lesser Caucasus. In late summer west winds bring sukhovey foehns to the Kura Lowland as they descend the eastern slopes of the Surami Range. Relatively warm winter temperatures averaging approximately 0°C on the southeastern slopes of the Great Caucasus attest to the frequency of foehns produced by northerly air finding its way southward through the passes of the Great Caucasus.

When a high pressure occupies the upper portions of the Great Caucasus foehns may occur on both the northeastern and southwestern slopes. The Kazbegi High Mountain Station averages 52 days of foehn per year, mostly associated with anticyclonic build-up over the mountains. The foehns are most frequent in valleys that are oriented directly downslope. Thaws associated with them may cause disastrous avalanches. Foehns also occur on the Armenian Plateau primarily in winter in the lower lying basins on the lee sides of the Lesser Caucasus and at the bases of some of the higher volcanic cones. These descending winds generally form rather strong temperature inversions somewhere along the lower slopes of the mountains and overlying the lower basins and plains. This is particularly true in winter when cold air usually occupies depressions. On the North Caucasian Foreland during winter temperature inversions frequently lie at about 500–600 m above sea level and thus intersect the ground over the Stavropol' bulge in the middle portion of the plain. Cold air circulates into the area from the east around the southern periphery of the western extension of the Asiatic high underneath the inversion. This air has circulated over the frozen northern end of the Caspian Sea, is relatively cold, and reaches its condensation level soon as it rises up the eastern slope of the Stavropol' Plateau. There it produces much low cloudiness, fog, mist, rime and glaze ice, and hoarfrost, for which this area is particularly known throughout the Soviet Union. In contrast, the western slope of the Stavropol' Plateau experiences warm sunny weather as the easterly flow of air descends the slope and warms adiabatically.

Most of the cyclonic storms that enter the western part of the North Caucasian Foreland from the Black Sea during winter veer off to the northeast around the northern end of the Stavropol' Plateau and do not affect the southeastern portion of the plain. The plateau thus marks an important circulation divide in spite of its relatively low elevation. It is not alone responsible for the determination of routes of cyclonic storms, since it is natural for the storms to move from southwest to northeast in this area, as they indeed do over much of the European plain to the north. They are able to cross the low northwestern end of the Great Caucasus into the western portion of the North Caucasian Foreland, but farther east where the range is higher they are dissipated on the southern slopes. Thus, it is the character of the Great Caucasus that bears much of the responsibility for the differences in climate between west and east in the North Caucasian Foreland, but nevertheless the Stavropol' Plateau marks this division, and it figures significantly in the formation of such surface phenomena as glaze and hoarfrost under the inversion on its eastern slope.

Although temperature inversions are common in winter west of the Stavropol' Plateau as well as east of it, the cyclonic storms in the west tend to break them up frequently. Therefore, the weather at higher elevations is not as consistent in the west as in the east. But everywhere above the inversions winter weather tends to be sunny and relatively warm for the altitude. This has been conducive to the establishment of a number of winter sanatoria along the northern slopes of the Caucasus at intermediate elevations.

South of the eastern Great Caucasus in the Kura Lowland during winter easterly flow comes into the area relatively cold and dry so that the lower Kura Valley next to the

Caspian experiences a great deal of clear dry weather. But as the air gradually rises up the valley to the west cloudiness and fog develop similar to the case along the northern slopes of the Great Caucasus.

The dry easterly flow of air enters the Armenian Plateau and combines with locally derived air from the high pressure system that is prevalent in winter over the plateau. This dry clear air is conducive to strong terrestrial radiation over a continuous snow cover during winter which at night reduces temperatures to as low as −40°C in the northwestern part of Armenia. This is by far the most continental part of the Transcaucasus and compares to parts of the Eurasian plain much farther north. During winter the Armenian Plateau is more affected by easterly flow than it is by westerly flow, and therefore these cold dry conditions persist a great deal of the time. Westerly flow does penetrate the area occasionally and causes foehn effects which break up inversions and cause considerable warming in sheltered areas. During times of extreme cold on the Armenian Plateau cold air drainage down some of the steeper slopes may produce bora conditions.

Probably the most famous cases of bora winds in the Soviet Union occur at either end of the Great Caucasus when an exceptionally strong push of cold air in the western extension of the Asiatic high lodges against the northern slopes of the range. Under such conditions the cold air spills across the low passes of the northwestern extension of the Great Caucasus and southeastward along the narrow coastal plain between the southeastern end of the Great Caucasus and the Caspian Sea.

In the west the bora reaches its greatest strength around Novorossiysk where average wind speeds exceed 20 m/sec during such conditions. 40% of the cases experience winds greater than 30 m/sec, and separate gusts reach as much as 70 m/sec. Temperatures in Novorossiysk during the bora may fall to −15° or −20°C. Such conditions generally last a day or so and sometimes as much as two or three days. They may even last up to 15 days. They are most frequent in winter, particularly in November and December, but they can occur at any time of year. During the average winter bora of various strengths and prolongations are observed about five times between November and March. On the average they blow more than 30 days per year.

The bora at Novorossiysk is accentuated by the presence of a depression over the relatively warm Black Sea which sharpens the baric gradient with the cold air on the northern slopes of the mountains. Often the bora begins very suddenly with passage of a cold front and then persists as the western nose of the Asiatic high moves into the area. The air funnels into the city through Markhotskiy Pass to the north which lies at an elevation of 436 m. In this pass more than 120 days per year have storm winds. From November through March the winds in the pass average 10 m/sec. They frequently rise to 60 m/sec. As the winds cross the coast they whip up the waves of the Black Sea and cause rime ice to form along all the objects on the water front, the piers, electric wires, and so forth. Glaze may appear on the coastal roads causing great hazards to transport. The high winds and choppy seas present hazards to sea and rail transport as well. The wind dies quickly out to sea, generally extending no more than 10 km beyond the coast.

Strong northeast winds also blow through Markhotskiy Pass during summer, but at this time of year they reach Novorossiysk as warm, dry foehns.

As the height of the Great Caucasus increases southeastward, the bora diminishes and finally ceases to exist south of Sochi. However, cold air may penetrate further southeastward parallel to the coast if the isobars over European U.S.S.R. are oriented essentially north–south across the plain. Under such conditions Arctic air may be brought into the Black Sea coastal area in fairly fresh condition. This type of flow brings the coldest weather to the western Transcaucasus during the winter. Northwest of Sochi the coldest weather is associated with bora winds. All along the Black Sea coast the coldest temperatures in winter are produced by advection processes, which is in contrast to the North Caucasian Plain where the coldest temperatures are associated with intense terrestrial radiation at night under clear calm conditions.

Bora type winds also occur on the Apsheron Peninsula at the eastern end of the Great Caucasus where the air funnels in from the north between the Great Caucasus and the Caspian Sea. They may occur anytime during the year but are somewhat more prevalent during winter when they take on true bora characteristics. In the summer they are hot and dry and blow a great deal of dust into the city of Baku. They are generally known as "Bakinskiy nord". During the winter these winds are usually related to the northwestern sectors of passing cyclones, and therefore often follow closely on the heels of warm foehns which have come in from the south in the warm sectors ahead of cold fronts. Under such conditions temperatures often fall as much as 20°C during a 24-h period. Minimum winter temperatures are experienced along the Caspian coast during such occurrences.

Also famous on the Apsheron Peninsula and adjacent Caspian during winter are foehn-like conditions brought in by occasional southwesterly storms known as "moryana". During such occurrences the temperature rises, humidity falls, and clouds dissipate.

The Caspian Sea coast actually has more high winds than does the Black Sea coast. Makhachkala has an average of 69.8 days per year with wind speeds greater than 15 m/sec. Baku has 66.5. Novorossiysk on the Black Sea has 56.7. The high winds of Novorossiysk are almost all of a bora type and have by far the greatest seasonality of the three stations. At Novorossiysk the high winds appear primarily in winter. Makhachkala has a less pronounced winter maximum, and Baku to the south has a summer maximum.

Mountain–valley circulations predominate in certain closed basins, such as the Yerevan Basin and the basin of Lake Sevan, and in portions of the Colchis and Kura lowlands due to differential heating on the valley floors and mountain slopes. These are particularly prevalent in summer when the general circulation is weak. These local circulations may go to heights of 2,500 to 3,000 m in the western Transcaucasus and to more than 3,000 m in the eastern Transcaucasus. Generally they are gentle circulations which produce little more than ventilation during extreme temperature conditions, but occasionally they may become strong enough to produce convective showers. Convection is most intense in late spring when insolation has become intense on the lower slopes while snow still lies on the intermediate and upper slopes and thereby retards the rise of air temperatures because of the consumption of latent heat from melting. Under these conditions the greatest differential heating with altitude takes place and hence the greatest instability is effected. Throughout most of the Transcaucasian area, except for the Black and Caspian coastal areas, May is the month of maximum thundershower activity.

Precipitation

The Caucasian region has the most complex precipitation distribution, in both space and time, of any region in the U.S.S.R. Totals generally are greatest along the Black Sea littoral where they increase from 688 mm per year at Novorossiysk in the northwest to 2,504 mm at Batumi in the southeast. They reach their highest recorded value anywhere in the country at Achishko at an elevation of 1,880 m which has an annual average of 2,617 mm. In both the North Caucasian Foreland and the lowlands of the Transcaucasus amounts decrease rapidly eastward. Along the western shore of the Caspian, Derbent north of the Great Caucasus has 428 mm per year and Baku south of the Great Caucasus has 238 mm. South of Baku precipitation picks up again in the Talysh area in the very southeastern corner of Azerbaydzhan where Astara has 1,252 mm. In the central portion of the Transcaucasus the Armenian Plateau and the Lesser Caucasus receive annual totals between 250 and 800 mm.

The Caucasus region is often characterized as having a winter maximum of precipitation associated with cyclonic storms that traverse the eastern Mediterranean–Black Sea–Caspian area at this time of year. But this is generally true only along the immediate shore of the eastern part of the Black Sea, and even there it is a great oversimplification. In the northwestern part of the coast at Novorossiysk the primary maximum falls in December, but it is not an outstanding maximum. Precipitation is fairly evenly distributed through the year and bounces up and down repeatedly from one month to another. A secondary maximum occurs in July but the amount then is about the same as in November and in January. Minima occur in August and May. It is obvious that precipitation resulting from winter cyclonic storms and from summer thunderstorms fairly well even out during the year. Perhaps the most outstanding feature in this rather nondescript regime at Novorossiysk is the secondary maximum in July, which with 70 mm of rain stands well above 54 mm in June and 40 mm in August. Thundershower activity here is rather heavily concentrated in midsummer.

Southeastward along the Black Sea coast the summer maximum shifts to August and then September as the winter maximum diminishes and finally disappears. At Batumi in the southeast there is almost four times as much annual precipitation as at Novorossiysk, and a single maximum occurs in September and a single minimum in May. It appears that thundershower activity becomes greater southeastward and its greatest intensity during the year lags from midsummer into early fall. Precipitation from winter cyclonic storms increases southeastward too, but not quite as much as that resulting from summer thundershowers. Batumi shows a secondary maximum in November, after a sharp drop in October, which is almost as high as the primary maximum in September. There appears to be a precipitation decrease in October between the September peak of thundershower activity and the November oncoming of winter cyclonic storms. A very pronounced minimum occurs in May at Batumi. This May minimum persists all along the Black Sea coast, and generally there is a secondary minimum which occurs immediately after the summer maximum. In the north this secondary minimum occurs in October after the single maximum in September. The shift in the summer maximum from July in the northwest to September in the southeast does not seem to be attributable to changes in surface–air temperature contrasts since maximum surface temperatures occur everywhere in August.

Very quickly inland the precipitation regime changes drastically as primary control shifts to summer thunderstorms. Tkibuli in the hills northeast of the Colchis Lowland has a primary maximum in June and a primary minimum in February with a secondary maximum in December and a secondary minimum in August. June and July stand out sharply above May and August, and December stands out sharply above November and January. Thus, there is a two-month primary maximum in early summer and a one-month secondary maximum in early winter. In both cases minima follow closely on the heels of maxima. Early winter seems to be the peak period for cyclonic storms throughout the Black Sea coast, the Colchis Lowland, and the western part of the Great Caucasus. December is the month of most frequent precipitation maxima and September is second. Nowhere is there a maximum during the period January–May.

The seasonal distribution of precipitation continues to change farther inland. Much of the Armenian Plateau, eastern Georgia, and western Azerbaydzhan have May maxima of precipitation. Tbilisi and Kirovabad both have primary maxima in May and secondary maxima in September. Both have primary minima in January and secondary minima in August. The Armenina Plateau generally shows a single maximum and a single minimum. Leninakan, Sevan, and Yerevan all have May maxima. Maximum thunderstorm activity throughout this area falls during May.

Farther east, at Baku, the primary maximum is in November and the secondary maximum is in April, while the primary minimum is in August and the secondary minimum is in February. Of course, there is greatly reduced precipitation throughout the year.

There seems to be a general time progression of thunderstorm activity from east to west across the Transcaucasus, the eastern part getting April and May activity and the western part getting September activity. This is represented by the secondary maximum at Baku in April, the primary maximum at Kirovabad, Tbilisi, and on the Armenian Plateau in May, the primary maximum at Tkibuli in June, and the September maximum along the southern part of the eastern Black Sea coast. Almost the opposite trend obtains north of the Great Caucasus. Here there is a general tendency for maxima to appear later and later eastward. The west-central portion of the North Caucasian Foreland as far east as Groznyy has a June maximum, while Makhachkala on the Caspian coast has a September maximum and Derbent southeastward down the coast has a November maximum similar to that at Baku farther southeast.

Astara, which has the highest precipitation in the Talysh region in the very southeastern corner of the Transcaucasus next to the Caspian Sea, has a primary maximum in October, a secondary maximum in March, a broad primary minimum in June and July, and a secondary minimum in January. It appears that the farther south one proceeds the more the winter precipitation maximum is delayed into early spring. The March secondary maximum at Astara matches the March maximum eastward across the Caspian Sea throughout the entire plains region of Central Asia. March and October might well mark the times of extreme differences in air masses between cold air protruding from the north and local air in southeastern Transcaucasia. October appears to be the time of greatest frequency of cold fronts in this sheltered area.

Sunshine, clouds, and precipitation frequencies

Among other things, the Caucasus are known for their sunshine. However, this varies according to elevation and exposure to moisture bearing winds. In general the intermediate elevations on the mountain slopes experience greater cloud frequencies than do either the lowland or the higher peaks. Everywhere summer is much sunnier than winter. This is due not only to the greater length of daylight period during the summer but also due to less cloudiness in the summer. On the northeastern Black Sea coast, Novorossiysk receives 75% of possible sunlight during July, August, and September, while it receives only about 40% during December, January, and February. Makhachkala on the Caspian shore receives about 70% of possible sunshine during June, July, August, while it receives only about 25% during January and February. The seasonality of sunshine is similar in the Transcaucasus. Only the high mountain zones show a relatively even distribution of possible sunshine during the year. This is because in winter they are generally above the clouds.

Everywhere in the Caucasus, except for some of the higher mountain zones, winters are much cloudier than summers. Winter cyclonic storms and the prevalence of temperature inversions at intermediate elevations during winter make for great amounts of stratus type clouds. In the North Caucasus, Pyatigorsk at 13h00 records an average of approximately 8 tenths sky cover during the winter months, while during the summer months it records approximately 5 tenths. Gagra on the Black Sea coast records approximately 8 tenths sky cover during winter and 5 tenths or less during summer. In the central Transcaucasus, Tbilisi records about 0.75 during winter and 0.41 in August. On the Caspian shore Baku has 14.4 days of overcast during January and only 2.2 during August.

Yerevan has the greatest number of clear days throughout the entire Caucasus, 106 per year. Most stations range from 50 to 80 days per year. The greatest number of overcast days occur at the high rainfall stations of Batumi and Astara along the southeast coast of the Black Sea and the western coast of the Caspian Sea, respectively. Batumi experiences 159 days with overcast per year and Astara 156. On the eastern slope of the Stavropol' Plateau in the North Caucasian Foreland Ordzhonikidze has 150 overcast days per year and Groznyy has 149. On the Caspian coast north of the Great Caucasus Makhachkala and Derbent both have 140 overcast days per year, as does the high rainfall station of Achishko in the western Great Caucasus. Along the northeastern coast of the Black Sea Sukhumi has 134 days overcast per year and Sochi has 119.

In most places in the Caucasus amounts of cloudiness, frequencies of precipitation, and amounts of precipitation show similar seasonal regimes. However, south of Sochi along the Black Sea coast and in portions of the western North Caucasian Plain there is almost an opposite seasonality between the number of rainy days and amounts of rainfall. Poti at the eastern end of the Black Sea experiences the greatest number of days with more than 0.1 mm of rain during March and April, while September is the month of maximum rainfall. The March–April period of high frequency must signify a great number of prolonged foggy, drizzly days associated with high atmospheric stability and cool surface water temperatures in the Black Sea.

In moving along the eastern Black Sea coast southeastward from Novorossiysk to Batumi the total amount of rainfall increases significantly, the number of days having more than 0.1 mm of rain become more evenly distributed through the year and much more frequent during the entire year, and the time of maximum rainfall backs up from December or January in the northwest to September in the southeast. Actually, there isn't a consistent shift in the month of maximum rainfall, but rather a jumping back and forth in the middle portion of the coast between double maxima of December and September until finally there is a single maximum in September in the southern part of the coast. At Novorossiysk there is a total of 125.4 days of precipitation of more than 0.1 mm, while at Batumi there are 176.5 days. Rainy days at Novorossiysk are at a maximum of 14 in January and a minimum of 6.3 in August. Batumi has a much more even distribution of rainy days, with a maximum of 15.6 in March and a minimum of 12.5 in October.

Things change rapidly inland. Along the northeast edge of the Colchis Lowland Tkibuli shows a rather even distribution of rainy days during the year, with a slight maximum in July. June is the month of maximum precipitation. Farther east, Tbilisi, Kirovabad, and Yerevan have May maxima of rainy days, rainfall amounts, and 24-h rainfall. These coincide with the time of greatest convective activity.

On the Caspian shore Baku shows a November maximum for rainy days, rainfall amounts, and 24-h rainfalls. Farther south, Lenkoran and Astara show October maxima for all these things. Among those stations recording 24-h rainfall maxima, Poti holds the record with 268 mm during a June day. However, many stations do not have records of 24-h rainfalls. This is true of Achishko, which holds the highest recorded annual rainfall in the country.

Fog, snow, glaze, and thunderstorms

Most of the fog in the Caucasus is of an advection type, produced either by warm moist air moving over a colder surface or by upslope action along mountain fronts. Its frequency varies in both amounts and regimes primarily according to elevation. The upper slopes have much greater frequencies of fog than do the lower basins. The greatest fog frequency is recorded on Mount Elbrus at an elevation of 4,250 m where 220 days per year experience fog.

In the lower elevations almost everywhere the annual fog regime is practically the opposite of that of the precipitation regime. In the high mountains, where fog is really clouds on the ground, the two regimes coincide. The Black Sea shore, which has relatively little fog, everywhere has the maximum in April and May, which is the period of greatest atmospheric stability and least precipitation. Over the Stavropol' Plateau on the North Caucasian Foreland the foggiest month is generally December or January. The upper portions of the eastern slope of the Stavropol' Plateau have outstandingly high numbers of days per year with fog due to easterly flows of surface air rising up the eastern slope underneath the inversion in winter. Ordzhonikidze in this region has 103 days of fog per year. The central and eastern Transcaucasus have infrequent fog. Tbilisi has only 10 days per year, most in December, January, or February. Yerevan averages 26 days, with maxima in January and February.

The number of days with snowstorms in the Caucasus varies almost directly with elevation. In the high mountains there are frequently more than 100 days per year with snowstorms, but on the plains it is generally only a few days or even a fraction of a day. Sevan at an elevation of approximately 2,000 m gets 12 days per year with snow, while Yerevan at approximately 1,000 m gets 0.9 days. Many of the lowland stations do not receive snow every year.

Generally snow cover is very light in the Caucasus. It is usually nil along either sea coast. The eastern Transcaucasus has the lightest snow cover away from the sea coasts. Kirovabad in the Kura Lowland averages only 2 cm during the 10 days of greatest snow depth. This is true at Baku also. Yerevan on the Armenian Plateau has a maximum snow depth of 19 cm, and Tkibuli in the foothills northeast of the Colchis Lowland has 29 cm. On the North Caucasian Plain Ordzhonikidze has a maximum snow depth of 10 cm. In the western Great Caucasus Achishko averages 470 cm on the ground during the ten days of maximum snow cover.

The number of days with snow cover varies from less than 10 days along parts of the Black and Caspian sea coasts to 288 days at the Kazbegi High Altitude Station. In the Surami Range Tbilisi has 16 days of snow cover, and on the Armenian Plateau Sevan records 106 days.

Ice storms occur occasionally in the Caucasus. They are very infrequent in the Transcaucasus, except at elevations above 1,000 m. However, the eastern slope of the Stavropol' Plateau on the North Caucasian Foreland is known for its frequency of glaze. This occurs primarily with overcast drizzly weather that moves up the eastern slope of the Stavropol' Plateau underneath a temperature inversion in winter. Groznyy at an elevation of 126 meters records the most glaze, 20 days per year. Farther west Ordzhonikidze at an elevation of 696 m records only 10 days per year. Apparently Ordzhonikidze frequently is within the temperature inversion while Groznyy is almost always below it.

The Caucasus region experiences the highest frequency of thunderstorm occurrence in the U.S.S.R. Throughout much of the Caucasus thunderstorms occur from 20 to 40 days per year. They are particularly frequent in the high mountain zones and the Armenian Plateau where they commonly occur more than 50 days per year. Along the eastern Black Sea coast thunderstorms average about 30 days per year, and along the Caspian coast 15 or less. There are only 7 thunderstorms per year at Baku. May through August is the season of maximum occurrence almost everywhere.

The high frequency of thunderstorms over the Caucasus is well known because of the hail damage to such crops as citrus and vineyards. Hail occurs from one to four days per year in lowland areas and more than 8 days per year at elevations of 2,000–3,000 m. The phenomena of atmospheric turbulence and hail have been under intensive investigation at the Vysokogornyy Geofizicheskiy Institut (High Altitude Geophysics Institute) in the Great Caucasus with the intent to develop effective controls on hail damage.

Radiation and temperature

Radiation amounts in the Caucasus are some of the greatest in the Soviet Union, being exceeded only by those in portions of Central Asia. Amounts are greatest in the high mountains where the air is usually free of pollutants and water vapor and clouds are at a minimum. Most of the Great Caucasus measures a global radiation of 140–150 kcal./cm^2 per year. It rises to 170 kcal. on Mount Aragats on the Armenian Plateau. The Kura Valley on the other hand averages 120–130 kcal., and the humid Colchis Lowland averages less than 120.

Temperatures in the Caucasus generally are milder even than in Central Asia (Fig.7-6). Because the Transcaucasus is protected from northerly winds in winter, minimum temperatures average well above those in Central Asia which is wide open to Siberian air to the north. But even in the warmer parts of the Transcaucasus hard freezes occur every year which are great hazards to subtropical crops such as citrus (Fig.7-7). In one

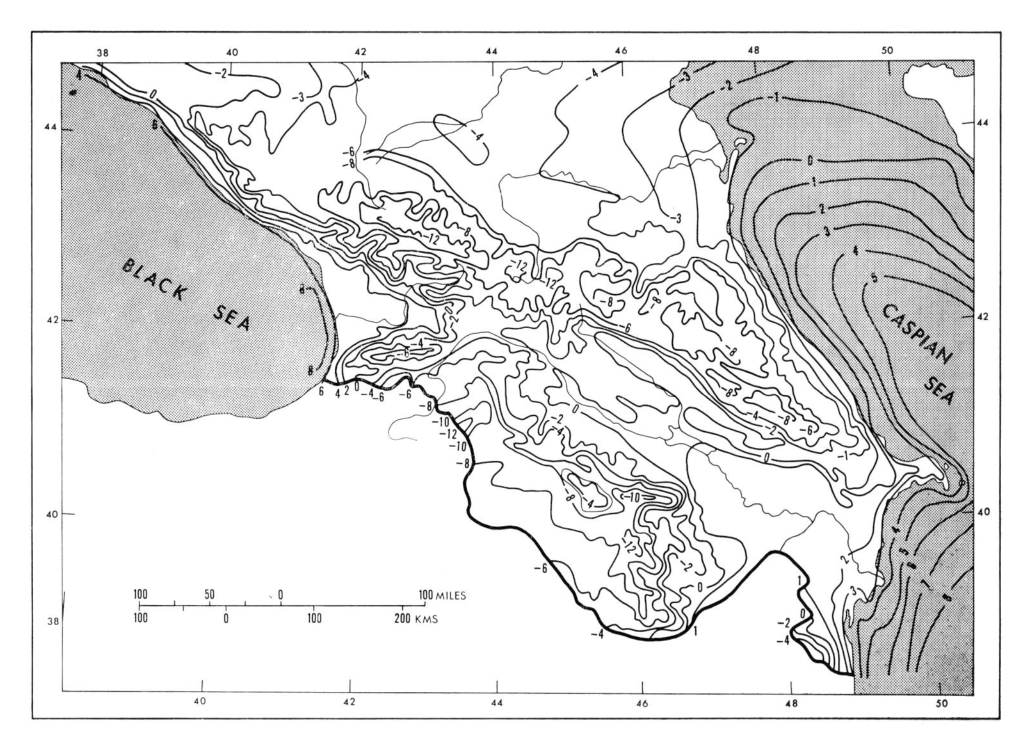

Fig.7-6. Average temperature (°C) of the surface air in January. (After ZANINA, 1961.)

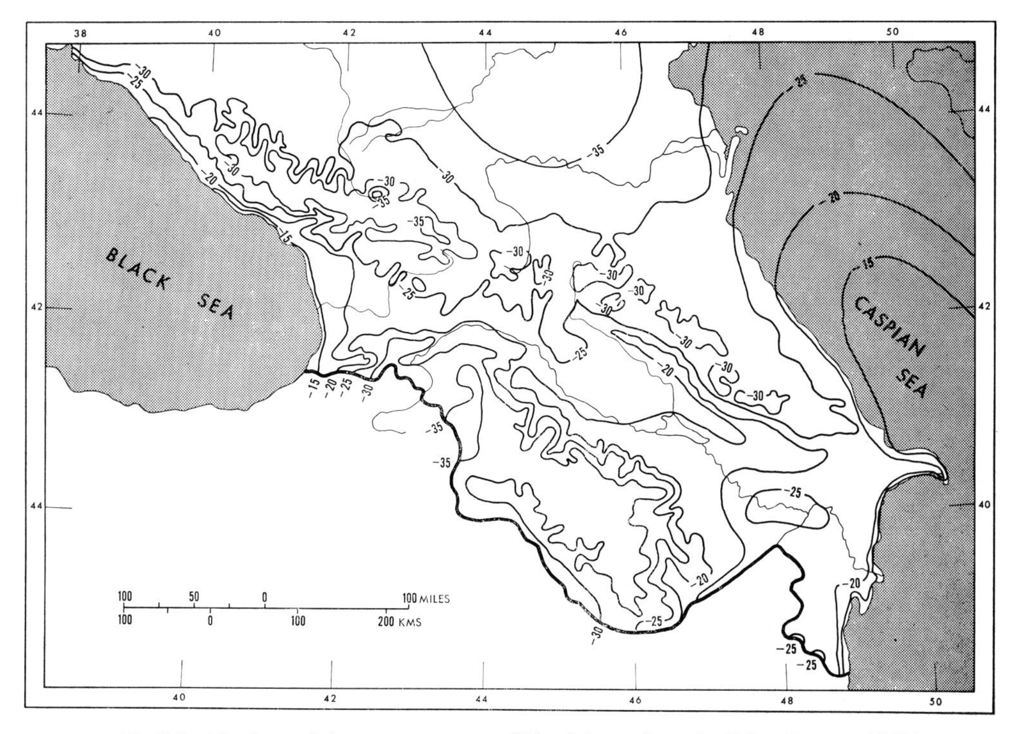

Fig.7-7. Absolute minimum temperature (°C) of the surface air. (After ZANINA, 1961.)

of the most sheltered areas along the Black Sea coast Batumi has experienced below freezing temperatures from November through April and it has experienced −8°C minima in January and February. The average annual frost free period in Batumi is 304 days. Farther northwest, Novorossiysk has experienced a minimum of −24°C and averages a frost-free period of only 231 days. In the Colchis Lowland, Kutaisi has experienced minima of −17°C and has a frost-free period of 288 days.

Temperatures, of course, are lower at higher elevations, but they do not necessarily decrease immediately upon ascent from sea coasts and lowland basins because of temperature inversions that persist in many areas particularly during the coldest parts of winter. These are enhanced by the many foehn winds that flow down most of the slopes at one time or another. Along the Black Sea coast of the Great Caucasus in winter there is generally a temperature inversion which causes the air to average several degrees warmer 100 m or so above the sea than it does at sea level. Gagra is the driest and warmest place in western Transcaucasia because the mountains approach the coast very closely and the air is heated by frequent foehns. Temperatures at Gagra can reach 20°C in winter under such conditions, accompanied by very low relative humidities.

Temperature extremes increase eastward in the Transcaucasus because of increasing continentality and increasing elevation. Tbilisi has experienced extreme values of −20°C and +38°C. Yerevan has experienced −27°C and +40°C. Leninakan at an elevation of 1,528 m has experienced an absolute minimum of −35°C. Lake Sevan on the Armenian Plateau does not freeze in winter and keeps the temperature along its immediate shore 2°C higher than further inland. The mean January temperature along the lake shore is about −6°C.

Because of the saucer-like shape of the Armenian Plateau cold air settles into the enclosed basins and acts as a source region for high pressure and cold continental air during the winter which makes the Armenian Plateau the coldest part of the Caucasus. At similar elevations the Great Caucasus generally experiences temperatures several degrees warmer than the Armenian Plateau in winter. Along the Caspian Sea shore, Baku has recorded an absolute minimum of −14°C and an absolute maximum of +38°C. The Black Sea has a much greater influence on temperatures in winter than does the Caspian Sea. Average January temperatures are about 3°C higher along the Black Sea than along the Caspian. In summer the Caspian coast averages about 25°C while the Black Sea coast averages about 23°C (Fig.7-8). Baku and Yerevan have the highest average monthly temperatures in summer, each having an average of 25.5°C in August.

The North Caucasian Foreland, of course, is open to northerly winds in winter and therefore averages considerably colder than the Transcaucasus. Pyatigorsk averages −4°C in January and has experienced temperatures as low as −33°C. Freezing temperatures occur from September through May, and the frost free period averages 175 days per year. Pyatigorsk averages only 21.6°C in July, but has experienced a temperature of 41°C.

Climatic regions of the Caucasus

The Caucasus region is generally divided into six climatic regions which owe their distinctiveness primarily to topography and location with respect to the Black and Caspian seas (Fig. 7-1).

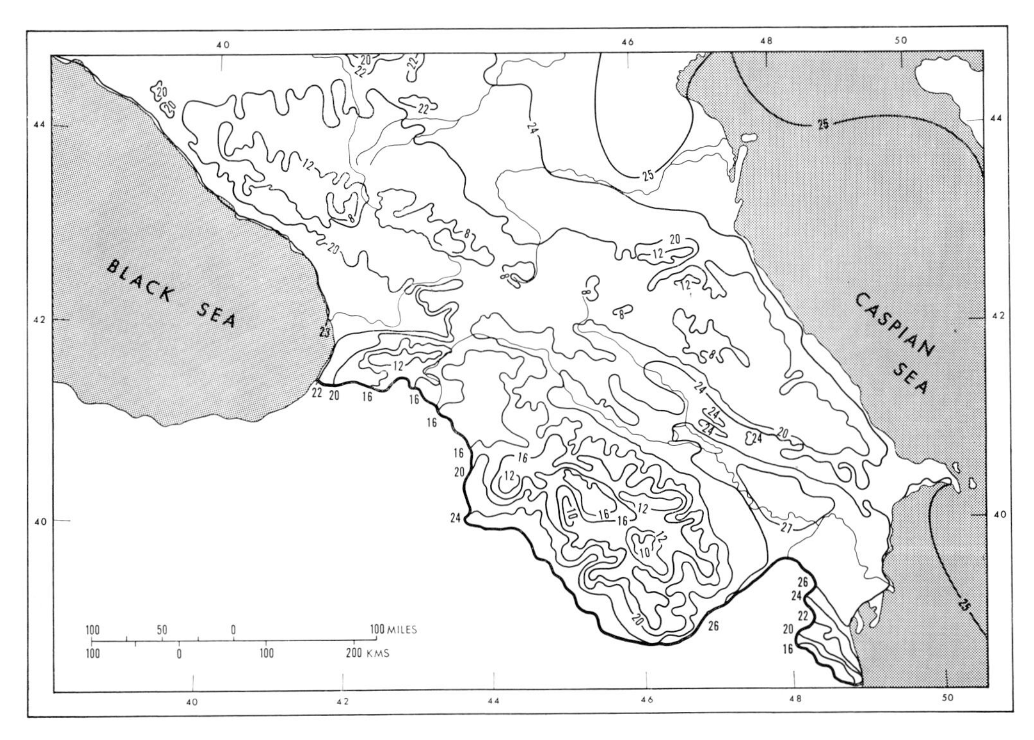

Fig.7-8. Average temperature (°C) of the surface air in July. (After ZANINA, 1961.)

Northern Caucasus

Except for the Black Sea coast, the Northern Caucasus reflects a high degree of continentality which produces cold winters and hot summers and a pronounced summer maximum of precipitation. Cyclonic activity is related mainly to European cyclones. Except in the northwest along the Black Sea coast, Mediterranean cyclones do not penetrate the height of the Great Caucasus very well. Arctic air pushing against the northern slopes of the Caucasus may drop temperatures to below −25°C, and form a strong temperature inversion along the mountain slopes. Winter temperatures decrease from west to east. Average January temperatures in the Kuban region in the northwest are around −2°C, while in the Terek Valley to the southeast they are approximately −5°C. Absolute minimum temperatures range from −20°C on the Black Sea to −30°C on the Stavropol' Plateau and −25°C on the Caspian Coast. Thaws occur rarely in winter, generally with foehn winds coming down the northern slopes of the mountains which may raise maximum daytime temperatures to 15°C on the plain. Snow reaches a depth of 20 cm in the western part of the Stavropol' Plateau and usually remains until the middle of March.

Precipitation diminishes from west to east. Maximum precipitation is in winter along the Black Sea coast and in summer throughout much of the rest of the North Caucasus. During fall the amount of precipitation gradually decreases in the central part of the North Caucasus and increases on both sea coasts. The increase takes place gradually from summer to winter along the Black Sea, but it increases very abruptly along the Caspian, which is a peculiarity of that coast in general, not only of its northern part.

Fall and early winter are the times of maximum precipitation along the Caspian coast of the Caucasus.

Cloudiness is generally greatest everywhere in winter and decreases rapidly in spring, except in the high mountains where it is the other way around. There summers experience the maximum amount of cloudiness associated with cumulus build-up. This change in annual regime of cloudiness takes place at an elevation of about 1,000 m.

The weather of the interior part of the North Caucasian plain is well exemplified by the Fedorov diagram for Pyatigorsk (Fig.7-9). Here it can be seen that the summers are relatively sunny and warm about 50% of the time. Midsummer experiences relatively low atmospheric humidities, and 2 or 3% of the time from early July until late September the region experiences temperatures greater than 22°C coupled with relative humidities of less than 40%. The other 50% of the time during summer is occupied by various types of cloud conditions and precipitation. About 25% of the time during midsummer convective activity causes cumulus cloud formations during the day which dissipate at night. Overcast skies occur approximately 20% of the time in summer, and rain occurs 5 to 10% of all days. The greatest rainfall amount occurs in June, and the total amount for the year is not great, 482 mm.

Winter in Pyatigorsk is mildly cold. About 50% of the days experience temperature changes across 0°C. Therefore there is a lot of freezing and thawing. The other 50% of the time the temperatures are generally below freezing and about 10% of the time they are below −15°C. Absolute minima reach −30°C and lower.

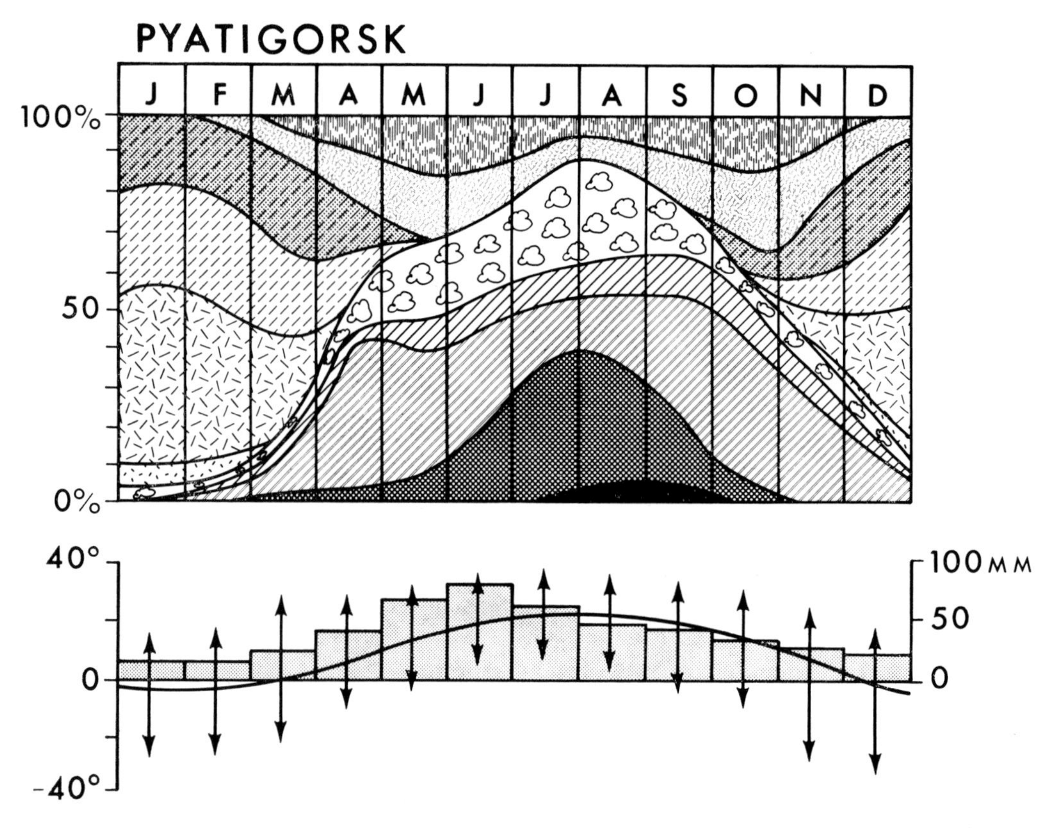

Fig.7-9. Frequencies of weather types, Pyatigorsk. (After BAYBAKOVOY and NEVRAYEVA, 1963.) For legend Fig.3-25.

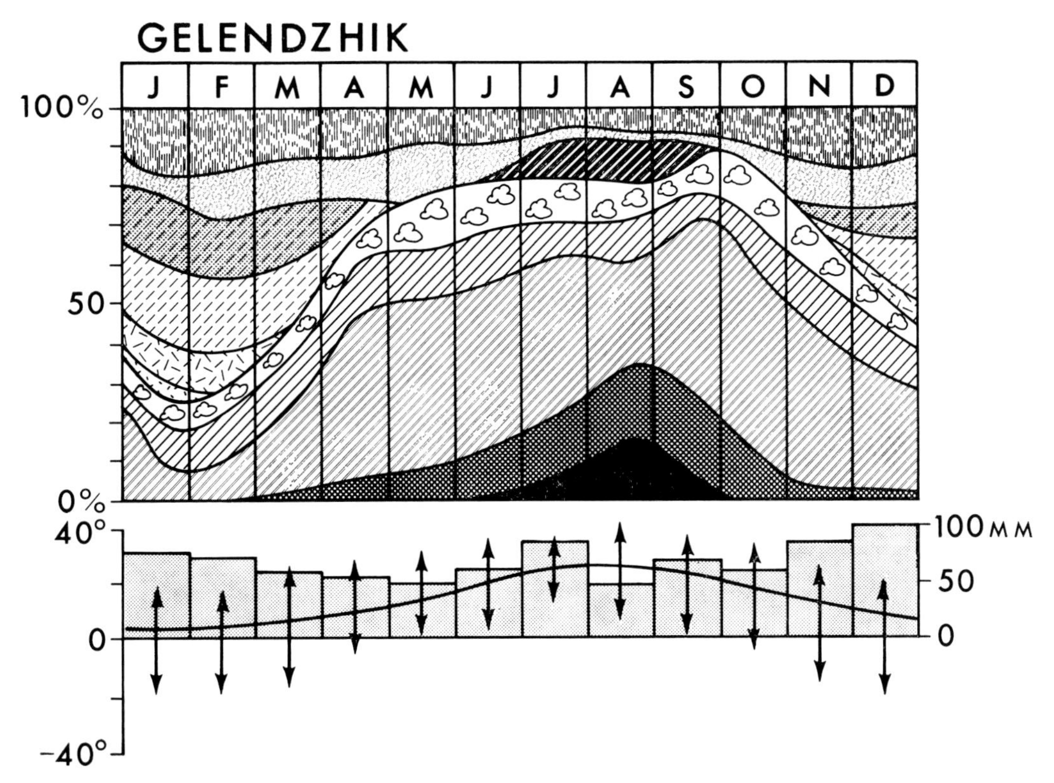

Fig.7-10. Frequencies of weather types, Gelendzhik, southeast of Novorossiysk. (After BAYBAKOVOY and NEVRAYEVA, 1963.) For legend see Fig.3-25.

The Black Sea coast of the Northern Caucasus is illustrated by the diagram for Gelendzhik, southeast of Novorossiysk (Fig.7-10). Here it can be seen that the winters are considerably milder than they are at Pyatigorsk and the summers are somewhat warmer, longer, and more humid. The rainfall regime is reversed from that at Pyatigorsk. There is more rainfall here, it is rather evenly distributed through the year, and it has a primary maximum in December, which reflects cyclonic activity, and a secondary maximum in July which reflects the build-up of thundershowers. All months at Gelendzhik average temperatures above freezing and there is much less freezing and thawing than at Pyatigorsk. The annual amplitude of temperature is less and the variation from year to year is less at Gelendzhik. Maximum frequency of cloudiness and precipitation takes place during the winter months. There is much less daytime cloudiness due to cumulus build-up during summer at Gelendzhik than at Pyatigorsk.

The high mountain zone of the Great Caucasus

The influences of elevation can be seen in the diagram for Kazbegi on the northern slope of the central Great Caucasus (Fig.7-11). The annual temperature regime shows a much reduced amplitude over that of the plains, which is caused primarily by the reduction of summer temperatures since winter temperatures do not vary much with elevation. Summers are cool and short, and winters are long and cold. Much of the cloudiness takes place during summer with the build-up of cumulus activity, and much of the precipitation is concentrated in the months May through August and comes in the form of showers.

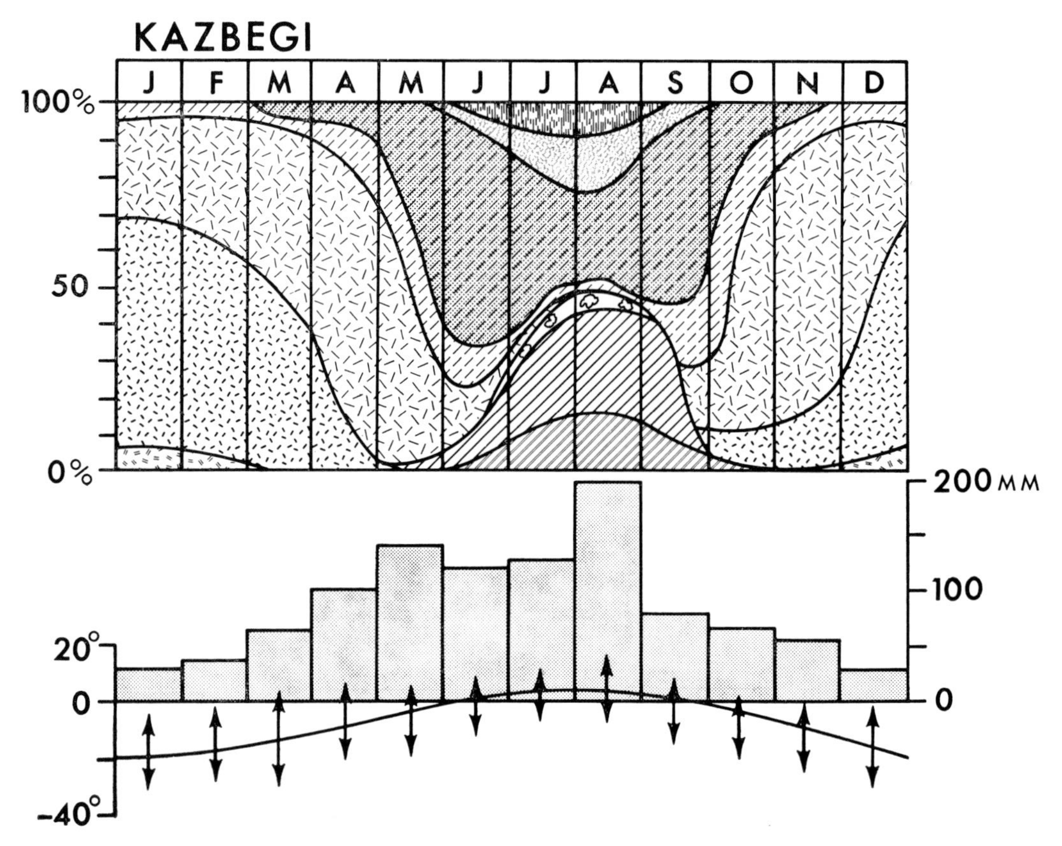

Fig.7-11. Frequencies of weather types, Kazbegi. (After GERASIMOV, 1964.) For legend see Fig.3-25.

August has perceptibly more precipitation than any other month during the year. More than 50% of the days during summer the temperature crosses the freezing point. In winter it is almost always below freezing. Winters are sunny since most of the stratus clouds lie below this altitude.

The western part of the Great Caucasus has much more precipitation than the eastern part, and the southwestern slopes, which are exposed to most of the moisture bearing winds, are much wetter than the northeastern slopes. Southwestern slopes of the western Great Caucasus may receive 3,000 mm of precipitation or more during the year, while the northeastern slopes probably do not exceed 1,000 mm. Differences can also be seen in the elevation of the snow line in the mountains, which on the average is 800 m higher on northeastern slopes than on southwestern ones.

A sunny winter is one of the biggest individualities of the climate in the high mountains, which is conducive to the location of winter sanatoria. Because of temperature inversions and the slow decrease of temperature with altitude during winter, temperatures in the mountains are not much different from what they are on the plains below. However, the high mountains are more subject to sharp weather changes, and can be subjected to severe blizzards and strong winds. On the other hand foehns may produce very warm dry days. Although they bring welcome breaks in the winter weather, the resultant thawing may bring on disastrous avalanches of snow in the mountains. Critical points along mountain highways must be protected with snow sheds built over the roads.

Spring is not very well distinguished in the high mountains because of the great amount of heat that has to be spent on melting of snow and glaciers. Summers are cool. The average July temperature at a height of 2,000 m is approximately 14°C. During summer mountain–valley circulations are well developed and a diurnal regime of cloudiness is sharply portrayed, with cumulus build-up during the day and dissipation at night. Torrential summer rains are common. The convectional rains are augmented by cyclonic precipitation during the summer, which is most active in June. The heaviest precipitation falls on slopes of intermediate elevation. There is generally a decrease above 3,000 m because of the low moisture holding capacity of the cold air at that altitude. Cloudiness generally remains relatively great in the high mountains except on the very highest summits.

Autumn is generally the best time of year in the Great Caucasus. Convective activity weakens and cloudiness decreases. Insolation is still fairly high, and the air is relatively warm over large areas, which causes small pressure gradients and weak winds. The average daily temperature decreases below freezing during the second half of November at an elevation of about 2,000 m and a snow cover becomes established soon afterward.

Western Transcaucasus

The warmest winters in the U.S.S.R. are observed in the western Transcaucasus. This region has the greatest protection from cold air intrusions from the north, it has the greatest influence of the advection of warm air in the warm sectors of cyclones passing inland from the Black Sea, and it has a great number of foehn winds which increase temperatures and lower relative humidities. This area also has the greatest amounts of precipitation in the country. The Black Sea littoral almost everywhere has a pronounced winter maximum of precipitation, but in spite of this winter insolation is fairly great. Periods between cyclonic storms are generally sunny.

Average January temperatures along the coast are approximately 6°C. Minima generally do not fall below −10°C. Coldest temperatures are associated with the intrusion of Arctic air from the northwest along the coast around the northwestern end of the Great Caucasus.

As can be seen in Figs.7-10, 7-12, and 7-13, as one progresses southeastward along the Black Sea coast from Gelendzhik to Kobuleti winters become much milder, total precipitation becomes much greater, and the annual regime of weather types becomes much more consistent through the year. Farther inland, at the eastern edge of the Colchis Lowland, Tskhaltubo, northwest of Kutaisi, shows greater seasonality in winter types but an evener distribution of amounts of precipitation through the year as increased thundershower activity in the interior during the summer makes up for diminishing rain from the winter cyclonic storms (Fig.7-14).

The enclosed nature of the Colchis Lowland east of the Black Sea makes for weak pressure gradients and wind circulations in summer and high relative humidity which generally is around 75 to 80%. This is considerably higher than in winter which is more frequented by dry foehn winds descending the western slopes of the Surami Range from the east. Although July temperatures in the Colchis Lowland average only around 23°C with maxima only slightly above 30°C, the high relative humidity creates an oppressively hot condition during the summer. Thus, much of the western Transcaucasus is truly subtropical in character. However, severe frosts occur in most places every year.

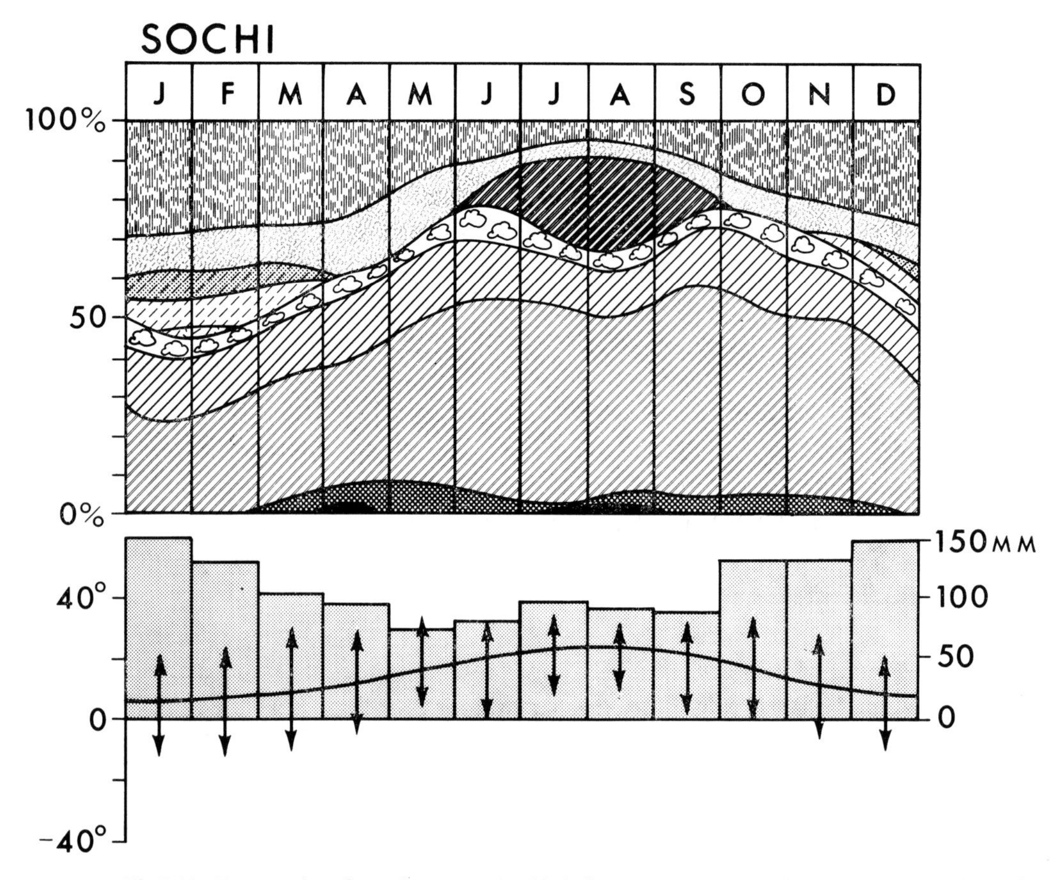

Fig.7-12. Frequencies of weather types, Sochi. (After BAYBAKOVOY and NEVRAYEVA, 1963.) For legend see Fig.3-25.

Eastern Transcaucasus

Eastern Transcaucasus is significantly drier than western Transcaucasus. The moist air which enters the Colchis Lowland from the Black Sea loses much of its moisture over the Colchis Lowland and the western slopes of the Surami Range and descends as a dry flow of air into the Kura Valley to the east. Most of the Kura Valley is semi-arid or arid. The slopes of the mountains on either side also are perceptibly drier than they are further west. Summers are a little warmer and winters are a little colder in the Kura Valley than they are in the Colchis Lowland. This is because of the increased continentality of the region. Cold northerly intrusions are drawn southward in winter along the western coast of the Caspian in the rear of passing cyclones and push their way into the Kura Valley bending around the southeastern end of the Great Caucasus. During such intrusions minimum temperatures may fall to —20°C even along the Caspian, except in the far southern corner known as the Lenkoran Lowland.

Throughout much of the eastern Transcaucasus, except along the Caspian shore, precipitation is greatest in May, associated with the time of maximum convective activity. During May the mid Kura Valley receives about 80 mm of precipitation which is about equal to that in the Colchis Lowland at the same time. Summers are hot and dry, with average temperatures in July ranging from 25° to 28°C. Relative humidities during the

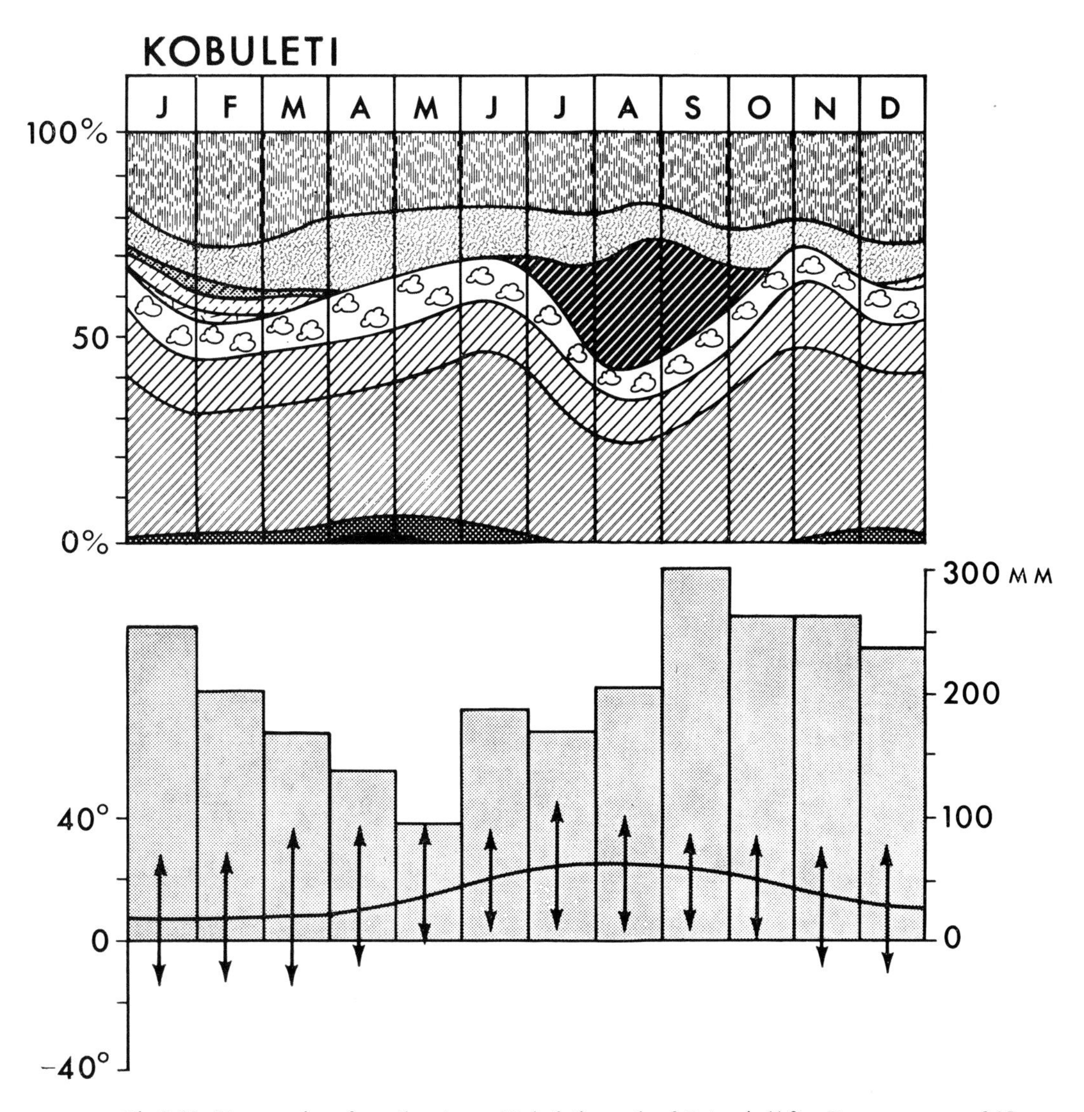

Fig.7-13. Frequencies of weather types, Kobuleti, north of Batumi. (After BAYBAKOVOY and NEVRAYEVA, 1963.) For legend see Fig.3-25.

day generally are no more than 40%, and moisture reserves in the soil become depleted so that irrigation is absolutely necessary for agriculture throughout much of the Kura Valley.

The diagram for Mardakyany, on the northeast coast of the Apsheron Peninsula, illustrates one of the drier parts of this region (Fig.7-15). Summers are long and warm, and winters are mild. However, there is a significant amount of freezing and thawing, a phenomenon which is relatively rare on the Black Sea coast of the Colchis Lowland.

The Talysh region

In southeastern Azerbaydzhan rainfall increases once more as air flow from the northeast approaches the Talysh Mountains that run northwest–southeast along the Soviet–Iranian border. In places yearly precipitation totals more than 1,000 mm. Thus, this restricted area is comparable to the Colchis Lowland with its warm humid subtropical

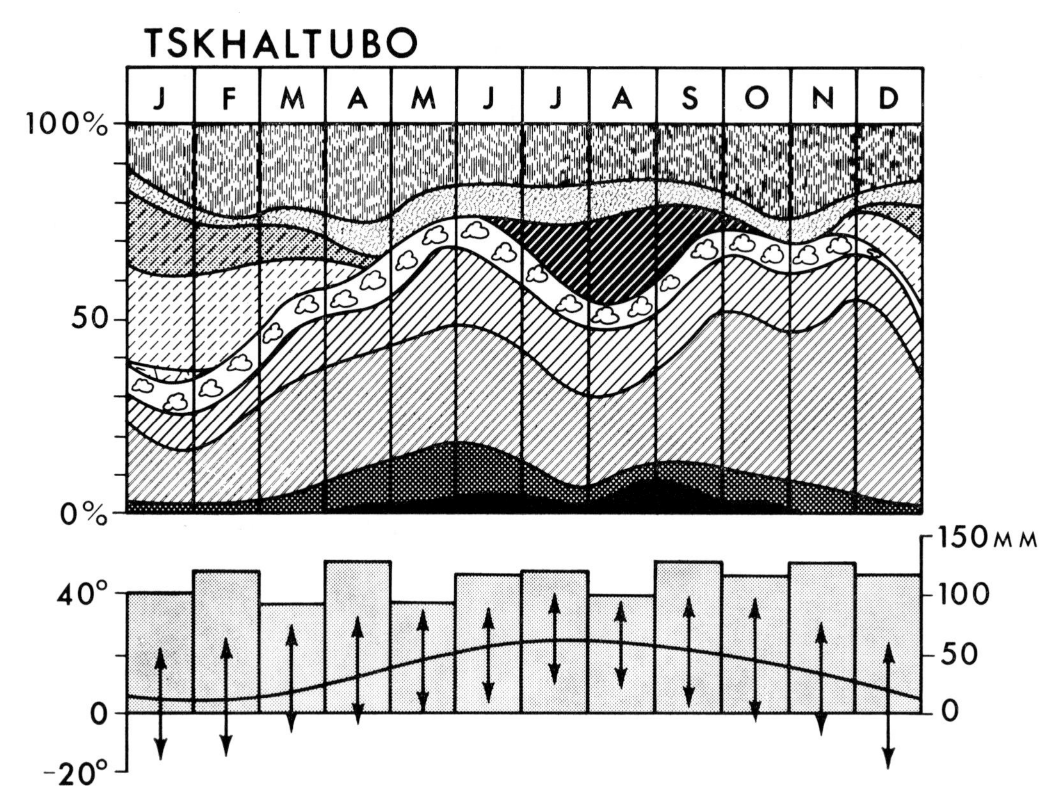

Fig.7-14. Frequencies of weather types, Tskhaltubo, northwest of Kutaisi. (After BAYBAKOVOY and NEVRAYEVA, 1963.) For legend see Fig.3-25.

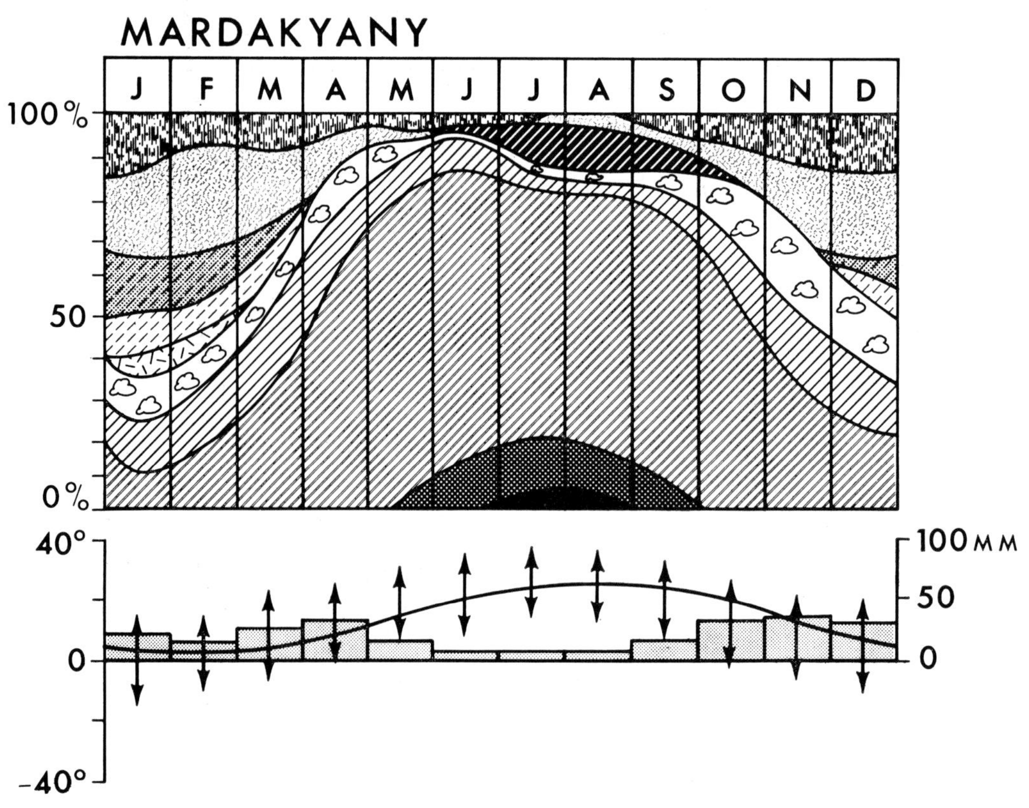

Fig.7-15. Frequencies of weather types, Mardakyany, northeast of Baku. (After BAYBAKOVOY and NEVRAYEVA, 1963.) For legend see Fig.3-25.

climate. However, here the maximum precipitation comes in the fall. There is a very rapid increase in September after dry July and August, and the maximum usually falls in October (Fig.7-16). Relative humidity at this time of the year averages about 80%. The transitional seasons seem to be the times of most frequent frontal activity in this area. Although spring is much drier than fall, it is wetter than either winter or summer. Moisture bearing winds are often drawn in from the east across the Caspian coast in the northeast sector of cyclonic storms.

The Armenian Plateau and the Lesser Caucasus

The Armenian Plateau is the most continental part of the Caucasus. It has cold and in some places severe winters for the latitude. Average January temperatures generally are below freezing, and minima may fall to −30°C or lower. The cause of the severe weather is not height alone but is also the plateau character of the region and the dryness of the air. At similar altitudes the Great Caucasus is considerably warmer.
The distribution of air temperature on the plateau varies greatly from place to place depending upon the land form. Lowest temperatures are observed on the high enclosed basins of the plateau, but on their slopes where cold air can drain away temperatures average relatively high. For instance, Leninakan at a height of 1,500 m in an enclosed basin on the plateau has an average January temperature of −12°C, lower than it is in

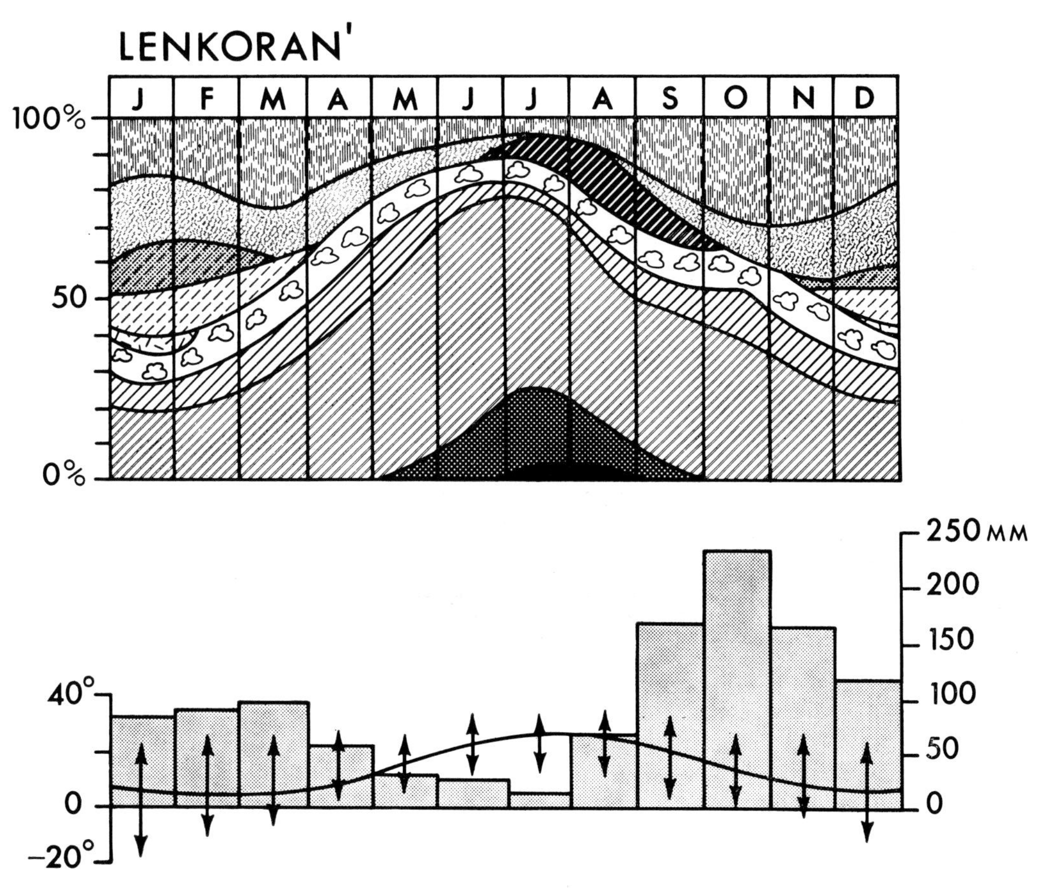

Fig.7-16. Frequencies of weather types, Lenkoran'. (After GERASIMOV, 1964.) For legend see Fig.3-25.

Moscow. At the same elevation on the eastern Karabach summit the January temperature averages −2°C, the same as in Crimea. Winters are relatively dry and sunny with the development of a local high pressure over the cold surface. Average cloudiness is only 50% during winter and precipitation amounts are small. The snow cover is unstable and winds are weak.

Temperatures rise rapidly during spring, and the precipitation increases rapidly to a maximum in May with the high development of convective activity. Precipitation decreases rapidly from May to June and reaches an absolute minimum in August. Although the high surface temperatures in summer make for relatively low surface air pressures, the surface low changes to a high aloft and much of the plateau is overlain by subsidence inversions during the summer which limit convection and precipitation. Thus the summers are relatively hot and dry. The northwestern slopes of the mountains facing the air flow from the Black Sea are an exception to this. They receive considerable precipitation in summer.

As was the case over much of eastern Transcaucasia, the transitional seasons are generally the rainiest on the Armenian Plateau. Winter and summer receive the least precipitation. The degree of aridity depends upon the topography. Driest areas are mainly on the eastern slopes of the Lesser Caucasus and adjacent areas. The driest portions are in southeastern Armenia, the Yerevan Basin, and the Araks Valley. Cloudiness is small in summer and insolation is high. In comparison to the Great Caucasus, which were warmer than the Armenian Plateau at similar elevations in winter, July temperatures in some of the mountain basins of the Armenian Plateau average more than 25°C which is considerably warmer than the Great Caucasus at this time of the year.

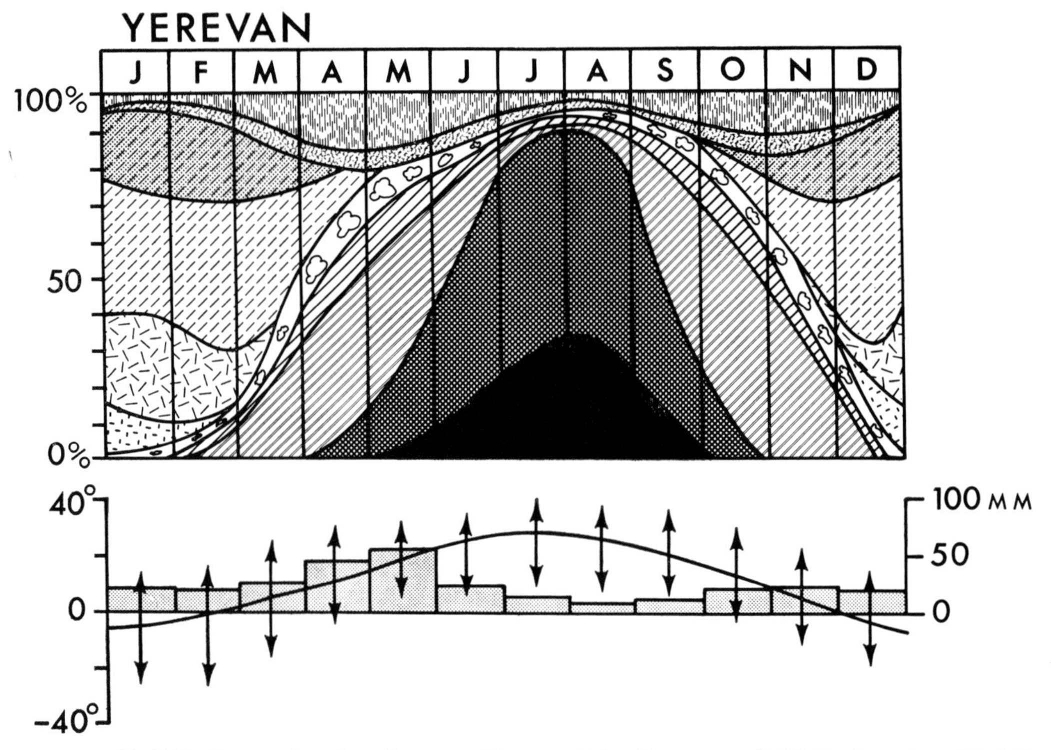

Fig.7-17. Frequencies of weather types, Yerevan. (After GERASIMOV, 1964.) For legend see Fig.3-25.

Weather in one of the lower basins of the Armenian Plateau is typified by the diagram for Yerevan which lies at an elevation of 907 m above sea level. The city sits in a basin which is almost completely encircled by higher land except in the south where the basin opens up to the Araks River lowland. Summers here are quite hot and sunny with stable air which is capped by an upper level inversion that traps much haze and atmospheric pollution. Winters are relatively cold with temperatures sometimes dropping to less than −20°C. The absolute range of temperature is from +40°C in summer to −27°C in winter. Winter experiences frequent passages of temperatures through the freezing point. At higher elevations on the plateau, of course, winter temperatures become even more severe and summers are cooler. The general regime of weather however remains about the same.

References and further reading

ALEKSANDRIAN, G., 1971. *Atmosfernye osadki v Armyanskoy SSR* (Precipitation in the Armenian SSR). Yerevan.

ALISOV, B. P., 1956. *Klimat SSSR* (Climate of the USSR). Moscow Univ., Moscow, 127 pp.

ANAPOL'SKAYA, L. E. and SUKHISHVILI, E. V., 1969. Raschetnyye skorosti vetra na kavkaze (Categorization of wind speeds in the Caucasus). *Gl. Geofiz. Obs. Tr.*, 246.

ANONYMOUS, 1968. *Gidrometeorologicheskiy rezhim Kolkhidskoy Nizmennosti* (Hydrometeorological regime of the Colchis Lowland). Tr. Zakavk. Nauchnoissled. Gidrometeorol. In-ta., Leningrad, 28: 239 pp.

ANONYMOUS, 1969. *Atlas SSSR* (Atlas of the USSR). Gl. Upr. Geod. Kartogr., Moscow, 2nd ed., 199 pp.

ANONYMOUS, 1971. *Klimat i Klimaticheskie Resursy Gruzii* (Climate and climatic resources of Georgia). Tr. Zakavk. Nauchnoissled. Gidrometeorol. In-ta., Leningrad, 44, 383 pp.

BAGDASARYAN, A. B., 1958. *Klimat Armyanskoy SSR.* (Climate of the Armenian SSR). Yerevan, 140 pp.

BAYBAKOVOY, E. M. and NEVRAYEVA, G. A. (Editors), 1963. *Ocherki po Klimatologii Kurortov* (Essays on the climatology of resorts). Tr. Tsentr. Nauchnoissled. Inst. Kurortol. Fizioter., Moscow. (Foldout map showing graphs of weather types at resorts in the USSR.)

BORISOV, A. A., 1970. *Klimatografiya Sovetskogo Soyuza* (Climatography of the Soviet Union). Leningrad Univ., Leningrad, 225 pp.

BUDYKO, M. I., 1971. *Klimat i Zhizn* (Climate and life). Gidrometeoizdat, Leningrad.

BURMAN, E. A., 1969. *Mestnyye Vetry* (Local winds). Gidrometeoizdat, Moscow, 340 pp.

GERASIMOV, I. P. (Editor), 1964. Fiziko-Geograficheskiy Atlas Mira (Physical Geographical Atlas of the World). Akad. Nauk SSSR, Moscow, 298 pp.

GONCHAROVA, E. F., 1940. O sinopticheskikh usloviyakh sil'nykh vetrov v Sochi (On the synoptic conditions associated with strong winds in Sochi). *Meteorol. Gridrol.*, 3: 36–41.

KAMINSKIY, A. A., 1932. *Klimat SSSR, Chast II, Davlenie vozdukha i veter v SSSR* (Climate of the USSR, Part II, Air pressure and winds in the USSR). Gl. Geofiz. Obs., Leningrad.

KUZNETSOV, N. N., 1940. Tipy fenov na zapadnom Kavkaze (Types of foehns in the western Caucasus). *Meteorol. Gidrol.*, 12: 19–32.

MALIK, S. A., 1939. Maskirovka frontov kavkazom (Camouflaged fronts in the Caucasus). *Meteorol. Gidrol.*, 6: 34–45.

MALIK, S. A., 1940. Sinopticheskie usloviya sil'nykh vetrov Chernomorskogo poberezh'ya kavkaza ot Anapy do Adlera (Synoptic conditions producing strong winds along the Black Sea Coast of the Caucasus from Anapa to Adler). *Meteorol. Gidrol.*, 3: 27–35.

MASTERSKIKH, M. A., 1968. K prognozy frontal'noy bory v rayone Novorossiyska (Toward the prognosis of frontal bora in Novorossiysk region). *Gl. Upr. Gidrometeorol. Sluzby Tr.*, 22: 43–46.

NAPETVARIDZE, E. A. and PAPINASHVILI, K. I., 1939. Sinopticheskaya kharakteristika tsentral'noy chasti glavnogo kavkazskogo khrebta (Synoptic characteristics of the central part of the main Caucasus range). *Meteorol. Gidrol.*, 10–11: 32–45.

SHIKHLINSKIY, E. M., 1967. The thermal balance of the Caucasus. In: M. I. BUDYKO (Editor), *Modern Problems of Climatology*. Translation Division, Wright-Patterson Air Force Base, Ohio, pp.137–155.

STEPANOVA, N. A., 1960. *Annotated Bibliography of the Climate of the Caucasus*. U.S. Weather Bureau, Foreign Area Section, Office of Climatology, Washington, D.C., 43 pp.

TSUTSKIRIDZE, YA. A., 1967. *Radiatsionnyy i Termicheskiy Rezhimy Territorii Gruzii* (Radiation and thermal regimes in Georgia). Gidrometeoizdat, Leningrad, 162 pp.

VORONTSOV, P. A., 1940. Prizemnye inversii temperatury pri sukhoveyakh v zapadnoy Gruzii (Surface temperature inversions during sukhovey in western Georgia). *Meteorol. Gidrol.*, 11: 44–49.

ZAMORSKIY, A. D., 1938. Sinopticheskie usloviya sil'nykh dozhdey na Chernomorskom poberezh'e kavkaze ot Anapy do Adlera (Synoptic conditions with heavy rain along the Black Sea Coast of the Caucasus from Anapa to Adler). *Meteorol. Gidrol.*, 9–10: 39–52.

ZANINA, A. A., 1961. *Klimat SSSR, Vypusk 2, Kavkaz* (Climate of the USSR, volume 2, the Caucasus). Gidrometeoizdat, Leningrad, 290 pp.

ZVEREV, A. S., 1957. *Sinopticheskaya Meteorologiya* (Synoptic meteorology). Gidrometeoizdat, Leningrad, 559 pp. pp.318, 388.

Chapter 8

The Thermal Factor

Heat energy is the driving mechanism for many earth systems. And the ultimate source of all heat energy received at the surface of the earth is the radiation from the sun. The amount of radiation, both direct and scattered, the so-called "global" radiation, that is received at any point determines the heat budget that that point on the earth's surface has to work with, although, of course, heat may be advected in or out of a region in the form of both sensible heat and latent heat. Some of the incoming radiation is reflected and lost as far as any use is concerned. The rest is absorbed at the surface and either effects a temperature rise or evaporation of water. A temperature rise at the earth's surface induces convective currents in the lower atmosphere, and a turbulent heat transfer takes place from the earth to the atmosphere. At the same time evaporated moisture is carried upward into the air. Thus, heat energy initiates the atmospheric motion and provides the driving force for the hydrologic cycle. Therefore, heat energy becomes the basic element in all climatological processes. A knowledge of its magnitudes and modes of exchange is of primary importance for the understanding of climatological processes that are taking place in any area.

Russian climatologists were among the first to recognize the basic need for understanding the various elements of the heat budget of the earth, and Soviet climatologists have led the world during the last two or three decades in research and publication on this aspect of climate. Foremost at the present time is undoubtedly the work of the Main Geophysical Observatory in Leningrad and its director, M. I. Budyko. They have been particularly concerned with the quantification of the heat budget as a means for understanding completely its many-faceted role in the growth of plants, the development of soils, and its pervasion of all man's activities. They look upon the heat budget as holding more possibilities for ultimately solving more relationships than can be done through the study of temperature since heat can be expressed as quantities of energy whereas temperature can only be expressed as energy levels.

Unfortunately at the present time practically all components of the heat budget are lacking adequate records of observational data. Not until the organization of the International Geophysical Year in the second half of the 1950's was a world-wide actinometric network set up to measure the total solar (global) radiation at several hundred stations. In most areas of the world such stations are still so sparsely scattered that they serve only to provide some observational check for computed values of the heat budget which use formulas that relate components of heat to commonly measured elements such as temperature, humidity, cloudiness, and wind. Components of the heat budget other than incoming radiation are almost always derived from long recorded data on traditional weather elements whose relationships to the heat budget are still so tenuous as to make the results of computations highly suspect. Nevertheless, Budyko and his co-workers laid the basis for such computations in his renowned monograph, *The Heat Balance of the Earth's Surface*, which was published in 1956 and translated into English in 1958. Since then methods have been refined, and the Main Geophysical Observatory, under the editorship of M. I. Budyko, has published a second edition of the *Atlas of Heat Balance of the Earth* in 1963, which, together with BARASHKOVA et al. (1961), provides the basis for most of the heat budget studies of the U.S.S.R. More recent publications by Budyko, Pivovarova, and others have added to these two basic sources.

Observational data on heat budget components are increasing rapidly, and heat budget theory is changing accordingly. This has injected much inconsistency in maps which are supposed to be depicting the same thing in publications that are spread only two or three years apart. This is true between the two basic sources cited, as well as between them and some of the more recent publications. It is quite evident that information on heat budget components is still fragmentary and mostly computed rather than observed. Frustrating as this might be, a discussion of the heat budget components will be presented here in as up-to-date form as is possible. Consistency within any one source will be sacrificed in order to utilize what appears to be the latest, most accurate information available on any given element. In all cases, it must be kept in mind that the small scale maps utilized here depict conditions at low elevations, generally below 500 m above sea level, and ignore wide local variations of such things as albedos. At any given point on the earth's surface, heat components could differ markedly from regional characteristics presented on the maps. This would be true particularly in mountainous areas, where it has been estimated that barring excessive cloudiness, incoming radiation increases by about 10% for each kilometer rise through the first 3 km of elevation.

Radiation components

Global radiation

The mean annual amounts of total incoming radiation, or global radiation, reflect the high latitudes of the Soviet Union (Fig. 8-1). The country shows generally low total amounts of radiation and only modest variations from one place to another. Moscow receives about 88 kcal./cm^2 per year, which is similar to Churchill in the central part of Canada. Leningrad with only 68 kcal./cm^2 compares to Great Slave Lake in northwestern Canada. In the far southwest, Odessa, with 118 kcal./cm^2 receives a little less than New York City. And Vladivostok, the southernmost city in the Soviet Far East, with 110 kcal. per year, compares to Halifax, Nova Scotia. The greatest amount of incoming radiation during the year is received in the southern part of Soviet Central Asia where computations indicate that a considerable area receives as much as 160 kcal. per year, although no recording station shows more than 140. This compares to more than 200 kcal./cm^2 in the deserts of southwestern United States and northwestern Mexico and more than 220 kcal. in the eastern Sahara and western Arabia.

By-and-large, incoming radiation in the Soviet Union compares latitudinally with North

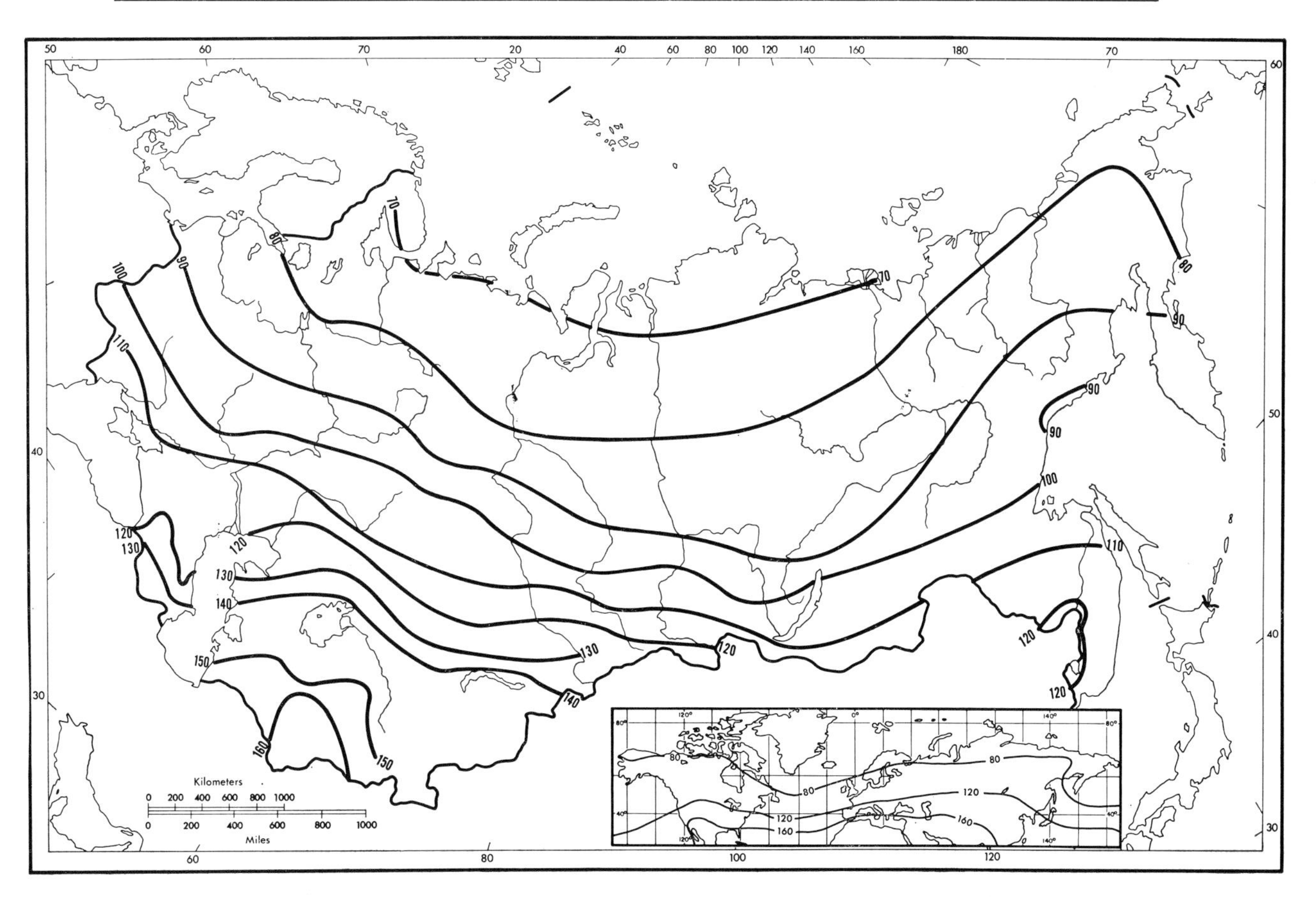

Fig.8-1. Mean annual global radiation (kcal./cm²). (After PIVOVAROVA, 1966; inset from BUDYKO, 1963.)

America. However, the latitude in the Soviet Union in most cases is quite high, so that most comparisons must be made with Canada rather than with the United States. There are some compensating factors, though, at these high latitudes, for yearly averages here hide much greater seasonal swings in radiational receipt than is the case in lower latitudes. High latitudes benefit greatly from long summer days. The Arctic coast of the Soviet Union, beyond the Arctic Circle, in June compares very favorably with northeastern United States, while over the entire year it receives only about 60% as much radiation as northeastern United States. Farther north, over the Arctic Ocean, June radiation becomes even greater and would compare favorably with the deserts of southwestern United States or the deserts of Soviet Central Asia.

In both Eurasia and North America the June sum of solar radiation decreases from south to north to about the 65th or 70th parallel, after which it increases again. Minimum values range from 14 to 16 kcal./cm² in the vicinity of the Arctic coast of the Soviet Union, which is comparable to the Great Lakes region in North America. Further north in the Soviet Arctic they increase to more than 20 kcal. which is comparable to the Central Asian deserts (Fig.8-2). Definite northward bulges of the isolines in the interiors of both Eurasia and North America attest to the transparency of the atmosphere in interior locations, which is due primarily to reduced moisture in the air. Thus, the largeness of the land mass of the Soviet Union is also a contributing factor to anomalously high values of radiation during the summer.

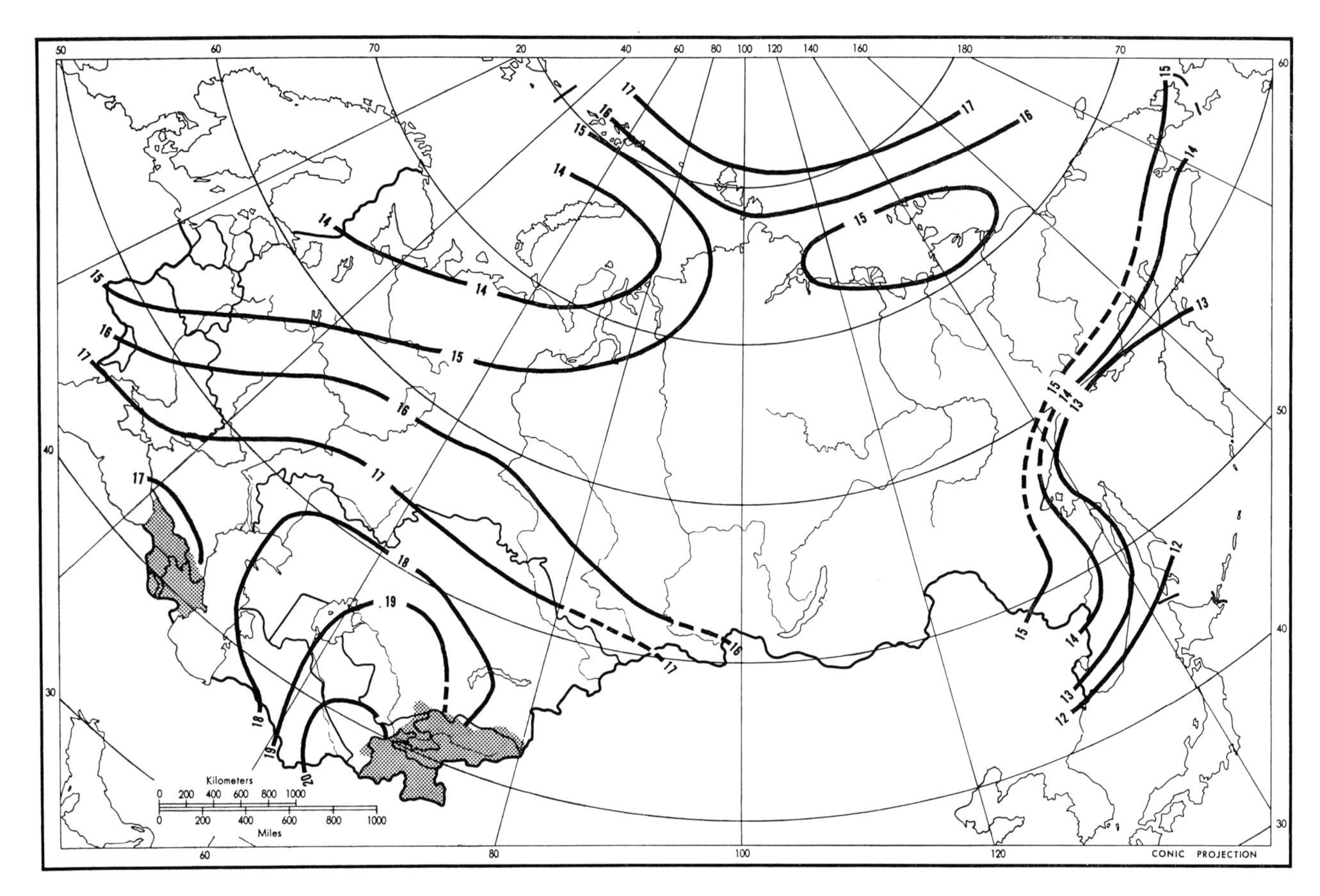

Fig.8-2. Mean global radiation (kcal./cm²) in June. (After BARASHKOVA et al., 1961.)

Radiation totals in June vary in a northeasterly direction from about 20 kcal. in the Central Asian desert to 15 kcal. around the mouth of the Lena River on the Arctic coast. To the east and to the west of this maximum axis radiation decreases as the coasts are approached. The Soviet Far East in June receives the least radiation in the country. This is no doubt due to the monsoon flow of air inland from the Pacific at this time of year which is accompanied by much cloudiness and precipitation. The isolines of radiation are almost parallel to the coast all the way from the Bering Straits southward to Korea. To a lesser degree, the Barents Sea and its fringes in the northwestern part of the Soviet Union also experience reduced radiation due to the high degree of cloudiness.

The distribution of radiation is arranged much more latitudinally in December than in June (Fig.8-3). Thus it is quite obvious that latitude is a more important control over radiation in winter than in summer. This is just the opposite of the case with temperature distribution in the Soviet Union. While latitude is the primary control over radiation during both seasons, in summer the state of the atmosphere becomes very important also. Atmospheric humidity and clouds exert much greater influences in summer when relatively large amounts of insolation are being received than they do in winter when insolation is very low over much of the territory anyway. North of about 69° latitude in December no radiation is received at all.

There is about the same radiational difference between south and north across the Soviet Union in June as there is in December. But the distribution of radiation is very different

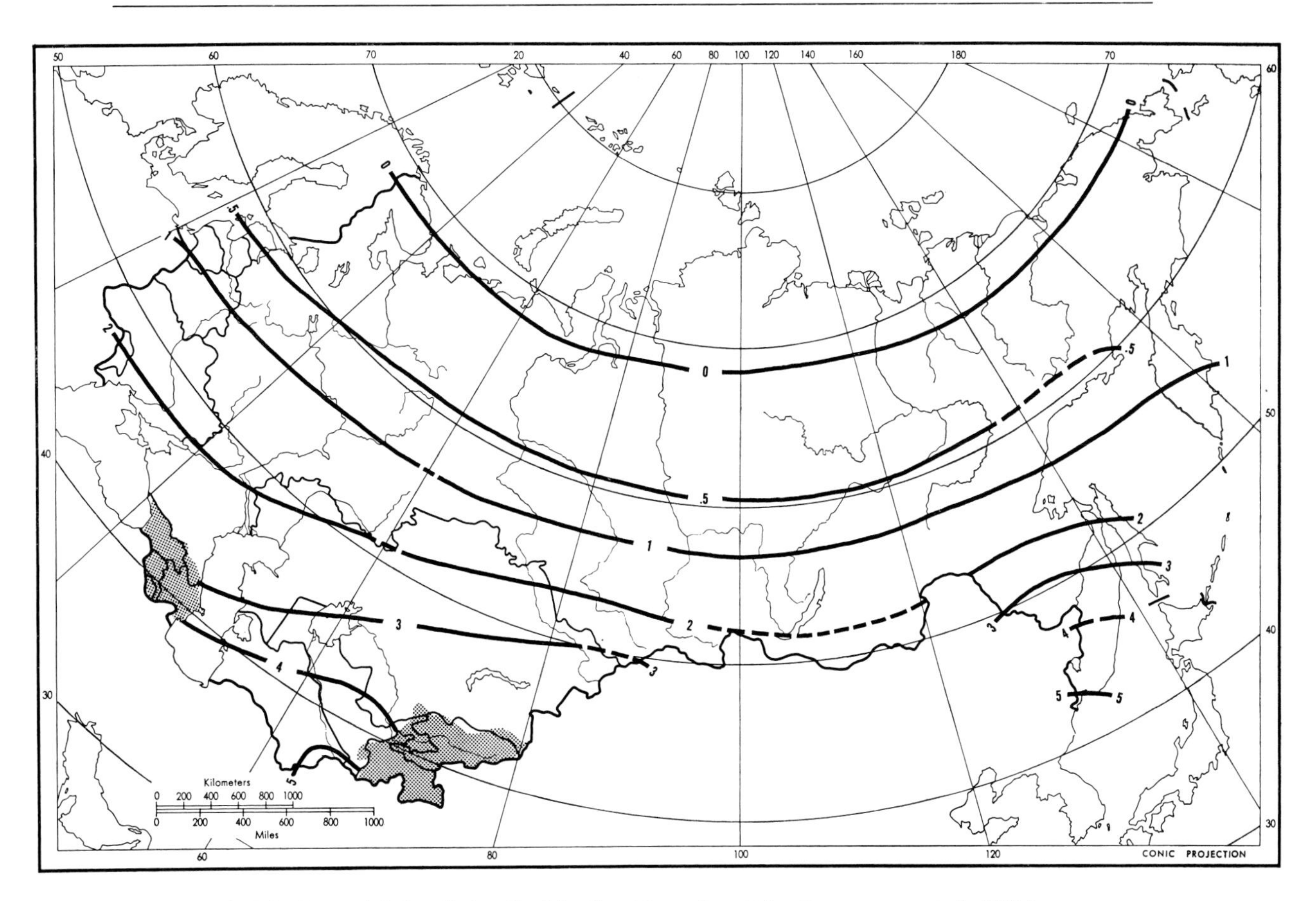

Fig.8-3. Mean global radiation (kcal./cm²) in December. (After BARASHKOVA et al., 1961.)

in the two months. During June the radiation decreases northward to about 70° latitude after which it increases again, while in December there is a gradual decrease northward with a minimum in the Arctic. The pronounced minimum over the Barents Sea in June disappears in December, as does the minimum in the Far East. The Pacific coast of Maritime Kray obviously is quite clear during the winter, since the southward bend of the isolines of radiation does not take place until over the water off shore. This relative clearness in winter is of course caused by the persistence of the Asiatic high over the area. Thus, Vladivostok in December receives about the same amount of radiation as does the southern fringe of Soviet Central Asia and about $2\frac{1}{2}$ times as much as comparable latitudes in the western part of the country.

An essentially latitudinal arrangement of incoming radiation exists from September through March. However, already in March the isolines begin to deviate from a simple zonal pattern, and by April there is very little semblance of a latitudinal distribution at all. A large bulge of high radiation values has developed over the eastern part of the country clear into the Chukchi Peninsula, and a region of low values has begun to develop over the northwestern part of the country. In Central Asia the highest values of all are beginning to show along the southern borders of the country. Isolines run from southwest to northeast all the way from the northern coast of the Black Sea to the Arctic Coast of the Chukchi Peninsula. These developments in April are accentuated in May when the southwest-to-northeast trend itself is lost and the area of low values over the northwestern part of the country spreads eastward throughout the central part of the country as far east as Lake Baykal. The high-value area of Central Asia intensifies, and a separate closed high-value area appears over the Arctic coast east of the Taymyr Peninsula. In the Far East, isolines, which began their southward deflection as early as February, now almost entirely parallel the coast of the Sea of Okhotsk and Maritime Kray. The June map (Fig.8-2) shows a pattern similar to May, but the elongated tongue of low values emanating from the northwest, which occupies central portions of the country during May, has receded back over the Barents Sea in June and lost some of its intensity. Lowest values over the Barents Sea in June are about 14 kcal./cm^2, whereas in May over northwestern European Russia they average about 12 kcal. At the same time, the area of high values along the southern border of Soviet Central Asia shows about 20 kcal. in both months. Therefore, gradients from south to north are somewhat weaker in June than in May. The closed area of high values along the northeastern coast in May has shrunken in size by June, and the gradient has increased across the Pacific coast, as radiation values along the coast fall from 13–14 in May to 12–13 in June.

The changing patterns from May to June may be a function largely of the increasing turbidity as the summer season progresses. Generally the distribution of radiation through the years is asymmetrical with respect to seasons, and spring receives considerably more radiation than fall. This is due to the fact that in spring the surface of the earth is cooler than in fall and convective currents are not as strong; therefore the moisture content in the air is not as high in spring as it is later in the summer, and neither are the concentrations of dust particles and other types of pollution. Since the absolute humidity of the air is the main determinant of its transmissivity to radiation, the most pronounced reduction of transmissivity during summer takes place in the southwestern part of the country where warm moist air intrudes from the Balkans and the Black Sea (Fig.8-4). And this region of low transmissivity extends eastward into southwestern Siberia and northern Kazakhstan along a favored route of cyclonic storms during this time of year. Rather low values of transmissivity are also experienced throughout much of Central Asia in conjunction with the great amount of dust particles in the air during the summer. The highest transmissivity by far is in the Arctic where cool temperatures preclude much convective activity and strictly limit the air's capacity to hold moisture, and wet or vegetated surface conditions reduce the number of particulates that are available for transportation upward into the atmosphere.

Direct and scattered radiation

The annual map of total radiation reflects the influences of direct radiation much more than those of scattered radiation because over the course of the entire year direct radiation is distributed in a latitudinal pattern whereas scattered radiation does not form a very distinguishable pattern. The magnitude of scattered radiation depends primarily upon the state of the atmosphere and the underlying surface and over the course of a year it does not vary much from one part of the country to another (Figs.8-5 and 8-6). The accumulation of scattered radiation through the year amounts to less than half that of direct radiation in the southern part of the country but in the northern part it is somewhat greater than direct radiation. This results from the changing angle of the sun's rays with latitude and the increase in general cloudiness toward the Arctic fringe. There

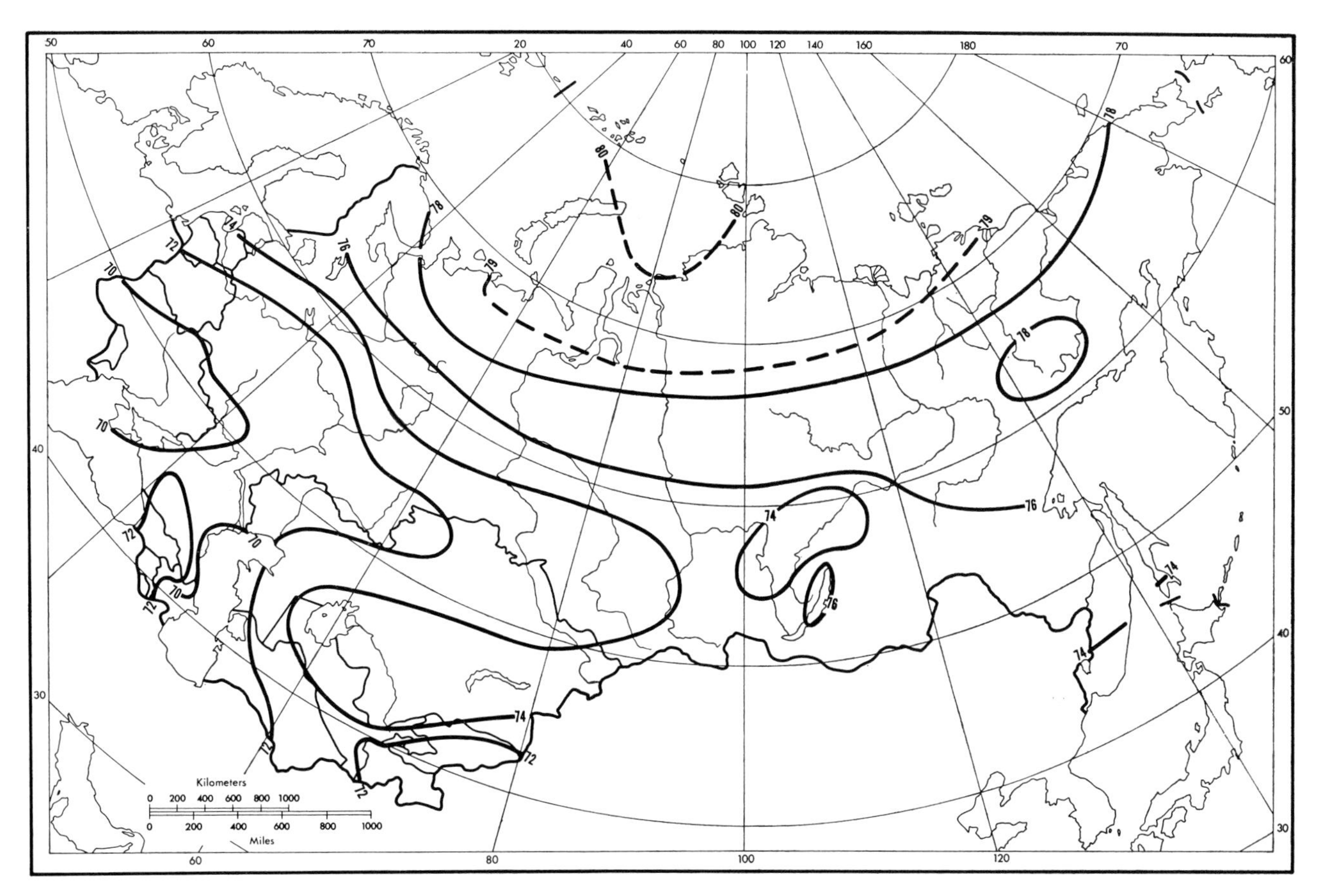

Fig.8-4. Coefficient of transmissivity of the atmosphere in June: $P_2 \cdot 10^{-2}$. (After PIVOVAROVA, 1966.)

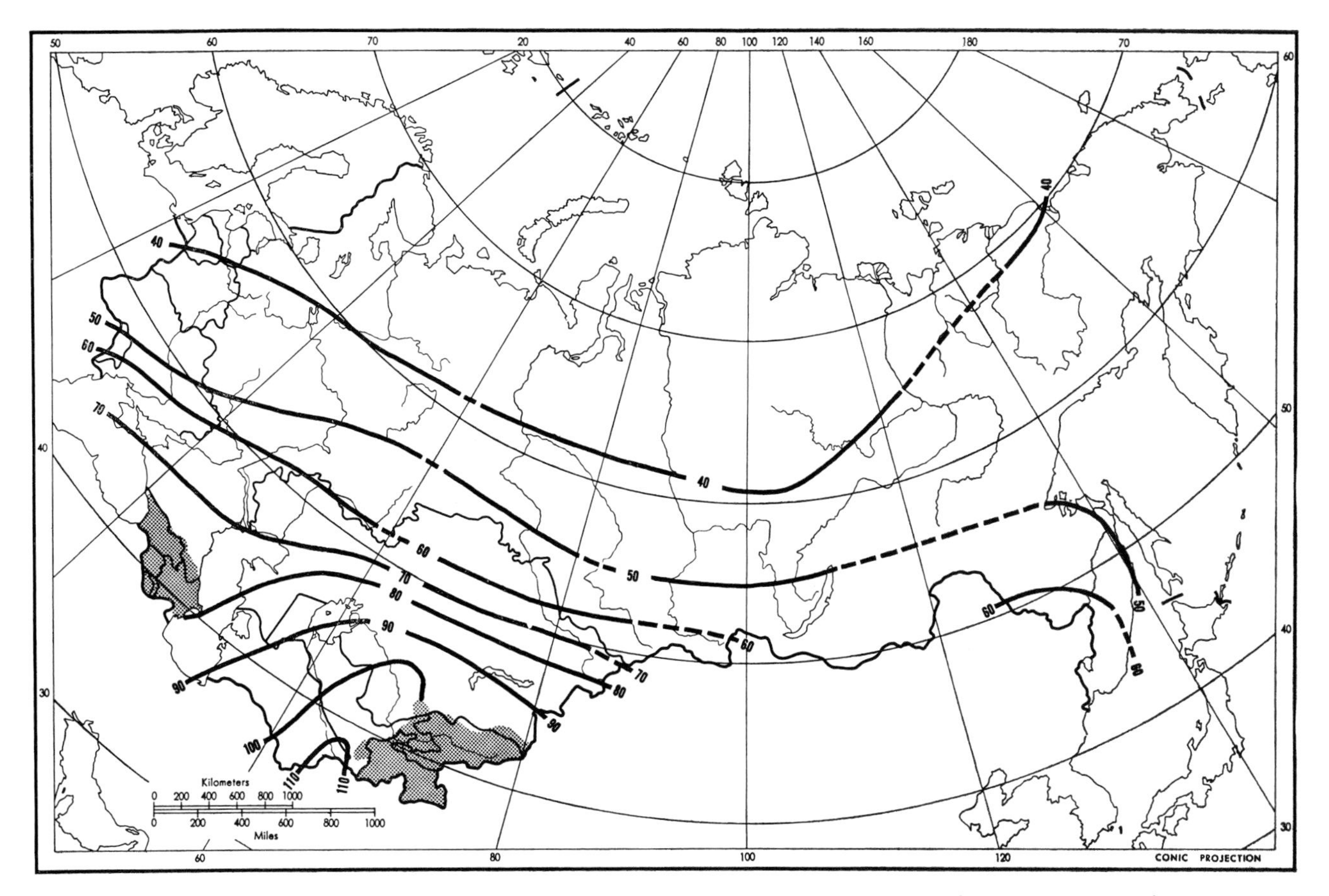

Fig.8-5. Mean annual direct radiation (kcal./cm²) on a horizontal surface. (After BARASHKOVA et al., 1961.)

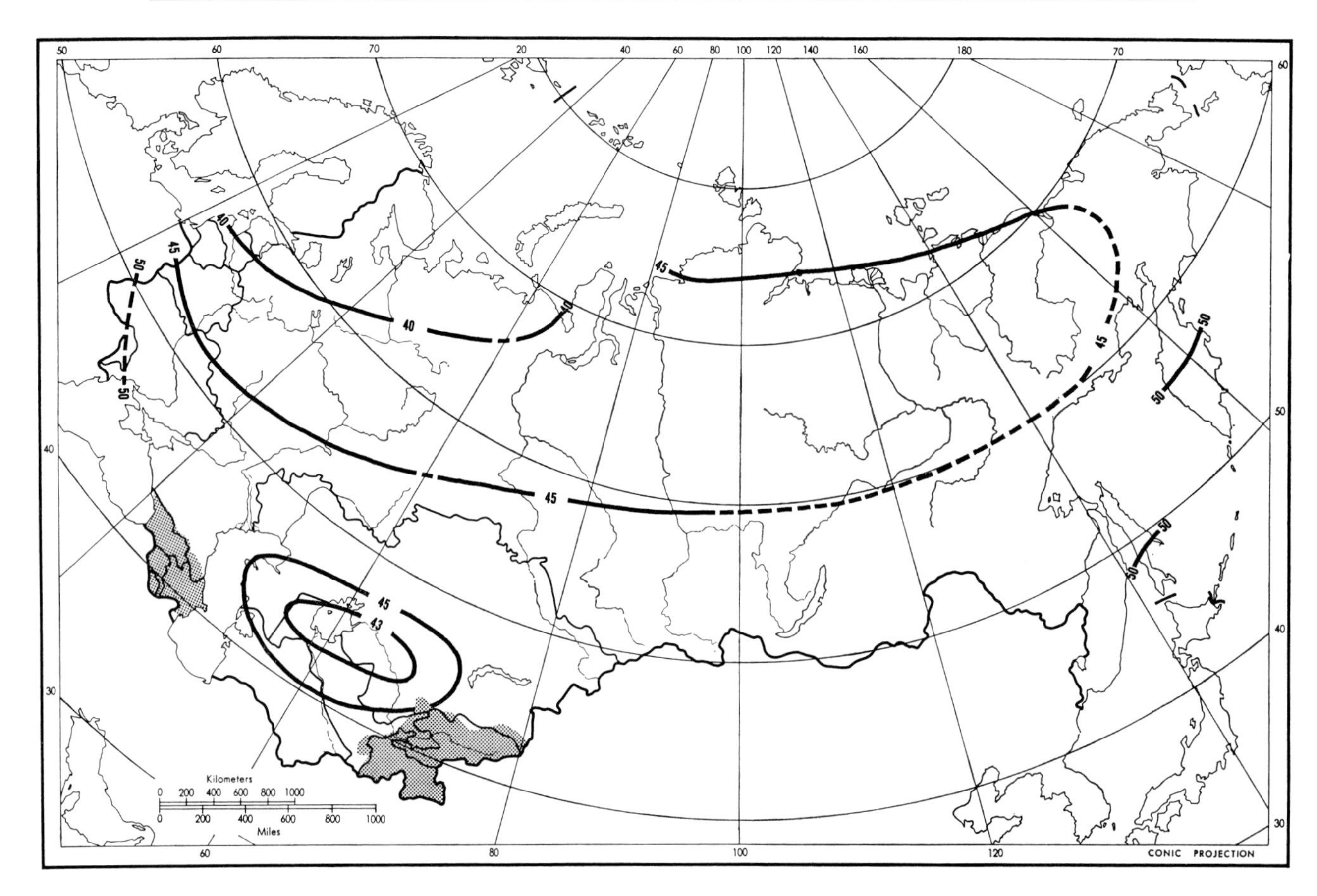

Fig.8-6. Mean annual scattered radiation (kcal./cm²) on a horizontal surface. (After BARASHKOVA et al., 1961.)

is a tendency for the two types of radiation to vary oppositely, with scattered radiation showing maximum amounts where direct radiation shows minima. The degree of cloudiness of course is the primary cause for this.

The annual distribution of direct radiation, although very latitudinal, shows a tendency to form an enclosed maximum in the southeastern part of Central Asia, which attests to the dryness of the air during much of the year in that region. In the Far East the isolines undergo a sharp southward bend as they approach the Pacific coast, which attests to the high amounts of humidity and clouds in the air in that region, particularly during the summer. The tendency for a Pacific coast minimum is accentuated by the fact that inland the northeastern part of the country experiences quite high values of direct radiation for the latitude, which again is a result of the dryness and clearness of the atmosphere in that region throughout much of the year.

Across the south-central half of European U.S.S.R. the isolines run east-northeastward rather than straight eastward, which indicates that the eastern part of the plain gets somewhat greater radiation than the western part does. This would be expected because of the decreasing cloudiness and the decreasing relative humidity toward the east. However, from the Urals foreland eastward across western Siberia and even into eastern Siberia the isolines run almost exactly parallel to the latitudes. This indicates that there is considerable atmospheric homogeneity throughout this broad area. Perhaps an increase of cyclonic activity in the northern part of western Siberia in summer, as compared to the eastern part of the European plain, might advect additional moisture from the southwest into western Siberia which would compensate for the tendency for moisture gradually to decrease from west to east across the plain. Therefore humidity and cloud conditions might be quite uniform all the way from Syktyvkar to the Yenisey River.

Annual totals of scattered radiation form a distribution pattern that is anything but latitudinal. The distribution depends very much upon cloud conditions and also upon the state of the underlying surface. Scattered radiation increases wherever snow is lying, and it especially increases at low elevations of the sun. At Pavlovsk outside of Leningrad experiments have shown that when the sun is low on the horizon indirect radiation increases by about 65% in the presence of a snow cover, whereas at a solar elevation of 50° it increases by only 12%. Thus, the greatest scattering of radiation is in the Arctic where direct radiation comprises only 24% of the total yearly radiation and scattered radiation comprises 76% of the total. At the Karadag observatory in the Crimea, on the other hand, the ratio is almost reversed, with direct radiation making up 68% of the total and the indirect radiation only 32%. The greatest totals of scattered radiation are in the Far East where the southern parts of Kamchatka and Sakhalin Island receive more than 50 kcal./cm^2 during the year. These high values of scattered radiation are due to the great amount of cloudiness during the summer when much solar radiation is being received. Relatively high values are also found in the Arctic, which has cloudiness throughout much of the year, plus a snow-covered surface at least 8 or 9 months of the year. The rest of the year has a surface of watery wastes on top of the ice or interspersed with the ice which show considerably reduced albedos, but albedos that nevertheless are still higher than a vegetated surface such as a mixed forest. There also tends to be an increase in scattered radiation along certain foothill areas where upslope motion produces an abnormal amount of cloudiness through the year. Good examples of this are the Stavropol' Plateau north of the Caucasus and the northern foothills of the Carpathians in western Ukraine.

The distribution of scattered radiation varies greatly throughout the year. December's pattern is quite latitudinal (Fig. 8-7), and this is typical from October through February. But in March an oval-shaped high-value area begins to appear over northern Kazakhstan elongated in an east–west direction, and by April the pattern has completely lost its zonal characteristics (Fig.8-8). The latitudinal pattern is somewhat re-established through May and June, but with gradients reversed from what they had been in winter (Fig.8-9). At this time of year there is a strong increase in scattered radiation northward toward the Arctic where absolute values become twice as great as they were in preceding months. In March scattered radiation decreases northward from about 5 kcal./cm^2 in Kazakhstan to less than 2 along the Arctic coast. In May on the other hand it increases northward from about 6 in Kazakhstan to more than 11 along the Arctic coast. In between, April shows an amorphous pattern that bears no resemblance to latitude, and obviously represents the transition between the winter and summer situations when radiation gradients are reversing themselves.

July, August, and September also deviate greatly from a latitudinal pattern, although there is some semblance of zonality. The July map, for instance, shows a value of around 7 kcal./cm^2 in the far north which reduces southward to less than 5 in Central Asia where an enclosed area of minimum values exists between the Aral Sea and Lake Bal-

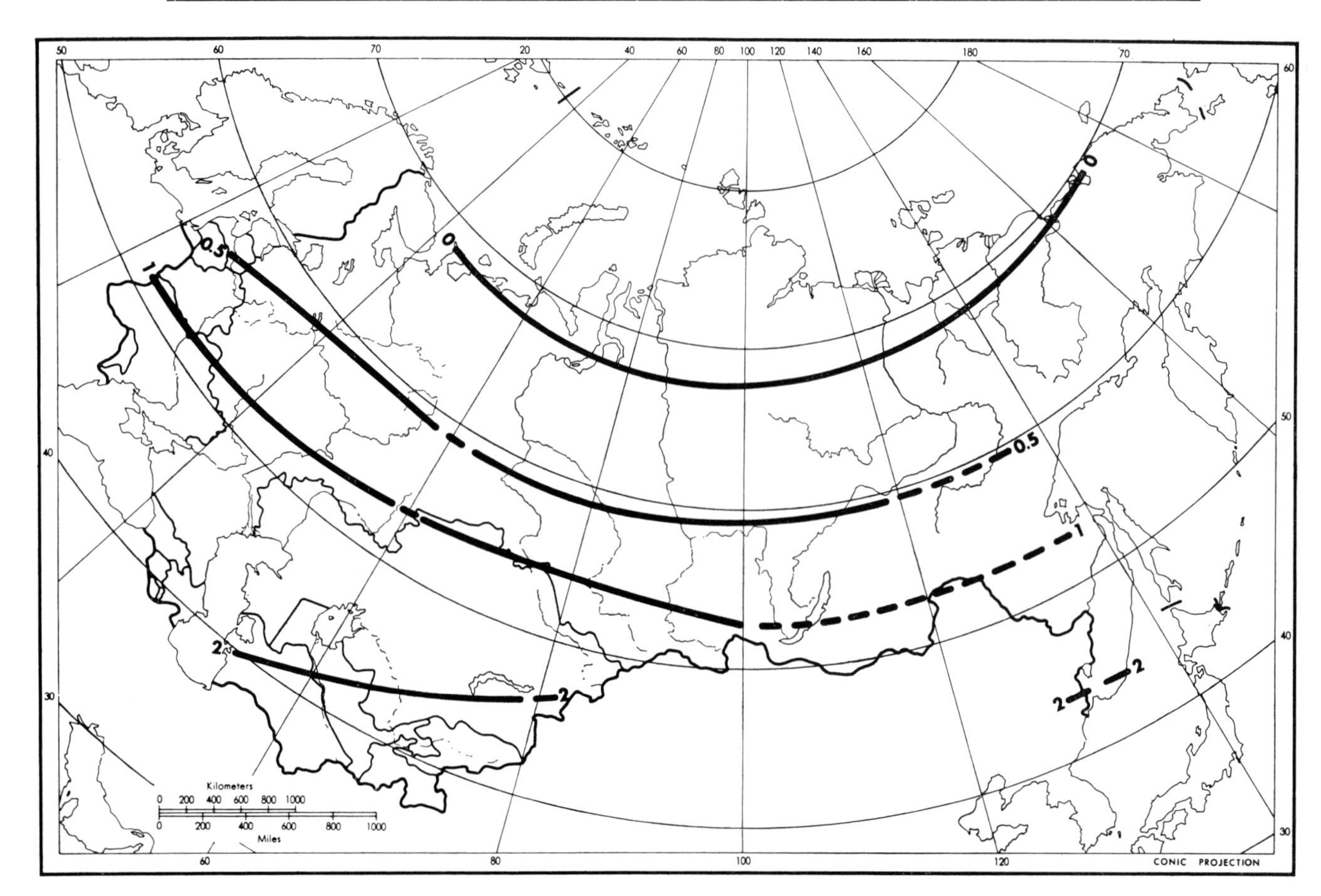

Fig.8-7. Mean scattered radiation (kcal./cm²) in December. (After BARASHKOVA et al., 1961.)

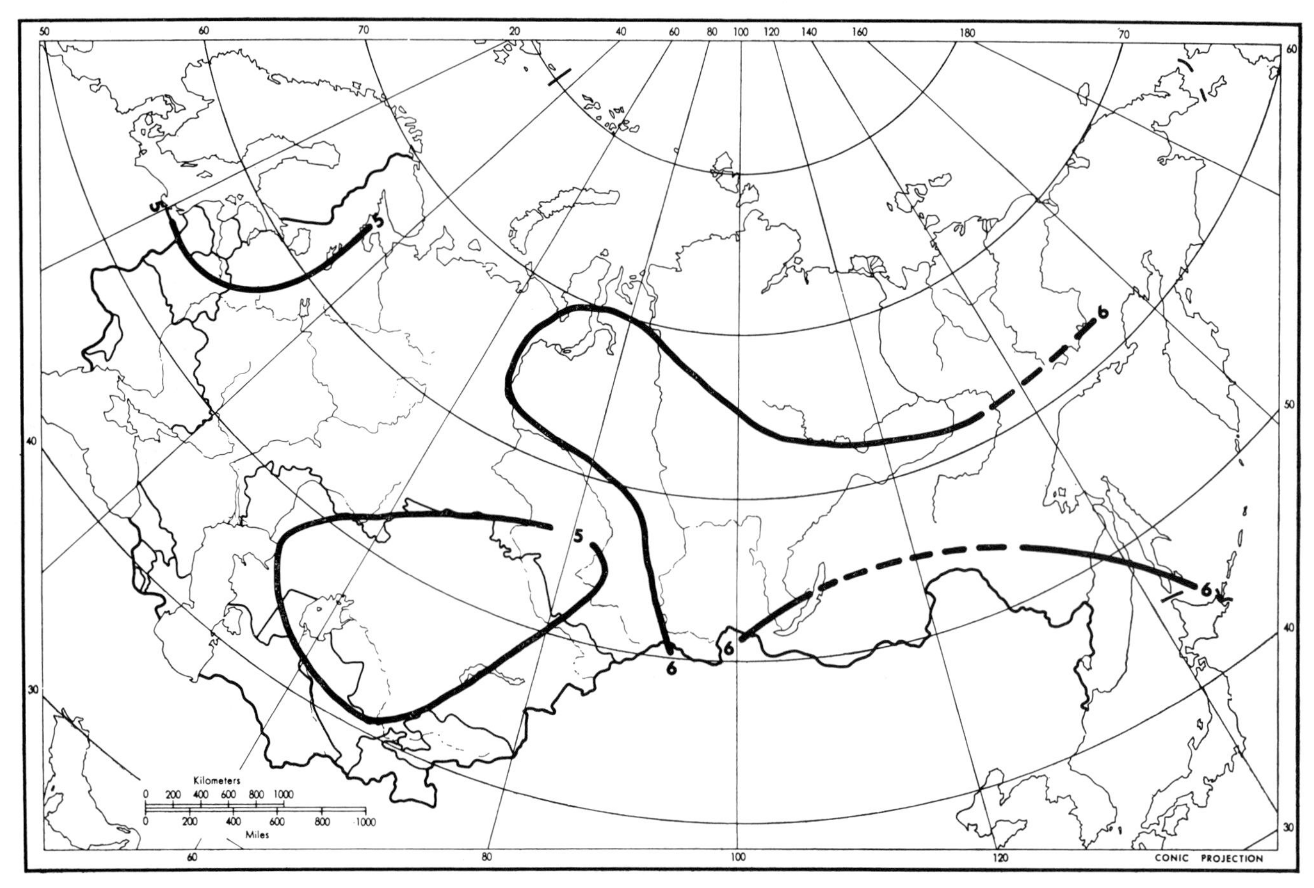

Fig.8-8. Mean scattered radiation (kcal./cm²) in April. (After BARASHKOVA et al., 1961.)

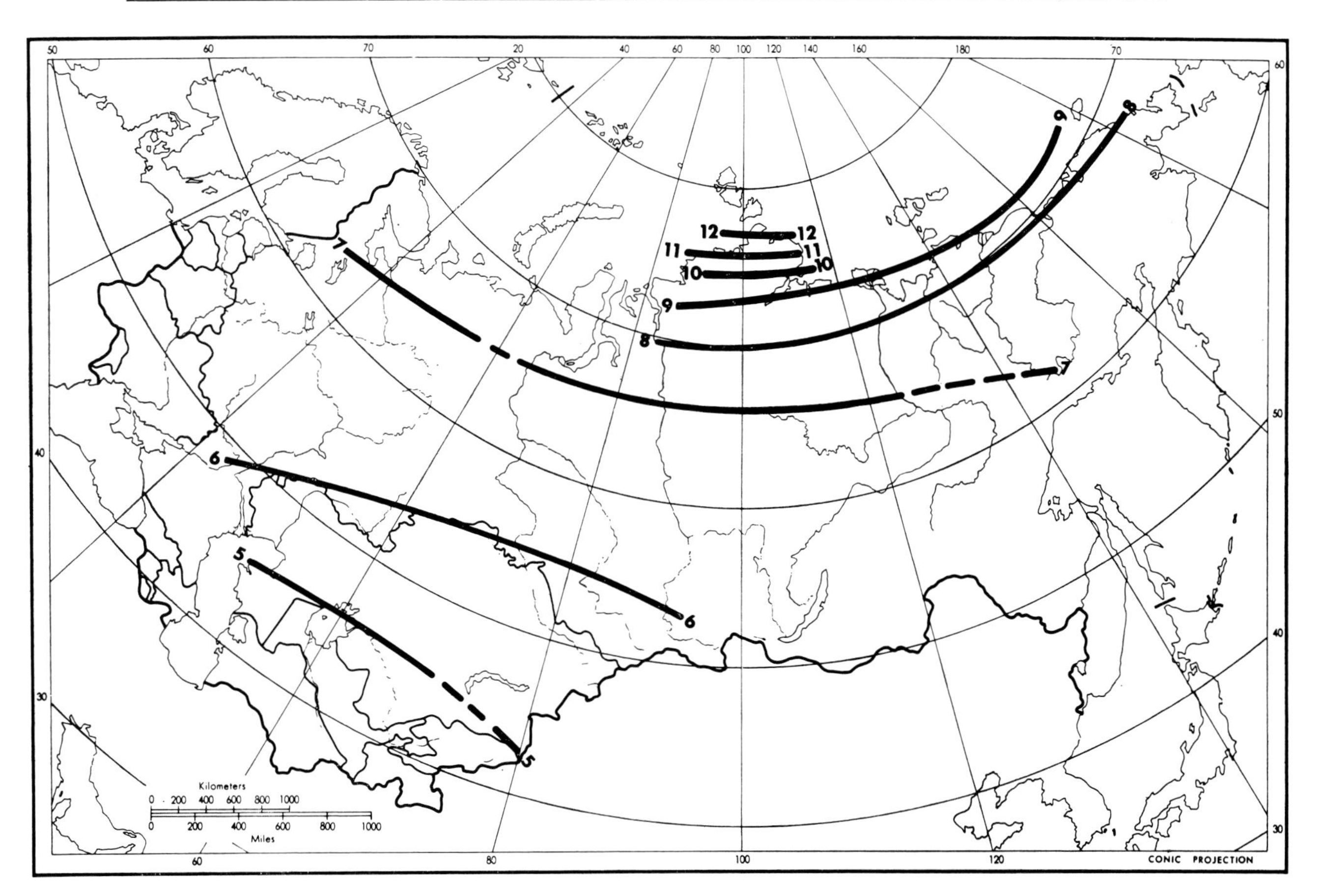

Fig.8-9. Mean scattered radiation (kcal./cm²) in June. (After BARASHKOVA et al., 1961.)

khash. By August the values in the Arctic have dropped to less than 5 and a broad extensive ridge of high values has enveloped most of the country. In the south the isoline of 5 runs through Kazakhstan. An enclosed isoline of 4 encloses minimum values between the Aral Sea and Lake Balkhash. This pattern continues into September, when values drop below 3 in the Arctic and much of the country registers somewhat above 3. Minimum values occur again in Central Asia where an enclosed area of less than 3 exists. Thus, the high absolute values of radiation in the Arctic during the June solstice, which make for very high scattered radiation values there, decrease through the autumn equinox and eventually bring on the winter pattern of distribution with the Arctic receiving little or no radiation and relatively higher values in the south, although the absolute magnitudes everywhere are less than they were during summer.

The most extreme gradient of scattered radiation at anytime occurs in June as values increase rapidly northward through the Taymyr Peninsula. This is probably due to the fact that snow is still lying on the northern portion of the continent at this time of year, and the high reflectivity of the snow causes very large amounts of scattered radiation to be received during the summer solstice when total radiation is at its peak.

Reflectivity, absorption, and terrestrial radiation

The albedo varies greatly in the Soviet Union from place to place and from season to season as landscapes change from deserts to steppes to forests to tundra and winter brings on an extensive snow cover (Table 8–I). In January much of the country has some snow cover and therefore albedos are universally high (Fig.8-10). In the north where the snow is relatively clean 80% or more of the meager radiation is reflected as soon as it hits the surface. This decreases southward, first slowly, and then more rapidly as the edge of the consistent snow cover is crossed, until in Central Asia only about 30% is reflected. The pattern remains very similar in February, but by March the snow cover is beginning to melt in the southern portions of the country and the zone of transition broadens. Much of Siberia still shows 80% reflectivity, but the Black Sea Steppes and North Caucasus now have only about 20%.

April shows a definite shift northward of the southern edge of the snow line, as evidenced by the zone of rapid transition of albedo across the central part of the country (Fig.8-11).

TABLE 8-I

ALBEDO OF NATURAL SURFACES
(After BUDYKO, 1963)

Type of surface	Albedo
Stable snow cover in high latitudes (above 60°)	0.80
Stable snow cover in temperate latitudes (below 60°)	0.70
Forest with stable snow cover	0.45
Unstable snow cover in spring	0.38
Forest with unstable snow cover in spring	0.25
Unstable snow cover in autumn	0.50
Forest with unstable snow cover in autumn	0.30
Steppe and forest during the period between the disappearance of snow cover and the transition of mean diurnal temperature above 10°C	0.13
Tundra during the period between the disappearance of snow cover and the transition of the diurnal temperature above 10°C	0.18
Tundra, steppe, deciduous forest in the period from the spring transition of temperature above 10°C until the occurrence of snow cover	p.18
Coniferous forest during the period of the spring transition of temperature above 10°C until the occurrence of snow cover	0.14
Forest, shedding leaves during the dry period of the year, savanna, semi-desert during the dry period of the year	0.24
The same for the wet period of the year	0.18
Desert	0.28

TABLE 8-II

ALBEDO OF WATER SURFACE

Latitude	Jan.	Feb.	Mar.	Apr.	May	June	July	Aug.	Sept.	Oct.	Nov.	Dec.
70°	–	0.23	0.16	0.11	0.09	0.09	0.09	0.10	0.13	0.15	–	–
60°	0.20	0.16	0.11	0.08	0.08	0.07	0.08	0.09	0.10	0.14	0.19	0.21
50°	0.16	0.12	0.09	0.07	0.07	0.06	0.07	0.07	0.08	0.11	0.14	0.16
40°	0.11	0.09	0.08	0.07	0.06	0.06	0.06	0.06	0.07	0.08	0.11	0.12
30°	0.09	0.08	0.07	0.06	0.06	0.06	0.06	0.06	0.06	0.07	0.08	0.09
20°	0.07	0.07	0.06	0.06	0.06	0.06	0.06	0.06	0.06	0.06	0.07	0.07
10°	0.06	0.06	0.06	0.06	0.06	0.06	0.06	0.06	0.06	0.06	0.06	0.07
0°	0.06	0.06	0.06	0.06	0.06	0.06	0.06	0.06	0.06	0.06	0.06	0.06

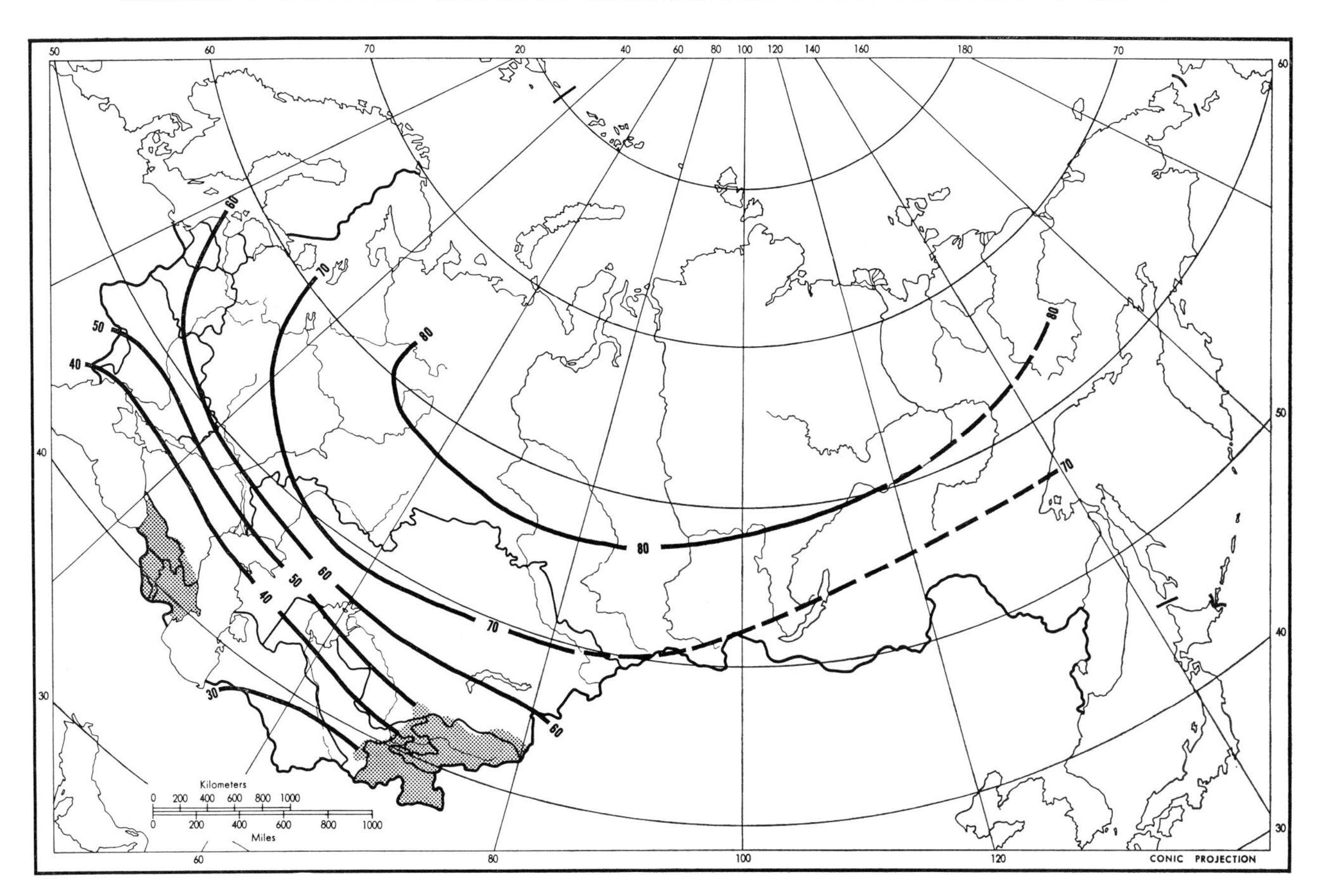

Fig.8-10. Albedo in January (%). (After BARASHKOVA et al., 1961.)

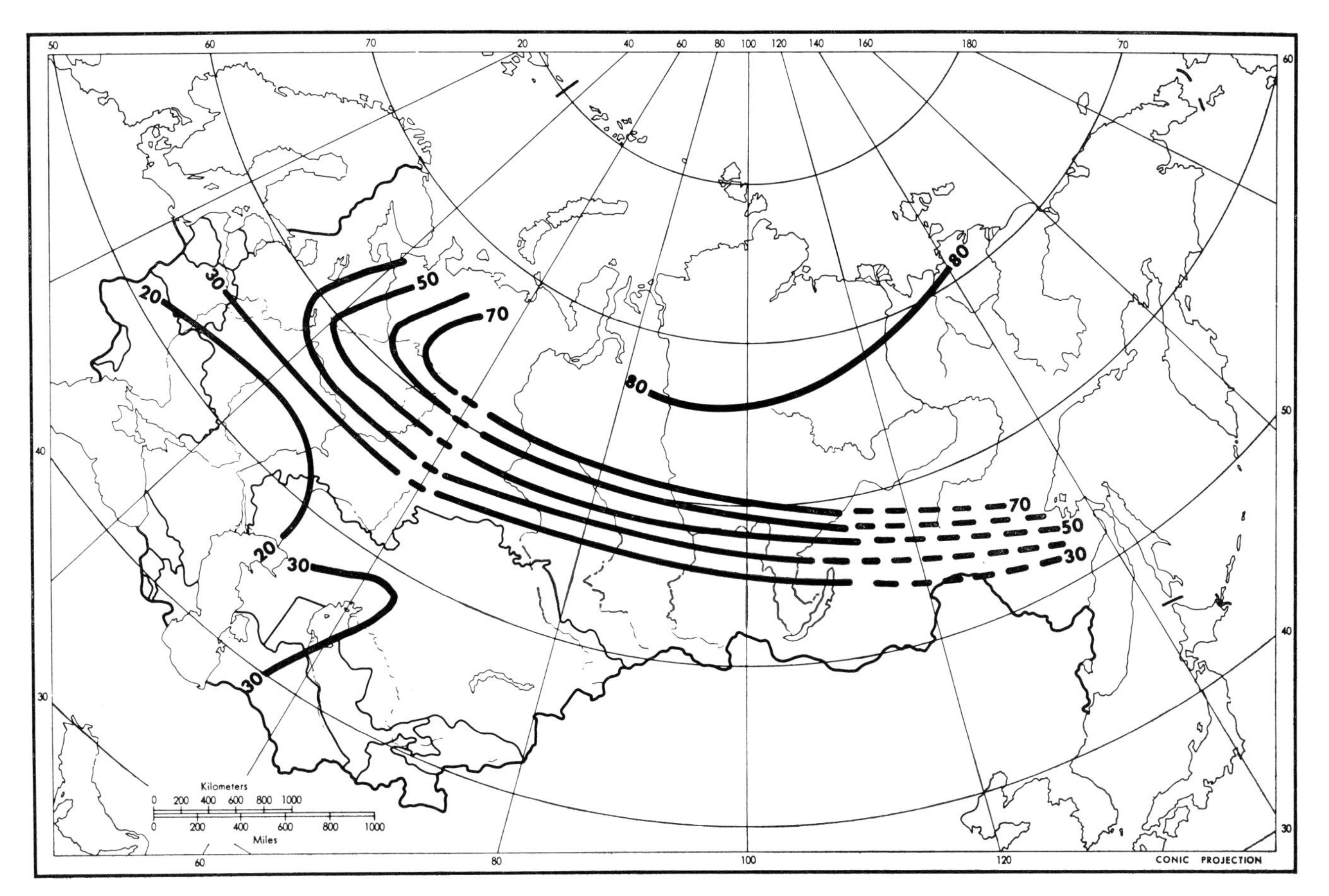

Fig.8-11. Albedo in April (%). (After BARASHKOVA et al., 1961.)

This zone of transition in April shows the most intense gradient of reflectivity of any time of year. Only a small part of north-central Siberia now has more than 80% reflectivity, and a large portion of the southern European plain has less than 20%. Much of Central Asia continues to have an albedo of about 30%. Central Asia is very consistent through the year, since it gains neither a significant snow cover in winter nor a significant vegetative cover in summer. The reflectivity of the desert surfaces does not vary much through the year.

By mid-June the snow is essentially all gone even in the northern part of the Soviet Union, and albedos of 20 to 30% cover much of the country. This continues through July, August, and September in much the same pattern that is shown on the July map (Fig.8-12). Much of the forest and tundra regions during the summer shows an albedo of less than 20%, much of the steppe region shows less than 20% since most of this area during the summer is in agricultural crops, and the desert area east of the Caspian shows around 30%. During autumn the winter situation is gradually reestablished with its highly reflective snow cover in the north, but October shows very significant differences from April (Fig.8-13). The snow cover is still not well established in October and is largely limited to the Arctic fringe of Siberia and the mountains east of the Lena River. Therefore, albedos along the Arctic are somewhat lower than they are in April, and the zone of transition through the central part of the country is much more gradual than in April and much farther north in the Far East where the water bodies along the Pacific coast

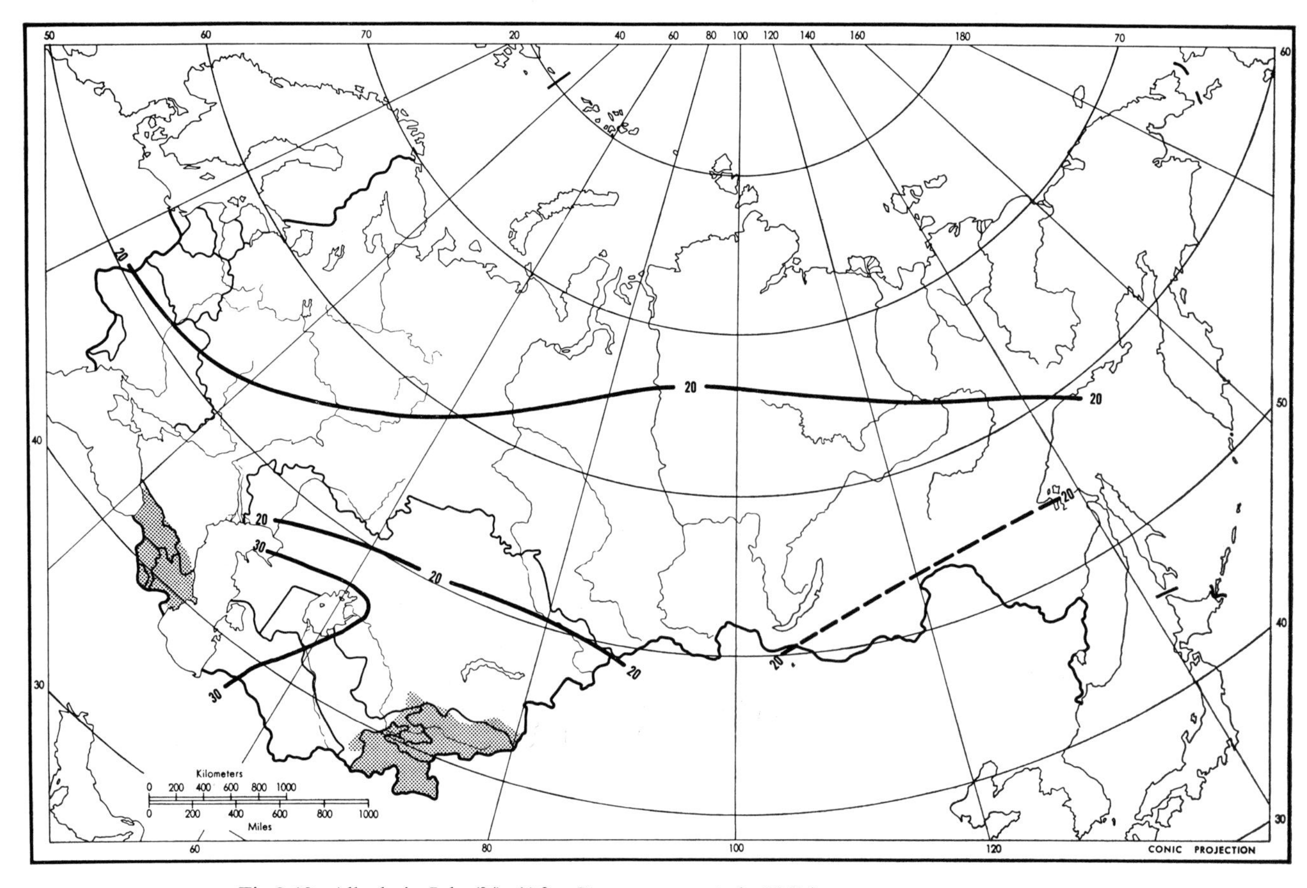

Fig.8-12. Albedo in July (%). (After BARASHKOVA et al., 1961.)

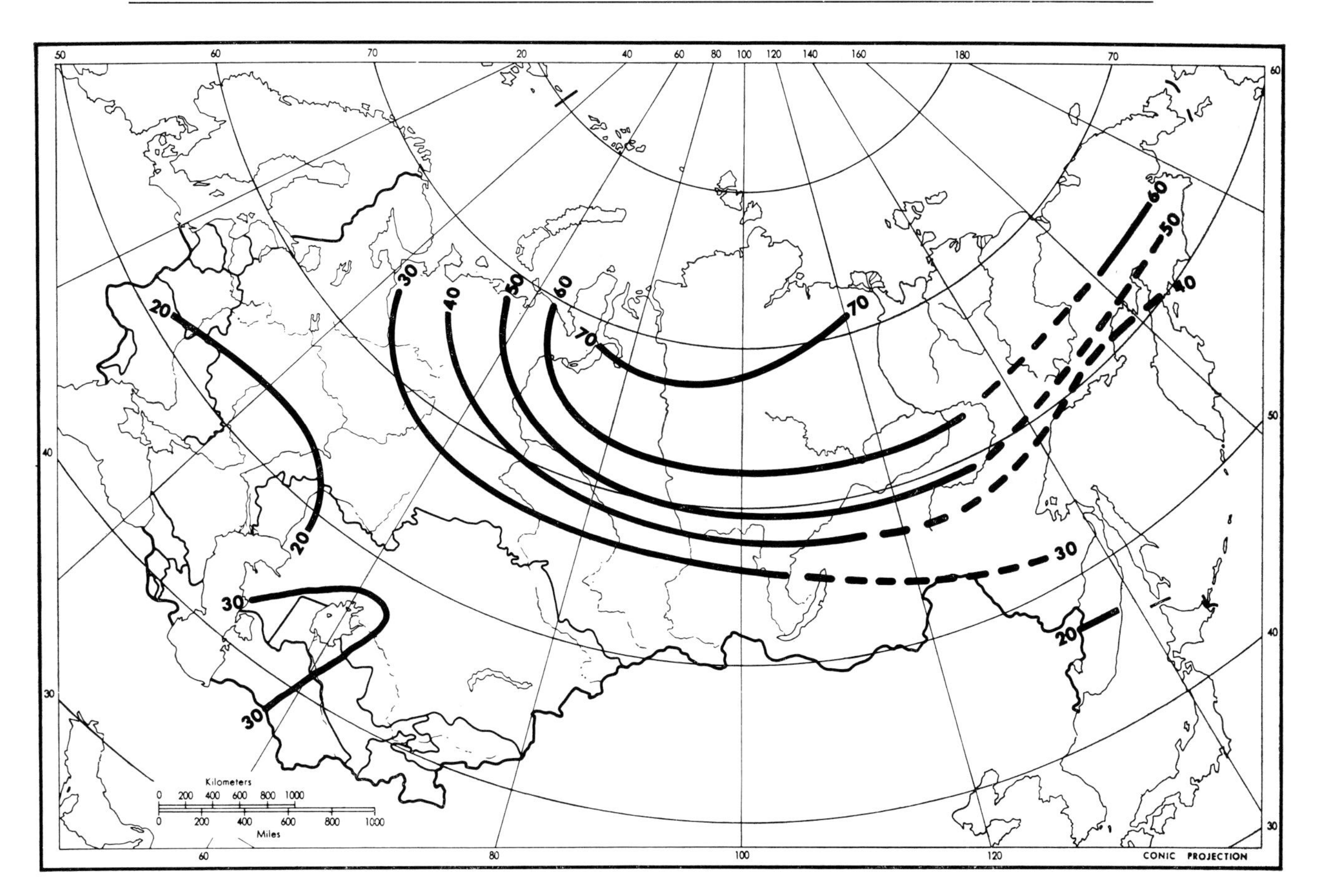

Fig.8-13. Albedo in October (%). (After BARASHKOVA et al., 1961.)

remain unfrozen and delay the coming of winter. In April, the same water bodies with their slowly thawing surfaces delay the coming of summer and maintain albedos higher than during corresponding times in autumn.

The high reflectivity of the Arctic in winter is somewhat inconsequential, since little, if any, radiation is received during that time of year. Therefore summer becomes all important in the heating of this region, and the high values of incoming radiation coupled with relatively low albedos make for a high degree of absorption which boosts the yearly total disproportionately high with respect to radiation absorption farther south. Whereas the difference in annual radiation between the south and north is about 90–100 kcal./cm^2, the difference in amounts actually absorbed is only about 40–50 kcal./cm^2 (Fig.8-14). Roughly 2/3 of the incident radiation is absorbed in both the Arctic fringe and in Central Asia. But Central Asia maintains this ratio throughout the year while the Arctic coast ranges from less than 20% in winter to about 80% in summer.

The net outgoing radiation from the earth's surface during the year describes almost a latitudinal pattern, although there is a tendency for the outgoing radiation to be a little greater in the eastern part of the country than in the west (Fig.8-15). This again attests to the reduced humidity and cloudiness in the east as compared to the west. Data are insufficient along the Pacific coast to indicate exactly what happens there, but probably the outgoing radiation is considerably less right along the coast because of the humidity and cloudiness of that region, especially during summer when the absolute amounts of

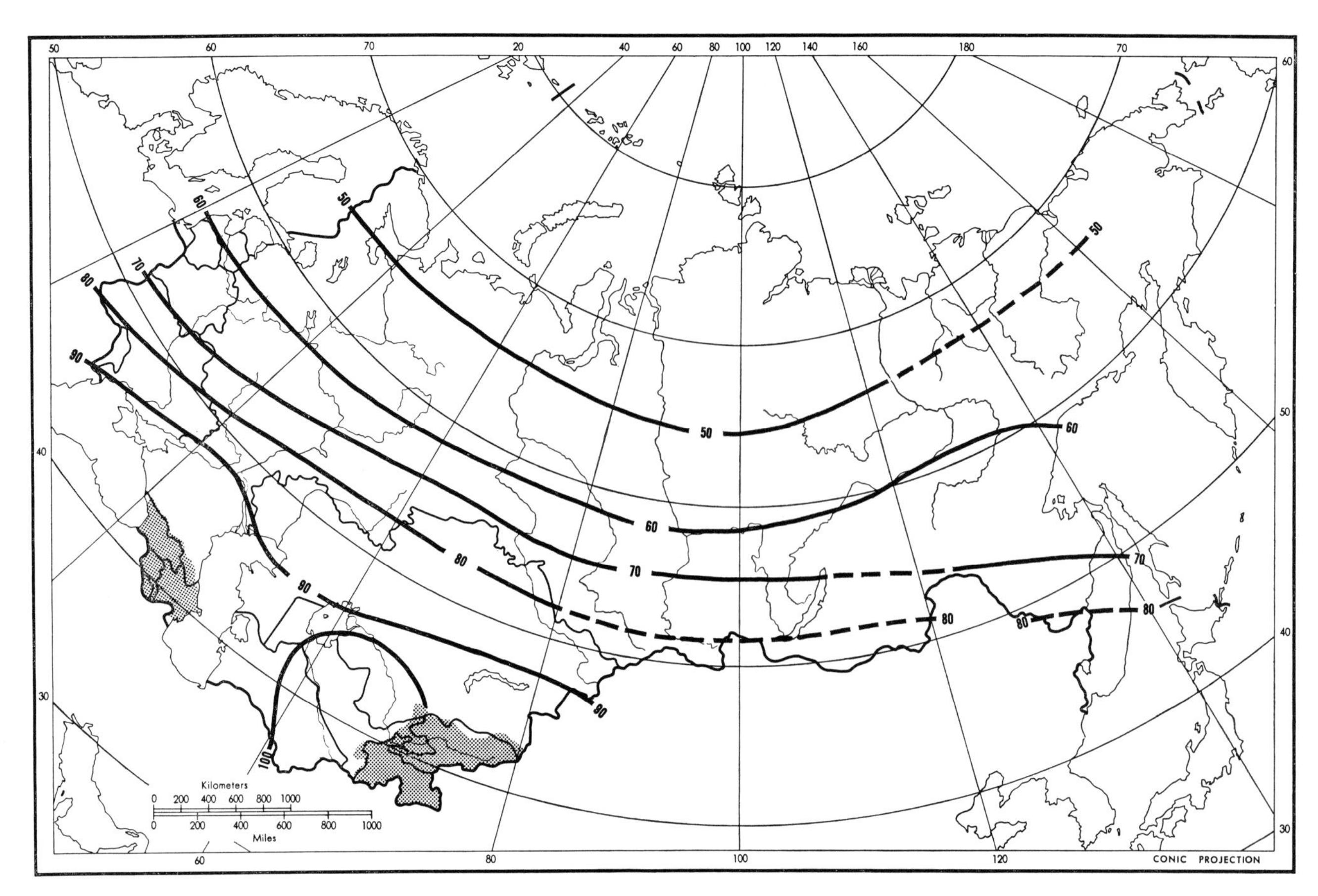

Fig.8-14. Mean annual radiation (kcal./cm²) absorbed at the earth's surface. (After BARASHKOVA et al., 1961.)

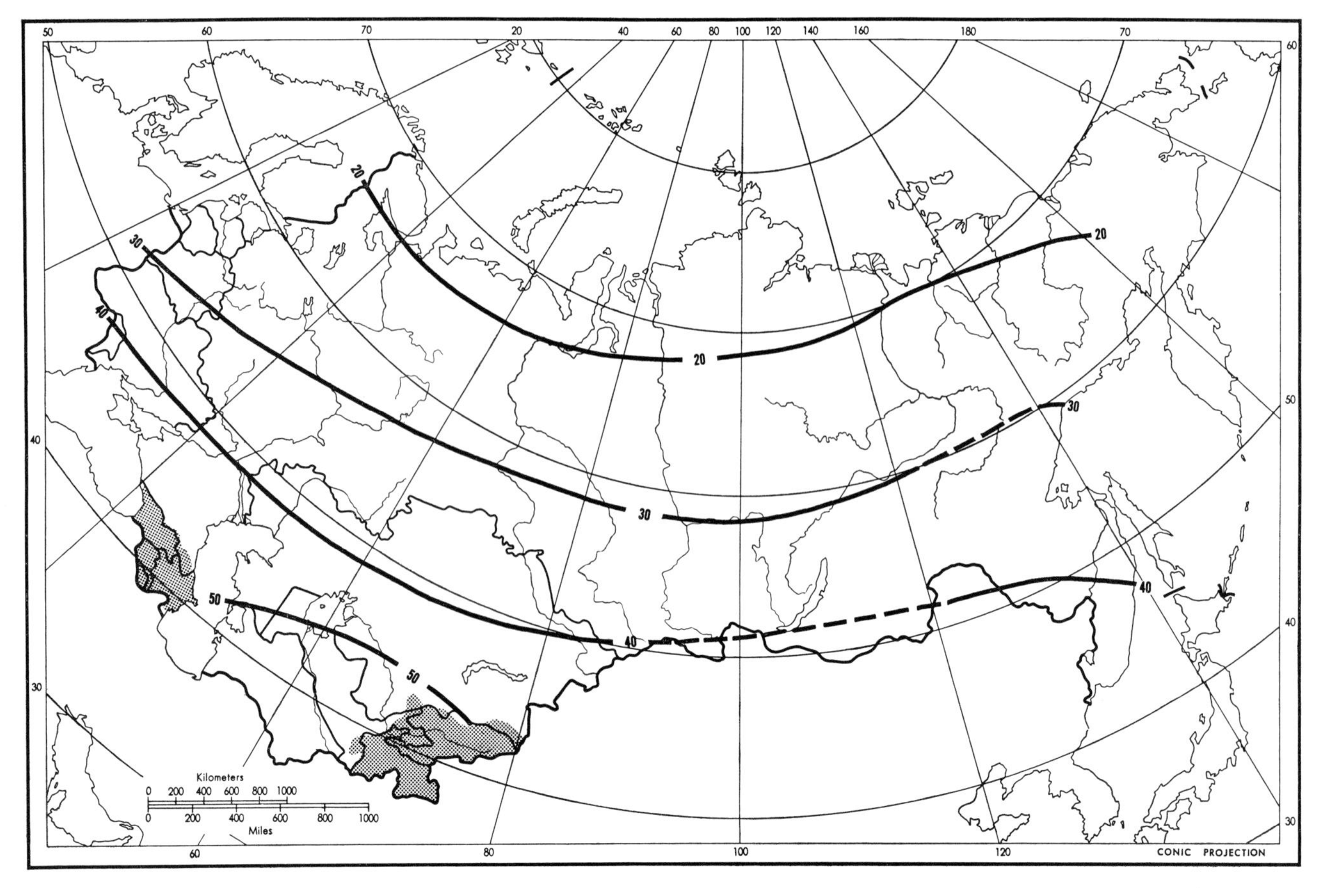

Fig.8-15. Mean annual net outgoing terrestrial radiation (kcal./cm²). (After BARASHKOVA et al., 1961.)

outgoing radiation are greatest. It appears that throughout much of the U.S.S.R. the net outgoing radiation is approximately equal to 1/3 of the incoming radiation and about 1/2 of the radiation absorbed at the earth's surface.

The pattern of outgoing terrestrial radiation over European U.S.S.R. varies significantly by season (Fig.8-16). During winter the isolines of net outgoing radiation describe parabolas open to the north, while during summer they run fairly parallel to the latitudes in the northern part of the plain and from southwest to northeast in the southern part of the plain. This seasonal change can be explained by seasonally different combinations of longitudinal distributions of temperature and absolute humidity that are superimposed on the general backdrop throughout the year of decreasing temperature and decreasing absolute humidity with increasing latitude. During winter at any given latitude temperature and absolute humidity decrease from west to east across European U.S.S.R. The decrease of temperature eastward tends to decrease outgoing terrestrial radiation eastward, but the decrease of absolute humidity eastward tends to increase outgoing radiation eastward. The result is higher outgoing radiation in the western and eastern parts of European U.S.S.R. than along a central meridian at about 40°E longitude. During summer, on the other hand, along a given latitude temperatures tend to increase from west to east while absolute humidity continues to decrease from west to east. The effect of both the temperature and humidity gradients is to increase outgoing radiation eastward. Therefore, during summer isolines of outgoing radiation trend southwest–northeast.

No seasonal data are available for Asiatic U.S.S.R., but it would appear likely that most

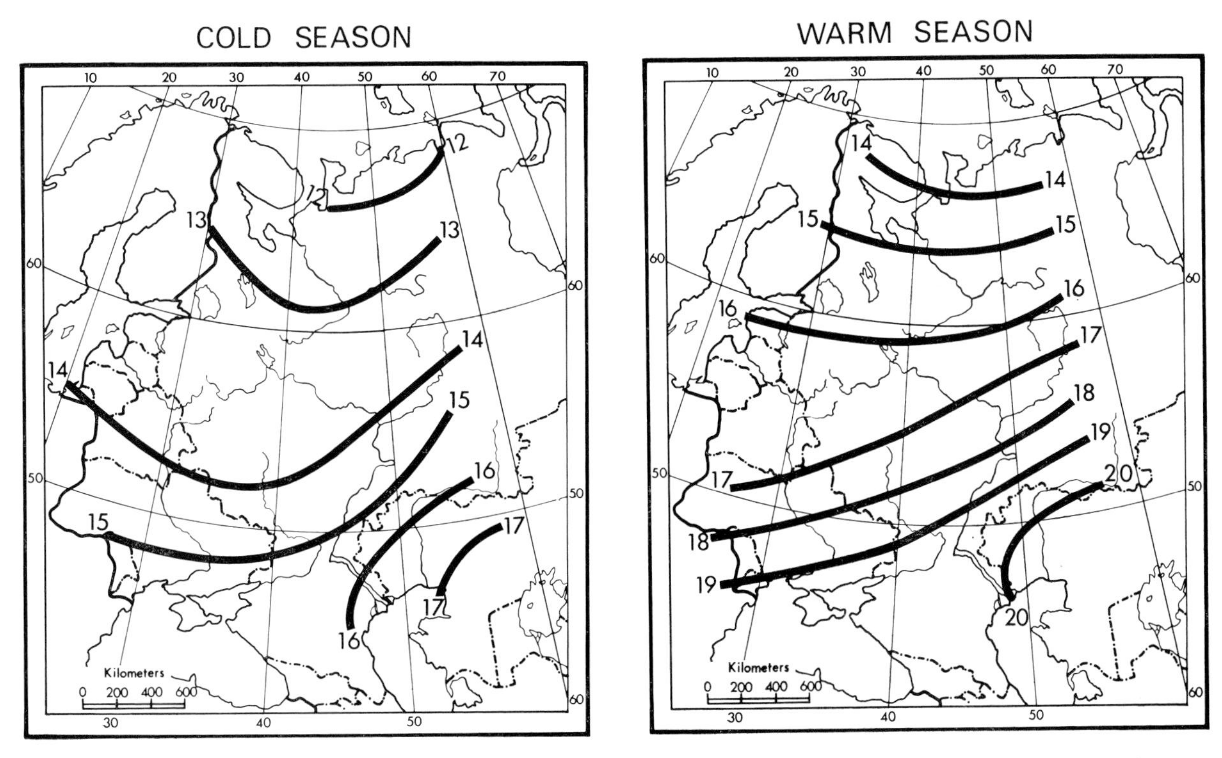

Fig.8-16. Net outgoing radiation (kcal./cm²) from European USSR during cold (October–March) and warm (April–September) seasons. (After DANILOV, 1963.)

of the outgoing radiation would take place during summer when the earth's surface is warmer and temperature inversions are not so prevalent in the lower atmosphere. On many occasions in winter, particularly in eastern Siberia, when very intense temperature inversions exist at or near the surface, there is actually a net radiation of heat downward to the earth's surface by the atmosphere, and in all cases temperature inversions would greatly impede the upward transfer of heat.

Heat balance at the earth's surface

Radiation balance

The sum or net effect of all the foregoing components of radiation are represented by radiation balance at the earth's surface, which for most purposes is the single most important quantity that one can know about the radiation of a region. It is one of the main controls on the transformation of air masses that pass over the surface, and it is the primary factor that controls the evaporation of moisture from the earth's surface. The magnitude of the radiation balance seems to be related intimately to various plant growth processes and many other processes that take place at the earth's surface, including those of soil formation, and because of this it has become a useful parameter to be used in the formulation of climatic classification schemes and so forth.

Yearly values of radiation balance by latitude are very similar in the Soviet Union and in North America. In general the Soviet Union shows slightly higher values over the western part of the country and slightly lower values over the eastern part of the country than obtained at comparable latitudes in North America. For instance, much of the Ukraine has a slight edge on similar locations in North America, although again it must be remembered, that the latitudinal comparison is with the prairie provinces of Canada, not with the corn belt of the United States. The very southern portions of the U.S.S.R. in Central Asia and the Transcaucasus have heat balances comparable to those in northern Iowa and southern Wisconsin. These areas experience radiation balances of as much as 50 kcal./cm^2 per year. But nothing in the Soviet Union compares to the southeastern United States which has a balance of 70–80 kcal./cm^2 per year.

The tendency for an east–west reversal between the two countries relates primarily to the different directions in moisture flux across the two areas. Eastern North America derives a great deal of moisture from a southerly flow of air out of the Gulf of Mexico, whereas eastern U.S.S.R. does not have such a source of moisture, and this leaves the western plains of European U.S.S.R. with the highest humidities, which are derived from the westerly flow of Atlantic air.

The east–west distribution of radiation balance in the U.S.S.R. is almost opposite to that of the radiation received (Figs.8-1 and 8-17). Whereas the eastern part of the country generally receives a little more radiation than the western part, because of less cloudiness and lower humidity, there generally is a lower radiational balance in the eastern part, for the same reasons. The clearer air in the eastern part of the country allows for greater radiational heat losses, and reduced evaporation of moisture from the earth's surface in the east allows for a greater portion of the absorbed heat to be returned to the atmosphere as back radiation. It is the lack of moisture exchange between the earth's surface

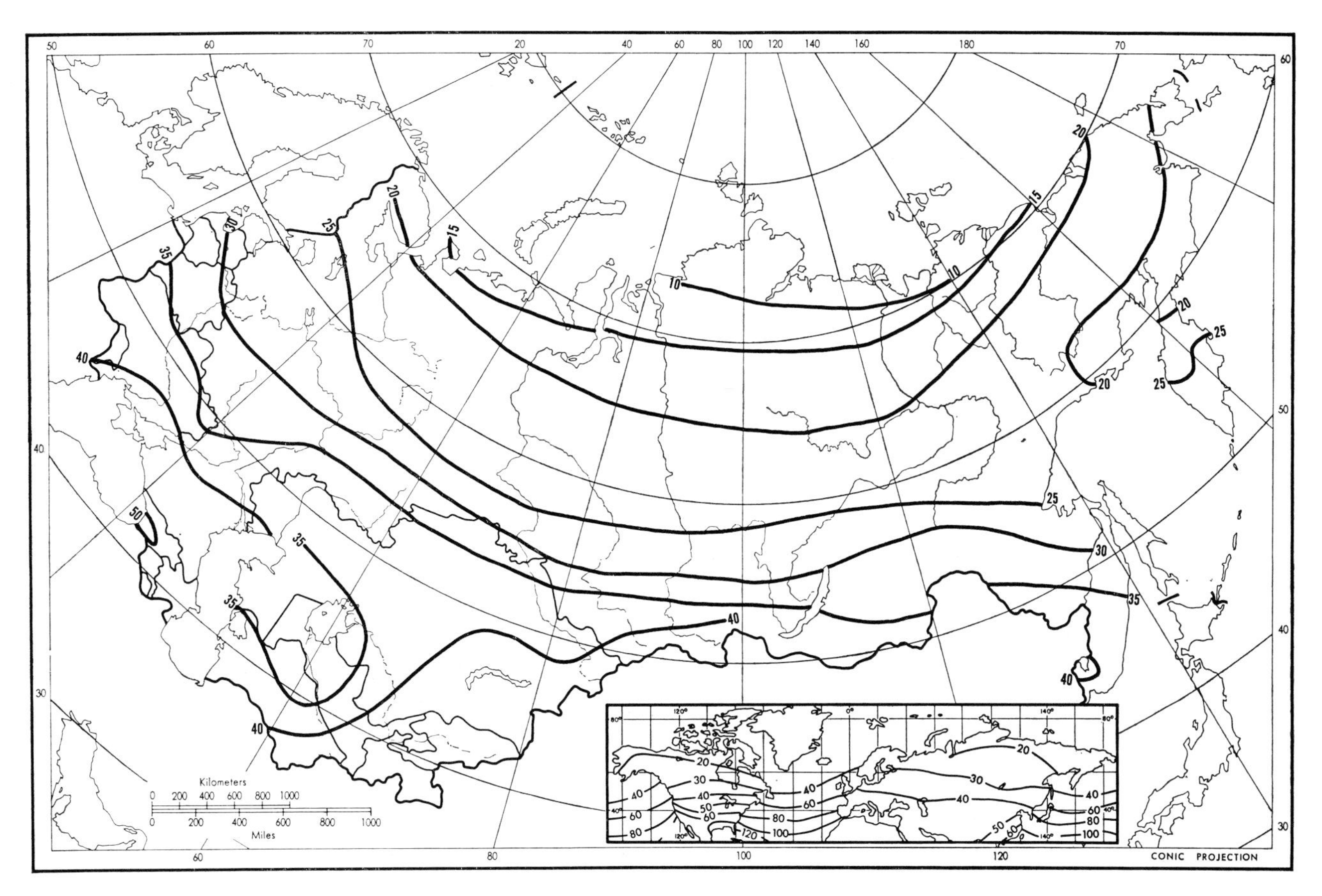

Fig.8-17. Mean annual radiation balance (kcal./cm²) at the earth's surface. (After PIVOVAROVA, 1966; inset from BUDYKO, 1963.)

and the atmosphere that largely allows for the seemingly contradictory coincidence of regions with low radiational heat balance and high surface air temperatures during the summer.

Most heat that is lost by the earth's surface to the air is done so through sensible temperature exchanges and not through latent heat exchanges in the evaporation processes. Thus, the radiation balance during summer is less in Central Asia than it is in southern Ukraine, because of the drier atmosphere in Central Asia and the greater amount of outgoing radiation which is commensurate with the much higher surface temperatures in Central Asia. Much of the desert of Central Asia ends up with about the same annual heat balance as does Belorussia, but summer temperatures in Central Asia average about 28°C, while those in Belorussia average about 18°C.

For the latitude, the Far East tends to experience the lowest values of annual radiational balance in the Soviet Union, because of an unfavorable seasonal distribution of clouds. Great cloudiness in summer reduces incoming radiation when it is at its peak, and clear weather in winter allows for large radiational heat losses from the earth's surface.

During January almost the entire Soviet Union has a negative radiation balance (Fig. 8-18). The lack of differentiation across the country is striking. In winter the northern areas receive little or no radiation and lose very little, and the balance is about the same as it is farther south where more heat is received and more is lost. In March there is a

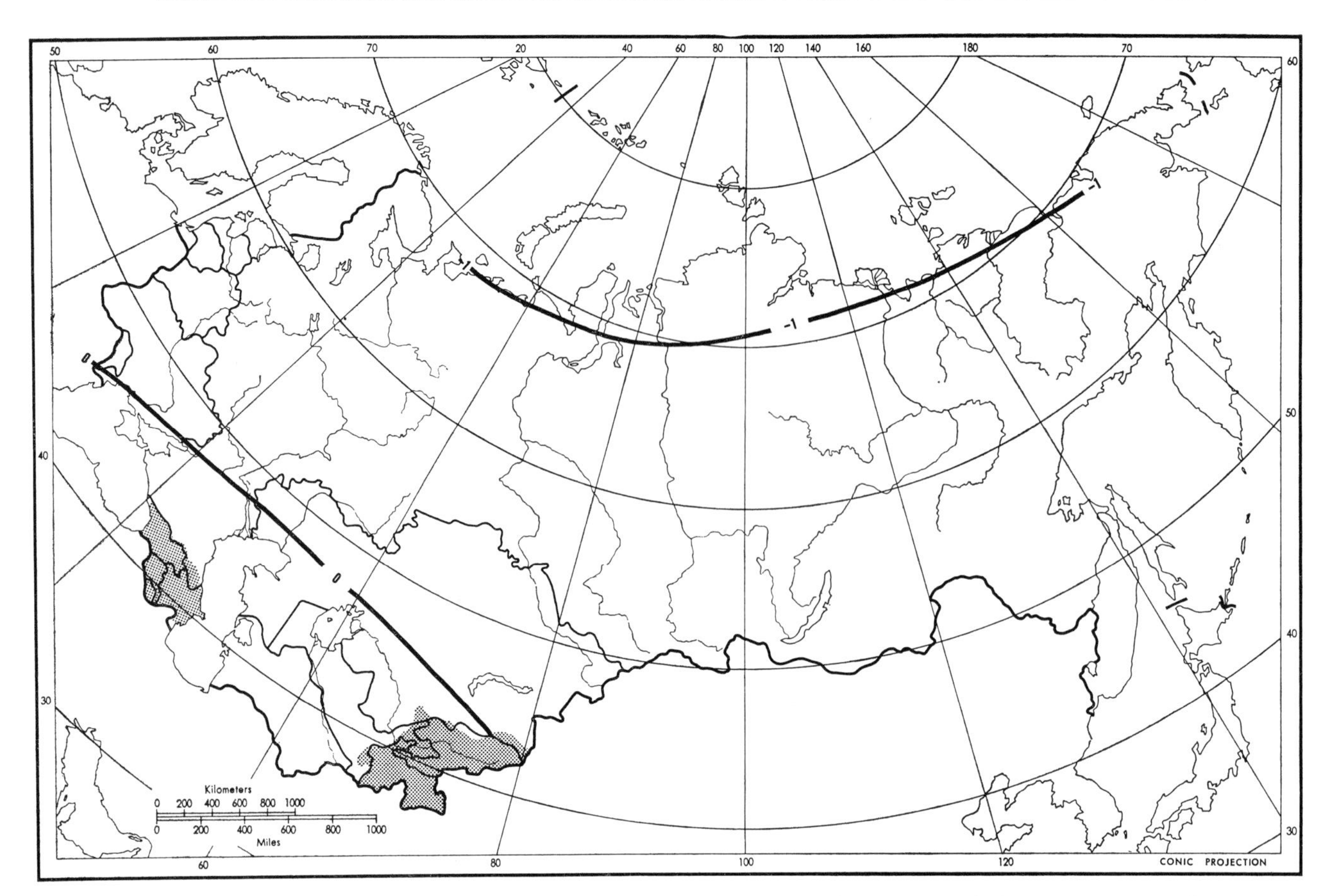

Fig.8-18. Mean radiation balance (kcal./cm²) at the earth's surface, January. (After BARASHKOVA et al., 1961.)

rapid increase in radiation balance across the country, and the zero line moves northward to the White Sea in European Russia and to central Yakutia in eastern Siberia. A broad dip in values still is pronounced in western Siberia, which might be associated with deep snow cover there that hasn't melted yet.

By April the entire Soviet Union has a positive radiation balance (Fig.8-19), and April and May are the months with the greatest differences across the country. The zone of rapid transition in April parallels the edge of the snow cover through central Siberia. By May this zone has moved northward to about the Arctic Circle. By June the north-central part of the country has an as great or greater radiation balance as does Soviet Central Asia, and in July the balance is practically the same everywhere (Fig.8-20). August also shows about the same radiation balance everywhere, but by then the magnitude has dropped from about 8 kcal./cm² in July to 6 kcal./cm² in August. A zonal distribution becomes re-established in September, and by October a winter distribution has definitely set in, with the zero line running through the center of the country (Fig.8-21).

The fall distribution is much different from spring. The radiation balance is much less in fall and there is no zone of rapid transition such as that which borders the front of the snow belt in April. It will be recalled that spring generally has more incoming radiation than fall because of the greater transmissivity of the atmosphere in spring. In addition, the outgoing radiation is higher in fall than in spring because the earth is much warmer in fall.

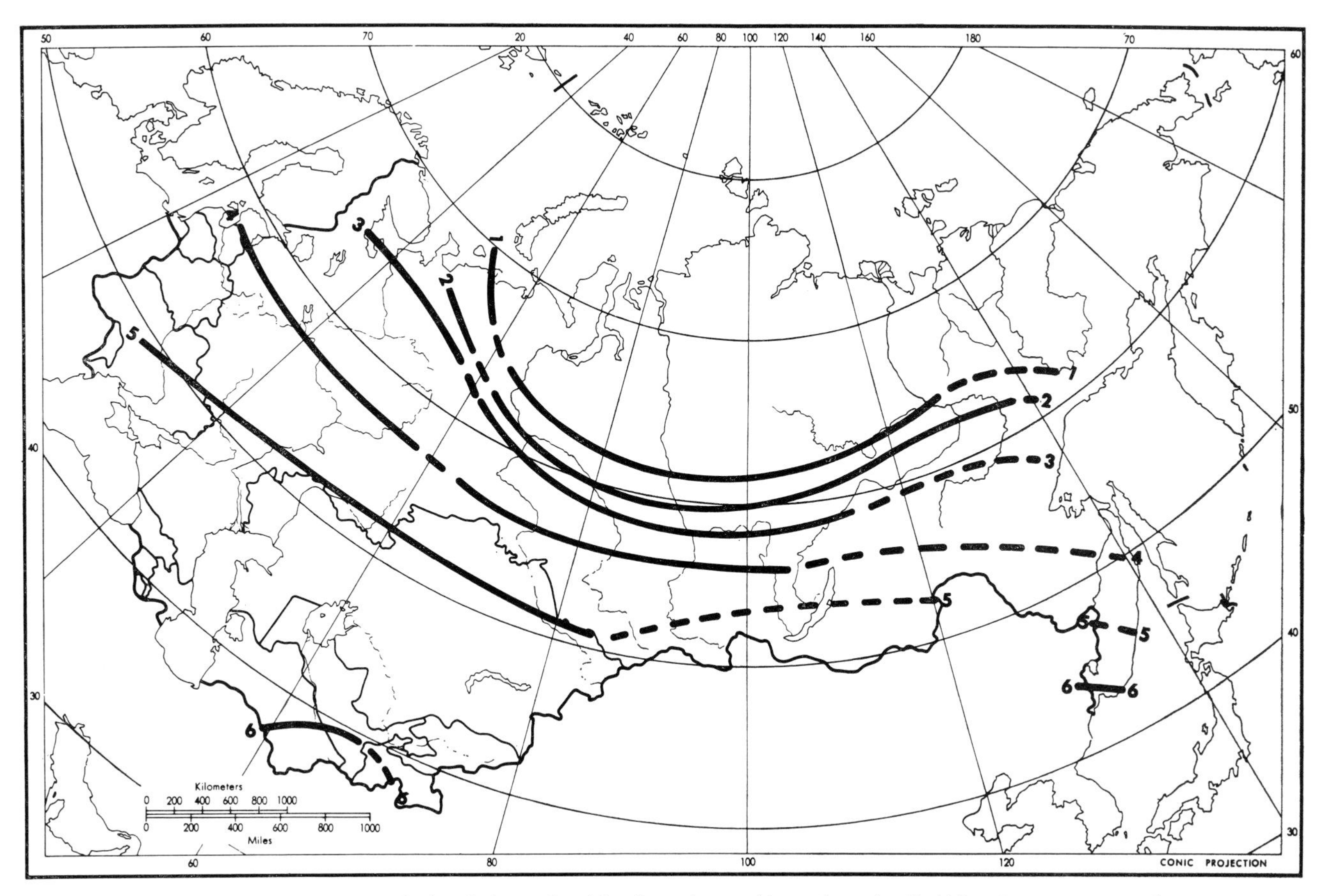

Fig.8-19. Mean radiation balance (kcal./cm²) at the earth's surface, April. (After BARASHKOVA et al., 1961.)

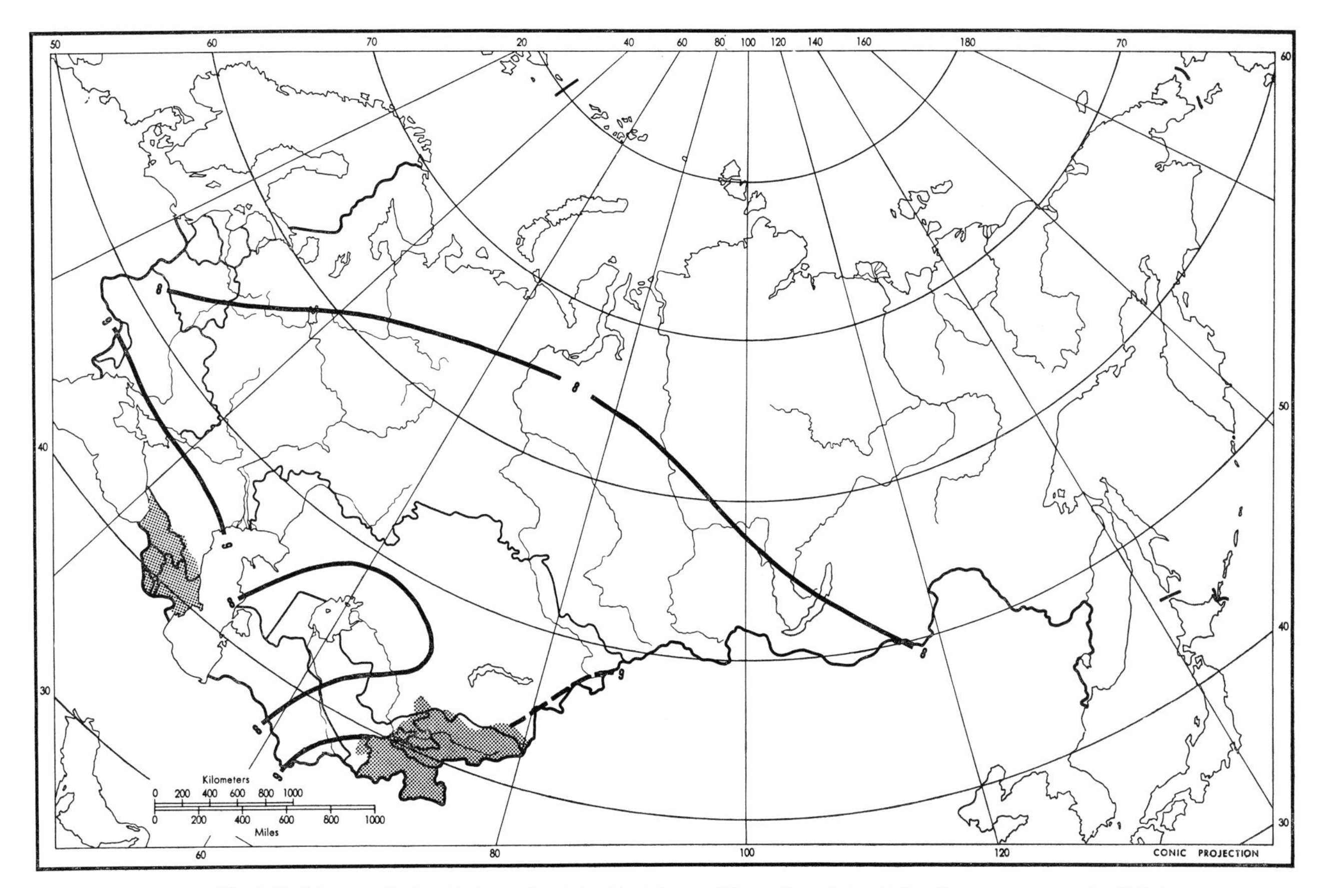

Fig.8-20. Mean radiation balance (kcal./cm²) at the earth's surface, July. (After BARASHKOVA et al., 1961.)

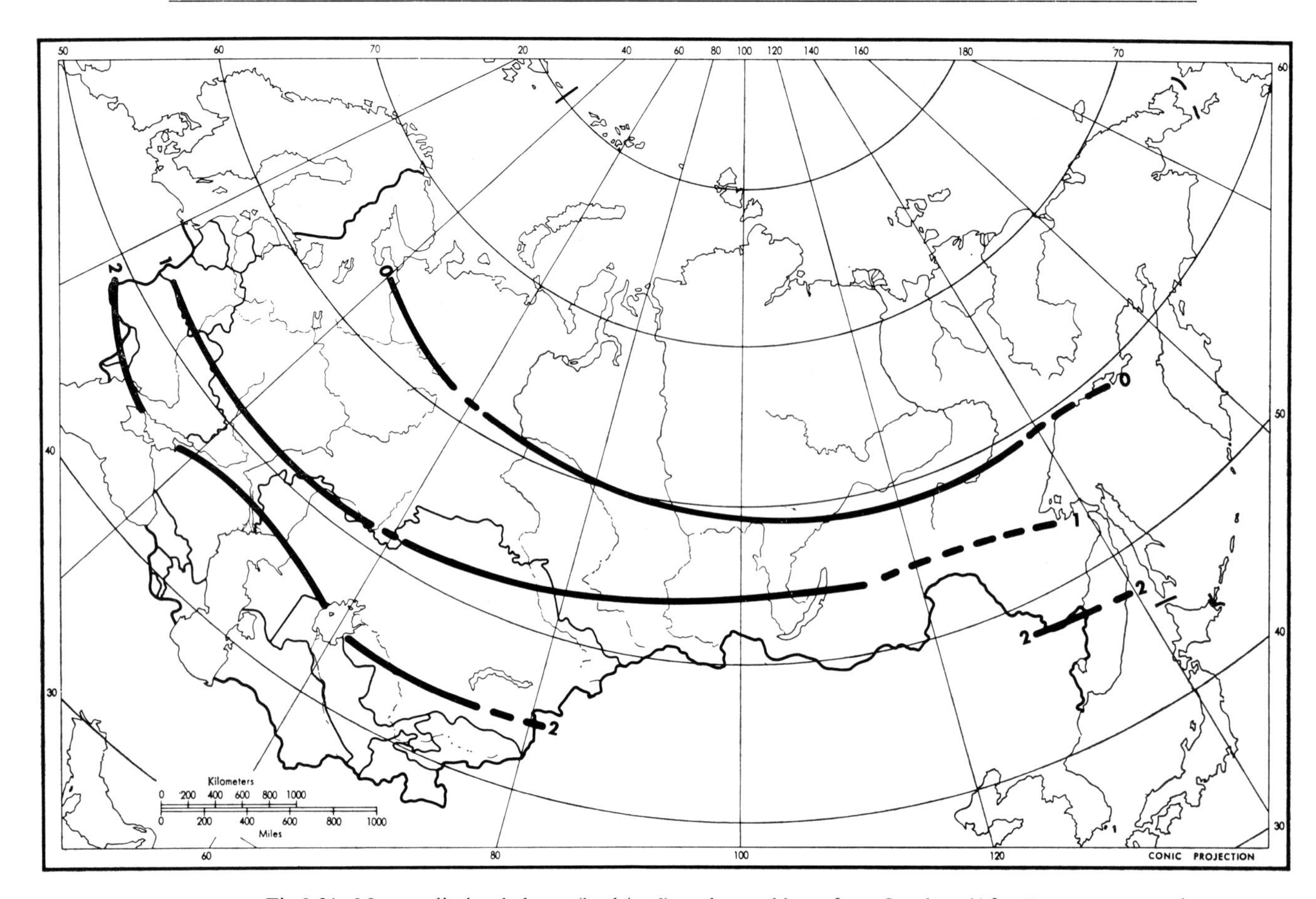

Fig.8-21. Mean radiation balance (kcal./cm²) at the earth's surface, October. (After BARASHKOVA et al., 1961.)

There seems to be a marked relationship between radiation balance and snow cover. The period with negative radiation balance generally is 15 to 20 days shorter than the period of snow cover. The transition from a positive to a negative balance in fall generally occurs a few days before the establishment of a snow cover, and the transition from a negative to a positive radiation balance in spring generally occurs 20–30 days before the disappearance of the snow cover. Thus in spring the balance becomes positive with higher albedo than in autumn. The greater radiation balance under comparable conditions in spring is primarily due to the fact that the snow is melting at that time of year and much of the radiation that strikes the surface is used in the melting process. Hence there is little temperature rise at the surface and little outgoing radiation from the surface. During autumn this is not the case. In fact, in autumn when the ground is freezing and a snow cover is forming, latent heat of fusion provides additional heat for back radiation and thereby reduces the radiation balance.

Snow cover also affects the diurnal regime of radiation balance. Without a snow cover, radiation balance commonly passes through zero about 1 h after sunrise and 1 h 10 min before sunset. With a snow cover, radiation balance passes through zero about 1.5 h after sunrise and 1.5 h before sunset.

At coastal stations in autumn the transition toward a negative radiation balance occurs later than in continental interiors, and conversely in spring the transition toward positive radiation balance occurs earlier near coasts than in interiors. One of the striking features of the world maps of radiation balance in the *Atlas of Heat Balance* is the discontinuity of lines at all coasts. Ocean areas obviously have much higher radiation balances than do continental areas adjacent to them. It is very important then whether there is a general flow of air into a continent or out away from it, since this will determine the direction of heat advection. The fact that high mountains rim the entire southern fringe of the Soviet Union and block the advection of warm moist air from any southern seas, as well as the fact that there is little penetration of Pacific air across the coast during winter, produce a tremendous deficit of heat across much of the Soviet Union as compared to many land areas such as eastern United States.

Heat expended on evaporation

During the course of a year much of the Soviet Union loses between 10 and 20 kcal./cm^2 of heat to the evaporation of moisture from the surface (Fig.8-22). These modest amounts compare to those throughout much of Canada. But they do not compare very favorably to much of western Europe which utilizes 20–40 kcal./cm^2, eastern United States which utilizes 30–60 kcal./cm^2, or vast Asia south of the Soviet Union which utilizes 30–60 kcal./cm^2. The highest values in the vicinity of the Soviet Union are over some of the bordering seas. The Black and Caspian seas in the south experience losses of 40–60

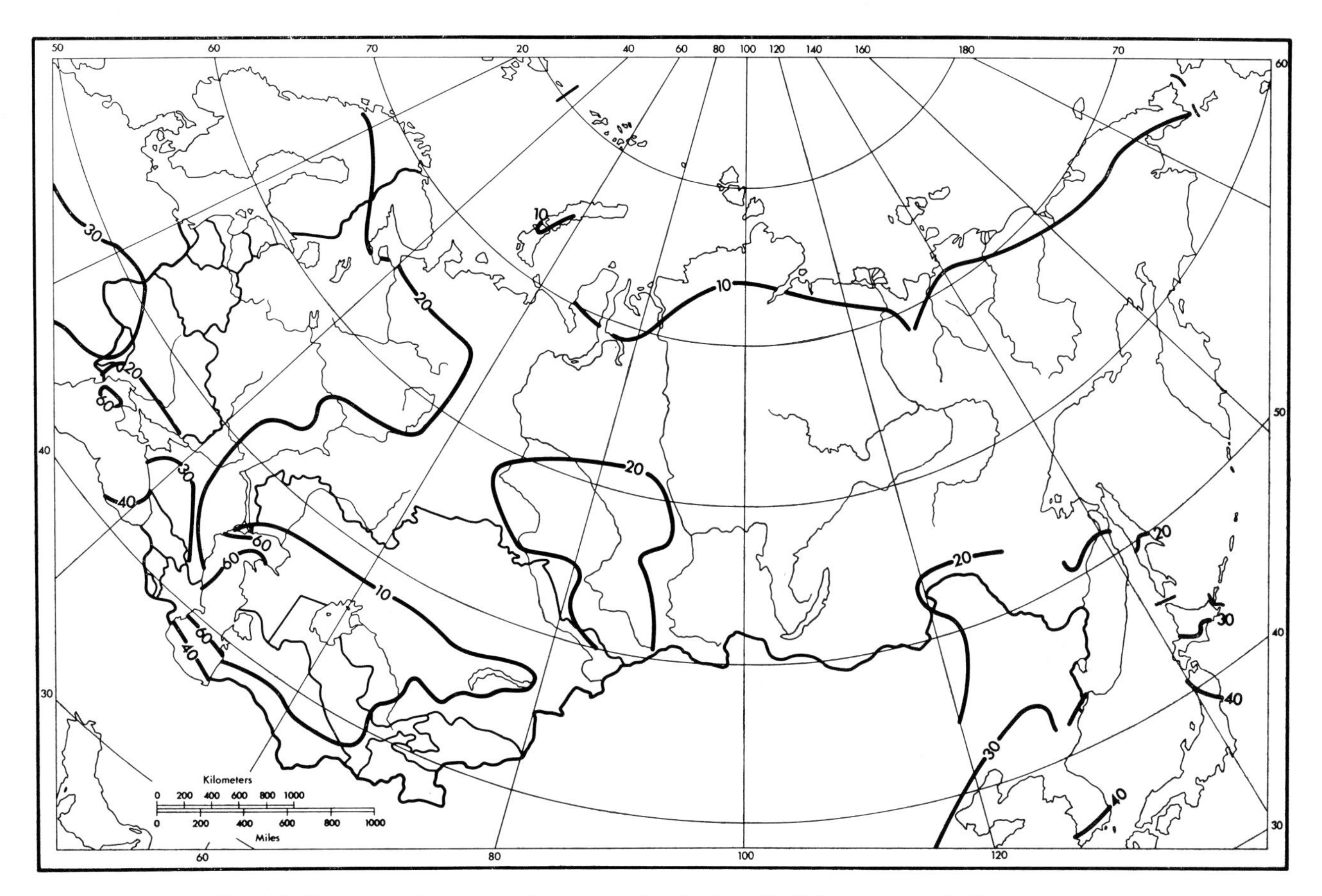

Fig.8-22. Mean annual heat used for evaporation (kcal./cm^2). (After BUDYKO, 1963.)

kcal./cm² to evaporation, as does much of the Sea of Japan in the Far East. But no water bodies adjacent to the Soviet Union experience the great evaporation that takes place off the east coast of the United States where more than 180 kcal./cm² are utilized. Thus, once again it can be seen that the Soviet Union has no source of moisture and latent heat commensurate to that of eastern United States. The annual evaporation of moisture over the land in southeastern United States is as much as that over the water surfaces of the Black and Caspian seas.

The percent of annual radiation balance of the earth's surface which is utilized in evaporation varies from less than 30% in Central Asia to more than 70% over the northern half of the European part of the country, much of Siberia, and the Far East (Fig.8-23). The low values in Central Asia are due to the lack of moisture available for evaporation rather than lack of heat.

The same factors determine the annual course of evaporation in all regions. Maximum evaporation throughout much of the country seems to take place in June, with July second, May third, and August a poor fourth. There is a large drop-off from July to August. Apparently this regime is a function of soil moisture which is rather abundant after spring thaws and the beginning of early summer rains, but which is very meager later in the summer after evaporation has expended most of the soil water.

In the middle of summer the greatest evaporation takes place primarily over the forested

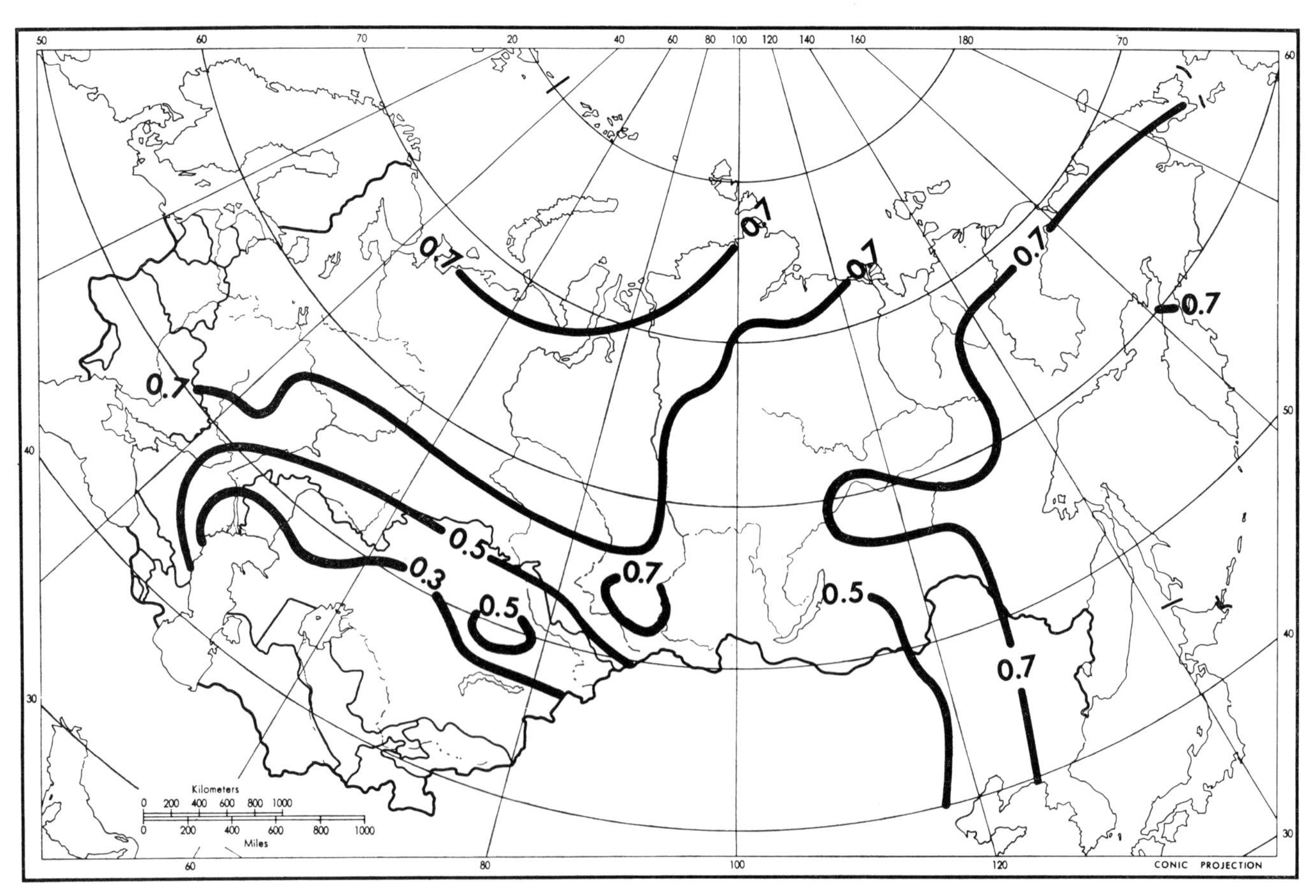

Fig.8-23. Mean annual heat used for evaporation as a portion of the heat balance at the earth's surface. (After ZUBENOK, 1966.)

regions of the country, where soil moisture is fairly available (Fig.8-24). In this broad zone heat expended on evaporation in July amounts to about 4 kcal./cm^2. Higher values are found only in the south, primarily over the Black and Caspian seas, and to a certain extent over the warm humid western portion of the North Caucasian Foreland.

The entire Soviet Union expends less than 1 kcal./cm^2 on evaporation during each month from November through February. And only fringe areas in the east, west, and south experience more than this during March and October. Therefore there is no reason to show a map for winter conditions since there is no differentiation across the country.

Turbulent heat exchange

Turbulent heat exchange is usually determined as a residual of the heat balance equation, and therefore is subject to more error than the other components, since it incorporates the sum of the errors committed in computing the other components. However, monthly and yearly maps of turbulent heat exchange have been presented in the *Atlas of Heat Balance*, and they are somewhat instructive.

Overall, the Soviet Union shows a rather high turbulent heat exchange for its latitude, which is to be expected because of a general lack of moisture over much of its territory. However, where areas are wet the turbulent heat exchange decreases accordingly. Outstandingly low-value areas are the Colchis Lowland, the Polesya, all of the Far East, and

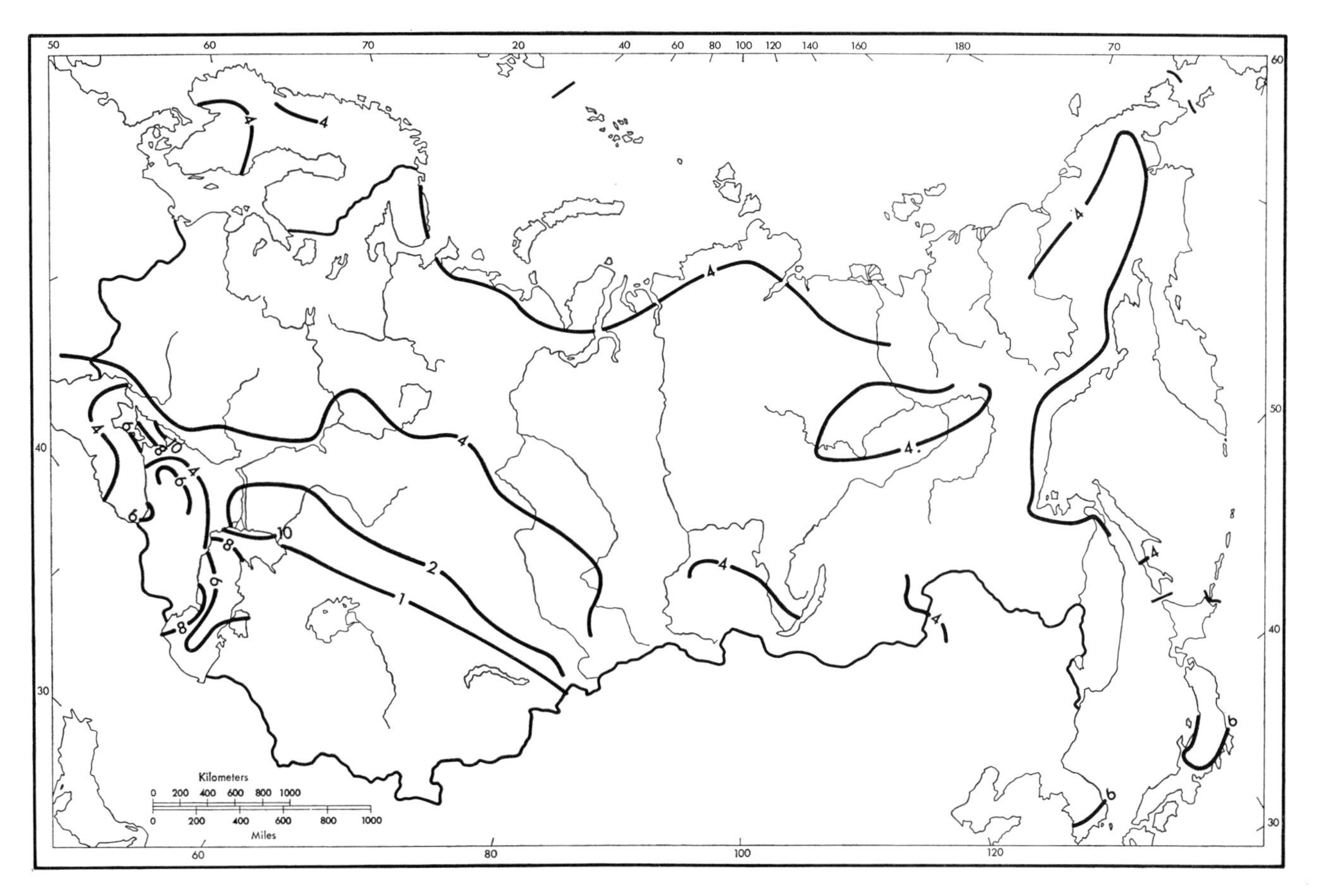

Fig.8-24. Mean heat used for evaporation (kcal./cm^2), July. (After BUDYKO, 1963.)

much of the northern half of European Russia and Siberia, except the Arctic fringe in summer. Whereas most of these areas experience a turbulent heat exchange of less than 10 kcal./cm^2 during the year, and extensive parts of these areas experience less than 5 kcal., the deserts of Central Asia experience more than 40 kcal. over wide regions. The yearly totals in Central Asia are similar to those in southwestern United States and northwestern Mexico which lie generally 5°–10° latitude farther south.

As one would expect, within the Soviet Union the pattern of turbulent heat exchange is somewhat opposite to that of the heat used in evaporation processes (Fig.8-25). Where soil moisture is lacking for evaporation purposes, turbulent heat exchange increases. Values are highest by far in the Central Asian desert, and intermediate values show a significant bulge northeastward into the Lena Valley, which is noted for its subhumid conditions at this high latitude. Low values of turbulent heat exchange show a significant southward projecting lobe in western European U.S.S.R. down the Dnieper Valley approximately to Kiev. This is obviously due to the wetness of the surface in this area and the high humidity of the atmosphere.

The yearly pattern of turbulent heat exchange in the U.S.S.R. is almost entirely a function of summer, since there is essentially no turbulent heat exchange over almost all the Soviet Union from December through February. The January map shows only Central Asia having somewhere between 0 and 1 kcal./cm^2. The rest of the country is entirely nil. By March the 0-line in the south has moved northward to about the 50th parallel so that much of the Ukraine, the North Caucasus, Central Asia, and the southern half of Kazakhstan experience some turbulent heat exchange in early spring. Central Asia shows around 2 kcal./cm^2 over much of its area. A significant area has already developed over the middle and upper Lena Valley where 0 to 1 kcal./cm^2 are expended in turbulent heat exchange in March. These tendencies intensify and spread northward through April, May, and June, and maximum amounts of turbulent heat exchange are experienced over most of the country in July (Fig.8-26). During the one month of July turbulent heat exchange along the Arctic fringes reaches higher values than farther south in the forested zone. Values along the Arctic in July equal those throughout much of the mixed forest, forest steppe, and steppe regions of European U.S.S.R. and Siberia. This is no doubt due to the very high incoming radiation along the Arctic at this time of year and the generally low absolute humidity of the air in the region. The regions of notably low turbulent heat exchange during July are the Polesya in the upper reaches of the Dnieper River in the west European part of the country, in the Colchis Lowland along the eastern end of the Black Sea in western Georgia, all of the Far East, and much of the tayga zone of northern European U.S.S.R. and Siberia.

It is instructive to compare Figs.8-14, 8-15, 8-23, and 8-25 to determine over the course of a year the disposition of the heat that is absorbed at the earth's surface in the Soviet Union. Of the 100 kcal./cm^2 of radiation that is absorbed at the earth's surface throughout much of Central Asia during the year, about 50 kcal./cm^2 escape as net outgoing radiation, 40 kcal. are transferred upward into the atmosphere by turbulence and only about 10 kcal. are transferred into the air through the evaporation of moisture. Over the European plain around Moscow, the earth absorbs about 60 kcal./cm^2 during the year, and loses approximately 30 kcal. through net outgoing radiation, 10 kcal. through turbulent heat exchange, and 20 kcal. through evaporation. The Arctic fringe of Siberia receives about 40 kcal./cm^2 per year, and loses about 20 through net outgoing radiation, 10

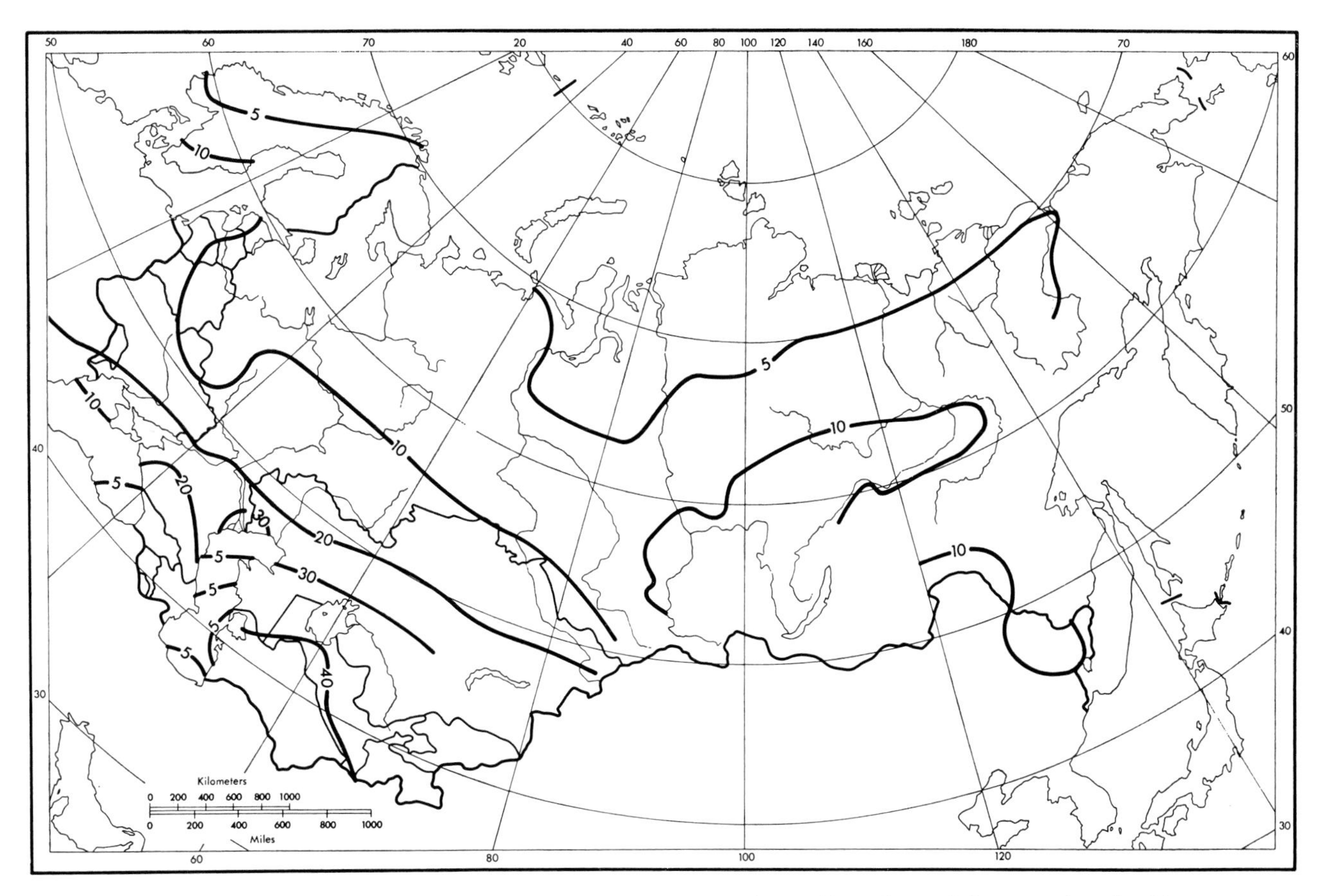

Fig.8-25. Mean annual turbulent heat exchange between the earth's surface and the atmosphere (kcal./cm²). (After BUDYKO, 1963.)

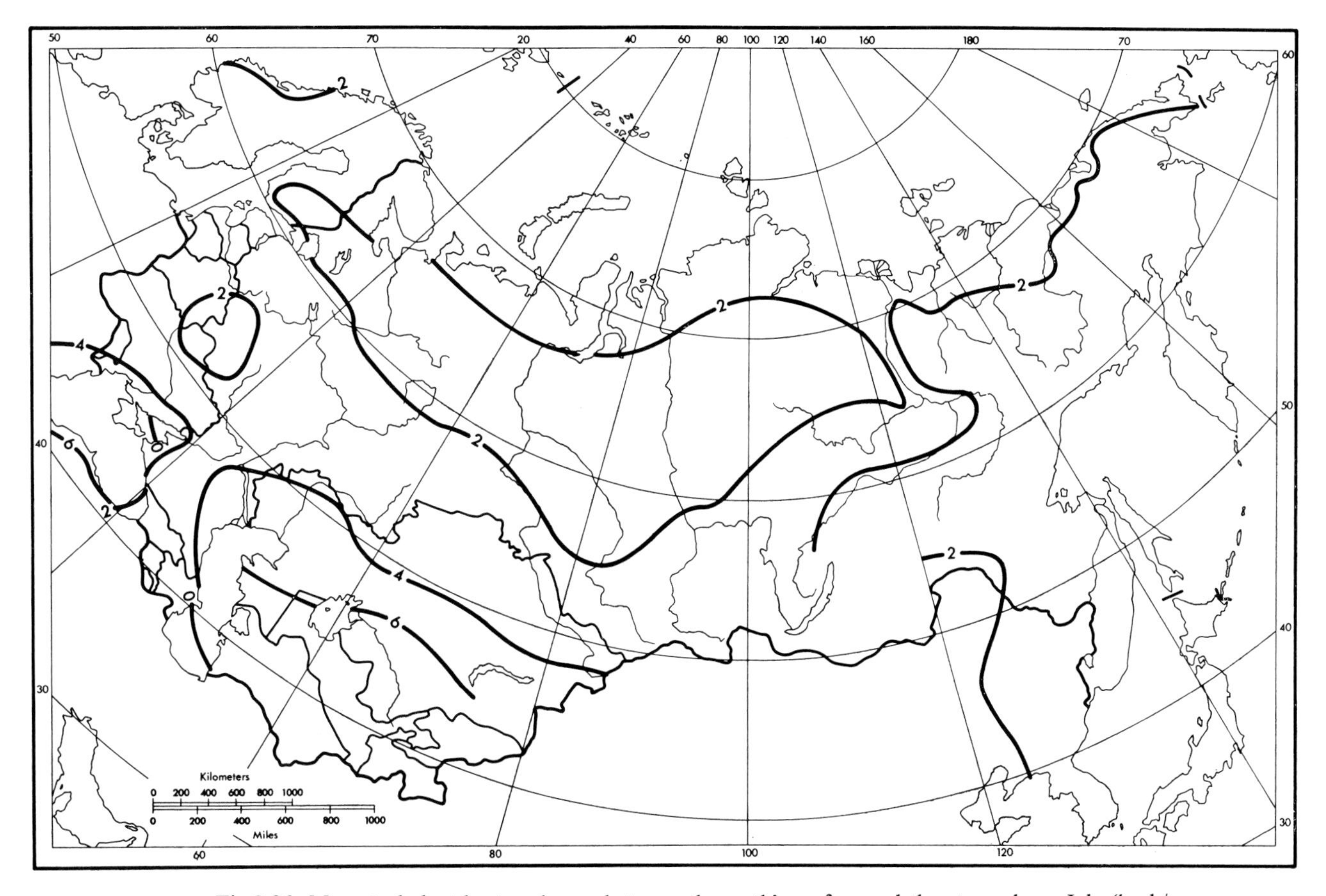

Fig.8-26. Mean turbulent heat exchange between the earth's surface and the atmosphere, July (kcal./cm²). (After BUDYKO, 1963.)

through turbulent heat exchange, and 10 through evaporation. Maritime Kray in the Far East receives about 80 kcal./cm^2 per year, and loses about 40 through net outgoing radiation, 15 through turbulent heat exchange, and 25 through evaporation. It appears that throughout much of the Soviet Union about half the heat that is absorbed at the earth's surface is lost as net outgoing radiation. But the ratio between turbulent heat exchange and evaporation varies greatly from place to place depending primarily on atmospheric and soil moisture conditions. In the deserts of Central Asia turbulent heat exchange amounts to about four times as much as the heat used in evaporation, but in the mixed forests around Moscow, evaporation accounts for more than twice as much heat as turbulent exchange does, and in the southern part of the Soviet Far East evaporation accounts for about 1.6 times as much heat as turbulent exchange does.

The relative magnitudes of the heat components at the earth's surface can be changed significantly in dry areas by irrigation. Irrigation reduces surface temperatures and increases surface atmospheric humidities. It thereby reduces net outgoing radiation and turbulent heat exchange and increases the radiation balance and the heat used for evaporation. Significant amounts of heat are thus shifted from radiation losses and sensible temperature increases to the evaporation process, which, if the irrigated area is being cropped, is largely accomplished in the form of useful transpiration through plants.

Heat advection

Latent heat

Except in the southern part of its Far Eastern territory, the Soviet Union suffers a deficit of heat advected into the country more than any other developed area of the world. Heat gained from the condensation of water vapor varies from less than 10 kcal./cm^2 per year in Central Asia to between 40 and 60 kcal./cm^2 over Maritime Kray and the southern halves of Sakhalin Island and Kamchatka (Fig.8-27). Over almost all the country the values range from 10 to slightly more than 20 kcal./cm^2. This compares poorly indeed to the eastern half of the United States and southeastern Canada which receive anywhere from 40 to well over 60 kcal./cm^2 per year. The highest values in the world, of course, are associated with the very high rainfall areas of the tropics, particularly some of the monsoon lands such as western Burma, part of the Guinea coast of Africa, and the west coast of Colombia, which receive over 180 kcal./cm^2 per year. Much of central and western Europe receive between 40 and 60 kcal./cm^2.

To determine net heat gain or loss through the hydrologic cycle for any given area, one would of course have to balance heat gained through condensation against heat lost through evaporation. Much of the Soviet Union shows little net effect. Western and central Siberia and the monsoon fringe of the Far East show a little gain, and much of the rest of the country shows almost an exact balance between gain and loss. Everywhere in the Soviet Union the hydrologic cycle operates at a fairly low ebb.

Sensible heat

Throughout the entire Soviet Union, except for a small area in the Far East, horizontal movements of air result in a net loss of sensible heat energy during the year (Fig. 8-28).

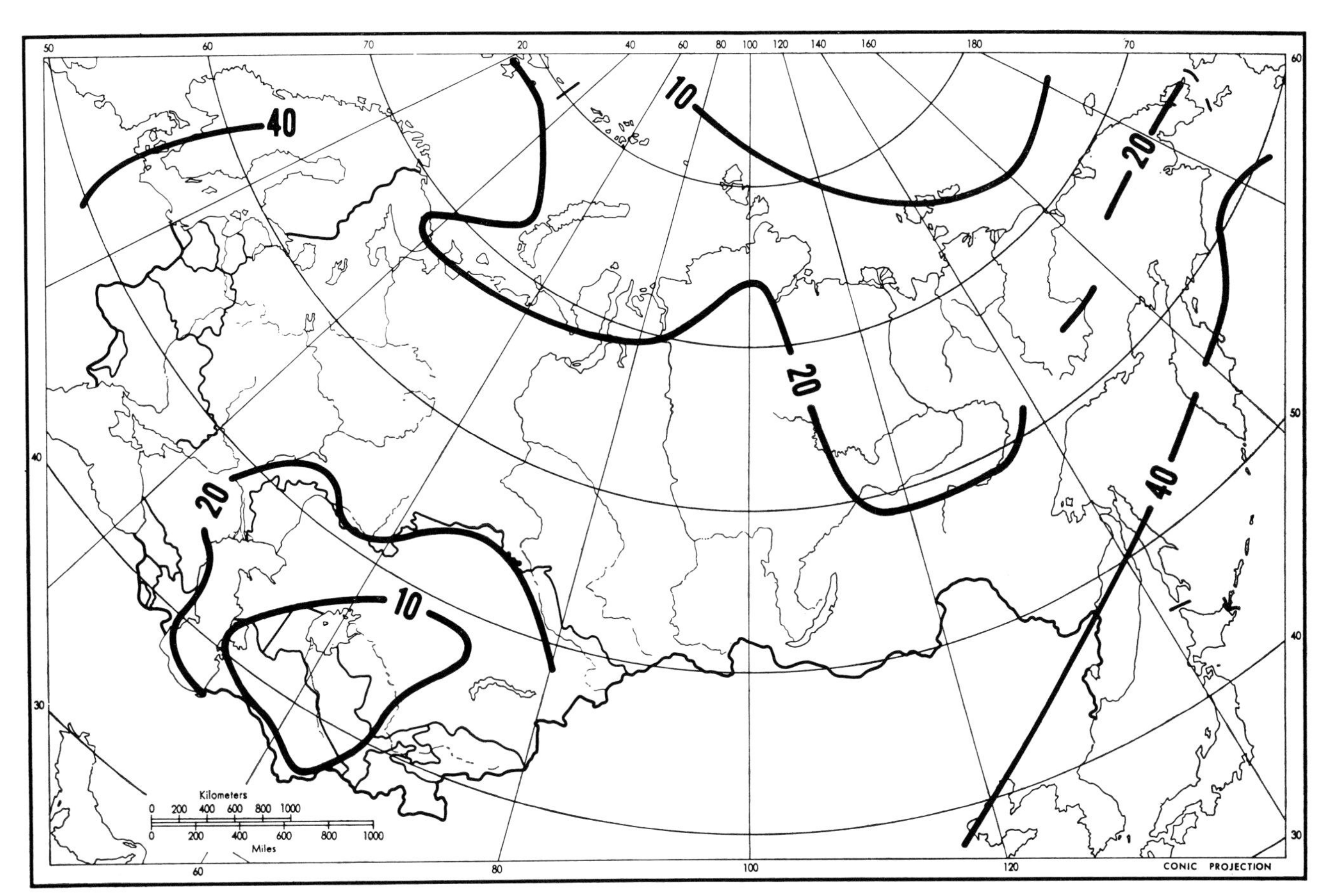

Fig.8-27. Mean annual heat gain through condensation of water vapor (kcal./cm²). (After BUDYKO, 1963.)

The world map of advection of sensible heat in the *Atlas of Heat Balance* shows that in general in low and lower middle latitudes heat is advected from oceanic areas toward interiors of land masses, while in higher middle and higher latitudes heat is generally advected out of continental areas toward oceanic areas. Thus, the Soviet Union, lying in higher middle and higher latitudes, derives a maximum climatic damage from this sort of relationship. Its heat deficit because of its high latitude is due to more than just astronomical relationships which determine the angles of sunrays and lengths of days. The high latitude works equally against the advective processes. Much of the Soviet Union experiences a heat loss through the advection of sensible heat that amounts to somewhat less than 20 kcal./cm² in the south and more than 40 kcal./cm² in the north. In sharp contrast, practically all of the United States, except for some of the mountainous west, receives a net gain of heat through sensible heat advection which amounts to as much as 40 kcal./cm² or more throughout the entire area of southeastern United States. Even southeastern Canada shows a significant heat gain, while comparable latitudes in the Soviet Union show significant losses. Figs. 8-27 and 8-28, perhaps more than any other maps, illustrate the basic climatic dilemma that faces the Soviet Union—a general lack of heat. Not only does the country receive a meager supply of heat through the radiation processes, because of its high latitude, but also it gains little, if any, help from the outside, and if anything contributes more heat to its surroundings through the advective processes than it gains.

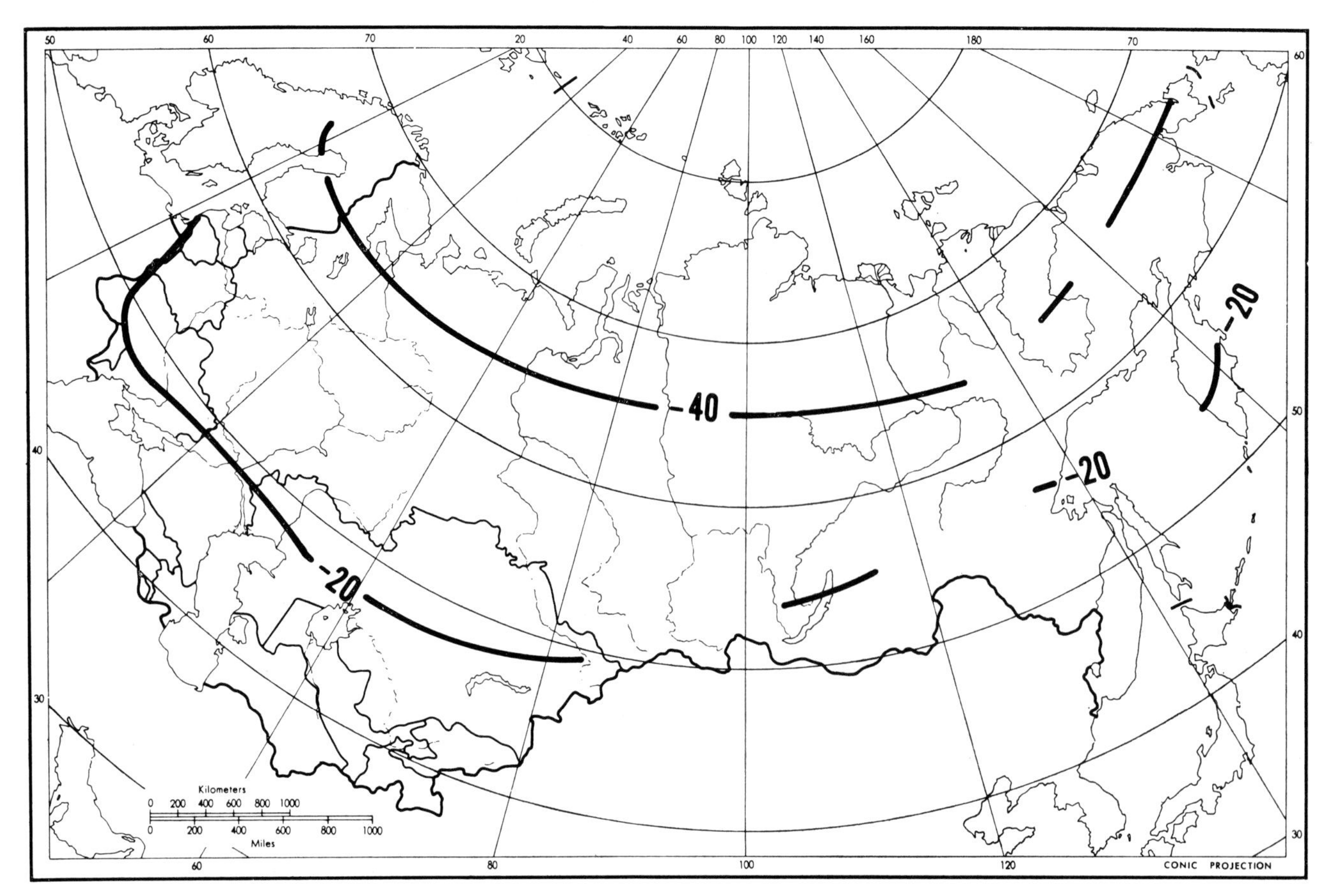

Fig.8-28. Mean annual advection of heat in the atmosphere (kcal./cm²). (After BUDYKO, 1963.)

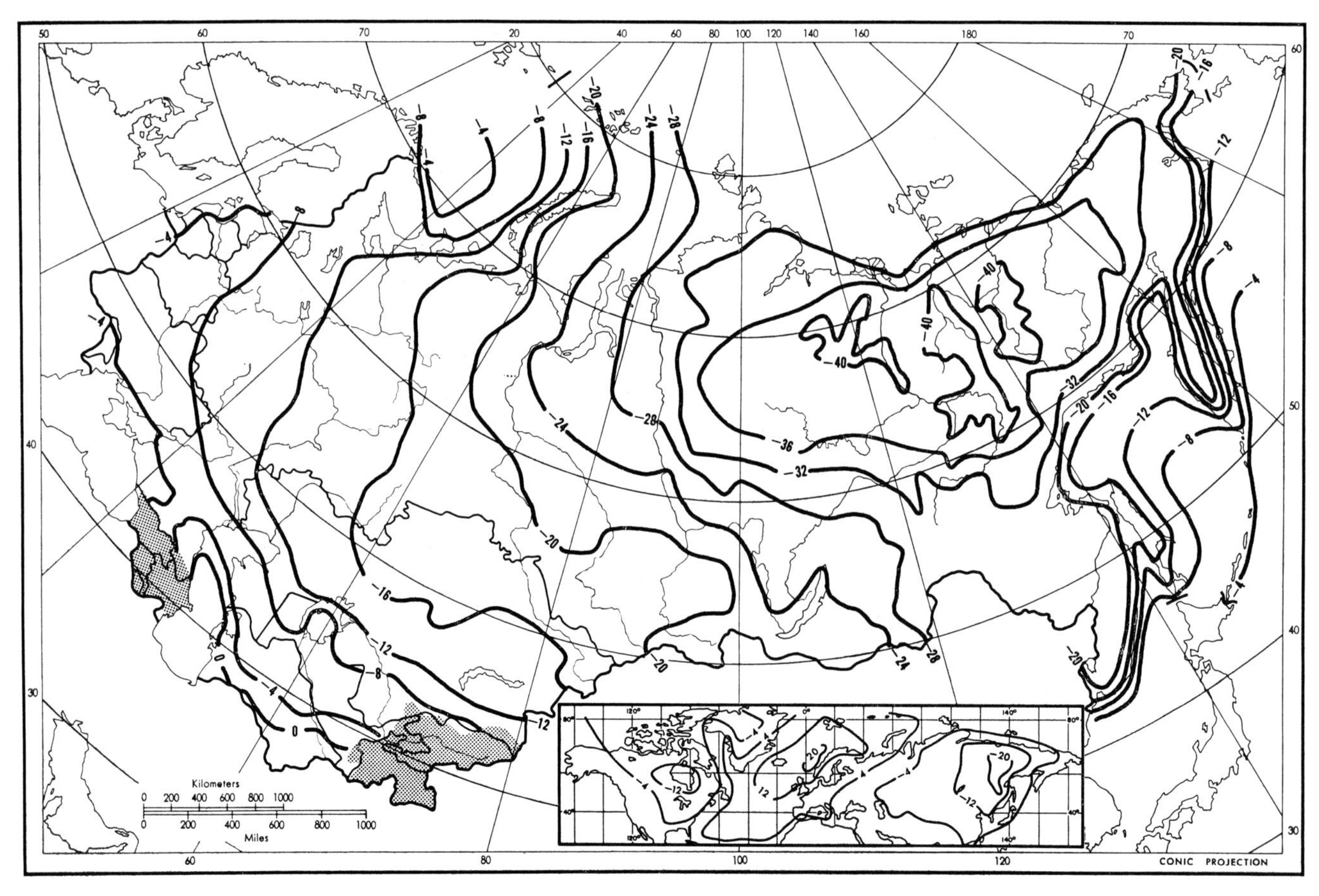

Fig.8-29. Mean surface air temperature, January (°C). Inset: mean temperature deviations from latitudinal normals, January. (After GERASIMOV, 1964.)

Temperature

Temperatures in the Soviet Union reflect the high latitude and high degree of continentality in most of the country. Generally temperatures are relatively cool, and seasonal variations are great. Even throughout much of the better agricultural regions in the southwestern part of the country summers are relatively cool compared to regions such as the corn belt of the United States. For the latitude, temperatures in the western parts of the country generally are not low, and in some cases are even higher than normal, particularly during winter, but latitudes are high compared to most other major agricultural countries of the world.

Temperature means

Mean surface air temperatures in January vary from a little above 0°C along the southern boundaries of Soviet Central Asia to —48.9°C at Verkhoyansk and —50.1°C at Oymyakon in the northeastern part of the country (Fig.8-29). The distribution of isotherms at this time of year reflects the great differences in temperature between sea and land, particularly in the west and eastern portions of the country. All the way from the western borders of the country eastward to beyond the Yenisey River isotherms run almost north–south in the northern half or two-thirds of the territory. This pattern reflects the great influence of the Atlantic at this time of year abetted by active cyclonic movements along the northwestern edge of the Asiatic high. It also reflects the insignificant differentiation of latitudes at this time of year at latitudes from 50° to 80° north, where little insolation is being received and where the snow cover reflects much of what is being received. Only south of 50° latitude do the isotherms take on a quasi zonal pattern. In the Far East the isotherms almost exactly parallel the coastline and their close spacing shows the extreme temperature gradients that exist across much of this coast during winter when the Asiatic high dominates the region and prevents any significant penetration of oceanic air. Even along the Arctic coast of Siberia there is close correspondence of isothermal patterns to coastal configuration. In spite of the fact that the Arctic is frozen over at this time of the year right up to the coast, the air over the land mass to the south is considerably colder than that over the frozen water surface, since the atmosphere over the frozen Arctic still receives considerable amounts of heat from the unfrozen water underneath the ice. Thus, colder winter temperatures are found over the land mass of northeastern U.S.S.R. than anywhere in the Arctic Basin. And in fact, colder temperatures are found in the semi-enclosed river valleys between the mountain ranges of this part of the U.S.S.R. than at similar elevations anywhere else in the world. The extremely cold temperatures observed in the Antarctic have been observed at elevations of around 3,000 m or more, whereas the cold temperatures observed at Verkhoyansk are at only 137 m above sea level. Temperatures in northeastern U.S.S.R. during January are as much as 24°–26°C below the average for their latitudes. At similar latitudes in central Canada, temperature anomalies do not amount to more than —15° to 16°C (inset of Fig.8-29).

Temperature anomalies

The contrast in the primary controls of air flow between the territory east of the Yenisey

River and that to the west is reflected very clearly in the temperature anomaly distribution shown in the inset in Fig.8-29. The western part of the country has significant positive temperature anomalies while the bulk of the country to the east of the Caspian and western Siberia experiences extreme negative temperature anomalies. The zero line runs from southwest to northeast from about the mid-section of the northern coast of the Black Sea to the Ob' Gulf and approximates the average position of the northwestern edge of the Asiatic high. Obviously air masses which have originated in the Atlantic Ocean or the Barents Sea bring a good deal of warmth to the western parts of the Soviet Union at this time of the year, while air masses associated with the Asiatic high extract a great deal of warmth from the eastern 3/4 of the country.

During January the western part of the Soviet Union receives much more warmth from the outside than does eastern North America. Temperature anomalies go as high as +10°C in Karelia. In North America, all of the continent east of the Rockies, except for the Florida Peninsula shows negative temperature anomalies at this time of year which reach values as great as −8°C throughout the entire upper Mississippi Valley. The different magnitudes and orientations of the isolines of temperature anomalies in Eurasia and North America conform to the different air mass situations over the two continents during winter. Whereas North America is open to incursions of fresh continental Polar air from the northwest, European Russia receives maritime Polar air from the northwesterly direction. The coldest air to reach European U.S.S.R. during winter comes from the southeast around the southern edge of the western extension of the Asiatic high. It has traversed a long southerly trajectory from its source region and is considerably modified, and therefore it does not exhibit the characteristics of fresh continental Polar air.

It is a different story in eastern Siberia in January. Here large negative anomalies indicate that this region is the coldest in the world for its latitude. Whereas the highest negative temperature anomalies in North America are approximately −15°C, in eastern Siberia they reach −24°C. Thus, much of European Russia is rather anomalously warm during the winter and much of Asiatic Russia is anomalously cold. The air mass control of temperature in the Soviet Union is much more important in winter than is the latitudinal control.

Moscow compares latitudinally with the southern part of Hudson Bay in North America, but January temperatures compare with northern Iowa. Kiev, which compares latitudinally with Winnipeg, Canada, has January temperatures comparable to Lincoln, Nebraska. However, because of the high latitude of the Soviet Union, there is nothing in the country that compares very well with southeastern United States south of the Ohio River. And certainly there is no counterpart during the winter in the Soviet Union to Florida or California. The southern coast of Crimea, with its exalted "Mediterranean" climate, experiences below freezing temperatures 20% of the time during January. On the average Sevastopol experiences frost 53 days per year, and Yalta 37.

Fig.8-30 shows that during January practically all of the Soviet Union east of the Ural Mountains experiences below freezing temperatures more than 99% of the time. Even the Maritime Province of the Far East receives below freezing temperatures more than 95% of the time in January, as does half of European U.S.S.R. West and south of Moscow a rather steep gradient of frequencies occurs. Ukraine experiences below freezing temperatures about 70% of the time in the north and about 40% of the time in the

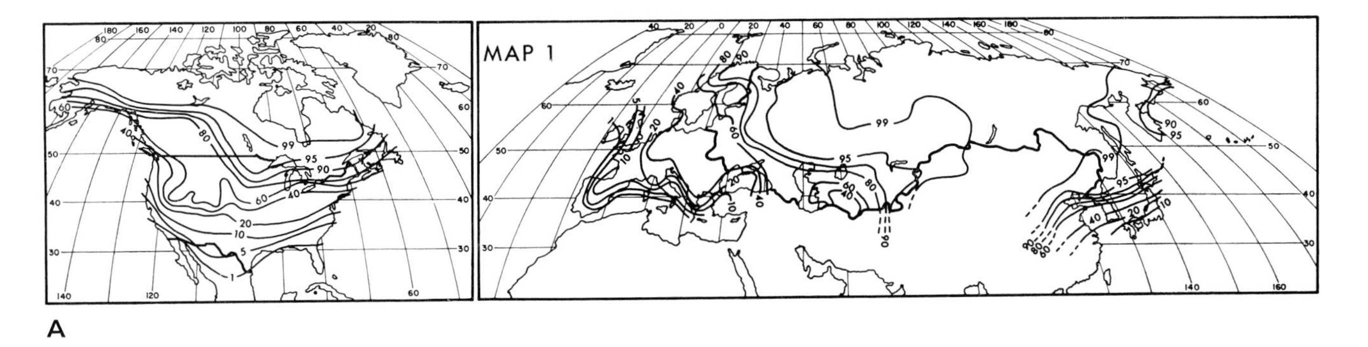

A

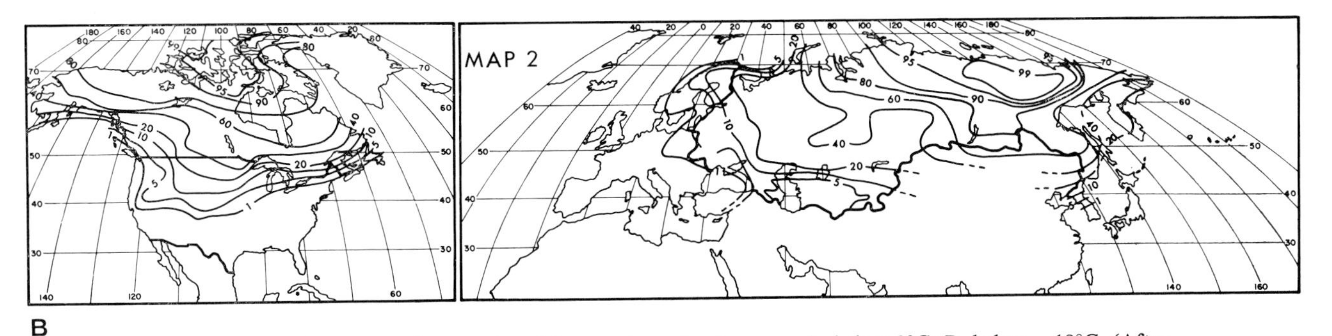

B

Fig.8-30. Percentage frequencies of temperatures for January: A. below 0°C; B. below −18°C. (After RAYNER, 1961.)

Black Sea Steppes. At similar latitudes in central North America, Minnesota averages below freezing about 90% of the time in January. So, at comparable latitudes central North America receives much more frequent freezing temperatures than does European U.S.S.R. However, the United States, lying at lower latitudes, has about half its territory receiving freezing temperatures less than 50% of the time in winter, and the Gulf Coast of the United States experiences freezing temperatures less than 1% of the time.

The January distribution of temperature in the U.S.S.R. prevails through much of November–March. In March there is a beginning of a wave of warmth in the southwestern part of the country that during April sweeps across the country. As far northeastward as a line joining Archangel with Omsk mean temperatures during April are above freezing (Fig.8-31). By May subzero temperatures are squeezed into the northeastern part of the country north of about 65° latitude where snow is still lying on the ground, and by June mean monthly averages of less than freezing appear only on the Arctic fringes of the Taymyr Peninsula and adjacent peninsulas and islands in the Arctic. Temperatures are warmest throughout most of the country in July, when they vary from an average of about 32°C in the southern part of Central Asia to near the freezing point along the Arctic coast of the Taymyr Peninsula and on island groups in the Arctic Ocean (Fig.8-32).

As can be seen on the inset in Fig.8-32, almost the entire land mass of the Soviet Union shows positive temperature anomalies during midsummer. In European U.S.S.R. the positive anomalies are not quite as great as they are in that area during winter, but they are still considerably higher than in eastern North America which shows negative anomalies throughout the eastern fringes of the continent. Moscow, which compared with northern Iowa in winter, during summer compares with Winnipeg Canada, about 400 miles Equator-ward from Moscow but considerably north of Moscow's winter temperature analogs in northern Iowa. Kiev, which had temperatures comparable to

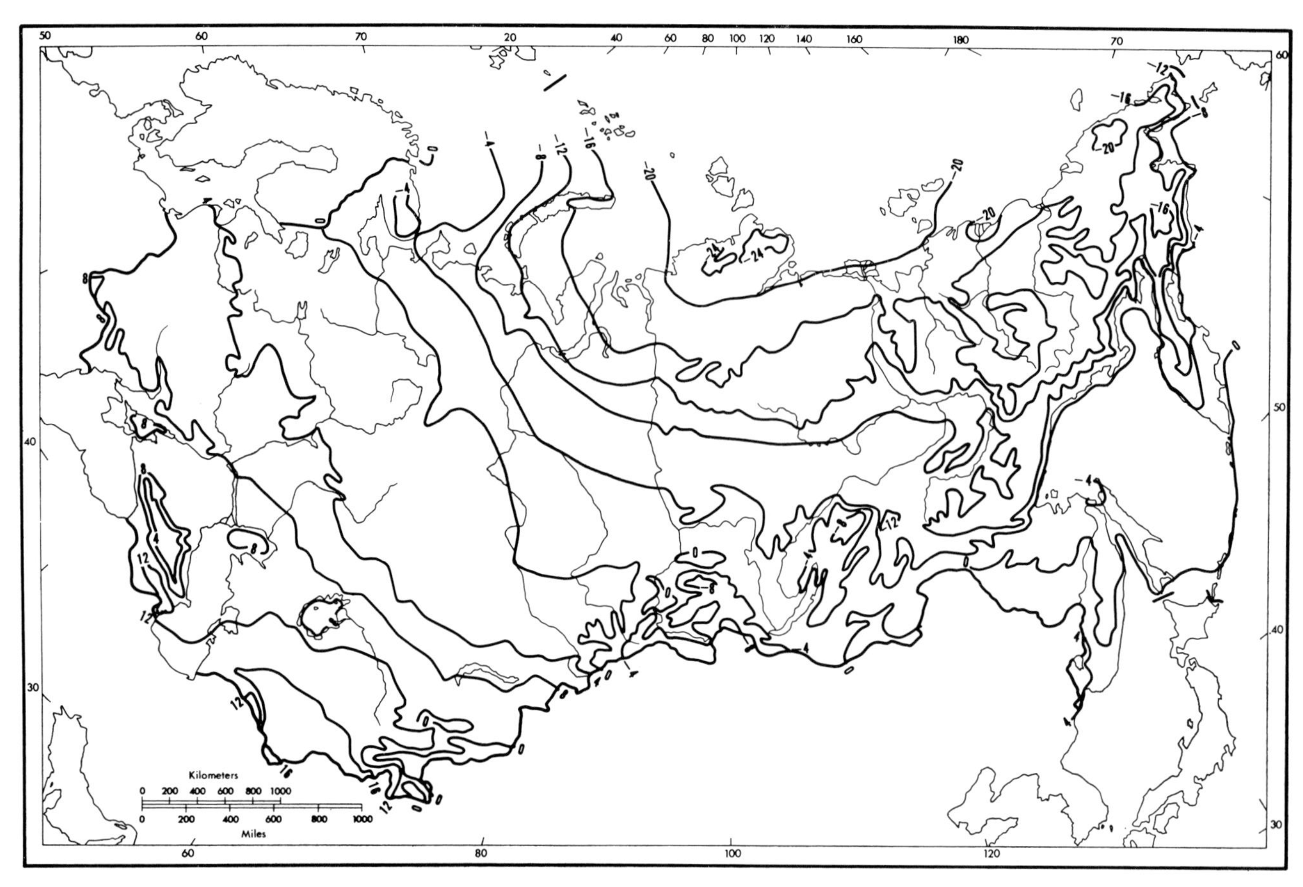

Fig.8-31. Mean surface air temperature (°C) in April. (After GERASIMOV, 1964.)

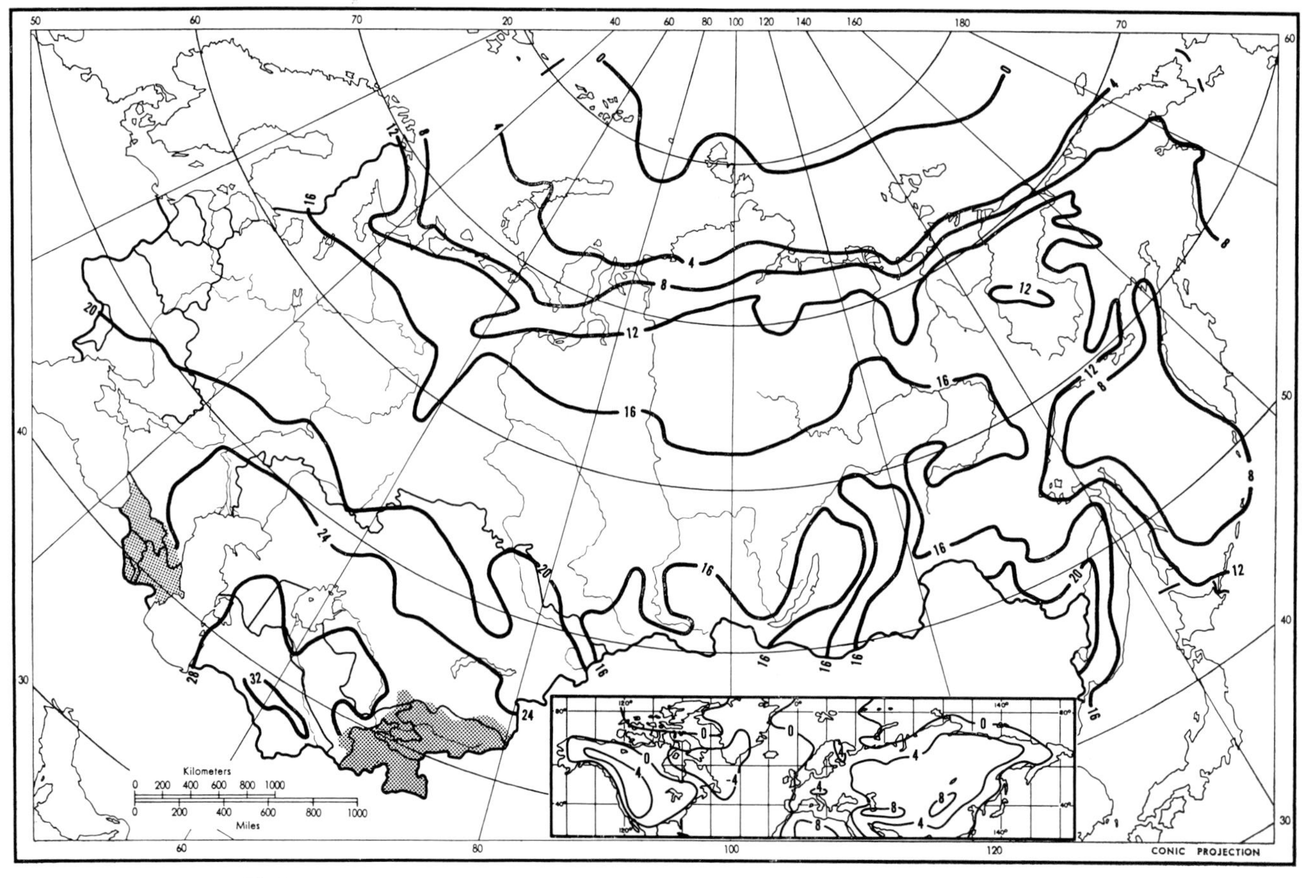

Fig.8-32. Mean surface air temperature (°C) in July. Inset: mean temperature deviations from latitudinal normals, July. (After GERASIMOV, 1964.)

Lincoln, Nebraska in winter, during summer has temperatures comparable to Bismark, North Dakota. Thus, European U.S.S.R. benefits from positive temperature anomalies year round, but more so in winter than in summer, which helps to ameliorate the winters somewhat but is not as advantageous as it might be for agriculture during summer. Asiatic U.S.S.R., on the other hand, which during winter showed very large negative temperature anomalies, during summer shows positive temperature anomalies that are considerably higher than those in European U.S.S.R. and are way above those in eastern North America at comparable latitudes. The same conditions that are conducive to extreme cooling during winter—extreme continentality, air stagnation in mountain basins, and low absolute humidity—produce unusually high temperatures during summer. The isothermal pattern in July is a much more zonal one than that in January. The air mass control of temperature in the Soviet Union is much less important in summer than in winter. Asiatic U.S.S.R. which is a source of extremely cold air in winter, during summer becomes warmer than European U.S.S.R. Yakutsk has a July mean temperature that is 1°C above that of Moscow about 700 km further south. Frequencies of temperatures above 30°C are considerably higher throughout the U.S.S.R. during July than they are at comparable latitudes in Canada (Fig.8-33). However, European U.S.S.R. would not compare to eastern United States in frequencies of high summer temperatures. Thus, once again, for the latitude the Soviet Union fares quite well, but in comparison to countries with comparable economic developments it exhibits definite climatic drawbacks because of its high latitude.

By October freezing temperatures have reestablished themselves over much of Siberia and the Far East, and a cold wave is gradually progressing from northeast to southwest across the country (Fig.8-34). The pattern of isotherms looks much as it did in April, but temperatures in the north are considerably higher in October than in April. Mean

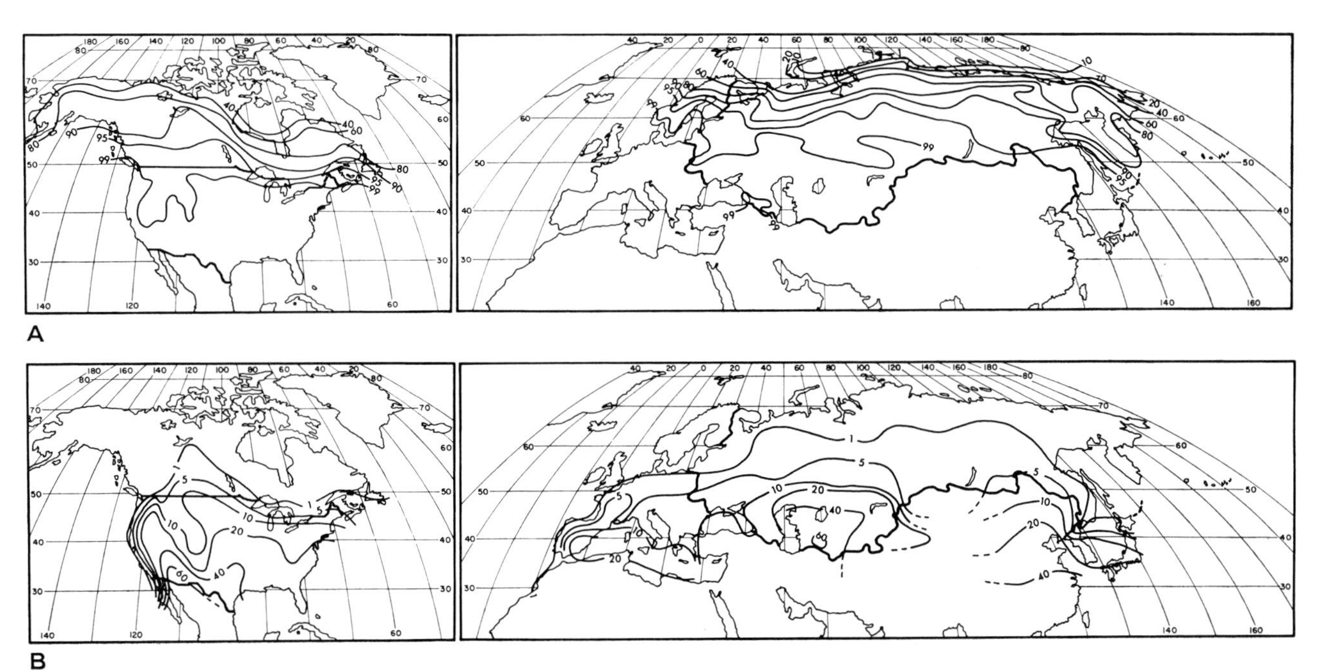

Fig.8-33. Percentage frequencies of temperatures for July: A. above 10°C; B. above 30°C. (After Rayner, 1961.)

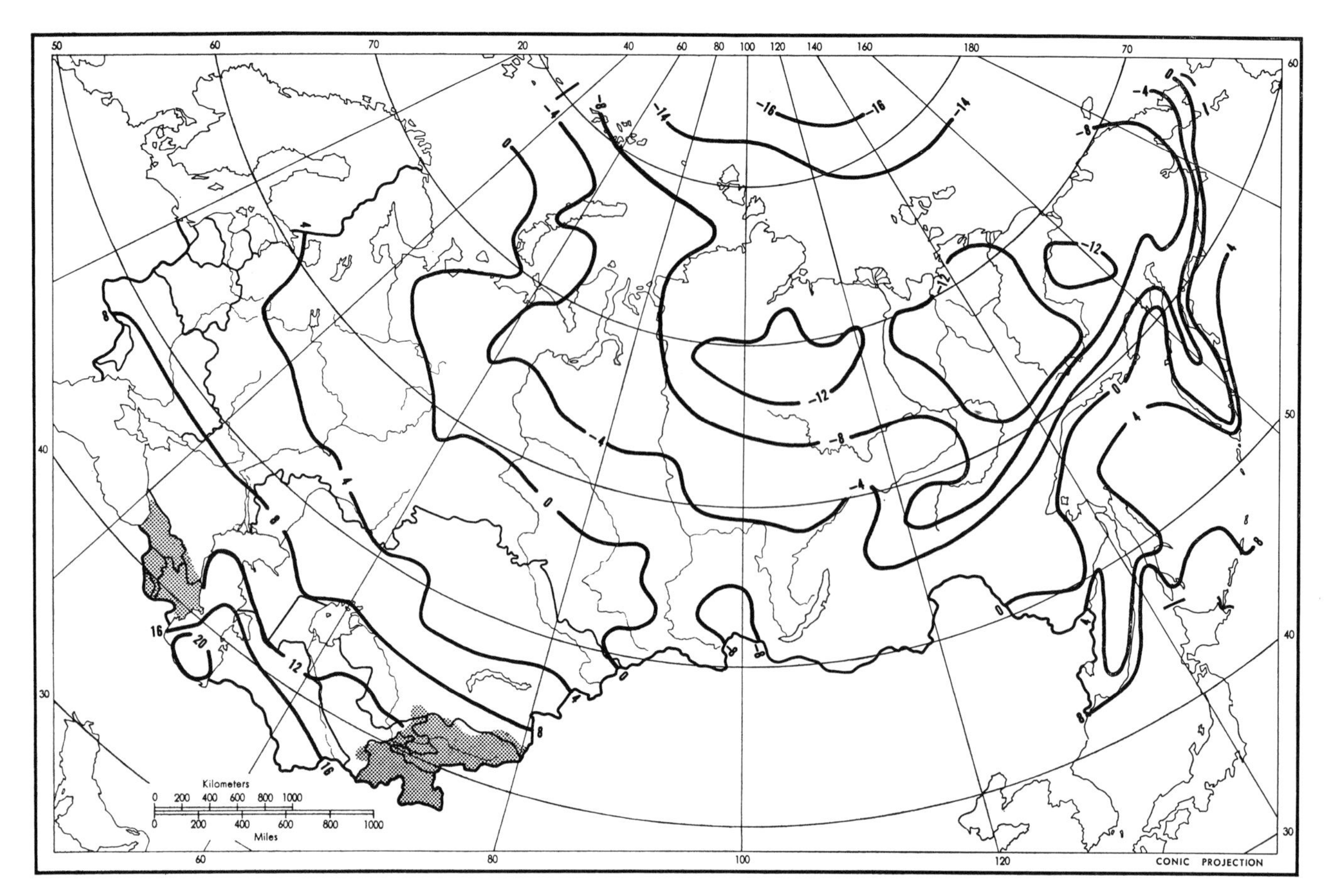

Fig.8-34. Mean surface air temperature in October (°C). (After GERASIMOV, 1964.)

monthly temperatures along the coast of the Taymyr Peninsula in October are about −10°C whereas in April they were below -22°C. This is because in October the latent heat of fusion is being given up to the air as the sea and land freeze and a snow cover forms, while in April the latent heat of fusion is being extracted from the air as the snow cover, the sea surface, and ground are thawing. This counterbalances the influences of incoming radiation which is more than five times as great in April as in October along the Arctic coast of Siberia. The discrepancy between October and April mean temperatures decreases southwestward until there is little difference between the two months south and west of the middle Urals. Most of the Pacific fringe of the Soviet Far East as well as the lower Amur Valley experience temperatures in October that are about 6°C higher than those in April, which relates to latent heat differences similar to those along the Arctic coast, but also reflects the monsoonal influx of Pacific air during late summer and fall.

Temperature variation

One of the most outstanding characteristics of the temperature of the Soviet Union is the high range and rapidity of change, both from time to time and from place to place. Mean annual ranges of temperature reach as much as 64.2°C at Verkhoyansk where January averages −48.9°C and July averages +15.3°C. This is the greatest mean annual temperature range on earth. The absolute temperature range at Verkhoyansk is 103°C (Fig.8-35). Oymyakon, at a higher elevation in the same general area, has an absolute range of 104°C. Temperature ranges decrease outward from this area and reach a minimum along the Black Sea coast where Batumi has an absolute annual range of only 48°C.

Characteristically, much of the Soviet Union, as is true of most high-latitude land masses, experiences very rapid rises and falls of temperatures during the transitional seasons. Spring and autumn are very short. This is particularly true of spring when initially air temperatures are retarded by the melting snow and land, after which they shoot up rapidly under the influences of intense insolation and long daylight hours of the approaching solstice.

The areal pattern of distribution of annual temperature variations is due more to winter conditions than to summer conditions. The absolute minimum temperature varies from −8°C at Batumi to −71°C at Oymyakon, while the absolute maximum varies only from 21°C at Uelen in the extreme northeast of the country along the Bering Straits to 50°C at Termez near the southern border of Soviet Central Asia. Thus, the range of absolute minimum temperatures across the country is more than twice that of the absolute maximum temperatures. During summer the almost continuous daylight in the northern part of the country produces maximum temperatures in enclosed mountain basins that are almost as great as those on the southern plains, and only the very dry regions of Central Asia experience temperatures much higher than the rest of the country (Figs.8-36 and 8-37).

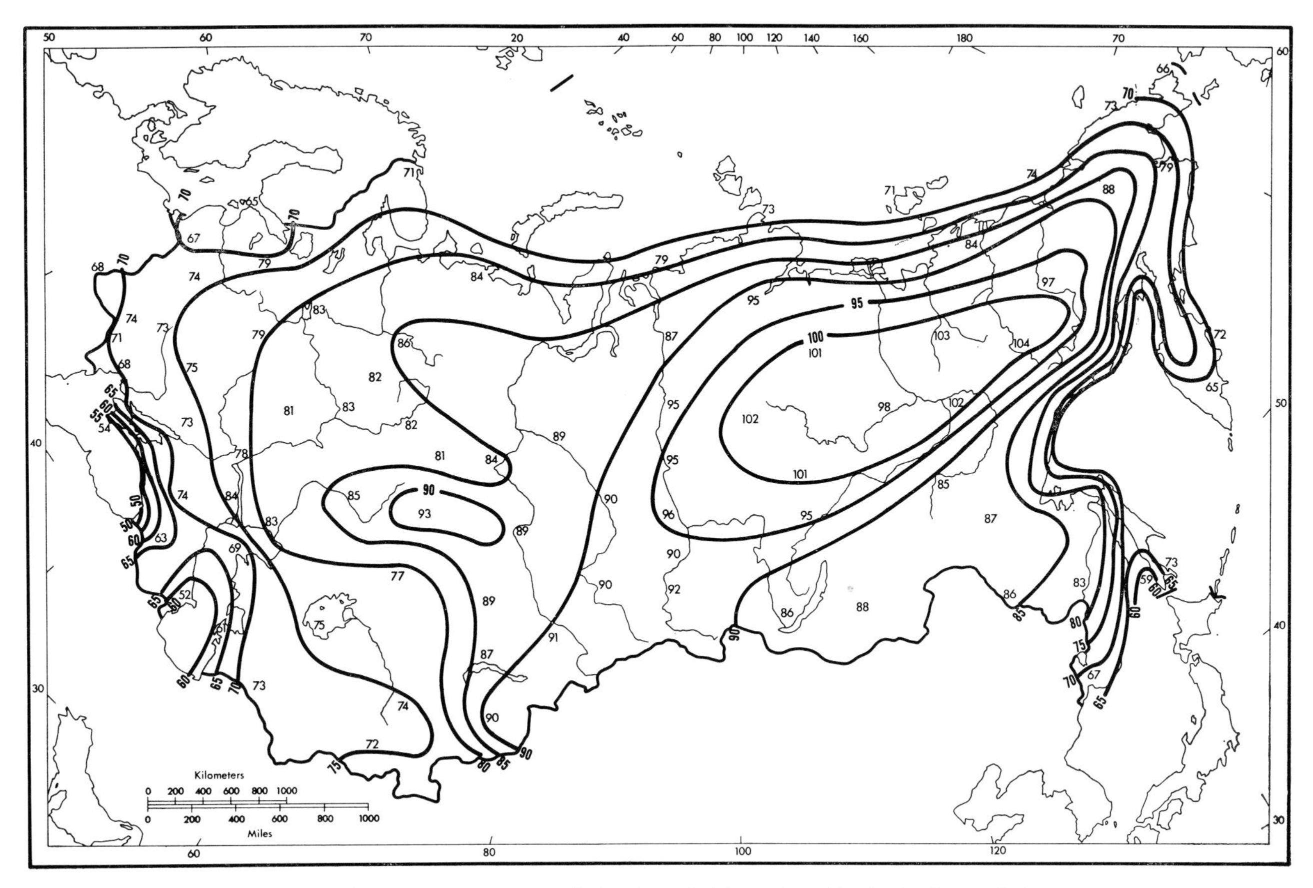

Fig.8-35. Absolute temperature range (°C). (Compiled from the tables in the Appendix.)

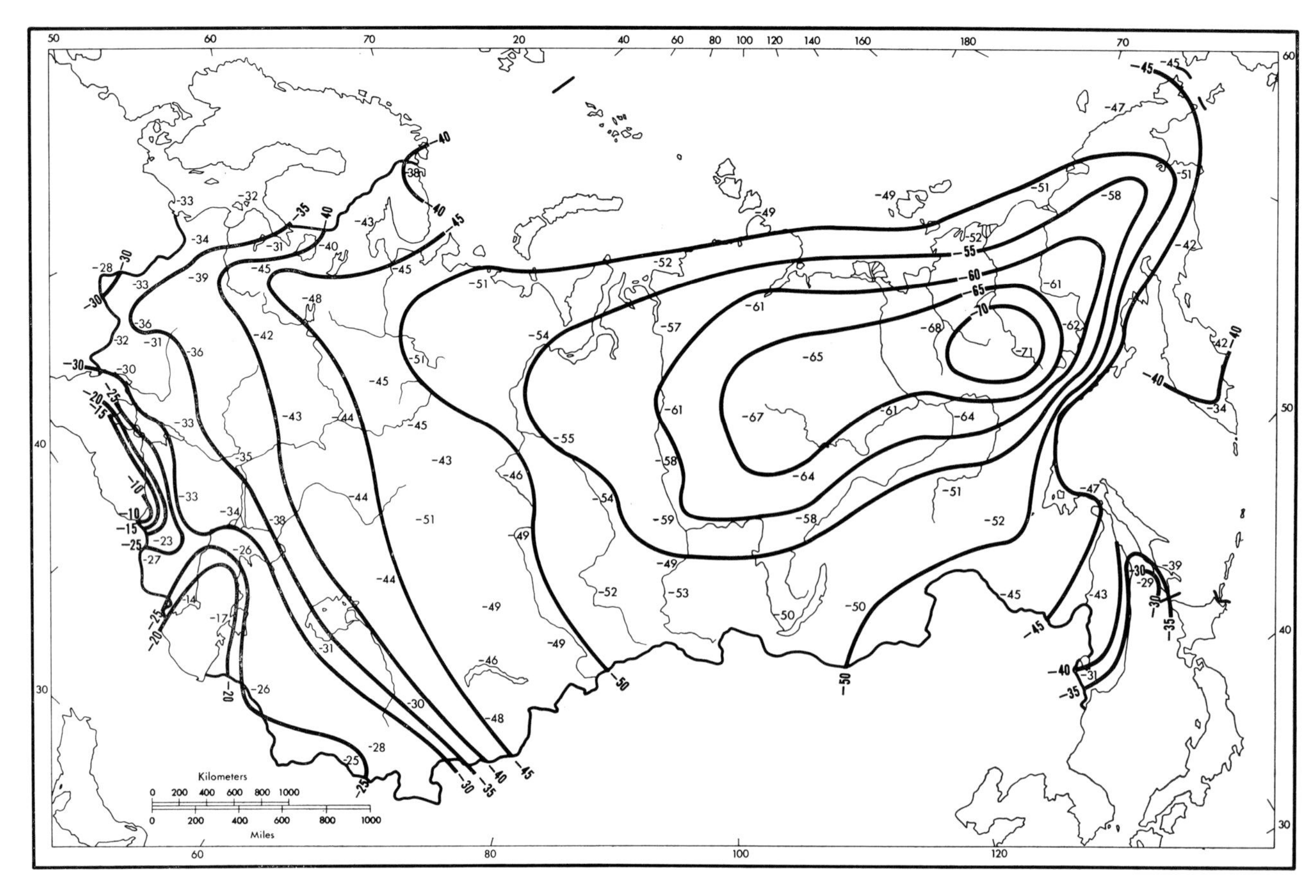

Fig.8-36. Absolute minimum temperature (°C). (Compiled from the tables in the Appendix.)

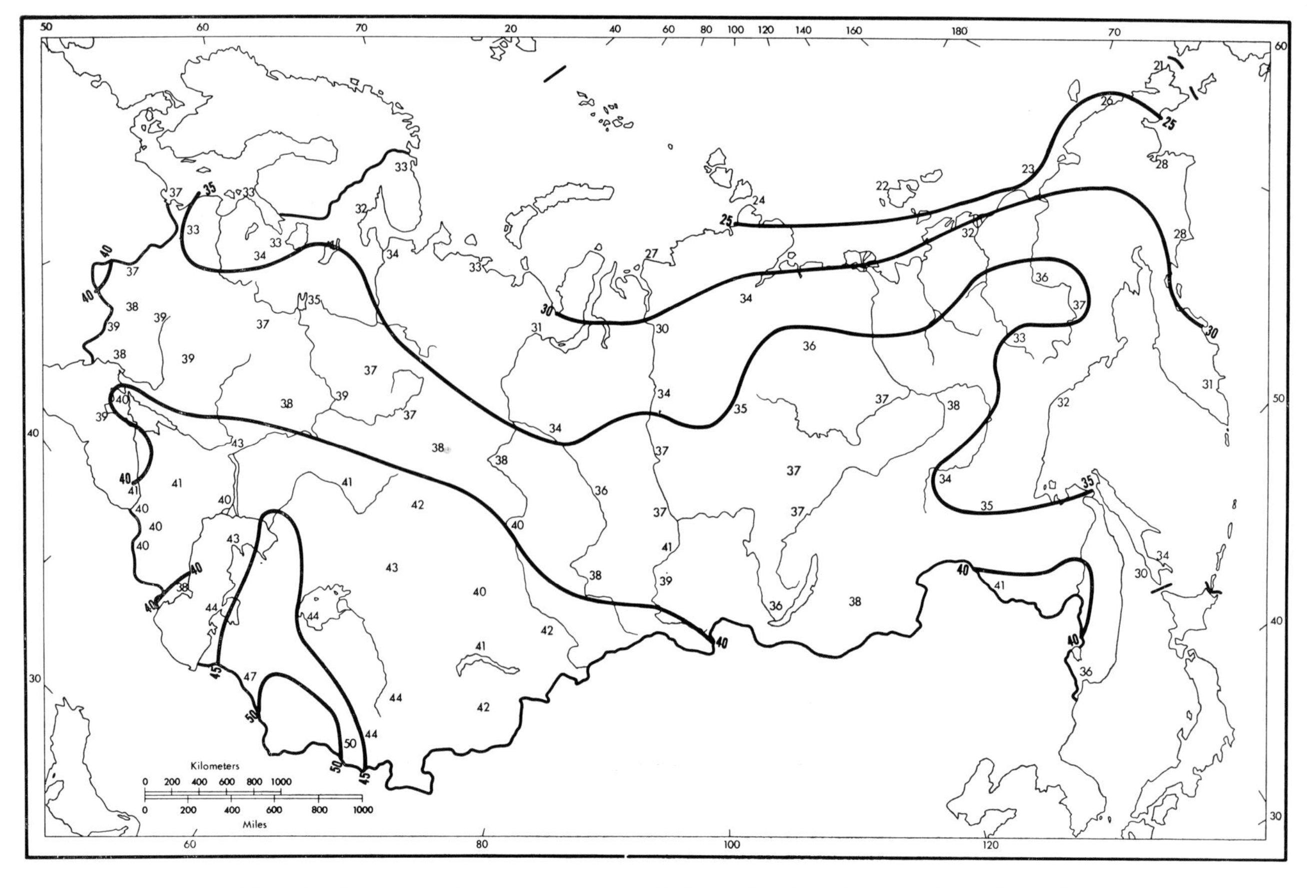

Fig.8-37. Absolute maximum temperature (°C). (Compiled from the tables in the Appendix.)

Since absolute temperatures depend upon the length of record, mean annual absolute temperatures are more meaningful. Generally for most purposes, particularly those of agriculture, absolute maximum temperatures are not as critical as absolute minimum temperatures. Soviet agroclimatologists have utilized average annual absolute minimum temperatures as one of their primary indexes of climate for agricultural purposes. The phenomenon of winter kill has been related to this index. Fig.8-38 shows the distribution of average annual absolute minimum temperatures across the country. These vary from about −5°C along the sea coasts on the eastern and western end of the Transcaucasus to less than −60°C in the upper Yana and Indigirka River Valleys in the northeastern part of the country. The Far East experiences very low minima because of the persistence of the Asiatic high during winter which sends cold continental air southeastward across the coast and largely prevents the intrusion of maritime air from the Pacific. Only very narrow coastal strips along the Pacific experience absolute minima comparable to those in the western portions of European U.S.S.R., but the Pacific littoral experiences such high winds during its coldest periods that effective temperatures are much colder than they are in European U.S.S.R. (Fig.8-39). In fact, wind chill is much more severe throughout the Soviet Far East, even in its southern portions, than it is in much of eastern Siberia where the coldest temperatures are recorded. The core area of the Asiatic high in winter stands out as a region of relatively mild wind chill compared to the Soviet Far East or even to western Siberia and the northern half of Kazakhstan. This is because of the calm

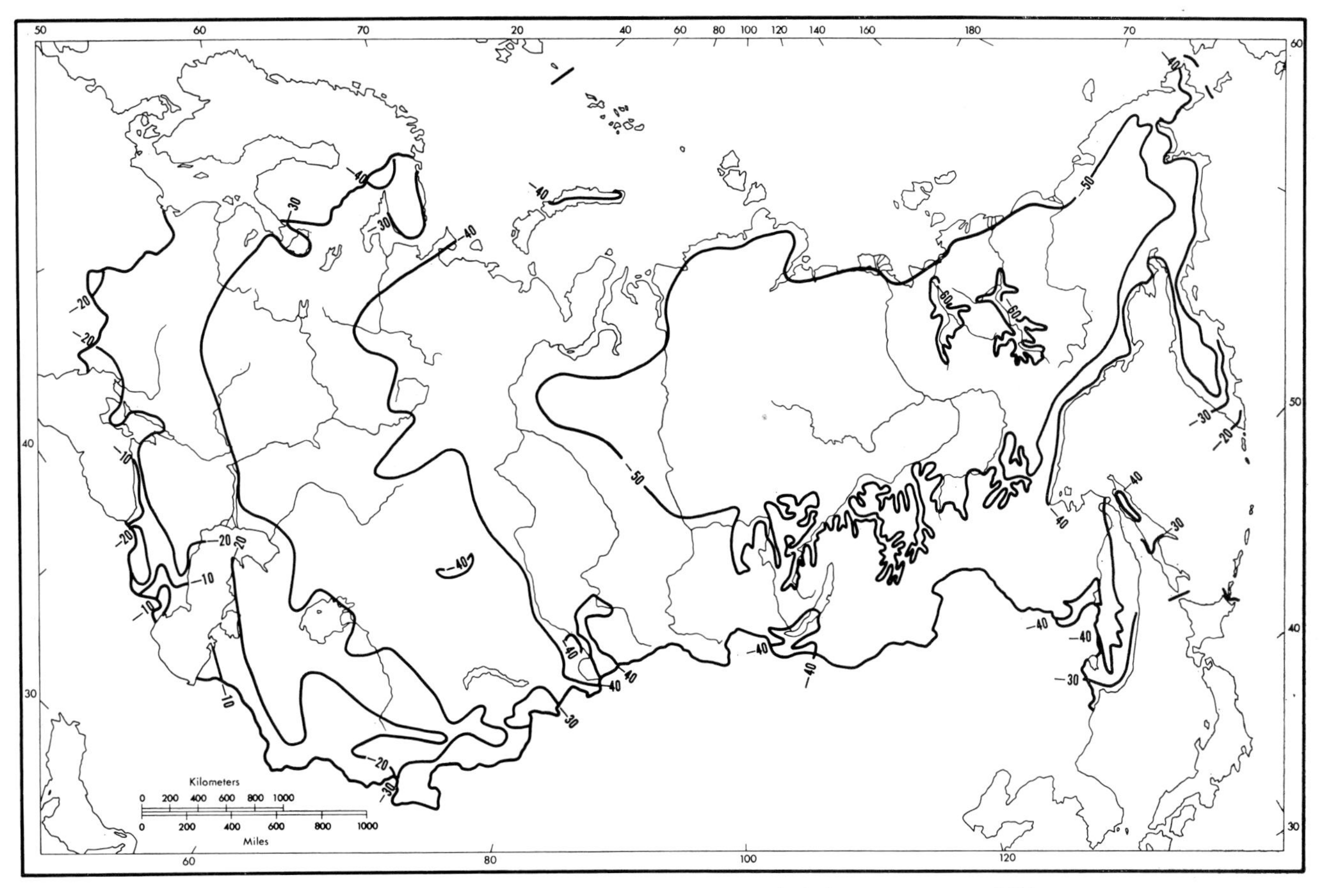

Fig.8-38. Mean annual absolute minimum temperature (°C). (After ANONYMOUS, 1960.)

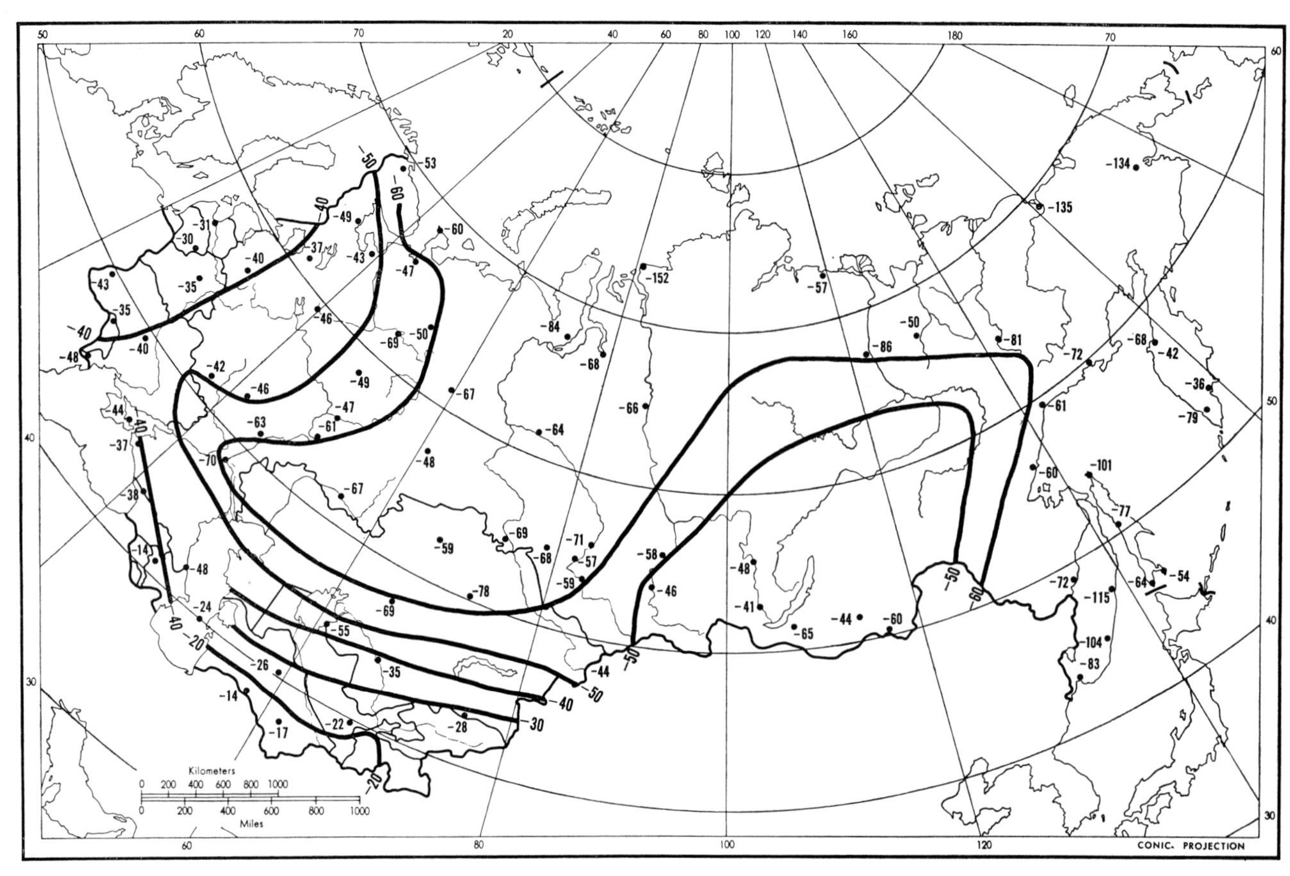

Fig.8-39. Effective minimum temperature for calculating heat loss from buildings, based on atmospheric temperature and wind conditions (°C). Probability once per year. (After ANAPOL'SKAYA and GANDIN, 1966.)

clear conditions much of the time in the core area of the Asiatic high. Effective minimum temperatures at Yakutsk are about the same as those along the middle Volga and even higher than those along the lower Volga. From Fig.8-39 it can be seen that the most severe conditions to be endured during the winter occur on exposed windy capes along the Arctic coast and throughout much of the Pacific margins of the Far East. Effective temperatures vary greatly along these coasts depending upon topography and exposure. In places there are extreme gradients of effective temperatures, particularly where mountain ranges come into close proximity with the sea.

Diurnal ranges of temperature are rather large throughout much of the Soviet Union, and in general they are much larger in summer than in winter (Figs.8-40 and 8-41). In January they vary from about 6°C around the Black and Caspian seas to more than 14°C in the southern part of Central Asia and Transbaykalia. In July diurnal temperature ranges vary from 8°C in the Baltic, Black Sea, and Caspian areas to about 20°C in central Siberia. Much of the European plain experiences diurnal ranges of about 8°C in January and 12°—14°C in July. These seasonal differences are caused by seasonal changes in lengths of daylight period, surface cover, and so forth. In winter much of the land is covered by snow and temperature inversions occur throughout the lower layers of the atmosphere much of the time. Such conditions, combined with low inputs of insolation, make for little change between day and night as compared to summer when daylight

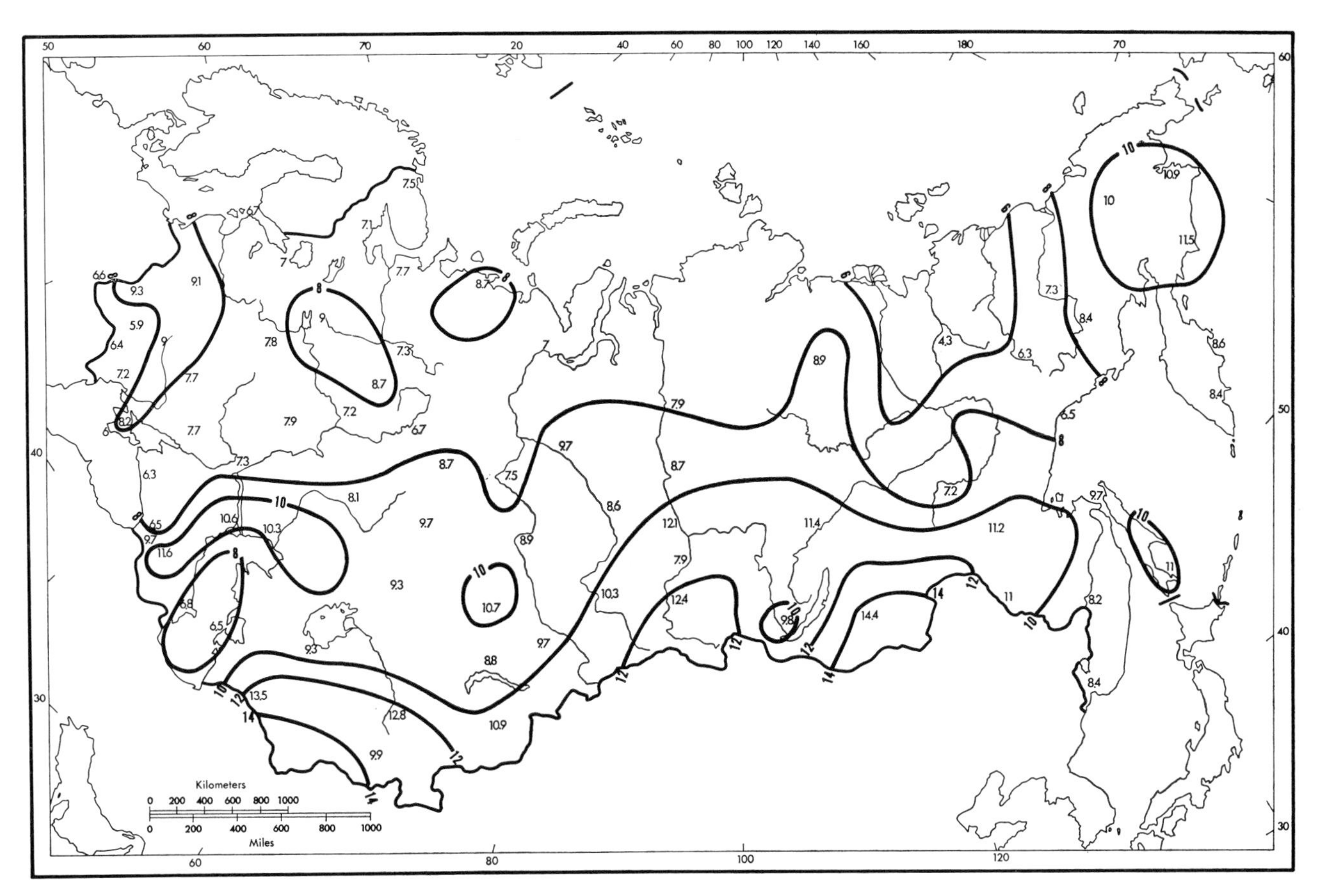

Fig.8-40. Mean diurnal range of temperature in January (°C). (Compiled from the tables in the Appendix.)

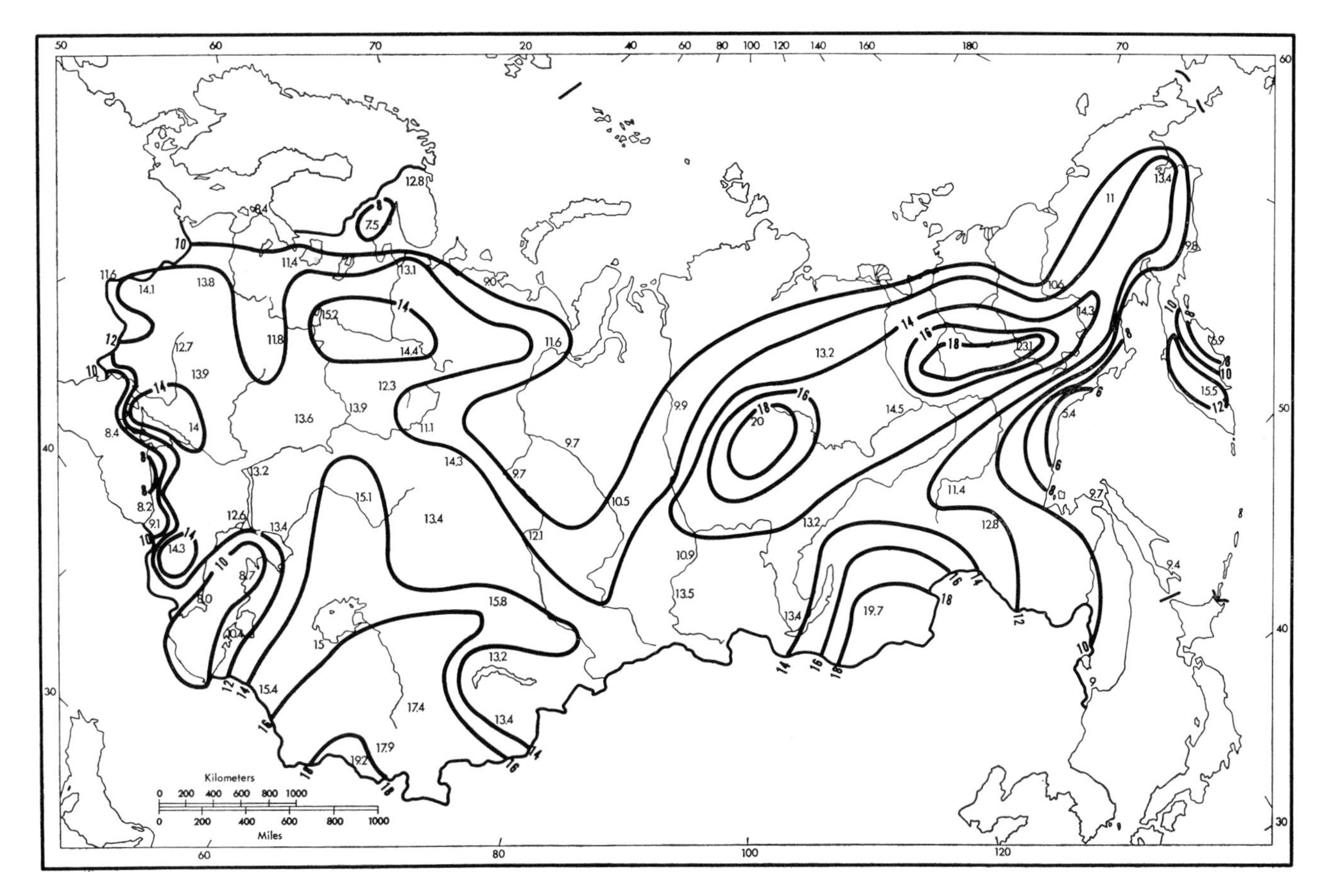

Fig.8-41. Mean diurnal range of temperature in July (°C). (Compiled from the tables in the Appendix.)

periods are very long and the high continentality of the interior of the country induces great increases in surface temperatures during the day. Enclosed basins between the complex mountains of the northeastern part of the country particularly benefit from air stagnation and almost continuous daylight during the summer solstice. The least temperature fluctuations take place in the northeastern part of the country during the Polar night. Under overcast skies the Arctic fringe of Siberia and the Far East experiences less than 1°C temperature variation during a 24-h period, and even under clear skies the variation is frequently less than 2°C.

Critical temperatures

The foregoing discussion of temperature has been based on data taken at standard weather shelter height, about 2 m above the ground. Temperatures vary considerably at times between this screen level and the ground surface. This is particularly true in summer when mean differences between daytime temperatures of the active surface and the air may amount to as much as 16°C in Central Asia (Fig.8-42). The difference is very little in the far north in May when the surface is still covered with snow but in July and August the whole northern half of the country including the Arctic coast shows a difference of as much as 6°C. Thus, temperature records compiled at screen level are

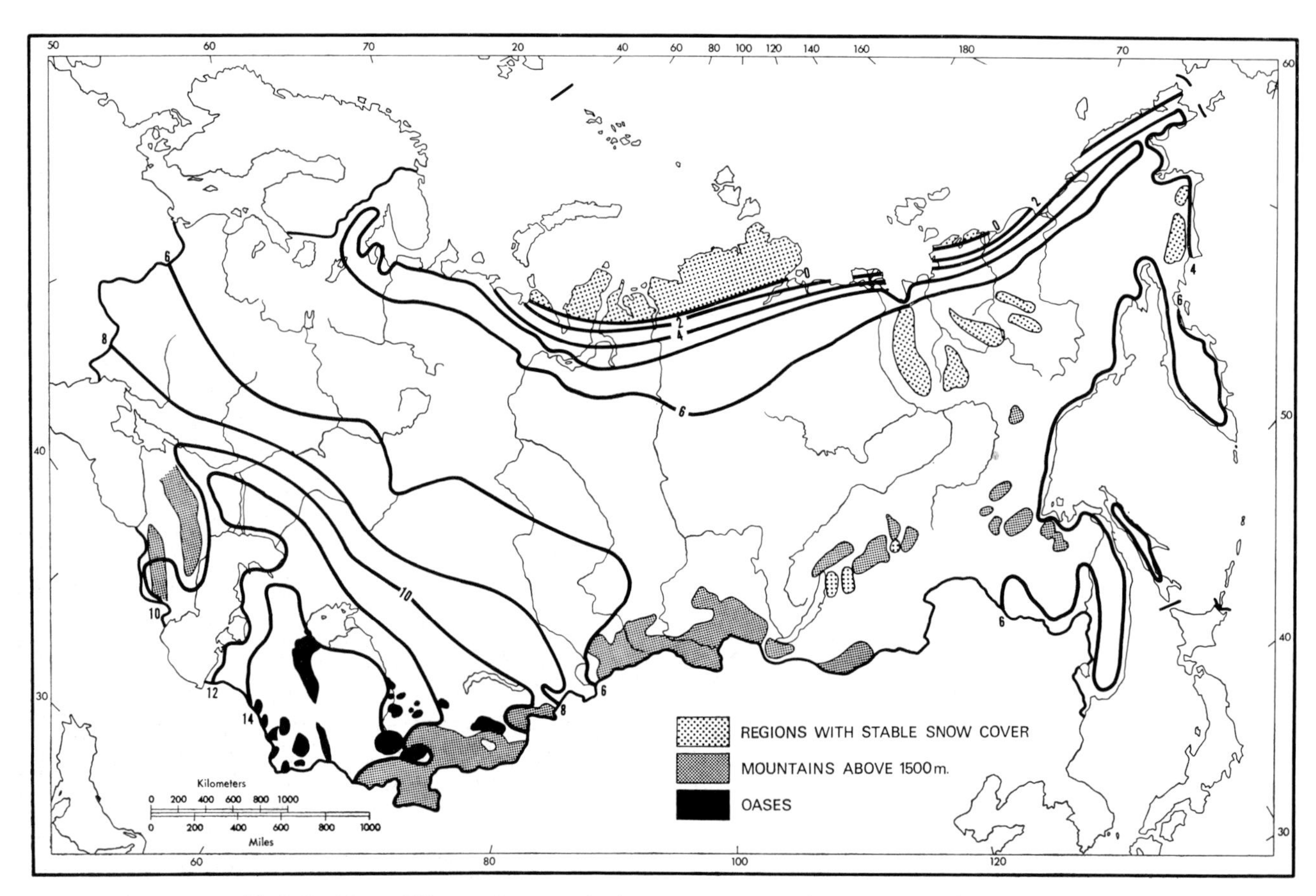

Fig.8-42. Mean difference between daytime temperatures of the active surface and the air in June. (After GOL'TSBERG, 1967.)

not too useful for many practical purposes, particularly those that emphasize certain critical temperatures.

Certain temperature levels are critical for various reasons. The freezing point is one of the more critical temperatures for many things. The change in state of water, which involves so much heat, exercises pronounced controls over many aspects of climate. Particularly for agriculture, it is important to know the length of the frost-free period. Fig.8-43 shows the mean dates of the last killing frosts in spring and Fig.8-44 the mean dates of the first killing frosts in fall, taking into account microclimatic influences. The length in days of the frost-free period varies from more than 300 in parts of the Transcaucasus to as little as 2 or even none at all on some of the more exposed capes and islands along the Arctic coast (Fig.8-45). Much of the agricultural zone of European U.S.S.R. has a frost-free period varying from 120 days north of Moscow to 180 days along the northern coast of the Black Sea. This is not an exceedingly long growing period, and, coupled with generally cool temperatures during the period, it sets many limitations on the choice of agricultural crops in this region. Even in the southern part of the agricultural region the limited heat resources during the growing season usually come into play before the limited moisture resources in determining which crops can be grown.

Frosts are quite haphazardly occurring, and their dates of appearance and disappearance vary greatly from one year to another. Fig.8-46 and 8-47 show the last possible dates of frost in spring and earliest possible dates of frost in autumn, and Fig.8-48 shows the

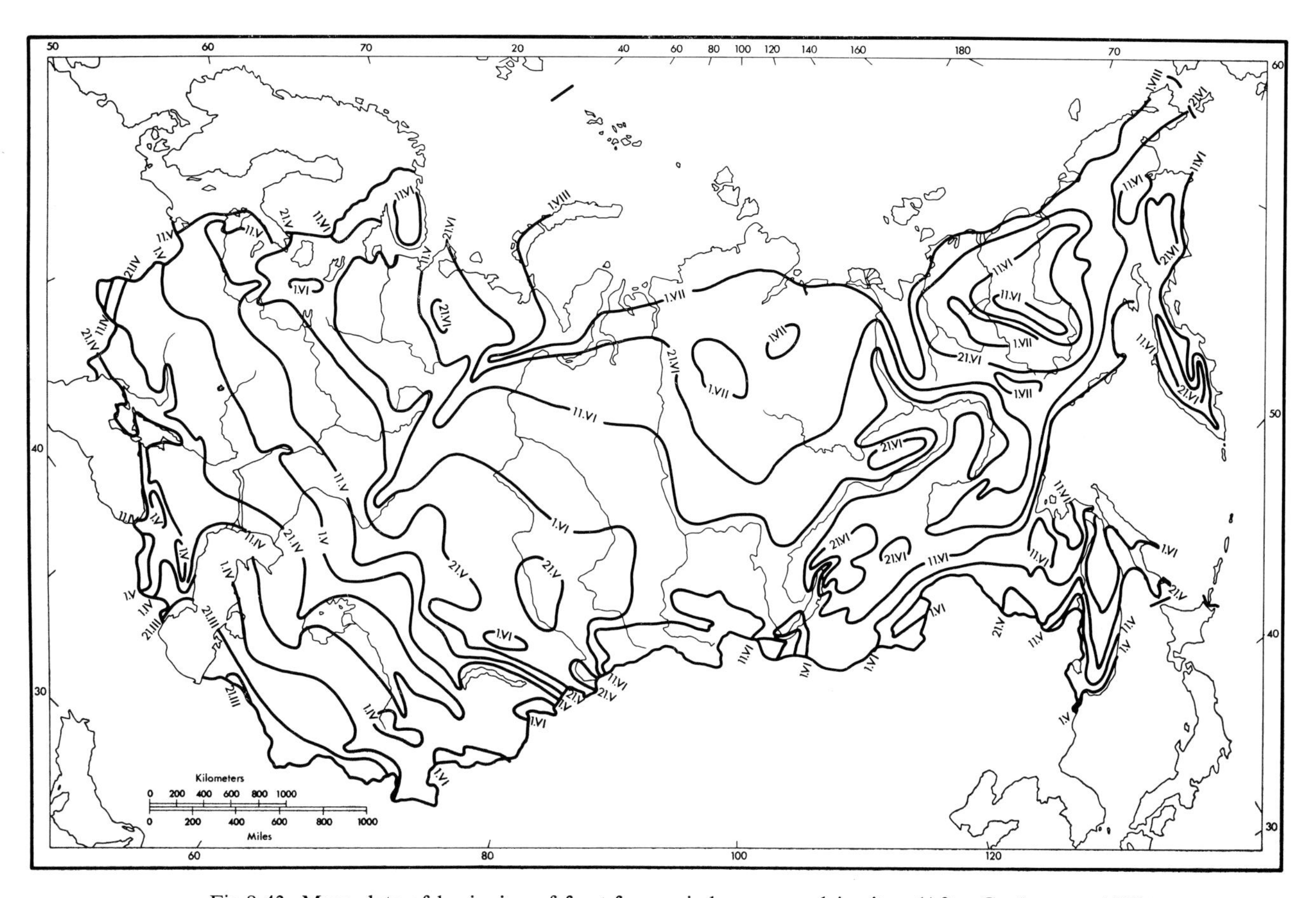

Fig.8-43. Mean date of beginning of frost-free period on open plain sites. (After GOL'TSBERG, 1969.)

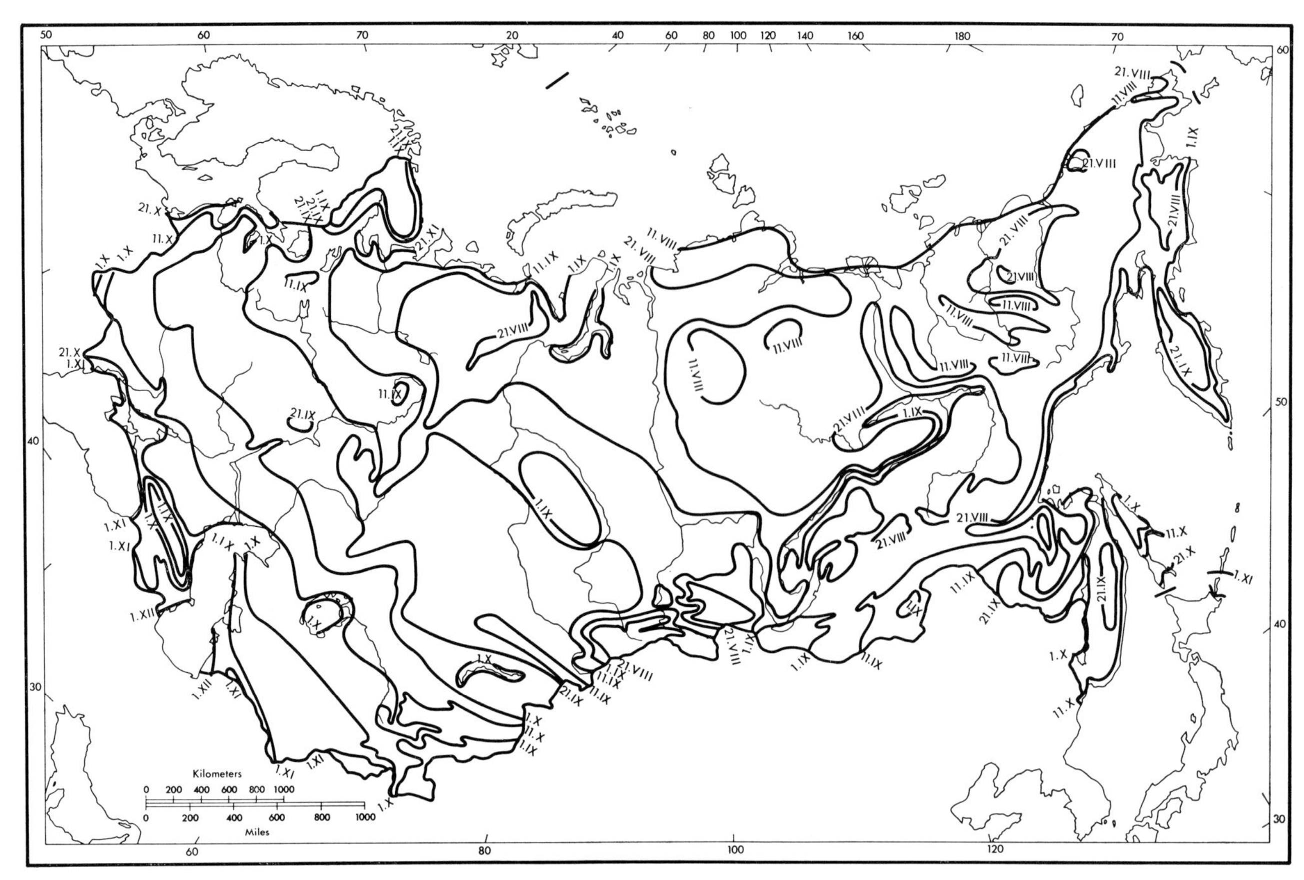

Fig.8-44. Mean date of end of frost-free period on open plain sites. (After GOL'TSBERG, 1969.)

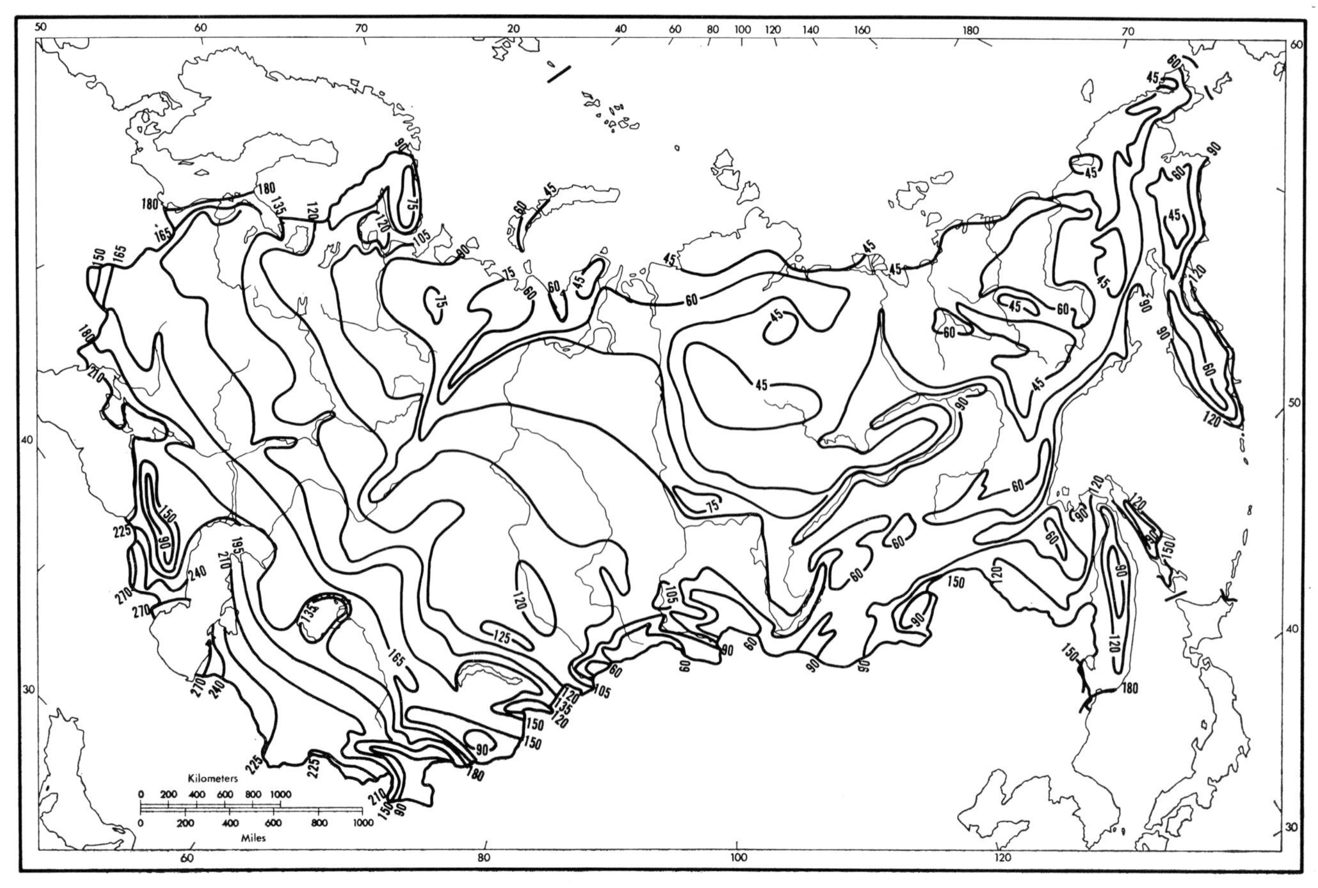

Fig.8-45. Average length of frost-free period on open plain sites (in days). (After GOL'TSBERG, 1969.)

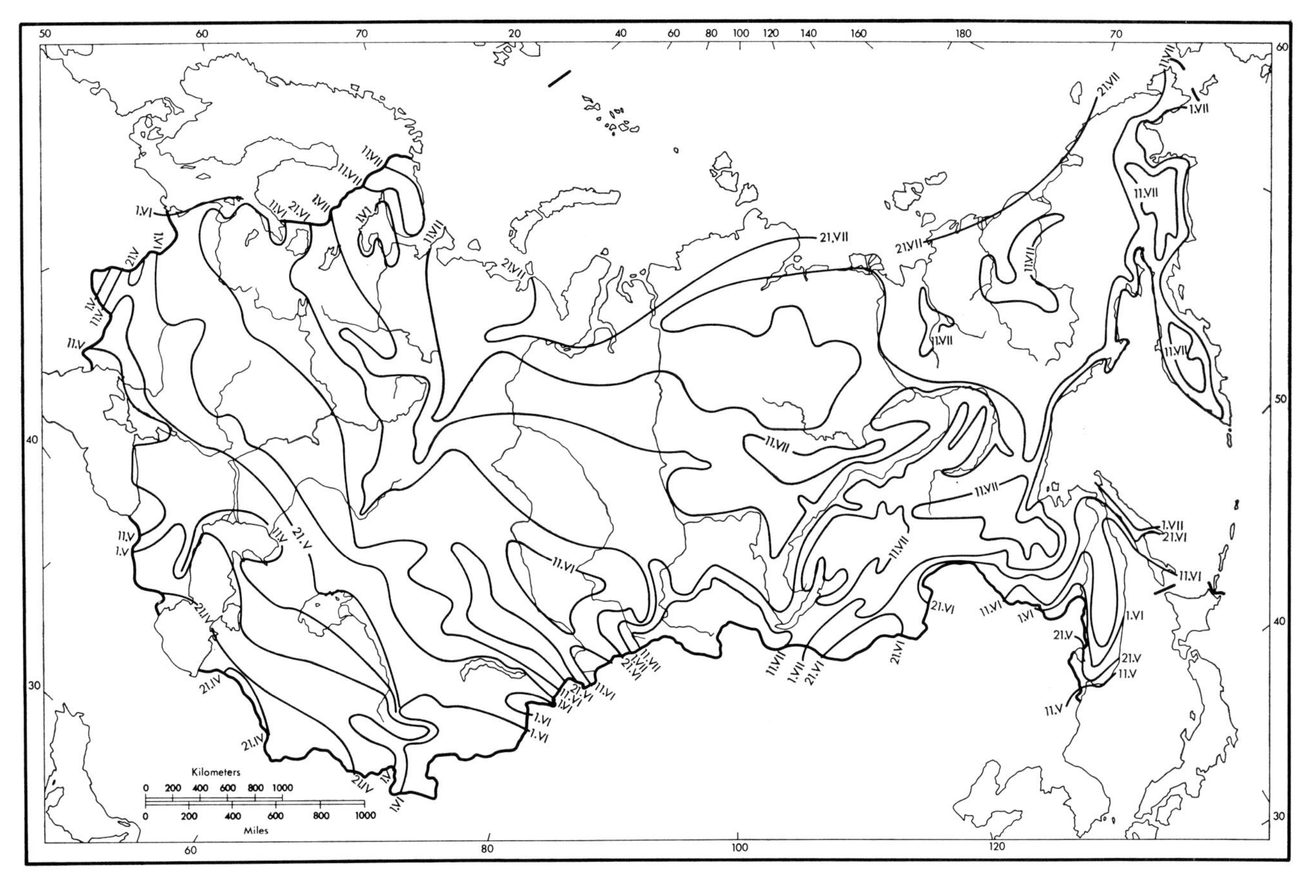

Fig.8-46. Last possible date of frost in spring. (After GOL'TSBERG, 1961.)

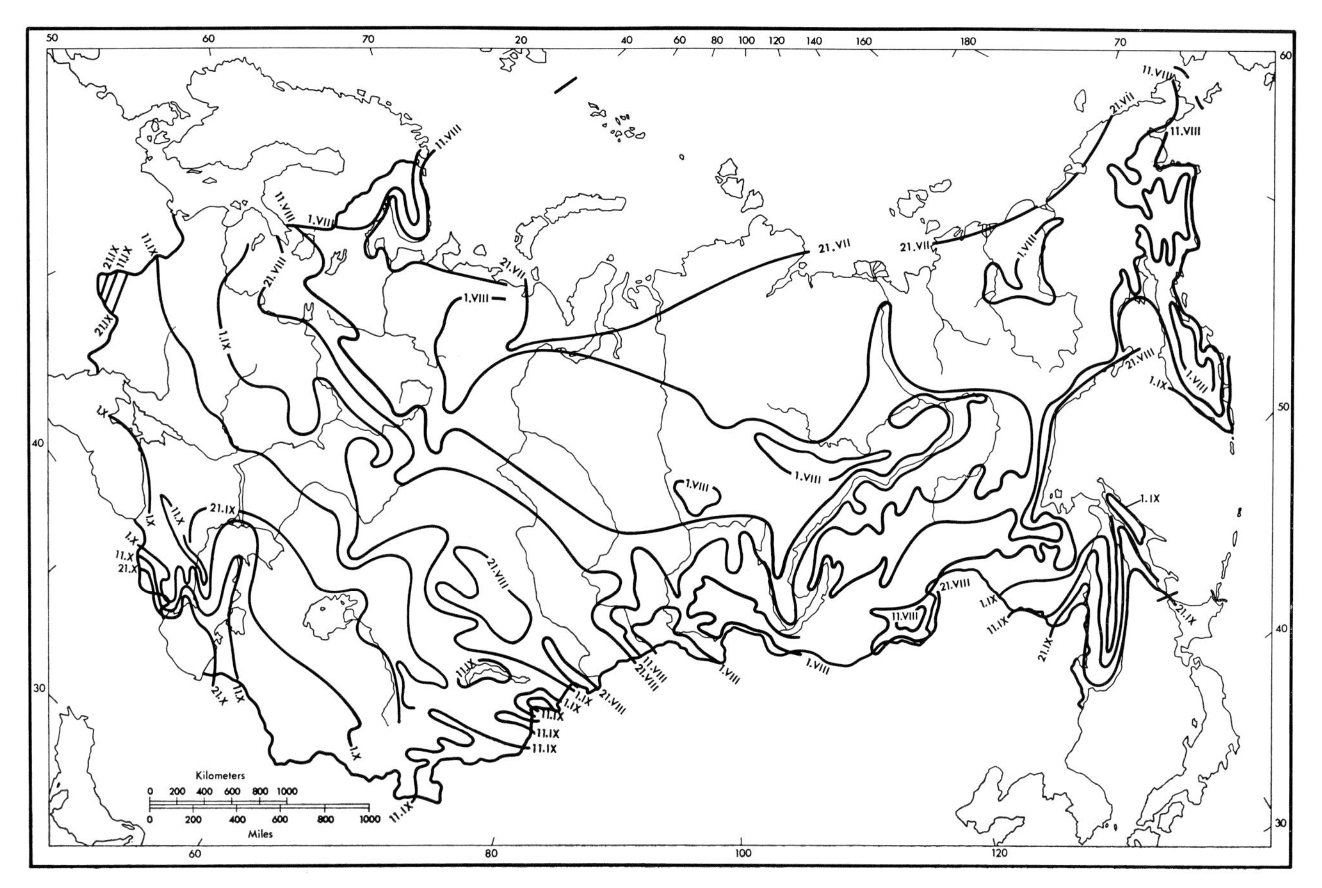

Fig.8-47. Earliest possible date of frost in autumn. (After GOL'TSBERG, 1961.)

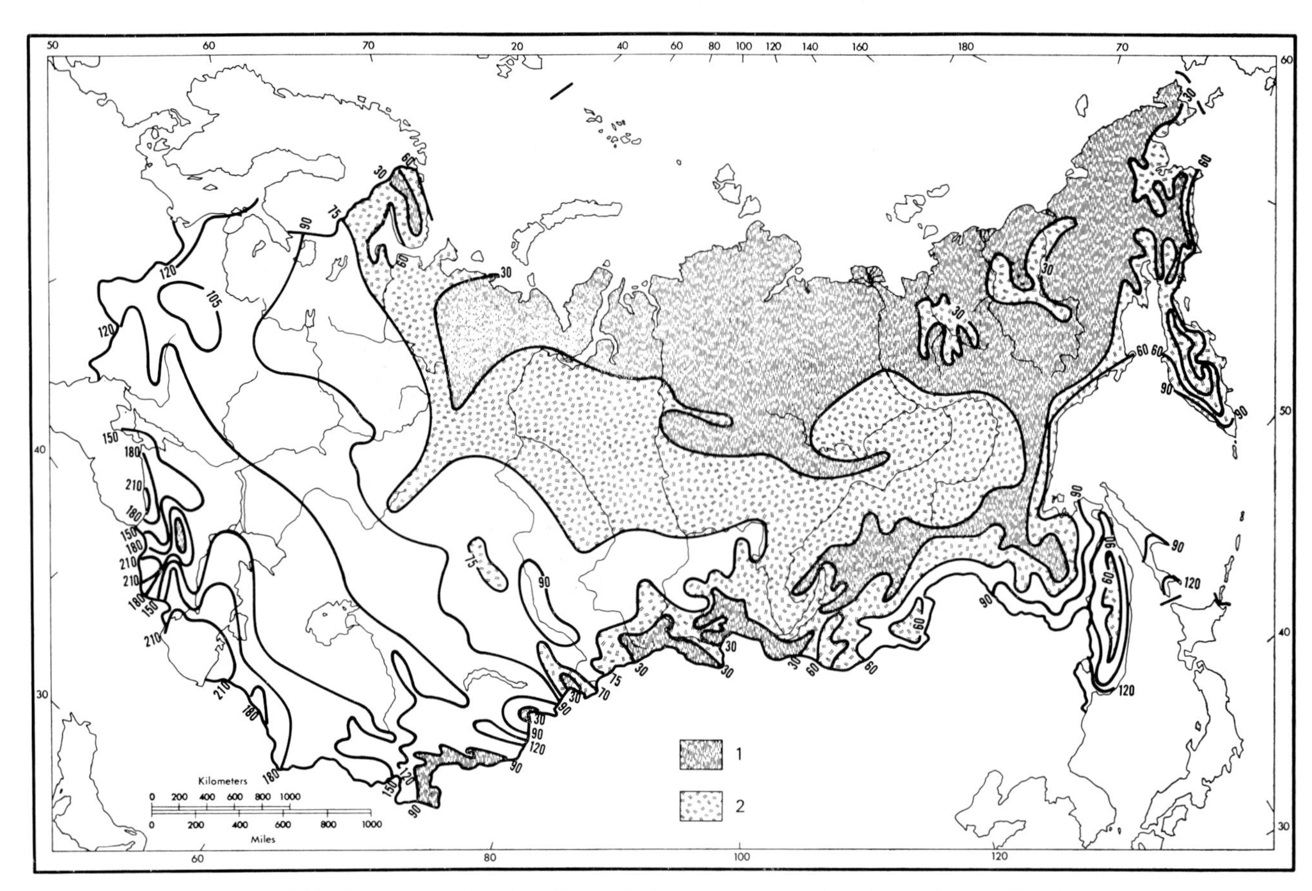

Fig.8-48. Shortest possible frost-free period on open plain sites (in days). *1* = Freezing air temperatures possible during entire summer; *2* = in July and early August the surface of the soil may experience frosts. (After GOL'TSBERG, 1961.)

shortest possible frost-free period on open plain sites in the Soviet Union. It can be seen that the frost-free period may actually be as much as 60 days shorter in individual years than it is on the average over a long period. The northern Black Sea Steppes average about 180 consecutive days per year without frost, but some years have no more than 120. In the Moscow region the average frost-free period is about 130 days, while the minimum frost-free period is only 80-85. Thus, frost presents a great hazard to agriculture, and on individual years it can cut short the growing season for most of the crops that are being grown.

Data given in the tables at the end of the book may be quite misleading for the larger cities because frosts often occur in the countrysides when they do not occur in urban heat islands. Frosts are not unknown, even in July and August, over large portions of the better agricultural lands of the Soviet Union. Below freezing temperatures can occur in the surface air during late August as far south as the 50th parallel in the Don and Volga River regions and even further south in Kazakhstan. Frosts are possible any time during the summer north of about 60° latitude, except for the western portions of the country which are somewhat protected by marine air from the Atlantic (Fig.8-49). The danger of summer frosts are even greater in the soil.

Major streams generally freeze up around the third week in December in the southwestern part of the country and as early as the first of October along the Arctic coast of

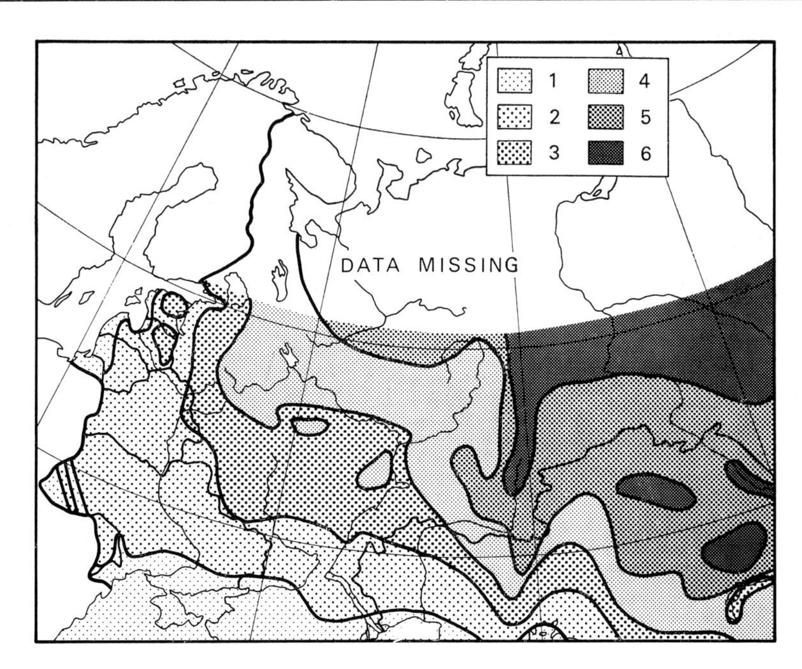

Fig.8-49. Frequencies of frosts in July and August in open plain sites. (After GOL'TSBERG, 1961.) At the soil surface: *1* = no frosts in July and August; *2* = 20–30/VIII from 5 to 20% of years; *3* = early August 5–20%, late August 20–30%; *4* = in July 10–30%, early August 10–40%, and late August up to 60%; *5* = late August every year; *6* = possible every summer. In surface air: *1* and *2* = no frosts in July and August; *3* = possible in late August; *4* = possible scattered frosts in July and early August; *5* = scattered frosts every year; *6* = frosts possible all summer.

Siberia (Fig.8-50). They thaw out again in spring around the first of March in the southwest and the second week of June in the extreme north (Fig.8-51). Hence they are frozen about 80 days in the southwest and up to 240 days in the extreme north (Fig.8-52).

The western part of the country experiences some thawing during the winter, but east of the Urals little thawing occurs anytime from December through February north of 50° latitude (Fig.8-53). On the average during these three months the western border of the country can expect about 50 days when temperatures rise above the freezing point, but the frequency decreases rapidly eastward and Moscow can expect no more than about 13 such days during the period December–February. Thawing occurs much less frequently when a good snow cover is lying on the ground. During such winters no more than 15 days of thawing can be expected along the western border and only about 1 day in Moscow. From the orientation of the isolines in Fig.8-53, it is quite obvious that thawing is controlled very significantly by the conflict of air masses coming from the west and the east during the winter. The cold Asiatic high hardly ever allows for any thawing, while air from the Atlantic induces frequent thaws.

Much of the eastern part of the country is underlain by permafrost (Fig.8-54). Much of this appears to be inherited from the ice ages, but the present climate is conducive to its slow rate of thawing, if any. This phenomenon presents many hazards to land development, both agricultural and urban, since man's disturbance of the top soil alters the permafrost conditions and induces thawing to greater depths in an uneven fashion which causes fields to become hummocky and buildings and roadways to sag.

Other temperature levels have been determined to be critical for various purposes by

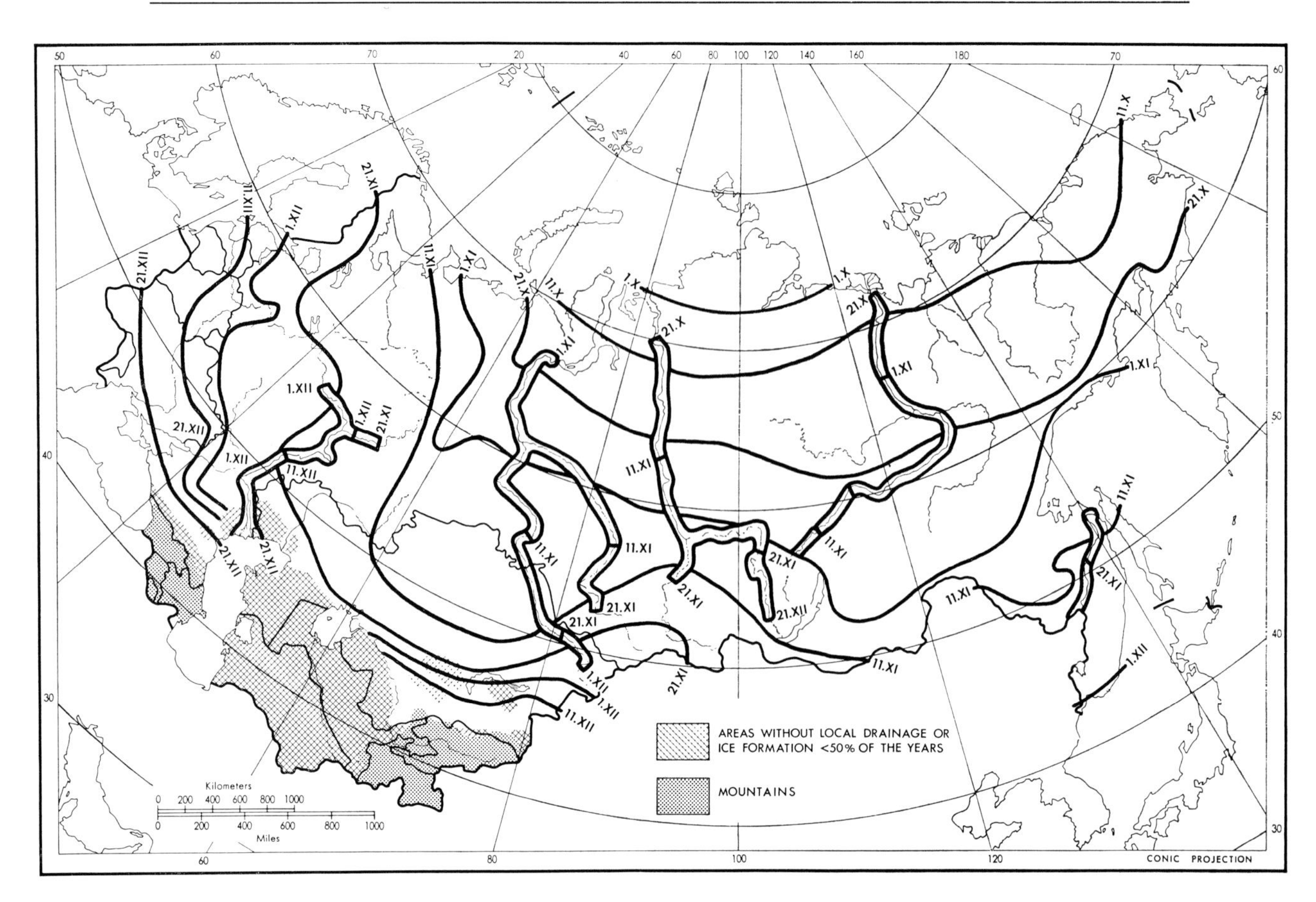

Fig.8-50. Mean dates of freezing of streams in autumn. (After GERASIMOV, 1964.)

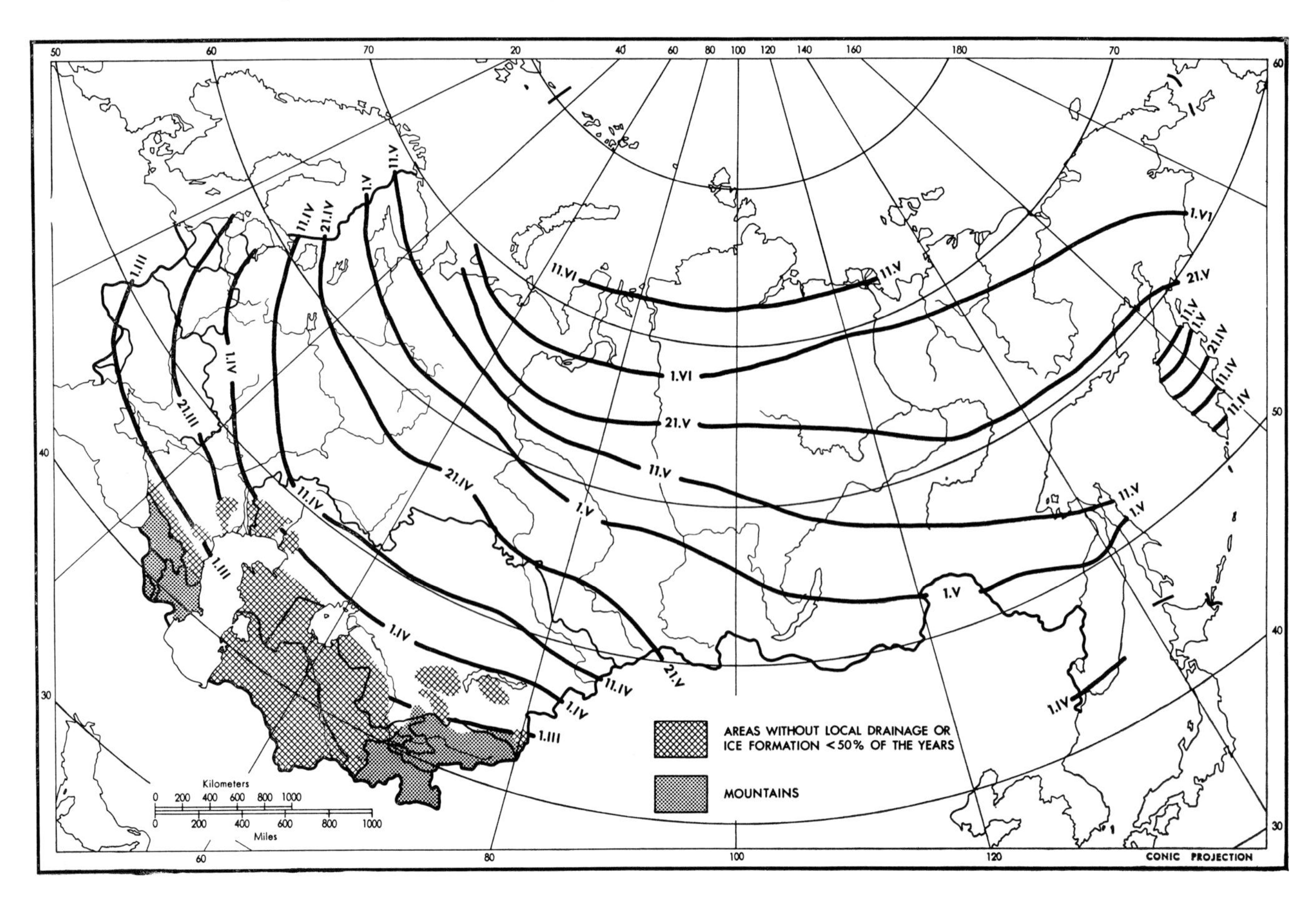

Fig.8-51. Mean dates of thawing of streams in spring. (After GERASIMOV, 1964.)

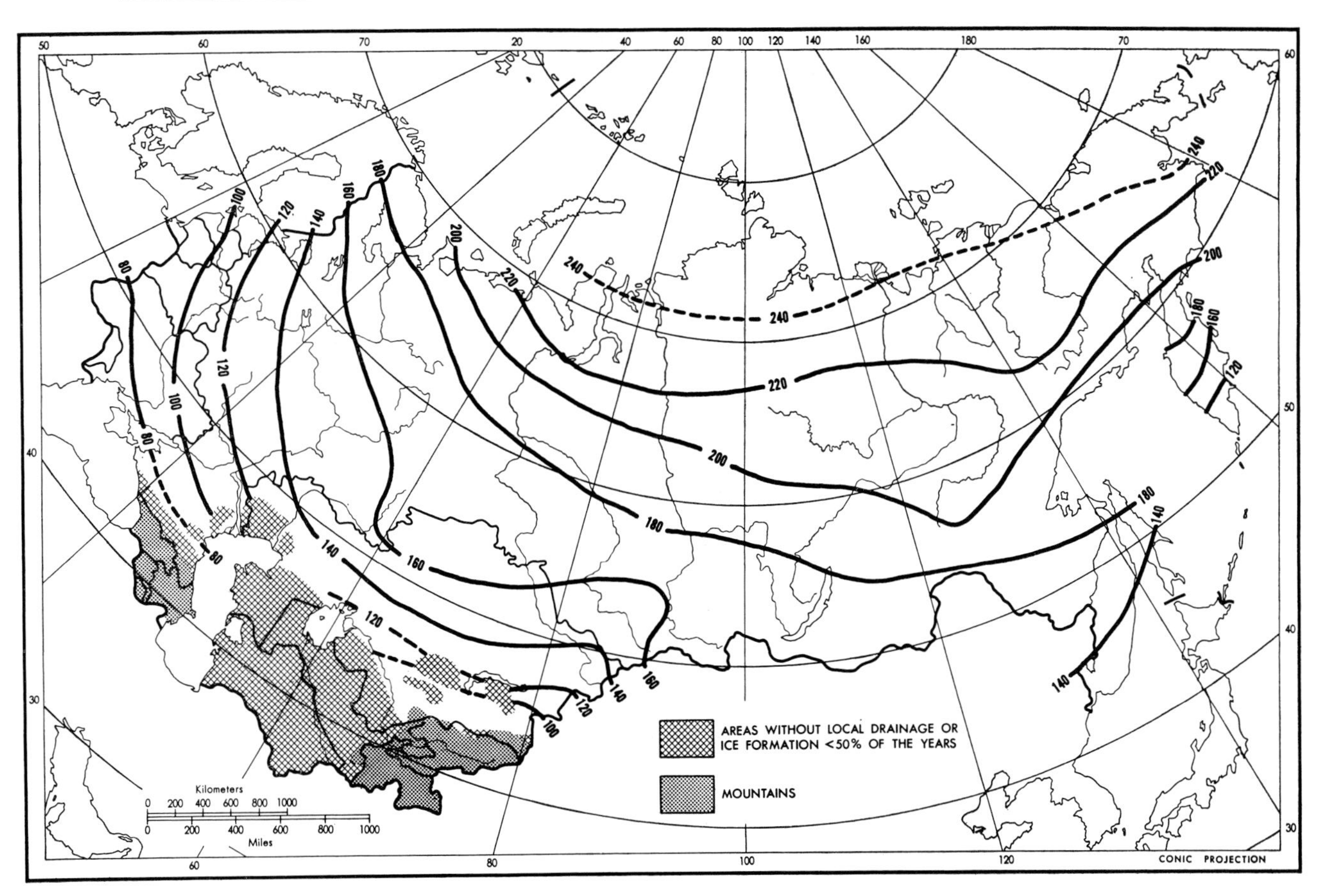

Fig.8-52. Mean length of period of frozen streams (in days). (After GERASIMOV, 1964.)

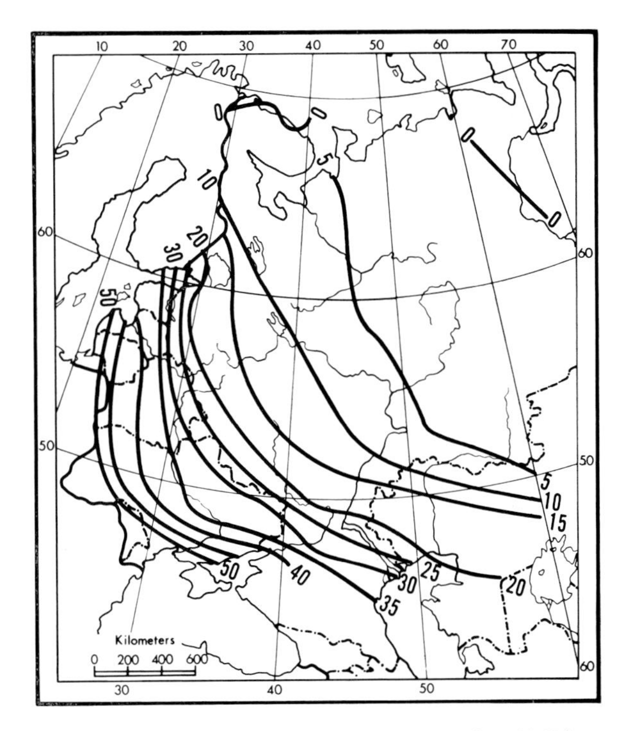

Fig.8-53. Number of days with thaw, 1 December–29 February. (After KOTLYAKOV, 1968.)

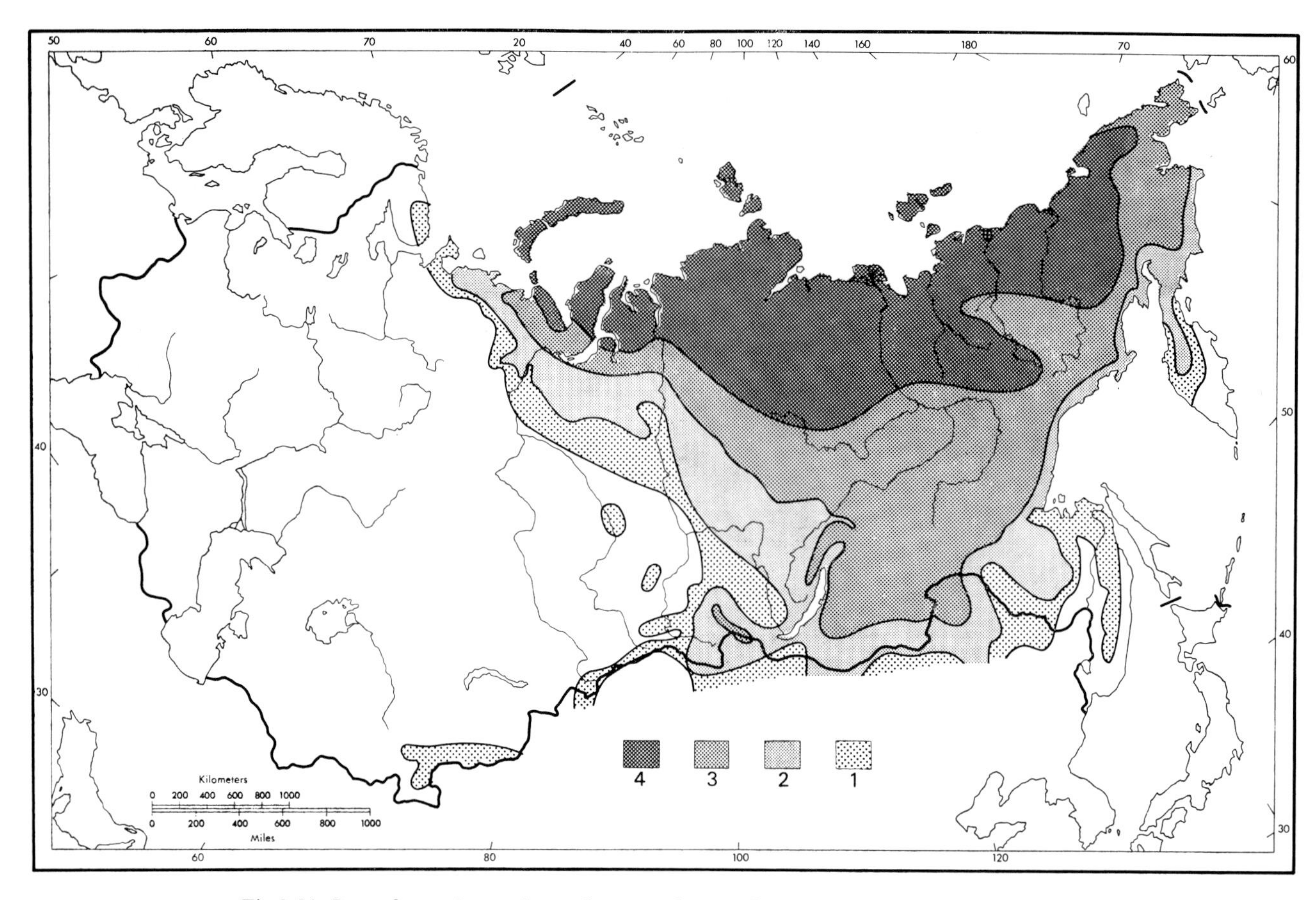

Fig.8-54. Permafrost: *1* = regions of scattered permafrost with thicknesses of less than 15 m; *2* = regions of widely distributed permafrost with thicknesses up to 60 m; *3* = thickness 60–250 m; *4* = thickness more than 250 m. (After SHCHERBAKOVA, 1961.)

Soviet climatologists. For example, many indices in agricultural climatology have used the mean diurnal temperature of 10°C, which seems to be the threshold of growth for many domesticated plants. Sums of temperatures during the period when the mean diurnal temperature remains above 10°C have been computed and utilized by agroclimatologists to determine growth possibilities, much as the degree-day concept has been used in Western Europe and the United States. The *Agricultural Atlas of the U.S.S.R.* published in 1960 includes a large number of maps showing dates at which critical temperature levels are passed in spring and autumn and sums of temperatures during periods when temperatures remain above these levels. For our purposes here we will present only those maps dealing with the 10°C level.

The date at which the mean diurnal temperature rises above 10°C in spring varies from around the third week in April in the southwest to mid-July in the north (Fig.8-55). Mean diurnal temperatures fall below 10°C in autumn around early August in the far north and around mid-October in the southwest (Fig.8-56). Hence, there is hardly any period of growth in the far north. It increases southward until along the northern Black Sea coast it amounts to about 170 days. In some of the protected lowlands of the Transcaucasus it is considerably longer than that, but data are lacking. During this growth period temperature sums total as much as 5,200°C in the southern part of Central Asia

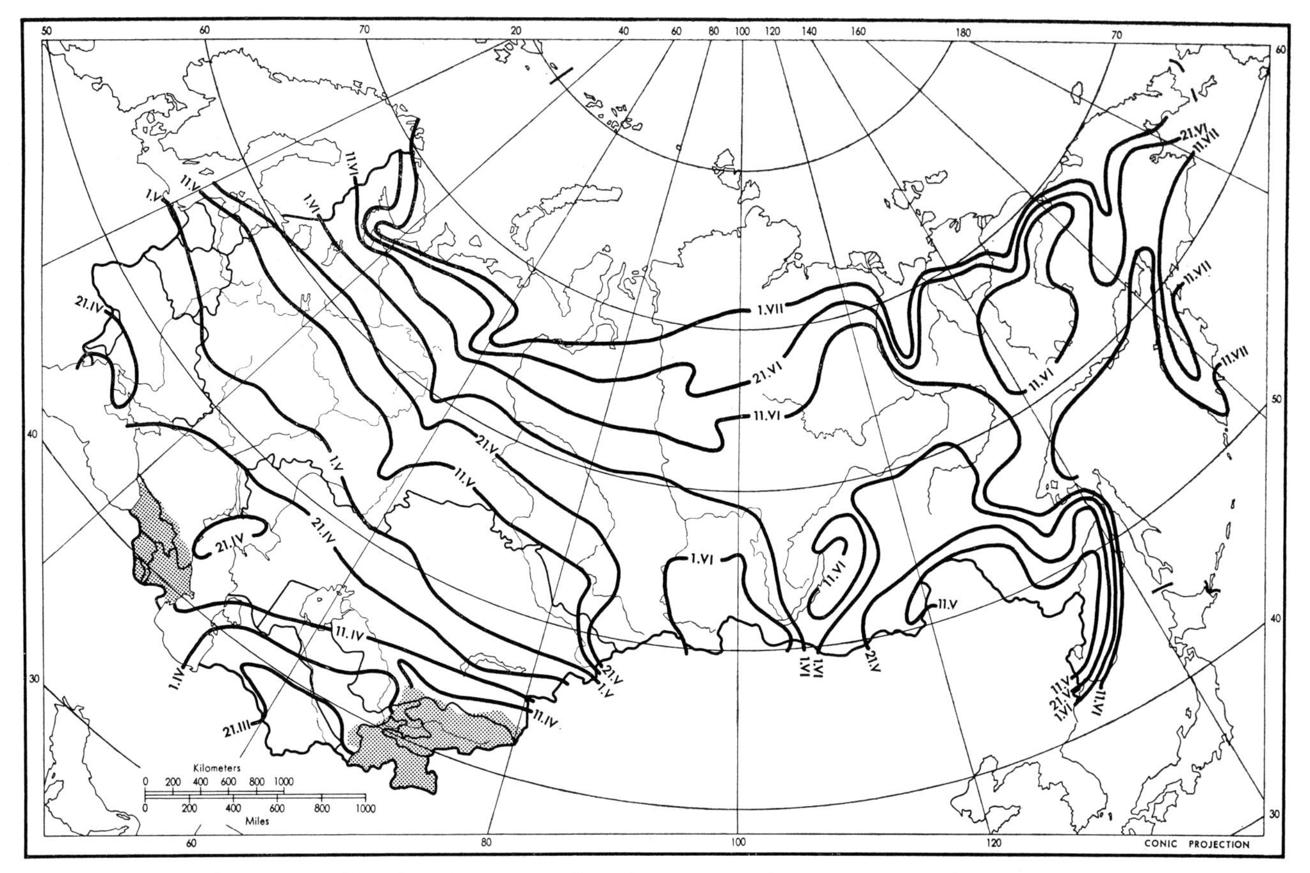

Fig.8-55. Date in spring when average diurnal temperature rises above 10°C. (After GERASIMOV, 1964.)

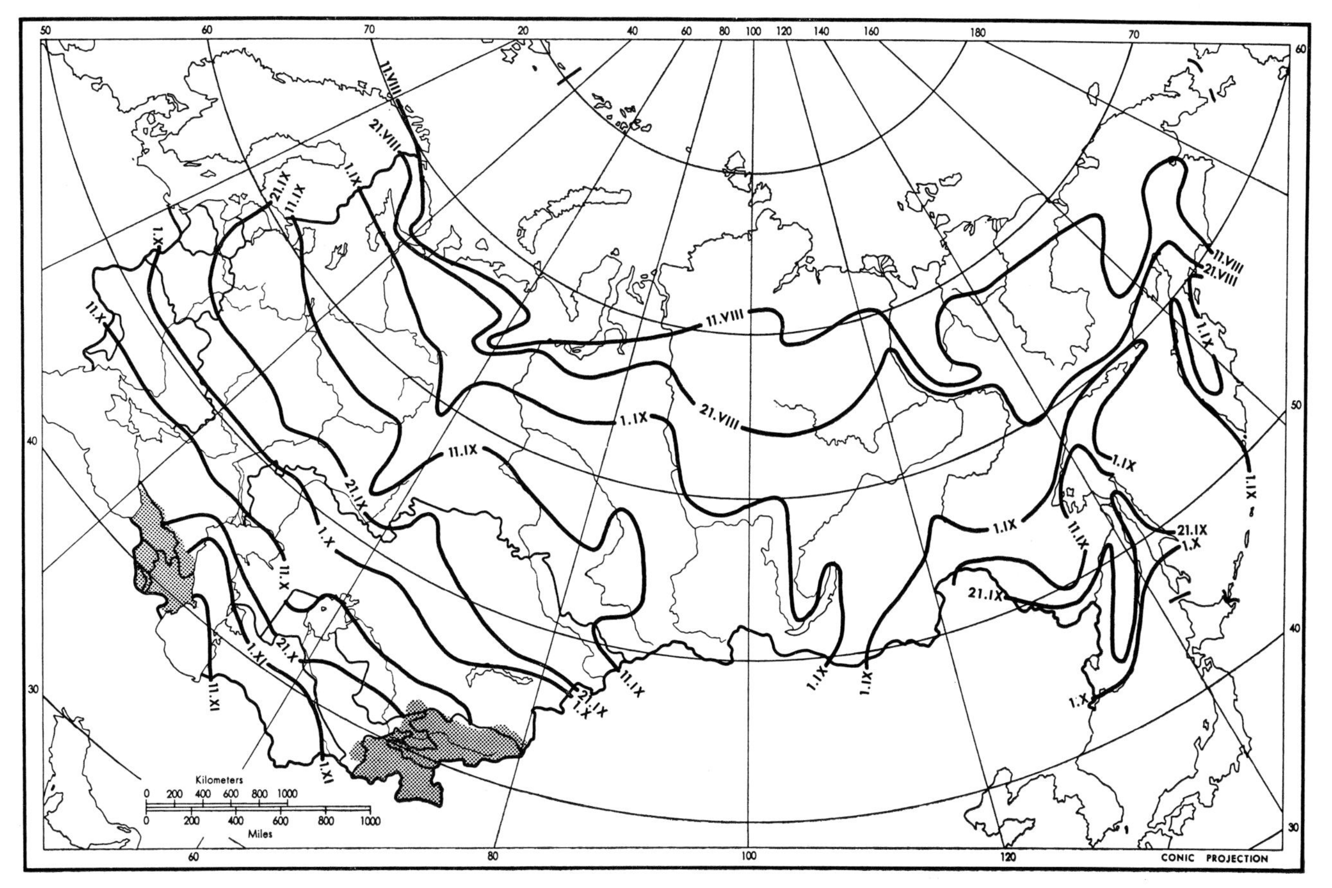

Fig.8-56. Date in autumn when average diurnal temperature falls below 10°C. (After GERASIMOV, 1964.)

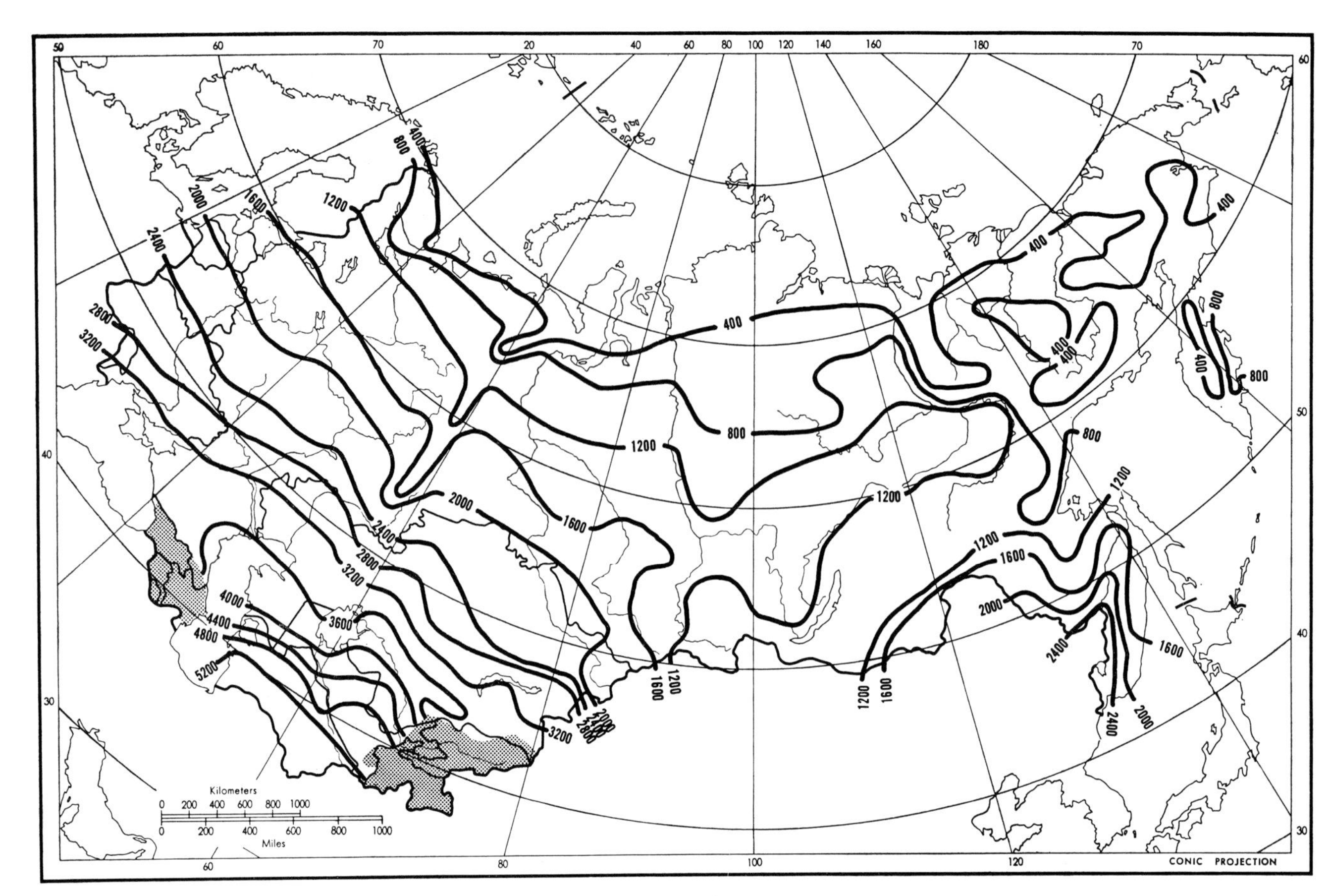

Fig.8-57. Sums of mean diurnal temperatures during period when mean diurnal temperatures are above 10°C. (After GERASIMOV, 1964.)

and less than 400°C in the far north (Fig.8-57). Such sums of temperatures have been correlated with growth phases of various types of crops and have therefore been given real significance by Soviet agroclimatologists. They have also been correlated with radiation balance at the earth's surface, thus allowing their use in lieu of radiation data which are largely lacking. These concepts will be made use of in Chapter 11 in determining climatic regions in the U.S.S.R. that have practical significance for agriculture and other human endeavors.

References and further reading

ANAPOL'SKAYA, L. E. and GANDIN, L. S., 1966. Teoreticheskie osnovy stroitel'noy klimatologii (The theoretical basis for construction climatology). In: M. I. BUDYKO (Editor), *Sovremennye Problemy Klimatologii*. Gidrometeoizdat, Leningrad, pp.263–279.

ANONYMOUS, 1960. *Atlas Sel'skogo Khozyaystva SSSR* (Atlas of Agriculture of the USSR). Gl. Upr. Geod. Kartogr., Moscow.

BARASHKOVA, E. P., GAEVSKIY, V. L., L'YACHENKO, L. N., LYGINA, K. M., and PIVOVAROVA, Z. I., 1961. *Radiatsionnyy Rezhim Territorii SSSR* (Radiation regime on the territory of the USSR). Gidrometeoizdat, Leningrad, 528 pp.

BUDYKO, M. I. (Editor), 1963. *Atlas Teplovogo Balansa Zemnogo Shara* (Atlas of Heat Balance of the Earth). Gl. Geofiz. Obs., Moscow, 2nd ed., 69 pp. (Translated supplement, U.S. Weather Bureau, Washington, 1964, 21 pp.)

BUDYKO, M. I. (Editor), 1966. *Sovremennye Problemy Klimatologii* (Contemporary problems in climatology). Gidrometeoizdat, Leningrad. (Translated as *Modern Problems of Climatology*. Foreign Technology Division, Wright-Patterson Air Force Base, Ohio, 29 November 1967, AD 670 893.)

BUDYKO, M. I., 1971. *Klimat i Zhizn'* (Climate and life). Gidrometeoizdat, Leningrad. (Translated as a volume of the Int. Geophys. Ser., Academic Press, New York, 1974.)

DANILOV, N. A., 1963. Effective radiation in European USSR: Its annual distribution and year-to-year variability. *Tr., All-Union Sci. Meteorol. Conf.*, 6: 358–365. (Translated by Environmental Sciences Services Administration.)

GERASIMOV, I. P., 1964. *Fiziko-Geograficheskiy Atlas Mira* (Physical geographical atlas of the world). Akad. Nauk, Moscow, 298 pp. (All legends and textual materials translated by Th. Shabad and published in: *Sov. Geogr. Rev. Transl.*, May–June, 1965, 403 pp.)

GOL'TSBERG, I. A., 1961. *Agroklimaticheskaya Kharakteristika Zamorozkov v SSSR i Metody Bor'by s Nimi* (Agroclimatic characteristics of frosts in the USSR and methods of combatting them). Gidrometeoizdat, Leningrad, 197 pp.

GOL'TSBERG, I. A. (Editor), 1967. *Mikroklimat SSSR*. Gidrometeoizdat, Leningrad. (Translated as *Microclimate of the USSR*. Isr. Progr. Sci. Transl., No.5345, Jerusalem, 1969, pp.62–212.)

GOL'TSBERG, I. A., 1969. Karty srednikh dat zamorozkov i dlitel'nosti bezmoroznogo perioda dlya SSSR (Maps of average dates of frosts and lengths of frost free periods in the USSR). *Gl. Geofiz. Obs. Tr.*, 247: 4–6.

KASHIN, K. I. and PAGAVA, S. T., 1959. *Issledovanie Znachitel'nykh Mesyachnykh Anomalii Temperatury Vozdukha na Evropeyskoy Territorii SSSR*. Gidrometeoizdat, Moscow. (Translated as *Significant Monthly Anomalies of Air Temperature over the European Territory of the USSR*. Isr. Progr. Sci. Transl., No.699, Jerusalem, 1962, 116 pp.)

KOTLYAKOV, V. M., 1968. *Snezhnyy Pokrov Zemli i Ledniki* (Snow cover on the earth and glaciers). Gidrometeoizdat, Leningrad.

PED', D. A. and TURKETTI, E. L., 1962. *Atlas Sutochnykh Amplitud Temperatury Vozdukha v SSSR* (Atlas of diurnal ranges of temperatures in the USSR). Gidrometeoizdat, Leningrad, 103 pp.

PIVOVAROVA, Z. I., 1960. Osnovnye kharakteristiki radiatsionnogo rezhima evropeyskoy territorii SSSR. *Gl. Geofiz. Obs. Tr.*, 115: 77–94. (Published as: *The Chief Characteristics of the Radiation Regime of the European Territory of the USSR*. U.S. Weather Bureau, Washington, 1961, 20 pp.)

PIVOVAROVA, Z. I., 1966. Izuchenie rezhima solnechnoy radiatsii v SSSR (Studying the regime of solar radiation in the USSR). In: M. I. BUDYKO (Editor), *Sovremennye Problemy Klimatologii*. Gidrometeoizdat, Leningrad, pp.41–56.

RAYNER, J. N., 1961. *Atlas of Surface Temperature Frequencies for North America and Greenland and Atlas of Surface Temperature Frequencies for Eurasia*. Arctic Meteorol. Res. Group, McGill Univ., Publications in Meteorology Nos.33 and 39, Montreal.

SHCHERBAKOVA, E. YA., 1961. *Klimat SSSR, 5, Vostochnaya Sibir'* (Climate of the USSR, 5. Eastern Siberia). Gidrometeoizdat, Leningrad, 300 pp.

VENTSKEVICH, G. Z., 1958. *Agrometeorologiya* (Agrometeorology). Translated by Isr. Progr. Sci. Transl., No.487, Jerusalem, 1961, 300 pp.

ZUBENOK, L. I., 1966. Rol' ispareniya v teplovom balance sushi (Role of evaporation in the heat balance of land areas). In: M. I. BUDYKO (Editor), *Sovremennye Problemy Klimatologii*. Gidrometeoizdat, Leningrad, pp.57–66.

Chapter 9

The Moisture Factor

A complete understanding of the hydrologic cycle is still beyond our grasp for any area of the earth, and the Soviet Union is no exception, although more efforts have been made to compute unknown values there than in most other parts of the world. Measured statistics exist in relative abundance for precipitation in most of its forms and for surface atmospheric humidity, both relative and absolute, over a network of stations that generally is dense enough to represent adequately the variations across the country. Measured statistics exist for atmospheric humidity content at various levels above the surface for a somewhat sparser network of stations, and relatively accurate estimates comprise long records of such things as cloud cover, fog frequencies, snow accumulation, and the like. To a certain extent, the disposition of precipitation in the form of surface and underground runoff has been determined for most parts of the Soviet Union as well. But the very important factor of evaporation and transpiration of water vapor from the earth's surface into the atmosphere still relies for its statistics primarily on computed values derived from other parameters. And the question of moisture flux in the atmosphere, which holds the promise of more complete understanding of the origins of precipitation and its related phenomena, has only recently been tackled in a few case studies that generally are based on no more than one to four years of observations of such things as humidity and wind readings at various levels.

It is hardly necessary to stress the importance of knowing all one can about moisture in all of its phases, for much of man's activities depend to a great extent upon the balance between precipitation and precipitation losses at the earth's surface, and his physiological functions and mental attitudes respond sensitively to such things as atmospheric humidity and cloud cover. The amount, frequency, and form of precipitation has profound effects on man's life, and only the amount is very well quantified. Even long-standing records on precipitation amounts are being questioned these days as it has been discovered that measuring devices might have misrepresented actual fall of hydrometeors by as much as 25% in some instances. During the past few years the Soviets have been in the process of correcting upwards most of their precipitation data, and therefore at the present time it is hard to tell whether one has truly representative samples or even comparable samples, since in most data tables that have been made available no notation has been given regarding data collection and collation. In most cases even the period of record is unknown, and therefore one might be trying to compare records of different lengths and different times.

Nevertheless, most of the data on the various facets of moisture are as good or better in the Soviet Union than in the rest of the world, and therefore they will be considered in detail here. Some understanding has already been gained about the mechanisms of precipitation and moisture balance in the various regions of the country. The discussion here will serve to tie together on a broader scale the causes behind the distributions and the country-wide distributions themselves which relate intimately to the look of the land and its use for agriculture and other economic endeavors. Use will be made of what information there is to work logically through the hydrologic cycle from its origins as vapor in the atmosphere through the various forms of condensation and precipitation to the disposition of precipitation at the surface of the earth and its influences on soil moisture and plant growth.

Atmospheric moisture content and moisture flux

Fortunately, there is now available a monograph which brings together the various case studies on atmospheric moisture content and moisture flux across the Soviet Union and other parts of the world (DROZDOV and GRIGOR'EVA, 1963). This embodies a number of maps of the U.S.S.R. and the world which are the first of their kind to depict exchanges of atmospheric moisture. But as the authors point out, time and energy have limited them to computations of only a small amount of the data available, and since atmospheric moisture content, moisture flux, evaporation, and so forth fluctuate greatly from year to year, even over the entire earth, a study based on only one year, or a few years at best, is not necessarily very representative of all time. The authors computed moisture content and moisture flux at four levels in the atmosphere across European U.S.S.R. for January, April, and July of 1951, 1952, 1954, and 1956 and for the entire U.S.S.R. for only January and July of 1956. Since 1956 was not a completely normal year it will be necessary to utilize some of the work done on European U.S.S.R. as well as that done on the entire country. In addition, some comparisons can be made of studies of moisture flux in North America and of computations of moisture content in the atmosphere over the entire earth.

The generally low moisture content of the air over the Soviet Union again reflects the high latitude and relatively small amounts of heat available to the Soviet Union.During July on the average in the first 5 km of the atmosphere most of the Soviet Union has 15–25 kg of moisture per square meter. This compares to the situation over Canada. Only in limited areas in the south, primarily around the Black and Caspian seas, do values reach 30–40 kg/m^2, which is similar to the eastern half of the United States. The values along the Arctic coast of Siberia during summer are similar to those in some of the subtropical deserts of the world. In January, of course, the atmospheric moisture content of the Northern Hemisphere decreases greatly. Over much of the Soviet Union it ranges from 2.5 to 5.0 kg/m^2. Again this compares to much of Canada, while eastern United States ranges from 7.5 to 20.

Fig.9-1 shows the distribution of moisture over the entire Soviet Union in the first 5 km of the atmosphere during January, 1956. During that month western Europe and a large part of European U.S.S.R. were under the influence of a low pressure which was considerably lower than normal, and the precipitation over the area was higher than normal. Nevertheless, the distribution shown in Fig.9-1 seems to agree fairly well with some other

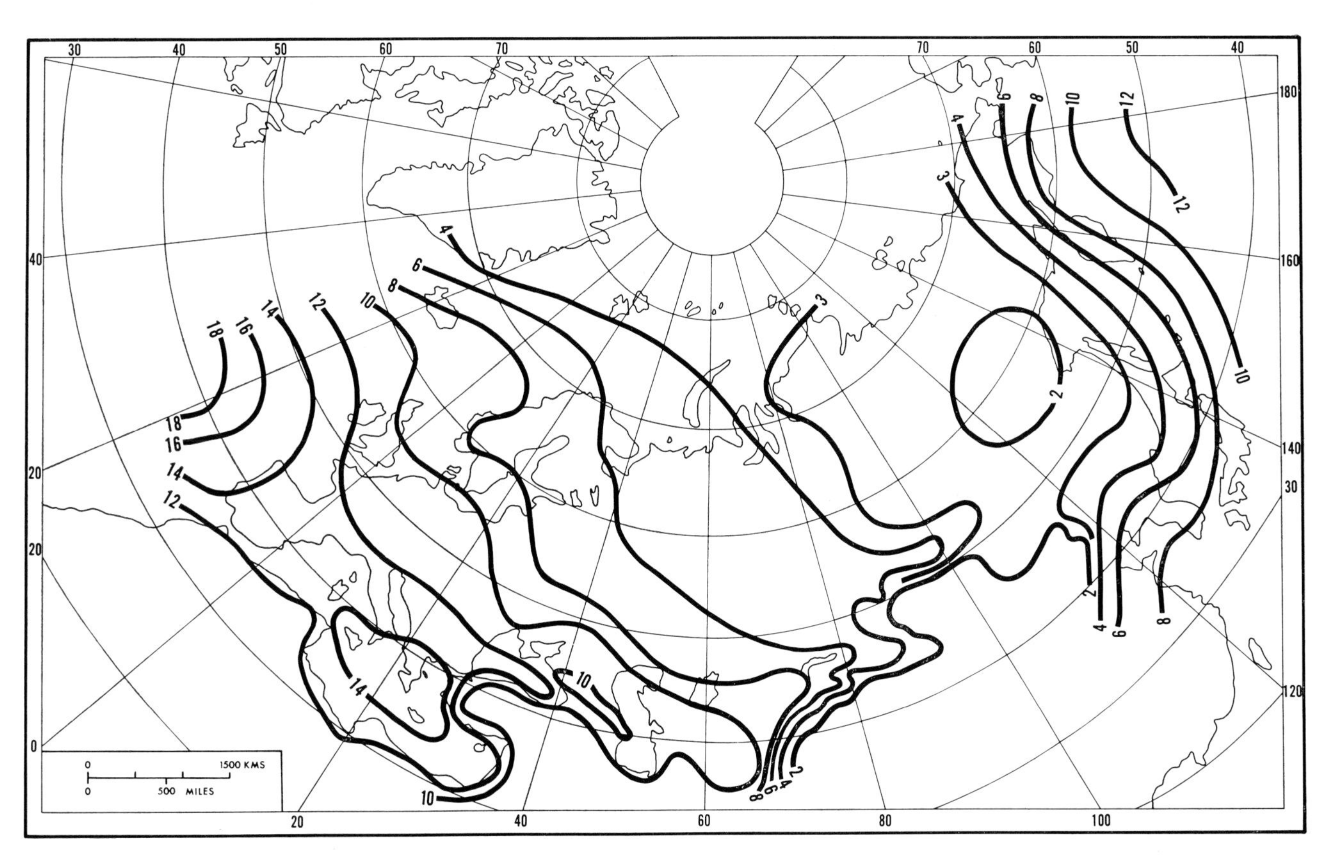

Fig.9-1. Atmospheric moisture content (kg/cm²) in the first 5 km of air, January 1956. (After DROZDOV and GRIGOR'EVA, 1963.)

studies of longer periods over individual parts of the Soviet Union. It can be seen that by far the greatest moisture content is in the western part of the country, and the driest air is in eastern Siberia over the middle Lena Valley. Moisture picks up again across the Pacific coast, but the air over the continental land mass even along the coast at this time of year is quite dry. This distribution is to be expected because of the consistent dominance of the Asiatic high during winter which rules out penetration of air from any direction except the west and causes the east Siberian area to be particularly void of intrusions of marine air.

This is brought out very clearly on the moisture flux map which shows a net westerly flux of moisture across the entire country clear to the Pacific coast at this time of year (Fig.9-2). Only in the far northeast, in the Chukchi and Kamchatka peninsulas, is there any net flux of air from the Pacific during January. Most of the moisture during winter is advected into the country from the North Atlantic and to a certain extent from the Mediterranean as the air drifts generally east-noıtheastward across the country along the eastern limb of the upper level trough over European U.S.S.R. and the northwestern periphery of the Asiatic high. In the Soviet Far East moisture transport along the coast sometimes occurs from the south and southeast in a layer up to 1.5–2 km thick, but higher up the transport in January is always from the west.

In July, of course, everywhere within the Soviet Union the atmosphere contains a great deal more moisture than it does in January (Fig.9-3). The distribution over the country is different also, which signifies changes in the distribution of moisture flux (Fig.9-4).

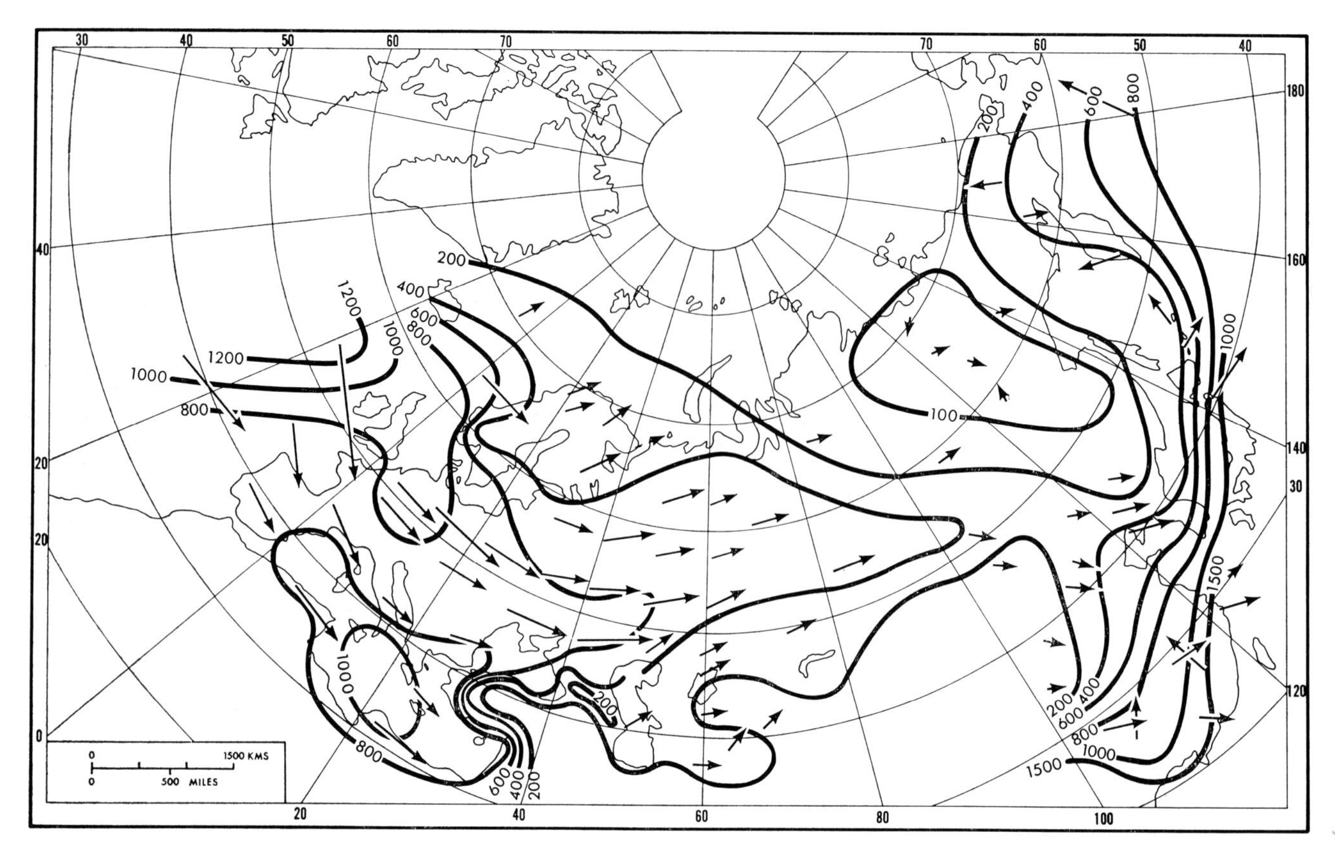

Fig.9-2. Total atmospheric moisture flux (g/cm per second) from 1,000 to 500 mbar, January 1956. (After DROZDOV and GRIGOR'EVA, 1963.)

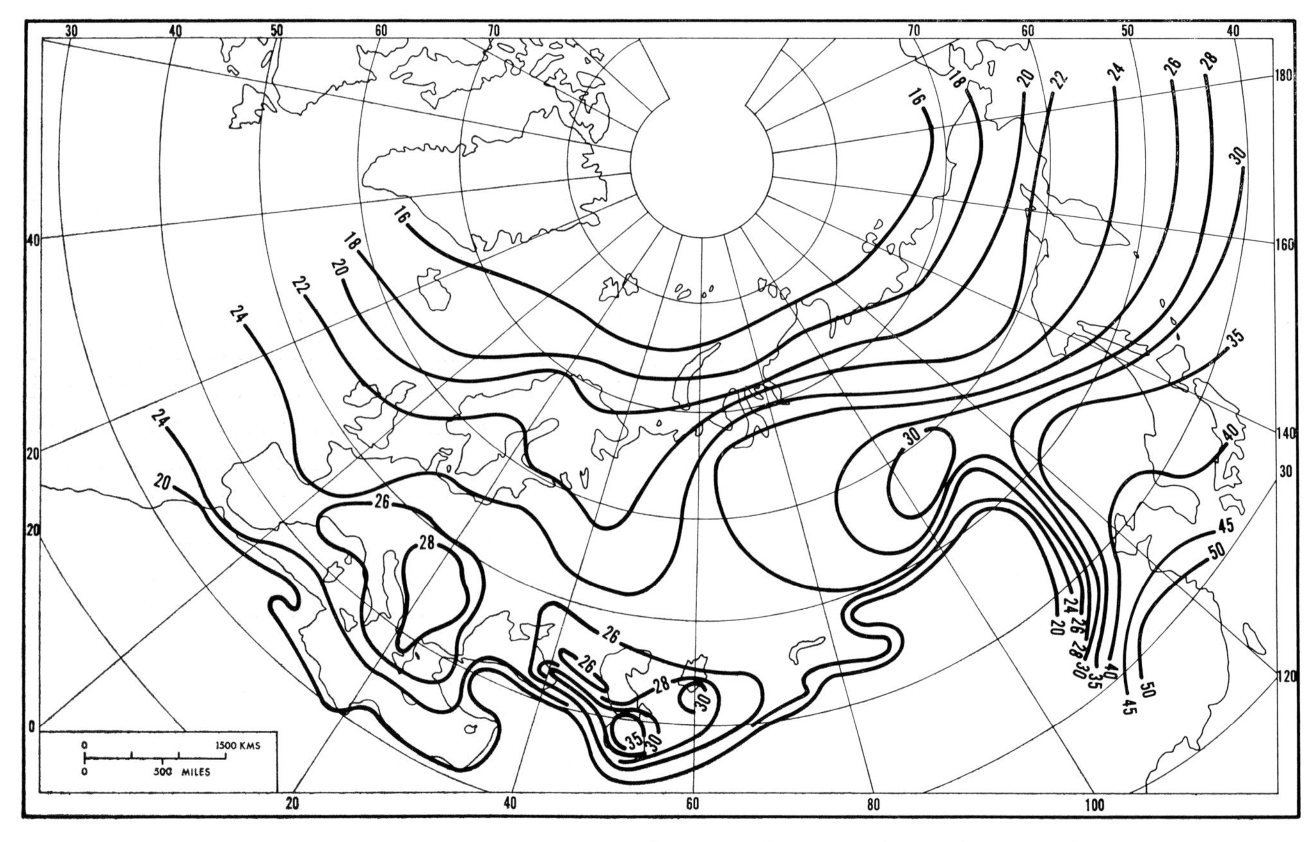

Fig.9-3. Atmospheric moisture content (kg/m²) in the first 5 km of air, July 1956. (After DROZDOV and GRIGOR'EVA, 1963.)

The flux over most of the country continues to be from west to east, but in northern European U.S.S.R. there is a significant movement from northeast to southwest, and in the Far East the Pacific monsoon now penetrates far up the Amur Valley. In addition, much of the marine air moving into the land picks up moisture through evaporation from the surface over which it is traveling, and since this land surface in summer is usually considerably warmer than the air moving over it convection and turbulent currents carry this evaporated moisture upward into the air rapidly. Therefore, the total moisture content in the first 5 km of air tends to remain fairly constant as the air sweeps across broad sections of the land from west to east; turbulent mixing upward of evaporated moisture compensates for loss of moisture through precipitation. Therefore, in summer, particularly across the western half of the country, there is not the general west-to-east decrease in atmospheric moisture content that there is in the winter, and neither is there such a rapid fall-off of precipitation eastward during summer.

Apparently there is a significant amount of precipitation additionally stimulated by the local effects of evaporation over the land during summer. This comes about not only through the addition of moisture into the air from the surface but also through the creation of greater conditional instability by continual addition of moisture to the lower layers of air, which then leads to a good deal of convective shower activity over much of the country during the summer. It is impossible to measure just what portion of the precipitation that falls in a given area is due to the influences of local evaporation proces-

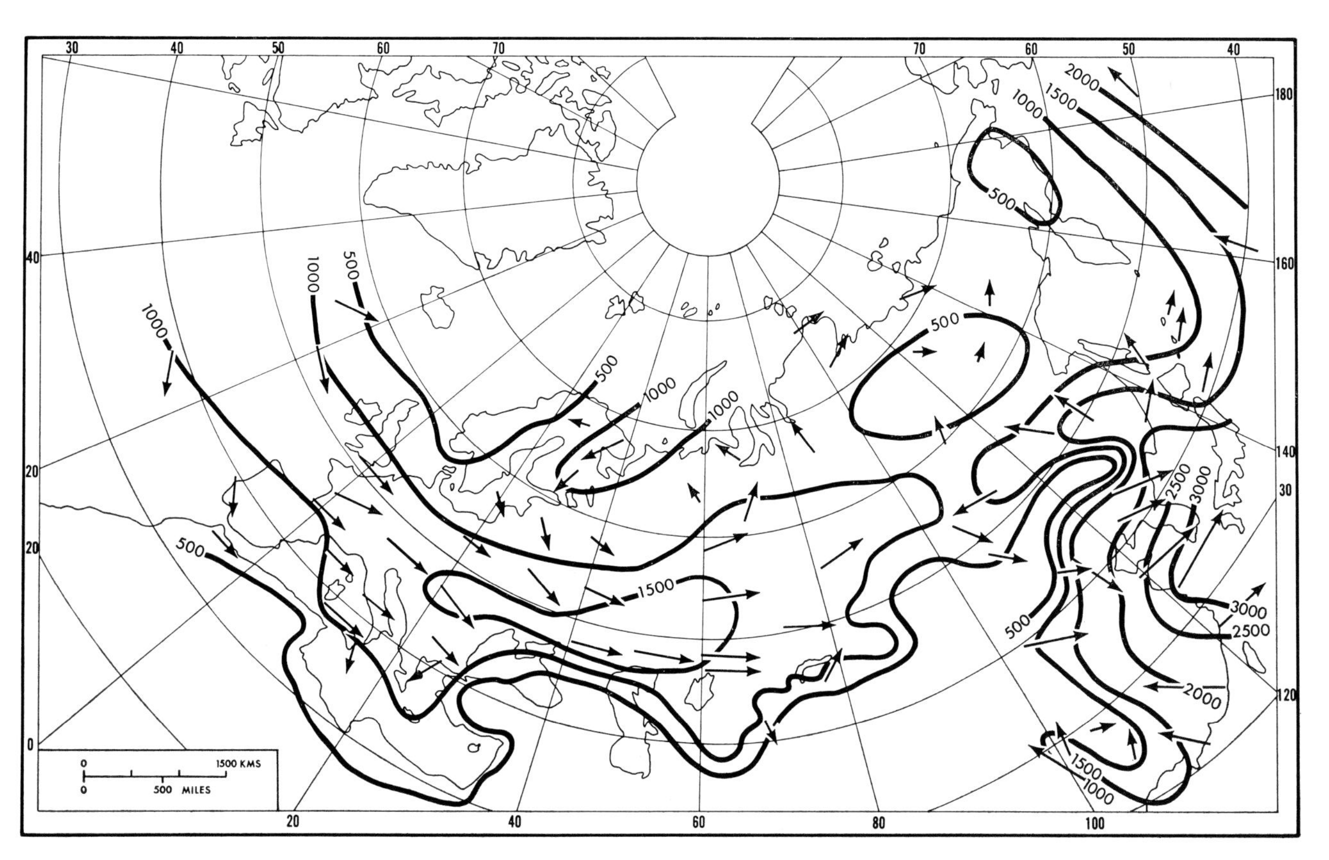

Fig.9-4. Total atmospheric moisture flux (g/cm per second) from 1,000 to 500 mbar, July 1956. (After DROZDOV and GRIGOR'EVA, 1963.)

ses and mixing upward of moisture from the surface, particularly since the precipitation in the Soviet Union shows a distribution which is more related to cyclonic tracks and other synoptic features than it is to atmospheric moisture content, but it has been estimated that in certain instances 10–15% of the precipitation that falls in an area during the summer might have been due either directly or indirectly to evaporation processes in the local area. Downstream from large bodies of water in dry areas evaporation might even effect a 25% increase of precipitation during the summer.

The moisture flux in July is three times as great as in January over European U.S.S.R., and five to six times as great over eastern Siberia. The moisture content over eastern Siberia is ten to twelve times higher in July than in January. During July the highest moisture flux is over the southern steppes in an east–west oval-shaped area from the Carpathian Mountains eastward to the northern shore of the Aral Sea. There it averages more than 1,500 g/cm per second. It decreases northward due to decreasing air temperatures and decreases southward due to decreasing wind speeds. Another area of maximum flux appears in the southern part of the Soviet Far East where the Pacific monsoon causes a flux of as much as 2,000 g/cm per second over Vladivostok. Lowest values occur over the central portion of eastern Siberia and the extreme northeastern part of the U.S.S.R. where generally stagnant air motion is associated with a broad shallow thermal low pressure area at this season of the year.

Atmospheric moisture content in July is highest in the extreme south, Central Asia and the Caucasus, and in the monsoon portion of the Far East. The areas of maximum moisture content in Central Asia are quite localized and appear to relate closely to the major water bodies in that region, the Caspian and Aral seas.

Moisture content and moisture flux over European U.S.S.R.

Moisture flux was computed by DROZDOV and GRIGOR'EVA (1963) for the European U.S.S.R. for January, April, and July of 1951, 1952, 1954, and 1956. Moisture fluxes were calculated at 1,000-, 850-, 700-, and 500-mbar surfaces and an integral flux was calculated for the entire thickness from 1,000 to 500 mbar. The total flux was made up of the average flux which was computed from observed predominant winds and moisture contents and the turbulent flux which is a function of the movement of eddies in the atmosphere which range in size from a few millimeters in diameter to cyclonic and anticyclonic proportions. Moisture in the form of liquid and solid states was not taken into account in any of the calculations since they generally make up less than 10% of the total moisture flux. Roughly half the moisture flux takes place in the layer of air below the 850-mbar level, and this is the most difficult to calculate since the wind is most variable in this lower layer. It was found that the air above the 500-mbar level contained only about 15% of the entire moisture flux of the atmosphere, and therefore the study did not include heights above 500 mbar.

Moisture content in different parts of European U.S.S.R. are due mainly to unequal frequencies of currents of air coming from different regions and depend little on local conditions. During the course of a year the most moisture-rich air reaches European U.S.S.R. from the south and southwest. In winter the air with the highest moisture content, about 18 kg/m^2 in the 500-mbar thickness, comes from the southeastern Mediterranean, while in summer it comes from Iran via the Caspian Sea and contains as much as 40 kg/m^2. The air which is poorest in water vapor naturally arrives from the Arctic. During February air from the Kara Sea contains only about 1 kg/m^2. Air coming into European U.S.S.R. from the east from the central part of the continent during winter contains 5–6 kg/m^2. During summer the moisture entering European U.S.S.R. from the Greenland Sea contains about 17 kg/m^2 and that coming from the middle latitudes of the Atlantic about 25 kg/m^2, which is about the same as the central Asian air that occasionally invades the southeastern part of European U.S.S.R. During a study in August 1951, Bashtan traced air transport into European U.S.S.R. all the way from the Far East across northern Eurasia around the northern edge of the slowly circulating Asiatic low. He found that this air also arrived with a moisture content of about 25 kg/m^2. Just how frequently this type of thing occurs is not known (DROZDOV and GRIGOR'EVA, 1963, p.204). During winter the contrast in moisture contents of different air masses is extremely high, amounting to as much as ratios of 15 to 1, while in summer differences are much less, with extreme ratios of approximately 2 to 1. In spring and autumn moisture contents differ by maximum ratios of about 4 to 1.

On the average in January the atmospheric moisture content in the layer between 1,000 and 500 mbar is low and decreases across European U.S.S.R. from southwest, where it is 9–10 kg/m^2 to northeast where it is 5–5.5 kg/m^2. In southwestern European U.S.S.R. the absolute humidity lapses with height from 3.5 to 4 g/m^3 at the 1000-mbar surface to 2.5 g/m^3 at the 850-mbar surface. In the central and northern parts of the plain the humidity does not vary up to the level of the 850-mbar surface, and in southeastern European U.S.S.R. it increases from 1.5 g/m^3 at the 1,000-mbar surface to 2 g/m^3 at the 850-mbar surface. Above 850 mbar the absolute humidity decreases with altitude over the entire territory.

At the earth's surface the moisture flux in January is small, and slightly more moisture is transported in a meridional direction than in a zonal one. With increasing height the role of the zonal moisture flux increases. At 850 mbar the average zonal flux over European U.S.S.R. is approximately double the meridional flux, at the 700-mbar surface it is 5 times as great, and at the 500-mbar surface 8 times as great. Of the four levels considered, the maximum flux is observed at the 850-mbar surface, since at this level the absolute humidity differs little from that at the earth's surface, and wind velocities are considerably higher than they are at the surface. Probably the maximum moisture flux is somewhat below the 850-mbar surface, but data are not available to confirm this. The average meridional moisture flux over European U.S.S.R. during January is everywhere directed from south to north.

The average moisture flux during January at the 850-, 700-, and 500-mbar surfaces is at a minimum in the northern part of European U.S.S.R., due to both the decrease in humidity there and the reduction of westerly velocities at these heights in comparison with the central region (Fig.9-5). A second minimum occurs in the southeastern part of European U.S.S.R. due partly to a decrease in moisture content but mainly due to the result of diminishing resultant westerly transport as a heavy transfer arrives from the east along the southern periphery of the Asiatic high. At the 500-mbar surface westerly transport prevails, and so the minimum in the southeast practically disappears at this level.

In July, in addition to moisture received from the atmosphere coming from oceanic areas, a considerable amount is contributed by evaporation from underlying land. Total moisture flux in July is approximately double that in January, the ratio being somewhat

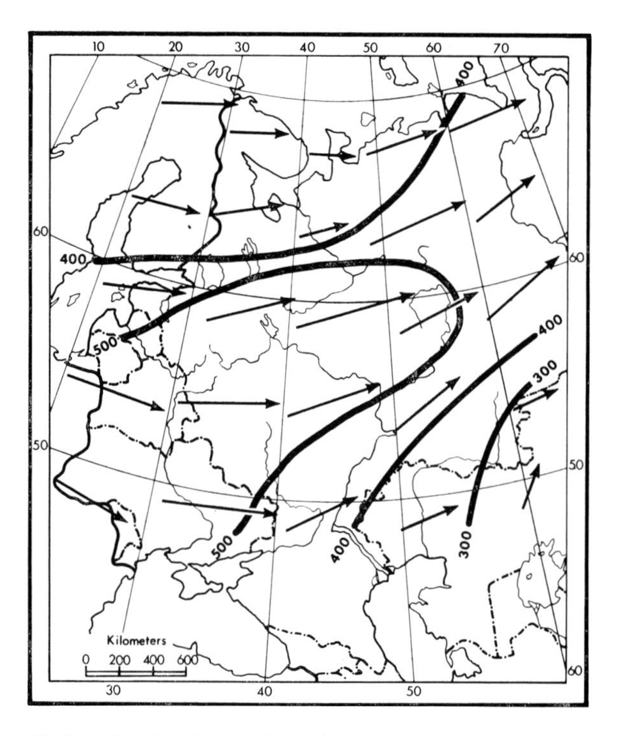

Fig.9-5. Total moisture flux (g/cm per second) over European USSR for January. (After DROZDOV and GRIGOR'EVA, 1963.)

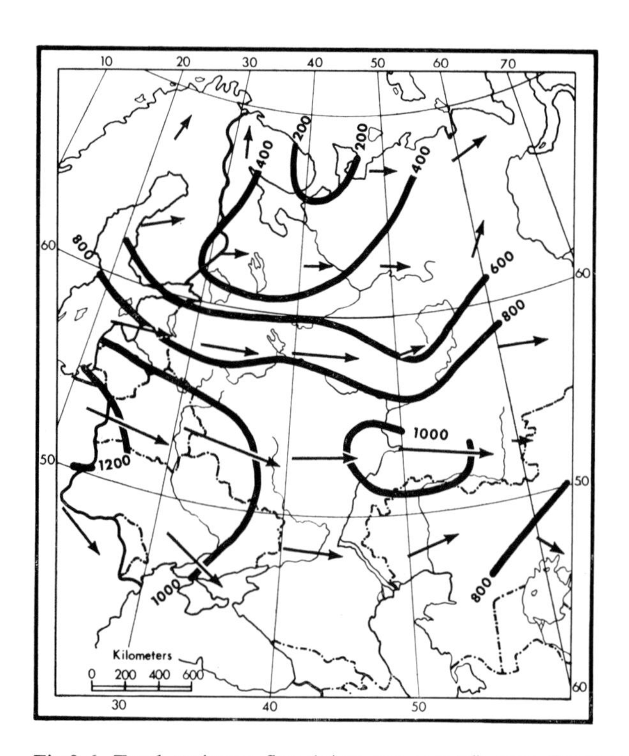

Fig.9-6. Total moisture flux (g/cm per second) over European USSR, July. (After DROZDOV and GRIGOR'EVA, 1963.)

less in the north. The maximum flux during July is over the western part of the plain, particularly in the Baltic and Belorussian areas where it reaches values of 1,100–1,200 g/cm per second (Fig.9-6). The minimum is along the White and Barents seas where values drop to 200 or less.

Moisture evaporated from the land is not transferred immediately into the upper layers of the air, and therefore the vertical distribution of moisture in summer is less uniform than it is in winter when practically all the moisture over European U.S.S.R. is of oceanic origin. Over all of European U.S.S.R. the absolute humidity decreases with height in summer. It lapses from 11 or 12 g/m^3 at the 1,000-mbar surface to 1.2 g/m^3 at the 500-mbar surface in the central and southern parts of the plain, and from 8.5 g/m^3 to 1 g/m^3 in the northern part. Resultant wind velocities at low altitudes in summer are weak but they increase rapidly with height. Therefore, in spite of the fact that in July the water vapor below 850 mbar constitutes a larger portion of the total than it does in winter, on the average the proportion of moisture transport at lower altitudes is less in summer than in winter. This of course varies drastically from time to time and place to place.

The average zonal flux at the earth's surface in July is directed from west to east south of 60° north latitude and from east to west north of 60°. It is obvious that during summer European U.S.S.R. derives a considerable amount of moisture from the Barents and Kara seas in the Arctic. The average zonal flux at the earth's surface is greatest in the southwestern part of European U.S.S.R. where it amounts to more than 15 g/cm per second while in the north the easterly flux has a relatively low magnitude, less than 5 g/cm per second. The largest amount of moisture is transported at the 850-mbar level. Above 850-mbar, the zonal flux everywhere is directed from west to east.

In July the average meridional moisture flux is less than the average zonal moisture flux at all levels. In the first 5 km of the atmosphere the meridional flux averages only 10–20% as much as the zonal flux in the western part of European U.S.S.R., about half as much in southeast European U.S.S.R., and somewhat more than the zonal flux in the northwest. At the earth's surface in July the meridional flux over a considerable region is directed from north to south, but with increasing height the area occupied by moisture flux from the north decreases.

The moisture flux differs greatly in magnitude and direction from one year to another, depending on the type of circulation that prevails. In January 1952 the westerly type of circulation prevailed and the maximum moisture flux throughout the central portion of the plain reached 800 g/cm per second. In 1954 a deep trough prevailed over European U.S.S.R. and the circulation was predominantly meridional. The average moisture flux in January that year was only 15–20% as much as it was in 1952. However, in the northern part of the plain moisture flux was greater in 1954 than in 1952.

In July 1954 westerly and easterly circulations were equally prevalent, and meridional circulation did not occur at all. The European U.S.S.R. was under the influence of the eastern periphery of a trough which was situated in the North Atlantic and the western periphery of a ridge situated over eastern European U.S.S.R. and western Siberia. The moisture flux over the entire European U.S.S.R. was southwesterly and it diminished from west to east. Values ranged from 1,500 south of the Baltic to 500 in the east. Isolines ran almost north–south through much of the plain, except along the Arctic coast where values of 1,000 prevailed. In July 1956, on the other hand, a deep trough was centered over European U.S.S.R. and the circulation was predominantly meridional. The largest moisture flux occurred in the south where it was more than 1,500 g/cm per second. It decreased in the center and increased again in the north to more than 1,000 g/cm per second associated with stable northeasterly winds.

Magnitudes of moisture flux tend to differ from year to year more in winter than in summer. During predominantly western circulation the Icelandic low is well developed, a strong pressure gradient is directed from south to north, and cyclonic circulations move rapidly from west to east and bring warm humid masses of air from the ocean to the continent which cause positive anomalies of temperature and atmospheric humidity over European U.S.S.R. The atmospheric moisture content and its gradient eastward are greatest during western type circulations. With this type of flow the Scandinavian Mountains have a significant influence in the north, reducing moisture content and moisture flux on the eastern side. They have no noticeable effect during meridional circulation.

When meridional circulation prevails there is a slackening of the average transport and an intensification of the turbulent transport of moisture. Particularly in winter and spring the meridional circulation is associated with a much reduced moisture flux, and the ratio of the turbulent flux to the total flux is higher. This type of circulation generally occurs when a deep trough is located over European U.S.S.R. At these times air masses move into western European U.S.S.R. from the north along the western periphery of the trough and bring low moisture contents to the western part of the plain.

When easterly circulation predominates, there is a large decrease in moisture content southeastward across European U.S.S.R. because of high frequencies of cold dry air masses that move into the southern part of the plain along the southern periphery of the Asiatic high. When the Asian anticyclone is well developed and extends across European U.S.S.R. into western Europe the moisture content is probably lower than even during a meridional circulation. However, generally the lowest moisture content over European U.S.S.R. is associated with meridional flow.

Humidity

Within the U.S.S.R. in July the vapor pressure in the surface air is highest in the Black Sea–Caspian area and in the lower Amur Valley of the Far East (Fig.9-7). Rather surprisingly high values also appear in the upper Yenisey and Ob River basins. Much of Central Asia and Kazakhstan are predictably low for their latitudes, and in the northern half of the country the vapor pressure generally decreases with latitude.

The highest values in the entire country are in the Colchis Lowland in the Transcaucasus at the eastern end of the Black Sea. High values continue eastward across the great Caucasus and into the North Caucasian Foreland and the Caspian Basin. The high values over the Caspian attest to the great amount of evaporation that takes place from the surface of the sea during the summer and do not signify moist conditions. Relative humidities in that area are around 50–60% and there is little precipitation over the northern and eastern parts of the basin.

The orientation of the vapor pressure isolines in the southern part of the Soviet Far East corroborates what has been said about the direction of moisture flux in this region during middle and late summer when monsoon air swings around the southwestern side

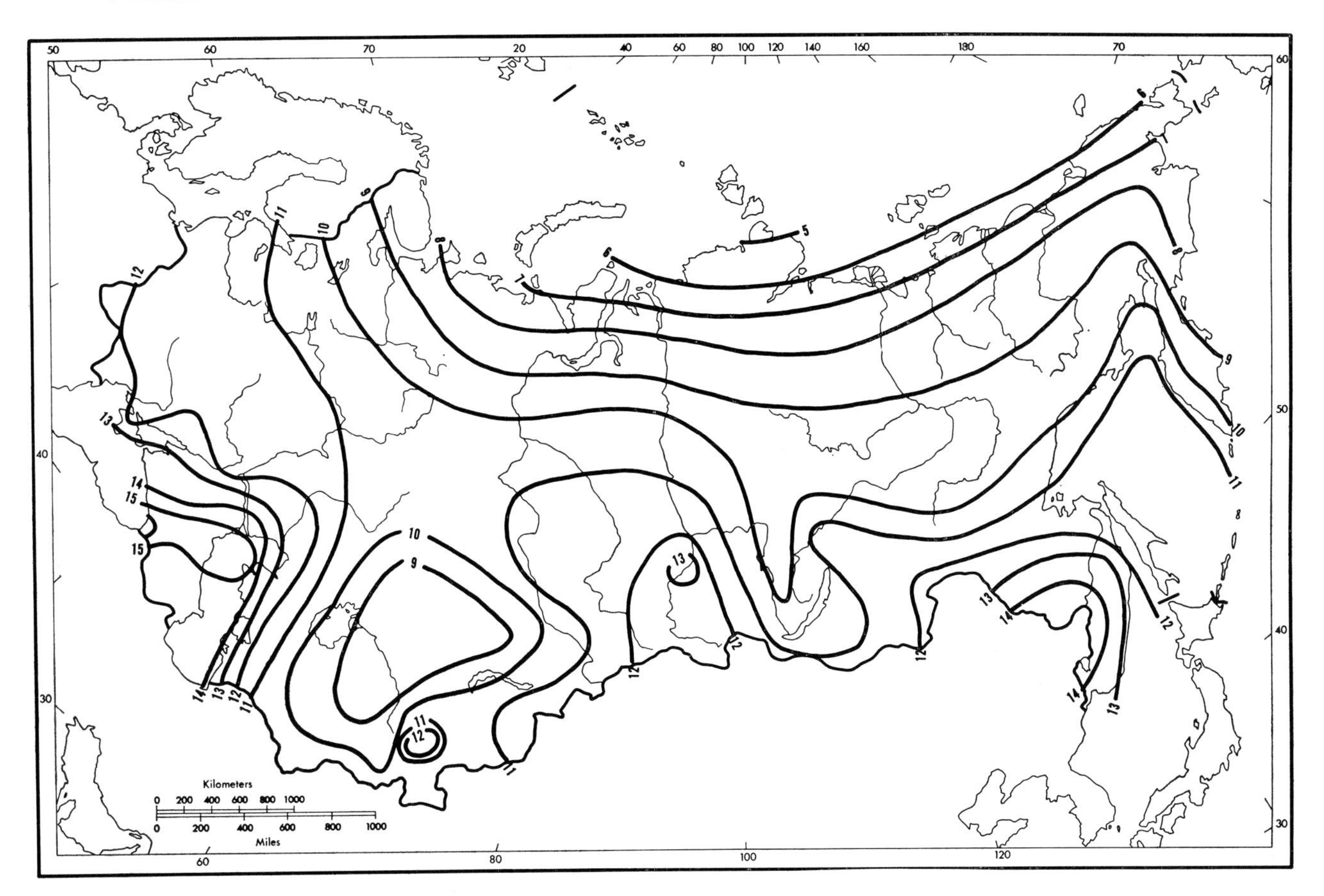

Fig.9-7. Mean vapor pressure, July, in millimeters of mercury. (Compiled from the tables in the Appendix.)

of the subtropical high in the Pacific and comes into the Amur Basin from the south and southwest through northern China. The vapor pressure is higher in the Amur–Ussuri valleys than it is along the eastern coast.

Relative humidity of the surface air is depicted in Figs.9-8 and 9-9. Over practically the entire country it is highest in January, since temperatures are much lower at that time of year and the capacity of the air to hold moisture is thereby reduced. Highest values in winter generally occur in the western and northern portions of the country. The high values in European U.S.S.R. reflect the high amount of moisture that is being advected into the region from the west with strong atmospheric circulation at this time of year, and the high values along the Arctic reflect the very cold temperatures. High values also exist in coastal and island areas along the Pacific fringe of the country. Everywhere in January relative humidity generally averages between 70 and 90%. The only regions that fall below this are the southern portions of Maritime Kray and the fringes of the Sea of Okhotsk. Such low values in these marine locations no doubt result from the continuous outflow of cold air from the Asiatic high during winter which, upon coming into contact with marine influence along the coast, experiences an immediate drop in relative humidity, as heat is injected into the surface air faster than moisture is.

Surface relative humidities in July generally reflect the degree of continentality. High values exist along the periphery of the country, particularly along the Pacific and Arctic

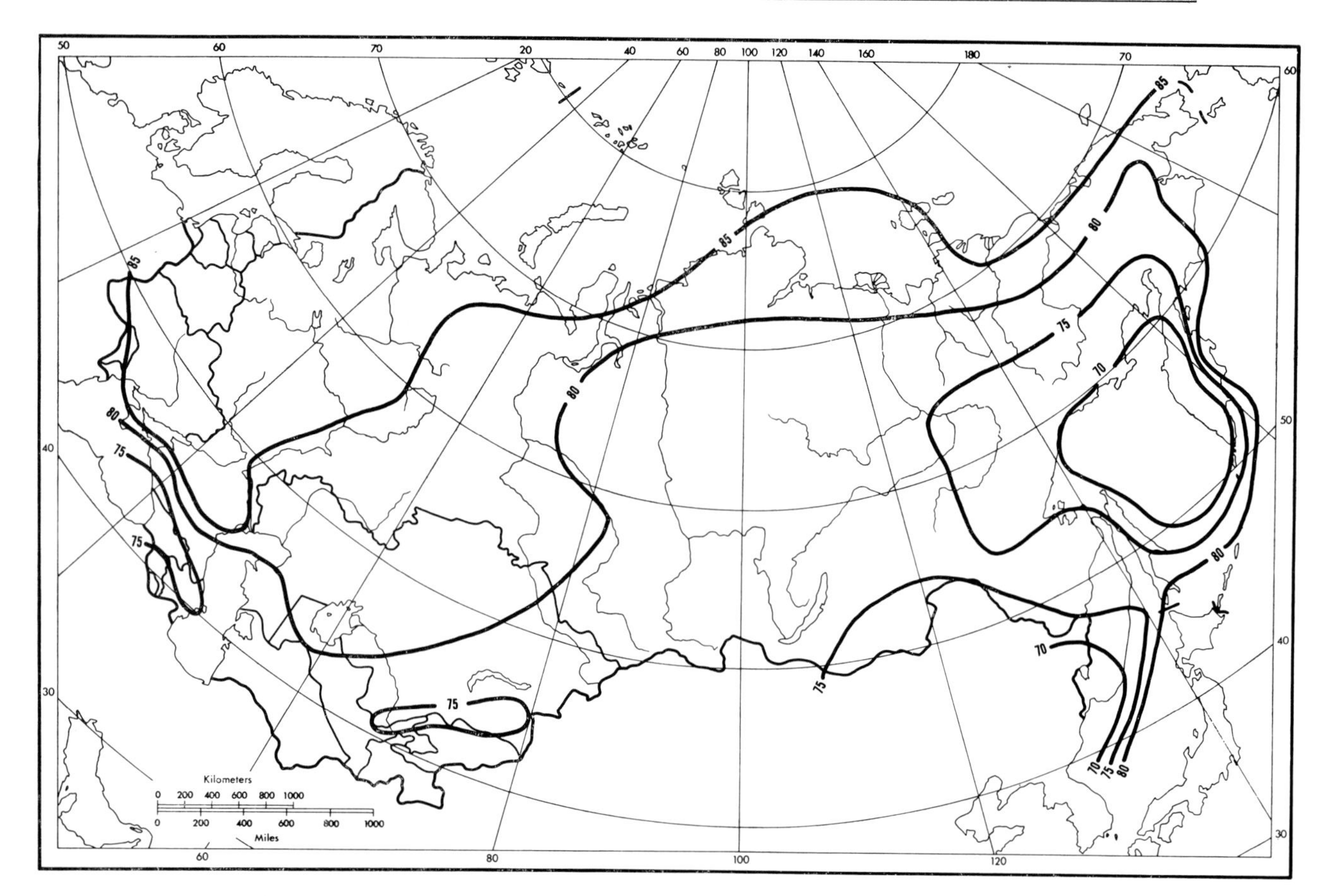

Fig.9-8. Mean relative humidity (%) of the surface air for January. (Compiled from the tables in the Appendix.)

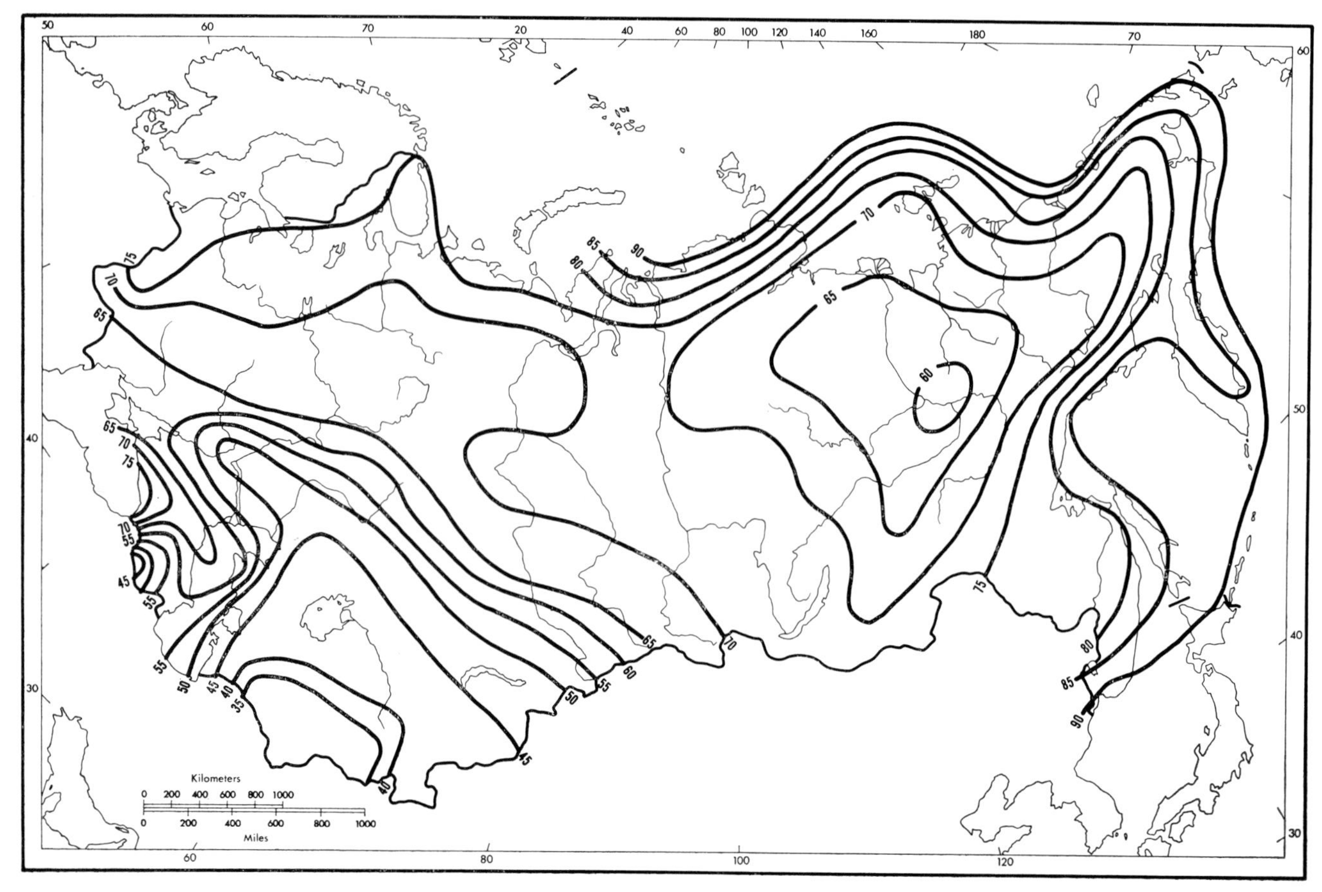

Fig.9-9. Mean relative humidity of the surface air, July, in percent. (Compiled from the tables in the Appendix.)

coasts. The very place that had the lowest value in winter, Vladivostok, has one of the highest values in summer. The western Transcaucasus also have values somewhat higher than they do in winter. The variation of relative humidity across the country is much greater than it was during the winter, and steep gradients occur in places such as the Transcaucasus, Central Asia, and the coastal areas along the Arctic and the Pacific. A great difference exists between the western and eastern Transcaucasus which corresponds to the general climatic contrast between these two areas. The Colchis Lowland in summer, as in winter, is moist and rainy, while the Kura Lowland to the east is rather arid. The Armenian Plateau in the southern part of the region has very dry sunny conditions.

Much of the European plain has relatively low humidities at this time of year, generally between 60 and 70%. The gradient is directed from northwest to southeast, and relative humidities reach their lowest values of all in the southern part of Central Asia. Relatively low values also exist over much of eastern Siberia. The extreme continentality of this area is conducive to very rapid heating of the surface in the summer so that the capacity of the air to hold moisture rises faster than the humidity content derived through evaporation from the surface.

Over most of the Soviet Union during summer relative humidities on the average are fairly comfortable. Only the western Transcaucasus and the southern part of the Soviet Far East experience prolonged periods of oppressively high relative humidities together with relatively high temperatures. Of course, on individual days influxes of warm humid air can affect almost any part of European U.S.S.R. and Siberia. Central Asia, on the other hand, experiences desiccatingly low relative humidities, which tend to ameliorate the high summer temperatures of that region but have telling effects on anyone who remains exposed to these desert conditions for prolonged periods of time.

Relative humidities are generally lowest over the country during spring when absolute humidity contents are still relatively low and the intense heating of the land is rapidly increasing the capacity of the surface air to hold moisture. This has already been mentioned with respect to the high transmissivity of the air to radiation at this time of year. Generally May is the month of lowest relative humidities, except for parts of Central Asia and the eastern Transcaucasus.

Clouds and fog

The distributions of clouds and fog are probably the hardest things to depict on small scale maps because they vary so abruptly locally according to topography, type of surface, and exposure to wind currents. It is probably significant that no such maps occur in any Soviet publications. Monthly maps showing frequencies of cloudy days were included in the climatic atlas of the U.S.S.R. published in 1933, but these were based on data taken between 1896 and 1915, which understandably were rather sparse, particularly in the eastern parts of the country. Isolines have been drawn very schematically to represent what seems to be logical, but they do not fit more recent data on cloud cover that have been made available in various statistical compilations. In 1962 SCHLOSS constructed winter and summer maps of cloudiness for the U.S.S.R. showing average percent of sky cover from December to February and from June to August. These maps were based on

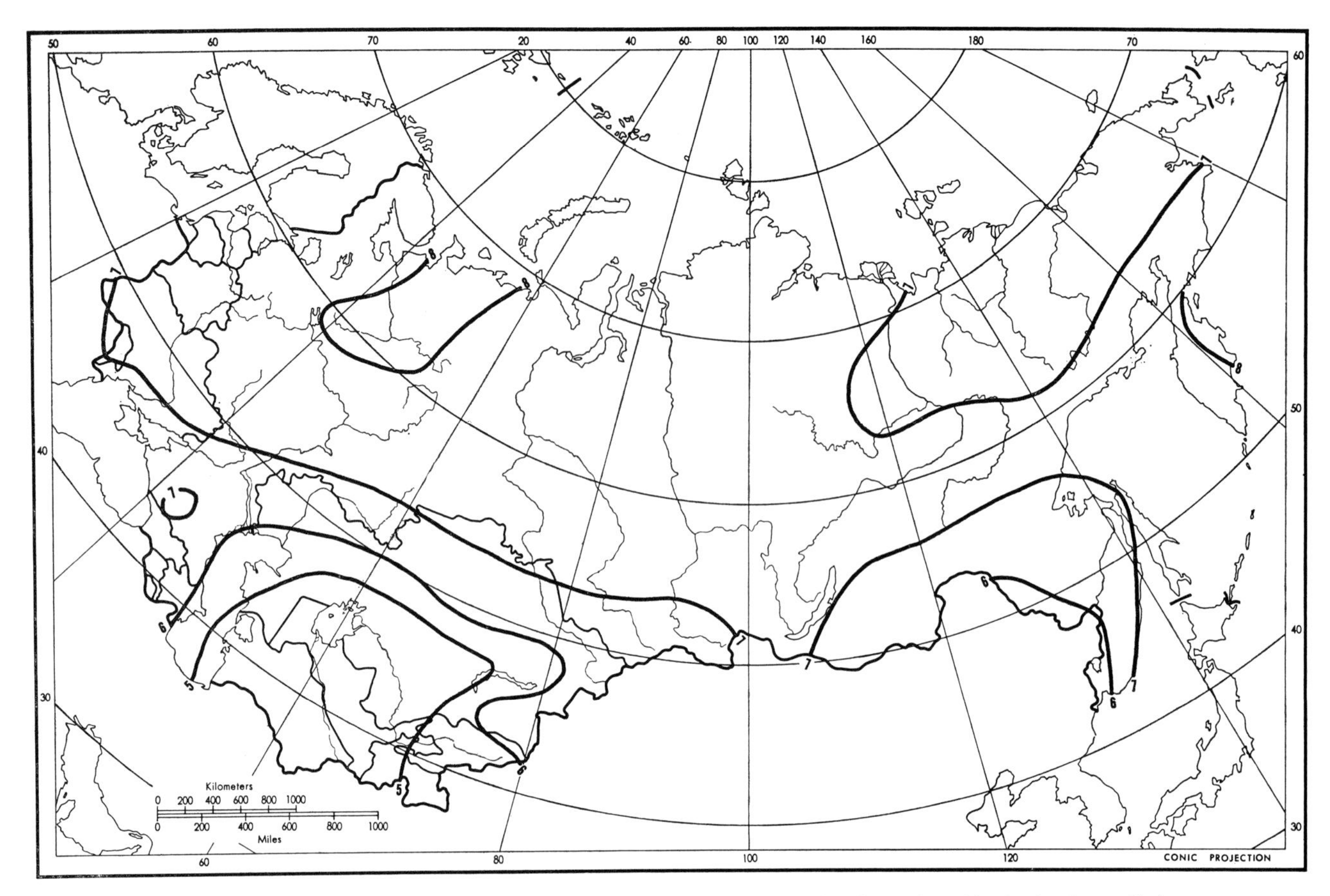

Fig.9-10. Mean annual tenths of sky cover at 13h00. (Compiled from the tables in the Appendix.)

data which were collected mainly before 1937, and again the northern and eastern parts of the country were very poorly represented. There were no data at all along either the Pacific or Arctic coasts where cloud conditions change rapidly.

Most of the more recent data that have appeared in statistical compilations show numbers of days with sky coverage between 8 and 10 tenths. Therefore, the data are not entirely satisfactory since they probably stress stratus-type clouds associated with fronts and cool stable atmospheric conditions at the expense of cumulus-type clouds associated with atmospheric instability and local convection.

Figs.9-10, 9-11, and 9-12 showing tenths of sky cover have been compiled from recent data collected from various sources and presented in the statistical tables at the end of this book. Although in certain instances more data could still be desired, a general pattern of cloudiness over the U.S.S.R. does emerge, partly expected and partly unexpected. During the course of the year cloud cover is greatest in the northwest and least in Central Asia, as would be expected. It is relatively great throughout European U.S.S.R. and Western Siberia, generally decreasing from the Barents Sea southward to the Black Sea and eastward across central Siberia. It increases again as the Pacific coast is approached.

Perhaps the most unexpected feature on the annual map is the low amount of cloudiness in the middle Amur Valley in the southern part of the Soviet Far East. Even during the summer monsoon period the region is not perceptibly cloudier than is the northern half of

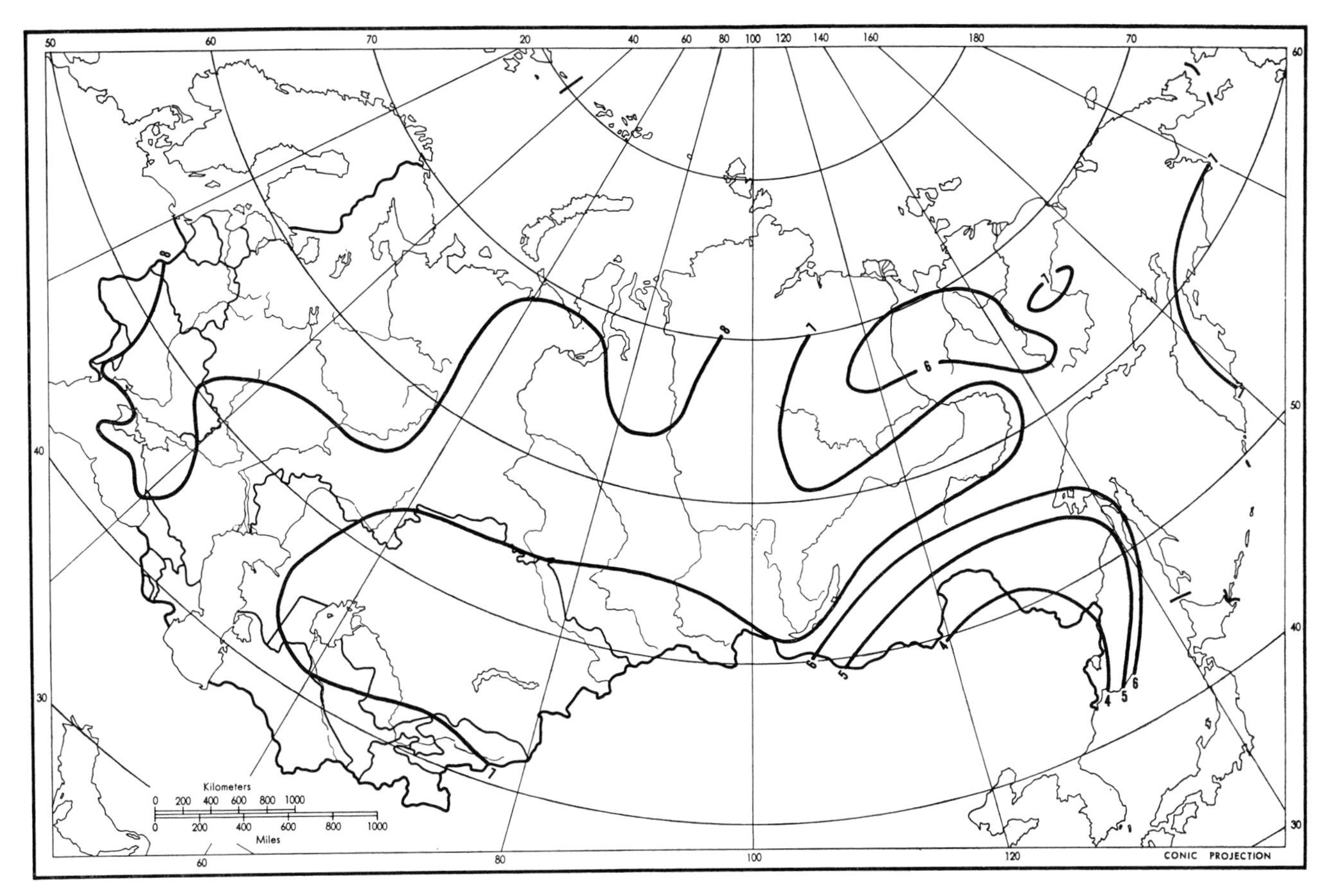

Fig.9-11. Mean January tenths of sky cover at 13h00. (Compiled from the tables in the Appendix.)

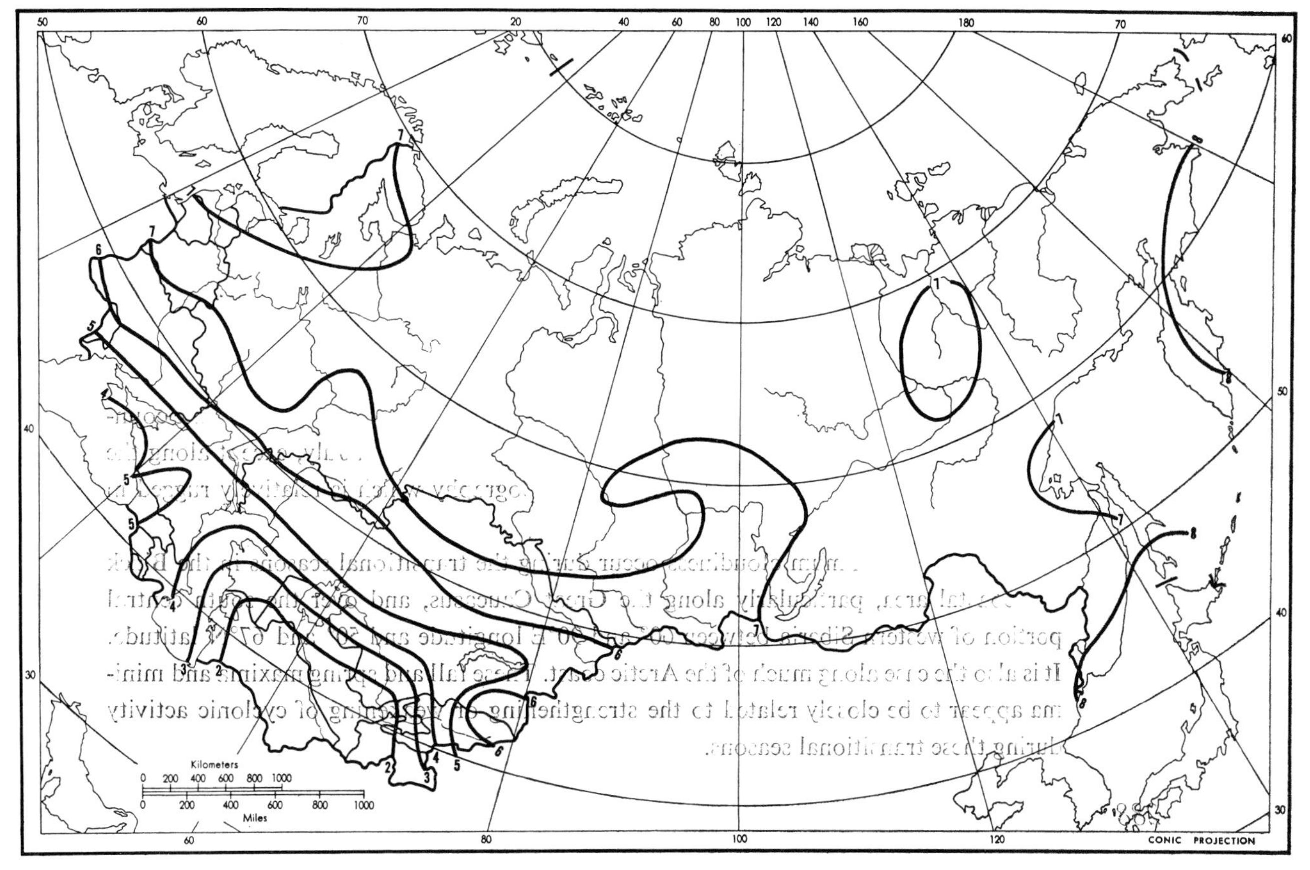

Fig.9-12. Mean July tenths of sky cover at 13h00. (Compiled from the tables in the Appendix.)

European U.S.S.R. and much of Siberia. In winter parts of the region have less than 4 tenths sky cover under the influence of the eastern edge of the main core of the Asiatic high. The area experiences particularly little overcast. Khabarovsk records only 18 days per year with more than 8 tenths sky coverage. (See the climatic tables in the Appendix.) There appear to be no days at all during January when the middle Amur Valley has more than 8 tenths sky coverage, and only 4 days in July. This seems to be rather unexpected in view of the summer monsoon in this area. Apparently the clouds associated with the monsoon and its precipitation in this region must usually constitute less than 8 tenths sky coverage. During July Khabarovsk on the average has more cloud cover than does Leningrad in the northwestern part of the country, but Leningrad has more than twice as many days with overcast skies as Khabarovsk does.

Over the course of a year a west–east corridor of greater cloudiness appears across the Aldan Plateau of southeastern Siberia between the clearer Amur Valley to the south and the relatively clearer northeastern part of Siberia and the Far East. This pattern is accentuated in winter when cyclonic storms frequent this corridor between the Mongolian and northeastern cores of the Asiatic high.

Throughout European U.S.S.R., western Siberia, and Central Asia cloudiness is significantly greater in winter than in summer. The contrast is greatest in Central Asia where the summers are dominated by hot dry air and the winters experience frequent incursions of cyclonic storms. In most of European U.S.S.R. winter has 2 to 3 times as many days with more than 8 tenths sky cover as summer has. The ratio diminishes eastward, but winter still has a maximum of overcast as far east as the Urals and beyond. Great variability of cloudiness is typical of this broad region. Both the winter maximum and the variability are related to the heavy cyclonic activity throughout much of European U.S.S.R. and the West Siberian Lowland. They are also related to the relatively high moisture flux that crosses this area from the Atlantic.

In eastern Siberia and the Far East the seasonality is just the reverse. Maximum cloudiness occurs in summer with the monsoonal influx of moisture from the Pacific and the Arctic and greater convective activity over the land mass. In winter most of eastern Siberia and the Far East have very low amounts of cloudiness due to the persistence of the Asiatic high. Generally in winter there are strong gradients along the eastern coasts as cloudiness increases rapidly to sea. This is not true along the Arctic coast of eastern Siberia which is frozen right up to the shore at this time of year. The most exposed fringes of the Arctic coast have very high amounts of cloudiness in summer when the cold but unfrozen water along the shore produces large amounts of fog and low cloudiness, and convective activity in the cool marine air as it penetrates the warmer mainland causes large amounts of cumulus clouds to form. The northeastern extremity of the country appears to have almost as much cloudiness in January as in July, except along the coast itself. Perhaps this is a function of the topography which is relatively rugged in this area.

Maximum and minimum cloudiness occur during the transitional seasons in the Black Sea coastal area, particularly along the Great Caucasus, and over the south central portion of western Siberia between 60° and 90°E longitude and 50° and 67°N latitude. It is also the case along much of the Arctic coast. These fall and spring maxima and minima appear to be closely related to the strengthening or weakening of cyclonic activity during these transitional seasons.

Higher mountain areas generally have maximum cloudiness in May or June and a minimum cloudiness in midwinter. During winter low stratus clouds do not reach the high mountain summits, such as the Caucasus or the mountains of Central Asia or the Altay, but in summer the heating of the land daily produces ascending air currents on the higher slopes which generally envelop the peaks in shrouds of clouds. This, of course, does not show up on such small scale maps as Figs.9-10, 9-11 and 9-12.

Over the country as a whole, except for the western Transcaucasus and portions of the Far East, gradients of sky cover are generally perpendicular to each other between winter and summer. This seasonal change in the direction of the cloudiness gradient is related to the seasonal change of the cause for much of the cloudiness over the country. During winter most of the cloudiness is of a stratiform type associated with the advection of moisture into the country, primarily from the west, and with the formation of cyclonic storms along portions of the Arctic and Polar fronts, again particularly in the west. During summer on the other hand, much of the cloudiness is due to convective activity initiated by the warming land mass as cool moist air is advected over the surface. The contrast between the temperatures of the land and the temperatures of the surface air is greatest in the north where air is moving onto the land across the Arctic coast from melt waters of the Arctic ice pack that are consistently around the freezing point. Generally, cloudiness in summer is not so great right along a flattish coast, such as the coastal area in the Baltic republics, but increases rapidly inland as the warming effect takes place. Along more rugged coast lines, such as at Cape Chelyuskin on the Arctic coast orographic uplift adds to the thermal effect and produces great cloudiness right along the coast.

The amounts and types of cloudiness are very dependent upon the relationship between the air mass that is dominating an area at a given time and the characteristics of the underlying surface (Table 10-I). For instance, European U.S.S.R. is affected by Arctic, continental temperate, maritime temperate, and tropical air masses. During the winter cloudiness is greatest with tropical air masses as they move northward across the plain and are cooled by the underlying surface. Condensation is reached easily because the

TABLE 9-I

FREQUENCIES OF OVERCAST (8–10 tenths) AND CLEAR (0–2 tenths) DURING DAYS WITH VARIOUS AIR MASSES IN EUROPEAN USSR (%)
(After BORISOV, 1967)

Air mass	Overcast	Clear
Winter		
Arctic	62	27
Continental Polar	80	15
Maritime Polar	93	4
Tropical	100	0
Summer		
Arctic	55	28
Continental Polar	51	28
Maritime Polar	30	46
Tropical	13	68

absolute humidity is high. Conversely, cloudiness is least in winter with Arctic air which has low absolute humidity. The situation is just the reverse in summer. Arctic air produces the most cloudiness because it is most affected by the heating of the land mass as it moves across it. The greatest instability at the surface is induced in the Arctic air and the greatest amount of cumuloform clouds are produced. Tropical air, on the other hand, during summer experiences little modification as it moves across the land mass, and therefore cloudiness is much reduced. During winter, then, when the land is cold, cloudiness depends primarily on the moisture of the air, but in summer cloudiness depends primarily upon the amount of convective activity induced in the air that is being advected over the heated land surface, which is a function of the temperature difference between the land surface and the air moving across it.

The diurnal march of cloudiness varies with area and season, and also with synoptic situation. In the European part of the country during the warm season there is a tendency for two maxima to occur, one during the daytime with the formation of cumulus as a result of convection, and the other at night with the formation of stratus clouds as a result of cooling under appropriate conditions. During the cold season, European U.S.S.R. generally experiences one maximum of cloudiness during a 24-h period, either at day or night. Throughout much of Asiatic U.S.S.R., summer usually experiences its greatest cloudiness during the daylight hours and winter during the evening hours. In mountainous regions there is usually a daytime maximum during the summer due to convective activity and a nighttime maximum during the winter due to condensation brought about by surface cooling processes.

Fog

If it is hazardous to draw isolines of cloudiness on a small-scale map, it is absolutely foolhardy to try to do so for fog distribution, for, as is well known, the formation of fog depends upon the most delicate balances among humidity, temperature, and wind speed and varies most abruptly over short horizontal and vertical distances. Therefore, stations next to each other on a small-scale map might show drastically different values, and if one attempts to draw isolines taking into account every value the resultant pattern will be almost a hodge-podge. Nevertheless, since it appears to be useful to show some simplified pattern, over the plain-portions of the country at least, a map has been compiled from the data available (Fig.9-13). This map, of course, does not represent the complex differences between summits and valleys or windward and leeward slopes in mountainous regions or drastic differences among separate segments of ragged coasts. It probably does not even represent the broad open plains of the country very well, since so many of the recording stations are in cities which are located in river valleys. Both the city itself and the topographical location might cause conditions to be quite different than they are on unurbanized interfluves. In early winter particularly observation stations along major streams may show high occurrences of fog because the water in the streams, which is still unfrozen, is so much warmer than the wintry blasts of air which are already descending upon the region. At this time of year steam fog rises from the rivers practically every day in the early morning hours. The exceptionally high occurrence of fog at Irkutsk at the southwest end of Lake Baykal is undoubtedly due to this phenomenon. Here the river remains unfrozen all winter since it has such a strong current issuing from

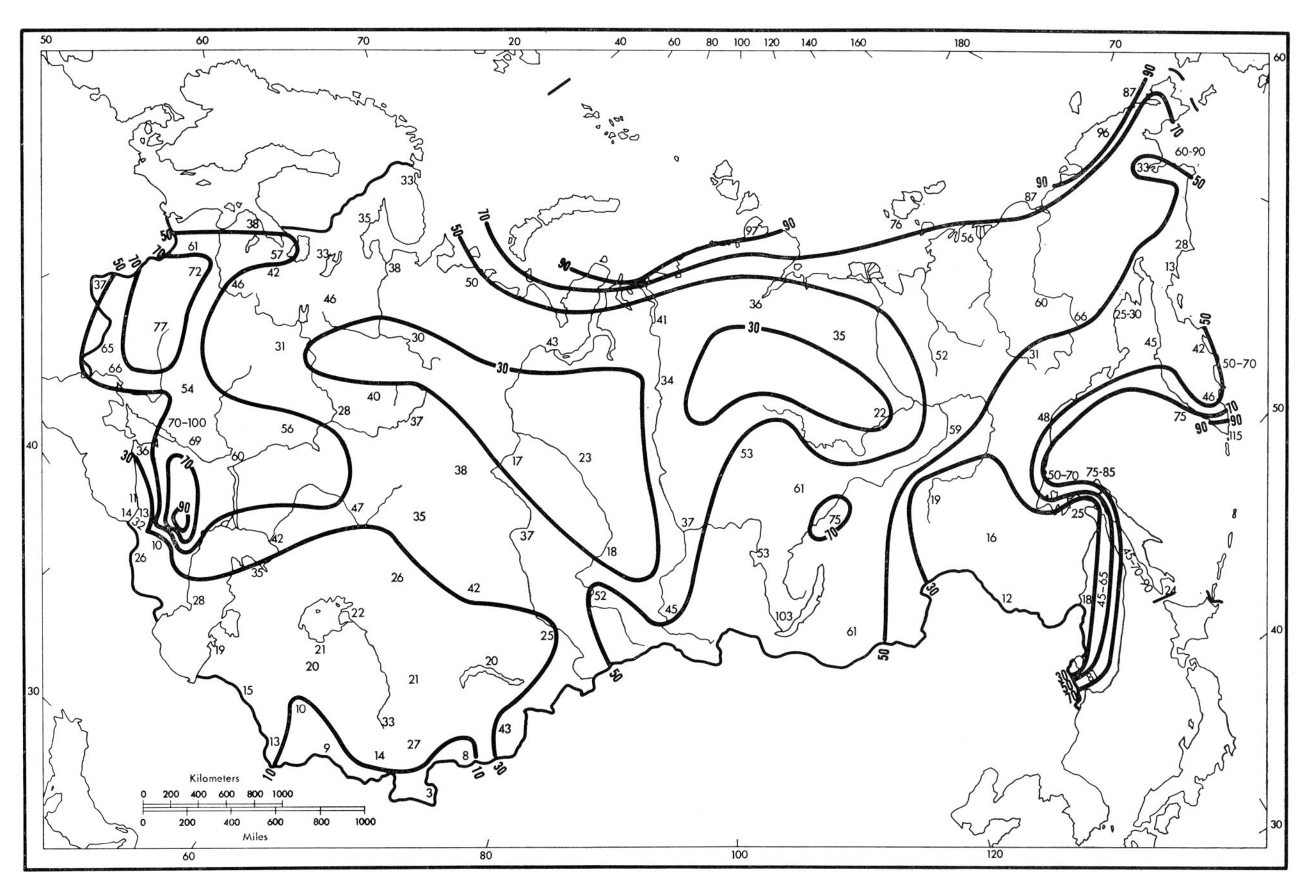

Fig.9-13. Mean annual number of days with fog. (Compiled from the tables in the Appendix, and information contained in ANONYMOUS, 1958–1963.)

Lake Baykal. It would be meaningless to draw isolines around this single point because the phenomenon is a point phenomenon. To a lesser degree the same is probably true of many high values along major streams, particularly in the eastern portion of the country where the winters are very cold.

Except for mountain areas, the most difficult region to depict is the Pacific coastal fringe, which not only varies greatly between sea and land because of generally ragged topography, but which also varies drastically from one segment of the shoreline to another along this deeply embayed coast with its many large peninsulas and islands which vary the exposure of a given point to moisture-laden air currents. In general along this coastline it appears that the deep embayments experience much less fog than do exposed headlands. And since most of the recording points are ports, which usually are situated in embayments in mouths of rivers, the observed data probably underestimate fog conditions in most cases. Therefore the observed data of this region have been augmented by estimates given in *Klimat SSSR*.

It appears that generally the greatest frequencies of fog are found along the immediate vicinities of the Arctic and Pacific coasts. Such conditions are to be expected, with the advection of relatively warm air over cold ocean waters. Frequencies in most of these areas are highest by far in late summer when the greatest cooling effect takes place from the sea. Most of this fog is advection fog, much of which forms over the sea and moves inland with the sea breezes when the land is being heated on long summer days. Generally it is dissipated fairly rapidly inland as the heating from the land increases the capacity of the air to hold moisture, and therefore exposed capes and islands show the greatest fog frequencies. Much of the Asiatic interior shows greatly reduced frequencies of fog, and in most places there are winter maxima associated with radiational heat losses at night through the calm clear air of the Asiatic high. Frequencies generally increase westward across the Eurasian plain as atmospheric moisture increases and radiation fog is joined by significant amounts of advection fog and prefrontal fog associated with cyclonic situations that are most frequent in the west.

To these three types of fog is added upslope fog in some portions of the plain where the surface is significantly bulged upward. This is particularly obvious over the broad gentle incline of the Stavropol' Plateau on the North Caucasian Foreland and over the Donets Ridge in the Eastern Ukraine, where in both instances frequencies as high as 100 days per year are observed. Fog also forms more frequently than in surrounding areas over the Smolensk–Moscow Ridge associated with air coming from the north, over the Volyn–Podolian Plateau with air coming from the south, and over the Central Russian Uplands with air coming from the east. From October to January in these regions fog forms on about ⅓ of all days. This type of fog, produced by adiabatic cooling of slowly rising air, is, of course, also prevalent in all mountain regions and accounts for very high fog frequencies on certain mountain peaks where the clouds produced by rising air lie at the surface. The higher peaks of the humid western part of the Great Caucasus experience some of the highest fog frequencies in the country. For instance, on Mount El'brus at a height of 4,250 m fog is observed 220 days per year on the average.

Karabi-Yayla in the Crimean Mountains experiences 150 days per year with fog, 22 of which are in December, and only 5 of which are in July. This annual regime corresponds to that of cyclonic occurrences in the area. The highest peaks in the northern and southern Urals have fog 80–160 days per year. The Urals get more fog on their western slopes than on their eastern slopes, which results from the general west-to-east movement of air over the region.

The other type of fog is steam fog which is produced by injection of moisture into already saturated cool air by a much warmer underlying water surface. This occurs along most streams during early winter before they freeze over, and, of course, continues later into the winter along the larger streams where ice forms latest. As has been pointed out before, since so many of the larger cities are located along major streams, many observation points represent these conditions only and not the interfluvial areas between streams. Steam fog is common also over some lakes and along coastal areas where water surfaces are frequently much warmer than the overlying air. This is particularly true along the western fringes of the Barents Sea where water temperatures in the extension of the Gulf Stream around northern Scandinavia contrast sharply with air temperatures during winter. Between 1945 and 1953 Murmansk experienced 439 days with steam fog during October–April. The maximum occurred in January, and the other winter months were arranged symmetrically around it.

Lowest fog frequencies are found in the southern part of Soviet Central Asia. This is to be expected in an area which generally has low relative humidities. Only a few days during late winter or early spring experience any fog at all in this region. Fog frequencies are generally highest in March, as are precipitation amounts. Almost equally free of fogs is the southern portion of the Soviet Far East, except for the Sikhota-Alin Mountains and their seaward slopes. Amur region generally experiences 10–20 days per year and the Khanka–Ussuri Lowland 15–30 days. There is a decided late summer maximum in this area, associated with influxes of very moist air at this time of year.

Fig.9-13, of course, does not tell anything about density of fog or duration of it during any one day. In most cases in European U.S.S.R., fog during the cold season lasts 1½ to 2 times as long as it does during the warm season. On the plains during the winter occurrences of fog generally last from 4½ to 7½ h, whereas in summer they last only from 2½ to 5½ h. In mountain regions fog generally persists longer, from 12 to 16 h during winter and from 10 to 12 h during summer. During the cold season the sum of all durations of fog on the plains of European U.S.S.R. amounts to about 150–300 h, which is 3–6% of all time. In cloud-shrouded mountains of the Urals it amounts to 1,500–2,400 h, or 30–50% of all time. During the warm season the plains total 30–100 h of fog, or 0.6–2% of the time, and the mountains total 650–1,400 h, or 15–30% of the time. Throughout much of European U.S.S.R. there is a fall and early winter maximum of fog and a June or July minimum. In the north the maximum falls generally between October and December and in the south during December or later.

On the plains of European U.S.S.R. during the warm period fog is most frequent in late night and early morning hours. More than 90% of all cases occur between midnight and noon. In mountains this is reduced to 60–65%, since much of the fog is upslope fog associated with rising air currents that reach their maximum development after noon. During the cold season fog is not so concentrated between midnight and noon but it still occurs then 60–70% of the time on the plains and 55–60% of the time in the mountains.

In the Transcaucasus the lowlands and plateaus generally experience maximum frequencies of fog in winter during the period October–March. Much of this fog is advection or upslope fog. In the higher mountains there is a well defined summer maximum due to cumulus build-up. The frequencies during all months are much higher in the mountains than they are in the lowlands. Generally, the lowlands and seacoasts at either end of the Caucasus have quite low frequencies of fog, about 10–35 days per year, whereas many of the mountain peaks and high passes experience 150–200 days per year.

In western Siberia, except in the extreme north and south, there are generally two maxima of fog frequencies per year, at the end of summer in August and September and in the second half of winter in February–March. In the southern part there is generally a winter maximum from December through March associated with high frequencies of anticyclonic weather. In the forests and eastern parts of the forest steppe zone fog occurs as frequently in August and September as it does in winter. On the shores of large rivers and lakes there is a summer–fall maximum. Along the shore of the Kara Sea in the north there is a broad summer maximum from June to August, with July having most frequent fogs.

In western Siberia during winter in 40–50% of the cases fog lasts 5–7 h at a time, whereas in summer in 50–70% of the cases it does not last more than 3 h. In the north summer fog occurs 60% of the time during night or morning hours, whereas in winter it occurs during the day and early evening hours. In more continental regions during winter fog occurs most frequently between 06h00 and 18h00, and in summer during late night and early morning hours.

The Altay Mountains do not seem to have very high fog frequencies even in the higher portions; 30–40 days a year is about the average. The yearly regime is much the same as in the steppes to the west, with winter and late summer–fall maxima. In the valleys of the Altay which are open to the west fog occurs about 30 days per year, but in the central Altay and on the eastern slopes valleys receive only 15–20 days per year.

Winter fog in eastern Siberia often occurs with temperatures between $-25°$ and $-45°C$, and relative humidities between 70 and 80%. Fog generally occurs with higher temperatures and higher humidities in the west than in the east. At Yeniseysk the greatest number of winter fogs occur with temperatures between $-40°$ and $-45°C$ and relative humidities between 70 and 85%. At Turukhansk fog has occurred with relative humidities as low as 65%. From May to October, fog occurs with relative humidities between 95 and 100%. The fog occurrence with such low relative humidities in winter must be due to urban influences, combustion and so forth, which put water vapor and hygroscopic particulants into the air. It may also have to do with the fact that much of the fog contains ice crystals which serve as nuclei for further condensation. Ice crystal fog is very common in central Yakutia. Some of the river valleys experience 50 to 65 days of such fog during the cold season. This kind of fog is not very dense and is very shallow. It often hangs over living quarters of settlements.

For some reason fog frequencies around the shores of Lake Baykal are rather low. In the southern part there are no more than 7–15 days per year and in the northern part 20–25. Perhaps there is too much wind much of the time right along the steep slopes of the lake shore. No explanation has been offered in Soviet literature. One would think that there would be considerable steam fog in early winter. This is certainly true in the Angara River where Irkutsk experiences 103 days per year. It is also true in the upper Lena and Vitim rivers which have 80–85 days of fog per year. There have been reports of pronounced amounts of steam rising from Lake Baykal in early morning hours during calm weather in November and December.

The Irkutsk fogs from the Angara River occur particularly in December and January. They may last from a few hours to all day and generally reach thicknesses of 100–300 m. The Angara at this point usually does not freeze until the middle of January and sometimes not until the end of January. In December the water temperature is generally 1° or 2°C while air temperatures above it are much lower. Fogs in Irkutsk cause a great handicap to aviation and other types of transportation. During the winter of 1955–56 the Main Geophysical Observatory conducted investigations on the Angara fogs. December experienced 210 h of fog and January 170 h. Generally during fog winds were weak, 25% of the time calm. Average wind speeds were 1–2 m/sec and maxima were 5–6 m/sec. Wind directions were prevailingly from the southeast and east and rarely from the north and northwest.

The Arctic coast generally has little fog during the winter time, about 20 days along the east Siberian Sea and only 10 days along the Laptev Sea. Cape Chelyuskin, which has 97 days of fog per year, has only 1 to 2 days in December. It has a very pronounced late summer maximum with about 22 days in August. But this regime changes rapidly inland, and at Dudinka at the head of the Yenisey estuary fog occurrence is almost the same every month of the year and there are only 41 days per year with fog. Farther toward the interior of the continent a winter maximum becomes pronounced.

In the Far East, Cape Lopatka at the southern tip of Kamchatka Peninsula has the greatest recorded fog frequency, 115 days per year. Surprisingly low frequencies occur in some of the embayments of the coast. In general frequencies seem to decrease from north to south across Magadan Oblast so that the northern coast of the Sea of Okhotsk has lower frequencies than anywhere on the mainland to the north. But apparently farther out to sea they increase again, particularly on exposures along Kamchatka Peninsula and Sakhalin Island. In the Tatar Strait frequencies run only 15 to 25 days per year but they increase rather rapidly on either side.

Along most of the shore zone of the Far East fog occurs with marine air blowing inland from the south and southeast, sometimes from the east or southwest. Along the coast there is universally a pronounced summer maximum of fog associated with the maritime monsoon at this time of the year. In almost all cases July is the month of maximum fog and June is second. In a few cases June is first. There seems to be a fairly rapid decrease after July, which seems odd in view of the fact that the monsoon flow inland strengthens in late summer. Perhaps the ocean waters are not so cold later in summer so the air being advected over them does not cool as much as it does earlier. The fog does not decrease so rapidly in late summer farther to sea. Across the Sea of Okhotsk the Kuril Islands experience fog 20–26 days per month during June, July, and August.

Along the shore fog is generally a night and early-morning phenomenon and it usually disappears after the sun comes up. In the open seas at the same time the sky may be clear and without fog. Along the more northerly parts of the coast overcast sky at night might represent high fog which descends to the water surface by morning. Much of the Far East experiences fog most frequently during night and early morning hours.

Inland in many places the seasonality changes to a pronounced winter maximum. Much of this winter fog is of a radiation type, augmented by steam fog along streams. Along the Amur River both the coastal summer and inland winter regimes are reflected, although total values are so low in this region that there is little variation through the year. Nevertheless, Blagoveshchensk shows a primary maximum in July and August, which would be associated with the monsoon, and a secondary maximum in November and December, which one can assume is primarily due to local radiation. Winter fog is rather rare in the southern portions of the shore zone, and some winters it does not appear at all.

The desert plains of Central Asia have only winter fog, except along the shores of some of the larger bodies of water. There is a summer maximum around Kara-Bogaz-Gol and Cheleken Peninsula on the east side of the Caspian. The occasional summer fogs in Kara-Bogaz-Gol have been explained by the lowering of temperature across the strongly evaporating surface of the gulf caused by relatively cool northwest winds coming in from the Caspian.

In the high mountains frequency of fog rises rapidly. In the desert north of Alma-Ata 15 days per year are experienced. Nearby at about 1,500 m elevation on the northern slopes of the Zaili Alatau fog frequency rises to 40–50 days per year, and at an elevation of 2,500 m in the Kirgiz Range to the south the frequency rises to 70–75 days. The highest frequency of fog occurs in the high passes of the western parts of the mountain ranges. Shakhristanskom Pass, for instance, at an elevation of 3,200 m receives 91 days of fog per year, and Anzob Pass more than 100 days. In contrast, fog hardly ever occurs in the narrow high mountain valleys in the eastern parts of the ranges, such as the middle and upper courses of the Zeravshan River, the Alay Valley, the western part of the Issyk-Kul' Basin, and also in the very dry mountain regions of the eastern Pamirs. In most places on the desert plains the fog maximum occurs in March, as does the precipitation maximum.

Precipitation

Precipitation over most of the Soviet Union is only light to modest. This again reflects the high latitude of the country and attendant cool temperatures of the air with low capacity to hold moisture and the high continentality of much of the area which precludes the intrusion of maritime tropical air masses. The bulk of the country receives between 400 and 800 mm of water equivalent per year (Fig.9-14). Most of the mountains of the country and some small lowlands in the Transcaucasus receive more than this. And large areas, including all of Central Asia and much of Kazakhstan, as well as the eastern plains in the Caucasus region and some of the northeastern extremities of the country, receive well under 400 mm per year. Sections of Central Asia receive less than 100 mm per year. The highest rainfall total at a lowland recording station occurs at Batumi in the southern foothills of the Colchis Lowland on the eastern shore of the Black Sea where the yearly average is 2,504 mm. The greatest amount recorded anywhere is at Achishko, a mountain station in the western Great Caucasus, which at an elevation of 1,880 m receives an average annual precipitation of 2,617 mm. The absolute minimum in the country occurs in scattered areas of the Kara-Kum desert in the southern part of Central Asia and in the basin of Lake Kara-Kul near Bukhara, also in Central Asia, where approximately 30 mm per year falls.

Across the plain of European U.S.S.R. the greatest decrease in precipitation takes place in

Fig.9-14. Mean annual precipitation (mm). (From ANONYMOUS, 1967. Inset from GERASIMOV, 1964.)

a southeasterly direction from the Baltic Sea to the Caspian. There is also a general decrease from west to east across European U.S.S.R. and Siberia as far east as Yakutsk in the Lena Valley. East of there the precipitation varies greatly according to topographic situation and generally increases significantly as the Pacific coast is approached. The Pacific slopes of Maritime Kray, Sakhalin Island, and Kamchatka Peninsula are very wet, particularly during certain portions of the year, as are the Kuril Islands which have a completely marine controlled climate.

The Ural Mountains, though they rarely rise above 1,500–2,000 m, present an elongated obstruction oriented almost perpendicular to much of the air flow across the Eurasian plain and increase precipitation as much as 200–400 mm annually above that which is received on the surrounding plain. The southwestern slopes of the mountains generally are considerably wetter than are the northeastern slopes, which correspond to the southwest–northeast movement of most of the cyclonic storms that pass through the area. The Putoran Mountains which rise to heights of about 2,000 m east of the Yenisey River in north central Siberia have a similar effect. In contrast, the Khibiny Massif in the central part of the Kola Peninsula which rises to elevations of about 1,000 m has little effect on precipitation totals. The air flow usually circumvents the massif and therefore orographic effects are minimized. In the southern part of the European plain the Carpathian Mountains in places receive more than 1,500 mm of precipitation annually and the Crimean Mountains receive as much as 1,100 mm.

Precipitation varies drastically according to elevation and general exposure to moisture-bearing winds in all the high mountains along the southern fringes of the country. The maximum at Achishko in the western part of the Caucasus has already been mentioned. The eastern part of the Great Caucasus generally exhibits quite dry aspects in spite of their continuing high elevations. In Central Asia the Tyan Shans and Pamir-Alay ranges receive over 1,000 mm per year on some of the westward facing slopes at intermediate altitudes. But in the high eastern portions of the Pamirs, summits generally are above most storms and as little as 75 mm per year may fall in some portions of eastern Tadzhikistan, where the climate has a very dry aspect in spite of the fact that temperatures are always quite cold and evaporation rates are low. Much of the precipitation which falls in the Central Asian mountains falls in the winter in the form of snow, which is very important to irrigation usage on the plains to the northwest. If the winter precipitation fell as rain it would run off immediately and probably cause more harm than good, but as it is, the snow remains in the mountains through the winter and the greatest meltwater run-off occurs in midsummer when it is most needed by irrigated crops.

Farther east, the Altay Mountains have a very complex distribution of precipitation which varies from 120 to 2,000 mm. Generally the southwestern slopes of the mountains and southwestward-facing valleys catch the moisture-bearing winds while the northeastern exposures are much drier.

As can be seen in Figs.9-15 and 9-16, most of the Soviet Union experiences great seasonal contrasts in precipitation amounts. Although the warm season has been defined as a 7-month period and the cold season as only a 5-month period, it is quite obvious that over most of the country the warm season has a much greater amount of precipitation than does the cold season. Such a regime reflects the seasonal reversal of pressure pattern across the country, with the Asiatic high in winter and a shallow thermal low in summer. It also reflects the contrast in moisture-holding capacity of the air during the two seasons.

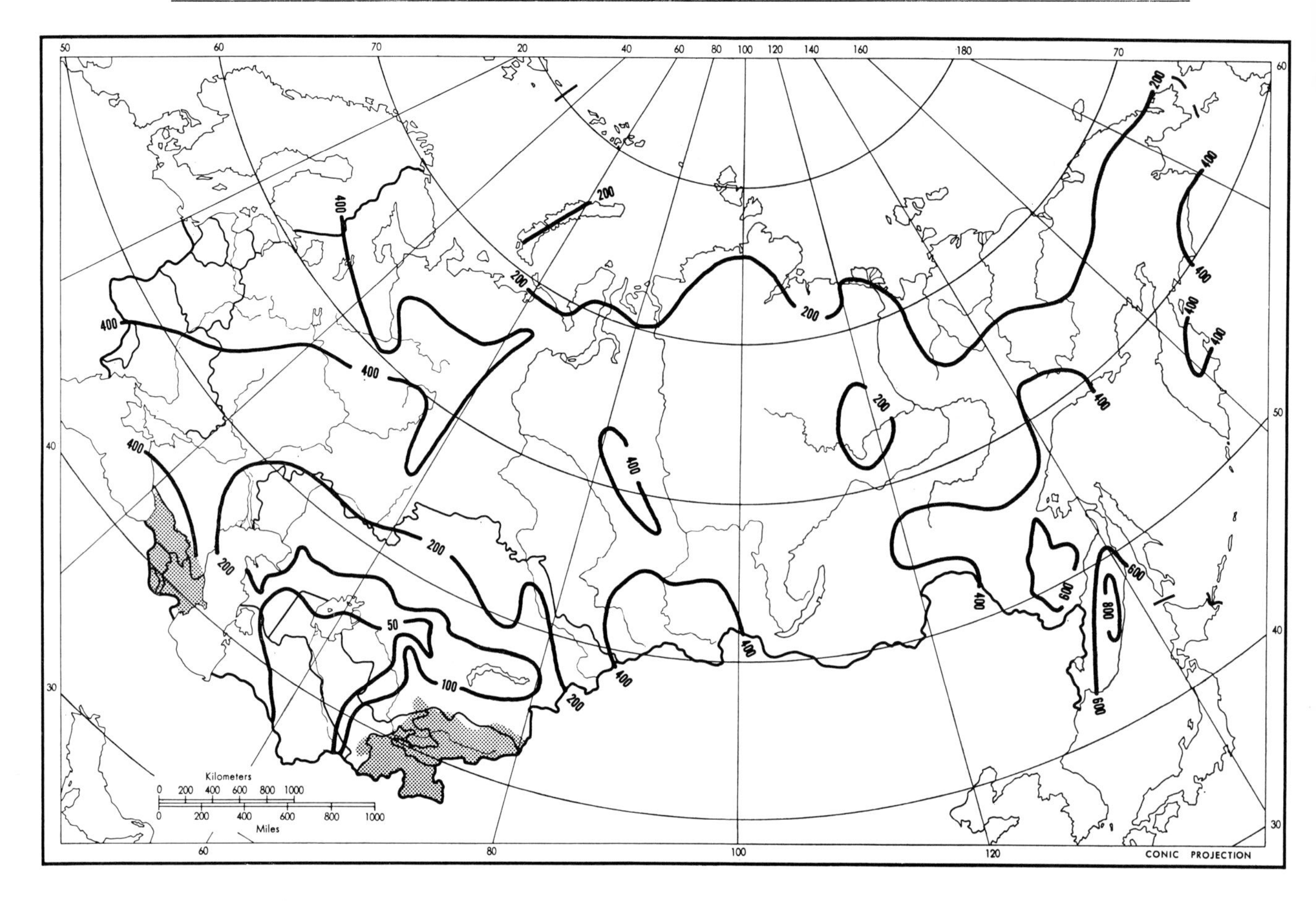

Fig.9-15. Mean warm season precipitation (mm), April–October. (After GERASIMOV, 1964.)

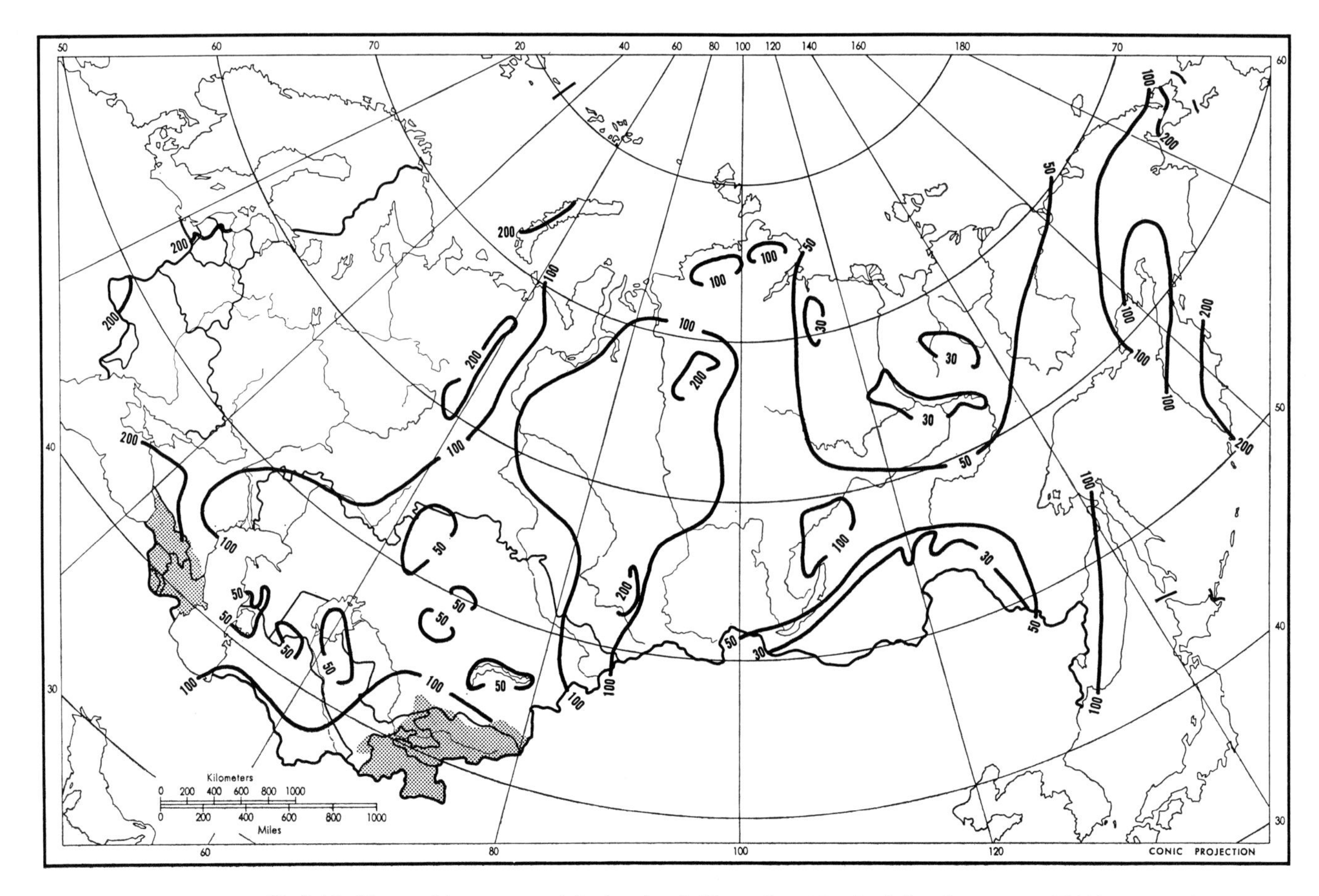

Fig.9-16. Mean cold season precipitation (mm), November–March. (After GERASIMOV, 1964.)

Only some peripheral areas, notably portions of the Caucasus and all of Central Asia, receive more precipitation during the cold season. But this is an oversimplification of relatively complex precipitation regimes in many parts of the country. Monthly maps of precipitation presented in the *Fiziko-Geograficheskiy Atlas Mira* show significantly different distributions from month to month. In some mountain regions and coastal areas regimes change over relatively short distances.

Throughout the broad central section of the plain extending across European U.S.S.R. and northeastward through Siberia and the interior portions of the Soviet Far East precipitation maxima occur rather regularly in July and there is a fairly symmetrical arrangement of precipitation around midsummer, although generally precipitation amounts are skewed a little toward autumn rather than spring (Fig.9-17). Minima in this region generally fall late in winter. The regime throughout this broad region is closely tied to annual regimes of atmospheric moisture content and convective activity, summer thundershowers generally accounting for more precipitation than winter cyclonic storms. Except for the Pacific fringe, as one proceeds northward from the central zone precipitation maxima generally lag into August and as one proceeds southward they advance into June. This shift in precipitation maxima from June in the south to July in the center to August in the north is related to the northward progression of convective activity as the heating of the land surface sweeps northward during the summer.

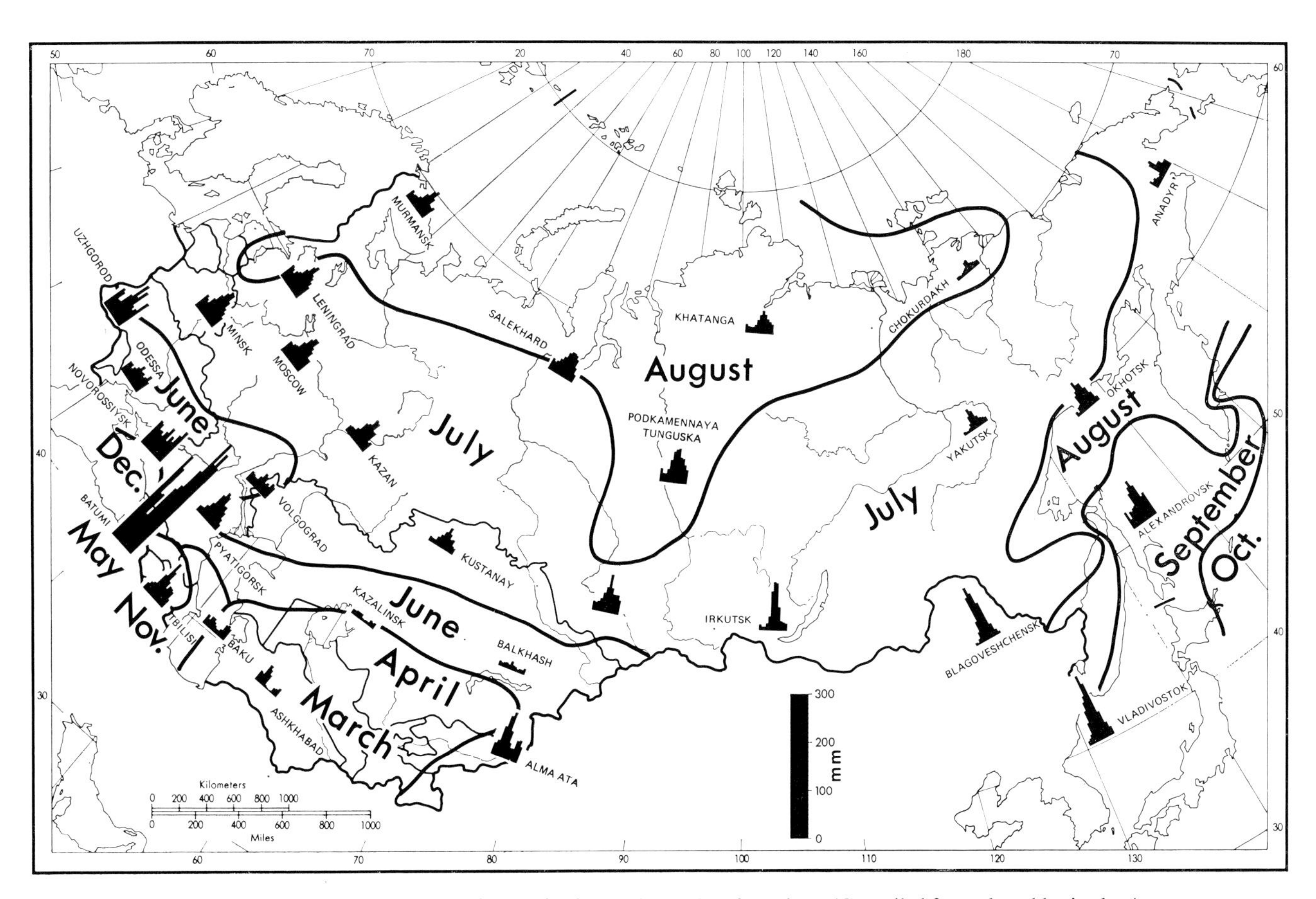

Fig. 9-17. Annual regimes of precipitation and months of maxima. (Compiled from the tables in the Appendix, and in ANONYMOUS, 1958–1963.)

The August maximum dips far southward in western Siberia and in the northwestern part of European U.S.S.R., which is a great hazard to harvesting in these areas which have rather short periods for harvest anyway because of temperature conditions.

The farther one proceeds in either direction from the center in the European plain the more complicated becomes the precipitation regime as additional precipitation controls alter the simple one-maximum–one-minimum regime that prevails over most of the plain. On the Arctic fringe Arkhangel'sk experiences a slight dip in precipitation during August with higher values on either side, the primary maximum falling in September. A broad minimum stretches from February through April. The autumn maximum is probably due to the maximum temperatures of the sea at this time of year and the oncoming intensification of winter cyclonic storms along the northern coast, and the late winter spring minimum coincides with minimum sea surface temperatures as well as cold temperatures over the snow-covered land to the south. In the south, Odessa on the northern Black Sea coast shows a primary maximum in June and almost as high a maximum in November with inconsistent variations in between. The June maximum relates to the prevailing conditions over the continent just to the north which is receiving its greatest convective activity at this time of year, while the November maximum relates to the oncoming of winter cyclonic storms following the southern track from the Mediterranean through the Black Sea into the Caucasus and Central Asia.

In the continental portions of the Soviet Far East, particularly in the south, the summer maximum is more pronounced than anywhere else in the country. The second half of summer receives considerably more than the first half and is related to the monsoonal influx of Pacific air at this time. As one proceeds to sea, the maximum lags later and later into autumn and the regime becomes more complicated. Much of the coast of the mainland facing the Sea of Japan and all of Sakhalin Island except the northern tip experience October and November maxima, as do the Komandorskiy Islands farther north and Petropavlovsk-Kamchatskiy on the southeastern coast of Kamchatka Peninsula. The regime at Petropavlovsk-Kamchatskiy is not a simple one; March receives almost as much precipitation as November does. The primary minimum falls in June, and there is a very pronounced secondary minimum in February immediately preceding the March secondary maximum. March receives almost twice as much precipitation as February does.

The seaward fringe of the Soviet Far East from Kamchatka southward receives significant amounts of precipitation in autumn from typhoons which at this time of year are following a route northeastward from Japan. They do not come close enough to the Soviet Union to produce wind damage, but they do cause perturbations in the airflow over the mainland coast which bring about prolonged precipitation over wide areas.

Throughout the southern plains of Central Asia March is the month of maximum precipitation. This coincides with the time of greatest cyclonic activity. In most places on the plains in Central Asia there is little or no precipitation from June through September. Only an occasional weak shower may provide some measurable precipitation. Both summer and winter precipitation increase from the plains into the foothill areas. In Alma-Ata at an elevation 848 m the driest month is August with 23 mm, and the wettest month is April with 99. The yearly total here is 581 mm. In the mountains these values increase still more. In the higher basins of the eastern Pamirs and Tyan Shans summer thundershowers become more important than winter cyclonic storms and the maximum shifts to May or June.

Proceeding northward in the Central Asian plains the spring maximum generally comes later, and a tendency develops for double maxima in spring and fall which is reminiscent of portions of the Caucasus to the west. In an east–west zone which includes the Aral Sea maxima generally fall in April, but around the Aral Sea there is also a secondary maximum in October. North of the Aral Sea and Lake Balkhash in a broad zone through central Kazakhstan the maximum usually falls in May or June, and this grades into June or July in northern Kazakhstan. It is evident that central Kazakhstan is the transition area between the predominance of the late winter and early spring cyclonic storms in Central Asia and the predominance of early and midsummer thunderstorms in northern Kazakhstan and much of western Siberia.

The Caucasian region exhibits the greatest complexities in precipitation of any part of the Soviet Union. Annual totals in low elevations vary from 2,504 mm at Batumi to 238 mm at Baku. In the higher mountains there are great variations related to altitude and exposure. One of the most characteristic features of the rainfall regime of the area is the frequency and amount of precipitation in winter associated with cyclonic storms. However, in many places the heaviest precipitation occurs in thunderstorms usually during the summer time, and many parts of the Transcaucasus have double maxima and minima which may fall during equinoxial periods as well as during solstices. A detailed analysis of the distribution of precipitation, in both space and time, has been included in the chapter on the Caucasus.

Frequency, intensity, and duration

Amounts of precipitation do not particularly reflect frequencies or total durations of precipitation. Generally overcast skies and low-intensity prolonged precipitation are often associated with cool water sources and high cyclonic frequency which are not particularly conducive to heavy precipitation. In the Soviet Union the most frequent precipitation of all intensities occurs in an elongated area that extends inland from the Barents Sea southeastward across the northeastern part of European U.S.S.R. and across the northern Urals into the mid-section of the West Siberian Lowland and the middle Yenisey Valley (Fig.9-18). This is somewhat coextensive with the number of overcast days which are great in northeastern European U.S.S.R. but in the west Siberian area the frequency of precipitation is not as closely linked with total cloudiness. Here it is probably related more to annual cyclonic frequencies and to convective activity during the summer.

Some precipitation falls on almost half of all days throughout a large portion of European U.S.S.R. and western Siberia, except in the south. Relatively high frequencies are also found in coastal areas of the Soviet Far East and the western portion of the Transcaucasus. The lowest frequency of all is found in the desert plains of Central Asia.

The distribution of frequencies of heavier precipitation is more similar to the distribution of the annual totals of precipitation in the Soviet Union (Fig.9-19). Here the outstanding feature is the wedge of generally high frequencies that engulfs the width of the plain along the western borders of the country and narrows eastward into Siberia. Frequencies pick up again in the mountains surrounding Lake Baykal, and spots of very high frequency occur in other mountainous areas around the country, particularly in the western portion of the Great Caucasus where the greatest annual totals of precipitation are found.

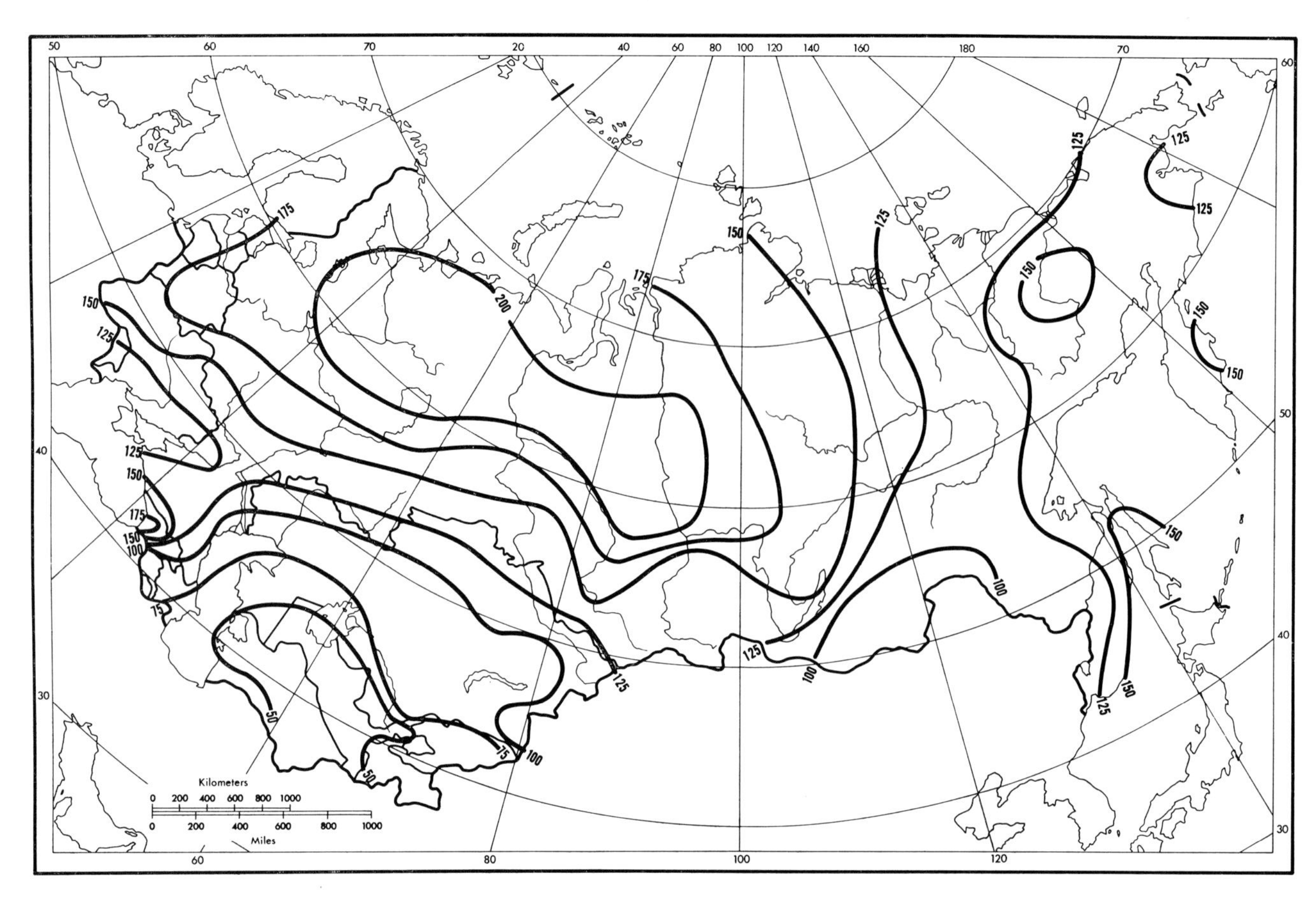

Fig.9-18. Mean annual number of days with precipitation ≥0.1 mm. (Compiled from the tables in the Appendix.)

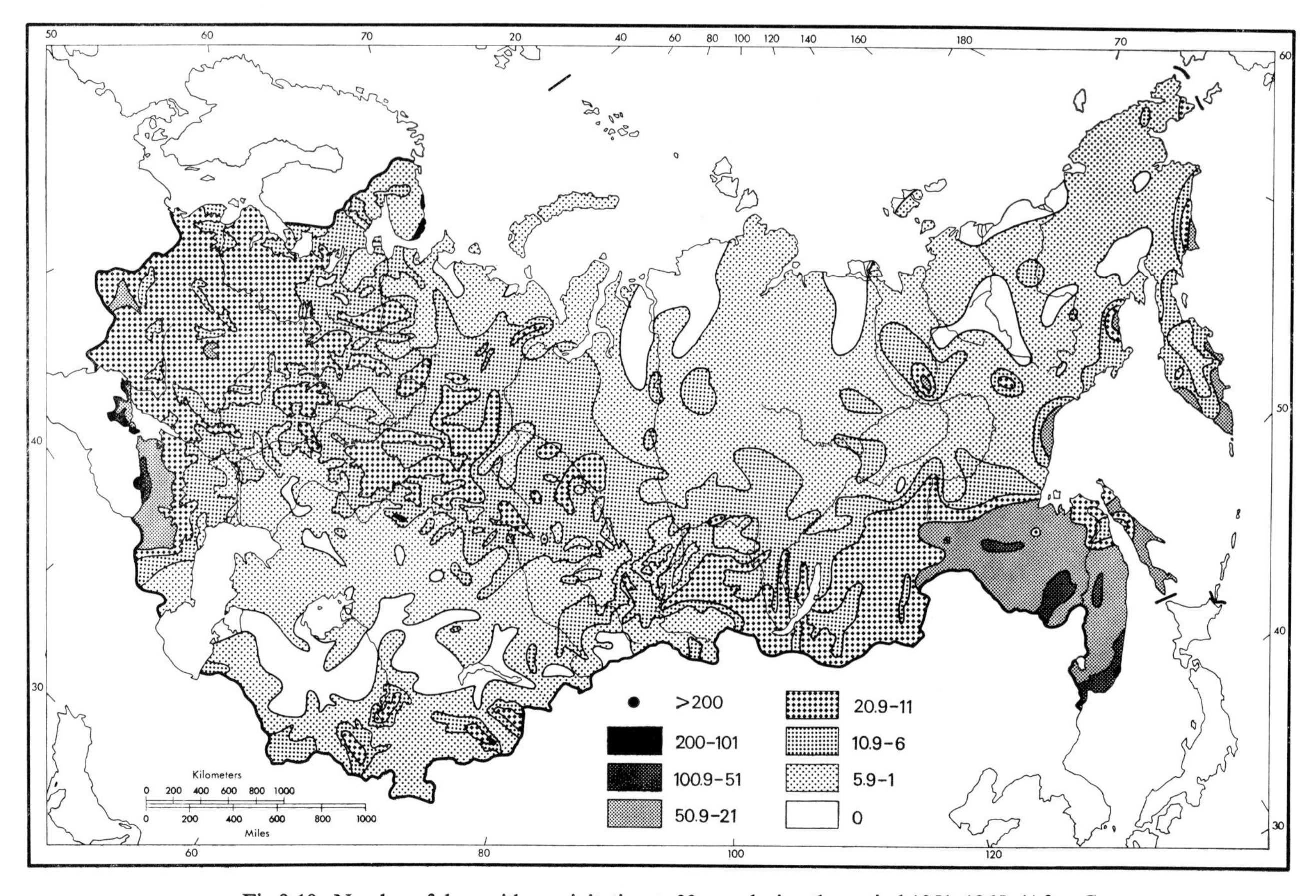

Fig.9-19. Number of days with precipitation ≥30 mm during the period 1951–1965. (After CHISTYAKOV et al., 1968.)

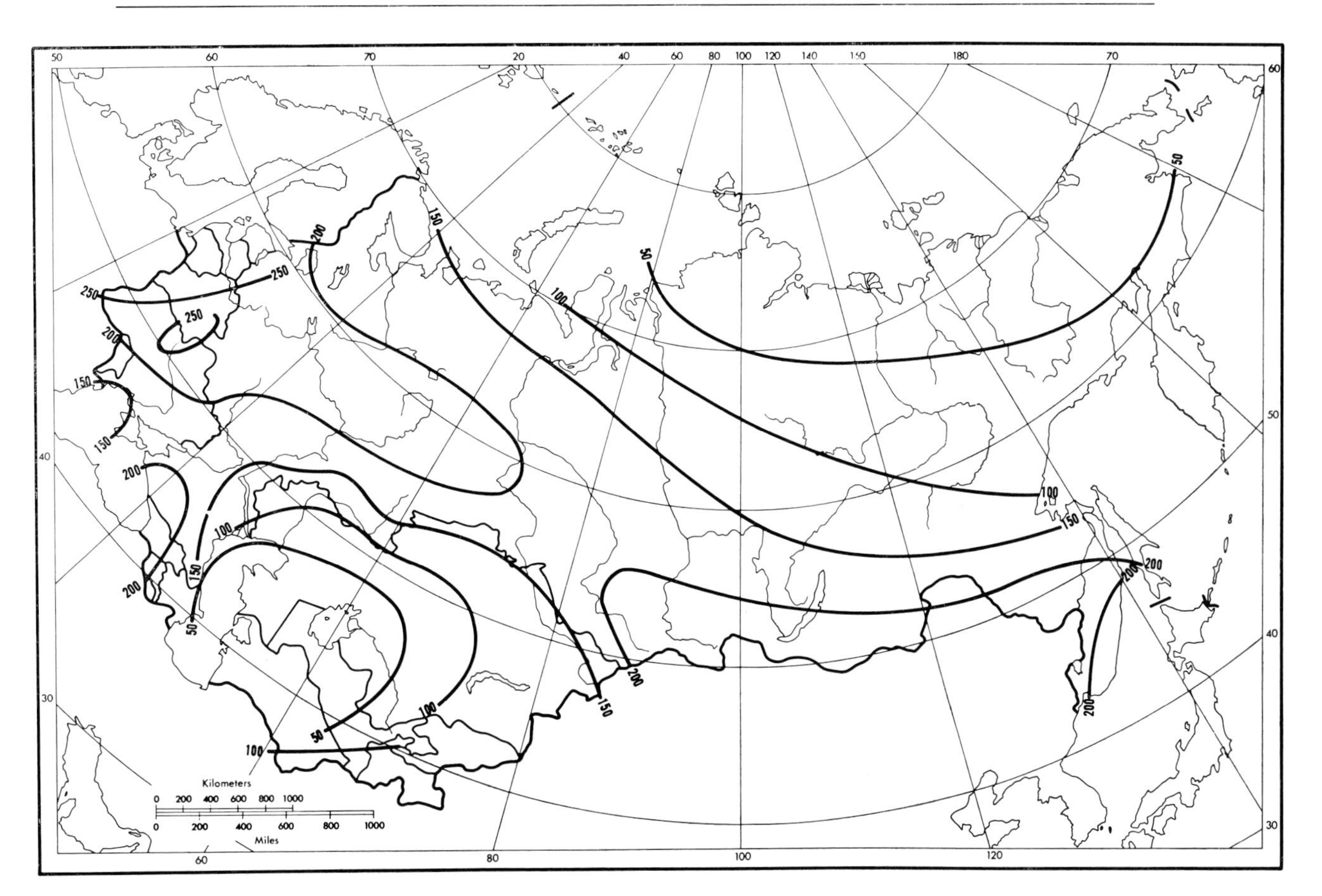

Fig.9-20. Mean annual precipitation (mm) falling in heavy showers. (After EVSEEV, 1958.)

There is little correspondence between the pattern of distribution of downpours and the pattern of distribution of prolonged precipitation. Downpours account for the greatest amounts of precipitation in the northwestern part of European U.S.S.R. and for the least amounts in Central Asia and along much of the Arctic coast of the country. Rather light amounts of this type of precipitation fall throughout much of eastern Siberia and the Soviet Far East (Fig.9-20). Prolonged precipitation accounts for the greatest amounts of accumulation in much of the Soviet Far East (Fig.9-21). Relatively high values also form a broad zone from west to east across the northcentral portion of European U.S.S.R. and western Siberia, where it swings southward into the mountains of southern Siberia. However, in this zone prolonged precipitation usually does not account for as great a portion of the total precipitation as does showery type precipitation.

Prolonged forms of precipitation generally account for about three times as much as showery precipitation throughout the Arctic fringe of Siberia and the Soviet Far East and about twice as much as showery precipitation in Soviet Central Asia. It is quite evident that most of the sparse precipitation that does come to Central Asia comes during winter as a result of cyclonic storm passages and is of a continuous, prolonged type associated with overcast skies. Farther north in the forest zone where summer temperatures are not so far above the dew points convective showers account for as much as 60% of the annual precipitation.

Except for the extreme north and south, European U.S.S.R. generally shows annual

Fig.9-21. Mean annual precipitation (mm) falling over a continuous prolonged period. (After EVSEEV, 1958.)

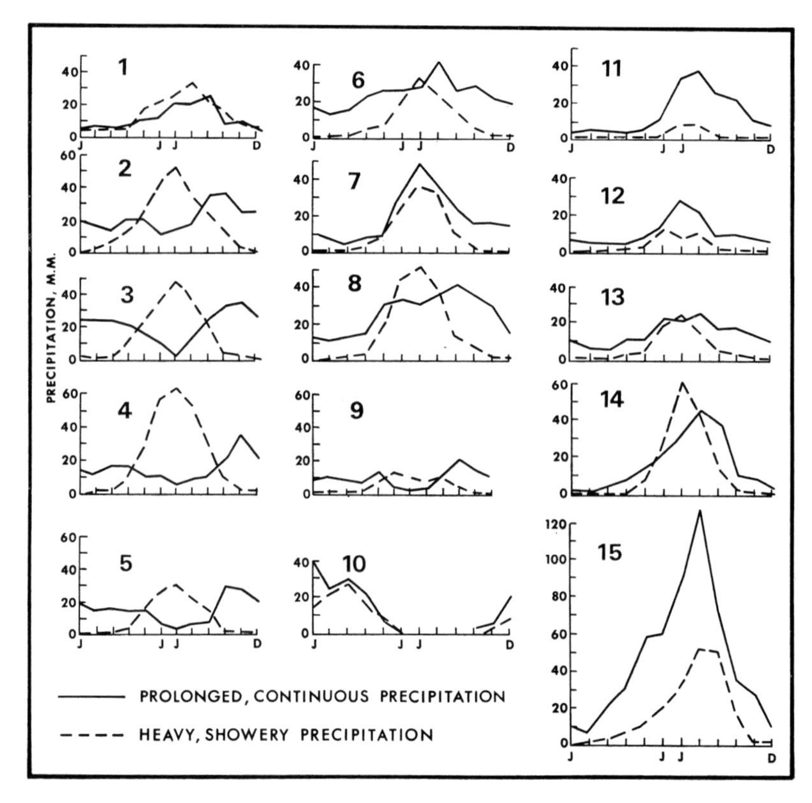

Fig.9-22. Yearly regimes of prolonged and showery precipitation at selected stations. *1* = Murmansk; *2* = Syktyvkar; *3* = Tambov; *4* = Orel; *5* = Kamyshin; *6* = Salekhard; *7* = Bratsk; *8* = Tomsk; *9* = Astrakhan; *10* = Kushka; *11* = Anadyr'; *12* = Verkhoyansk; *13* = Vilyuysk; *14* = Chita; *15* = Vladivostok. (After EVSEEV, 1958.)

regimes of showery and prolonged precipitation that are almost the opposites of one another (Fig.9-22). For example, Tambov and Orel show very definite peaked maxima of showery precipitation in July and less pronounced maxima of prolonged precipitation in autumn. Minima for showery precipitation fall in midwinter, and minima for prolonged precipitation in early- or midsummer. Farther east there is a tendency for the two regimes to become more in phase with one another as cyclonic activity and associated overcast skies appear during approximately the same seasons as does maximum convective activity. Western Siberia still retains a tendency for a midsummer maximum of showery precipitation and an autumn maximum of prolonged precipitation, but eastern Siberia shows middle- or late-summer maxima for both, and Vladivostok in the Far East shows a very pronounced August-September maximum of prolonged precipitation associated with the late summer–autumn monsoon when showery activity is at its height also and is also associated with the monsoon flow of air. The tendency throughout much of the European part of the country and western Siberia to have autumn maxima of the prolonged type of precipitation presents a great hazard to agricultural harvesting in these regions which is hampered by relatively high amounts of precipitation at this time of year. The prolonged form of the precipitation compounds the problem.

During the warm season of the year, April–October, there is an approximately zonal distribution of total duration of precipitation, with high values in the north and low values in the south (Fig.9-23). This reflects the generally high amounts of cloudiness in the northern parts of the country at this time of year which are due to three major factors. During summer the Arctic coast, particularly east of the Urals, is a fairly active route of cyclonic storms. In addition temperature differences between land and sea generally induce intrusions of marine air into the continent at this time of the year which carry considerable moisture and in fact may advect large areas of stratoform clouds onto the continent from their origins over the sea. This inflowing air experiences rapid warming at the surface, and a great deal of cumulus clouds builds up as it moves inland.

In the northern part of the Soviet Far East similar conditions exist with the influx of Pacific air into the continent. This is true to a certain extent over the southern part of the Soviet Far East as well, although here cloud systems and precipitation are more associated with cyclonic storms and convective activity along the Mongolian section of the Polar front. This imparts an interrupted character to the cloud cover which is in considerable contrast to the rather continuous overcast misty conditions along coastal areas in the north.

The total duration of precipitation is greatest during the winter months throughout most of the country except for the southern part of eastern Siberia and most of the Soviet Far East whose regimes are so intimately connected with the monsoonal inflow of moist air from the Pacific during summer (Fig.9-24). Extremely long durations of precipitation occur during the winter on the average in the middle Yenisey Valley and occasionally on the windward slopes of the Urals, as well as in some of the mountainous regions of the northeastern part of the country. None of these areas have particularly large amounts of precipitation during the winter, but are characterized by prolonged overcast skies associated with light precipitation.

Maximum amounts of precipitation that have been observed to fall during any one 24-h period are scattered widely across the country and depend to a great extent upon local conditions, but generally the very highest amounts are found in the south where as much

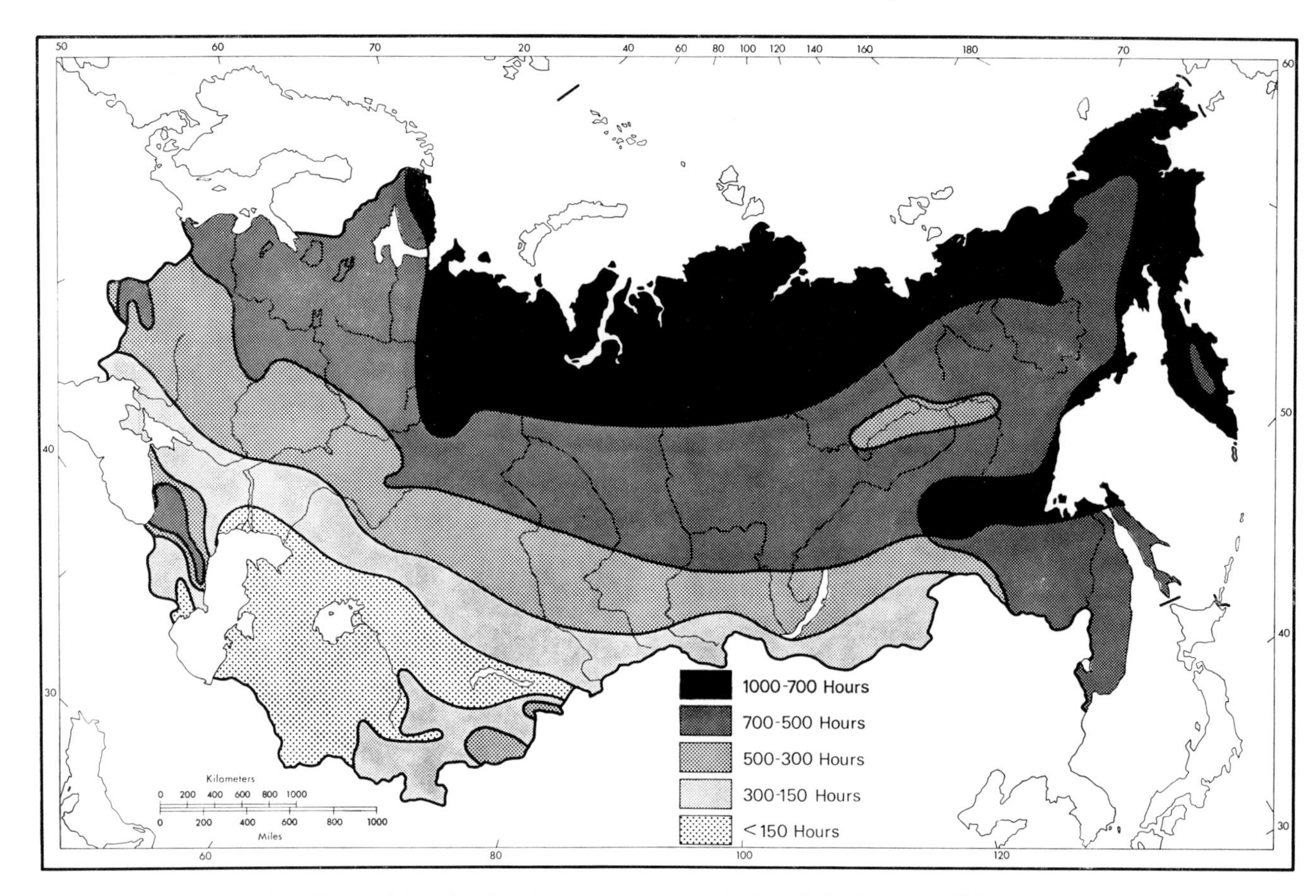

Fig.9-23. Total duration (h) of warm season precipitation. (After LEDEBEV, 1964.)

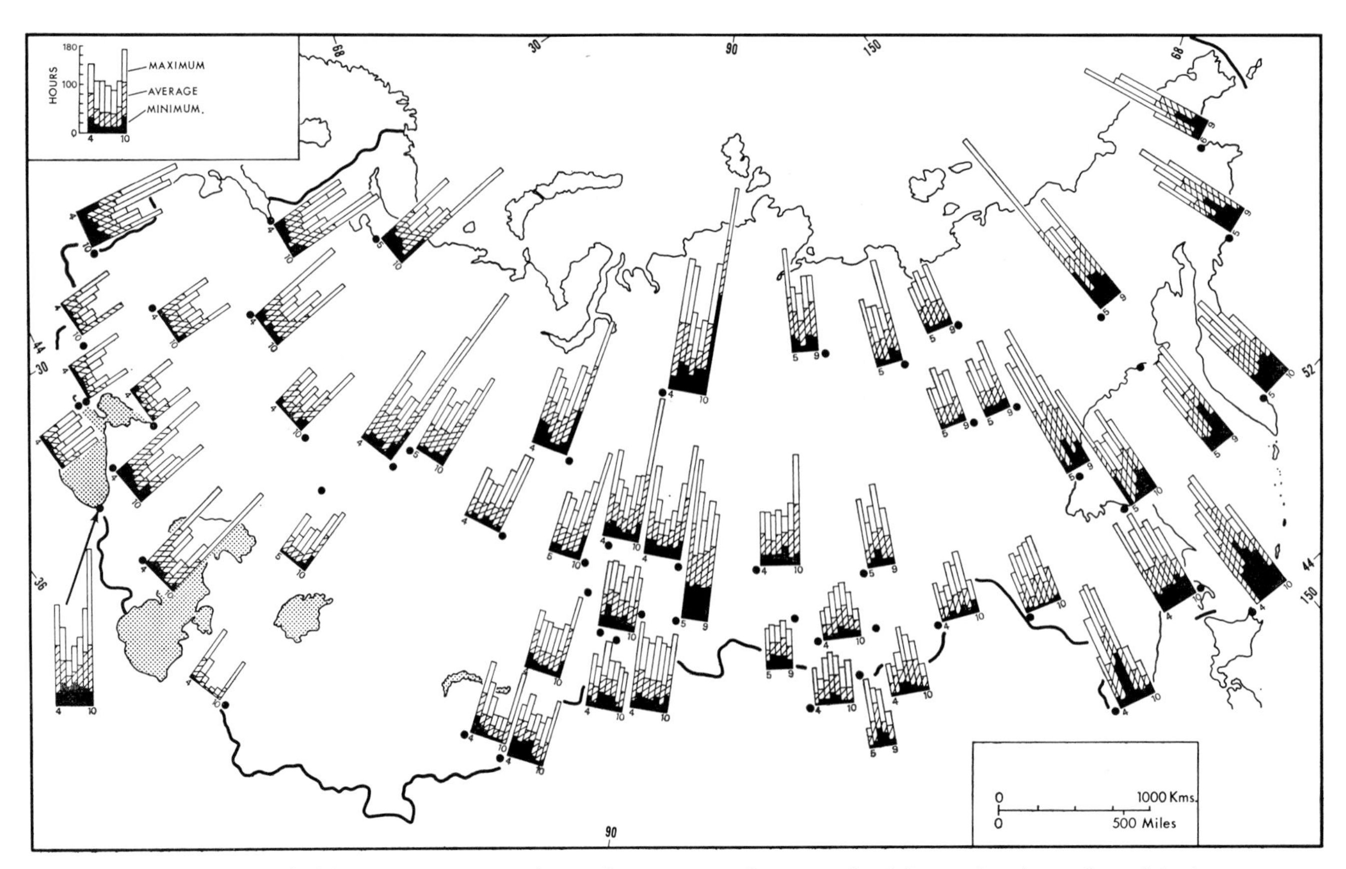

Fig.9-24. Warm season regimes of average, maximum, and minimum durations of precipitation. Numbers at bases of graphs refer to months. (After LEBEDEV, 1964.)

as 200 mm have fallen during a 24-h period in spots along the northern slopes of the Carpathians, the Black Sea slopes of the Caucasus, and the windward slopes of the Sikhota-Alin Mountains of the Soviet Far East (Fig.9-25 and Table 9-II). Similar amounts have also occurred in the mountains around Lake Baykal and on the Kuril Islands chain. The highest 24-h rainfall in the entire U.S.S.R. during the period 1951–1965 was 333.8 mm at Bilyasar in the Azerbaydzhan S.S.R. (Table 9-III).

The lowest maxima generally occur in the north and eastern parts of the country which usually experience small moisture fluxes in the atmosphere. Low precipitation maxima also occur in some of the desert regions of Central Asia where summer temperatures are usually well above the dew point.

Variability

As might be expected, the portions of the Eurasian plain with the highest amounts of yearly precipitation generally also enjoy the greatest constancy of this precipitation from one year to another. Annual variations are lowest along the western borders of Belorussia and the Baltic republics and extend eastward across the forest zone into the middle Ob Valley (Fig.9-26). The variation increases markedly southward into Soviet Central Asia

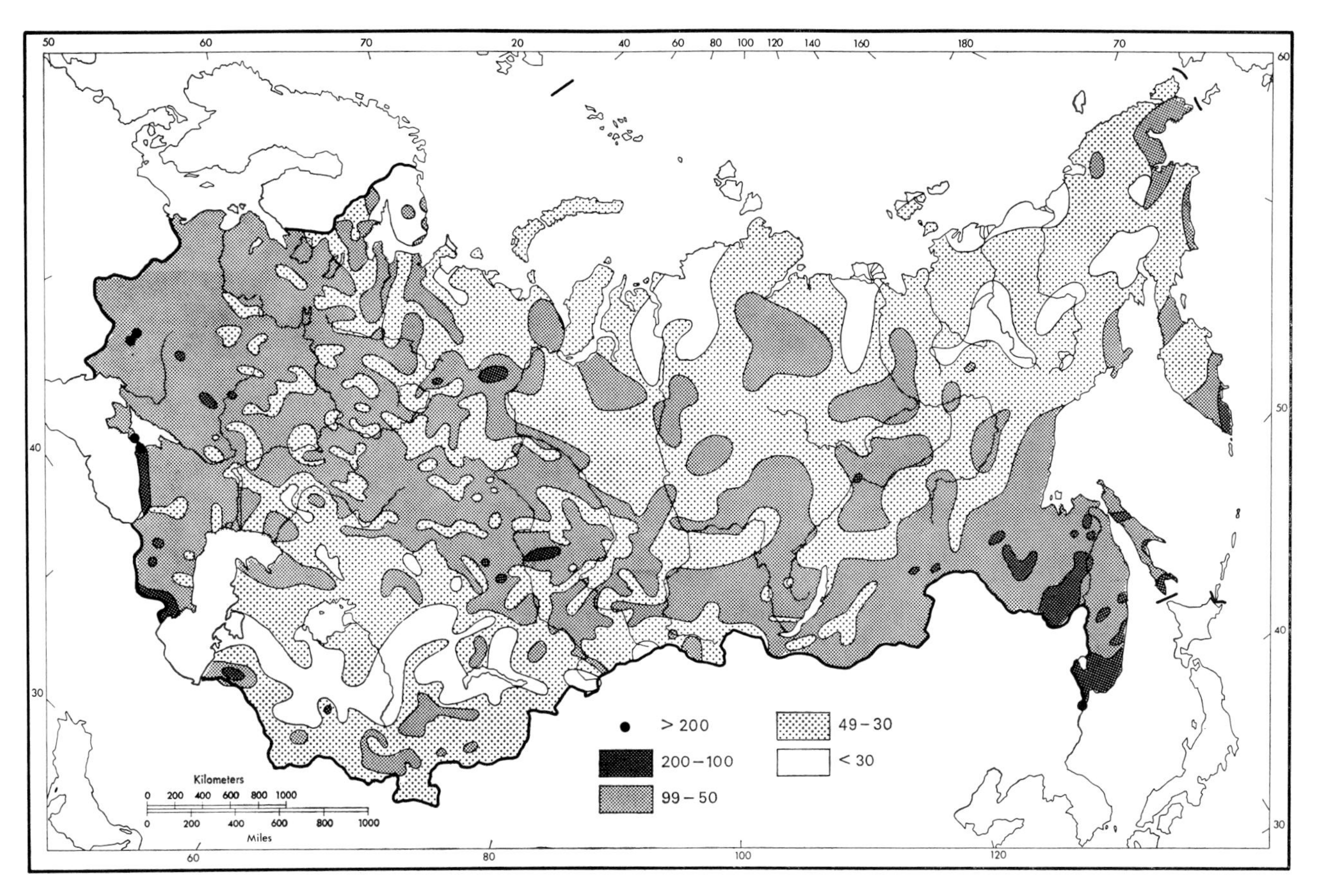

Fig.9-25. Maximum precipitation (mm) during a 24-h period, 1951–1965. (After CHISTYAKOV et al., 1968.)

TABLE 9-II

FREQUENCIES OF OCCURRENCE OF CATEGORIES OF 24-H RAINFALLS IN NUMBERS OF CASES DURING THE PERIOD 1951–1965; NUMBERS IN PARENTHESES REPRESENT NUMBERS OF CASES PER STATION
(After CHISTYAKOV et al., 1968)

Region	Number of Stations	≥30 mm	≥70 mm	≥110 mm	≥150 mm	Maximum in 24 h (mm)
Murmansk	42	162 (3.9)	5 (0.1)			96.8
Northern	292	1,442 (4.9)	22 (0.07)	1 (0.003)		133.6
Northwestern	213	2,631 (11.4)	40 (0.14)	5 (0.012)		140
Karelian ASSR	150	1,000 (9.4)	17 (0.11)	1 (0.006)		116.1
Estonian SSR	98	1,260 (12.8)	19 (0.2)	1 (0.01)		112.4
Latvian SSSR	151	1,391 (9.2)	63 (0.4)	3 (0.02)		161.2
Lithuanian SSR	80	959 (12.0)	36 (0.45)	1 (0.01)		130.7
Belorussian SSR	141	1.924 (13.6)	41 (0.3)	8 (0.05)		110.2
Ukrainian SSR	322	5,317 (16.6)	235 (0.7)	27 (0.08)	3 (0.009)	281.6
Moldavian SSR	73	1,224 (16.8)	46 (0.6)	3 (0.04)		124.3
Central Chernozem oblasts	68	980 (14.4)	27 (0.4)	1 (0.01)		122.4
Central oblasts	289	3,313 (11.5)	70 (0.2)	3 (0.01)		134.3
Upper Volga	405	2,375 (5.9)	52 (0.1)	2 (0.005)		117.2
Volga	256	1,646 (6.5)	59 (0.2)	6 (0.02)		144.4
North Caucasus	295	7,161 (24.2)	518 (1.7)	72 (0.1)	13 (0.04)	298.2
Dagestan ASSR	69	426 (6.2)	15 (0.2)	1 (0.01)		109.9
Georgian SSR	264	22,343 (85.3)	2,668 (10.1)	464 (1.3)	94 (0.3)	238.7
Azerbaydzhan SSR	196	4,609 (23.6)	166 (2.2)	98 (0.5)	26 (0.1)	333.8
Armenian SSR	193	1,090 (5.6)	21 (0.1)	2 (0.01)	2 (0.01)	176.0
Ural	256	2,484 (9.7)	64 (0.2)	1 (0.004)		137.2
Omsk	83	582 (7.0)	15 (0.2)	2 (0.02)		121.9
West Siberia	472	2,518 (5.3)	79 (0.16)	7 (0.01)		148.2
Krasnoyarsk	370	993 (2.5)	18 (0.05)	3 (0.008)		138.5
Transbaykal	174	2,078 (11.9)	47 (0.27)	1 (0.006)		149.5
Irkutsk	155	1,819 (11.7)	86 (0.55)	10 (0.06)	3 (0.02)	207.9
Yakutsk	234	630 (2.7)	15 (0.06)			105.9
Far East	165	3,998 (42.3)	241 (1.4)	17 (0.1)	1 (0.006)	227.2
Maritime Kray	59	3,130 (52.9)	337 (4.6)	79 (1.3)	17 (0.3)	278.5
Sakhalin	41	1,603 (39.0)	141 (3.4)	25 (0.6)	3 (0.07)	206.2
Kamchatka	76	1,015 (13.4)	15 (0.6)	8 (0.1)	4 (0.05)	186.4
Kolyma	126	413 (3.3)	6 (0.04)			108.0
Kazakh SSR	817	3,595 (4.4)	104 (0.1)	3 (0.003)		131.4
Turkmen SSR	96	252 (2.6)	10 (0.1)	1 (0.01)		123.6
Uzbek SSR	183	1,575 (8.6)	27 (0.1)	3 (0.003)	2 (0.01)	166.9
Tadzhik SSR	163	2,340 (14.4)	18 (0.3)	1 (0.006)		144.0
Kirgiz SSR	208	1,559 (7.4)	21 (0.1)			100.3
Total USSR	7,274	91,748	5,724	864	169	

and it increases a little toward the Arctic coast. Unfortunately, no similar data have been worked out for the area east of the Ob Basin.

During the warm season coefficients of variation form a pattern similar to that of the yearly one, with a somewhat greater range between minimum and maximum values. The variation is quite high during summer in Central Asia, but this is relatively meaningless since the total amounts of precipitation during the warm period in this area are almost nil (Fig.9-27). The pattern shifts considerably during the cold part of the year (Fig.9-28).

TABLE 9-III

CATASTROPHIC RAINFALL (>200 MM/24 H) OBSERVED IN THE USSR DURING THE PERIOD 1951–1965 (After CHISTYAKOV et al., 1968)

Station	Max. precip. (mm)	Date
Georgian SSR		
Dzhvari	227.4	2/VIII 1962
Akhuti	233.7	20/X 1954
Kheta	221.6	25/VII 1953
Dabla-Tsikhe	213.7	23/VII 1956
Shroma	203.3	28/VIII 1960
Tsikhisdziri	230.6	2/VIII 1963
Chakva (agro)	234.7	2/VIII 1963
Makhindzhauri	205.0	10/VIII 1956
Batumi	238.7	12/IX 1962
Charnali	213.2	12/IX 1962
Darcheli	207.0	25/VII 1953
Kvemo-Kheta	226.5	9/X 1962
Vakidzhvari	206.5	9/X 1962
Gudava	221.2	9/VII 1962
Dagva	230.5	2/VIII 1963
Sindieti	213.8	12/IX 1962
Maradidi	209.2	12/IX 1962
Azerbaydzhan SSR		
Lerik 30-th km	263.2	16/VIII 1955
Byursyulyum	227.0	16/VIII 1955
Bilyasar	333.8	16/VIII 1955
Lenkoran', zonal	264.4	24/VIII 1962
Lenkoran', aerological	215.8	20/X 1951
Astara	228.3	8/XI 1955
Ukrainian SSR		
Podgaytsy	281.6	3/VI 1957
Opasnoe	206.3	17/VII 1956
Chernovtsy	222.0	6/VI 1965
Krasnodarskiy Kray		
Temryuk	200.8	19/VIII 1953
Achishkho	298.2	26/VI 1956
Kepsh	227.6	26/VI 1956
Irkutskaya Oblast		
Khamar-Daban	207.9	20/VI 1960
Kuril Islands		
Simushir Island	206.2	8/VIII 1957
Primorskiy Kray		
Askol'd	226.9	3/VIII 1962
Gamov	278.5	3/VIII 1962
Khabarovskiy Kray		
Bolon'	227.2	9/VIII 1958

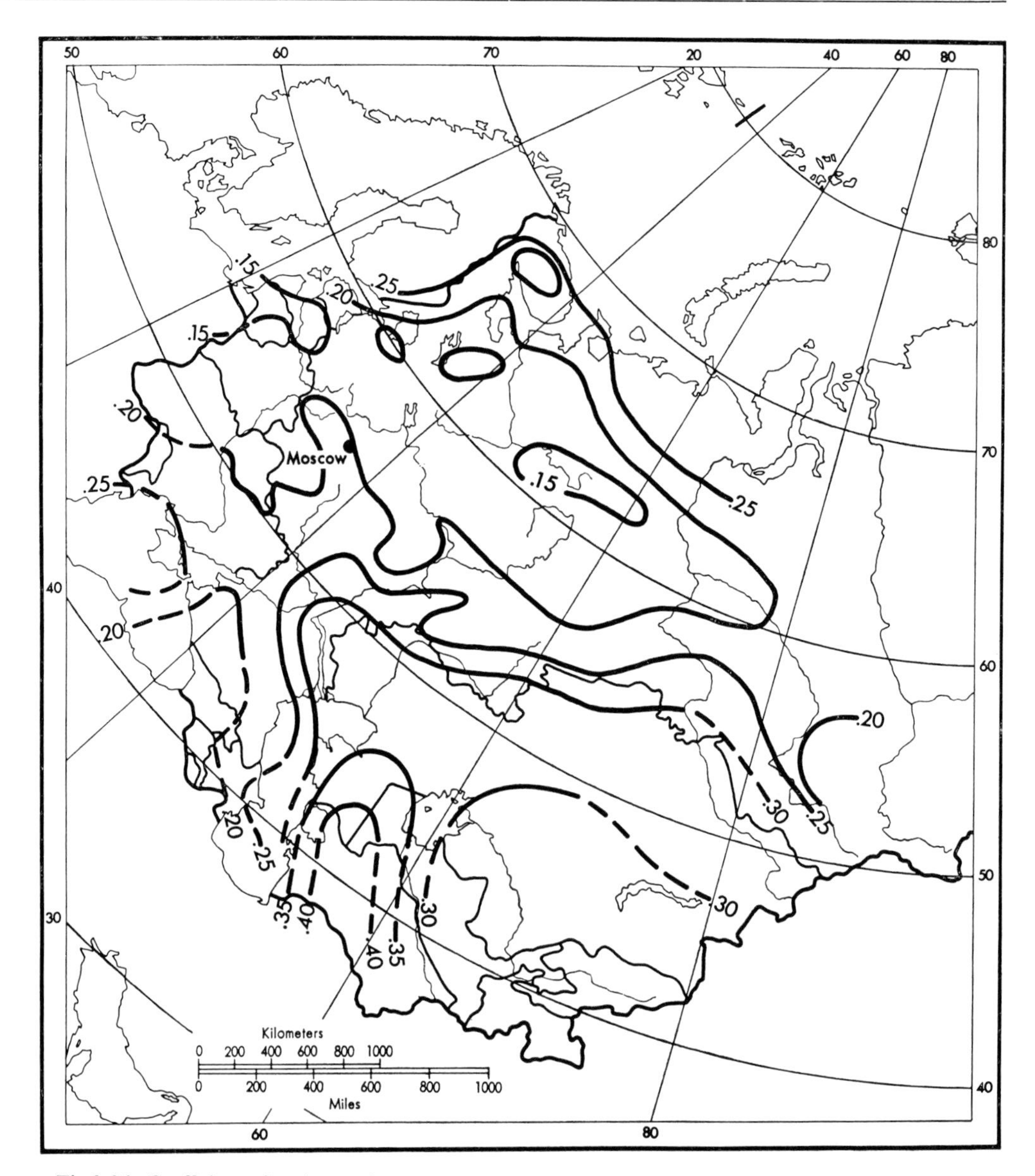

Fig.9-26. Coefficient of variance of annual precipitation. (After BATTALOV, 1968.)

Most significant are the higher values along the Arctic coast which must be related to rather wide year-to-year fluctuations in the numbers and intensities of winter cyclonic storms in this area. Winter precipitation here has about the same reliability as it does in Central Asia, which at this time of the year is also relying upon cyclonic storm passages. Winter precipitation is less reliable than summer precipitation over much of the middle portions of the plain, but this part of the country generally has the most reliable precipitation during all seasons. The western portion of the North Caucasus also enjoys fairly reliable precipitation all during the year.

Probabilities of receiving various amounts of precipitation during the year and during the warm and cold seasons are presented in Figs.9-29–9-34. Again, the north-central portions of the plain, which receive the greatest average annual precipitation, show the greatest probabilities of having higher values of precipitation. Central Asia never receives

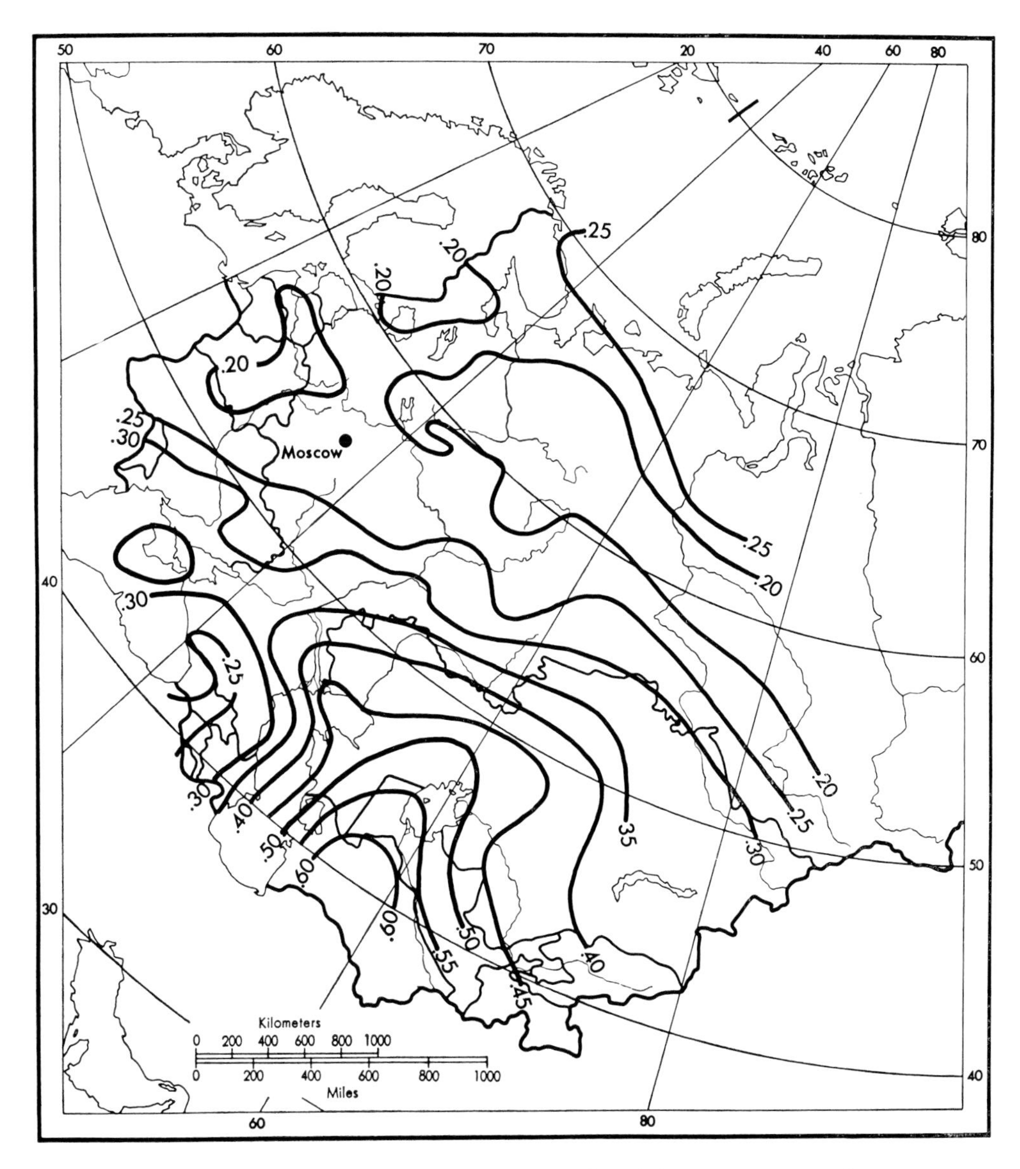

Fig.9-27. Coefficient of variance of warm season precipitation. (After BATTALOV, 1968.)

as much as 400 mm of precipitation per year, and the Arctic coast seldom receives 600. The outstanding feature on the annual maps is probably the consistency of the middle Urals and a localized region about 350 miles northeast of Moscow to receive more than 600 mm of precipitation per year. Another characteristic that should be noted on the 600 mm map is the rapid decrease of probability both north and south along the western border of the country from a maximum value around Kaliningrad. Kaliningrad can expect 600 mm of precipitation per year about 70% of the time, whereas only a short distance up the Baltic coast Tallin can expect it only 30% of the time.

The 400-mm map shows rather surprisingly high probabilities over the mouth of the Dnieper, which on the 600-mm map shows a very low probability. This area is relatively dry and seldom receives as much as 600 mm of precipitation per year, but apparently it has very consistent annual precipitation totals of around 450 mm. The reliability of the

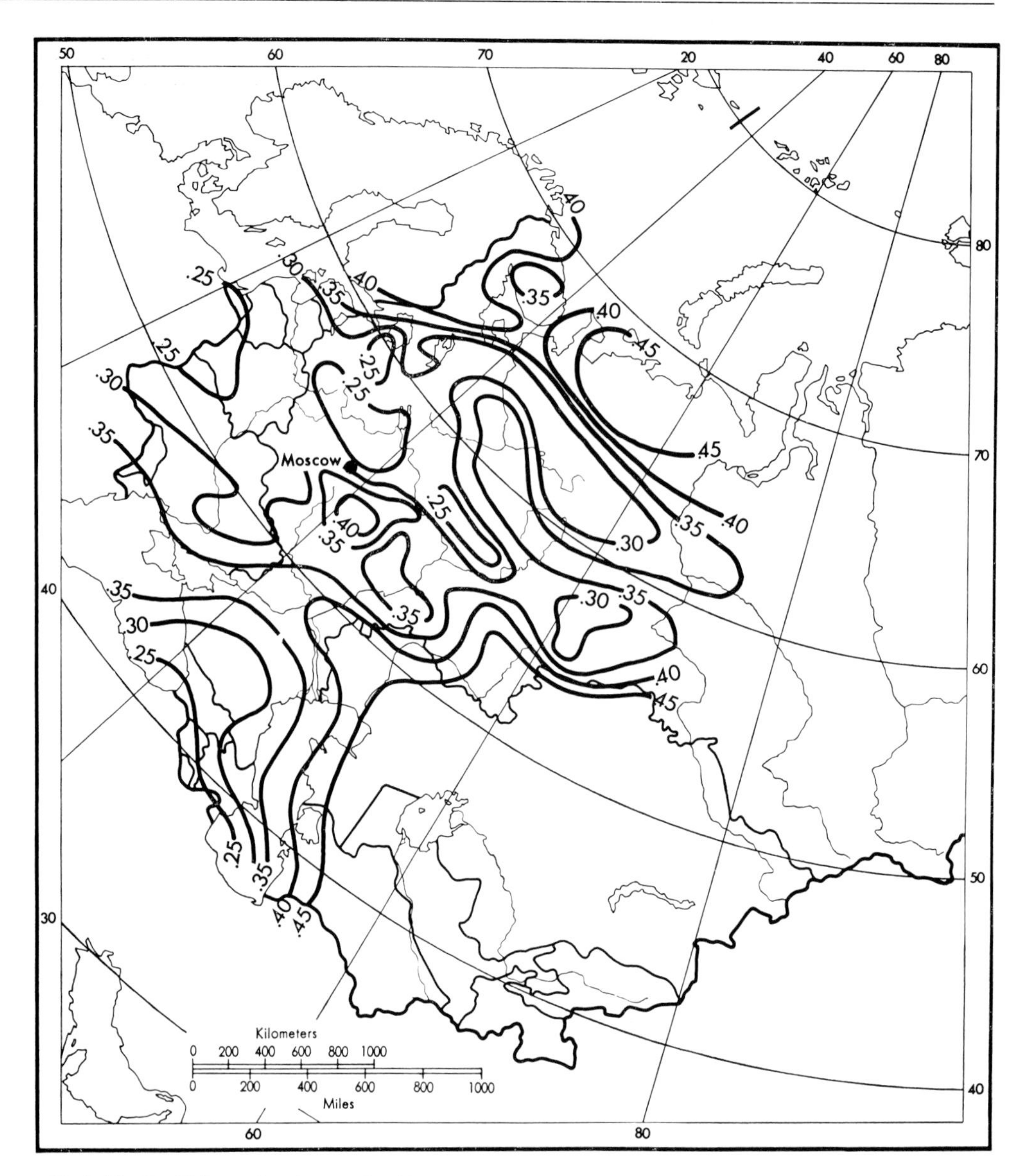

Fig.9-28. Coefficient of variance of cold season precipitation. (After BATTALOV, 1968.)

annual rainfall in this dryish area can be a significant factor in the development of agriculture in this region which is now undergoing intensification with the use of irrigation water from the Kakhovka Reservoir. The probability of receiving 400 mm of precipitation is also very high along the western portions of the north Caucasus, which is of great significance to the rich agricultural development of Krasnodar Kray.

The distributions of probabilities during the warm season are very similar to those during the entire year. The wetter north-central portion of the European plain and the western Caucasus show the highest values. The dryer zone running from the northern coast of the Black Sea northeastward across the middle Volga Valley shows lower probabilities, and Central Asia shows the lowest of all. The cold season patterns are quite different. Much of the plain averages less than 200 mm during the months November–March, and therefore the higher probabilities for this amount are concentrated

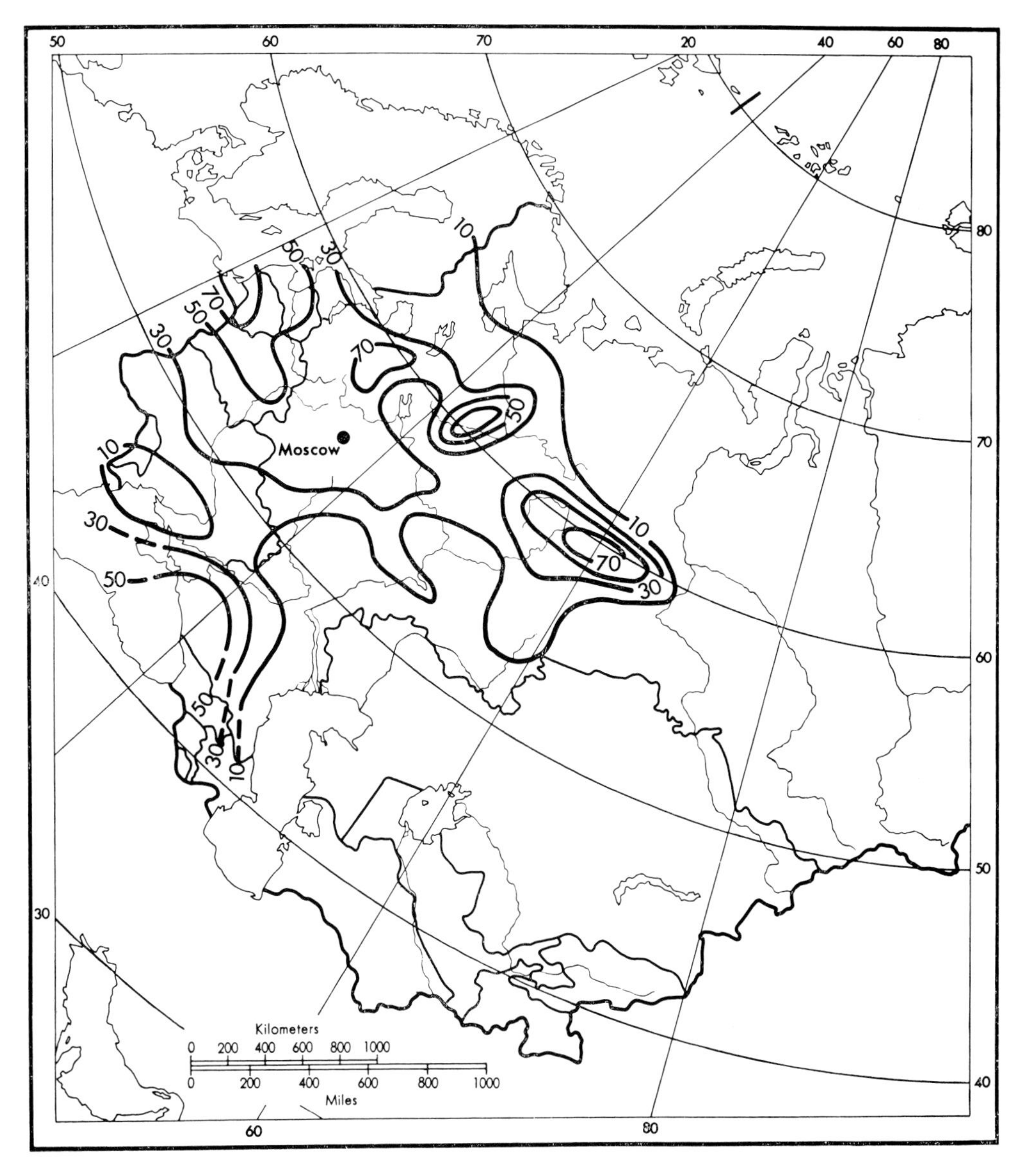

Fig.9-29. Probabilities of annual precipitation exceeding 600 mm, in percents of all years. (After BATTALOV, 1968.)

around the Black Sea. However, most of the map, except for Central Asia and the Arctic coast, shows high probabilities for receiving at least 100 mm of precipitation during the five-month cold season.

A study by DROZDOV and GRIGOR'EVA (1969) shows that precipitation varies more during the cold part of the year than during the warm part, and during both seasons it varies differently in different parts of the country (Fig.9-35). During the period November–March, over periods longer than 50 years precipitation varies by as much as 40% in southeast European U.S.S.R., in western Siberia, and in eastern Siberia north of Lake Baykal. The precipitation during the months April–October generally varies less than 15%, except in the southern part of the Soviet Far East where it varies as much as 20%. It appears that precipitation variations tend to occur in cycles of different lengths in different parts of the Soviet Union. The same study identified 20–30 year cycles over

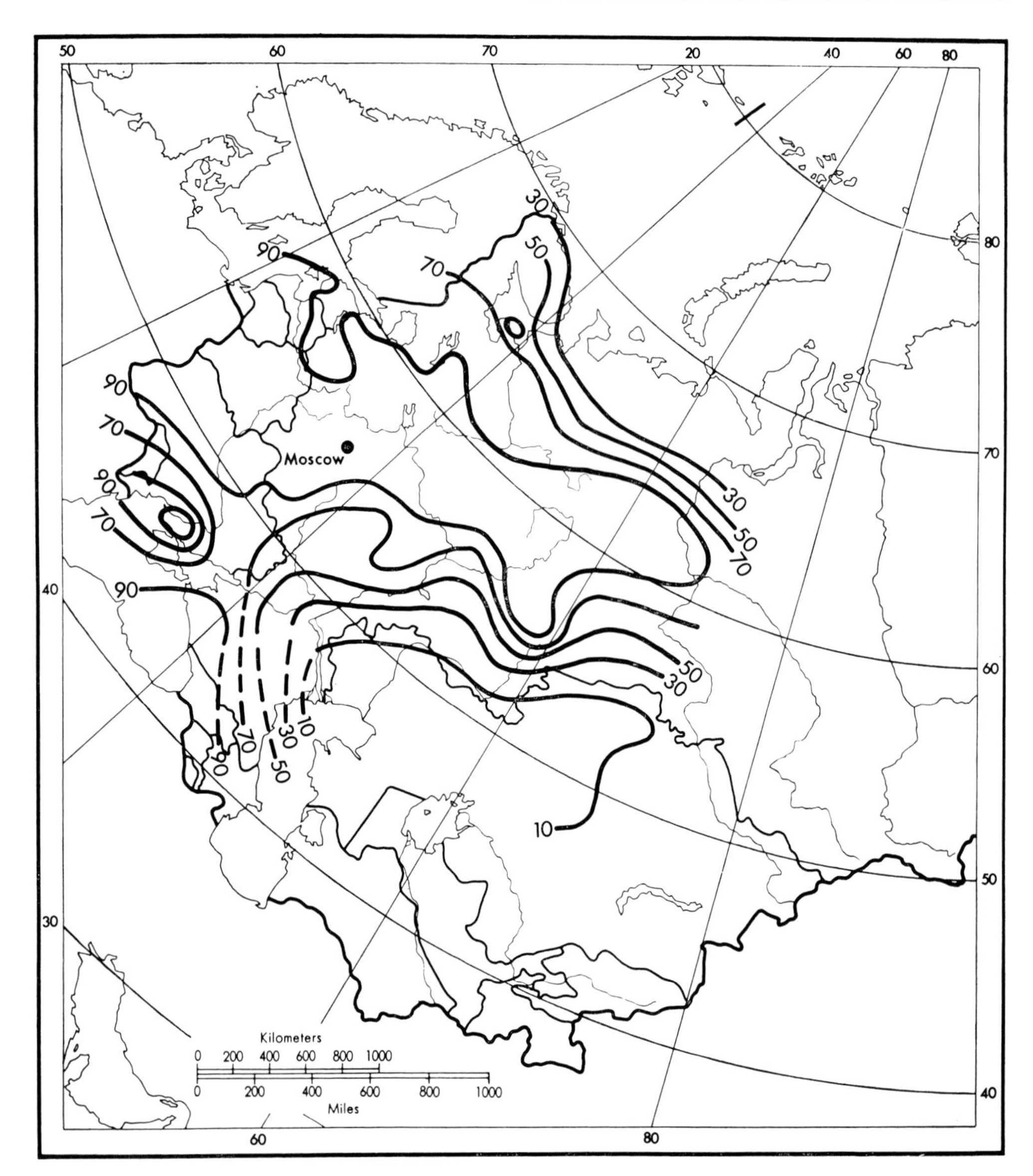

Fig.9-30. Probabilities of annual precipitation exceeding 400 mm, in percents of all years. (After BATTALOV, 1968.)

much of European U.S.S.R. and parts of western Siberia. It identified periods of 10–13 years over the northeastern part of the Soviet Far East, over the southern part of eastern Siberia, over the central portion of Kazakhstan and the southern Urals, over the extreme northwestern part of the country, and over the Black Sea steppes and North Caucasian Foreland. 5–6 year cycles were distinguished in limited areas in the central part of the European plain of the U.S.S.R., over the central part of the Ob' Basin, over southeastern Central Asia, and over a limited area around Lake Baykal.

Precipitation type

Thunderstorms and hail

Thunderstorms occur throughout the Soviet Union, but they are relatively rare along the

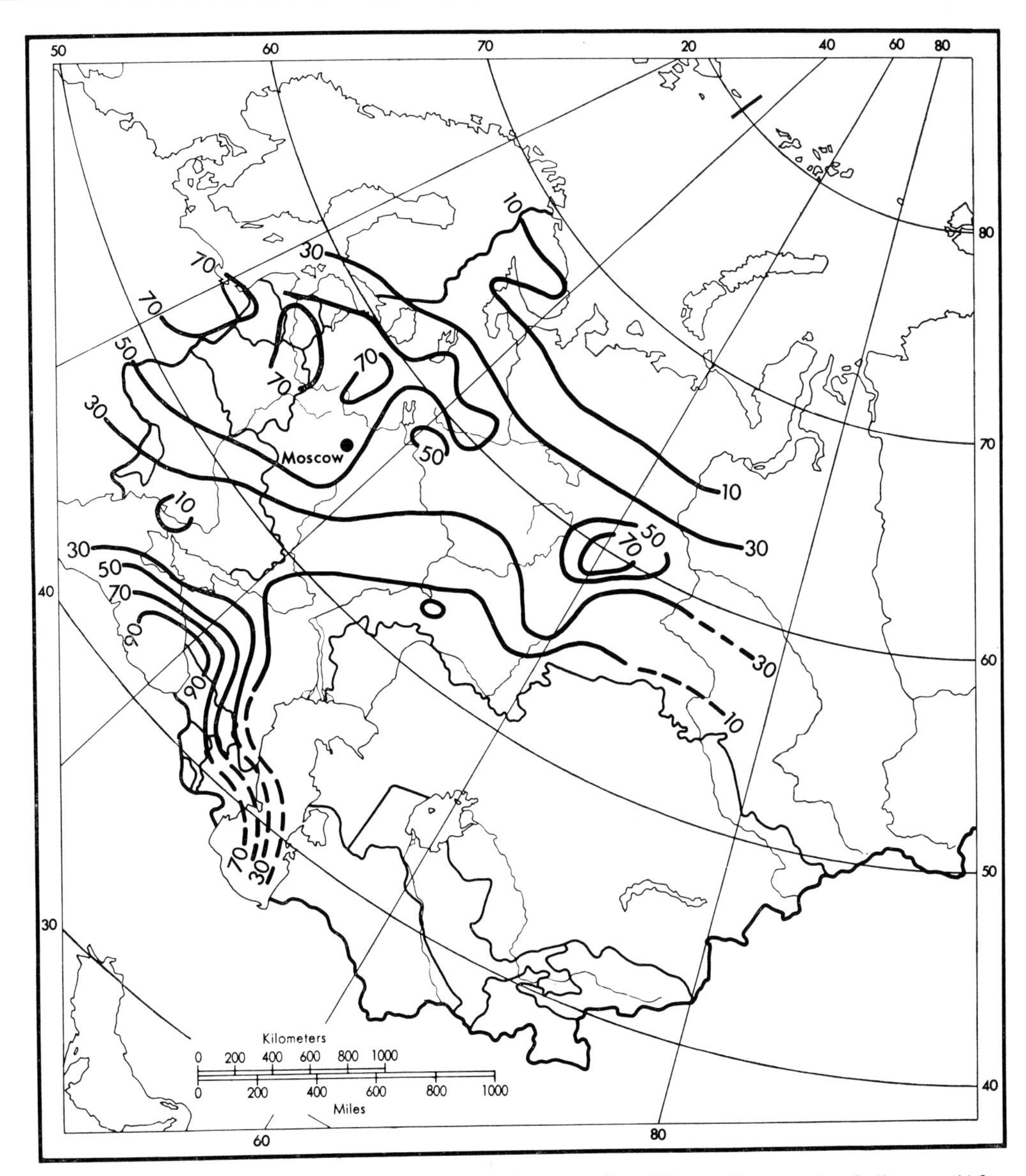

Fig.9-31. Probabilities of warm season precipitation exceeding 400 mm, in percents of all years. (After BATTALOV, 1968.)

Arctic and Pacific coasts (Fig.9-36). They generally occur from 40–80 h per year over the central and southern portions of European U.S.S.R. as far east as the Volga River, as they do in the central and southern Urals and an area extending eastward through the southern part of western Siberia and northern Kazakhstan. Frequencies are highest in the mountains along the southwestern border of the country where they occur more than 100 h per year in eastern Georgia, Armenia, and Carpathian Ukraine.

Everywhere, thunderstorms are primarily a summer phenomenon. In Kaliningrad Oblast and western Lithuania as many as 60% of all winters may experience at least one thunderstorm, but this frequency diminishes rapidly eastward to only 15% of all winters along a line from Archangel to Perm to Volgograd. East of this line only very rare occurrences of thunderstorms are observed in winter. Much of Siberia, the Far East, and Central Asia receive no winter thunderstorms at all.

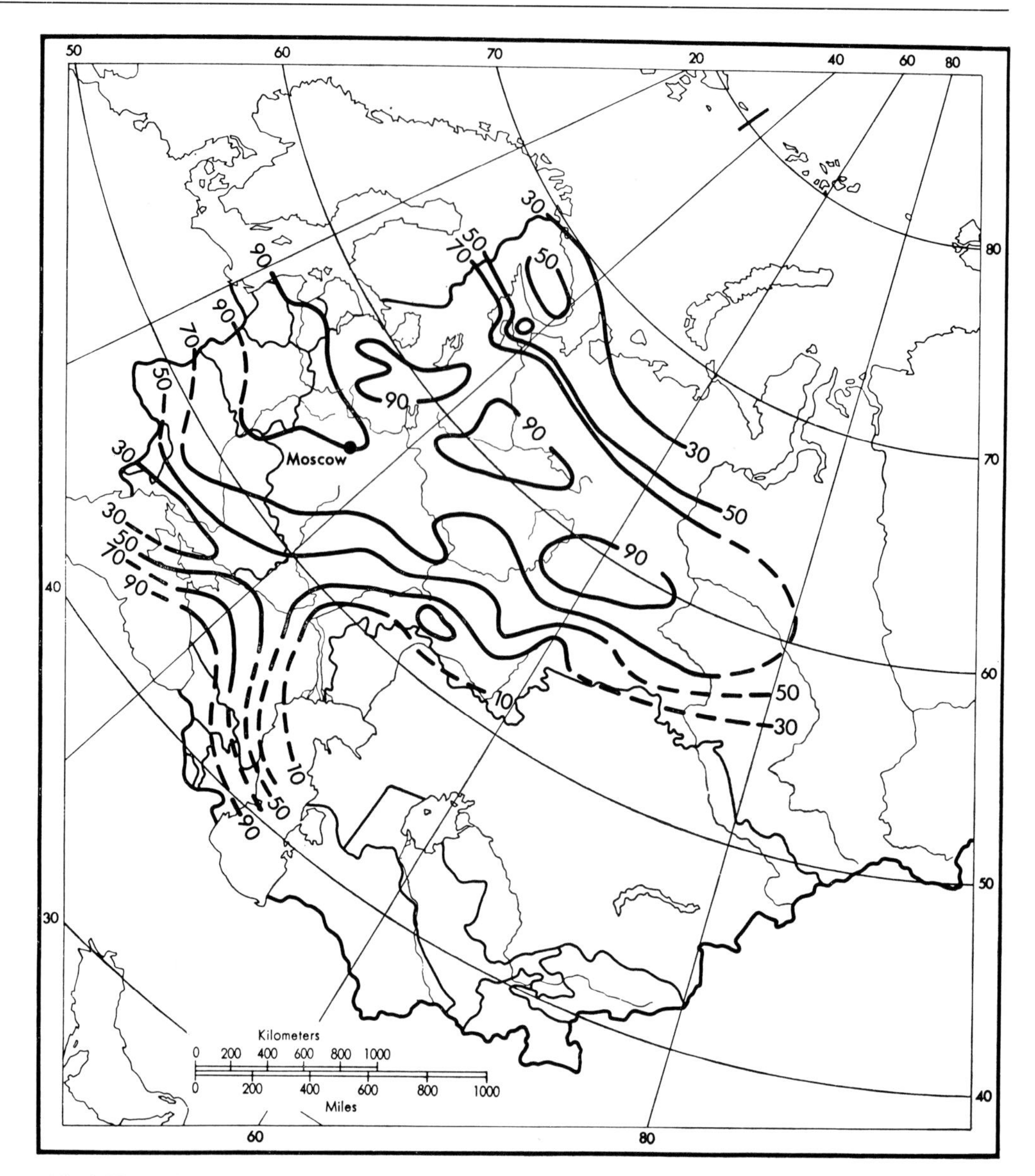

Fig.9-32. Probabilities of warm season precipitation exceeding 300 mm, in percents of all years. (After BATTALOV, 1968.)

Summer thunderstorms are both frontal and air-mass in character and, of course, in the mountainous areas are triggered by orographic uplift. A study by TORBINA (1937) showed that on the average throughout the European part of the country thunderstorms were associated with cold fronts 50% of the time, with warm fronts 22% of the time, and with single air masses 28% of the time. During midsummer they are most frequent over the mountainous areas of the southern part of the Far East and over the Caucasus (Fig.9-37). They are least along the Arctic and Pacific coastal areas and in Central Asia.

Season of maximum thunderstorm activity varies from place to place. In the northwestern part of European U.S.S.R. they are generally at a maximum in July. Much of the mid-section of the country experiences six to eight thunderstorms during this month. Maximum frequency shifts to June southward and eastward across the plain, although many stations there still report July maxima. Much of the Transcaucasus and Central Asia

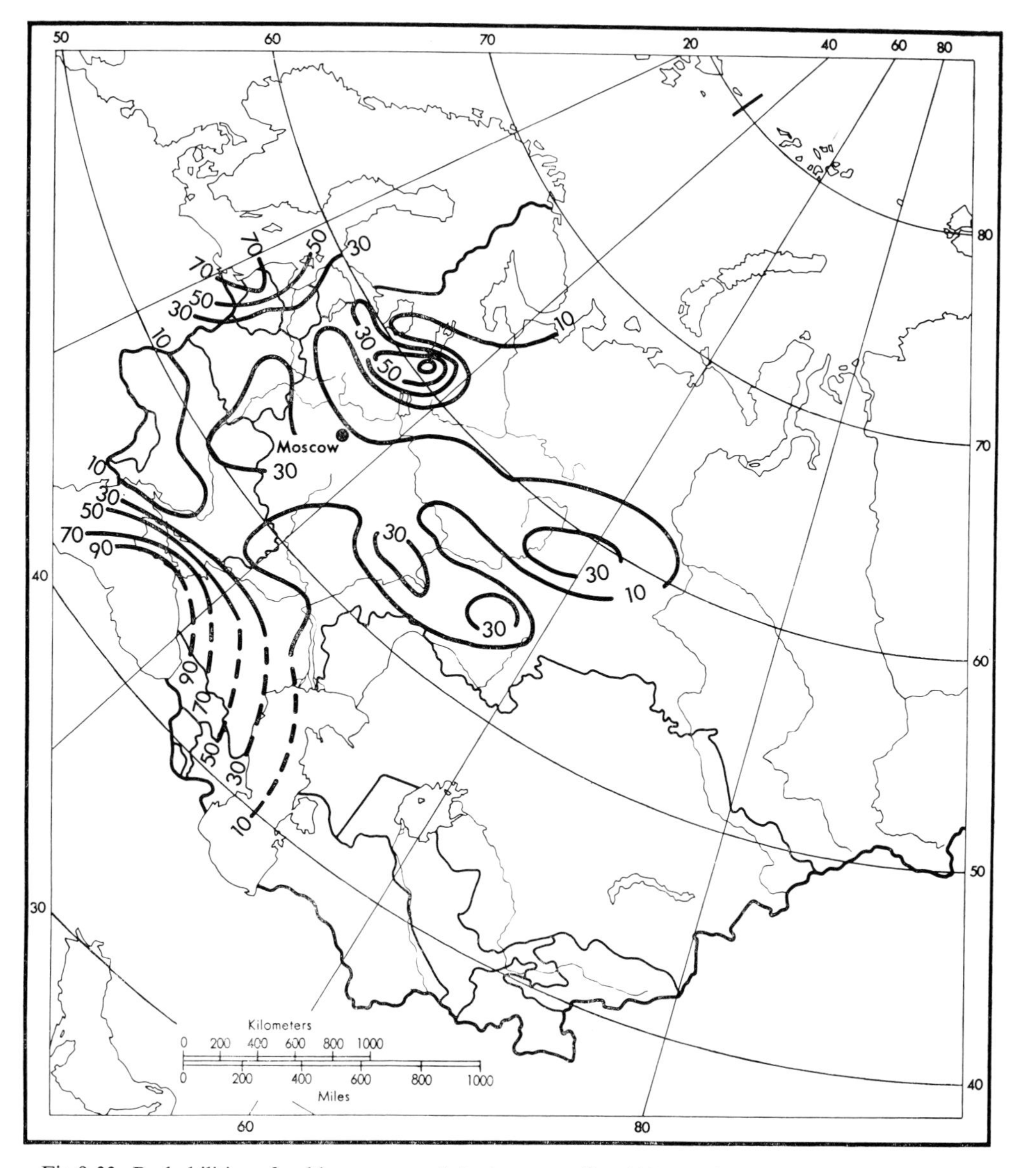

Fig.9-33. Probabilities of cold season precipitation exceeding 200 mm, in percents of all years. (After BATTALOV, 1968.)

show a May maximum of thunderstorms. Typically there is an abrupt rise from April to May and then a more gradual decline through June, July, August, and September. The northern foothills of the Tyan Shans in Central Asia have a summer maximum centered on July. Much of Siberia has a summer maximum rather symmetrically centered on July, although in places there is a tendency toward a summer–fall maximum. In the Far East thunderstorms occur infrequently and no pronounced monthly maximum exists. Thunderstorms are scattered from May through September. On Sakhalin Island the maximum is in September. On Kamchatka Peninsula coastal stations experience practically no thunderstorms, but Mt. Klyuchevski has a rather high total with a pronounced maximum centered on July.

Hail, of course, is associated with severe thunderstorms, and therefore its temporal and spatial distributions can be expected to be similar to those of thunderstorms. Its occur-

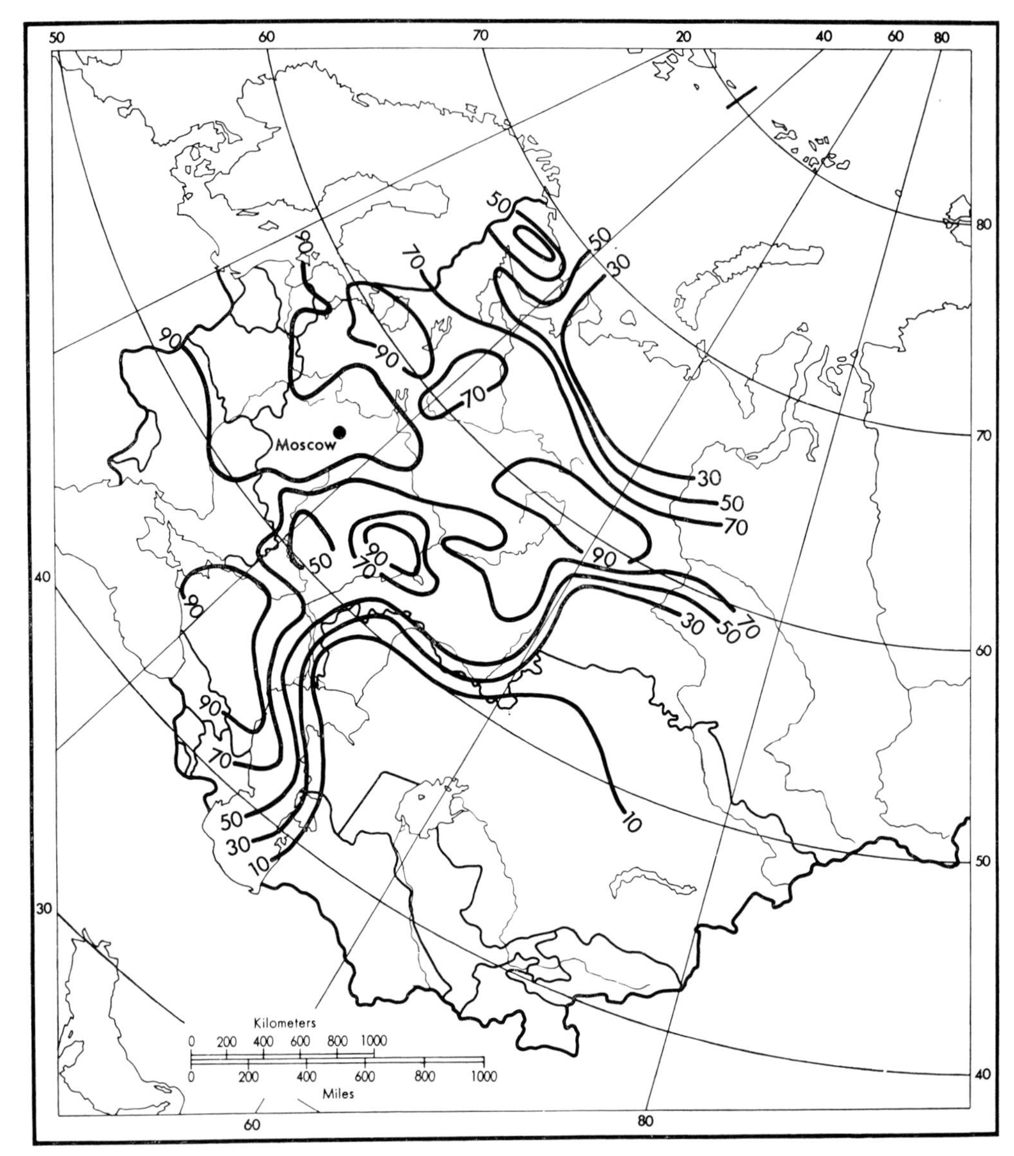

Fig.9-34. Probabilities of cold season precipitation exceeding 100 mm, in percents of all years. (After BATTALOV, 1968.)

rence is generally lowest along the Arctic and Pacific margins and in the desert plains of Central Asia, whereas maximum frquencies occur in the mountains of the south, particularly in the Caucasus (Fig.9-38). Everywhere the occurrence of hail is influenced greatly by topography. For instance, in the Valday upland northwest of Moscow, Staraya Russa at an elevation of 26 m gets an average of 1.4 days per year with hail, while Valday at an elevation of 220 m gets 3.5 days per year. Similarly, in the Central Russian Uplands, which extend north–south across the mid-section of European U.S.S.R., a difference of only 80 m elevation may change the hail occurrence from 1.4 days per year to 3 days per year. In mountainous areas, of course, the influence of elevation is even greater. In the Caucasus, hail occurrence varies from 1–4 days per year in lowlands to more than 8 days per year at elevations of 2,000–3,000 m. The high frequency of hail in the Caucasus is well known for the damage that is done to such crops as grapes and citrus. The Soviets

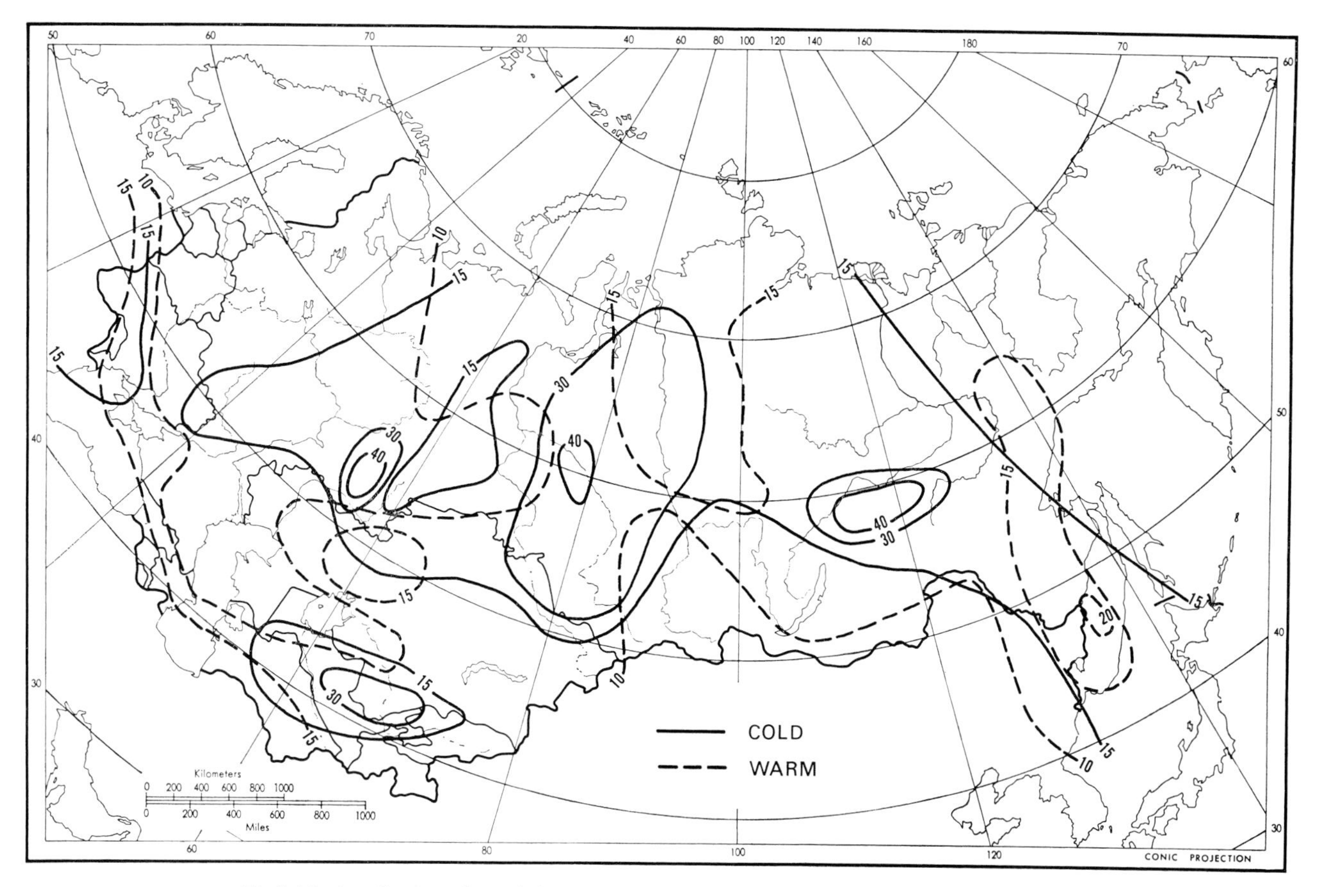

Fig.9-35. Amplitudes of precipitation variation (%) of cycles more than 50 years long, cold season and warm season. (After DROZDOV and GRIGOR'EVA, 1969.)

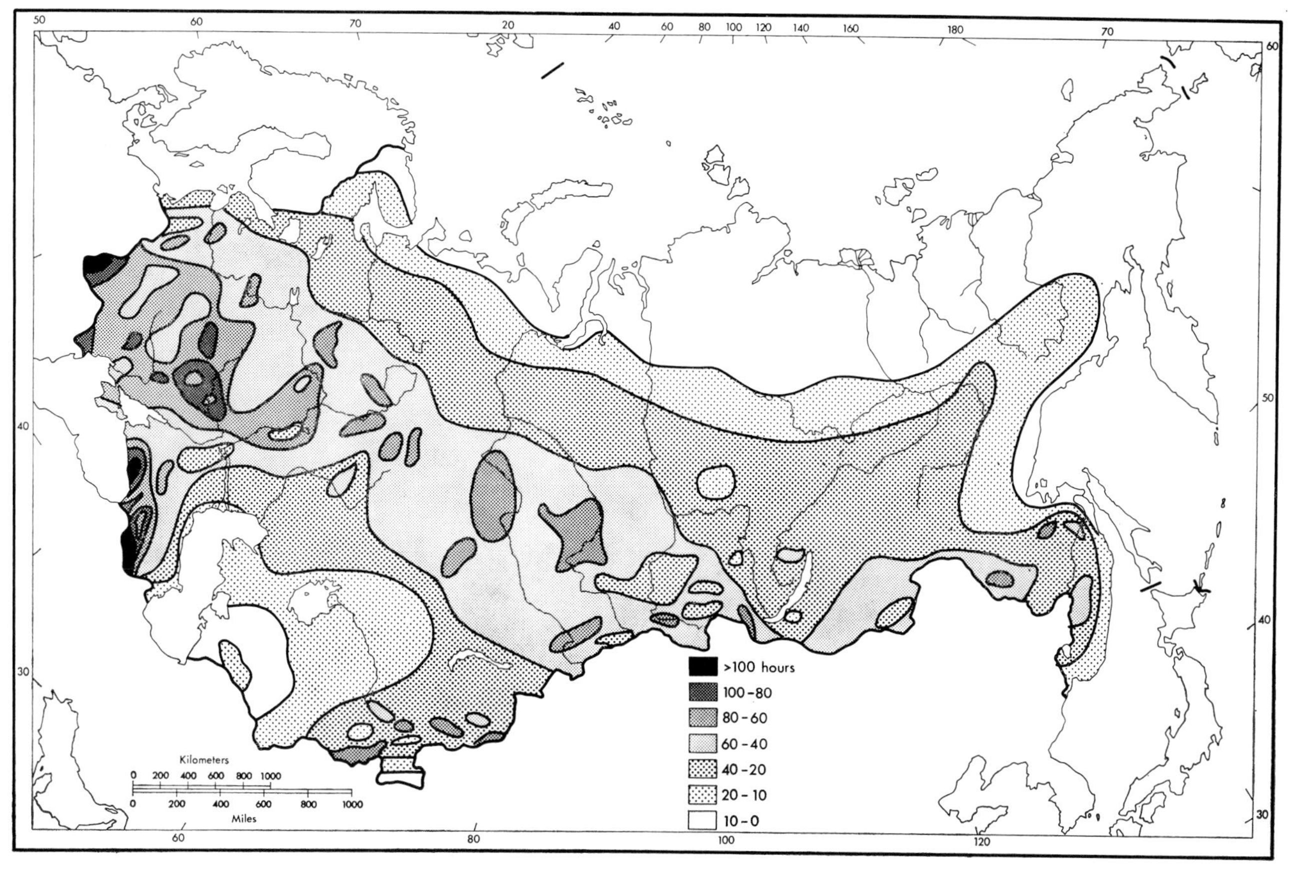

Fig.9-36. Mean annual duration of thunderstorms (h). (After LOBODIN, 1973.)

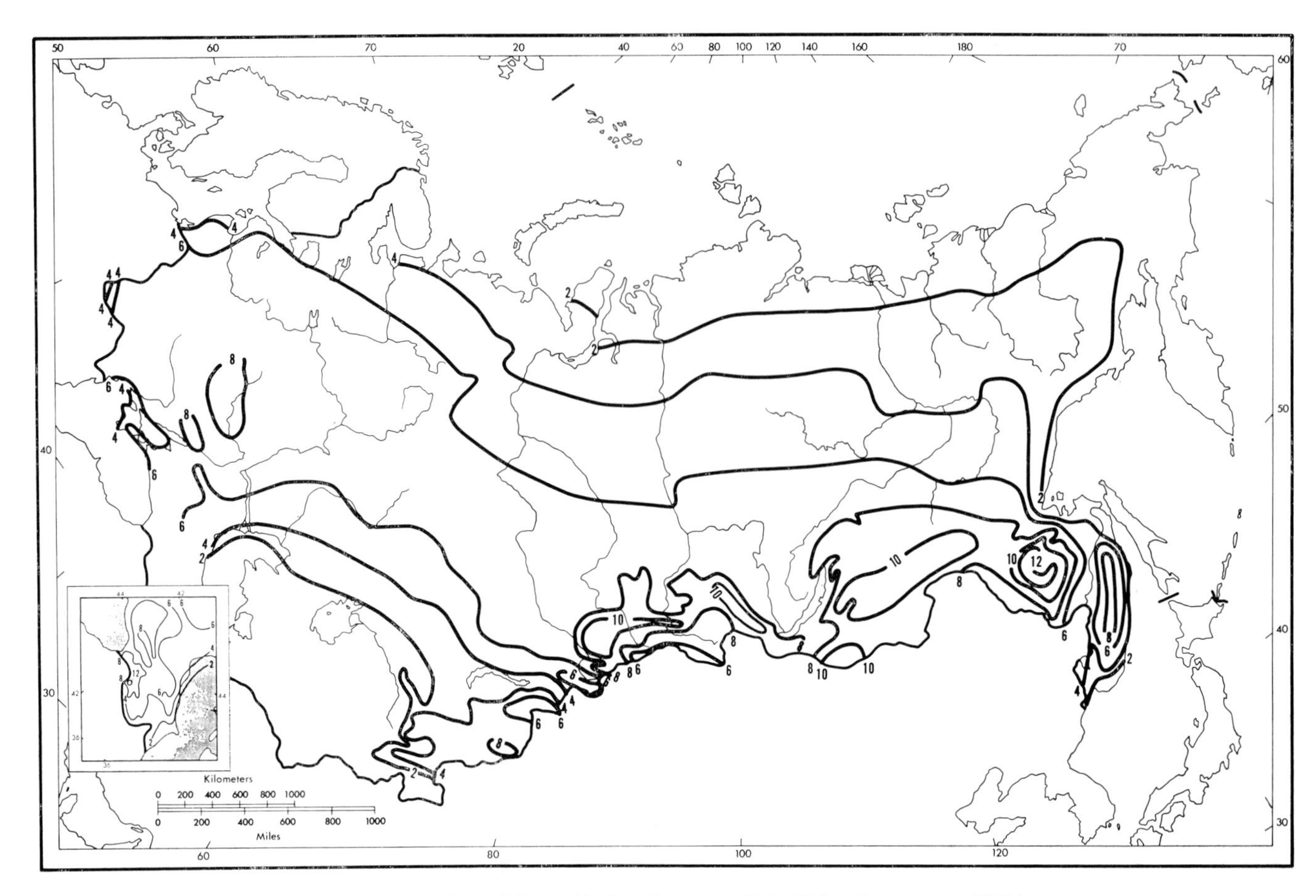

Fig.9-37. Mean number of days with thunderstorms, July. (After ARKHIPOVA, 1957.)

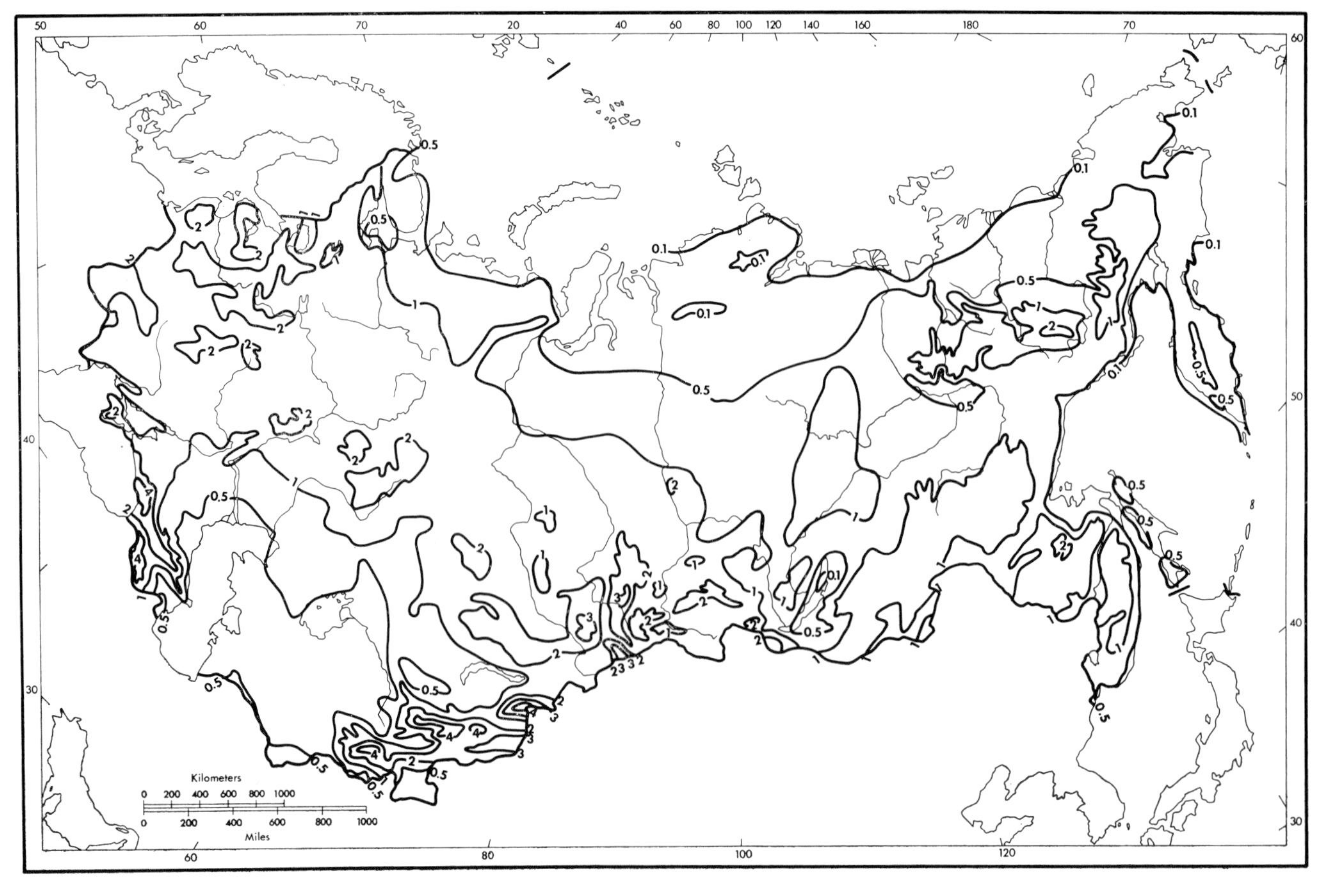

Fig.9-38. Mean annual number of days with hail. (After PASMUKH and SOKHRINA, 1957.)

have carried out many investigations in this area to study the possibilities of suppressing growth of cumulus clouds before they reach the hail stage.
Throughout much of the central and southern plain of European U.S.S.R. the maximum occurrence of hail is in May–June. Along the Baltic coast this shifts rather abruptly to September–November. Along the Black Sea coast of Georgia hail occurs most frequently in November–January, but throughout the interior of Georgia and eastward throughout the rest of the Transcaucasus it is generally May–June. May and June are also usually the months of maximum hail frequency throughout the Asiatic part of the country as far east as the Pacific fringe. Beyond there Sakhalin shows a maximum in October. Of course the annual frequency there is quite low.

Snow

One of the outstanding characteristics of the climate of the U.S.S.R. is the deep and prolonged snow cover over much of the country which exerts such important influences on other climatic factors. Once a snow cover is established, as much as 80–90% of the incident radiation is reflected and lost. This decreases to 50% or less for old and dirty snow, but even that albedo is higher than for any other type of surface. The snow also acts as almost a perfect radiator, so that heat loss from the surface is high and a large downward flux of heat takes place from the atmosphere to the snow surface which over a prolonged period of time produces extremely low surface air temperatures and strong surface temperature inversions. The snow insulates the ground from the air and thereby regulates the temperature of the underlying soil so that it is not greatly affected by the extremes of temperature taking place in overlying air masses. The air derives very little heat through the snow cover from the underlying ground, which is one of the major reasons why surface air temperatures become so cold.
If a snow cover is established early enough in autumn before a deep freeze has set in, the ground is not frozen to great depths during the winter and it is free to absorb waters from melting snow and increased rainfall in the spring. Thus a snow cover can be a great regulator of soil moisture as well as of soil temperature. At the end of winter, the melting of a deep snow cover will delay the coming of spring because any excess heat either radiated or advected into the area will be utilized for latent heat of fusion in the melting of the snow, and the surface air temperatures will not rise much above the freezing point until all the snow is gone. Once the snow is melted, there is still some delay in the sharp rise of air temperatures because of the evaporation that is taking place from the ground which is left very wet after the thawing period. Thus the rise in air temperature characteristic of spring is delayed by melting and evaporation processes until late spring when insolation is already great. Therefore, when the air temperatures do finally rise they rise very rapidly, and the season changes abruptly from winter to summer with little spring in between. It is characteristic of most high latitude areas to experience extremely brief transitional seasons, particularly in spring, and to a large extent this is due to the effects of snow cover.
The Soviet Union, more than any other large populous agricultural country, is affected by deep prolonged snow cover over much of its territory. It is only reasonable to expect that the influences of snow cover should have been studied in great detail first in Russia and that a continuous program of investigation should be manned to discover precisely

the relationships of snow cover to temperature, moisture, and other aspects of climate. The renowned Russian climatologist, A. I. VOYEYKOV, was the first to recognize the great significance of snow cover and the first to publish a lengthy monograph on it, published in 1899 by the Russian Geographical Society and republished in 1948 by the Academy of Sciences of the U.S.S.R. as part of his collected works. A half century later G. D. RIKHTER began publishing a long list of papers and monographs on snow cover, most of which are cited in his most recent effort as editor of the volume *Geography of Snow Cover* (in Russian) published by the Academy of Sciences of the U.S.S.R. in 1960. Outstanding are his monographs *Snow Cover, its Formation and Properties*, and *The Role of Snow Cover in Physical Geographical Processes* (1945 and 1948, respectively; both in Russian). Later publications by him and other authors have dealt particularly with the possibilities of altering the distribution of snow cover in order to put it to the greatest use for agriculture. Much theoretical and practical work on snow cover has been combined with some discussion of snow cover over the U.S.S.R. and other parts of the earth's surface in a recent compendium by KOTLYAKOV (1968).

Much of the Soviet Union receives only modest snowfall, but because of the long cold winters much of the country accumulates a relatively thick and persistent snow cover. Snowfall on the Eurasian plain during November–March varies from about 300 cm in the hillier parts of the western portion of European U.S.S.R. to less than 50 cm in much of the Soviet Far East except for the Pacific margins. The eastern half of eastern Siberia, the Arctic fringes of all of Siberia, and a portion of Central Asia between the Caspian and Aral seas also receive less than 50 cm (Fig.9-39). Greater amounts fall in some of the high mountains of the south, particularly in the western Great Caucasus. Snowfall data are not available there, but snow depths at Achishko at an elevation of 1,880 m reach as much as 400–500 cm during average winters and 700–800 cm during extreme winters. Snowfall varies greatly from one winter to another, as much as 150–200% in the northeastern part of the country and 400–450% in the west. During exceptionally cold winters very heavy snowfalls are associated with the passage of depressions, especially in the southern parts of the country, where even Gagra, Sukhumi, and Batumi along the eastern Black Sea coast suffer from abundant snows during such synoptic situations.

The number of days with snow storms during the winter varies over the European plain of the U.S.S.R. from 80–100 along the Arctic coast to 5 or less along the Black Sea coast (Fig.9-40). From a comparison of Figs.9-39 and 9-40, it is quite clear that individual snowfalls are much lighter in the north than in the south. Although the frequency of snowfall is 5 to 6 times greater in the north than in the south, the amount that falls during the winter is about the same in the two areas. And in the mid-section of the plain where the frequency is no more than $\frac{1}{3}$ as much as that along the Arctic coast, the total snowfall during the winter amounts to at least twice as much as along the Arctic coast. This is to be expected because of the great difference in moisture content between northern and southern air masses. It is quite obvious that over the entire plain the moisture source for snow comes from the south.

Generally snow accounts for 25–30% of the total annual precipitation through the broad mid-section of the country. Its share in total precipitation increases somewhat to the north and decreases rapidly southward. In Archangel it constitutes 31% of the total precipitation, in Moscow 28%, and in Kherson on the Black Sea coast 11%. Northeast of a line running from Leningrad southeastward to the mouth of the Ural River in the

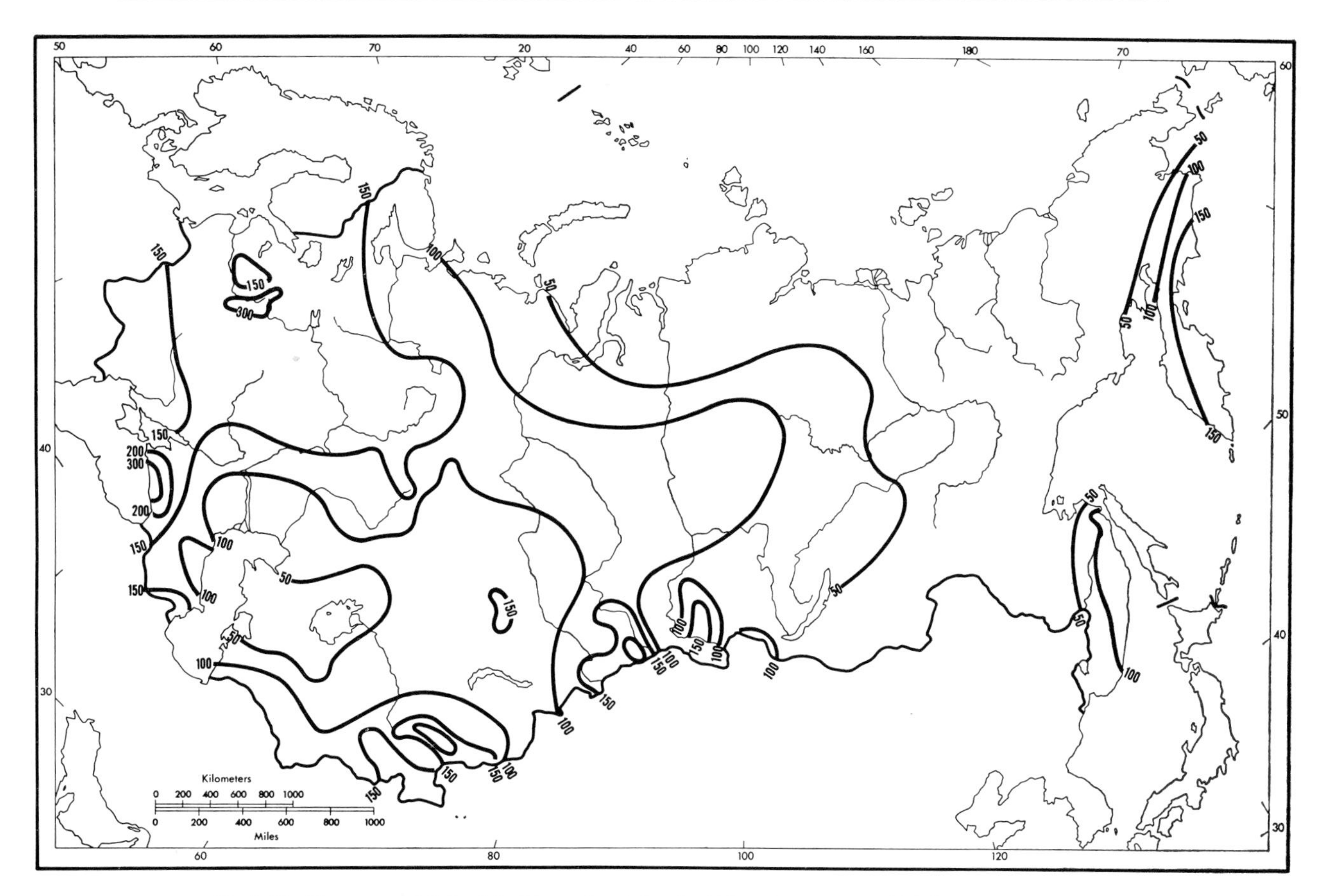

Fig.9-39. Mean snowfall (cm) in November–March. (After KOTLYAKOV, 1968.)

north Caspian, snow constitutes about 4% of the monthly precipitation in September, 40% in October, 80% in November, 97–100% in December–February, about 90% in March, 50% in April, 8% in May, and 1% in June. Southwest of this line snow generally does not fall at all from May to September. In October it amounts to about 10% of the monthly precipitation, in November about 40%, in December 70%, in January–February about 80%, in March 65%, and in April 20%.

The distribution of snow cover is not the same as that of snowfall because of the temperature influences on melting. Over the Eurasian plain the region of maximum snowfall in the west experiences enough thawing during the winter to keep maximum depths of snow cover at no more than 20 cm, except in the Khibiny Mountains on the Kola Peninsula where thicknesses can reach as much as 120 cm. Farther east, as the snowfall decreases the colder temperatures and associated lack of thawing allow accumulation to more than 80 cm on the western slopes of the Urals and in the middle Yenisey Valley (Fig.9-41). The greatest depths are generally found at latitudes around 63° or 64°. North of there snow cover diminishes because of diminishing snowfall, and south the cover diminishes because of increased thawing. The thinnest snow cover is in the south in a region stretching from the Black Sea steppes eastward into Central Asia where a snow cover is formed less than 50% of all years. It is also quite thin in the southern Lake Baykal region because of the extremely low snowfall during the winter in this core area of the Asiatic high. However, the mountains surrounding the basin in this area

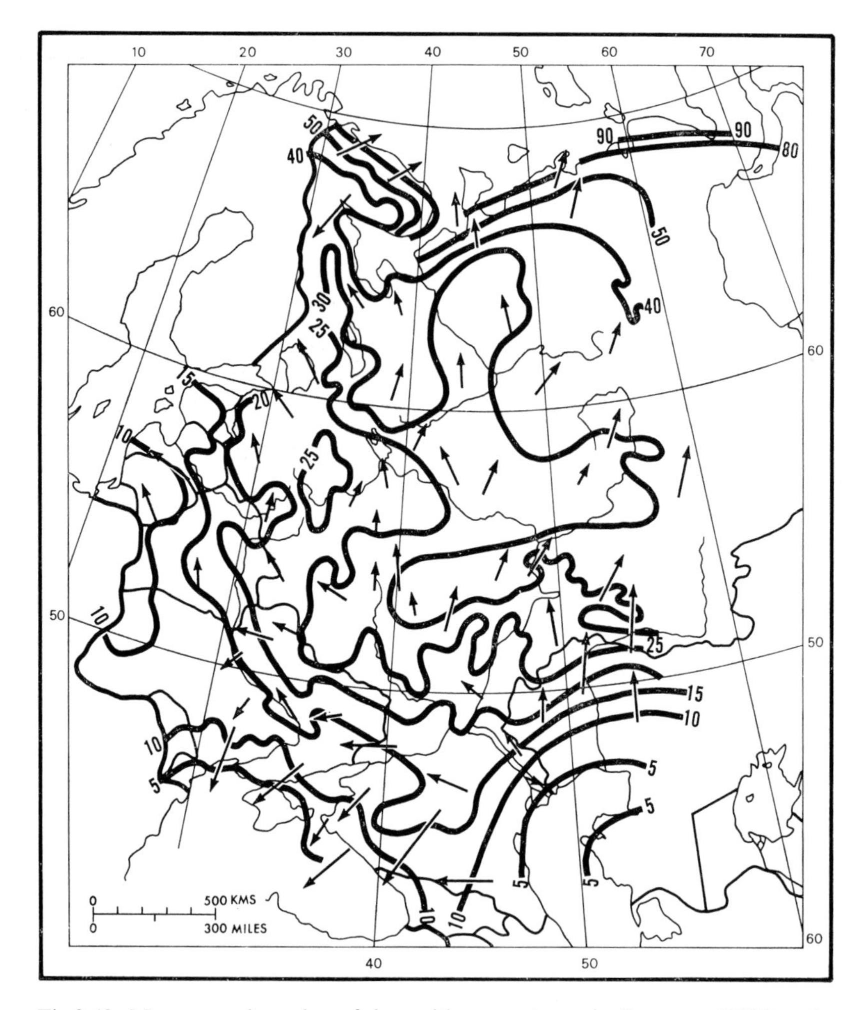

Fig.9-40. Mean annual number of days with snow storms in European USSR and prevailing surface winds during snow storms. (After KOTLYAKOV, 1968.)

may have snow depths up to 100 cm. Snow cover thickens rapidly as the Pacific coast is approached, and on the volcanic peaks of Kamchatka Peninsula it reaches depths of more than 160 cm.

Figs.9-42 and 9-43 show the dates of establishment and disappearance of snow cover and Fig.9-44 shows number of days with snow cover during the year. These three patterns are somewhat similar. In the western part of European U.S.S.R. the isolines run mainly north–south from Leningrad to eastern Belorussia and then turn east-southeastward as they cross the country in a rather zonal fashion. This is considerably different from the pattern of snow depth which shows enclosed areas of maximum values in the central portions of the country on the western slopes of the Urals and in the middle Yenisey. Apparently snow depth has little effect on snow duration. The interior parts of the country which accumulate great amounts of snow in the winter because of cold temperatures undergo rapid thawing in late spring as radiational heating, unhampered by marine influences, raises temperatures rapidly. In the western part of the country the marine influence from the Atlantic and adjacent seas greatly control thawing, but further eastward latitude is the primary control.

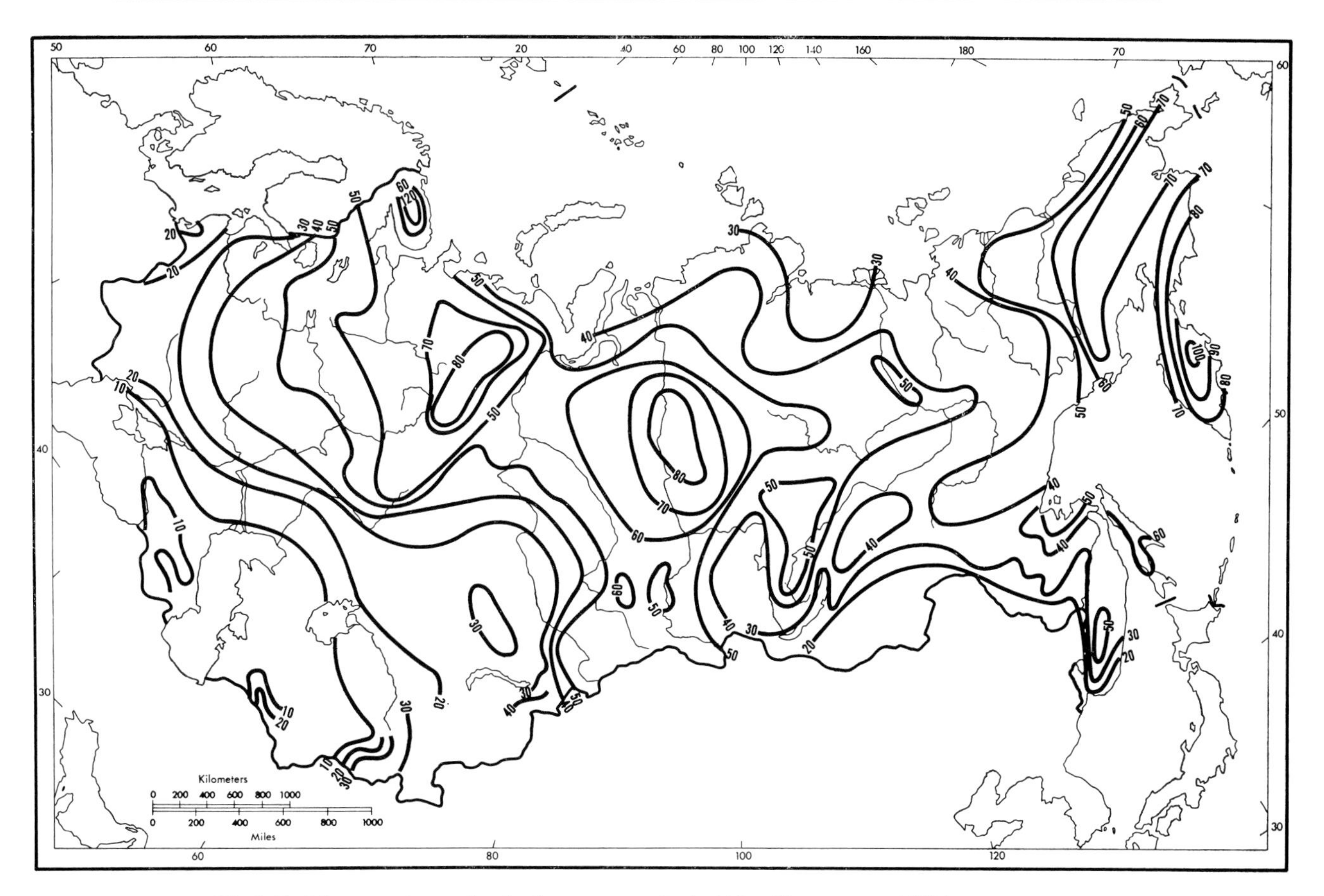

Fig.9-41. Mean maximum depth of snow cover (cm). (After KOTLYAKOV, 1968.)

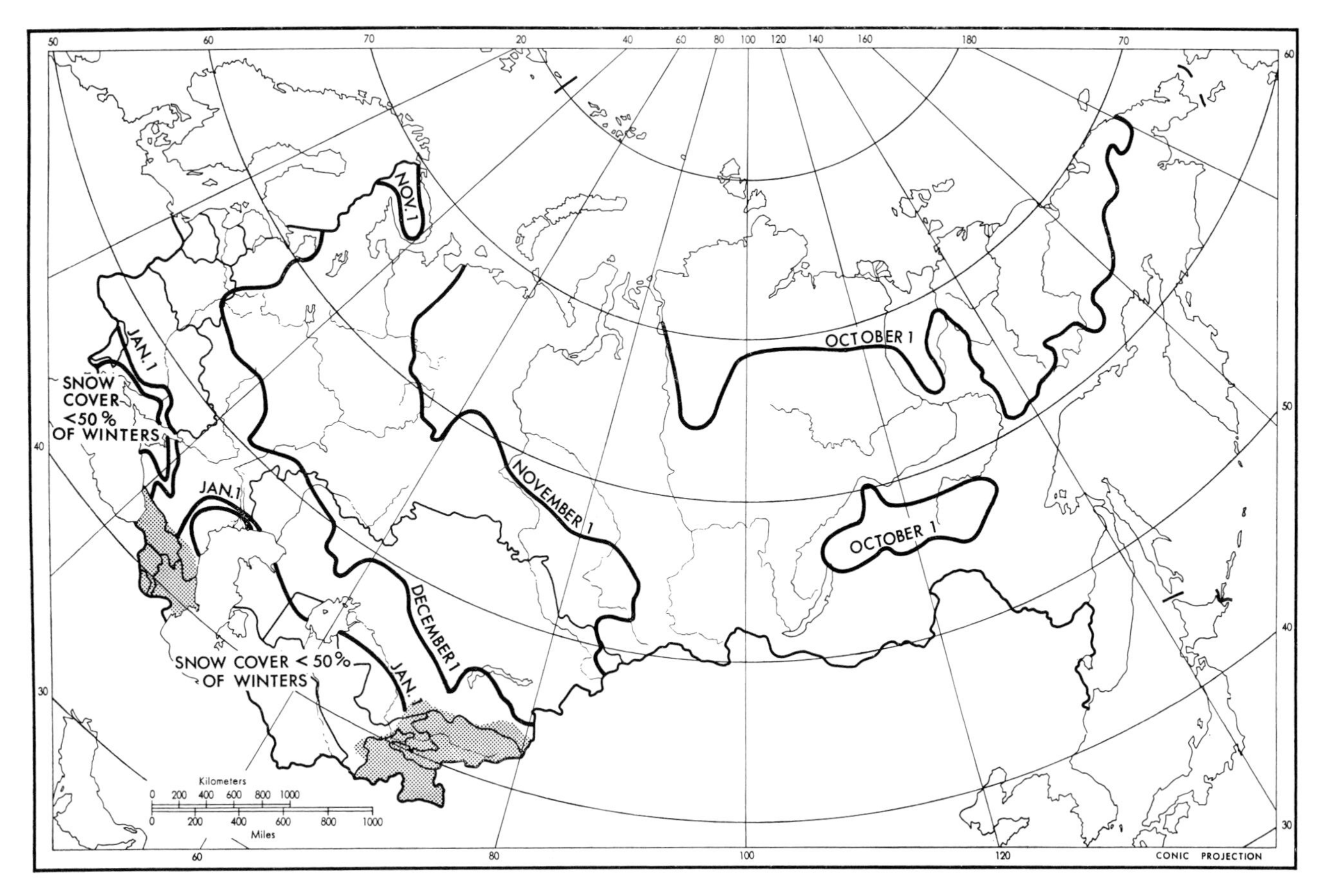

Fig.9-42. Mean date of establishment of snow cover. (After ANONYMOUS, 1960.)

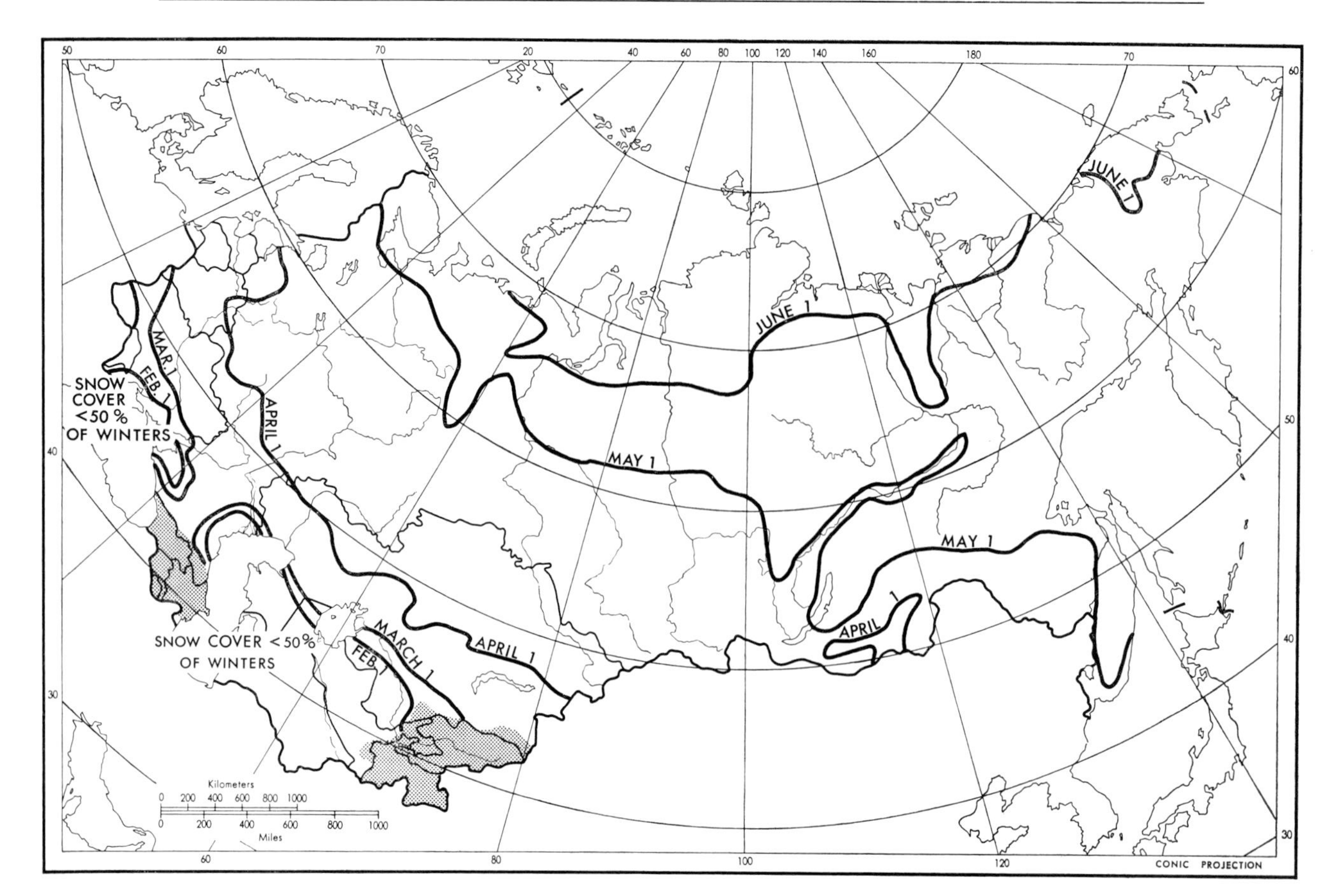

Fig.9-43. Mean date of disappearance of snow cover. (After ANONYMOUS, 1960.)

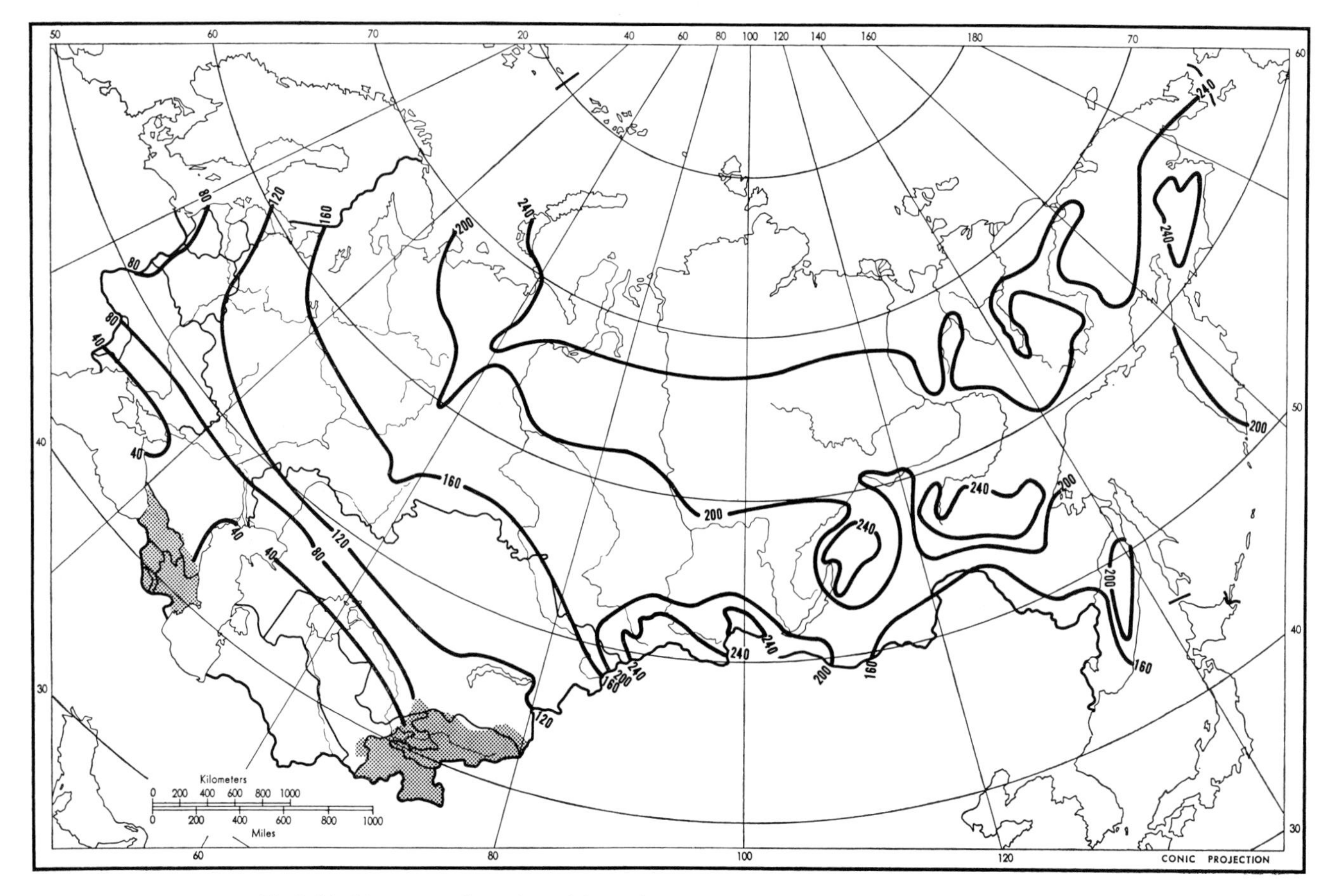

Fig.9-44. Mean annual number of days with snow cover. (After ANONYMOUS, 1960.)

The snow cover begins to form first in the northeastern part of the country and in the high mountains along the southern border of Siberia and the Far East where it begins to form in September. By the end of October almost all of Siberia and the Far East is covered, and by the end of November the edge of the snow cover lies well south of Moscow. During December it forms over much of the rest of the plain except the very southern fringes along the Black Sea coast and east of the Caspian Sea. Generally throughout much of the country persistent snow cover is preceded by two or more temporary snow covers in early winter. Occasionally snow may form in the south before it does in the north if an extremely heavy snowfall accompanies the southerly path of a cyclonic storm. The snow melts very irregularly in spring. The whole season of snow melting from the southwest to the northeast lasts approximately 50 days. Within the interior of the continent the snow cover frequently disappears rapidly under the influence of the sun, particularly during times of warm rain. The central part of Yakutia experiences the disappearance of the snow cover earlier than either the southern or northern parts. This is because the central part has less snow to begin with and it has a great deal of solar heating and great atmospheric transparency in spring.

The number of days during the year with snow cover varies from less than 20 in the south to as much as 280 along the Arctic fringes of Siberia. It lasts even longer than that on many of the islands in the Arctic Ocean as well as on some of the higher mountain peaks in the southern part of the country where snow fields and glaciers remain year round. The extreme western portions of the country experience less than 80 days of snow cover because of midwinter thawing in maritime air masses coming from the Atlantic. But eastward and northward the length of period of snow cover increases rapidly. Much of the central portion of the country experiences 120 to 160 days per year. This includes much of the agricultural land of the country. Northeast European U.S.S.R. and all of Siberia and the Far East except some of the southern basins experience more than 160 days per year and as much as 280 days per year.

Glaze and rime

The formation of ice encrustations on surfaces represents a great hazard to transportation and communication lines as well as to the wintering of livestock and crops in many parts of the country. Glaze ice usually is present from 5 to 45 days during the winter, depending upon locality and topographic position (Fig.9-45). Generally it occurs more frequently on uplands than on lowlands. The Timan Ridge in the European north, the Stavropol' Plateau in the north Caucasus, and the Donets Ridge in eastern Ukraine are particularly noted for this phenomenon. On transmission wires etc. glaze is formed by freezing rain, but on the ground it may be formed by either freezing rain or alternate freezing and thawing of snow. Since midwinter thaws are relatively frequent in the western part of the country and nearly absent in the east, much more glaze is found on the ground in the west than in the east.

Rime ice forms when saturated air is blown against objects with surface temperatures below the freezing point, and therefore it is usually found on windy promontories, particularly in uplands and on coasts.

Both types of ice can accumulate to thicknesses of more than 8 cm on electric and telephone lines and vegetation (Fig.9-46). Under these conditions the great weight might

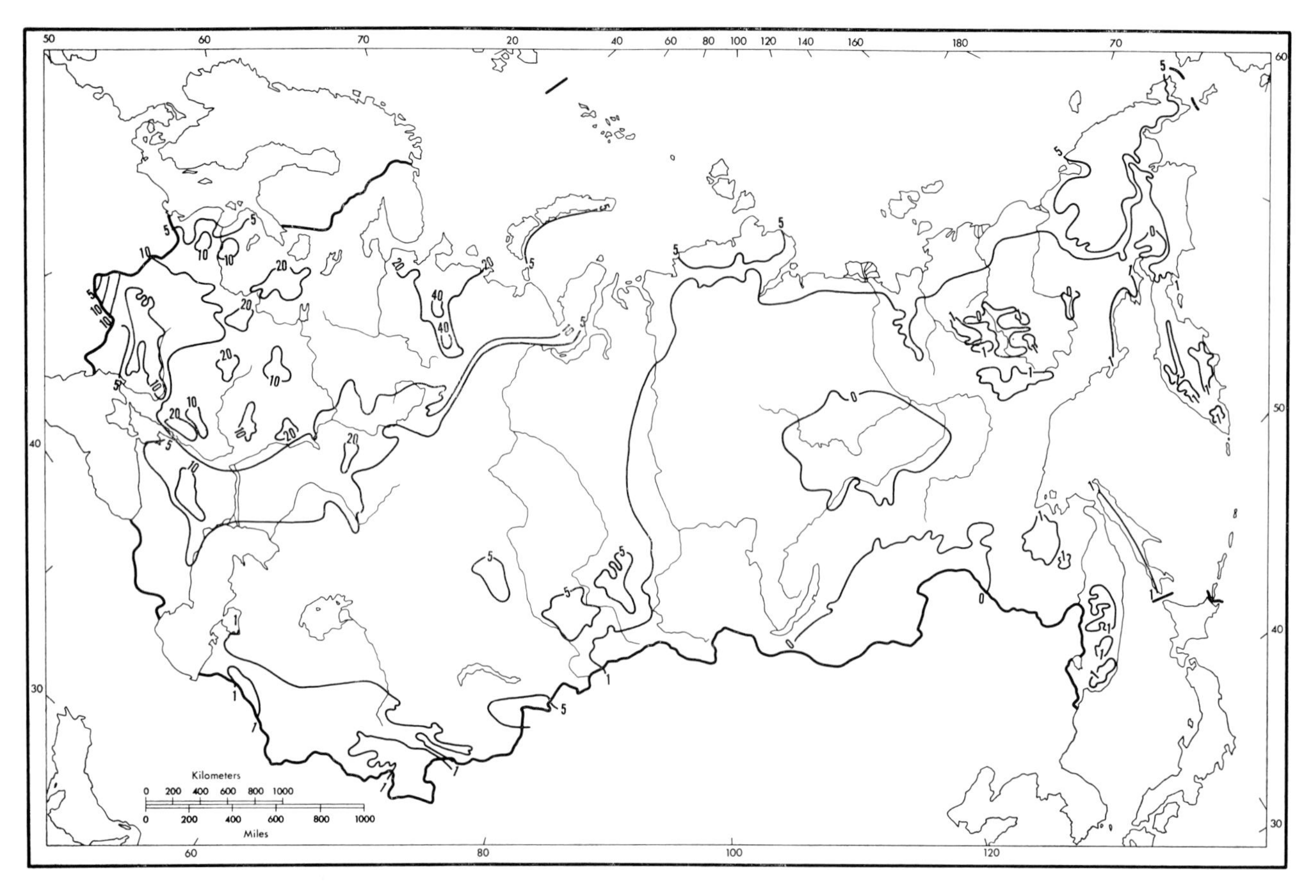

Fig.9-45. Mean annual number of days with glaze ice. (After RUDNEVA, 1961.)

break the object on which the ice has formed. Wet snow which clings to objects may be a similar hazard because of the weight involved. Generally these phenomena are more prevalent in the southwest where moist air is more frequently available during winter. They accumulate to maximum thicknesses in the Urals. In the dry air of the interior of Siberia and the Far East such phenomena are not observed every year.

The synoptic situation which seems to be most conducive to the formation of glaze and rime in European U.S.S.R. is southerly flow of air around the western periphery of the Asiatic high when its western protrusion occupies European U.S.S.R. The formation of glaze and rime is particularly intense in such a flow when a warm front lies just to the south in the Black Sea–Caucasus–Caspian area. This often occurs in association with pre-warm frontal fog.

Water losses

Evaporation

In order to be able to judge the moisture conditions of any given area, one must know the magnitudes of moisture losses, particularly those of evaporation and stream runoff so that these can be balanced against precipitation. Unfortunately, data for evaporation and

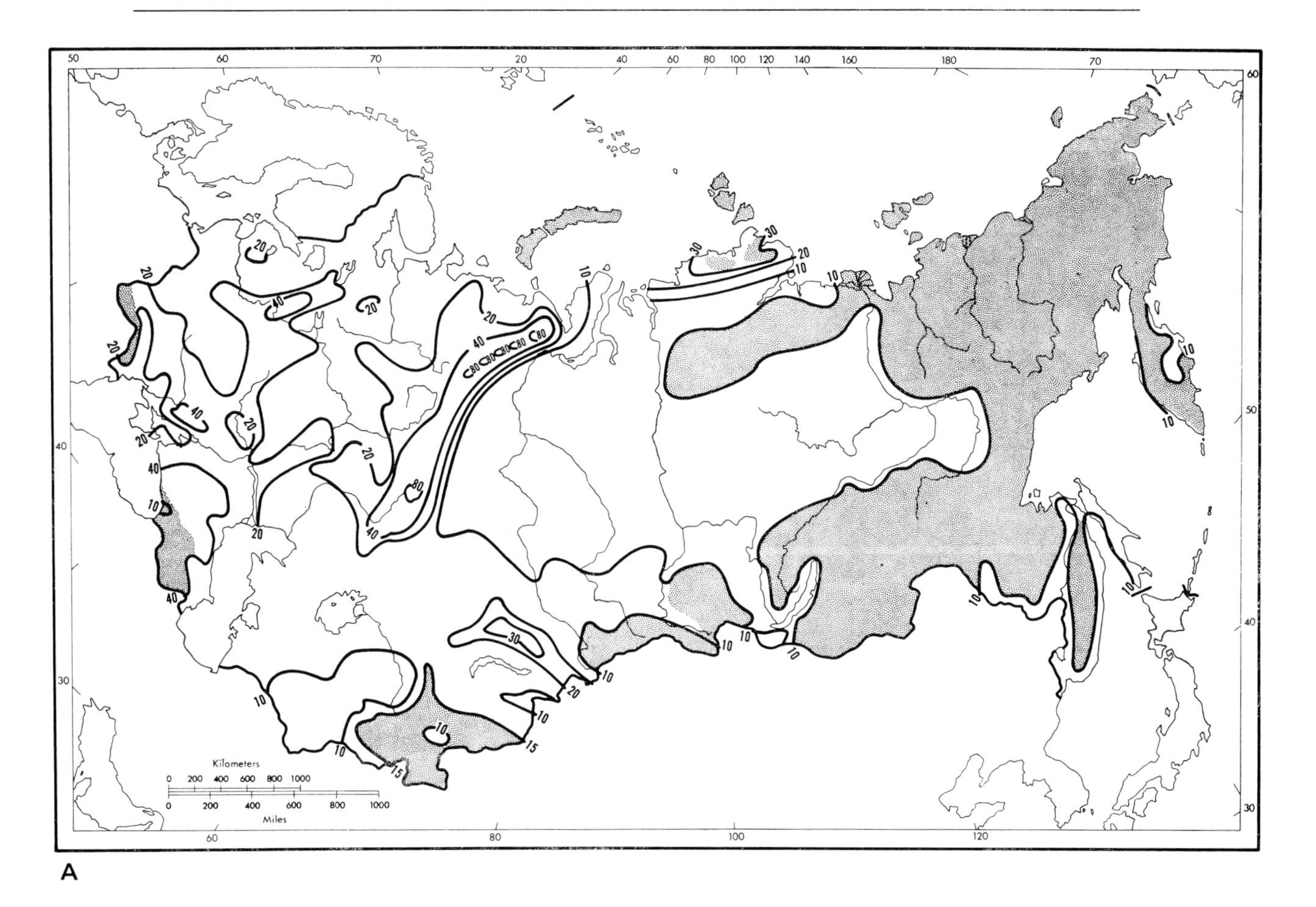

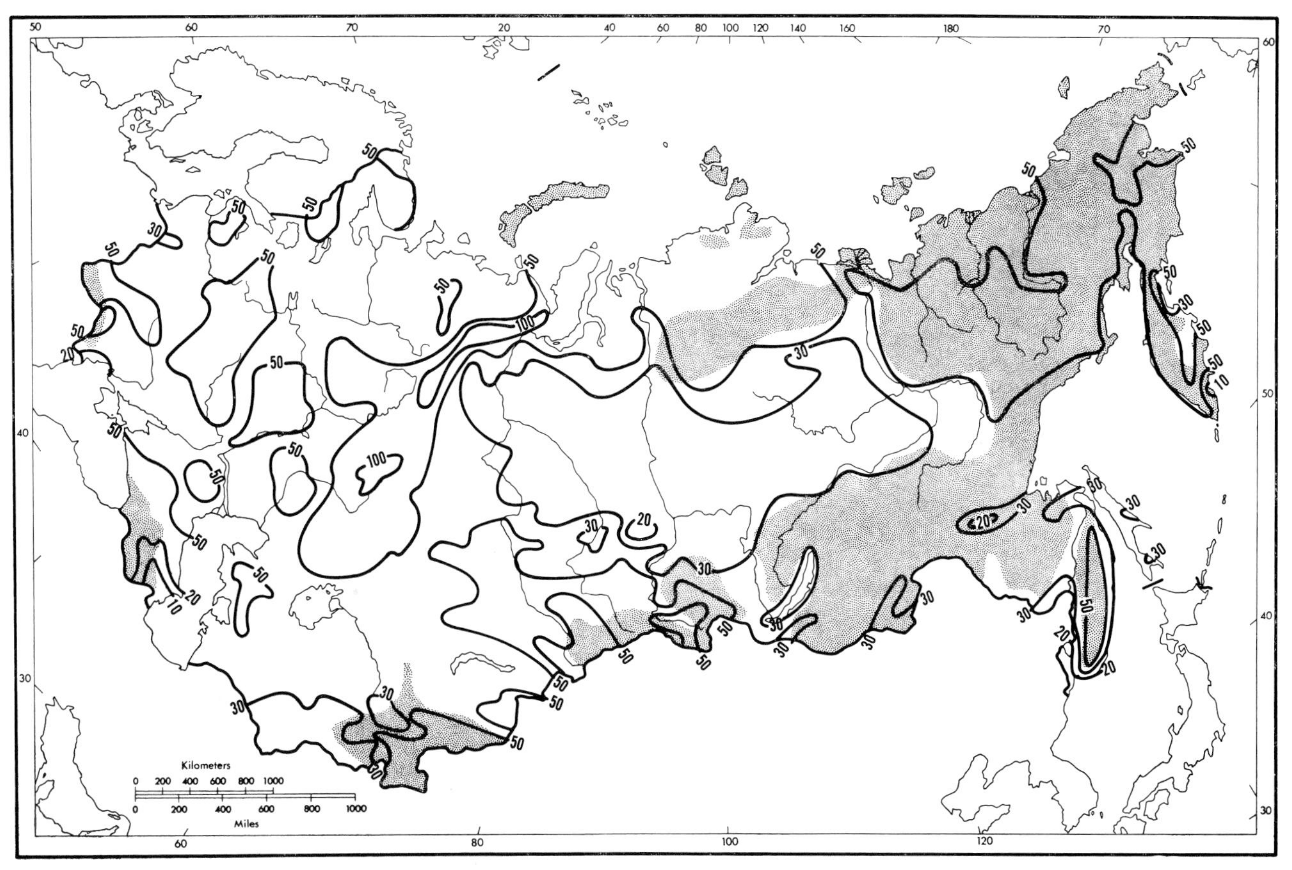

Fig.9-46. Maximum diameters of accumulation (mm) on transmission lines of freezing rain (A) and rime (B). Shading indicates mountains and other areas where few wires exist. (After RUDNEVA, 1961.)

stream runoff are not as good as those for precipitation, and most data that have been made available for evaporation have been computed rather than observed. Nevertheless, Soviet climatologists have been very concerned about the moisture loss factor and have computed values for evaporation that are as good as or better than those in any other part of the world, and therefore we will make use of them here.

Because of its high latitude and associated cool temperatures, the Soviet Union shows only low to moderate values for both potential evaporation and actual evaporation. The potential evaporation, as computed from mean monthly air temperature and mean monthly relative humidity, varies from a low of less than 125 mm along the Arctic coast to a high of somewhat more than 2,250 mm in the southern part of Central Asia (Fig. 9-47). The pattern is quite latitudinal although there are deviations because of topography and water bodies, particularly in the Far East where high atmospheric humidity and cloudiness along the Pacific margins contrast greatly with the dry clear conditions of the interior.

Actual evaporation shows a very different distribution, since it is limited in the north by low temperatures and in the south by low moisture availability (Fig.9-48). The greatest values are found in the southwestern part of the country where 500–600 mm per year are evaporated in the Pripyat Marsh and adjacent regions. A tongue of high values protrudes eastward across the mid-section of European U.S.S.R. and western Siberia. Values decrease eastward toward the Urals but pick up again in the south-central portion of the west Siberian plain where very swampy conditions prevail. All of eastern Siberia and much of the Far East, except the southern portion, shows only very modest values of actual evaporation, and the isolines, in contrast to those for potential evaporation, cut right across the Pacific coast, indicating that about the same amount of actual evaporation takes place in the interior as over water bodies offshore. The dry sunny weather of the interior, coupled with some deficit of soil moisture, produces about the same amount of evaporation as the cool cloudy conditions over the sea with unlimited moisture supply. In Maritime Kray and the lower Amur Valley, as well as on portions of Sakhalin Island and Kamchatka Peninsula, evaporation reaches 450–500 mm, which is almost as much as the central portion of European U.S.S.R. Most of this evaporation takes place during summer when a monsoon flow of air brings warm moist conditions to these eastern regions.

Other relatively high values of evaporation are found in some of the more well moistened mountain slopes of southern Siberia, Central Asia, and the Caucasus. In fact, the highest measured evaporation in the entire country lies in a small area along the southern slopes of the wet western portion of the Great Caucasus, where values of more than 700 mm have been recorded. The lowest values in the country are found along the Arctic coast of eastern Siberia and the Far East which evaporates less than 100 mm per year. Other low values are found in the central portions of Central Asia and Kazakhstan where less than 150 mm per year are evaporated. Such values of course do not represent the water bodies in Central Asia which evaporate more than 1,000 mm per year.

The difference between potential evaporation and actual evaporation, of course, is greatest over Central Asia, where surface temperature and atmospheric conditions are conducive to great amounts of evaporation but soil moisture is inadequate for this evaporation to take place (Fig.9-49). The difference between the two varies from more than 1,200 mm in the central part of Kara-Kum in Central Asia to less than 100

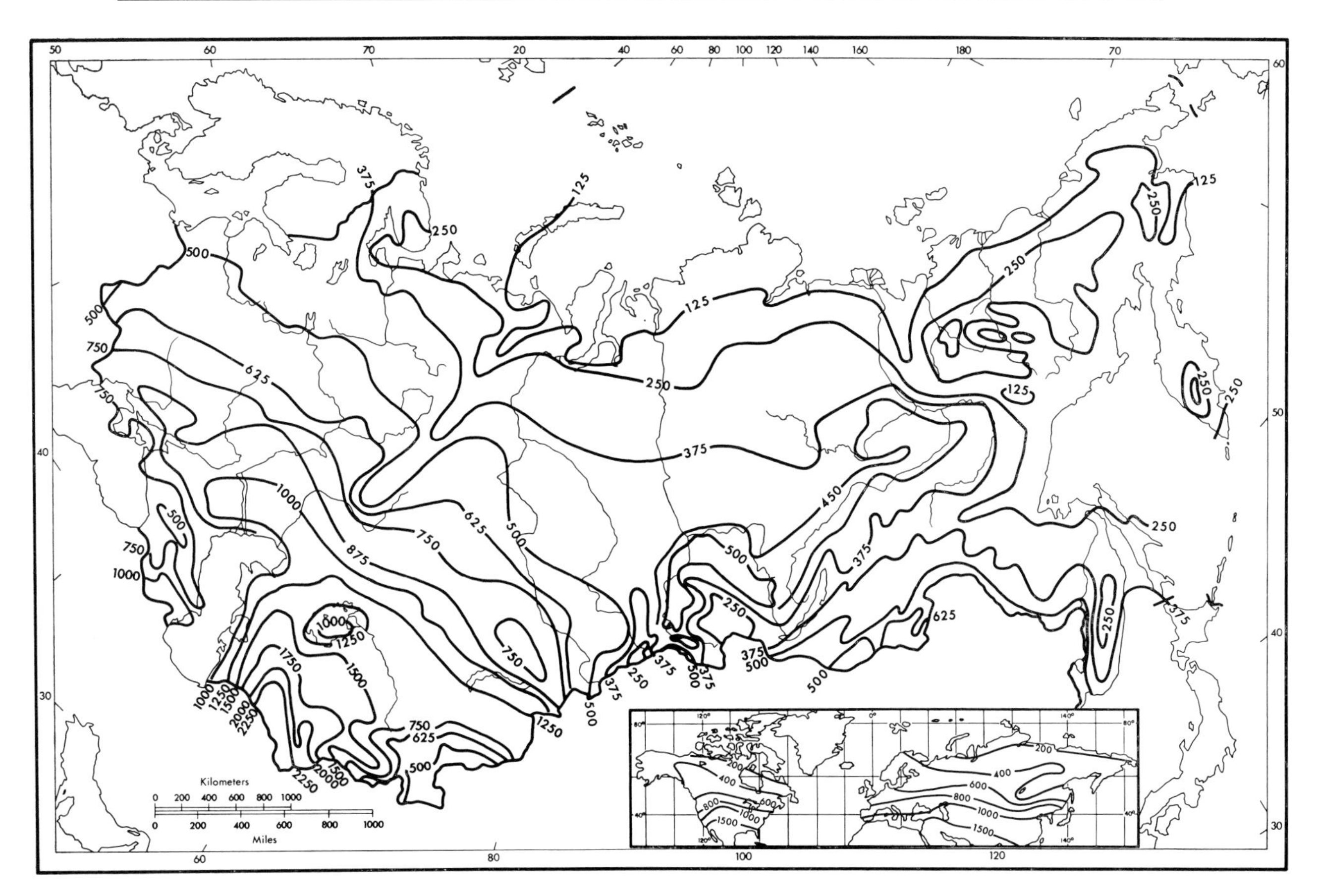

Fig.9-47. Mean annual potential evaporation (mm). (After BORISOV, 1967; inset from GERASIMOV, 1964.)

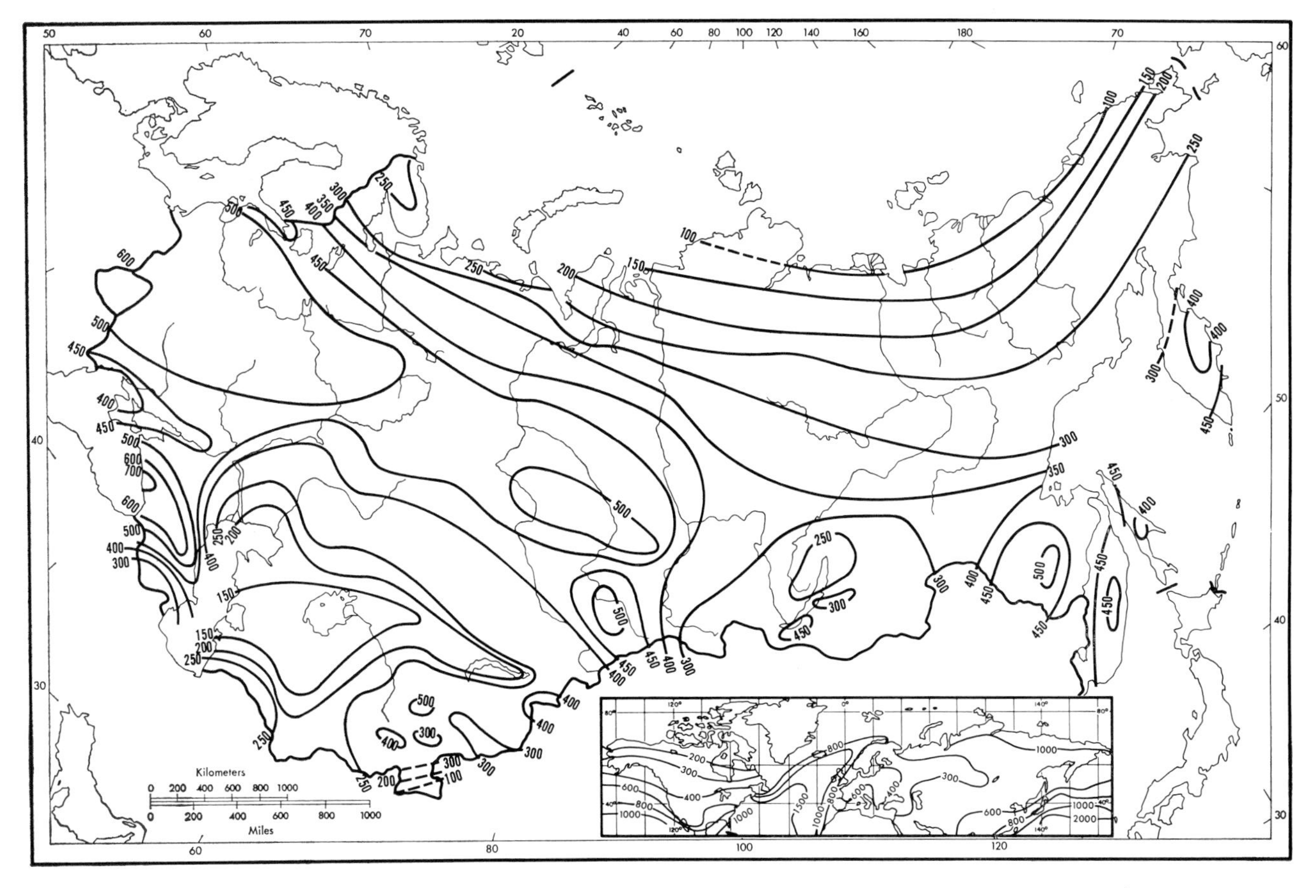

Fig.9-48. Mean annual evaporation (mm). (From ANONYMOUS, 1967; inset from GERASIMOV, 1964.)

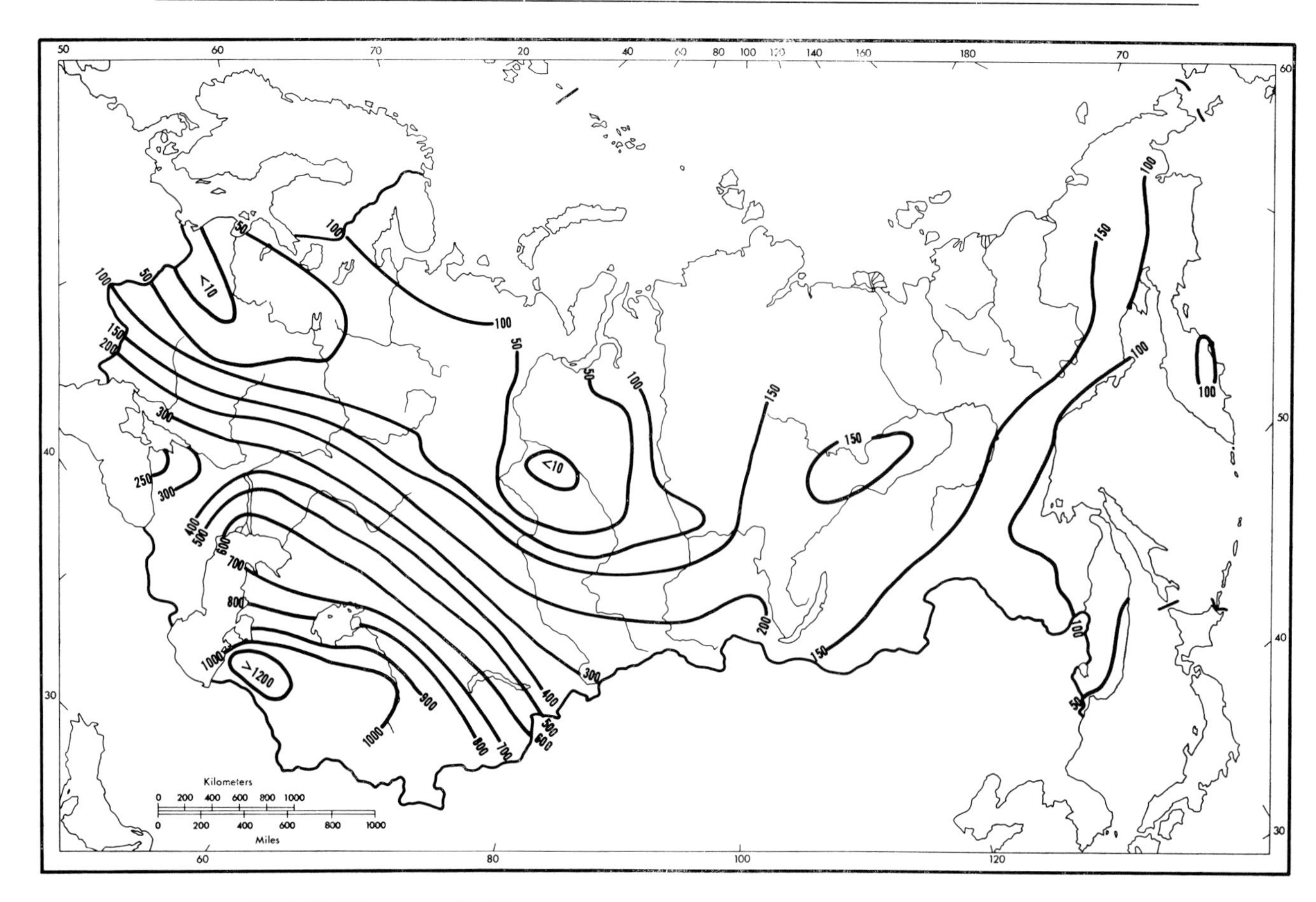

Fig.9-49. Mean annual difference (mm) between potential and actual evaporation. (After DAVITAYA, 1959.)

throughout much of the middle and northern portions of the country. In the western part of the country and in the central Ob' Basin differences dwindle to less than 10 mm. Over the course of a year, there is not much moisture deficit throughout broad sections of the forested zone of the country.

More than 9/10 of potential evaporation is actually evaporated north of 60° latitude in European U.S.S.R. and western Siberia (Fig.9-50). This is true also in the western part of the Transcaucasus and in the southern portions of the Soviet Far East. On the other hand less than 1/10 of the potential evaporation is actually evaporated in the drier parts of Central Asia. It is interesting to note that much of eastern Siberia has smaller ratios of evaporation to potential evaporation than do similar latitudes farther west. This reflects the clear dry conditions that prevail over the middle Lena Valley much of the time, which make this region subhumid in character.

Actual evaporation is practically nil over the entire Soviet Union during winter. *The Atlas of Heat Balance* (BUDYKO, 1963) shows evaporation taking place only over the Black and Caspian seas in the south, the Barents Sea in the north, and seas adjacent to the Pacific coast during November–February. Over the continent during this period evaporation everywhere is less than 10 mm per month. By March much of Central Asia has increased to 10 mm or more as temperatures rise and late winter rains provide the soil moisture necessary for evaporation. Evaporation also becomes significant in March

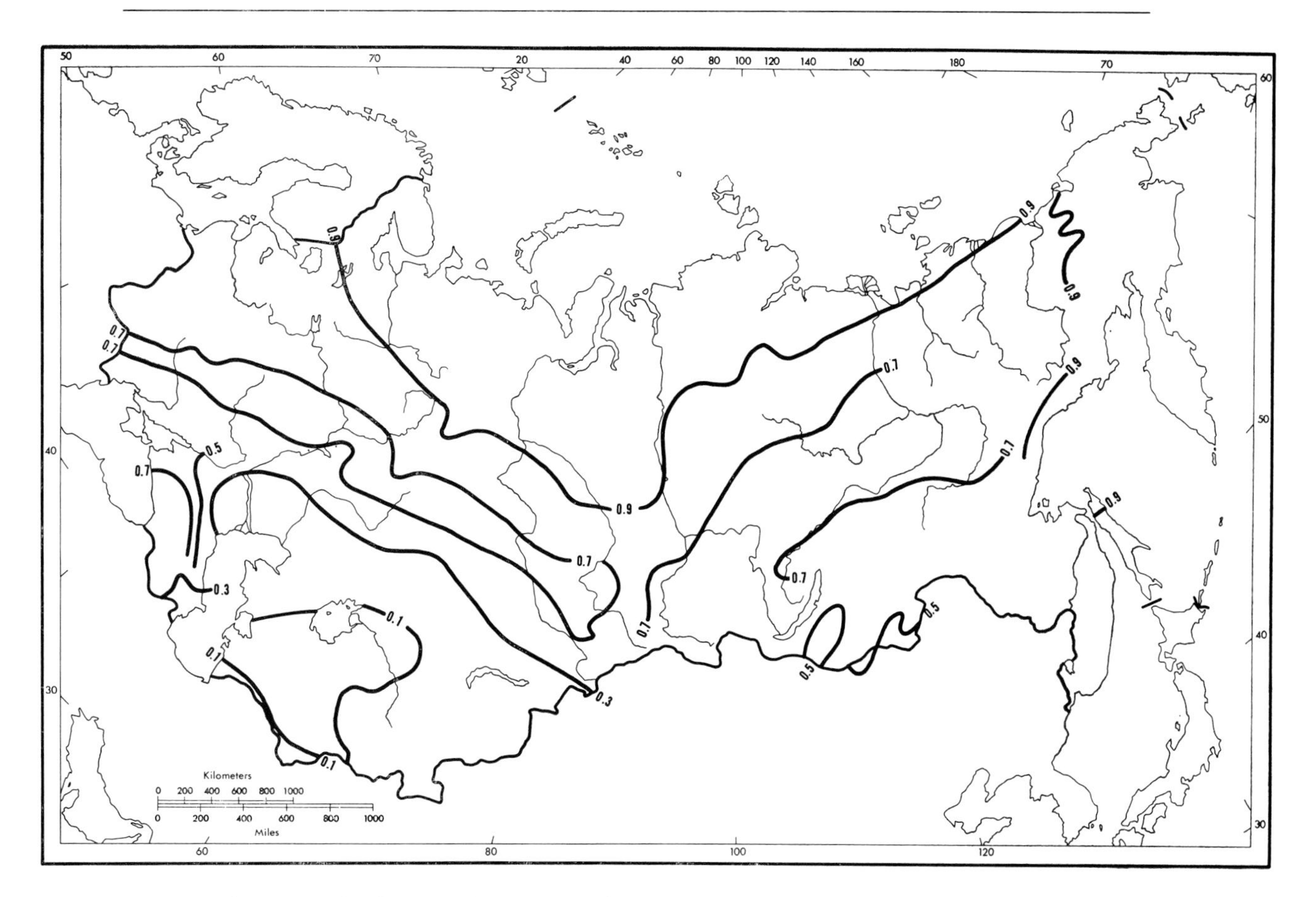

Fig.9-50. Ratio of evaporation to potential evaporation, annual. (After ZUBENOK, 1965.)

east of the Sea of Azov and in the southern portions of the Soviet Far East. Evaporation continues to increase throughout the country during April, and May shows the greatest differentiation of evaporation across the country.

During May the south-central part of the Soviet Union all the way from its western borders to the Yenisey River evaporates 70 mm or more of moisture from the surface (Fig.9-51). A rapid decrease takes place in the northern part of the country to values of less than 10 mm along the Arctic where snow is still lying on the ground. A moisture deficit has already set in in Central Asia which limits evaporation in that region to no more than 30 mm. Therefore, the yearly pattern, with maximum values in the middle of the country and decreasing values to the north and to the south, has become established by May.

This type of pattern continues throughout the summer when most of the evaporation during the year takes place. By far the most evaporation takes place from the Soviet Union during the month of June when most of the country shows values of 70 mm or more. There is a rapid decrease to less than 20 mm on the Taymyr Peninsula on the Arctic coast and to less than 10 mm in the southwestern portion of Central Asia. July shows similar values, but the broad central region of maximum values has shifted northward to engulf much of the Arctic coast at this time, and much of the country other than Central Asia evaporates 70 mm of moisture or more.

By August the sun has shifted southward again, and the turbidity of the atmosphere has

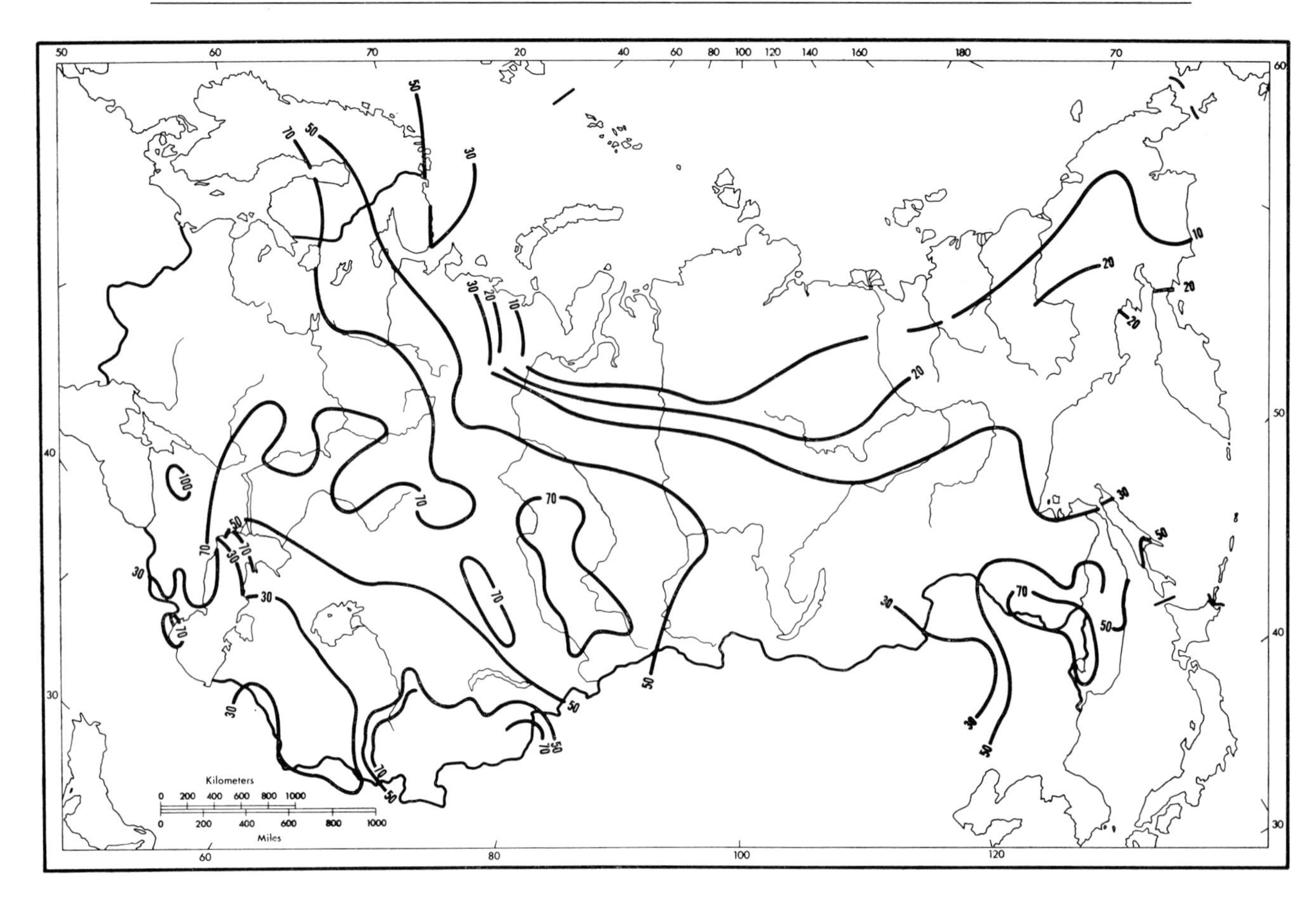

Fig.9-51. Mean evaporation (mm) in May. (After BUDYKO, 1963.)

increased to the point that evaporation suffers a significant drop over most of the country. Only the north-central portion of the plain west of the Yenisey River and the southern portion of the Far East show values of more than 70 mm during August. All of Central Asia and the southern half of Kazakhstan show less than 10 mm, and the Arctic coast shows about 30 mm. A sharp decrease takes place throughout the country during September, and by October a winter situation has definitely set in. Only the southwestern part of the plain, the southern portion of the Soviet Far East, and isolated spots around the Altay Mountains show more than 10 mm of evaporation during October.

Data from agroclimatic handbooks for oblasts in south-central European U.S.S.R. indicate that evaporation from the surfaces of ponds is greatest in June. May is almost as high. July is lower than either June or May. August shows a great decrease from July, and September is somewhat lower than April. This asymmetrical distribution of evaporation with regard to the temperature regime is probably due to two things. First, during spring and early summer the ground is being heated rapidly by increasing insolation. This produces persistent strong heating of the surface air which maintains a considerable saturation deficit in the surface air. Such processes are much reduced in late summer. Second, as the summer wears on, the turbidity of the air increases as more moisture, dust, and other particles are derived from the surface. Therefore, the air is less clear, and evaporation rates tend to be reduced.

Monthly maps for the warm period in ZUBENOK (1965) indicate that May is the month with the maximum ratio of evaporation to potential evaporation throughout much of the country except perhaps the southern portion of the Soviet Far East where the monsoon sets in later. May stands out above succeeding months particularly in the west-central portion of the European part of the country and in Central Asia. During May west-central European U.S.S.R. shows a ratio of more than 9/10, whereas in June it shows a ratio between 7/10 and 9/10 and during July, August, and September less than 7/10. Central Asia in May shows a ratio of about 3/10, whereas during June–September it is less than 1/10. The decrease after May in the ratio between actual evaporation and potential evaporation is probably due primarily to the increase in atmospheric turbidity in western European U.S.S.R. and to the decreasing supply of soil moisture in Central Asia. The soil might also be drying out a little in the better drained parts of the western portion of the country, but that would not seem to be a significant limitation on evaporation as early as May.

Stream runoff

Some measure of the water balance of an area can be gained from a study of stream runoff (Fig.9-52). The pattern is very different from that of evaporation. Generally the values are highest in mountainous areas, particularly where the snow pack is thickest in the winter time. The highest recorded values lie in the western Caucasus which receive

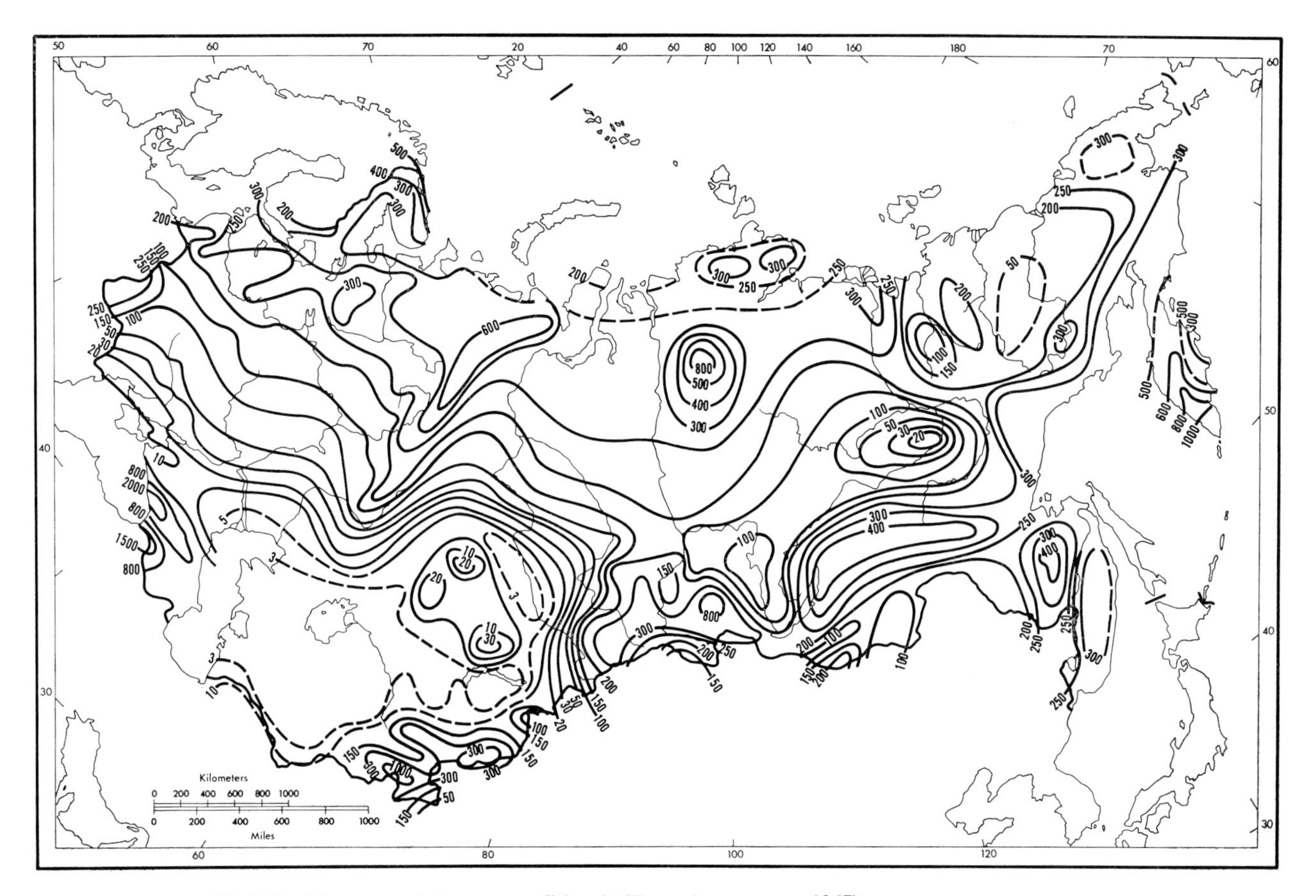

Fig.9-52. Mean annual stream runoff (mm). (From ANONYMOUS, 1967).

much precipitation all year round and heavy snowfall during the winter. The Urals stand out on this map perhaps more than on any other map of hydroclimatic elements. Great variations occur in eastern Siberia which generally is an area of some homogeneity. The Putoran Mountains in the northwestern portion of the Central Siberian Uplands show values of more than 800 mm per year, and the plateaus northeast of Lake Baykal show over 600 mm per year, while not too far to the north the central Lena Basin around Yakutsk shows less than 20 mm per year. Much of the European plain, which shows rather high values of evaporation, shows moderate to low values of runoff.

Generally, surface runoff is much greater than underground runoff (Figs.9-53 and 9-54). This is particularly true in mountainous regions where steep slopes are conducive to the immediate runoff of rainfall and frozen ground in spring and early summer retards the percolation of snow melt. Surface runoff is about five times as great as underground runoff in the desert regions of Central Asia, because of the intensive nature of much of the precipitation, although total runoff here is very small indeed.

Moisture balance, drought, and sukhovey

The moisture balance for the months May–July is presented in Fig.9-55 according to an index called the "hydrothermic coefficient" which was originated by G. T. Selyaninov in 1928 and has subsequently been widely adopted throughout the Soviet Union. The hydrothermic coefficient is defined as the precipitation in millimeters divided by the sum of temperatures in Celsius during the period when the temperature is above 10°C. According to Selyaninov categories of this index correspond to the following types of surface: less than 0.3, desert; 0.3–0.5, semidesert; 0.6–0.8, steppe; 0.8–1.0, forest steppe; and 1.1–1.3, forest. According to Fig.9-55, it can be seen that during the period May–July there is sufficient moisture to support forests north of the 1.0 line except for isolated pockets of drier areas in intermontain basins along the southern fringe of Siberia and in the Yana River Valley and a small area facing the Arctic coast in the northeastern part of the country. Much of the northern half of European U.S.S.R. and the western half of Siberia, as well as large areas in the southern half of the Far East, show a very humid moisture balance at this time. But the southern half of Ukraine, the lower Volga region, the eastern half of the Transcaucasus, the north Caucasus, and all of Kazakhstan and Central Asia show moisture deficits. Of course the deserts of Central Asia show by far the lowest moisture balance values.

Fig.9-56 shows the distribution of frequencies of moderate and severe droughts during the period 1885–1954 in percents of all the years. It can be seen that European U.S.S.R. and western Siberia northwest of a line from northern Moldavia northeastward through Gorkiy and then eastward across the central Urals to the middle Yenisey Valley and beyond seldom suffer from significant drought. But southeast of this line there is a rapid transition to the Caspian lowland where drought is commonplace. In the northeastern extremity of the country it is significant that river valleys such as the Yana and the middle Lena suffer drought 20–30% of all years.

Fig.9-57 gives an indication of soil moisture conditions at the beginning of the growing season in much of the agricultural land of the U.S.S.R. North of a line running north of Kiev and south of Moscow eastward to Tobolsk and Irkutsk there is little if any saturation deficit in the soil after the spring thawing period. But south of this line deficits

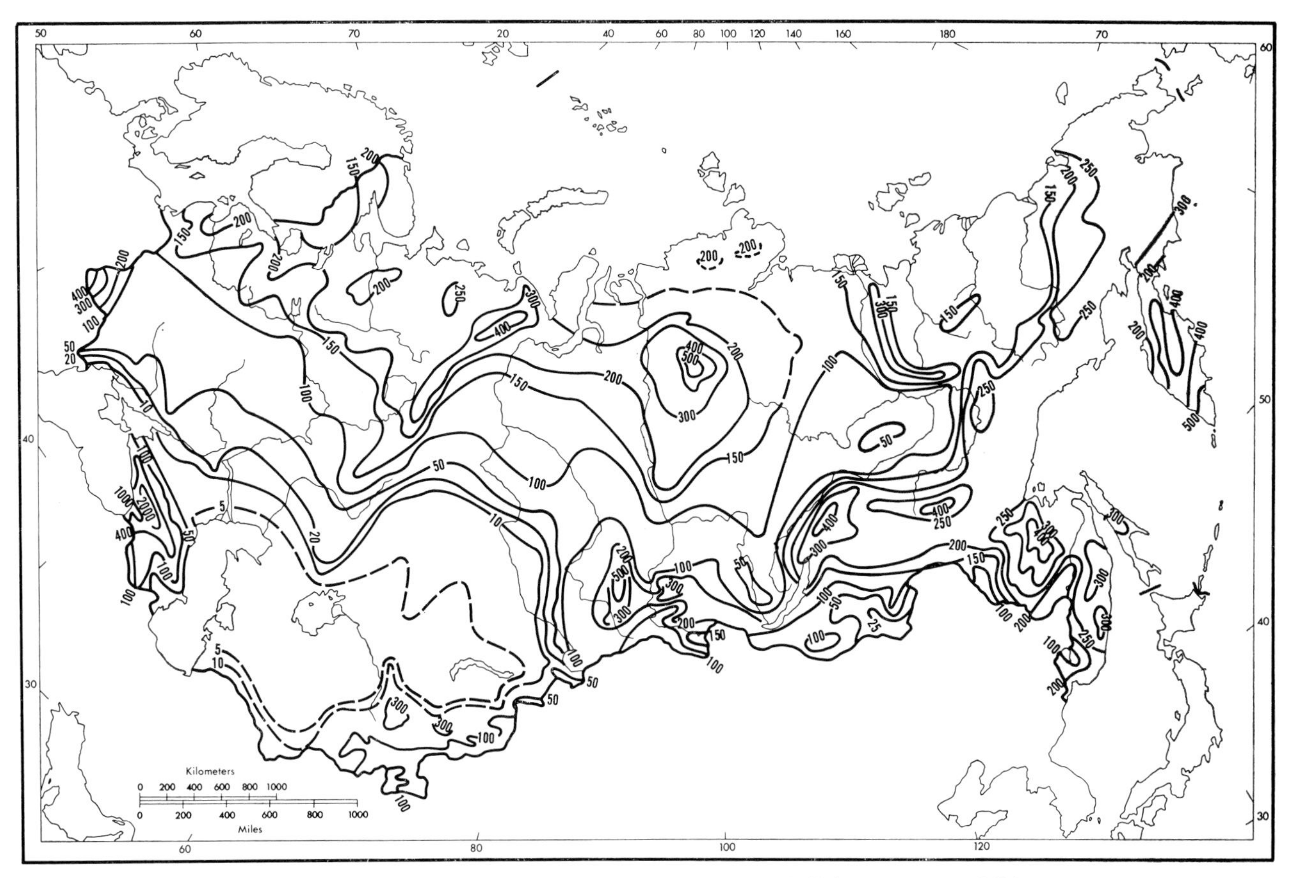

Fig.9-53. Mean annual stream runoff fed by surface water (mm). (After L'VOVICH, 1969.)

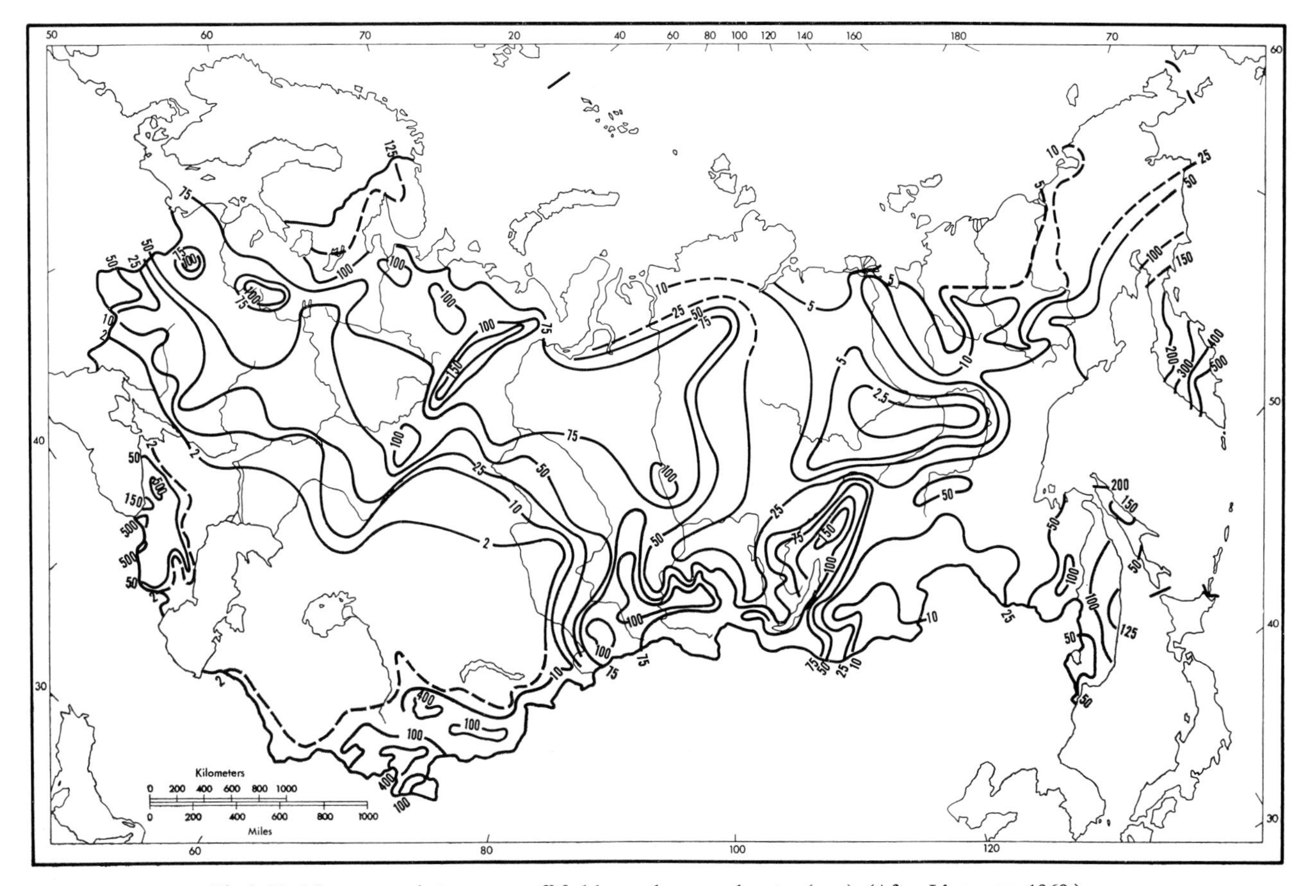

Fig.9-54. Mean annual stream runoff fed by underground water (mm). (After L'VOVICH, 1969.)

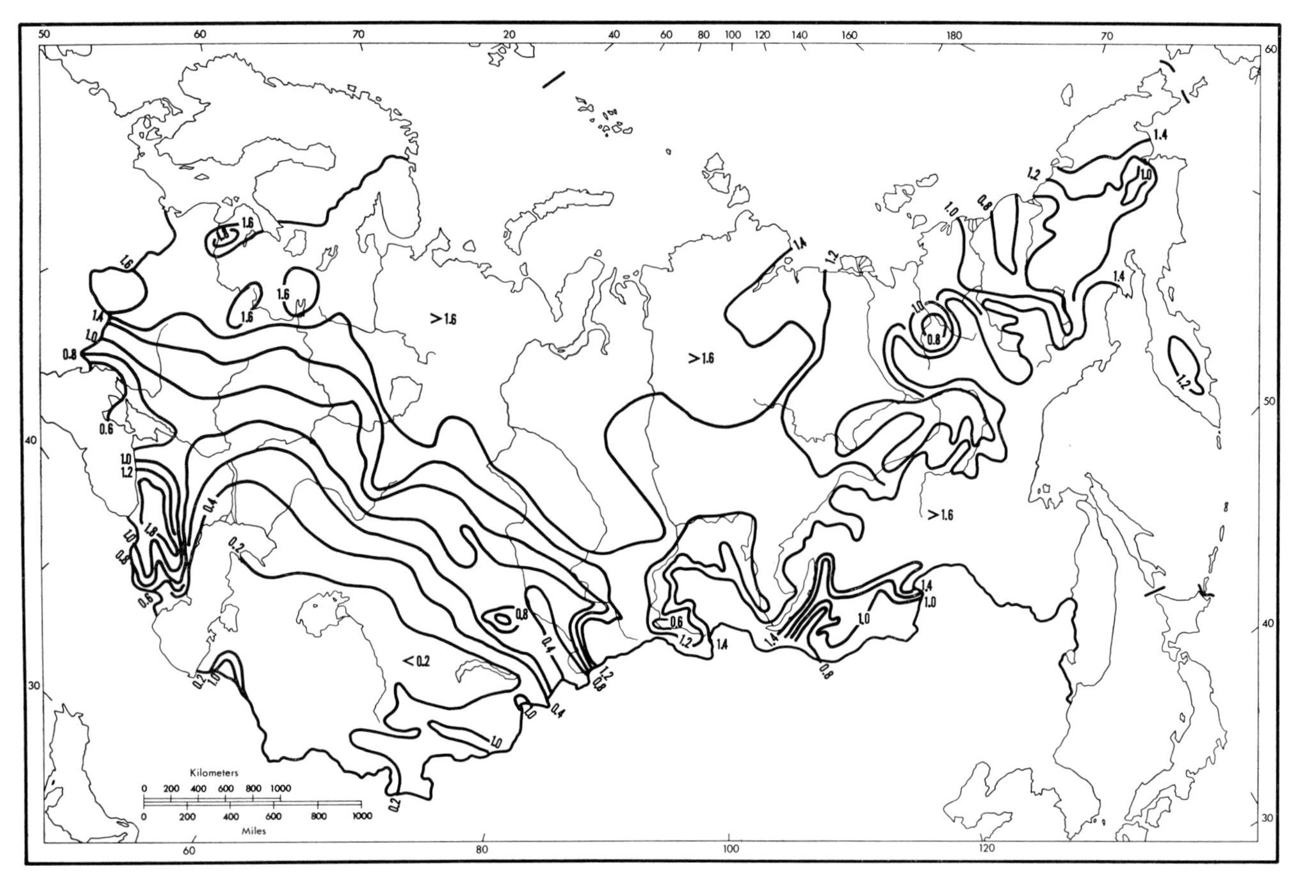

Fig.9-55. Moisture supply according to Selyaninov's hydrothermic coefficient, May–July. (After RUDENKO, 1958.)

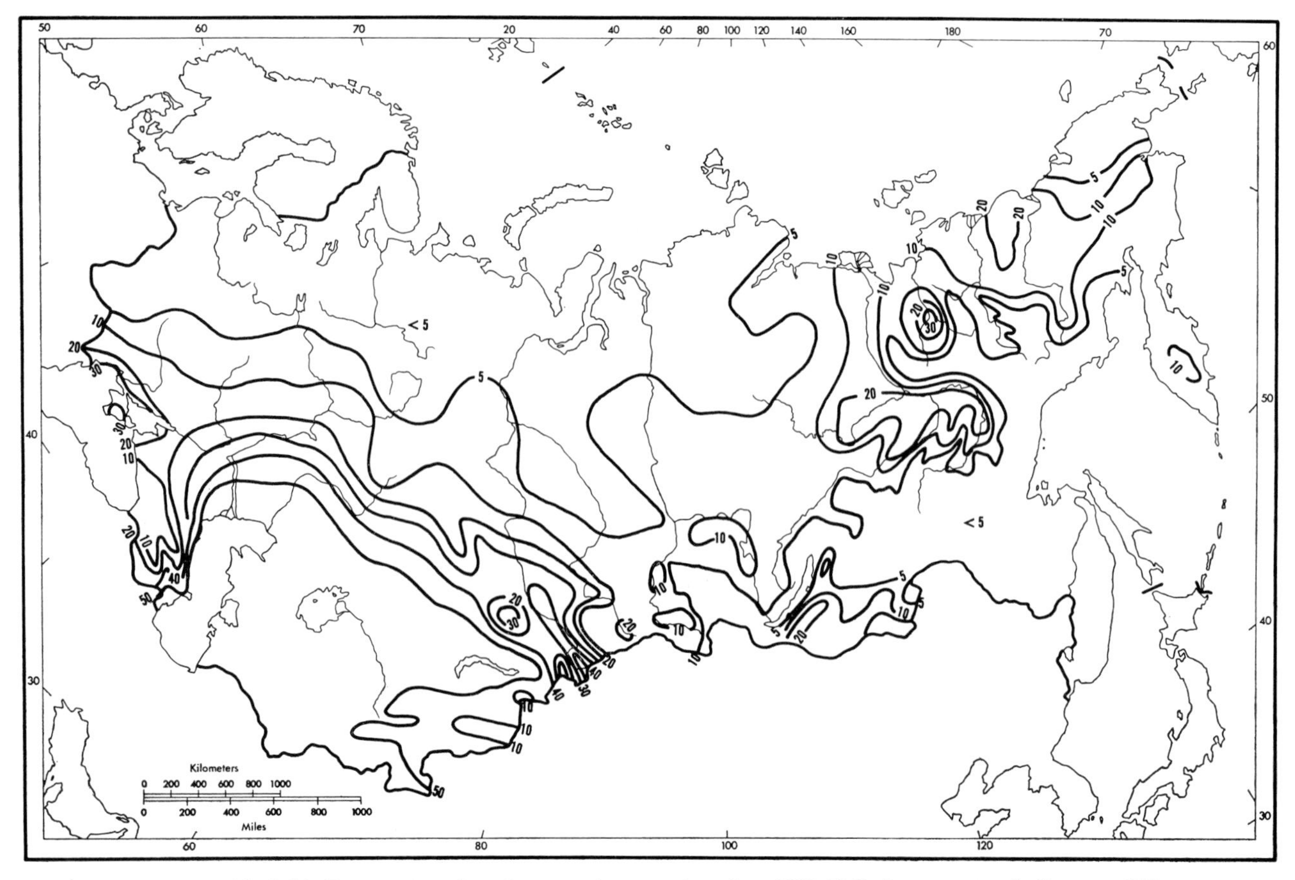

Fig.9-56. Frequencies of moderate and severe droughts, 1885–1954, in percents of all years. (After RUDENKO, 1958.)

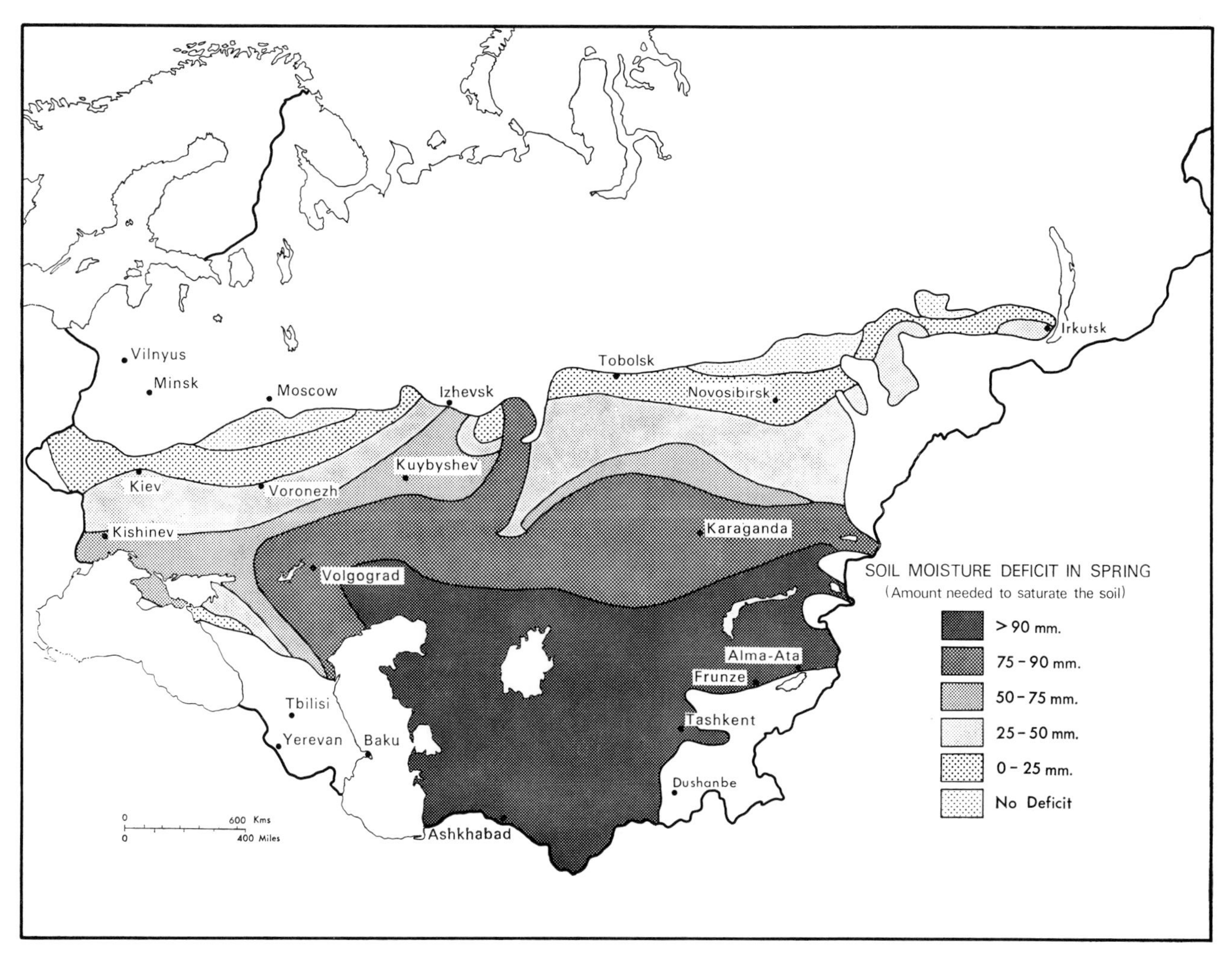

Fig.9-57. Degrees of drought as signified by soil moisture deficits in spring (amounts needed to saturate the soil, in mm). (After LYDOLPH, 1964.)

increase rapidly, particularly in southeastern European U.S.S.R., and throughout the Caspian Lowland and eastward through central and southern Kazakhstan and all of Central Asia the soil would be able to hold at least an additional 90 mm of moisture. Moisture deficits do not occur simultaneously throughout the agricultural land of the Soviet Union. According to DAVITAYA (1958) during 62 years drought occurred 26 times in Ukraine and 28 times in Kazakhstan, but only 3 times in both areas simultaneously. He explains the reason for this to be the distance of the separation of the two regions which is approximately equal to $\frac{1}{2}$ wavelength of the standing waves of the upper troposphere. Thus precipitation in the two areas tends to fluctuate in opposite directions. This was a significant factor in governmental decisions to plow up the virgin lands of northern Kazakhstan. It would ensure a good wheat crop in either Ukraine or Kazakhstan each year.

A phenomenon often associated with drought, and sometimes occurring even without drought as a background, is the so-called "sukhovey", which is a Russian term signifying very desiccating conditions for plants. This might be brought on by a combination of factors, such as soil moisture deficits, low atmospheric humidities, high temperatures, and high wind speeds, which in combination put a great strain on the plant when it tries to transpire moisture from its leaves faster than it is taking it up through its roots. This is usually a relatively sudden occurrence which may cause severe damage to plants in a very short time, but it can last for several days. There are few occurrences of this phenomenon as far northwest as Moscow, although occasionally it can happen (Fig.9-58). It is most devastating in the Black Sea steppes and in the North Caucasus where well developed agriculture normally relies on natural precipitation and uncontrolled weather conditions.

Another phenomenon which is a product of drought is the dust storm. The southern part of the country is famous for its dust storms. Occasionally dust may be carried in the air from southeast to northwest all the way across European U.S.S.R. clear into Scandinavia. Fig.9-59 shows that the greatest frequency of dust storms occurs in the lower Ural River Valley and eastward in northern Kazakhstan and southwestern Siberia, as well as in the north Caucasus, the Black Sea steppes of southern Ukraine, and in the northern part of the Crimean Peninsula. They are also relatively common in the eastern part of the Transcaucasus where they frequently blow into the large city of Baku and cause miserable conditions for the residents there. Severe deflation of the soil takes place in many locali-

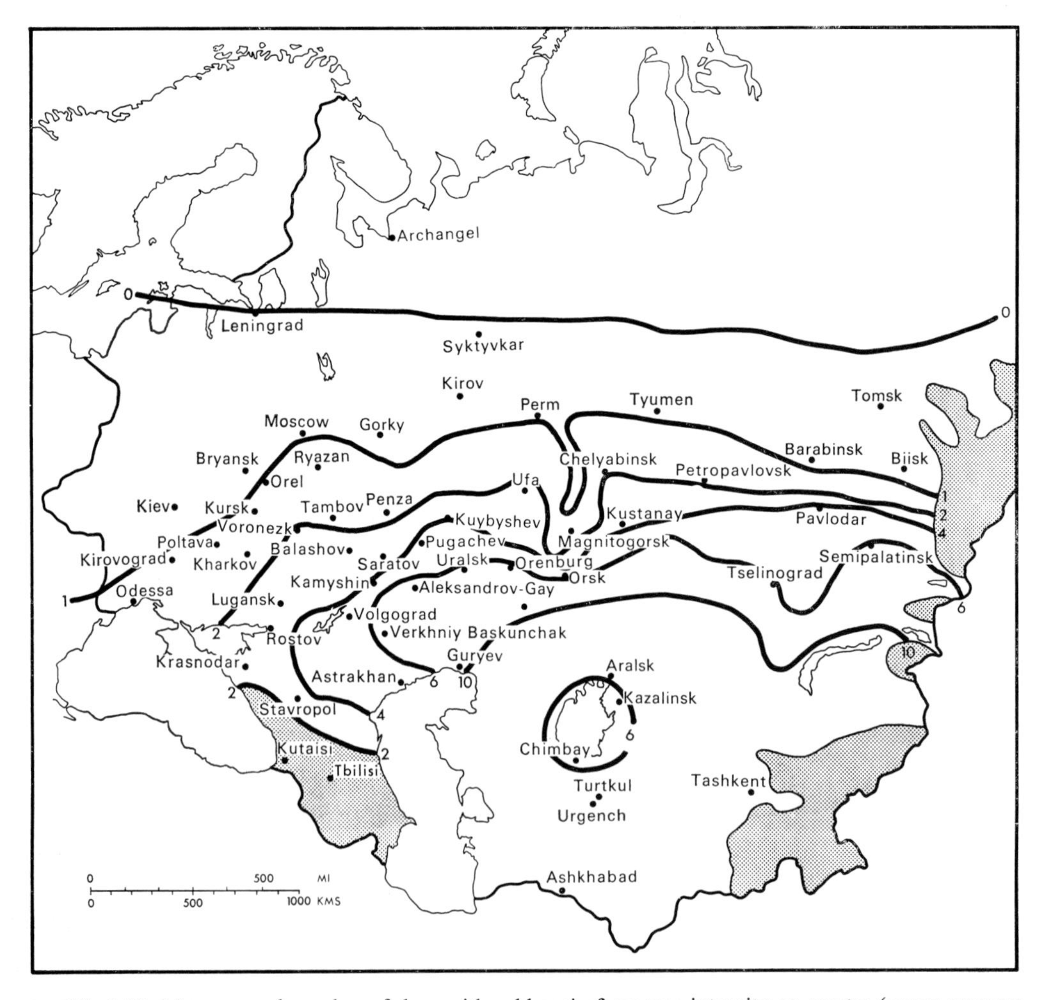

Fig.9-58. Mean annual number of days with sukhovei of average intensity or greater (vapor pressure deficit ≥ 20 mm) during the growing season of winter grains. (After LYDOLPH, 1964.)

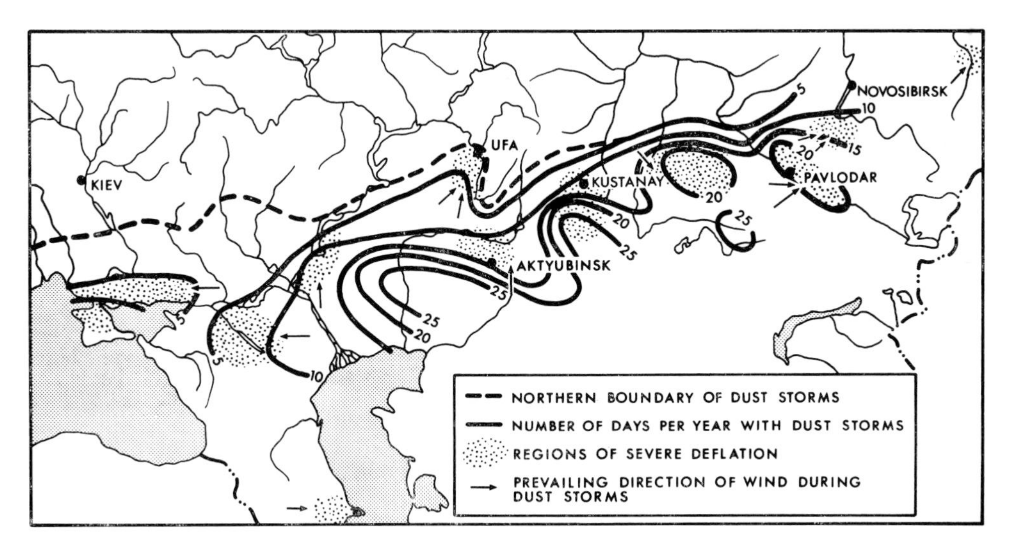

Fig.9-59. The distribution of dust storms in the USSR. (After ZAKHOROV, 1965.)

ties within this general belt. Deposits of wind-blown soil can be found over wide portions of adjacent areas to the north and west of this belt of dust storms. The mineral-rich loess, or wind-blown soil, has added a great deal of fertility to the surface soils throughout much of Ukraine and adjacent areas.

A very thick belt of loess deposits rings the northern foothills of the southern mountains in Soviet Central Asia. In this case the dust has been blown from the north and west southeastward out of the desert regions. A high quantity of dust is commonplace in the air throughout most of the settled oases of the mountain foothills of Central Asia. Written accounts by early travelers generally mention the haziness of the atmosphere in this region particularly over towns where the activity of man and animals constantly stir dust into the atmosphere. Frequently in this region dust rises to 2–3 km in the air, and sometimes even higher. Undoubtedly the dustiness of the atmosphere is a significant factor affecting heat and moisture balances.

References and further reading

ARKHIPOVA, E. P., 1957. Karty geograficheskogo raspredeleniya chisla dney s grozoy na territorii SSSR (Maps of geographical distribution of the number of days with thunderstorms in the USSR). *Gl. Geofiz. Obs. Tr.*, 74: 41–60.

ANONYMOUS, 1933. *Klimatologicheskiy Atlas SSSR* (Climatological atlas of the USSR). *Tr. Konf. razmeshcheniyu Proizvod. SSSR, vyp. III.* Izd. Gosp.ana SSSR, Leningrad, 72 pp.

ANONYMOUS, 1957. Grozy i grad na territorii SSSR (Thunderstorms and hail in the USSR). *Gl. Geofiz Obs. Tr.*, 74: 103 pp.

ANONYMOUS, 1958–1963. *Klimat SSSR.* (Climate of the USSR). Gidrometeoizdat, Leningrad, 6 regional volumes.

ANONYMOUS, 1960. *Atlas sel'skogo Khozyaystva SSSR* (Atlas of agriculture of the USSR). Gl. Upr. *Geod Kartogr.*, Moscow, pp.32–44.

ANONYMOUS, 1967. *Vodnye Resursy i Vodnyy Balans Territorii Sovetskogo Soyuza* (Water resources and the water balance of the USSR). Gidrometeoizdat, Leningrad, 199 pp.

ANONYMOUS, 1969. *Vodnyy Balans SSSR i yego Preobrazovaniye* (The water balance of the USSR and its transformation). Nauka, Moscow.

BATTALOV, F. Z., 1968. *Mnogoletnie Kolebaniya Atmosfernykh Osadkov i Vychislenie Norm Osadkov* (Long term variations in precipitation and the computation of precipitation norms). Gidrometeoizdat, Leningrad, 183 pp.

BORISOV, A. A., 1967. *Klimaty SSSR* (Climates of the USSR). Izd. Prosveshchenie, Moscow.

BUDYKO, M. I. (Editor), 1963. *Atlas Teplovogo Balansa Zemnogo Shara* (Atlas of heat balance of the earth). *Gl. Geofiz. Obs.*, Moscow, 69 pp.

CHISTYAKOV, A. D., SOROCHINSKIY, M. A., YUSHENKO, G. P. and SHUSHAKOVA, L. G., 1968. Raspredelenie po territorii sovetskogo soyuza sil'nykh dozhdey za period s 1951 do 1965 g. (The distribution of intense rainfall over the USSR during the period 1951–1965). *Tr., Ord. Lenina Gidrometeorol. Nauchnoizsled. Tsentr. SSSR*, 32: 3–15.

DAVITAYA, F. F., 1958. Issledovanie zasukh i sukhoveev (Studies of drought and sukhovei). *Izv. Akad. Nauk SSSR, Ser. Geogr.*, 5: 131–136.

DAVITAYA, F. F., 1959. Nauchnye osnovy bor'by s zasukhoy po prirodnym zonam SSSR (Scientific principles for combating drought according to natural zones in the USSR). *Izv. Akad. Nauk SSSR, Ser. Geogr.*, 1: 7–28.

DAVITAYA, F. F. and MEL'NIK, YU. S., 1970. *Problema Prognoza Isparyaemosti i Orositel'nykh Norm*, (Problems of predicting potential evaporation and irrigation norms). Gidrometeoizdat, Leningrad 72 pp.

DROZDOV, O. A. and GRIGOR'EVA, A. S., 1963. *Vlagooborot v Atmosfere*. Gidrometeoizdat, Leningrad. (Published in English as *The Hydrologic Cycle in the Atmosphere*. Isr. Progr. Sci. Transl., Jerusalem, 1965, No.1375, 282 pp.)

DROZDOV, O. A. and GRIGOR'EVA, A. S., 1969. O proyavlenii tsiklichnosti v khode kolichestva atmosfernykh osadkov na territorii SSSR (On the cyclic nature of the march of precipitation in the USSR). *Gl. Geofiz. Obs. Tr.*, 245: 3–13.

DZERDZEEVSKIY, B. L. (Editor), 1957. *Sukhovei ikh Proiskhozhdenie i Bor'ba s Nimi*. Akad. Nauk, Moscow. (Published in English as *Sukhoveis and Drought Control*. Isr. Progr. Sci. Transl., Jerusalem, 1963, No.890, 366 pp.)

DZERDZEEVSKIY, B. L. (Editor), 1960. *Gidroklimaticheskii Rezhim Lesostepnoi i Stepnoi zon SSSR v Zasusblivye i Vlazhnye Gody* (The hydroclimatic regimes in the forest–steppe and steppe zones of the USSR during dry and wet years). Akad. Nauk, Moscow, 169 pp.

EVSEEV, P. K., 1958. Raspredelenie livnevykh i oblozhnykh osadkov na territorii SSSR (The distribution of heavy showers and continuous prolonged precipitation in the USSR). *Meteorol. Gidrol.*, 3: 13–20.

GALAKHOV, N. N., 1959. *Izuchenie Struktury Klimaticheskikh Sezonov Goda* (A study of the structure of climatic seasons of the year). Akad. Nauk, Moscow, 182 pp.

GERASIMOV, I. P., 1964. *Fiziko-Geograficheskiy Atlas Mira* (Physical-geographic atlas of the world). Akad. Nauk, Moscow, 298 pp. (All legends and textual materials translated by Theodore Shabad and published in *Sov. Geogr. Rev. Rransl.* May–June, 1965, 403 pp.)

GLEBOVA, M. YA., 1958. Napravlenie metelnykh vetrov na evropeyskoy territorii SSSR (The directions of winds with snowstorms in European USSR). *Gl. Geofiz. Obs. Tr.*, 85: 73–80.

KAMINSKY, A. A., 1934. Types of droughts and dry winds of the plains in the USSR. *Gl. Geofiz. Obs. Tr.*, 1: 7–48.

KOTLYAKOV, V. M., 1968. *Snezhnyy Pokrov Zemli i Ledniki* (Snow cover on the earth and glaciers). Gidrometeoizdat, Leningrad, 479 pp.

LEBEDEV, A. N., 1964. *Prodolzhitel'nost' Dozhdey na Territorii SSSR* (The duration of precipitation in the USSR). Gidrometeoizdat, Leningrad, 510 pp.

LOBODIN, T. V., 1973. Duration of thunderstorms over the territory of the USSR. *Meteorol. Hydrol.*, 2: 85–90.

L'VOVICH, M. I. (Editor), 1969. *Vodnyy Balans SSSR i ego Preobrazovanie* (The water balance of the USSR and its transformation). Nauka, Moscow, 338 pp.

LYDOLPH, P. E., 1964. The Russian sukhovey. *Ann. Assoc. Am. Geogr.*, 54: 291–309.

MILLER, D. B. and FEDDES, R. G., 1971. *Global Atlas of Relative Cloud Cover, 1967–70, Based on Photographic Signals from Meteorological Satellites*. U.S. Dept. of Commerce and U.S. Air Force, Washington, D.C., 237 pp.

OBRUCHEV, V. A., 1951. Rol' i znachenie pyli v prirode (The role and significance of dust in nature). *Izv. Akad. Nauk SSSR, Ser. Geogr.*, 3: 15–27.

PASMUKH, V. P. and SOKHRINA, R. F., 1957. Grad na territorii SSSR (Hail in the USSR). *Gl. Geofiz. Obs. Tr.*, 74: 3–31.

PUPKOV, V. N., 1964. Formirovanie, raspredelinie i izmenchivost snezhnogo pokrova na aziatskoy territorii SSSR (Formation, distribution, and variability of snow cover over Asiatic USSR). *Meteorol. Gidrol.*, 8: 34–40.

RIKHTER, G. D. (Editor), 1960. *Geografiya Snezhnogo Pokrova* (Geography of snow cover). Akad. Nauk, Moscow, 222 pp.

RUDENKO, A. I. (Editor), 1958. *Zasukhi v SSSR: ikh Proiskhozhdenie, Povtoryaemost', i Vliyanie na Urozhay* (Droughts in the USSR: their causes, frequencies, and influences on crop yields). Gidrometeoizdat, Leningrad, 207 pp.

RUDNEVA, A. V., 1961. *Gololed i Obledenenie Provodov na Territorii SSSR* (Glaze and ice covered transmission lines in the USSR). Gidrometeoizdat, Leningrad, 174 pp.

SAVINA, S. S., 1963. *Gidrometeorologicheskii Pokazatel' Zasukhi i ego Raspredelenie na Territorii Evropeyskoy Chasti SSSR* (The hydrometeorological index of drought and its distribution over European USSR). Akad. Nauk, Moscow, 103 pp.

SCHLOSS, M., 1962. Cloud cover of the Soviet Union. *Geogr. Rev.*, 52: 389–399.

SLABKOVICH, G. I. (Editor), 1968. *Klimaticheskiy Atlas Ukrainskoy SSR* (Climatic atlas of Ukraine). Gidrometeoizdat, Leningrad, 232 pp.

SOROCHAN, O. G., 1966. Mussony Azii (Monsoon Asia). In: M. I. BUDYKO (Editor), *Sovremennye Problemy Klimatologii.* Gidrometeoizdat, Leningrad, pp.202–221. (Translated as *Modern Problems of Climatology*, Foreign Technology Division, Wright-Patterson Air Force Base, Ohio, 1967.)

STEPANOVA, N. A., 1967. *An Annotated Bibliography on Cloudiness in the USSR.* U.S. Dept. of Commerce, Washington, D.C., 66 pp.

TORBINA, N. V., 1937. Sinopticheskie usloviya groz na ETC (Synoptic conditions for thunderstorms in European USSR). *Meteorol. Gidrol.*, 2: 3–17.

TSINZERLING, V. V., 1952. Prirodnye vodooboroty i ikh vliyanie na klimat SSSR (The hydrologic cycle and its influence on climate in the USSR). *Izv. Akad. Nauk SSSR, Ser. Geogr.*, 5: 58–77.

TUSHINSKIY, G. K., 1963. *Ledniki, Snezhniki, Laviny Sovetskogo Soyuza* (Glaciers, snowfields, and avalanches in the Soviet Union). Gosudarstv. Izd. Geogr. Literatury, Moscow, 311 pp.

UTESHEV, A. S., 1972. *Atmosfernye Zasukhi i ikh Vliyanie na Prirodnye Yavleniya* (Atmospheric droughts and their influences on natural phenomena). Nauka Kazakhskoy SSR, Alma-Ata, 175 pp.

VENTSKEVICH, G. Z., 1958. *Agrometeorologiya.* Gidrometeoizdat, Leningrad. (Published in English as *Agricultural Meteorology*, Isr. Progr. Sci. Transl., Jerusalem, 1961, No.487, 300 pp.).

VOYEYKOV, A. I., 1889. Snezhnyy pokrov, ego vliyanie na pochvu, klimat i pogodu, i sposoby issledovaniya (Snow cover: its influence on soil, climate and weather, and methods of its investigation). *Zap. Russk. Geogr. ob-va po Obshchey Geografii*, T. 18, No.2. Also published in *Izbr. soch* (Collected works), T.2, Akad. Nauk, Moscow, 1948.

ZAKHOROV, P. S., 1965. *Pylnye Buri* (Dust storms). Gidrometeoizdat, Leningrad, 164 pp.

ZAVARINA, I. V., 1954. *Zasukha i Borba s Ney* (Drought and the struggle against it). Geografgiz, Moscow, 86 pp.

ZUBENOK, L. I., 1965. Kharakteristika uvlazhneniya territorii sovetskogo soyuza (Moisture characteristics of the Soviet Union). *Gl. Geofiz. Obs. Tr.*, 179: 161–171.

Chapter 10

Wind

Atmospheric circulation is one of the primary controls of climate. It conveys heat and moisture from one location to another and largely determines the motion of major perturbations such as cyclonic storms which determine so much of the weather of a region. But in addition the wind itself often becomes a significant climatic element and must be taken into consideration in a complete analysis of any region. And in this sense the magnitude, perhaps, becomes more important than the direction since it is the absolute movement of air across any given location on the earth's surface which determines such things as rates of evaporation, wind chill, deflation, and the like, and, if excessive, damage. Besides, the consequences of wind direction and the special characteristics of local winds have already been covered in preceding chapters. Wind speeds can also be equated to energy resources.

The Soviets are beginning to talk a great deal about wind as an energy source. It has been estimated that 20 trillion kilowatt hours of electrical power could be generated each year by wind energy across the country. To generate such an amount of electricity from thermal stations would require the burning of 10 billion tons of oil (FATEYEV, 1971). Unfortunately, about half of this lies along a narrow fringe of the northern coast and sections of the Pacific coast where winds are strongest and few people live. But significant amounts of wind energy are available in the heavily populated southern portion of the European plain as well as in portions of Kazakhstan, Central Asia, and west-central Siberia.

Even in the weaker wind zones of the country wind energy can be put to good use through the use of windmills to pump water for agriculture, particularly in areas which are so sparsely settled that the construction of transmission lines and electrical systems to pump water would be prohibitive in cost. Windmills have been used for a long time in the Soviet Union and in old Russia to pump water and to grind grain, but their numbers have never been nearly sufficient. For instance, at the present time there are about 300 windmills used for water pumping for cattle pastures in Kazakhstan, but it has been estimated that about 100,000 are needed. In 1960 the Soviet Union manufactured 40,000 windmills (FATEYEV, 1971).

Since much of the interior of the U.S.S.R. is dominated by a large high pressure system in winter and a shallow low pressure system in summer, pressure gradients are generally weak except along the peripheries of the Eurasian land mass where they are accentuated by land and sea differences. They may become exceptionally strong along mountainous coasts. This is particularly true in the east and north. During the year, wind speeds average from 6 to 9 m/sec along much of the Arctic coast and parts of the Pacific coast, as well as in scattered patches in the southern and western portions of the country. The northwestern and southeastern portions of European U.S.S.R. and much of Kazakhstan and Central Asia average 4–6 m/sec during the year, and north-central European U.S.S.R. and a huge region in south-central Siberia and the Far East away from the coasts average only 2.5–4 m/sec (Fig.10-1).

The most consistently high wind speeds are found along the north and east coasts where rather large stretches of land experience more than 5 days per year with wind speeds of more than 20 m/sec, and certain localities experience more than 20 such days per year. On the other hand, an extensive area in southern European U.S.S.R. and Central Asia generally experiences no more than one day per year with wind speeds of more than 20 m/sec, and the bulk of the country during the average year does not experience such winds at all (Fig.10-2).

The highest winds in the country appear to be along the coasts of the Chukchi and Bering seas, eastern Kamchatka, and the northeastern coast of the Sea of Okhotsk, and the seaward slopes of the Sikhote-Alin in Maritime Kray facing the Sea of Japan. In these regions it can be expected that wind speeds of at least 40 m/sec will be encountered every year and wind speeds equal to or greater than 50 m/sec will be encountered at least once in 20 years (Fig.10-3 and Table 10-I). Wind speeds diminish rapidly out to sea. The greatest wind speed gradients in the U.S.S.R. are found during winter along the Sikhote-

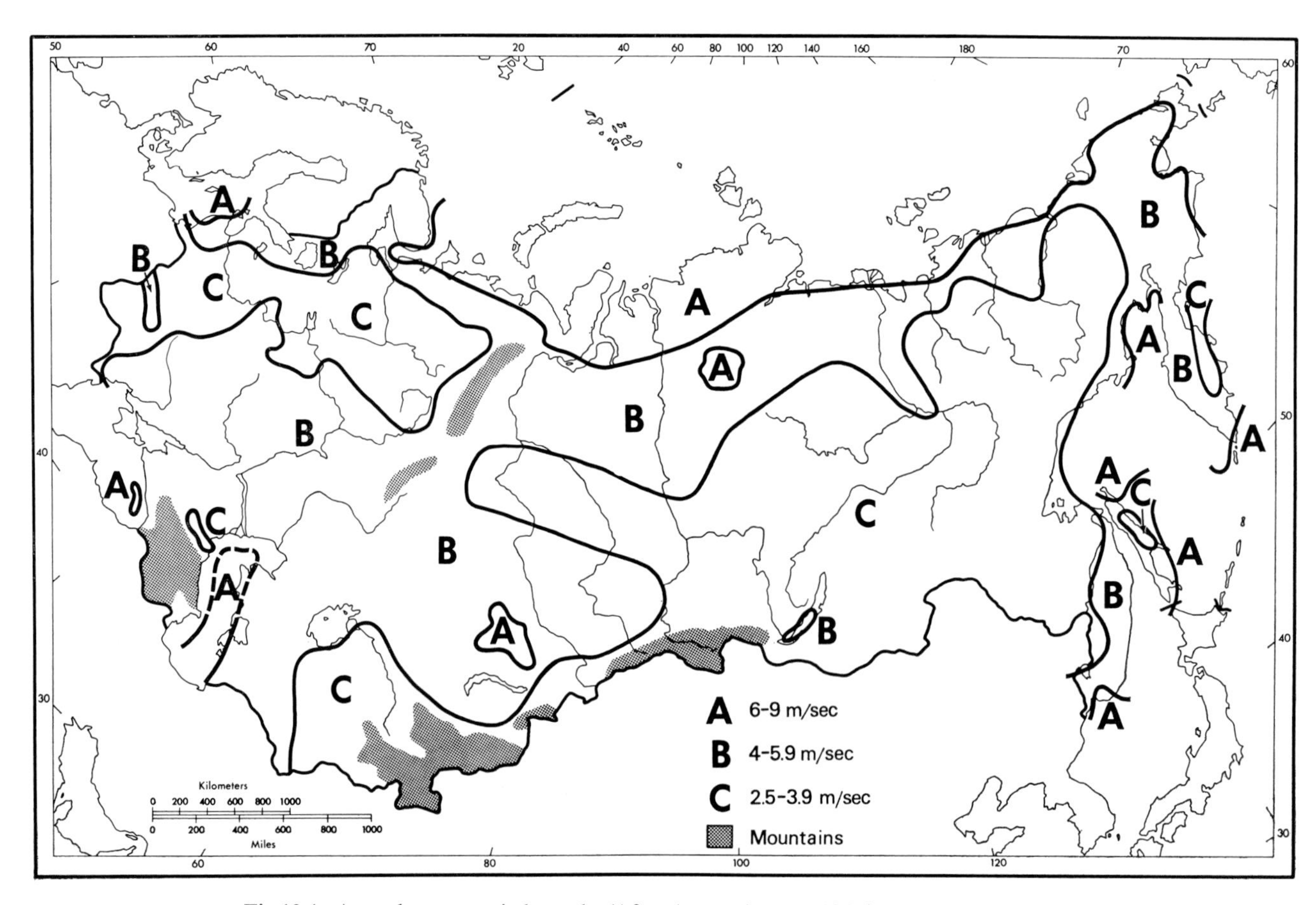

Fig.10-1. Annual average wind speeds. (After ANAPOL'SKAYA, 1961.)

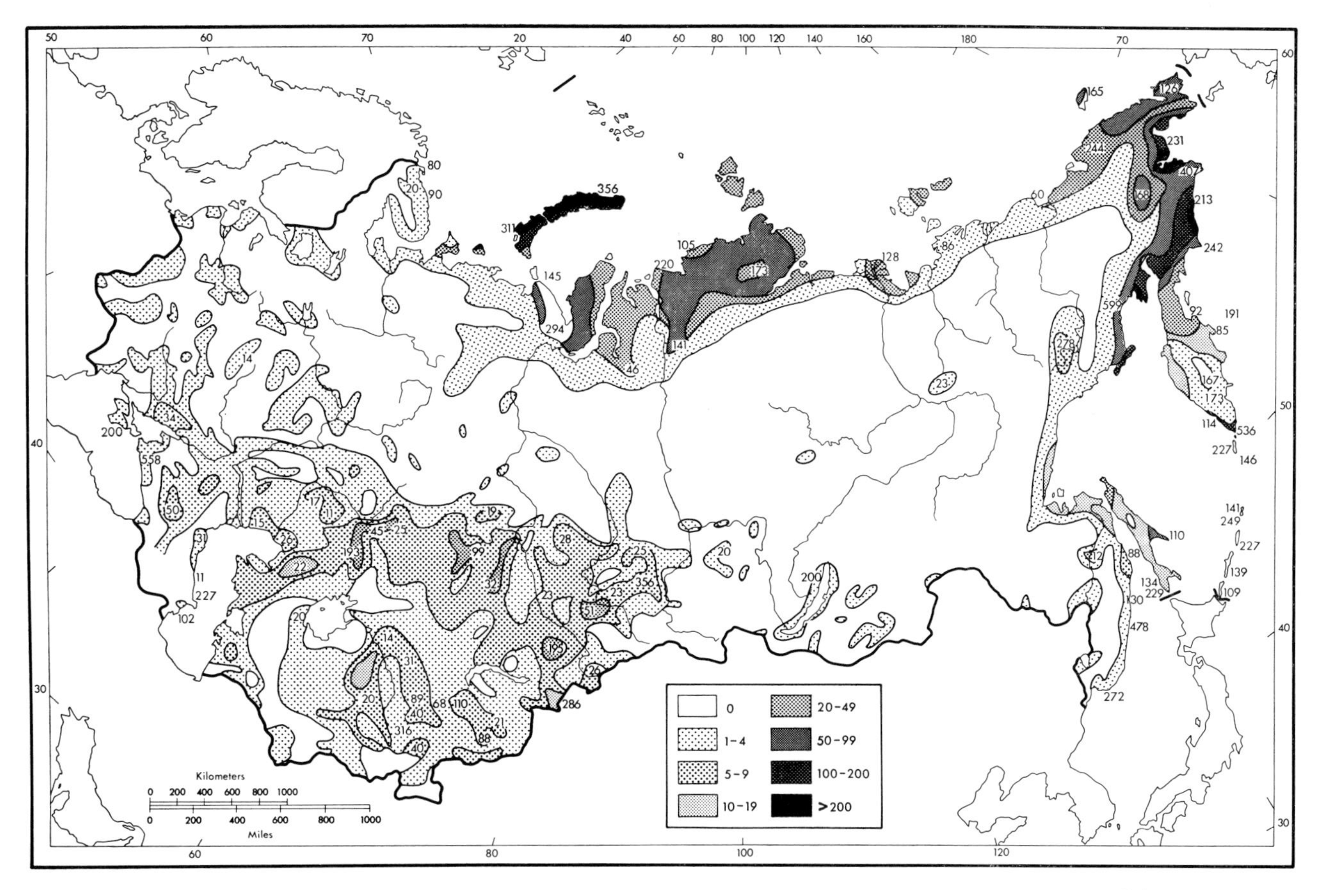

Fig.10-2. Number of days with wind speeds exceeding 20 m/sec during the period 1956–1965. (After SOROCHINSKIY et al., 1968.)

Alin. Winds between 60 and 70 m/sec have been observed on the eastern slopes, while to sea and on the western slopes of the mountains they were no more than half that amount. Wind speed gradients are also very great along the northern coast of the Sea of Okhotsk where steep air pressure gradients develop along the narrow Kolyma Range that separates the Sea of Okhotsk to the south from the extensive basin of the Kolyma River to the north. Passes in the Kolyma Range along the headwaters of the Kolyma River seem to be a major route of transit for air from the Arctic to the Pacific during winter.

Winds vary greatly along this mountainous, irregular Pacific coast. In general, exposed capes, islands, and summit areas of mountains experience the greatest wind velocities. The mountainous peninsula of Kamchatka appears to have very high wind speeds on the higher volcanic peaks, but no reliable records exist. Wind speeds are higher along the east coast of Kamchatka than along the west. The highest recorded wind speed on Kamchatka is on Cape Lopatka where every year speeds in excess of 50 m/sec are observed. On mountainous Sakhalin Island winds are stronger on the western side than on the east. They are particularly strong on the northern end where they have been observed as high as 44 m/sec at Cape Terpeniya. Wind speeds continue high all along the Arctic coast, although generally they are not quite as high as they are in places along the mountainous Pacific coast, except in such localities as Malye Karmaeuly, where a

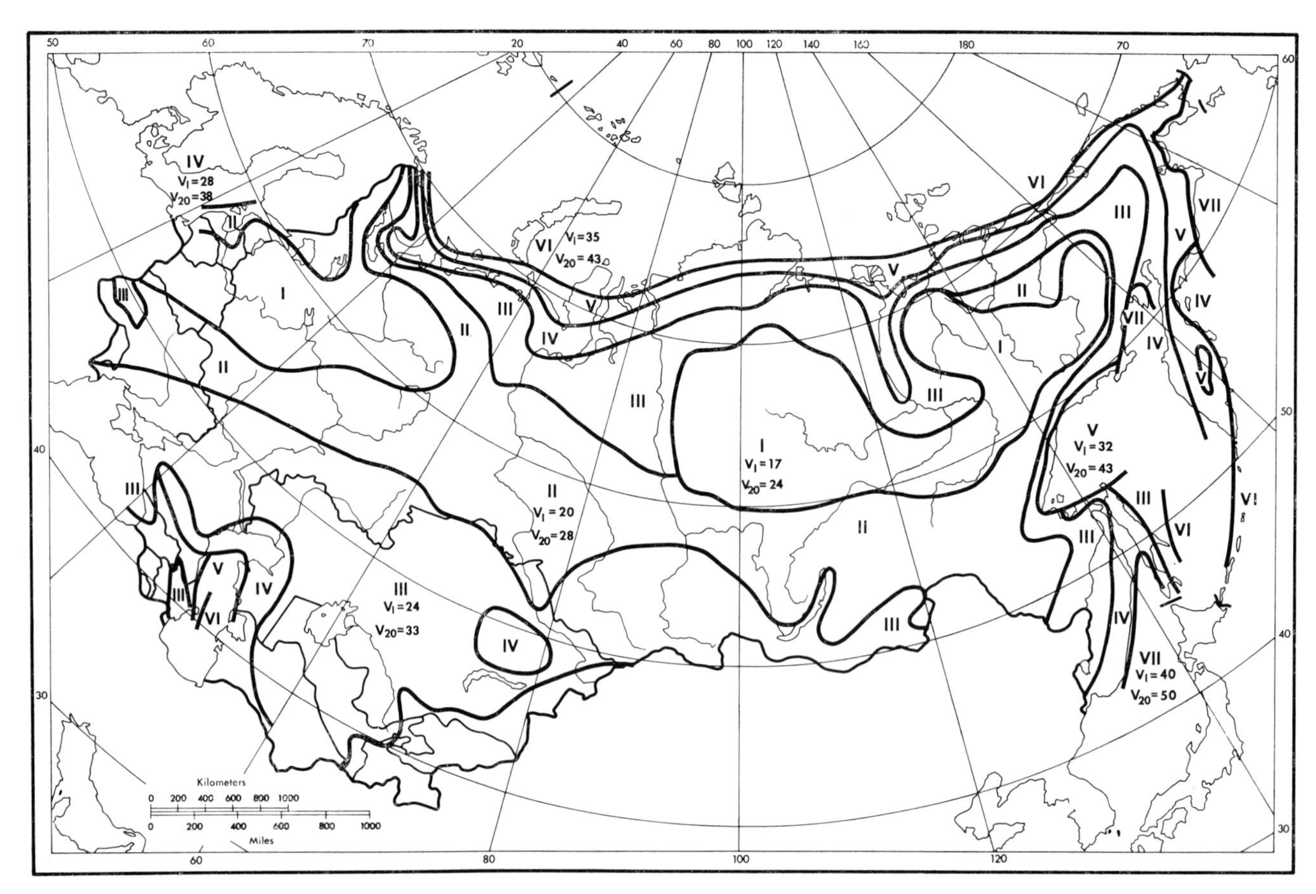

Fig.10-3. Maximum wind speed probabilities (m/sec), annually and once in twenty years. (After ANAPOL'SKAYA, 1961.)

Category	V_1 once per year	V_{20} once in 20 yrs.
I	16–18	22–26
II	19–21	27–30
III	22–25	31–35
IV	26–29	36–40
V	30–33	41–44
VI	34–36	45–49
VII	37–41	≥50

TABLE 10-I

FREQUENCIES OF OCCURRENCE OF WIND SPEEDS, IN PERCENT OF ALL CASES
(After ANAPOL'SKAYA, 1961)

Station	Geogr. site	Wind speed (m/sec) ≥4	≥10	≥16	≥24	≥34	≥40
European USSR							
Zhelaniya Mys	76°N 69°E	—	—	10.8	1.9	0.19	0.044
Khibiny	68°N 33°E	47.5	8.9	1.2	0.05	—	—
Arkhangel'sk	64°N 40°E	61.1	7.4	0.8	0.003	—	—
Vorkuta	68°N 64°E	57.1	10.0	1.3	0.08	0.01	—
Liepaya	56°N 21°E	62.8	19.5	5.8	0.4	0.03	—
Tallin	60°N 25°E	51.6	6.7	0.9	0.007	—	—
Kaunas	55°N 24°E	36.5	3.4	0.3	—	—	—
Minsk	54°N 27°E	56.5	4.1	0.2	—	—	—

TABLE 10-I (*continued*)

Station	Geogr. site	Wind speed (m/sec) ≥4	≥10	≥16	≥24	≥34	≥40
Arzamas	55°N 44°E	58.1	11.6	1.2	—	—	—
Kiev	50°N 31°E	44.4	4.0	0.3	—	—	—
Ay Petri	44°N 34°E	53.4	18.2	9.0	2.8	0.8	0.42
Caucasus							
Stavropol'	45°N 42°E	45.7	14.8	5.6	0.3	0.06	0.02
Makhachkala	43°N 47°E	64.6	22.7	6.7	—	—	0.4
Kirovabad	41°N 46°E	30.9	3.2	0.6	—	—	—
Sumgait	41°E 50°E	70.4	29.4	9.6	1.3	0.1	—
Central Asia							
Frunze	43°N 74°E	25.0	0.72	0.16	0.004	—	—
Urgench	42°N 61°E	39.2	3.6	0.8	—	—	—
Karakul'	40°N 64°E	32.1	1.7	0.4	—	—	—
Fedchenko Glacier	38°N 72°E	74.9	10.5	1.3	0.2	—	—
Dushanbe	38°N 69°E	7.33	0.22	0.07	0.006	—	—
Termez	37°N 67°E	20.2	4.5	1.18	0.077	—	—
Western Siberia and Kazakhstan							
Dikson	73°N 81°E	78.3	33.6	9.0	1.9	0.3	0.1
Dudinka	69°N 86°E	66.8	18.9	3.5	0.4	0.02	—
Salekhard	67°N 67°E	51.4	12.3	2.8	0.3	0.01	0.01
Surgut	61°N 74°E	52.4	4.9	0.9	—	—	—
Tobolsk	58°N 68°E	43.5	5.1	1.1	0.01	—	—
Tomsk	57°N 85°E	45.5	8.6	1.9	0.13	0.01	—
Krasnoyarsk	56°N 93°E	34.9	3.5	0.5	0.01	—	—
Chelyabinsk	55°N 62°E	45.9	3.6	0.3	—	—	—
Petropavlovsk	55°N 69°E	65.0	12.2	2.4	0.03	—	—
Novosibirsk	55°N 83°E	38.6	4.3	0.6	0.03	—	—
Pavlodar	52°N 77°E	64.0	13.5	3.1	0.1	0.03	—
Kemerovo	55°N 86°E	49.7	10.7	2.4	0.06	0.01	—
Karaganda	50°N 74°E	57.9	12.4	3.3	0.4	0.1	—
Gur'yev	47°N 52°E	53.2	8.4	1.3	0.01	—	—
Balkhash	47°N 75°E	63.6	9.3	0.8	0.0	—	—
Alma-Ata	43°N 77°E	14.3	0.3	0.05	—	—	—
Dzhambul	43°N 71°E	23.2	4.0	2.3	0.01	—	—
Eastern Siberia and the Far East							
Cape Chelyuskin	78°N 104°E	—	—	4.2	0.3	0.01	—
Tura	64°N 100°E	18.9	1.1	0.1	—	—	—
Anadyr'	65°N 177°E	69.2	22.2	8.0	—	—	—
Yakutsk	62°N 130°E	19.6	0.7	0.12	—	—	—
Oymyakon	64°N 143°E	10.5	0.3	—	—	—	—
Aldan	58°N 126°E	32.6	2.9	0.3	—	—	—
Okhotsk	59°N 143°E	45.0	5.6	1.0	—	—	—
Kyzyl	52°N 94°E	12.7	1.4	0.3	—	—	—
Chita	52°N 113°E	16.8	1.2	0.14	—	—	—
Ulan-Ude	52°N 108°E	26.5	3.7	0.9	—	—	—
Bomnak	55°N 129°E	18.5	0.5	0.07	—	—	—
Petropavlovsk-Kamchatskiy	53°N 158°E	70.0	27.8	12.6	—	—	—
Birobidzhan	49°N 133°E	37.6	3.5	0.4	—	—	—
Khabarovsk	48°N 135°E	59.0	12.1	1.7	—	—	—
Alexandrovsk	51°N 142°E	56.2	14.9	4.1	—	—	—
Yuzhno-Sakhalinsk	47°N 143°E	41.6	5.1	1.2	—	—	—
Vladivostok	43°N 132°E	—	—	6.5	0.4	0.06	—
Askol'dskiy Mayak	43°N 133°E	—	—	13.3	3.2	0.8	0.3

strong bora wind frequently blows through the narrow, steep-sided strait that divides Novaya Zemlya into two separate islands.

In the European part of the Soviet Union the greatest wind speeds are observed on the northern shore of the Kola Peninsula, on exposed capes and islands, and in the open sea. Tikhaya Bay on Franz Joseph Land can expect winds of 37 m/sec every year and 51 m/sec once every 20 years. Wind speeds diminish rapidly toward the interior of the Kola Peninsula. The immediate coastal area of the Baltic is quite windy, particularly in Latvia between Liepaya and Ventspils, where wind speeds of 39 m/sec can be expected once in 20 years. Winds diminish perceptibly only 30–60 km inland from the Baltic. Liepaya and Kaunas illustrate this nicely (Table 10-I). Sand dunes along the Baltic coast can decrease wind speeds by 1.5–3 m/sec. Also forests diminish speeds by several meters per second, as do urban areas. Leningrad port for instance experiences winds of 27 m/sec once in 20 years while the weather station at Leningrad experiences only 19 m/sec.

In the North Caucasian Foreland wind speeds once in 20 years reach 50 m/sec. This is true both on the Stavropol' Upland and in the major river valleys. Within the Great Caucasus, winds in narrow valleys reach 20–24 m/sec once in 20 years, but on open summits they reach 55–60 m/sec. In the Kura Valley in the eastern Transcaucasus on open steppes at heights of 300–400 m wind speeds once in 20 years reach 33 m/sec. In forested regions they reduce to 19–20 m/sec. Wind speeds increase significantly with height. In Tbilisi the observatory at 404 m observes 34 m/sec once in 20 years while a nearby station at 490 m experiences 50 m/sec. On the Armenian Plateau in open unvegetated areas winds reach 40–45 m/sec once in 20 years, while in gullies and valleys only 20–25 m/sec. The greatest wind speeds in Armenia are found on Mt. Aragats at an elevation of 3,229 m where once in 20 years the wind reaches 46 m/sec. Lake Sevan sits in a completely surrounded basin protected from strong winds. Once in 20 years it experiences 33 m/sec. The same is true of the valley of the Araks River.

At the eastern end of the Black Sea the Colchis Lowland experiences winds as high as 33 m/sec once in 20 years. Most of the higher speeds are associated with foehns that descend the western slopes of the low Surami Range during winter when a cold high builds up in the east. Mountain peaks and passes above the lowland experience much higher wind speeds. In the western part of the Great Caucasus the Gagrskiy Range experiences 51 m/sec once in 20 years. Farther northwest, Markhotskiy Pass experiences winds up to 60 m/sec during winter as the famous Novorossiysk bora blows from the cold high to the northeast of the mountains southwestward to the Black Sea coast where Novorossiysk on the coast experiences winds up to 50 m/sec once in 20 years. Still farther northwest, lowland Crimea experiences 33 m/sec once in 20 years, but Ay-Petri, the highest point along the limestone precipice in the Crimean Mountains overlooking the Black Sea, can expect winds of 53 m/sec once in 20 years.

The shores of the Caspian experience fairly high winds. The Apsheron Peninsula experiences the highest, with 48–50 m/sec once in 20 years. The strongest winds generally blow around the eastern end of the Great Caucasus from the north during winter when high pressure builds up in the cold Asiatic high on the north side of the mountains. Consequently, in the cities of Sumgait and Baku these winds are known as "nord".

East of the Caspian winds continue to be locally strong where topography and synoptic conditions cause constrictions in air flow. Most of these winds are of foehn or katabatic character. At Termez the 20-year wind speed probability is 40 m/sec in conjunction with the "afganets" or "harmsil" which blow from southerly directions ahead of cold fronts associated with eastward moving cyclones in winter. Farther east, in the Pamirs wind speeds of 25 m/sec in narrow valleys and 40–45 m/sec on mountain summits and open passes can be expected once in 20 years. To the north, in the Tyan Shans, Issyk-Kul' basin has a 20-year speed probability of 42 m/sec associated with westerly winds that pick up speed rapidly as they descend the mountains into the town of Rybach'ye at the west end of the lake. "Ursatsk" winds blow westward through the narrow neck of the Fergana Valley and reach speeds greater than 40 m/sec in the southeast part of the Golodnaya (Hungry) Steppe. And exceptionally strong winds, the "evgey" or "ebe", blow in winter northwestward through the Dzhungarian Gate east of Lake Balkhash where speeds as great as 70 m/sec have been recorded. In the Kazakh Folded Country to the north winds of 25 m/sec are experienced annually on open interfluves. Wind speeds vary 8–10 m/sec between upland and lowland in this area.

In the Altay and Sayan Mountains winds frequently reach 50 m/sec on the open summits. Foehn winds are common, particularly along the southern end of Teletsk Lake in the western Altay. And the katabatic winds that descend the many steep, narrow river valleys into Lake Baykal are of world renown. Most famous is the "sarma" which blows from the northwest down the Sarma Valley with speeds sometimes exceeding 40 m/sec. A study of the winds in this valley during the entire year, 1902, recorded 113 days with northwest or north-northwest winds with speeds of 15–40 m/sec. They are most frequent from October through December when the greatest temperature contrasts are found between the land and the lake, before the lake freezes over (BURMAN, 1969).

In general the central zone of the Soviet Union receives only moderate winds. The Ural Mountains experience speeds of more than 50 m/sec on some of the higher summits and significantly reduce wind speeds on their eastern sides, except during strong storms, but otherwise the terrain induces little variation in atmospheric circulation. A large region in central and eastern Siberia can expect wind speeds of $>$24m/sec no more than once in 20 years. The same is true of much of the forest areas of European U.S.S.R. Vegetation is second only to topography in producing local effects on wind speeds. The winds within wooded areas of northwestern European U.S.S.R. generally are 3–4 m/sec slower than those along the shores of lakes and reservoirs and on exposed uplands. On a zonal scale on the Russian plain wind speeds on the average increase southeastward from the forested region through the wooded steppe into the steppe and desert areas.

Within all the zones in the U.S.S.R. there are local winds produced by differences in relief, water bodies, and so forth, and often these dominate a given locality. They are particularly frequent in hilly mountainous regions, which include much of the Soviet Union east of the Yenisey River, the southern fringes of Central Asia, the Caucasus, the Urals, and major river valleys which have rather steep banks, as well as major water bodies such as the Caspian Sea, Aral Sea, Lake Balkhash, and Lake Baykal (Fig.10-4). Gentle gravity winds, particularly cold air drainage at night, can be significant even in gently rolling terrain if the general circulation is weak. Such air flows are particularly well developed in the drier portions of the country where diurnal heating differences along slopes are most intense. Thus, the frequency of occurrence of so-called "slope" winds is distributed in a rather zonal fashion (Fig.10-5). The formation of various types of local wind regimes results in a hodge-podge of directions of surface winds, especially in areas

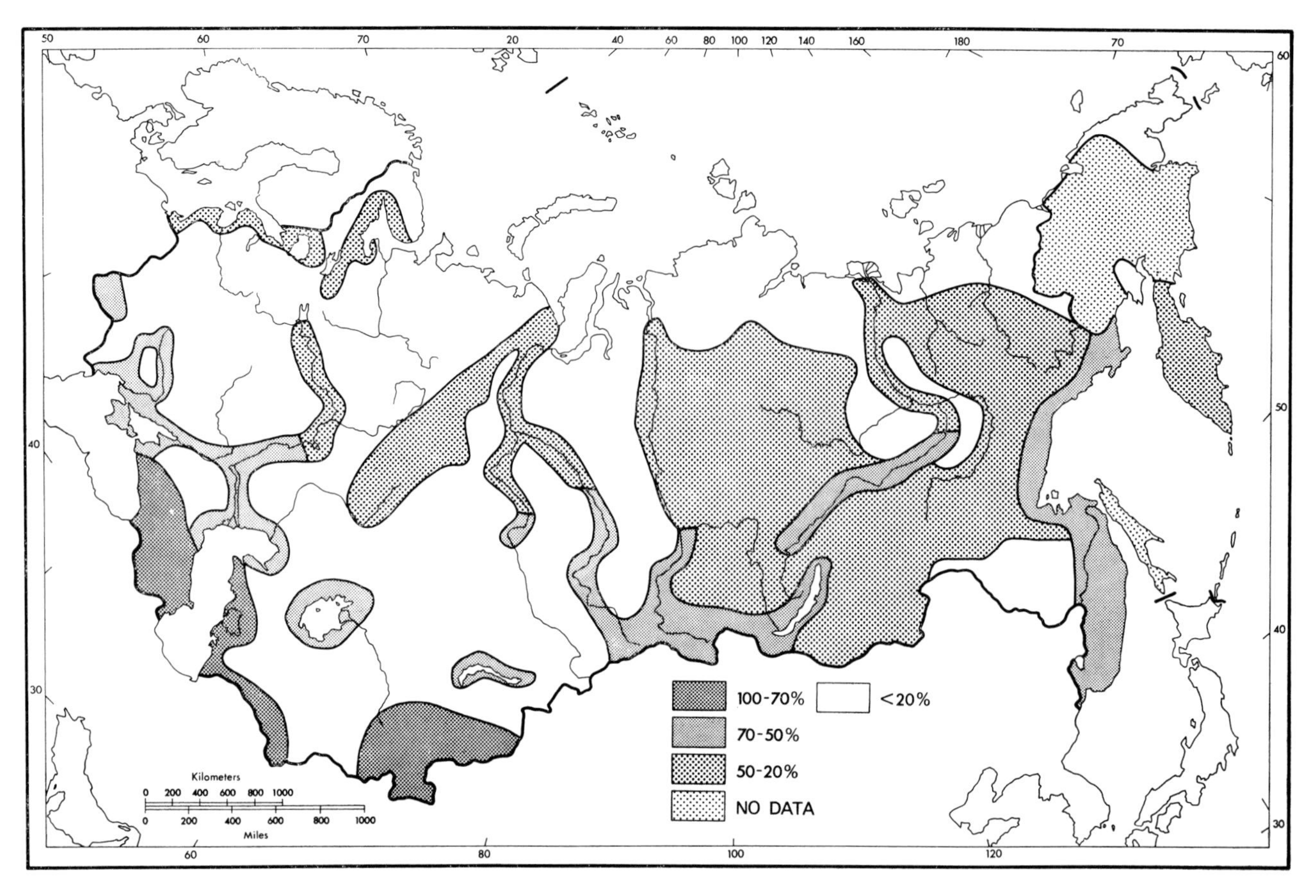

Fig.10-4. Frequencies of local wind regimes, in percents of all time, July. (After BURMAN, 1969.)

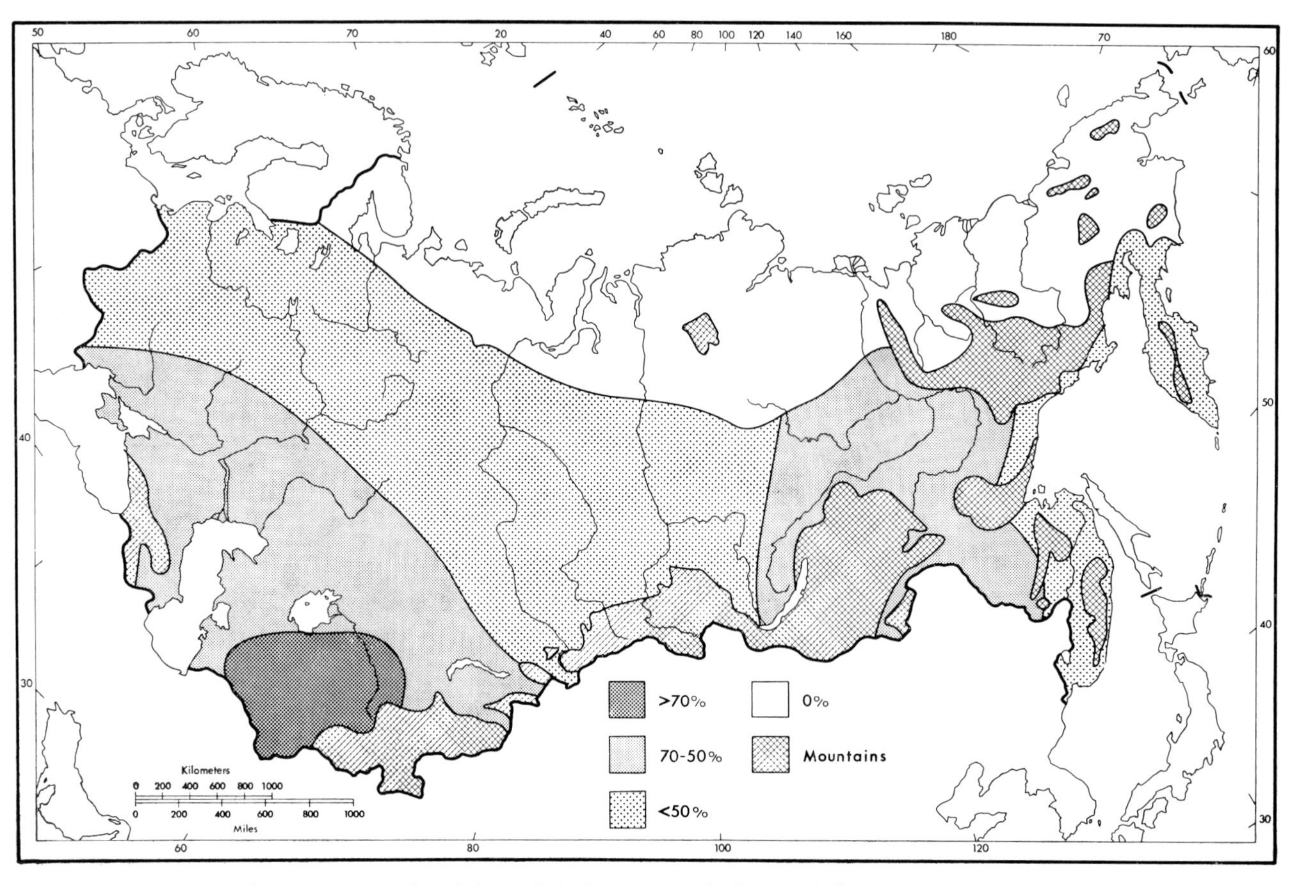

Fig.10-5. Frequencies of slope winds, in percents of all times. (After GOL'TSBERG, 1967.)

TABLE 10-II

OUTSTANDING WINDS IN THE USSR (1946–1958)
(After ANAPOL'SKAYA, 1961)

Location	Date	Speed (m/sec)	Characteristics
Moscow	25/VII/57	40–45	Severe thunderstorm, uprooted trees; blew off roofs
Ukraine–Central Urals	8/II/58		Passage of deep cyclone
Don and Central Volga Basins		30–40	
Kanin Nos	26/II/48	40	Blew down telephone lines and antennae
Archangel	17/VII/52	40	Uprooted trees; blew down telegraph and telephone lines
Golitsyno (near Kalinin)	24/VIII/56	?	Tornado; cut swath 100–200 m wide and 3 km long
Rostov (Yaroslavl O.)	24/VIII/53	recorded up to 60	Tornado; hurled 40-ton railroad freight cars off tracks
Stavropol' Plateau (North Caucasus)	6–12/I/50	35–40	Blizzard; northwest and west winds
Ay-Petri (Crimea)	20/XI/46	50	Blew down fire tower and enormous trees; blew off roofs
Ra-Iz (northern Urals)	4/I/48	>40	North winds lasted from 14h20 to 21h50
Ra-Iz	27/I/48	50	Northwest winds
West, North, and central Kazakhstan	18–22/XII/54		Prolonged, widespread winds
Aktyubinsk and Pavlodar	20–21/XII/54	30–40	Highest velocity during period
Norilsk	15–17/I/57	40	
Khatanga	28/III/53	>40	
Novosibirsk	21/X/54	34–40	
Tyukhtet (Half way between Tomsk and Krasnoyarsk)	8/VIII/48		Blew down 200 trees, many roofs, and communication lines
Nizmennyy Mayak (southeast coast of Maritime Kray)	14/IX/54	40	Typhoon passage; blew down roofs, telegraph poles, electric lines
Melkovodnaya (west coast of Gulf of Penzhina)	5/II/46	40	
Ostrov Zav'yalova (south of Magadan)	27/V/50	40	Strong burya
Bukte Nagaevo (near Magadan)	17/XI/55	>40	Blew down transmission lines and poles
Mys Sosunova	31/XII/50	40	Blizzard and snowfall; blew down transmission lines and blew fishing vessels up onto the shore to a distance of 2.5 m above sea level
Nakhodka	18/IX/46	40	
Kholmsk (southwest Sakhalin)	20/II/55	40	Blew down electric and telephone lines; heavy snowfall blocked railroad
Aleksandrovsk-Sakhalin	5/II/50	40	
Kuril'sk	23/X/52	30–33 gusts to 40	Northwest winds blew off many roofs

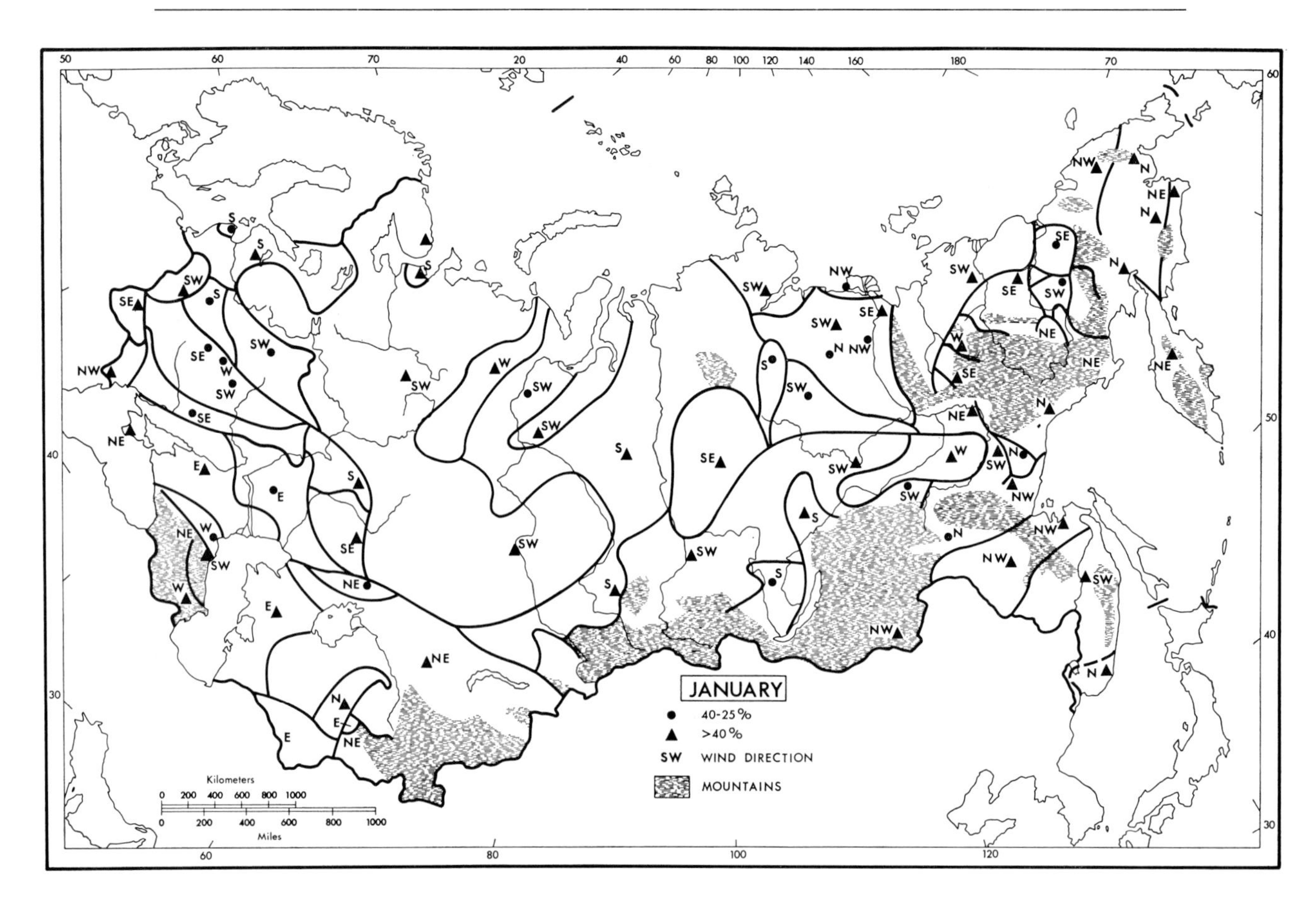

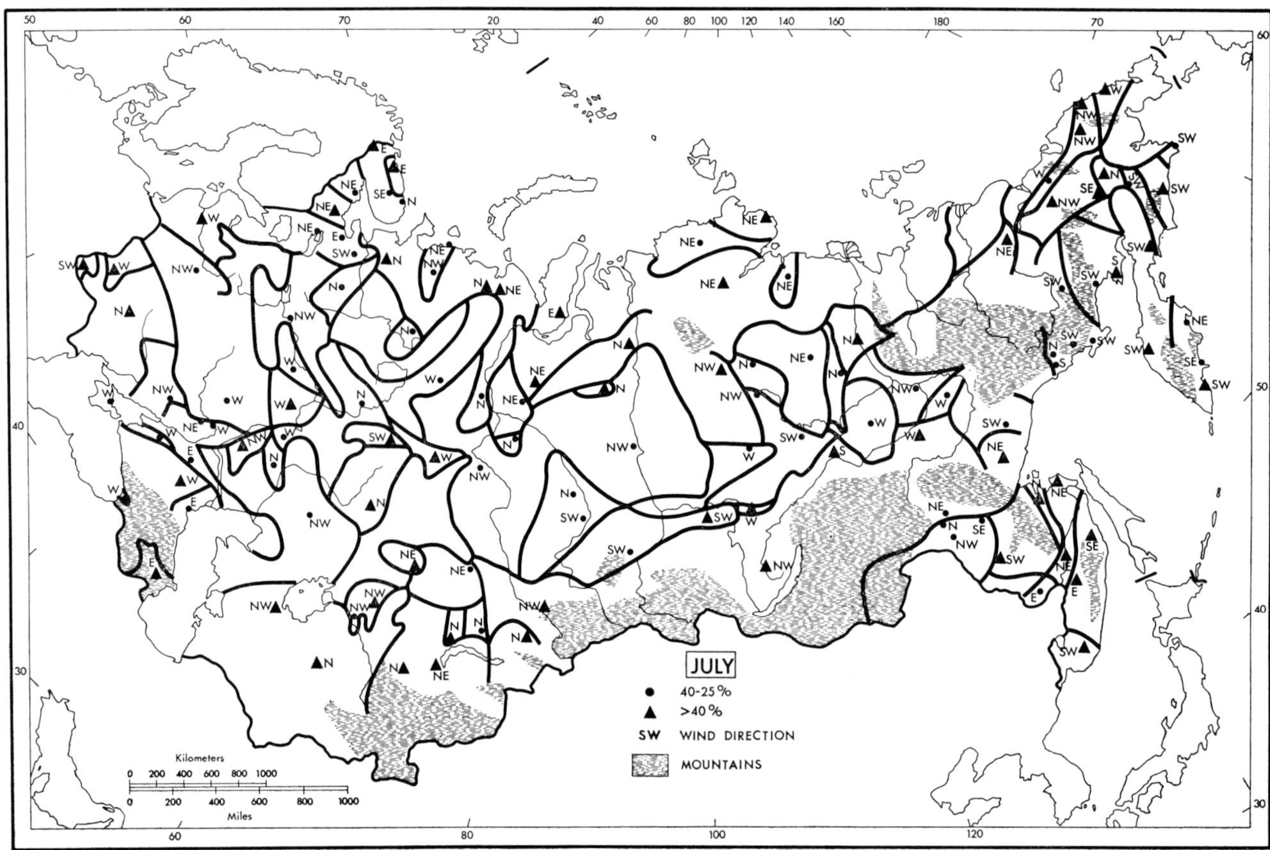

Fig.10-6. Prevailing surface winds, January and July. (After GOL'TSBERG, 1967.)

where pressure gradients associated with the general circulation are not great. This is particularly true in midsummer over a great deal of the country and in midwinter in central Siberia where the Asiatic high holds sway most of the time and pressure gradients are so weak that local topographic factors become the predominant control over the weak surface circulation (Fig.10-6).

Almost anywhere in the country, though, winds can occasionally reach extreme speeds if the right synoptic conditions develop. They are most likely to develop in areas already described for their high wind speeds, but they can also occur on the broad plains of European U.S.S.R. and western Siberia where traveling cyclonic storms are most prevalent (Table 10-II). One of three terms is usually applied to these occasions which are usually brought on by intensifications of pressure gradients with the approach of cyclonic storms. The general term in use in most parts of the country, but especially in the southwest, is "burya". This signifies simply high winds, with speeds of 20 m/sec or more. It might be translated most precisely as "tempest". On sea it connotes stormy waves with confused patterns. On land it usually connotes some destruction. In the dry southeastern Ukraine and surrounding areas it is the main cause of dust storms, "pyl'naya burya", or more commonly "chernaya burya" or black storm as distinguished from "metel' burya" or snow storm, a term that is little used.

Usually a snow storm is designated simply as "metel". But if it reaches blizzard proportions, with high winds and rapidly falling temperatures, it is usually designated as "buran" or "purga". Buran is the term commonly used in the steppes and deserts of the south; it is apparently derived from a Turkish language in the area. Purga is the term more commonly used in the forests and tundra of the north. It apparently stems from the Karelian language. Its usage appears to be restricted primarily to describe blinding swirling snow and numbing cold in unprotected non-wooded areas such as openings in the forest or the open tundra. Of course, over time, as people have moved from one part of the country to another, these terms have been carried with them and their usage has become more widespread and less precise.

Another term which has come into fairly common usage, particularly in the southwest, is "uragan" (hurricane). Of course there are no hurricane storms in this area. The term refers to winds of hurricane velocity. Usually the term is applied after the fact upon viewing of the resultant destruction, uprooted trees, destroyed houses, etc. Some of these extraordinarily high winds are no doubt associated with tornadoes. Although tornadoes are rare in the area, they are by no means completely absent. Although records are inadequate to make even an educated guess about distributions and frequencies, there are enough scattered accounts in the literature, some accompanied by photographs, to conclude that high winds with funnel clouds do occur perhaps on the order of once every five years in the Ukraine and adjoining European Russia at least as far north as Moscow.

References and further reading

Anapol'skaya, L. E., 1961. *Rezhim Skorostei vetra na Territorii SSSR* (Wind velocity regime in the USSR). Gidrometeoizdat, Leningrad, 199 pp.

Anapol'skaya, L. E. and Gandin, L. S., 1966. Teoreticheskie osnovy stroitel'noy klimatologii (The-

oretical basis for construction climatology). In: M. I. BUDYKO (Editor), *Sovremennye Problemy Klimatologii.* Gidrometeoizdat, Leningrad, pp.263–279.
ANAPOL'SKAYA, L. E. and TYUKTIK, V. V., 1969. Vetrovy rezhim gornykh rayonov vostochnoy sibiri, dal'nego vostoka, p-ova Kamchatka i o. Sakhalin (Wind regime in the mountain regions of eastern Siberia, the Far East, Kamchatka Peninsula, and Sakhalin Island). *Gl. Geofiz. Obs. Tr.*, 246: 15.
BURMAN, E. A., 1969. *Mestnye Vetry* (Local winds). Gidrometeoizdat, Leningrad, 341 pp.
FATEYEV, E. M., 1971. Wind Energy and Its Use. In: I. P. GERASIMOV, D. L. ARMAND, and K. M. YEFRON (Editors), *Natural Resources of the Soviet Union: Their Use and Renewal.* Freeman, San Francisco, Ill., pp.89–98.
GOLT'SBERG, I. A. (Editor), 1967. *Mikroklimat SSSR.* Gidrometeoizdat, Leningrad. (Translated as *Microclimate of the USSR*, Isr. Progr. Sci. Transl., Jerusalem, 1969, No.5345.
SOROCHINSKIY, M. A., KOSHEL'KOVA, G. A. and YUSHENKO, G. P., 1968. Rezhim shtormovogo vetra na territorii SSSR za period s 1956 po 1965 g. (Regime of storm winds over the USSR from 1956 to 1965). In: V. F. CHERNOVA (Editor), *Prognoz Baricheskogo Polia i Opasnykh Yavleniy pogody* (Prediction of the pressure field and hazardous weather phenomena). *Gidrometeorol. Nauchnoissled. tsentr. SSSR, Tr.*, 32: 23–34. (English abstract and two maps presented in: Guide to Soviet Literature Accessions in the Atmospheric Sciences Library and the Geophysical Sciences Library, July–August, 1969, pp.46–48.)

Chapter 11

Climate Distribution

Classification schemes

The climate of a place, of course, cannot be described by any one of its elements, nor even by separate descriptions of all its elements, for it is the combination of things interacting with one another that determines the potentials of the climate for the use and welfare of man. And it is the assessment of these potentials for man's use that is usually accepted as the criterion for evaluating and classifying the climate of an area.

Many different ways of expressing the climate of a region have been devised through the years by people in many countries for various purposes, some very general, others more specific. The Soviets themselves have been quite active in this field of classification. Several schemes in the past have caught the attention of scholars outside the Soviet Union, but many have remained relatively unknown to the rest of the world. A few of these have been summarized in BORISOV's *Climates of the U.S.S.R.*, and some have been presented in various issues of *Soviet Geography: Review and Translation*. A brief list of some of the more significant pieces of Soviet literature on climatic classification has been provided in LYDOLPH, 1971, pp.656–657. A summary of all these will not be attempted here, since as was stated at the outset, this book cannot hope to cover all the practical applications of climate. But in order to provide some integrated picture of the individual components of climate one Soviet scheme will be chosen to depict the climatic regions of the U.S.S.R. (Fig.11-1).

The Grigor'yev–Budyko scheme

This classification scheme was devised by A. A. GRIGOR'YEV and M. I. BUDYKO in the 1950's, and it immediately captured a great deal of attention in the Soviet Union because of both the eminence of the authors and the use of radiation values rather than temperature values as a thermal index, which was more in keeping with the current climatic thought in the Soviet Union. Dr. Grigor'yev has been director of the Institute of Geography in the Academy of Sciences until 1951, probably the top position in the field of geography in the Soviet Union, and Dr. Budyko was director of the Main Geophysical Observatory and is world-renowned for his work in radiation studies.

The climatic system devised by them has been calculated to approximate the observed distribution of natural vegetation and soils and the potentialities of the land for agricul-

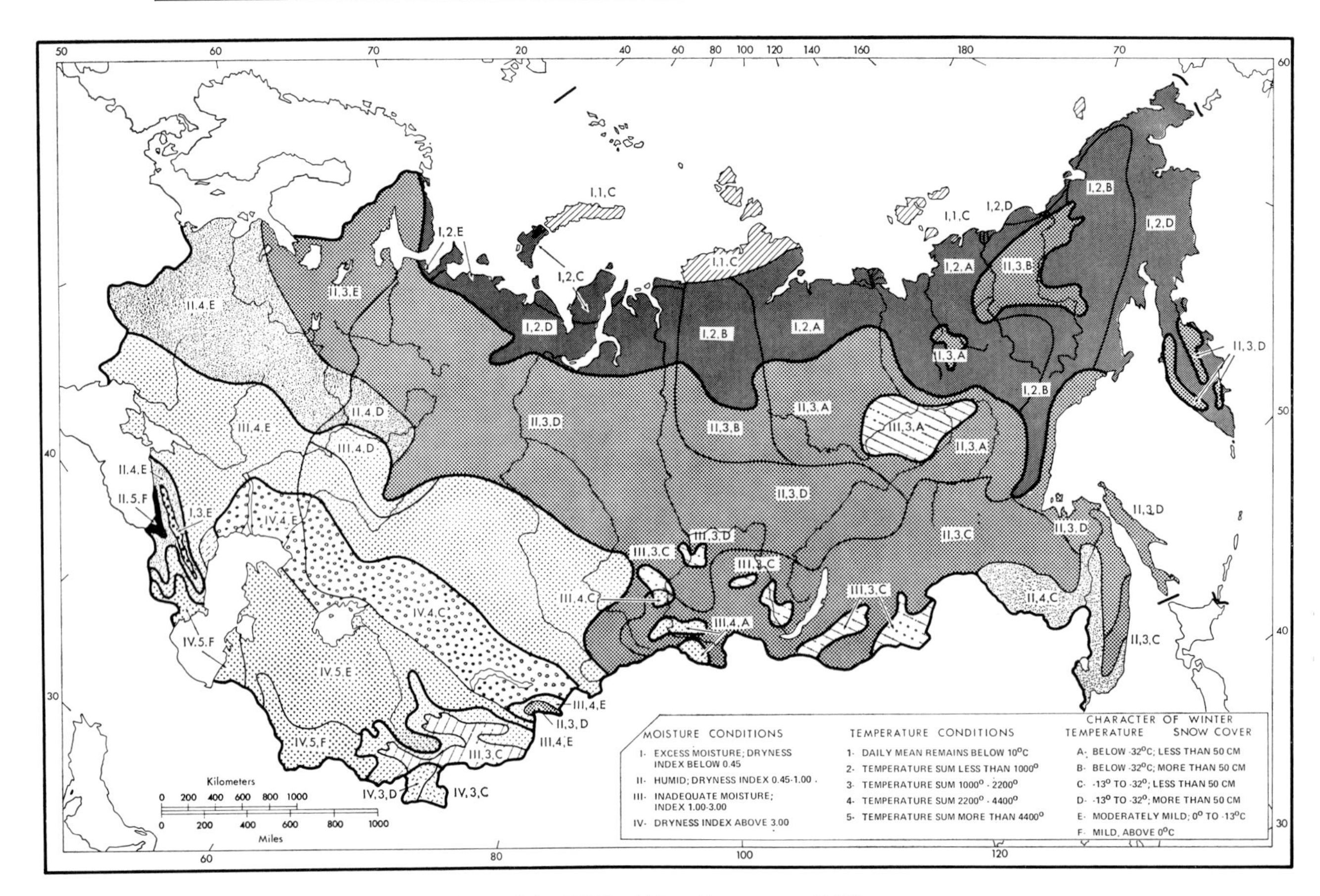

Fig.11-1. Climatic regions of the USSR. (After GERASIMOV, 1964.)

ture, as so many other climatic classifications also have been devised. The difference between this classification and those preceding it is the attempt to express the thermal parameter in terms of heat units rather than temperature levels and then to utilize the summation of heat units to balance against precipitation to form a moisture parameter. Thus, as is so commonly the case, the climatic classification is based on the two parameters, heat and moisture.

In this case the heat parameter is expressed as R, the radiation balance at the earth's surface, and the moisture parameter is expressed as R/Lr, the so-called "radiation index of aridity", where R is the radiation balance at the earth's surface, L the latent heat of condensation, and r the mean annual precipitation. Unfortunately, radiation balance data are largely lacking, so the value has been substituted by the sum of daily mean temperatures during the period when the daily mean temperature remains above 10°C. It had previously been illustrated by Budyko that the annual radiation balance at the earth's surface is linearly correlated with this sum of temperatures expression. Therefore, as can be seen in the legend of Fig.11-1, the primary parameters of the classification scheme become the sum of temperatures during the period when the mean diurnal temperature remains above 10°C and a moisture index based on a ratio between the sum of temperatures and the latent heat required to evaporate all the precipitation that falls during the year. Secondary parameters are winter (January mean) temperatures and depth of snow cover during the 10-day period of maximum snow cover.

All these parameters are closely tied to the agricultural potentials of the land. The sum of temperatures has become the standard index in the U.S.S.R. by which to judge stages of development in crop growth during the growing season, comparable to the use of the degree-day concept in many other parts of the world. The moisture index, which in this case is expressed the other way round, as degree of aridity, of course, relates to moisture availability in the soil. And mean annual absolute minimum temperatures and snow cover depths have been identified by the Soviets as prime agroclimatic indices.

Of course this classification does not take into consideration many facets of climate which might supply much additional information. But then this is true of all classification schemes. Their authors have found that they have to simplify drastically in order to achieve comprehensibility. Certainly there is no classification that comes close to depicting the totality of climate; that is why bookfulls of descriptive words have to be written to supplement maps of climatic regions based on classification schemes.

This classification scheme by GRIGOR'YEV and BUDYKO (1959) seems to be as comprehensive as any, and it holds considerable promise for achieving a considerable step forward in the understanding of heat and moisture relationships once adequate radiational data become available. The scheme has become widely used in the Soviet Union; it is the basis for the climate map of the U.S.S.R. which was included in the *Fiziko-Geograficheskiy Atlas Mira* (Physical-Geographic Atlas of the World). It is this wide acceptance among Soviet climatologists themselves which has prompted the use of it here. It does seem to fit the distribution of natural phenomena in the Soviet Union, such as vegetation and soils, better than any other climatic classification that has yet been devised. The scheme evolved from prior studies by Grigor'yev and Budyko on what they termed "the periodic law of geographic zonality" (GRIGOR'YEV and BUDYKO, 1956). In the study cited the authors claim to have found a periodically recurring order of natural vegetation and soils on the earth which correspond to thermal zones and moisture categories within zones. Therefore, their climatic regions theoretically have been devised to correspond closely to observed distributions of vegetation and soils.

Moisture distribution

As can be seen in Fig.11-1, the moisture categories tend to describe a zonal pattern in the Soviet Union with the most humid conditions along the Arctic coast and progressively drier conditions southward, except for some mountainous areas such as the Caucasus. However, there are some significant deviations from the zonal pattern which have been described many times with respect to vegetation and soil conditions but which have not become apparent on climatic maps before. Chief among these are the relatively drier areas in the northeast in a region which is generally considered to be humid throughout. The most significant of these areas is the Lena–Vilyuy–Aldan Lowland around Yakutsk, which shows an inadequate moisture index (category *III*). This is the first climatic map which shows this region to be somewhat droughty and thereby explains the steppe-like nature of some of the vegetation and soils of the region.

Thermal distribution

Sums of temperatures during the growing season also tend to show a zonal pattern, but winter temperatures deviate greatly from a latitudinal distribution. A significant boundary runs southward from the White Sea along the 40th meridian east and then swings south-southeastward into Central Asia where it finally trends latitudinally east of the Aral Sea. This separates a region on the west with moderately mild winters, cold month average between 0° and −13°C, from a region to the east which has a cold month average ranging from −13°C to −32°C throughout the eastern part of European U.S.S.R., western Siberia, and northern and central Kazakhstan. The northern part of this boundary running southward from Archangel is similar to boundary lines which have been drawn across this flat plain on a number of climatic maps using various classification schemes, but always rather arbitrarily until now. In spite of the fact that there is no significant topographic feature in this area, it appears that there is a significant climatic boundary here which during the winter separates a primarily marine-controlled region to the west from a primarily continentally controlled one to the east.

It is interesting that Moscow is placed in a region which is shown as having a moderately mild winter. Although to most non-Russians Moscow seems quite cold during winter, for its high latitude it is rather anomalously warm because of the marine influence from the west. During summer, on the other hand, this western portion of the plain suffers from the marine influence as far as temperatures are concerned. Sums of temperatures during the growing seasons are higher at similar latitudes in the east. Of course in the very northeastern part of the country mountainous topography reduces summer temperatures.

Climatic regions

Along the Arctic fringe, categories *I,1* and *I,2* are regions of excessive moisture. Precipitation in most cases is quite light, probably no more than 250 mm, if that, but the temperatures are so cool year round that evaporation rates are down, and when the surface of the soil thaws out during the short summer the region is generally water-logged. In many sections immediately along the coast, particularly on the Taymyr Peninsula, and on the islands off shore, daily mean temperatures never rise above 10°C. Most of the vegetation is tundra. According to GRIGOR'YEV and BUDYKO (1959) the lack of trees relates more to the excessive moisture conditions than to the coolness of the temperatures. The regions are subdivided according to winter temperatures, which range from moderately mild in the west around the fringes of the Barents Sea which remains open during winter to extremely cold in eastern Siberia where the regions expand far southward because of higher elevations. They are also subdivided according to depth of snow cover. Category *II,3*, which covers the broad north-central portion of the country and expands eastward to include much of eastern Siberia, is a region of humid conditions with temperature sums between 1,000° and 2,200°C. This is a region which is conducive to the growing of coniferous trees. Generally the region is not cultivated except in certain sheltered areas where probably the climate locally would not fit the category shown. This is particularly true in intermontain basins along the southern fringes of Siberia, where generally moisture conditions are category *III*, somewhat inadequate, which have been conducive to grass growth rather than tree growth and to the formation of fertile soils.

Category *II,4*, with humid conditions and temperature sums between 2,200° and 4,400°C corresponds fairly closely to the zone of mixed forest. Generally the region does not suffer from drought, and with its greater heat resources the region is adaptable to a considerable variety of crops. With proper agricultural practices and adequate fertilization this region can be maintained at a fairly high level of agricultural productivity. A similar region, but with colder winters, occupies much of the Amur Basin and the Ussuri–Khanka Lowland in the Far East. This is the most agriculturally productive part of the Soviet Far East.

Category *III* generally outlines the steppes with somewhat inadequate moisture supplies. This forms a broad zone in the southern part of the European U.S.S.R. which continues across the southern Urals into the southern part of western Siberia and adjacent northern Kazakhstan. This belt is generally considered to be the best farming region in the Soviet Union because of its larger resources of heat and its excellent soils which have developed under grass vegetation. However, the region universally suffers from drought. Nevertheless, most of the grain cultivation of the country is found here, as well as much of the sugar beets and practically all of the sunflowers, which respectively are the primary sources of sugar and vegetable oil for the country.

Corn, which is a relatively recent crop, since the mid-1950's, has been placed primarily in this region. The corn that is raised for dry grain is generally found in quite droughty portions of this zone in southeastern Ukraine and the north Caucasus. It is quite obvious that the Soviets have been willing to risk drought in order to gain heat for such crops which require relatively long growing seasons. Therefore, one might conclude that considerations of heat come into play before considerations of moisture in the decision-making process that determines crop distributions.

In the Lena–Vilyuy–Aldan Lowland in eastern Siberia is a large outlier of this subhumid type of climate, but it has lower heat resources and therefore much more restricted agricultural possibilities. Intermontane basins along the southern border of Siberia fall in a similar category, the largest areas of which are the Buryat region southeast of Lake Baykal and the Chita Basin farther east. West of Lake Baykal the Irkutsk, Krasnoyarsk, Minusinsk, Tuva, and some lesser basins provide the best agricultural potentials in Siberia east of the Ob River.

Category *IV* with its various subdivisions occupies the large territory in Soviet Central Asia between the Caspian Sea and the China border. Generally this area is too dry for agriculture without irrigation. But the southern half of the region has greater heat resources than any other part of the country, and this has induced irrigation agriculture in many places where it is possible to procure adequate water supplies, primarily from major river systems. Cotton is the big commercial crop in the warmest parts of this region where the growing season is long enough. Alfalfa is its rotation crop primarily to utilize accumulations of salts from irrigation water. And many varieties of fruits and vegetables are grown. In some of the foothills where the precipitation increases and the climate changes to a very warm version of subhumid category *III* dry farming of grains is practised. Category *III* climates continue into the mountains, but here it is too cool for much crop growth, and most of the mountains are used only for grazing and lumbering. Most of the lowland desert areas, where irrigation is not possible, are utilized for extensive grazing also. Much transhumance is still practiced, with extensive livestock drives into the mountains during the summer and back to the desert plains in the winter.

The Caucasus present a mixture of climatic types ranging from excessively wet, cool conditions along the crest of the Great Caucasus to humid, very warm conditions in the Colchis Lowland at the eastern end of the Black Sea to dry, hot conditions in the Kura and Araks river valleys in the east and south. The most distinctive characteristic of the climate in the Transcaucasus at low elevations is the mild winter which everywhere averages above freezing (subcategory *F*). Except for some sheltered spots in the very southern part of Soviet Central Asia and along the southeastern coast of the Crimean Peninsula, such mild winters do not exist anywhere else in the U.S.S.R. And these localities are not as protected from occasional intrusions of cold northerly air as the Transcaucasian lowlands are. It is the relative freedom from hard frosts that imparts the special agricultural character to portions of the Transcaucasus where crops such as citrus and tea are grown.

References and further reading

Borisov, A. A., 1965. *Climates of the USSR.* Aldine, Chicago, Ill., 225 pp.

Budyko, M. I., 1956. *Teplovoy Balans Zemnoy Poverkhnosti.* Gidrometeoizdat, Leningrad, 255 pp. (Translated by Nina A. Stepanova as *The Heat Balance of the Earth's Surface.* U.S. Weather Bureau, Washington, D.C., 1958, 259 pp.

Gerasimov, I. P. (Editor), 1964. Fiziko-Geograficheskiy Atlas Mira (Physical-geographical Atlas of the world). Akad. Nauk, Moscow, 298 pp.

Grigor'yev, A. A. and Budyko, M. I., 1956. O periodicheskom zakone geograficheskoy zonal'nosti (On the periodic law of geographic zonality). *Dokl. Akad. Nauk*, 110: 129–132. (Translated as: The heat and moisture balance and geographic zonality. *Sov. Geogr. Rev. Transl.*, 2(7): 3–16.)

Grigor'yev, A. A. and Budyko, M. I., 1959. Klassifikatsiya klimatov SSSR. *Izv. Akad. Nauk, Ser. Geogr.*, 3: 3–19. Translated as: Classification of the Climates of the USSR. *Sov. Geogr. Rev. Transl.*, 1(5): 3–24.

Lydolph, P. E., 1971. Soviet work and writing in climatology. *Sov. Geogr. Rev. Transl.*, 12: 637–666.

Appendix—Climatic Tables

Data for the climatic tables were derived from the following primary sources, as well as some other scattered sources.
(1) *Stroitelnye Normy i Pravila*, Chast'II, razdel A, glava 6, Stroitel'naya klimatologiya i geofizika, Moscow, 1963.
(2) The six volumes of *Klimat SSSR*.
(3) *CLINO Tables* of the World Meteorological Organization.
(4) Barashkova, E. P., Gaevskiy, V. L., D'yachenko, L. N., Lugina, K. M., and Pivovarova, Z. I., 1961. *Radiatsionnyy Rezhim Territorii SSSR*. Gidrometeoizdat, Leningrad.

Additional data were supplied from the records of the following agencies:
(1) Foreign Division, Main Administration of the Hydrometeorological Service of the Soviet Union. V. Maslow, Chief.
(2) Department of Climatology, Institute of Geography, Academy of Sciences of the USSR. The late B. L. Dzerdzeevskii, Head.
(3) Vakhushti Institute of Geography, Academy of Sciences of the Georgian SSR. F. F. Davitaya, Head.
The author is extremely grateful to Drs. Maslov, Dzerdzeevskii, and Davitaya for assigning their personnel the task of completing these data tables. Collaborators with Dr. Davitaya were D. G. Mumladze, Ts. D. Zhorzhikashvili and N. V. Gvasalia.
Note: Monthly values might not total to annual values if they have been derived from different sources. In some columns, such as minimum and maximum precipitation, there is no reason for them to do so.
A location map of the stations is given in Fig.A-1.

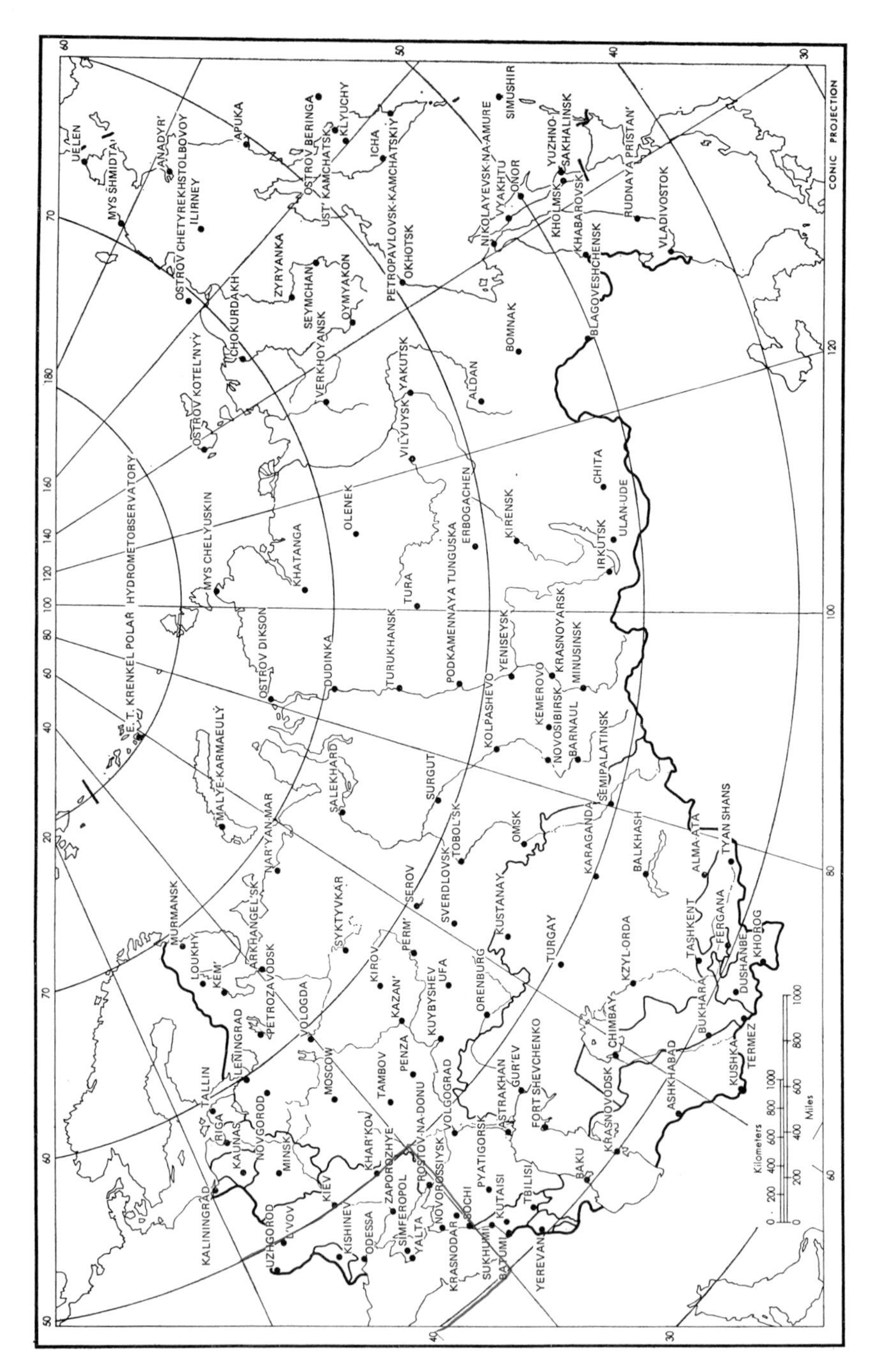

Fig.A-1. Station locations corresponding to the climatic tables.

TABLE I

CLIMATIC TABLE FOR ALDAN

Latitude 58°37′N longitude 125°22′E elevation 680 m

Month	Air press. (mbar) sea level	station	Wind preval. direct.	mean speed (m/s)	Global radiat. (kcal./cm²)	Temperature (°C) mean	mean daily range	max.	min.	Humidity vapor press. (mm)	relat. (%)	Precipitation (mm) mean	max.	min.	max. in 24 h
Jan.	1024.7		SW	2.9	1.1	−27.6	7.2	−2	−51	0.4	78	19	41	5	9
Feb.	1025.0		SW	2.6	2.9	−23.9	8.4	2	−50	0.6	76	15	38	2	6
Mar.	1022.6		SW	3.3	7.1	−16.1	10.2	6	−40	0.9	69	15	31	2	12
Apr.	1015.5		SW	3.5	13.0	−6.1	10.3	13	−33	2.0	63	27	84	8	26
May	1010.1		SW	3.2	14.8	3.5	11.4	30	−16	3.9	63	63	104	15	35
June	1006.2		SW	2.8	15.6	13.0	12.3	33	−4	7.3	60	73	189	16	75
July	1005.1		SW	2.6	13.2	17.1	11.4	34	0	9.7	67	88	191	23	56
Aug.	1008.7		SW	2.5	10.1	13.3	10.5	33	−2	8.6	73	83	217	28	64
Sept.	1005.5		SW, W	2.5	6.9	5.6	8.8	26	−14	5.2	75	70	139	18	37
Oct.	1019.2		SW	3.2	4.0	−6.4	7.6	16	−30	2.3	76	49	88	7	20
Nov.	1023.0		SW	3.2	1.6	−20.7	7.6	7	−44	0.8	78	28	38	7	10
Dec.	1023.2		SW	2.7	0.8	−26.8	7.2	−2	−48	0.5	79	21	50	6	11
Year	1015.7		SW	2.9	91.1	−6.3		34	−51		71	551	779	357	75

Month	Number of days wind (speed ≥15 m/s)	frost-free period	fog	precip. (≥0.1 mm)	thunder-storm	snow-cover	clouds (8–10 tenths)	Snow cover max. 10-day average depth (cm)	Cloudiness (tenths) 07h	13h	19h	Mean sun-shine (h)	Evaporation (mm) actual	potent.
Jan.	0.4		1	22.6			17.8		7.9	7.8	7.2	72	<10	
Feb.	0.5		0.3	18.0			17.1		7.4	7.3	6.4	106	<10	
Mar.	0.8		0.5	14.7			13.6		7.0	7.2	6.3	218	<10	
Apr.	1.2		0.1	13.7	0.1		14.0		7.2	7.6	7.3	225	>10	
May	0.7		1	14.0	0.7		16.6		7.7	8.1	7.7	232	>30	
June	0.7		1	12.9	5		12.0		6.5	7.8	7.1	283	70	
July	1.0		3	13.5	7		14.2		6.8	7.7	7.0	271	70	
Aug.	0.4		4	15.2	3		13.3		7.2	7.7	7.2	199	50	
Sept.	0.2		5	15.1	0.5		15.5		7.8	7.9	7.3	138	30	
Oct.	0.7		0.3	19.7			20.5		8.2	8.0	7.2	96	<10	
Nov.	0.5		0.5	19.9			16.2		7.9	8.0	7.1	54	<10	
Dec.	0.2		2	21.1			15.9		7.4	7.7	7.0	48	<10	
Year	7.3	99	19.	200.0	16.0	227	184.0	44	7.4	7.8	7.1	1942	300	400

TABLE II

CLIMATIC TABLE FOR ALMA-ATA

Latitude 43°14′N longitude 76°56′E elevation 848 m

Month	Air press. (mbar) sea level	station	Wind preval. direct.	mean speed (m/s)	Global radiat. (kcal./cm²)	Temperature (°C) mean	mean daily range	max.	min.	Humidity vapor press. (mm)	relat. (%)	Precipitation (mm) mean	max.	min.	max. in 24 h
Jan.	1029.9		SE	1.3	4.5	−8.8	10.9	11	−43	2.1	74	26	81	4	26
Feb.	1026.8		SE	1.6	6.0	−7.4	11.0	15	−48	2.3	74	32	68	1	33
Mar.	1026.4		SE	2.0	7.9	−0.1	11.9	27	−31	3.6	73	64	134	12	35
Apr.	1019.4		SE	2.2	11.2	9.9	13.7	33	−14	5.9	59	89	190	17	46
May	1015.0		SE	2.6	15.2	15.7	12.8	38	−9	8.3	55	99	209	13	62
June	1009.8		SE	2.7	16.4	19.8	12.8	40	−1	10.3	51	59	152	3	54
July	1004.2		SE	2.7	16.8	22.2	13.4	42	5	11.1	45	35	102	0	55
Aug.	1008.5		SE	2.2	14.6	21.0	13.8	41	0	9.7	44	23	75	0	33
Sept.	1015.6		SE	2.0	12.0	15.3	13.7	37	−5	7.3	45	25	70	0	44
Oct.	1022.6		SE	1.9	8.4	7.4	13.1	33	−18	5.0	55	46	139	0	38
Nov.	1028.3		SE	1.6	4.6	−0.8	11.3	24	−40	3.8	70	48	113	4	46
Dec.	1030.0		SE	1.4	3.5	−6.2	10.9	13	−39	2.7	74	35	87	4	33
Year	1019.8		SE	2.0	121.1	7.3		42	−48		60	581	923	296	62

Month	Number of days wind (speed ≥15 m/s)	frost-free period	fog	precip. (≥0.1 mm)	thunder-storm	snow-cover	clouds (8–10 tenths)	Snow cover max. 10-day average depth (cm)	Cloudiness (tenths) 07h	13h	19h	Mean sun-shine (h)	Evaporation (mm) actual	potent.
Jan.	0.2		9	8.4		31	11		6.2	6.2	4.7	115	<10	
Feb.	0.3		8	7.5		28	10		6.5	6.5	5.1	119	<10	
Mar.	0.6		6	10.8	0.1	11	13		7.0	7.1	6.3	140	10	
Apr.	1.0		1	11.8	2		12		6.6	7.2	7.3	195	30	
May	2.6		0.3	10.6	6		9		6.1	7.6	7.5	242	30	
June	2.9		0.3	9.9	7		6		5.5	7.5	7.6	282	<30	
July	3.2		0.04	8.6	7		4		4.8	6.8	6.8	316	30	
Aug.	1.9		0.03	5.9	5		3		4.3	6.0	5.9	296	<10	
Sept.	0.8		0.5	5.2	2		4		3.8	5.0	4.6	249	<10	
Oct.	0.5		2.0	6.9	0.4		7		5.1	5.5	4.3	198	<10	
Nov.	0.4		6	8.6			10		6.0	5.9	4.4	129	<10	
Dec.	0.1		10	8.7		28	13		6.1	6.0	4.6	111	<10	
Year	14.5	181	43	102.9	28	111	102	31	5.7	6.4	5.8	2392	200	800

TABLE III

CLIMATIC TABLE FOR ANADYR'

Latitude 64°47'N longitude 177°34'E elevation 61.6 m

Month	Air press. (mbar) sea level	Air press. (mbar) station	Wind preval. direct.	Wind mean speed (m/s)	Global radiat. (kcal./cm²)	Temperature (°C) mean	Temperature (°C) mean daily range	Temperature (°C) max.	Temperature (°C) min.	Humidity vapor press. (mm)	Humidity relat. (%)	Precipitation (mm) mean	Precipitation (mm) max.	Precipitation (mm) min.	Precipitation (mm) max. in 24 h
Jan.	1016.4		NW	7.3		–22.7	10.9	6	–51	1.0	82	21	124	0	
Feb.	1019.1		NW	8.2		–21.6	11.0	3	–51	0.9	81	14	54	0	
Mar.	1018.8		NW	6.8		–20.2	11.9	2	–41	1.0	81	16	61	0	
Apr.	1015.0		NW	6.6		–13.3	13.7	7	–36	1.6	85	13	60	0	
May	1015.6		NW	5.4		–3.5	12.8	10	–24	3.2	84	9	34	0	10
June	1013.4		SE	5.7		4.8	12.8	26	–8	5.4	81	13	62	0	26
July	1010.2		SE	6.0		10.5	13.4	28	0	7.7	81	39	84	7	46
Aug.	1010.2		SE	5.9		9.6	13.8	24	–4	7.4	81	46	101	5	40
Sept.	1012.9		NW	5.9		3.9	13.7	17	–11	5.1	81	26	73	5	29
Oct.	1010.3		NW	7.1		–5.0	13.1	12	–29	2.9	84	27	101	3	
Nov.	1013.3		NW	8.1		–14.2	11.3	5	–40	1.6	84	17	87	0	
Dec.	1014.6		NW	7.2		–20.9	10.9	4	–44	1.0	83	19	101	0	
Year	1014.1		NW	6.7		–7.7		28	–51		82	260	436	16	

Month	Number of days wind (speed ≥15 m/s)	Number of days frost-free period	Number of days fog	Number of days precip. (≥0.1 mm)	Number of days thunder-storm	Number of days snow-cover	Number of days clouds (8–10 tenths)	Snow cover max. 10-day average depth (cm)	Cloudiness (tenths) 07h	Cloudiness (tenths) 13h	Cloudiness (tenths) 19h	Mean sun-shine (h)	Evaporation (mm) actual	Evaporation (mm) potent.
Jan.	9.5		2	10		31	2		5.5	6.4	5.4	32	3	
Feb.	8.6		2	10		28	3		5.9	6.2	5.4	84	3	
Mar.	7.5		1	8		31	2		5.9	5.9	5.4	185	3	
Apr.	6.1		2	8		30	3		6.8	6.5	6.4	207	5	
May	3.6		5	8	0.0	31	5		7.4	6.8	6.9	232	5	
June	1.5		7	8	0.04	6	4		7.1	6.8	6.6	277	40	
July	2.4		3	12	0.3		7		7.8	7.7	7.5	278	<60	
Aug.	2.7		2	12	0.2		5		8.0	7.8	7.6	188	<40	
Sept.	3.1		2	11	0.0		5		7.6	7.6	7.3	149	10	
Oct.	6.7		2	10		12	7		7.3	7.4	6.6	110	5	
Nov.	9.8		3	11		30	5		6.5	7.2	6.0	57	3	
Dec.	7.6		2	10		31	3		5.1	6.2	5.1	21	3	
Year	69.1	82	33	118	0.5	230	51	53	6.7	6.9	6.3	1820	<200	200

TABLE IV

CLIMATIC TABLE FOR APUKA

Latitude 60°26'N longitude 169°40'E elevation 10 m

Month	Air press. (mbar) sea level	Air press. (mbar) station	Wind preval. direct.	Wind mean speed (m/s)	Global radiat. (kcal./cm²)	Temperature (°C) mean	Temperature (°C) mean daily range	Temperature (°C) max.	Temperature (°C) min.	Humidity vapor press. (mm)	Humidity relat. (%)	Precipitation (mm) mean	Precipitation (mm) max.	Precipitation (mm) min.	Precipitation (mm) max. in 24 h
Jan.	1007.8	1006.7	NE	5.9	1.0*	–11.5		3	–42	2.6	78	36			
Feb.	1009.0	1007.9	NE	6.6	2.6	–13.0		5	–42	2.2	76	20			
Mar.	1011.7	1010.6	NE	6.5	7.2	–12.0		6	–35	2.2	74	22			
Apr.	1009.9	1008.9	NE	4.5	10.8	–5.1		10	–29	3.5	81	27			
May	1012.7	1011.7	NE	3.0	13.8	0.7		15	–18	5.5	84	18	99	0	7
June	1012.4	1011.4	SE	3.1	13.6	6.1		20	–4	8.3	87	27	55	1	9
July	1009.0	1007.9	SE	3.2	12.2	10.0		28	–1	11.0	89	60	142	9	15
Aug.	1009.5	1008.5	W	3.3	9.9	10.6		23	–3	11.2	88	62	125	10	15
Sept.	1010.6	1009.5	NE	3.5	7.1	6.6		20	–10	8.2	83	53	192	2	13
Oct.	1007.5	1006.5	NE	4.4	3.5	–1.2		14	–22	4.6	78	47	123	2	14
Nov.	1007.4	1006.3	NE	6.1	1.4	–7.9		11	–32	3.0	76	35			
Dec.	1005.6	1004.6	NE	6.4	0.6	–10.5		2	–38	2.4	76	24			
Year	1009.3	1008.4	NE	4.7	83.7	–2.3		28	–42	5.4	81	431			

Month	Number of days wind (speed ≥15 m/s)	Number of days frost-free period	Number of days fog	Number of days precip. (≥0.1 mm)	Number of days thunder-storm	Number of days snow-cover	Number of days clouds (8–10 tenths)	Snow cover max. 10-day average depth (cm)	Cloudiness (tenths) 07h	Cloudiness (tenths) 13h	Cloudiness (tenths) 19h	Mean sun-shine (h)	Evaporation (mm) actual	Evaporation (mm) potent.
Jan.	12.8		0.	10.0		31	12.9		6.7	7.6	6.3	32	5	
Feb.	0.4		0.3	8.6		28	11.8		6.9	7.1	6.3	84	5	
Mar.	9.0		0.4	7.8		31	12.8		6.5	6.4	6.2	129	<10	
Apr.	4.7		0.6	11.1		30	16.8		7.3	7.1	7.3	160	<10	
May	2.3		5	8.5		21	19.5		7.9	7.7	7.8	182	20	
June	0.2		6	10.3			21.8		8.6	8.3	8.4	136	50	
July	0.4		7	14.8	0.2		24.2		8.8	8.6	8.8	125	50	
Aug.	0.5		6	13.5	0.0		21.2		8.3	8.2	8.3	142	50	
Sept.	0.9		2	12.4	0.1		15.3		7.4	7.4	7.3	157	30	
Oct.	4.9		0.3	12.0			13.0		7.2	7.1	6.4	107	<10	
Nov.	8.3		0.5	11.3		22	12.8		6.9	7.3	6.2	53	<10	
Dec.	11.8		0.3	12.3		31	12.2		6.4	7.2	6.1	45	5	
Year	56.2	96	28	132.6	0.3	198	194.	69	7.4	7.5	7.1	1362	>200	200

*for Korf

TABLE V

CLIMATIC TABLE FOR ARKHANGEL'SK

Latitude 64°30′N longitude 40°30′E elevation 3.9 m

Month	Air press. (mbar)		Wind		Global radiat.	Temperature (°C)				Humidity		Precipitation (mm)			
	sea level	station	preval. direct.	mean speed (m/s)	(kcal./cm²)	mean	mean daily range	max.	min.	vapor press. (mm)	relat. (%)	mean	max.	min.	max. in 24 h
Jan.	1011.6	1010.7	NW	4.6	0.2	−12.6	7.7	5	−45	1.9	88	33	64	7	21
Feb.	1012.7	1011.8	NW	4.5	1.3	−12.1	8.8	4	−41	1.8	86	28	66	8	16
Mar.	1011.6	1010.7	NE	4.4	5.3	−8.0	12.4	10	−37	2.2	82	28	85	7	16
Apr.	1014.2	1013.3	NW	4.1	9.6	−0.8	11.7	23	−27	3.4	76	28	66	6	15
May	1015.0	1014.2	SE	4.5	12.5	5.5	11.2	30	−14	5.2	71	39	92	4	43
June	1011.5	1010.7	SE	4.3	14.3	12.1	12.6	32	−4	7.4	69	59	140	5	45
July	1010.5	1009.7	SE	3.8	13.7	15.6	13.1	34	1	9.4	71	63	154	2	47
Aug.	1010.7	1009.9	NW	4.1	9.6	13.4	13.6	33	0	9.1	78	57	137	9	55
Sept.	1011.1	1010.3	NE	4.8	4.3	8.0	10.8	28	−7	6.8	85	66	132	13	49
Oct.	1011.8	1010.9	NE	5.0	1.6	1.3	8.3	17	−20	4.6	88	55	124	14	25
Nov.	1011.0	1010.1	NE	5.1	0.4	−5.1	5.8	1-	−36	3.2	89	44	91	6	18
Dec.	1012.6	1011.7	NW	4.5	0.1	−10.2	7.0	4	−43	2.2	89	39	103	8	22
Year	1012.0	1011.2		4.4	72.9	−0.6		34	−45		81	539	775	279	55

Month	Number of days							Snow cover	Cloudiness (tenths)			Mean sun-shine	Evaporation (mm)	
	wind (speed ≥15 m/s)	frost-free period	fog	precip. (≥0.1 mm)	thunder-storm	snow-cover	clouds (8–10 tenths)	max. 10-day average depth (cm)	07h	13h	19h	(h)	actual	potent.
Jan.	1.4		4	22		31	19.5		8.0	8.6	8.0	7	<10	
Feb.	1.0		4	19		28	16.9		8.4	8.3	7.5	40	<10	
Mar.	1.6		4	19		31	15.0		7.7	7.3	7.0	121	<10	
Apr.	1.0		3	15	0.1	23	14.3		7.5	7.4	7.2	187	10	
May	1.4		1	14	1		13.9		7.4	7.3	7.1	224	70	
June	1.3		0.7	14	3		12.0		6.8	7.2	6.7	283	>70	
July	0.6		0.9	12	4		10.9		6.4	6.7	6.3	302	70	
Aug.	0.9		2	14	2		14.7		7.4	7.4	6.8	230	>50	
Sept.	1.5		4	19	0.4		15.7		8.4	8.6	7.9	102	30	
Oct.	1.3		5	21			21.2		8.6	8.8	8.1	59	10	
Nov.	1.5		5	21		22	21.7		8.6	8.8	8.5	19	<10	
Dec.	1.1		4	23		31	22.0		8.2	8.7	8.0	2	<10	
Year	14.6	119	38	213	11	175	198	62	7.8	7.9	7.4	1576	>300	400

TABLE VI

CLIMATIC TABLE FOR ASHKHABAD

Latitude 37°58′N longitude 58°20′E elevation 219 m

Month	Air press. (mbar)		Wind		Global radiat.	Temperature (°C)				Humidity		Precipitation (mm)			
	sea level	station	preval. direct.	mean speed (m/s)	(kcal./cm²)	mean	mean daily range	max.	min.	vapor press. (mm)	relat. (%)	mean	max.	min.	max. in 24 h
Jan.	1023.8	995.4	E	1.7	4.9	2.1	13.5	29	−23	3.9	78	22	50	0	23
Feb.	1021.7	993.7	E	2.2	6.8	4.7	14.4	33	−26	4.7	71	21	61	2	22
Mar.	1019.2	991.7	NW	2.2	8.2	8.8	15.6	38	−21	6.2	70	44	106	10	55
Apr.	1015.1	988.5	NW	2.3	11.4	16.3	16.2	39	−4	8.1	57	38	128	0	47
May	1013.1	987.1	NW	2.4	15.8	23.3	15.8	46	5	9.2	45	28	71	3	29
June	1008.9	983.4	NW	2.6	17.0	28.6	15.4	46	8	10.1	34	6	50	0	49
July	1006.6	981.5	NW	2.4	18.0	31.2	15.4	47	12	10.5	32	2	35	0	25
Aug.	1009.8	983.9	NW	2.3	17.3	29.3	15.6	46	10	9.6	32	1	28	0	21
Sept.	1015.1	989.1	E	2.0	14.3	23.5	16.2	44	3	8.1	37	3	17	0	13
Oct.	1020.9	994.1	E	1.9	10.2	15.9	16.5	40	−5	6.9	48	11	77	0	24
Nov.	1023.1	995.5	E	1.7	5.8	7.7	14.8	33	−15	5.5	65	15	71	0	23
Dec.	1023.7	995.5	E	1.7	3.7	2.8	13.2	31	−22	4.9	77	19	55	0	24
Year	1016.7	990.0	NW	2.1	133.4	16.2		47	−26		54	210	377	136	55

Month	Number of days							Snow cover	Cloudiness (tenths)			Mean sun-shine	Evaporation (mm)	
	wind (speed ≥15 m/s)	frost-free period	fog	precip. (≥0.1 mm)	thunder-storm	snow-cover	clouds (8–10 tenths)	max. 10-day average depth (cm)	07h	13h	19h	(h)	actual	potent.
Jan.	0.4		5	10.0	0	31	13					111	<10	
Feb.	1.2		3	8.8	0.4	16	11					122	10	
Mar.	1.5		2	10.4	1		11					150	30	
Apr.	1.0		0.6	8.9	4		8					195	50	
May	1.3		0.2	6.6	6		4					289	30	
June	1.2		0	2.4	4		2					327	<10	
July	1.3		0	1.3	2		0.8					358	<10	
Aug.	0.6		0.05	1.1	1		0.5					353	<10	
Sept.	0.9		0	0.8	0.6		1					293	<10	
Oct.	0.9		0.5	3.6	0.9		4					244	<10	
Nov.	0.6		0.7	6.1	0	14	8					159	<10	
Dec.	0.4		3	9.0	0	18	12					111	<10	
Year	11.3	233	15	69.0	20	18	75	4				2712	200	1200

TABLE VII

CLIMATIC TABLE FOR ASTRAKHAN

Latitude 46°16′N longitude 48°02′E elevation 18 m

Month	Air press. (mbar) sea level	Air press. (mbar) station	Wind preval. direct.	Wind mean speed (m/s)	Global radiat. (kcal./cm²)	Temperature (°C) mean	Temperature (°C) mean daily range	Temperature (°C) max.	Temperature (°C) min.	Humidity vapor press. (mm)	Humidity relat. (%)	Precipitation (mm) mean	Precipitation (mm) max.	Precipitation (mm) min.	Precipitation (mm) max. in 24 h
Jan.	1021.2		E	5.2	2.5	−6.9	10.6	12	−32	2.7	86	16.3	45	0	21
Feb.	1020.8		E	5.4	4.8	−5.8	11.6	17	−32	2.8	83	10.7	70	0	17
Mar.	1020.5		E	5.7	7.3	0.2	12.7	24	−30	3.9	77	19.8	57	0	17
Apr.	1017.1		E	5.6	10.9	9.3	15.0	32	−10	6.0	62	9.9	56	0	43
May	1014.7		E	5.5	14.9	17.7	13.5	36	−2	9.5	58	23.3	95	0	73
June	1012.4		E	4.9	16.5	22.5	11.6	39	6	12.8	54	16.3	126	0	49
July	999.5		E	4.2	16.2	25.1	12.6	40	11	14.9	56	23.2	94	0	62
Aug.	1011.2		E	4.1	14.5	23.3	13.7	39	6	13.6	60	12.0	68	0	64
Sept.	1016.8		E	4.3	11.3	17.1	13.6	36	−1	10.1	66	16.0	66	0	60
Oct.	1020.7		E	4.5	6.7	9.9	13.5	30	−10	7.2	74	13.8	63	0	35
Nov.	1025.6		E	5.0	3.4	2.3	10.1	21	−26	5.1	82	11.4	55	0	22
Dec.	1022.9		E	5.0	2.0	−3.5	9.1	16	−34	3.5	86	17.4	64	0	15
Year	1017.8		E	5.0	111.0	9.3		40	−34		70	190.1	305	67	73

Month	Number of days wind (speed ⩾15 m/s)	frost-free period	fog	precip. (⩾0.1 mm)	thunder-storm	snow-cover	clouds (8–10 tenths)	Snow cover max. 10-day average depth (cm)	Cloudiness (tenths) 07h	13h	19h	Mean sunshine (h)	Evaporation (mm) actual	potent.
Jan.	0.9		7	9.1			16.0		7.7	7.9	7.0	67	<10	
Feb.	1.2		6	7.8			12.6		5.6	7.2	6.1	95	<10	
Mar.	2.1		6	6.8			12.2		7.1	7.2	6.3	161	<10	
Apr.	2.6		2	5.1			7.4		6.1	6.5	6.1	231	>20	
May	2.0		0.4	5.2	0.5		4.7		5.3	6.1	5.7	300	45	
June	1.6		0.3	5.0	1		2.5		4.6	5.4	5.1	333	40	
July	0.9		0.3	4.8	3		2.3		3.8	4.6	4.0	327	20	
Aug.	1.0		0.7	4.1	5		2.5		3.6	4.3	3.9	314	15	
Sept.	1.2		2	4.5	2		3.4		4.2	4.6	3.8	256	10	
Oct.	1.0		5	5.9	0.9		6.2		6.3	6.0	4.4	189	10	
Nov.	1.2		7	8.0	0.3		11.3		7.3	6.9	5.9	110	<10	
Dec.	0.9		8	10.7			16.3		7.9	7.8	7.2	58	<10	
Year	16.6	187	45	77	13	12	97		6.0	6.2	5.4	2441	200	900

TABLE VIII

CLIMATIC TABLE FOR BAKU (observatory)

Latitude 40°21′N longitude 49°50′E elevation 10 m

Month	Air press. (mbar) sea level	Air press. (mbar) station	Wind preval. direct.	Wind mean speed (m/s)	Global radiat. (kcal./cm²)	Temperature (°C) mean	Temperature (°C) mean daily range	Temperature (°C) max.	Temperature (°C) min.	Humidity vapor press. (mm)	Humidity relat. (%)	Precipitation (mm) mean	Precipitation (mm) max.	Precipitation (mm) min.	Precipitation (mm) max. in 24 h
Jan.				5.4		3.5	6.8			4.8	75	22			
Feb.				6.4		3.9	8.0			4.9	76	16			
Mar.				6.7		6.3	8.3			5.8	77	24			
Apr.				6.5		10.7	8.9			7.6	70	25			
May				5.8		17.5	8.4			11.0	65	17			
June				6.2		22.3	8.2			13.6	61	13			
July				6.7		25.3	8.0			15.9	60	8			
Aug.				6.4		25.5	7.7			17.2	64	7			
Sept.				6.0		21.7	6.5			15.0	68	15			
Oct.				5.7		16.8	6.7			11.2	73	31			
Nov.				5.7		11.0	6.9			8.2	77	33			
Dec.				5.9		6.6	7.5			6.0	77	27			
Year				6.1		14.3					70	238			

Month	Number of days wind (speed ⩾15 m/s)	frost-free period	fog	precip. (⩾0.1 mm)	thunder-storm	snow-cover	clouds (8–10 tenths)	Snow cover max. 10-day average depth (cm)
Jan.			4	8.7			14.4	
Feb.			5	7.0			14.1	
Mar.			5	7.3			14.7	
Apr.			3	6.0	0.5		9.5	
May			2	5.3	1		5.0	
June			0.6	3.3	2		2.3	
July			0.4	2.2	1		2.6	
Aug.			0.6	2.3	1		2.2	
Sept.			0.8	4.4	0.8		5.1	
Oct.			2	6.9	0.3		9.2	
Nov.			2	9.4	0.1		12.7	
Dec.			3	8.4	0.1		13.9	
Year		293	28	71.2	7	7	106	2

TABLE IX

CLIMATIC TABLE FOR BALKHASH

Latitude 46°54'N longitude 75°00'E elevation 423 m

Month	Air press. (mbar) sea level	Air press. (mbar) station	Wind preval. direct.	Wind mean speed (m/s)	Global radiat. (kcal./cm²)	Temperature (°C) mean	Temperature (°C) mean daily range	Temperature (°C) max.	Temperature (°C) min.	Humidity vapor press. (mm)	Humidity relat. (%)	Precipitation (mm) mean	Precipitation (mm) max.	Precipitation (mm) min.	Precipitation (mm) max. in 24 h
Jan.	1031.8		NE	5.7		−15.6	8.8	4	−46		78	10			
Feb.	1029.6		NE	5.6		−13.0	9.5	8	−40		78	8			
Mar.	1025.8		NE	5.2		−5.7	11.4	24	−30		76	10			
Apr.	1020.8		NE	5.0		6.2	14.0	28	−14		58	11			
May	1014.6		NE	5.2		16.1	14.8	34	−7		47	9			
June	1008.7		NE	4.9		21.3	13.6	41	1		44	19			
July	1005.1		NE	4.7		23.9	13.2	41	8		44	11			
Aug.	1008.7		NE	4.8		21.7	13.7	32	3		44	9			
Sept.	1016.3		NE	4.6		15.0	14.0	37	−5		48	4			
Oct.	1023.2		NE	4.5		6.5	13.7	28	−20		60	8			
Nov.	1030.4		NE	4.8		−5.4	10.8	17	−33		74	9			
Dec.	1032.6		NE	4.9		−11.6	8.9	7	−41		80	12			
Year	1020.6		NE	5.0		5.1		41	−46		61	115			

Month	Number of days: wind (speed ⩾15 m/s)	frost-free period	fog	precip. (⩾0.1 mm)	thunder-storm	snow-cover	clouds (8–10 tenths)	Snow cover max. 10-day average depth (cm)	Cloudiness (tenths) 07h	Cloudiness (tenths) 13h	Cloudiness (tenths) 19h	Mean sun-shine (h)	Evaporation (mm) actual	Evaporation (mm) potent.
Jan.	1.0		2	8.3		31	8.6		5.8	5.6	4.8		<10	
Feb.	1.3		3	5.8		28	7.5		5.4	5.3	4.5		<10	
Mar.	1.2		3	5.4		17	7.9		6.2	5.7	5.1		<10	
Apr.	1.3		0.7	4.6	0.7		6.1		5.2	5.5	5.4		<20	
May	2.3			4.1	2		3.9		4.3	5.1	5.2		<50	
June	2.4			4.2	3		2.7		3.8	4.8	4.8		30	
July	1.8			4.7	4		1.8		3.1	4.5	4.1		<20	
Aug.	1.7			3.3	4		1.5		2.8	3.6	3.4		10	
Sept.	1.3		1	2.4	0.3		1.5		3.2	3.4	3.0		10	
Oct.	1.1		0.4	4.4	0.1		4.7		4.8	4.7	3.7		<10	
Nov.	0.7		4	6.6			7.8		5.9	5.7	4.5		<10	
Dec.	1.0		6	8.5		25	10.0		6.2	6.0	5.2		<10	
Year	17.1	166	20	62.3	14	110	64	12	4.7	5.0	4.5		<200	800

TABLE X

CLIMATIC TABLE FOR BARNAUL

Latitude 53°20'N longitude 83°48'E elevation 158 m

Month	Air press. (mbar) sea level	Air press. (mbar) station	Wind preval. direct.	Wind mean speed (m/s)	Global radiat. (kcal./cm²)	Temperature (°C) mean	Temperature (°C) mean daily range	Temperature (°C) max.	Temperature (°C) min.	Humidity vapor press. (mm)	Humidity relat. (%)	Precipitation (mm) mean	Precipitation (mm) max.	Precipitation (mm) min.	Precipitation (mm) max. in 24 h
Jan.	1030.5	1009.0	SW	3.3		−17.7	10.3	5	−52	1.2	77	19	93	8	
Feb.	1030.0	1008.6	SW	3.6		−16.4	11.4	8	−49	1.3	76	18	83	1	
Mar.	1027.1	1006.3	SW	3.6		−9.7	11.3	13	−41	1.9	76	20	70	3	
Apr.	1021.6	1001.8	SW	3.4		1.2	10.9	27	−28	3.6	68	26	79	2	
May	1015.5	996.5	SW	3.6		11.0	13.2	35	−16	5.8	56	38	73	3	
June	1009.7	991.3	NE	3.2		17.6	12.8	37	−2	9.4	62	52	117	7	
July	1007.0	988.7	NE	2.8		19.6	11.8	38	3	11.6	68	79	187	17	
Aug.	1009.9	991.4	NE, SW	2.7		17.0	12.1	36	−1	10.1	76	52	139	2	
Sept.	1017.2	998.1	SW	3.0		10.7	12.3	34	−10	6.7	71	39	92	1	
Oct.	1022.7	1002.9	SW	3.5		2.3	9.3	25	−36	4.2	73	40	114	11	
Nov.	1027.1	1006.5	SW	3.9		−7.9	8.3	13	−45	2.5	79	40	112	2	
Dec.	1030.3	1009.0	SW	3.4		−15.1	9.1	7	−50	1.4	79	31	85	10	
Year	1020.7	1000.9	SW	3.3		1.0		38	−52		71	464	670	266	

Month	Number of days: wind (speed ⩾15 m/s)	frost-free period	fog	precip. (⩾0.1 mm)	thunder-storm	snow-cover	clouds (8–10 tenths)	Snow cover max. 10-day average depth (cm)	Cloudiness (tenths) 07h	Cloudiness (tenths) 13h	Cloudiness (tenths) 19h	Mean sun-shine (h)	Evaporation (mm) actual	Evaporation (mm) potent.
Jan.	3.6		7	15.1		31	3.4		7.0	6.7	5.6	58	<10	
Feb.	2.5		7	11.7		28	8.9		7.0	6.4	5.1	95	<10	
Mar.	4.2		8	11.2		31	8.9		7.2	6.8	5.9	146	<10	
Apr.	4.0		3	8.7	0.2	6	7.9		6.7	7.0	6.9	196	>10	
May	4.8		1	11.9	3		8.0		6.3	7.0	6.7	250	70	
June	3.0		0.5	12.0	6		6.2		6.0	7.1	6.5	284	70	
July	1.8		2	12.5	8		6.5		6.1	7.1	6.7	278	30	
Aug.	1.1		3	12.3	6		6.8		6.1	6.9	6.2	251	20	
Sept.	3.3		5	16.9	1		7.7		6.7	7.0	5.8	176	<20	
Oct.	4.9		5	13.5	0.2		14.2		7.9	8.0	6.6	105	10	
Nov.	6.1		4	16.4	0.02	23	16.6		7.9	7.7	6.8	55	<10	
Dec.	6.1		6	16.6		31	14.4		7.6	7.7	6.8	42	<10	
Year	45.4	116	52	152.8	24	164	120	67	6.9	7.1	6.3	1936	300	600

TABLE XI

CLIMATIC TABLE FOR BATUMI

Latitude 41°45′N longitude 41°40′E elevation 92 m

Month	Air press. (mbar)		Wind		Global radiat. (kcal./cm²)	Temperature (°C)				Humidity		Precipitation (mm)			
	sea level	station	preval. direct.	mean speed (m/s)		mean	mean daily range	max.	min.	vapor press. (mm)	relat. (%)	mean	max.	min.	max. in 24 h
Jan.	1019.2	1018.8	SW	2.4		6.4	9.9	25	−8	5.6	73	237	522	51	93
Feb.	1018.3	1017.9	SW	2.5		6.5	9.8	28	−8	5.6	77	205	490	29	73
Mar.	1016.6	1016.2	SW	2.0		8.4	10.4	32	−7	6.4	77	136	324	20	73
Apr.	1015.0	1014.6	SW	1.8		11.3	10.9	38	−1	8.0	80	138	238	13	74
May	1014.2	1013.8	SW	1.6		16.1	10.8	36	2	11.1	82	82	215	5	98
June	1012.7	1012.3	SW	1.5		20.3	9.6	38	9	14.2	79	165	524	40	157
July	1011.1	1010.7	SW	1.4		23.0	9.1	40	13	16.6	80	178	407	24	147
Aug.	1011.8	1011.4	SW	1.4		23.2	9.5	40	13	17.3	81	233	699	54	261
Sept.	1015.5	1015.1	SW	1.4		20.2	10.1	37	7	14.3	82	315	703	47	221
Oct.	1017.8	1017.6	SW	1.6		16.3	10.8	33	2	11.6	83	261	703	19	147
Nov.	1019.4	1019.0	SW	1.8		12.1	10.1	29	−4	8.4	80	294	859	20	130
Dec.	1019.2	1018.8	SW	1.9		9.2	9.5	28	−6	6.5	76	260	508	37	100
Year	1015.9	1015.5	SW	1.8		14.4		40	−8		79	2504	3943	1299	261

Month	Number of days							Snow cover max. 10-day average depth (cm)	Cloudiness (tenths)			Mean sunshine (h)	Evaporation (mm)	
	wind (speed ≥15 m/s)	frost-free period	fog	precip. (≥0.1 mm)	thunderstorm	snow-cover	clouds (8–10 tenths)		07h	13h	19h		actual	potent.
Jan.	1.9		0.4	14.6	0.7		14.0		7.5	7.7	6.9	99	20	
Feb.	2.0		1	15.5	0.2		13.9		7.9	7.7	7.3	105	30	
Mar.	1.9		3	15.6	0.1		14.6		8.0	7.9	7.8	126	50	
Apr.	2.1		4	15.1	0.6		13.0		7.7	7.5	7.6	148	50	
May	1.3		3	14.3	2		11.6		7.0	6.9	7.6	199	>90	
June	0.8		0.2	14.0	5		10.2		6.0	5.7	6.6	235	>90	
July	0.6		0.5	15.5	6		12.0		6.7	6.5	7.1	214	100	
Aug.	0.5		0.6	15.3	5		11.5		6.4	6.2	6.9	223	100	
Sept.	0.7		0.4	14.2	6		11.8		6.0	5.8	6.5	201	70	
Oct.	1.1		0.1	12.5	3		8.1		6.3	6.2	5.7	176	30	
Nov.	1.3		0.2	14.4	1		11.0		7.0	6.9	6.0	125	30	
Dec.	1.7		0.8	15.5	1		13.5		7.0	7.3	6.2	107	30	
Year	16.0	304	14	176.5	32	13	159	14	7.0	6.9	6.8	1958	<700	1000

TABLE XII

CLIMATIC TABLE FOR BLAGOVESHCHENSK

Latitude 50°16′N longitude 127°30′E elevation 132 m

Month	Air press. (mbar)		Wind		Global radiat. (kcal./cm²)	Temperature (°C)				Humidity		Precipitation (mm)			
	sea level	station	preval. direct.	mean speed (m/s)		mean	mean daily range	max.	min.	vapor press. (mm)	relat. (%)	mean	max.	min.	max. in 24 h
Jan.	1024.4	1004.5	NW	3.0	3.3	−24.2	11.0	−1	−42	0.5	72	5	19	0	6
Feb.	1022.4	1003.0	NW	3.2	6.3	−18.5	12.4	5	−45	0.7	70	3	15	0	11
Mar.	1018.1	999.4	NW	4.7	10.4	−9.5	13.8	14	−36	1.5	66	9	38	0	29
Apr.	1010.5	992.7	NW	5.8	11.8	2.4	12.5	28	−16	2.9	58	21	66	1	32
May	1005.9	989.4	N, NW	5.0	13.4	19.8	16.4	33	−8	5.2	58	45	143	0	47
June	1004.1	987.5	S	4.0	14.2	17.5	14.6	38	0	10.2	70	100	235	29	67
July	1004.5	988.1	S	3.4	12.6	21.5	12.8	41	8	13.9	76	120	284	2	113
Aug.	1006.7	990.2	NW	3.4	11.6	19.2	13.8	37	4	12.6	77	106	325	7	122
Sept.	1011.7	994.6	NW	3.6	8.5	12.2	14.5	34	−6	7.3	74	80	183	13	48
Oct.	1016.1	998.2	NW	4.9	6.3	1.9	13.6	26	−25	3.5	65	26	79	0	28
Nov.	1019.7	1000.9	NW	3.6	4.0	−11.3	10.5	11	−33	1.5	70	13	39	0	17
Dec.	1022.6	1002.9	NW	2.9	3.0	−21.7	11.0	4	−41	0.7	73	6	20	0	8
Year	1013.9	995.9	NW	4.0	105.4	0.0		41	−45		69	534	785	260	122

Month	Number of days							Snow cover max. 10-day average depth (cm)	Cloudiness (tenths)			Mean sunshine (h)	Evaporation (mm)	
	wind (speed ≥15 m/s)	frost-free period	fog	precip. (≥0.1 mm)	thunderstorm	snow-cover	clouds (8–10 tenths)		07h	13h	19h		actual	potent.
Jan.	1.2		1	6		31	0		3.2	3.9	2.6	151	<10	
Feb.	1.1		0.6	4		28	0		3.2	3.7	2.3	185	<10	
Mar.	3.3		0.4	4		23	0.6		4.7	4.8	3.9	215	10	
Apr.	6.0		0.3	7	0.2		0.9		5.8	6.9	6.2	199	20	
May	5.1		0.1	10	2		2		6.1	7.2	7.1	215	50	
June	2.6		0.7	15	7		4		6.4	7.0	7.1	228	70	
July	1.6		2	14	10		4		6.8	7.3	7.0	235	<70	
Aug.	1.4		2	14	5		3		6.3	6.9	6.5	220	>70	
Sept.	1.6		2	13	3		3		6.3	6.6	5.9	173	>50	
Oct.	3.5		0.7	6	0.2		1		5.3	5.5	4.3	170	30	
Nov.	1.7		0.8	7		20	0		4.4	4.4	3.3	148	<10	
Dec.	1.0		1	7		31	0		3.4	4.3	3.0	127	<10	
Year	30.1	144	12	107	27	147	18.5	21	5.2	5.7	4.9	2266	400	600

TABLE XIII

CLIMATIC TABLE FOR BOMNAK

Latitude 54°43'N longitude 128°56'E elevation 357 m

Month	Air press. (mbar)		Wind		Global radiat. (kcal./cm²)	Temperature (°C)				Humidity		Precipitation (mm)			
	sea level	station	preval. direct.	mean speed (m/s)		mean	mean daily range	max.	min.	vapor press. (mm)	relat. (%)	mean	max.	min.	max. in 24 h
Jan.	1025.9	975.5	NE	0.4		−32.4	11.2	−9	−51	0.4	74	4	17	0	5
Feb.	1023.5	974.7	NE	0.9		−24.4	14.7	2	−49	0.6	66	3	15	0	5
Mar.	1018.8	972.1	NE	1.7		−13.6	14.9	8	−40	1.4	58	9	32	0	21
Apr.	1011.1	966.7	E	2.6		−1.5	12.3	23	−28	3.0	53	26	80	2	29
May	1007.4	964.7	E	2.6		7.9	13.2	31	−11	5.7	55	45	134	9	45
June	1005.3	963.8	E	2.3		15.0	14.4	34	−2	11.0	67	82	182	13	56
July	1005.2	964.2	E	2.1		18.2	12.8	35	0	15.4	75	113	406	16	93
Aug.	1007.2	965.8	E	1.9		15.4	11.5	32	−1	14.2	81	130	304	26	101
Sept.	1012.3	969.5	E	2.0		8.6	12.1	29	−10	8.5	76	90	175	6	57
Oct.	1016.3	971.5	E	1.4		−2.7	11.6	21	−35	3.5	67	33	102	2	29
Nov.	1020.5	972.7	E	1.1		−20.0	11.2	6	−46	1.2	75	23	59	5	17
Dec.	1022.9	973.8	NE	0.6		−30.2	2.8	−1	−52	0.5	76	8	31	0	7
Year	1014.7	969.6	E	1.6		−4.9		35	−52	5.4	68	546	896	226	101

Month	Number of days							Snow cover max. 10-day average depth (cm)	Cloudiness (tenths)			Mean sunshine (h)	Evaporation (mm)	
	wind (speed ⩾15 m/s)	frost-free period	fog	precip. (⩾0.1 mm)	thunderstorm	snow-cover	clouds (8–10 tenths)		07h	13h	19h		actual	potent.
Jan.	0.03		0.1	6.3		31	3.1		3.5	5.1	2.8	111	5	
Feb.	0.2		0.03	4.0		28	3.1		4.1	4.5	2.6	154	5	
Mar.	0.5		0.1	4.8		31	6.0		5.5	5.4	4.4	209	5	
Apr.	1.2		0.3	7.2	0.03	26	9.6		6.3	6.7	6.6	204	<20	
May	0.8		0.7	9.8	2		12.9		6.9	8.0	7.5	205	70	
June	0.5		1	12.4	8		14.3		7.1	7.5	7.7	219	70	
July	0.3		3	12.9	9		14.5		7.4	7.2	7.4	228	<70	
Aug.	0.03		5	13.8	6		14.5		7.6	7.5	7.4	185	>70	
Sept.	0.1		4	12.3	2		11.9		7.3	7.2	6.5	152	>50	
Oct.	0.2		1	8.4	0.1	10	9.0		6.1	6.4	4.9	141	20	
Nov.	0.1		0.4	9.8		30	5.9		5.4	5.6	3.9	101	5	
Dec.	0.1		0.1	7.9		31	5.2		4.1	5.5	3.4	82	5	
Year	4.06	100	16	110	27	191	110		5.9	6.4	5.4	1991	<400	500

TABLE XIV

CLIMATIC TABLE FOR BUKHARA

Latitude 39°43'N longitude 64°37'E elevation

Month	Air press. (mbar)		Wind		Global radiat. (kcal./cm²)	Temperature (°C)				Humidity		Precipitation (mm)			
	sea level	station	preval. direct.	mean speed (m/s)		mean	mean daily range	max.	min.	vapor press. (mm)	relat. (%)	mean	max.	min.	max. in 24 h
Jan.			N	2.6	5.8					5.2	80	16			
Feb.			N	3.3	8.5					5.8	75	16			
Mar.			N	3.5	11.4					7.4	72	26			
Apr.			N	3.9	15.3					9.6	59	20			
May			N	4.4	19.5					11.5	46	11			
June			N	4.0	22.0					12.0	38	1			
July			N	4.0	21.9					13.9	40	1			
Aug.			N	3.9	20.0					13.5	44	0			
Sept.			N	3.3	16.4					10.5	48	0.1			
Oct.			N	2.5	12.0					7.8	56	3			
Nov.			N	2.9	7.5					5.5	64	8			
Dec.			N	2.2	5.6					5.3	79	11			
Year			N	3.4	165.9					9.0	58	113.1			

Month	Number of days							Snow cover max. 10-day average depth (cm)	Cloudiness (tenths)			Mean sunshine (h)	Evaporation (mm)	
	wind (speed ⩾15 m/s)	frost-free period	fog	precip. (⩾0.1 mm)	thunderstorm	snow-cover	clouds (8–10 tenths)		07h	13h	19h		actual	potent.
Jan.	1.4		6		–		9.1						10	
Feb.	3.0		3		0.1		7.4						10	
Mar.	5.4		1		0.6		9.9						12	
Apr.	5.6		0.4		2		9.2						19	
May	7.5				2		3.9						9	
June	4.6				1		0.5						0	
July	2.8				0.4		0.2						0	
Aug.	1.3				0.1		0.1						0	
Sept.	1.5				–		0.1						0	
Oct.	1.7		0.5		0.1		1.9						2	
Nov.	1.7		1.0		0.1		5.5						7	
Dec.	1.1		6		–		10.6						9	
Year	37.6		18		6.4		58.4						78	

TABLE XV

CLIMATIC TABLE FOR CHIMBAY

Latitude 42°57'N longitude 59°48'E elevation 66 m

Month	Air press. (mbar)		Wind		Global radiat.	Temperature (°C)				Humidity		Precipitation (mm)			
	sea level	station	preval. direct.	mean speed (m/s)	(kcal./cm²)	mean	mean daily range	max.	min.	vapor press. (mm)	relat. (%)	mean	max.	min.	max. in 24 h
Jan.	1026.4	1017.5		3.5		−6.2	9.3	17	−31	3.1	80	7	19	0	
Feb.	1024.3	1015.5		4.0		−4.1	10.8	24	−28	3.7	79	14	50	0	
Mar.	1021.6	1013.1		4.3		2.0	12.0	33	−24	5.5	71	13	46	0	
Apr.	1017.0	1008.9		4.3		12.0	15.0	39	−11	7.9	55	12	47	0	
May	1014.7	1006.7		4.1		19.6	15.6	42	−2	10.0	48	10	32	0	
June	1010.2	1002.3		3.7		24.2	15.7	44	4	12.9	49	6	37	0	
July	1008.2	1000.5		3.2		26.2	15.0	44	8	16.1	52	2	25	0	
Aug.	1011.4	1003.5		3.0		24.0	15.1	43	6	15.1	56	2	22	0	
Sept.	1017.5	1009.5		2.6		17.5	16.0	39	−3	10.9	59	3	18	0	
Oct.	1023.4	1015.2		2.7		9.7	15.1	34	−12	6.8	62	7	28	0	
Nov.	1025.9	1017.4		3.2		0.9	12.1	28	−28	6.1	69	5	21	0	
Dec.	1026.7	1017.9		3.3		−4.7	9.3	17	−28	4.3	80	8	32	0	
Year	1018.9	1010.7		3.5		10.1		44	−31	8.5	63	89	214	54	

Month	Number of days							Snow cover	Cloudiness (tenths)			Mean sun-shine	Evaporation (mm)	
	wind (speed ≥15 m/s)	frost-free period	fog	precip. (≥0.1 mm)	thunder-storm	snow-cover	clouds (8–10 tenths)	max. 10-day average depth (cm)	07h	13h	19h	(h)	actual	potent.
Jan.	0.2		5	5.6	0.03		12.3					106	<10	
Feb.	1.2		3	5.8			10.4					130	10	
Mar.	1.4		2	5.2	0.1		8.7					180	>10	
Apr.	1.4		0.6	4.8	0.4		5.9					242	30	
May	1.2		0.04	3.5	2		3.5					329	30	
June	1.1			2.5	2		1.6					370	10	
July	0.2			1.0	1		1.0					387	<10	
Aug.	0.4		0.1	1.1	0.6		0.0					364	<10	
Sept.	0.4		1	2.0	0.2		1.2					303	<10	
Oct.	0.5		1	2.8	0.1		3.4					245	<10	
Nov.	0.3		2	3.5			7.0					164	<10	
Dec.	0.04		6	5.6			13.0					107	<10	
Year	8.34	184	21	43.4	6	31	68	5				2927	>100	900

TABLE XVI

CLIMATIC TABLE FOR CHITA

Latitude 52°01'N longitude 113°20'E elevation 662 m

Month	Air press. (mbar)		Wind		Global radiat.	Temperature (°C)				Humidity		Precipitation (mm)			
	sea level	station	preval. direct.	mean speed (m/s)	(kcal./cm²)	mean	mean daily range	max.	min.	vapor press. (mm)	relat. (%)	mean	max.	min.	max. in 24 h
Jan.	1033.2	943.1	NW	1.1	2.8	−26.8	14.4	0	−51	0.5	74	2	10	0	9
Feb.	1030.8	942.6	NW	1.5	5.0	−21.8	18.2	7	−44	0.7	72	2	10	0	7
Mar.	1024.1	939.7	NW	2.0	9.5	−11.7	19.3	18	−40	1.4	64	4	36	0	22
Apr.	1014.9	934.7	NW	3.0	12.6	0.0	17.9	29	−25	2.5	50	7	39	0	22
May	1009.4	931.8	NW	2.6	15.1	7.8	20.9	34	−12	4.1	47	20	71	0	32
June	1005.1	929.9	NW	2.2	15.0	15.1	20.6	36	−3	7.7	59	51	157	12	34
July	1003.8	929.7	NW	1.8	13.4	18.5	19.7	38	−2	11.0	70	93	230	13	51
Aug.	1007.3	032.1	NW	1.6	11.2	15.2	19.2	41	−6	9.6	73	90	197	22	104
Sept.	1014.7	936.9	NW	1.7	8.6	8.4	19.8	28	−10	5.7	71	51	98	1	28
Oct.	1020.7	939.7	NW	1.8	5.8	−1.4	16.2	22	−29	3.0	67	12	43	0	19
Nov.	1025.9	940.6	NW	1.6	3.2	−14.3	14.4	10	−38	1.4	73	7	32	0	13
Dec.	1031.1	942.1	NW	1.2	1.9	−24.1	13.2	2	−50	0.6	78	4	10	0	5
Year	1018.4	936.9	NW	1.8	104.1	−2.9		41	−51		66	343	580	212	104

Month	Number of days							Snow cover	Cloudiness (tenths)			Mean sun-shine	Evaporation (mm)	
	wind (speed ≥15 m/s)	frost-free period	fog	precip. (≥0.1 mm)	thunder-storm	snow-cover	clouds (8–10 tenths)	max. 10-day average depth (cm)	07h	13h	19h	(h)	actual	potent.
Jan.	0.8		12	3.2		31	2		3.9	5.2	3.3	123	<10	
Feb.	1.0		6	2.5		28	2		4.2	5.0	3.1	170	<10	
Mar.	3.0		3	2.7	0.0	21	4		5.2	5.4	4.2	237	<10	
Apr.	5.0		1	3.8	0.1		6		6.1	7.3	6.2	237	>10	
May	10.0		1	7.0	0.9		8		5.9	7.6	7.2	271	>30	
June	2.4		2	10.1	5.0		8		6.7	7.6	7.2	255	<70	
July	0.6		5	14.0	8.0		10		6.9	7.5	7.4	234	<60	
Aug.	0.6		8	15.2	5.0		10		7.2	7.1	6.7	227	>50	
Sept.	1.4		5	8.8	0.6		7		7.1	6.9	6.1	191	<30	
Oct.	0.8		3	4.0	0.0		6		6.3	6.7	4.8	175	>10	
Nov.	1.8		5	4.2		17	5		5.6	6.2	4.2	131	<10	
Dec.	0.4		10	5.0		31	3		4.4	6.0	3.7	102	<10	
Year	27.8	99	61	80.5	20.0	145	71	11	5.8	6.5	5.3	2353	300	500

TABLE XVII

CLIMATIC TABLE FOR CHOKURDAKH

Latitude 70°37′N longitude 147°53′E elevation 20 m

Month	Air press. (mbar) sea level	Air press. (mbar) station	Wind preval. direct.	Wind mean speed (m/s)	Global radiat. (kcal./cm²)	Temperature (°C) mean	Temperature (°C) mean daily range	Temperature (°C) max.	Temperature (°C) min.	Humidity vapor press. (mm)	Humidity relat. (%)	Precipitation (mm) mean	Precipitation (mm) max.	Precipitation (mm) min.	Precipitation (mm) max. in 24 h
Jan.	1021.0		W	3.7		−36.2		−8	−52	0.3	86	7			
Feb.	1023.2		W	3.9		−34.1		−3	−50	0.4	86	8			
Mar.	1020.1		W	4.4		−29.2		−2	−46	0.5	86	6			
Apr.	1015.3		NE	4.9		−18.3		1	−40	1.2	85	8			
May	1013.5		E	5.5		−6.3		18	−30	3.4	84	10			
June	1008.5		NE	6.0		6.1		29	−8	7.2	79	13	40	1	
July	1007.6		NE	6.0		10.2		32	0	9.5	75	22	60	6	
Aug.	1008.9		NE	5.6		7.0		29	−4	8.2	81	24	65	5	
Sept.	1013.3		W	4.2		1.0		20	−14	5.9	84	15	48	2	
Oct.	1012.1		SW	4.1		−12.3		7	−34	2.4	87	12			
Nov.	1018.4		SW	3.9		−24.9		−4	−44	0.8	85	11			
Dec.	1022.0		SW	3.8		−32.8		−4	−48	0.4	85	9			
Year	1015.3		W	4.7		−14.2		32	−52	3.4	84	145			

Month	Number of days wind (speed ≥15 m/s)	frost-free period	fog	precip. (≥0.1 mm)	thunder-storm	snow-cover	clouds (8–10 tenths)	Snow cover max. 10-day average depth (cm)	Cloudiness (tenths) 07h	13h	19h	Mean sunshine (h)	Evaporation (mm) actual	potent.
Jan.	1.4		2	6.9		31	6					0	5	
Feb.	0.7		3	9.6		28	7					48	5	
Mar.	1.2		2	7.6		31	6					179	5	
Apr.	2.0		3	7.1		30	7					284	<10	
May	1.8		6	8.1		29	14					269	<10	
June	1.0		6	9.5	0.3		16					283	>50	
July	2.4		6	10.8	0.2		18					278	<60	
Aug.	1.4		7	12.5	0.1		18					134	<20	
Sept.	0.8		6	12.4		5	17					94	10	
Oct.	1.8		6	16.1		31	18					43	<10	
Nov.	1.4		5	11.4		30	11					4	5	
Dec.	1.6		4	9.0		31	9						5	
Year	17.5	60	56	121.0	0.6	240	147	25				1616	<200	

TABLE XVIII

CLIMATIC TABLE FOR DUDINKA

Latitude 69°24′N longitude 86°10′E elevation 20 m

Month	Air press. (mbar) sea level	Air press. (mbar) station	Wind preval. direct.	Wind mean speed (m/s)	Global radiat. (kcal./cm²)	Temperature (°C) mean	Temperature (°C) mean daily range	Temperature (°C) max.	Temperature (°C) min.	Humidity vapor press. (mm)	Humidity relat. (%)	Precipitation (mm) mean	Precipitation (mm) max.	Precipitation (mm) min.	Precipitation (mm) max. in 24 h
Jan.	1018.1	1015.1	SE	6.4		−29.5		−2	−57	0.5	78	12			
Feb.	1018.4	1015.4	SE	7.2		−25.7		−1	−55	0.7	79	11			
Mar.	1015.7	1012.8	ESE	6.6		−22.5		4	−52	0.8	79	9			
Apr.	1013.3	1010.5	E	7.5		−16.0		9	−42	1.5	80	10			
May	1011.2	1008.5	N	7.4		−6.4		16	−36	2.7	81	12			
June	1008.3	1005.7	NE	6.3		3.8		28	−15	5.4	79	29	86	6	
July	1007.5	1005.0	NE	6.2		12.0		30	−1	8.0	71	32	129	1	
Aug.	1008.6	1006.1	NE	6.2		10.4		30	−2	7.8	78	49	120	10	
Sept.	1010.4	1007.8	NE	6.7		3.2		24	−20	5.2	80	33	114	22	
Oct.	1009.9	1007.2	SE	7.1		−8.4		11	−40	2.3	86	28			
Nov.	1013.7	1010.8	SE	6.3		−21.8		3	−48	1.0	82	18			
Dec.	1017.1	1014.1	SE	6.4		−26.9		0	−54	0.6	79	13			
Year	1012.8	1008.9	SE	6.7		−10.7		30	−57		79	267			

Month	Number of days wind (speed ≥15 m/s)	frost-free period	fog	precip. (≥0.1 mm)	thunder-storm	snow-cover	clouds (8–10 tenths)	Snow cover max. 10-day average depth (cm)	Cloudiness (tenths) 07h	13h	19h	Mean sunshine (h)	Evaporation (mm) actual	potent.
Jan.	5.5		4	17.7		31	15					0	5	
Feb.	4.9		3	15.9		28	13					35	5	
Mar.	5.7		3	13.5		31	12					131	<10	
Apr.	6.6		3	14.1	0.0	30	13					204	<10	
May	6.3		4	12.2	0.03	29	19					238	<10	
June	3.6		4	14.7	0.6		18					243	70	
July	4.6		2	13.8	2.0		14					320	<60	
Aug.	4.3		3	15.2	2.0		19					219	<20	
Sept.	3.8		4	19.2	0.1	5	19					85	>10	
Oct.	5.3		4	19.2	0.0	31	22					38	<10	
Nov.	4.7		4	17.2		30	17					5	<10	
Dec.	5.7		3	15.5		31	15					0	<10	
Year	61	77	41	188.5	5	237	196	46				1518	200	300

TABLE XIX

CLIMATIC TABLE FOR DUSHANBE

Latitude 38°35'N longitude 68°47'E elevation 824 m

Month	Air press. (mbar)		Wind		Global radiat.	Temperature (°C)				Humidity		Precipitation (mm)			
	sea level	station	preval. direct.	mean speed (m/s)	(kcal./cm²)	mean	mean daily range	max.	min.	vapor press. (mm)	relat. (%)	mean	max.	min.	max. in 24 h
Jan.	1028.3	927.1	E	1.4	4.6	1.4	9.9	21	−28	3.6	75	79	141	5	36
Feb.	1026.7	926.3	E	1.8	7.4	3.2	10.4	23	−22	4.0	71	74	165	29	45
Mar.	1022.4	924.3	E	2.2	9.9	8.7	11.0	31	−17	5.5	67	108	248	30	58
Apr.	1017.4	921.9	E	2.0	12.3	14.9	12.8	36	−8	7.8	62	111	229	43	83
May	1013.9	920.3	E	1.7	17.5	19.6	14.0	38	2	9.5	57	73	203	5	68
June	1005.8	916.1	E	1.6	20.1	24.2	16.6	43	8	9.5	43	19	84		49
July	1001.3	913.0	E, SW	1.4	19.5	28.2	17.9	43	12	9.2	33	1	23		14
Aug.	1004.1	915.3	W	1.5	17.9	26.8	18.4	44	9	8.2	32	3	36		30
Sept.	1012.6	921.3	SW	1.6	14.0	21.6	19.6	41	3	6.3	35	1	12		7
Oct.	1020.5	926.5	SW	1.6	10.6	14.9	17.5	36	−6	5.4	46	19	144		43
Nov.	1024.1	927.9	E	1.6	6.3	9.5	13.6	32	−15	4.6	59	45	129		40
Dec.	1026.0	928.1	E	1.4	4.2	4.7	10.0	24	−23	4.1	70	71	144	19	33
Year	1016.9	922.3	E	1.6	144.3	14.8		44	−28		54	604	969	389	83

Month	Number of days							Snow cover	Cloudiness (tenths)			Mean sun-shine	Evaporation (mm)	
	wind (speed ⩾15 m/s)	frost-free period	fog	precip. (⩾0.1 mm)	thunder-storm	snow-cover	clouds (8–10 tenths)	max. 10-day average depth (cm)	07h	13h	19h	(h)	actual	potent.
Jan.	0.6		5	13.2	0.1		15		7.4	7.3	6.3	112	5	
Feb.	0.9		2	13.8	0.6		14		7.8	7.5	6.8	120	10	
Mar.	2.0		2	13.2	1		14		8.0	7.9	7.3	146	20	
Apr.	1.8		0.5	12.1	6		10		7.0	7.2	7.1	213	<40	
May	1.4		0.2	9.8	10		7		5.0	5.6	5.7	287	30	
June	1.2			3.4	6		2		2.3	3.1	3.4	332	5	
July	0.7			0.8	1		0.5		1.8	1.6	1.6	363	5	
Aug.	0.1			0.3	0.3		0.2		1.5	1.1	1.0	342	5	
Sept.	0.3			0.3	0.3		0.5		1.6	1.2	1.1	295	5	
Oct.	0.7			3.2	0.6		4		3.9	3.5	2.9	229	5	
Nov.	0.6		1	8.0	0.2		9		6.3	6.1	5.2	158	5	
Dec.	0.4		3	12.4	0.3		14		7.2	7.0	6.1	116	5	
Year	10.7	224	14	90.5	26	39	90	12	5.0	4.9	4.5	2713	>100	1100

TABLE XX

CLIMATIC TABLE FOR ERBOGACHEN

Latitude 61°16'N longitude 108°01'E elevation 287 m

Month	Air press. (mbar)		Wind		Global radiat.	Temperature (°C)				Humidity		Precipitation (mm)			
	sea level	station	preval. direct.	mean speed (m/s)	(kcal./cm²)	mean	mean daily range	max.	min.	vapor press. (mm)	relat. (%)	mean	max.	min.	max. in 24 h
Jan.	1023.5		SW	1.4	0.8	−31.2		0	−61		79	14	21	6	6
Feb.	1025.9		SW	1.5	2.5	−25.6		4	−58		78	10	23	2	6
Mar.	1021.8		SW	2.1	6.9	−15.8		11	−51		66	8	17	2	6
Apr.	1014.2		SW	2.8	10.4	−5.4		22	−42		59	16	27	5	8
May	1010.1		SW	3.0	12.0	4.8		32	−23	3.9	56	24	58	6	18
June	1006.1		SW	2.6	14.0	14.0		34	−8	7.1	60	42	127	22	71
July	1005.9		NE	2.2	14.1	17.0		37	−6	9.6	68	60	113	16	45
Aug.	1008.7		NE	2.1	9.6	13.0		35	−11	8.4	74	49	118	2	36
Sept.	1015.0		SW	2.1	5.8	5.6		27	−20	5.2	76	30	86	3	16
Oct.	1017.1		SW	2.4	3.2	−5.0		18	−44		79	32	63	14	16
Nov.	1021.5		SW	2.0	1.1	−21.6		4	−52		80	21	38	7	7
Dec.	1024.6		SW	1.7	0.3	−29.9		2	−64		80	17	26	6	8
Year	1016.2		SW	2.2	80.7	−6.7		37	−64		71	323	446	246	71

Month	Number of days							Snow cover	Cloudiness (tenths)			Mean sun-shine	Evaporation (mm)	
	wind (speed ⩾15 m/s)	frost-free period	fog	precip. (⩾0.1 mm)	thunder-storm	snow-cover	clouds (8–10 tenths)	max. 10-day average depth (cm)	07h	13h	19h	(h)	actual	potent.
Jan.	0.04		13	18.9		31	8.1		6.6	6.9	6.0	40	5	
Feb.	0.2		6	14.5		28	5.9		6.6	6.4	5.2	108	5	
Mar.	0.9		2	11.1		31	7.3		6.7	6.7	5.7	198	<10	
Apr.	2.1		0.1	10.3		30	10.2		6.6	7.4	6.8	228	>10	
May	1.8		1	11.9		4	12.6		7.3	8.1	7.8	228	>30	
June	1.0		3	12.1			10.9		6.6	7.8	7.2	289	>70	
July	0.7		6	12.1			10.6		6.4	7.3	6.7	298	>70	
Aug.	0.2		10	11.9			9.9		6.8	7.4	6.7	222	<30	
Sept.	0.3		5	13.1			12.0		7.9	7.9	7.1	127	<20	
Oct.	0.7		2	18.7		19	18.1		8.6	8.6	7.6	73	<10	
Nov.	0.5		3	18.9		30	10.9		7.1	7.5	6.4	50	5	
Dec.	0.3		10	20.0		31	10.7		6.9	7.5	6.3	16	5	
Year	8.74	67	61	174		205	127	46	7.0	7.5	6.6	1878	<300	400

TABLE XXI

CLIMATIC TABLE FOR E.T. KRENKEL HYDROMETOBSERVATORY

Latitude 80°37′N longitude 58°03′E elevation 20 m

Month	Air press. (mbar)		Wind		Global radiat. (kcal./cm²)	Temperature (°C)				Humidity		Precipitation (mm)			
	sea level	station	preval. direct.	mean speed (m/s)		mean	mean daily range	max.	min.	vapor press. (mm)	relat. (%)	mean	max.	min.	max. in 24 h
Jan.			NE	6.1	0.0							9			
Feb.			NE	6.3	0.0							7			
Mar.			NE	6.4	1.5							9			
Apr.			NE	5.8	8.0							8			
May			NE	5.3	15.8							10			
June			NE, SE	5.3	16.5							12			
July			W	4.9	10.8							21			
Aug.			NE	4.9	7.8							24			
Sept.			NE	5.6	2.0							18			
Oct.			NE	7.0	0.2							16			
Nov.			NE	5.8	0.0							12			
Dec.			NE	5.5	0.0							8			
Year			NE	5.7	62.6							154			

Month	Number of days		
	wind (speed ≥15 m/s)	frost-free period	fog
Jan.	9.6		2
Feb.	9.9		2
Mar.	8.3		2
Apr.	7.0		2
May	3.3		1
June	2.7		5
July	2.9		14
Aug.	3.0		11
Sept.	4.0		6
Oct.	7.4		2
Nov.	5.5		2
Dec.	6.5		2
Year	70.1		51

TABLE XXII

CLIMATIC TABLE FOR FERGANA

Latitude 40°23′N longitude 71°45′E elevation 578 m

Month	Air press. (mbar)		Wind		Global radiat. (kcal./cm²)	Temperature (°C)				Humidity		Precipitation (mm)			
	sea level	station	preval. direct.	mean speed (m/s)		mean	mean daily range	max.	min.	vapor press. (mm)	relat. (%)	mean	max.	min.	max. in 24 h
Jan.	1028.1	956.2	SE	1.2	4.9	–2.7	12.3	16	–28	3.2	48	20	73	1	24
Feb.	1025.6	954.7	SE	1.5	6.9	0.2	12.7	21	–24	3.9	83	16	74	0	32
Mar.	1021.2	952.3	SE	2.0	9.3	7.5	14.8	30	–24	5.4	71	27	116	0	46
Apr.	1015.7	949.1	W	2.0	13.8	15.2	16.2	35	–5	8.8	63	18	66	1	33
May	1012.4	947.1	W	2.2	16.9	20.4	16.3	41	2	9.2	52	19	49	0	51
June	1006.9	942.3	SE	1.8	19.0	24.5	16.8	43	6	10.5	49	10	40	0	22
July	1003.2	939.3	NW	1.8	18.8	26.4	16.8	42	10	12.5	49	5	48	0	18
Aug.	1005.9	942.3	SE	1.6	18.0	24.8	16.9	41	8	11.4	49	3	52	0	10
Sept.	1013.9	948.2	SE	1.4	14.1	19.0	17.8	38	2	8.5	52	2	26	0	24
Oct.	1022.2	954.3	SE	1.3	9.9	11.8	17.1	33	–10	6.4	61	12	51	0	37
Nov.	1026.8	957.0	SE	1.3	518	5.4	13.9	26	–23	4.7	73	19	155	0	67
Dec.	1028.1	957.1	SE	1.2	3.3	0.6	12.3	19	–28	3.7	80	18	80	0	19
Year	1017.5	950.1	SE	1.6	140.7	12.8		43	–28		63	169	342	68	67

Month	Number of days							Snow cover max. 10-day average depth (cm)	Cloudiness (tenths)			Mean sun-shine (h)	Evaporation (mm)	
	wind (speed ≥15 m/s)	frost-free period	fog	precip. (≥0.1 mm)	thunder-storm	snow-cover	clouds (8–10 tenths)		07h	13h	19h		actual	potent.
Jan.	0.1		9	5.5	0		15		7.3	7.3	6.2	112	<10	
Feb.	0.7		4	5.0	0		12		7.9	7.8	6.5	127	10	
Mar.	2.2		1	6.7	0.1		12		7.5	7.7	7.1	158	20	
Apr.	3.4		0.1	6.3	0.9		9		6.9	6.8	7.3	218	<50	
May	4.2		0.1	6.8	4		7		5.3	5.5	7.0	285	30	
June	4.6		0.1	3.9	6		3		3.8	3.5	5.7	331	<10	
July	4.2		0.1	2.1	3		1		3.0	2.3	4.1	362	<10	
Aug.	1.5		0	1	0.9		2		2.5	1.7	3.2	352	<10	
Sept.	0.8		0.2	1.2	0.2		2		2.2	1.9	3.0	298	<10	
Oct.	1.3		0.8	3.3	0.1		4		4.3	4.0	3.4	234	<10	
Nov.	0.8		2	5.0	0		9		6.3	6.1	4.9	152	<10	
Dec.	0.8		10	5.3	0		14		7.2	7.2	6.1	99	<10	
Year	24.6	213	27	52.2	15	30	88	9	5.4	5.2	5.4	2728	<200	1000

TABLE XXIII

CLIMATIC TABLE FOR FORT SHEVCHENKO

Latitude 44°33′N longitude 50°17′E elevation −23 m

Month	Air press. (mbar)		Wind		Global radiat.	Temperature (°C)				Humidity		Precipitation (mm)			
	sea level	station	preval. direct.	mean speed (m/s)	(kcal./cm²)	mean	mean daily range	max.	min.	vapor press. (mm)	relat. (%)	mean	max.	min.	max. in 24 h
Jan.	1021.4		SE	8.7	2.9	−3.6	7.9	14	−22	3.4	77	7	60	0	
Feb.	1019.9		SE	9.5	4.9	−2.7	8.3	18	−26	3.2	77	8	38	0	
Mar.	1019.8		SE	8.9	8.0	2.2	9.2	24	−19	4.3	73	8	33	0	
Apr.	1017.0		SE	7.9	12.5	10.0	10.0	32	−8	6.2	64	14	67	0	
May	1015.1		N, SE	7.4	16.5	17.8	9.0	38	2	9.5	62	11	61	0	
June	1011.8		N	6.7	16.7	22.7	8.6	40	6	18.4	63	17	87	0	
July	1009.3		N	6.4	16.2	25.8	8.7	43	13	15.9	63	15	71	0	
Aug.	1011.8		N	7.2	15.1	24.5	9.2	43	10	14.2	60	7	85	0	
Sept.	1016.7		SE	8.3	11.3	19.4	9.3	38	3	10.1	58	14	68	0	
Oct.	1020.7		SE	8.9	7.4	12.0	8.6	30	−3	7.3	63	12	58	0	
Nov.	1024.8		SE	9.6	3.7	4.9	8.2	22	−17	4.9	68	9	40	0	
Dec.	1023.4		SE	9.7	2.8	−0.2	7.2	18	−21	3.7	73	8	63	0	
Year	1017.6		SE	8.3	118.0	11.0		43	−26		67	130	290	57	

Month	Number of days							Snow cover max. 10-day average depth (cm)	Cloudiness (tenths)			Mean sunshine (h)	Evaporation (mm)	
	wind (speed ⩾15 m/s)	frost-free period	fog	precip. (⩾0.1 mm)	thunderstorm	snow-cover	clouds (8–10 tenths)		07h	13h	19h		actual	potent.
Jan.	5.3		4	8.6			14		7.8	7.9	6.7	69	<10	
Feb.	5.0		4	6.4			12		7.4	7.1	6.1	110	10	
Mar.	6.2		6	6.8	0.03		11		7.0	6.8	6.1	161	>10	
Apr.	4.5		5	5.6	0.02		8		6.1	5.8	5.8	222	<50	
May	2.8		2	5.4	0.7		4		5.3	5.1	5.2	318	30	
June	1.5		1	4.1	2		3		4.5	4.2	4.1	343	<10	
July	1.0		1	4.3	2		2		3.9	3.4	3.2	341	<10	
Aug.	1.6		1	4.4	2		3		3.5	3.2	3.0	337	<10	
Sept.	2.3		1	4.6	0.7		4		4.1	3.8	3.4	263	<10	
Oct.	3.5		3	5.6	0.3		5		5.8	5.4	4.4	198	<10	
Nov.	5.2		3	7.0	0.1		12		6.7	6.6	5.3	123	<10	
Dec.	5.6		4	9.7	0.02		16		7.2	7.5	6.4	78	<10	
Year	44.5	214	35	72.6	8	18	94	4	5.8	5.6	5.0	2563	<200	900

TABLE XXIV

CLIMATIC TABLE FOR GUR'EV

Latitude 47°01′N longitude 51°51′E elevation 22.8 m

Month	Air press. (mbar)		Wind		Global radiat.	Temperature (°C)				Humidity		Precipitation (mm)			
	sea level	station	preval. direct.	mean speed (m/s)	(kcal./cm²)	mean	mean daily range	max.	min.	vapor press. (mm)	relat. (%)	mean	max.	min.	max. in 24 h
Jan.	1022.7		SE	6.3	3.8	−10.4	10.3	7	−38	2.2	82	14	49	0	20
Feb.	1021.4		E	6.4	6.1	−9.4	10.8	15	−38	2.2	81	12	42	0	11
Mar.	1020.0		SE	6.4	10.0	−2.5	12.2	23	−32	3.5	78	12	36	0	17
Apr.	1017.5		E, SE	6.0	14.1	8.2	13.8	32	−12	5.6	65	12	56	0	30
May	1014.9		SE	6.1	18.4	17.7	13.3	37	−4	8.4	54	16	56	0	27
June	1011.8		W	5.4	19.3	22.6	13.2	42	4	11.0	52	19	134	0	36
July	1008.6		W	4.8	18.9	25.4	13.4	45	8	13.3	53	20	114	0	87
Aug.	1011.5		W	4.6	16.7	23.2	14.3	45	4	11.7	52	11	99	0	45
Sept.	1016.7		NW	4.5	12.6	16.2	14.3	37	−6	8.7	58	8	68	0	47
Oct.	1020.8		SE	4.9	7.9	8.2	12.4	29	−13	6.0	70	16	69	0	24
Nov.	1025.9		SE	5.5	4.5	0.2	10.6	19	−30	4.0	76	10	55	0	20
Dec.	1024.3		SE	5.6	2.9	−6.2	9.7	12	−38	2.8	84	14	64	1	18
Year	1018.0		SE	5.5	135.2	7.8		45	−38		67	164	321	74	87

Month	Number of days							Snow cover max. 10-day average depth (cm)	Cloudiness (tenths)			Mean sunshine (h)	Evaporation (mm)	
	wind (speed ⩾15 m/s)	frost-free period	fog	precip. (⩾0.1 mm)	thunderstorm	snow-cover	clouds (8–10 tenths)		07h	13h	19h		actual	potent.
Jan.	4.0		7	10.3		31	14.3		6.9	7.1	5.9	74	<10	
Feb.	4.5		5	7.6		28	10.6		7.0	6.7	5.3	107	<10	
Mar.	6.0		6	8.3	0.02	4	10.8		6.8	6.7	5.7	158	<10	
Apr.	5.6		2	5.6	0.2		8.0		5.7	6.0	5.6	241	>20	
May	6.7		0.7	4.9	2		7.2		5.0	5.8	5.6	325	40	
June	4.5		0.2	5.2	3		5.1		4.0	5.3	5.0	340	<30	
July	4.8		0.1	4.3	3		4.7		3.6	4.7	4.1	343	<15	
Aug.	3.0		0.9	3.6	2		3.5		3.2	4.3	3.7	338	<15	
Sept.	2.5		2	4.5	0.6		4.1		3.9	4.4	3.5	266	10	
Oct.	4.6		4	6.5	0.1		6.7		5.6	5.9	4.1	192	<10	
Nov.	3.7		7	8.5			11.4		6.0	6.2	4.8	127	<10	
Dec.	4.3		7	12.4		11	16.8		6.9	7.3	6.1	68	<10	
Year	54.2	176	42	81.7	11	73	10.3	13	5.4	5.9	5.0	2579	<200	800

TABLE XXV

CLIMATIC TABLE FOR ICHA

Latitude 55°42′N longitude 155°38′E elevation 4 m

Month	Air press. (mbar)		Wind		Global radiat. (kcal./cm²)	Temperature (°C)				Humidity		Precipitation (mm)			
	sea level	station	preval. direct.	mean speed (m/s)		mean	mean daily range	max.	min.	vapor press. (mm)	relat. (%)	mean	max.	min.	max. in 24 h
Jan.	1004.9	1003.9	NE	4.9		−13.2	8.0	7	−36	2.2	78	25			
Feb.	1005.4	1004.5	NE	4.6		−13.6	8.1	5	−40	1.9	78	15			
Mar.	1008.9	1007.9	NE	5.1		−9.7	8.4	9	−32	2.5	76	20			
Apr.	1009.1	1008.2	N	5.3		−3.2	6.9	13	−27	4.0	81	32			
May	1010.7	1009.8	N	4.9		2.0	5.7	24	−14	6.1	86	38			
June	1010.9	1009.9	S	4.6		6.2	5.2	24	−3	8.5	89	37			
July	1010.7	1009.8	S	4.6		10.3	5.0	29	1	11.5	92	69			
Aug.	1010.7	1009.8	S	4.3		11.4	5.3	25	0	12.7	91	84			
Sept.	1011.3	1010.3	S	4.6		8.9	6.3	22	−4	10.1	88	67			
Oct.	1009.1	1008.2	S	6.2		2.8	5.8	16	−12	6.4	83	117			
Nov.	1005.9	1005.0	SE	6.1		−4.2	6.5	13	−24	4.0	82	70			
Dec.	1013.0	1002.1	NE	5.6		−9.4	7.4	9	−34	2.5	78	46			
Year	1008.3	1007.4	N, NE, S	5.1		−1.0	–	29	−40	6.0	84	620			

Month	Number of days							Snow cover	Cloudiness (tenths)			Mean sun-shine (h)	Evaporation (mm)	
	wind (speed ⩾15 m/s)	frost-free period	fog	precip. (⩾0.1 mm)	thunder-storm	snow-cover	clouds (8–10 tenths)	max. 10-day average depth (cm)	07h	13h	19h		actual	potent.
Jan.	2.3		1	12.6			10.9		6.5	6.9	5.6	79	2	
Feb.	1.2		1	9.7			9.8		6.9	6.8	5.2	106	3.6	
Mar.	1.3		0.8	11.9			11.8		7.0	6.6	6.1	160	6.5	
Apr.	2.3		2	14.0			16.2		7.9	7.6	7.5	152	7.2	
May	1.5		7	10.7	0.03		18.9		8.5	8.1	7.9	153	25	
June	1.0		12	10.9	0.2		20.6		8.7	8.3	8.3	158	40	
July	1.6		14	14.9	0.4		21.4		9.1	8.4	8.3	165	42	
Aug.	0.9		12	14.5	0.3		20.0		8.7	7.9	7.9	163	40	
Sept.	1.1		5	14.2	0.07		16.1		8.0	7.6	7.3	145	20	
Oct.	3.7		1	21	0.03		18.2		8.6	8.4	7.4	81	10	
Nov.	3.8		0.4	19.0			16.6		7.9	8.2	7.1	51	7	
Dec.	3.0		0.6	16.2			13.1		7.2	7.8	6.5	39	1	
Year	23.7	121	57	170	1	187	194	32	7.9	7.7	7.1	1435	20.43	289.1

TABLE XXVI

CLIMATIC TABLE FOR ILIRNEY

Latitude 67°20′N longitude 168°11′E elevation 426 m

Month	Air press. (mbar)		Wind		Global radiat. (kcal./cm²)	Temperature (°C)				Humidity		Precipitation (mm)			
	sea level	station	preval. direct.	mean speed (m/s)		mean	mean daily range	max.	min.	vapor press. (mm)	relat. (%)	mean	max.	min.	max. in 24 h
Jan.	1019.6		W	3.2		−34.2	10.0	2	−57	0.8	76	12			
Feb.	1021.6		W	2.4		−32.2	9.4	−2	−58	0.6	74	9			
Mar.	1021.0		W	2.8		−28.3	12.0	0	−58	0.6	74	7			
Apr.	1017.5		W	2.6		−17.7	13.3	4	−46	1.5	75	6			
May	1015.5		E	3.2		−4.5	10.2	18	−37	3.6	75	7			
June	1010.1		E	3.1		6.9	11.3	27	−12	6.8	70	21			
July	1009.1		W	3.3		10.4	11.0	30	−2	9.1	72	48			
Aug.	1010.5		W	3.3		7.7	9.6	27	−7	8.2	75	40			
Sept.	1014.9		W	3.1		0.2	7.5	19	−17	5.3	80	18			
Oct.	1015.8		W	3.3		−12.5	7.6	13	−39	2.2	82	13			
Nov.	1018.3		W	3.0		−23.8	8.9	3	−54	0.8	79	14			
Dec.	1018.8		W	2.4		−31.4	8.7	4	−58	0.6	76	11			
Year	1016.1		W	3.0		−13.3		30	−58	3.3	76	206			

Month	Number of days							Snow cover	Cloudiness (tenths)			Mean sun-shine (h)	Evaporation (mm)	
	wind (speed ⩾15 m/s)	frost-free period	fog	precip. (⩾0.1 mm)	thunder-storm	snow-cover	clouds (8–10 tenths)	max. 10-day average depth (cm)	07h	13h	19h		actual	potent.
Jan.	1.8			12.3		31	11.6					2	5	
Feb.	0.4			9.4		28	8.9					76	5	
Mar.	0.9			8.8		31	7.1					203	5	
Apr.	0.3			8.5		30	9.6					263	5	
May	0.5			8.5		31	13.6					323	10	
June	0.2			11.7	1.0	1	13.3					307	60	
July	0.3			14.5	0.2		17.2					258	60	
Aug.	0.5			13.7			17.8					204	50	
Sept.	0.0			10.7			17.6					120	>10	
Oct.	1.1			13.6		29	14.9					87	5	
Nov.	0.5			12.5		30	12.0					24	5	
Dec.	1.3			10.4		31	10.1					0	5	
Year	7.8	44		135	1.0	246	154	23				1867	200	

TABLE XXVII

CLIMATIC TABLE FOR IRKUTSK

Latitude 52°16′N longitude 104°79′E elevation 468 m

Month	Air press. (mbar)		Wind		Global radiat. (kcal./cm²)	Temperature (°C)				Humidity		Precipitation (mm)			
	sea level	station	preval. direct.	mean speed (m/s)		mean	mean daily range	max.	min.	vapor press. (mm)	relat. (%)	mean	max.	min.	max. in 24 h
Jan.	1033.2		SE	2.1	2.0	-20.9	9.8	2	-50	0.8	80	12	29	2	9
Feb.	1032.6		SE	2.3	4.0	-18.5	12.6	6	-45	0.9	74	8	23	1	7
Mar.	1026.4		SE	2.7	8.5	-10.0	13.7	16	-37	1.6	68	9	34	1	11
Apr.	1019.6		NW	3.3	11.4	0.6	12.9	29	-32	2.8	59	15	46	2	14
May	1009.1		NW	3.4	14.5	8.1	15.3	33	-14	4.4	56	29	109	5	43
June	1008.9		NW	2.7	15.6	14.5	15.1	35	-4	7.7	66	83	303	11	49
July	1005.1		SE	2.2	15.3	17.5	13.4	36	0	10.7	74	102	226	13	68
Aug.	1010.2		SE	2.2	13.1	15.0	12.6	34	-3	9.6	78	99	173	20	76
Sept.	1017.8		NW	2.5	8.5	8.0	12.8	29	-12	6.0	78	49	140	10	34
Oct.	1022.7		SE	2.6	5.1	0.1	11.4	26	-30	3.5	75	20	73	2	21
Nov.	1028.6		NW	2.3	2.6	-10.7	10.1	14	-40	1.9	80	17	40	5	13
Dec.	1032.0		NW	1.8	1.4	-18.7	9.1	4	-46	1.1	85	15	71	2	30
Year	1020.5		SE	2.5	102.0	-1.2		36	-50		73	458	797	251	76

Month	Number of days							Snow cover max. 10-day average depth (cm)	Cloudiness (tenths)			Mean sunshine (h)	Evaporation (mm)	
	wind (speed ≥15 m/s)	frost-free period	fog	precip. (≥0.1 mm)	thunderstorm	snow-cover	clouds (8–10 tenths)		07h	13h	19h		actual	potent.
Jan.	0.7		20	11.8		31	10		7.4	7.5	5.9	86	<10	
Feb.	0.4		13	9.2		28	6		6.9	6.8	5.0	132	<10	
Mar.	2.0		6	7.1	0.02	31	7		7.0	6.6	5.7	205	<10	
Apr.	2.7		0.4	7.6	0.1		9		7.4	7.7	7.0	221	>10	
May	4.0		0.5	10.7	0.6		13		7.6	7.9	7.4	239	>30	
June	2.3		3	12.4	4.0		14		7.3	7.5	7.6	247	<60	
July	1.6		5	14.4	7.0		15		7.6	7.1	7.7	244	70	
Aug.	1.1		6	14.6	4.0		11		7.6	7.0	7.2	222	>50	
Sept.	1.0		8	12.0	0.8		10		7.7	7.0	6.1	178	>30	
Oct.	1.8		6	10.2	0.2		12		7.9	7.4	5.1	141	<10	
Nov.	2.1		12	12.2		28	13		8.3	7.9	6.1	83	<10	
Dec.	0.6		23	14.0		31	14		8.3	8.2	7.1	50	<10	
Year	20.3	94	103	136.2	17.0	162	134	28	7.6	7.4	6.5	2046	>300	400

TABLE XXVIII

CLIMATIC TABLE FOR KALININGRAD

Latitude 54°42′N longitude 20°37′E elevation 27 m

Month	Air press. (mbar)		Wind		Global radiat. (kcal./cm²)	Temperature (°C)				Humidity		Precipitation (mm)			
	sea level	station	preval. direct.	mean speed (m/s)		mean	mean daily range	max.	min.	vapor press. (mm)	relat. (%)	mean	max.	min.	max. in 24 h
Jan.	1016.4	1015.6	SW	5.1		-2.7			-33	3.5	86	56.4	116	11	21
Feb.	1015.8	1015.0	SW	5.0		-2.1			-31	3.6	84	39.2	109	4	24
Mar.	1013.9	1013.1	SE	4.8		0.8			-22	4.2	79	26.6	84	7	20
Apr.	1013.4	1013.6	SE	4.2		6.2			-10	5.4	76	43.3	91	4	35
May	1015.2	1014.5	W	4.2		11.8			-4	7.5	72	39.3	108	8	48
June	1013.5	1013.7	W	4.0		15.0			1	9.4	73	54.9	127	10	67
July	1012.5	1011.7	W	3.6		17.3			4	11.4	77	90.4	201	12	83
Aug.	1012.7	1012.1	SW, W	3.6		16.4			6	11.2	80	83.8	250	14	110
Sept.	1015.8	1015.0	SW	3.5		12.9			-1	9.2	83	83.0	179	12	51
Oct.	1015.1	1014.3	SW	4.0		7.7			-7	7.0	84	69.9	233	3	57
Nov.	1015.4	1014.6	SW	4.6		2.6			-14	5.1	88	52.4	183	8	41
Dec.	1014.6	1013.8	SW	5.0		-1.0			-22	1.0	87	58.8	149	12	25
Year	1014.6	1013.8	SW	4.3		7.1		37	-33		81	698.0	931	481	110

Month	Number of days							Snow cover max. 10-day average depth (cm)	Cloudiness (tenths)			Mean sunshine (h)	Evaporation (mm)	
	wind (speed ≥15 m/s)	frost-free period	fog	precip. (≥0.1 mm)	thunderstorm	snow-cover	clouds (8–10 tenths)		07h	13h	19h		actual	potent.
Jan.	1.5		8	18.0	0.1	31	19.5		7.8	8.0	7.6	42	<10	
Feb.	1.6		5	16.1	0.1	28	16.4		8.2	7.7	7.2	59	<10	
Mar.	1.3		5	15.0	0.1	11	14.5		7.1	6.8	6.3	130	10	
Apr.	0.4		2	13.9	1		11.1		7.1	6.9	6.3	184	>40	
May	0.5		2	12.6	3		7.9		6.2	6.5	5.7	256	>70	
June	0.3		1	12.7	3		8.2		6.2	6.6	5.7	276	>75	
July	0.4		1	14.5	4		8.3		6.7	7.2	6.0	252	70	
Aug.	0.6		1	16.0	4		8.0		6.6	7.3	6.3	230	50	
Sept.	0.7		3	15.5	2		8.0		6.7	7.0	5.8	180	>35	
Oct.	0.8		5	17.4	0.3		12.9		7.9	7.7	6.4	107	>10	
Nov.	0.6		8	18.1	0.1		18.0		8.5	8.5	8.0	39	<10	
Dec.	9		7	19.3	0.2	2	20.9		8.3	8.6	7.9	31	<10	
Year	17.7	182	48	189.1	18	73	154	10 cm	7.3	7.4	6.6	1786	400	600

TABLE XXIX

CLIMATIC TABLE FOR KARAGANDA

Latitude 49°48′N longitude 73°08′E elevation 537 m

Month	Air press. (mbar)		Wind		Global radiat.	Temperature (°C)				Humidity		Precipitation (mm)			
	sea level	station	preval. direct.	mean speed (m/s)	(kcal./cm²)	mean	mean daily range	max.	min.	vapor press. (mm)	relat. (%)	mean	max.	min.	max. in 24 h
Jan.	1032.2		SW	4.6		−15.2	10.7	7	−49	1.3	80	11			
Feb.	1029.1		SW	4.5		−14.0	11.8	7	−46	1.4	79	11			
Mar.	1026.4		SW	5.2		−8.9	12.5	22	−35	2.1	81	15			
Apr.	1022.6		SW	4.8		2.4	13.2	28	−25	4.0	70	22			
May	1015.4		SW	4.8		13.0	16.6	34	−10	5.8	55	28			
June	1010.2		NE	4.0		18.5	16.2	38	−2	7.7	53	41			
July	1006.5		NE	3.9		20.6	15.8	39	3	8.8	55	43			
Aug.	1010.3		NE	3.8		18.3	16.1	40	−1	7.7	55	28			
Sept.	1016.2		SW	3.5		11.8	16.0	35	−6	5.6	57	21			
Oct.	1023.6		SW	4.7		3.2	15.3	28	−25	4.1	69	24			
Nov.	1030.3		SW	5.0		−6.9	10.7	18	−38	2.7	81	15			
Dec.	1031.4		SW	5.0		−9.4	9.8	6	−43	16.	81	14			
Year	1021.2		SW	4.5		−7.2		40	−49		68	273			

Month	Number of days							Snow cover max. 10-day average depth (cm)	Cloudiness (tenths)			Mean sunshine (h)	Evaporation (mm)	
	wind (speed ≥15 m/s)	frost-free period	fog	precip. (≥0.1 mm)	thunderstorm	snow-cover	clouds (8–10 tenths)		07h	13h	19h		actual	potent.
Jan.	4.5		4	11.8		31	16		6.4	6.2	5.4	99	<10	
Feb.	4.7		6	10.3		28	11		6.3	5.9	4.8	127	<10	
Mar.	5.6		7	12.0		31	11		7.0	6.3	6.0	159	<10	
Apr.	4.4		4	7.2	0.2	4	7		6.3	6.5	6.1	217	<15	
May	5.4		1	9.8	3		5		5.3	6.6	6.0	281	<40	
June	4.4		0.5	9.8	5		6		5.0	6.5	5.7	302	<40	
July	3.0		1	10.8	8		4		4.8	6.7	5.4	303	20	
Aug.	2.5		1	7.5	4		3		4.6	6.1	5.1	291	<10	
Sept.	3.0		2	6.0	0.5		4		5.2	5.9	4.7	226	<10	
Oct.	3.6		2	10.6	0.1		11		6.9	6.9	5.6	145	<10	
Nov.	3.6		5	12.6		19	15		7.4	7.0	6.0	86	<10	
Dec.	4.9		8	14.8		31	16		6.9	6.9	6.1	66	<10	
Year	49.6	129	42	123.2	21	151	109	16	6.0	6.5	5.6	2302	<200	600

TABLE XXX

CLIMATIC TABLE FOR KAUNAS

Latitude 54°53′N longitude 23°53′E elevation 75 m

Month	Air press. (mbar)		Wind		Global radiat.	Temperature (°C)				Humidity		Precipitation (mm)			
	sea level	station	preval. direct.	mean speed (m/s)	(kcal./cm²)	mean	mean daily range	max.	min.	vapor press. (mm)	relat. (%)	mean	max.	min.	max. in 24 h
Jan.	1017.2	1006.3	SW	4.8	1.3	−4.7			−34	3.1	87	33	52	8	22
Feb.	1016.6	1005.7	SE	4.2	3.3	−4.0			−33	3.1	84	35	66	6	24
Mar.	1014.7	1003.9	SE	3.8	7.1	−0.2			−26	3.9	79	29	66	4	20
Apr.	1014.1	1003.5	SE	3.6	8.1	6.3			−12	5.3	75	40	91	5	29
May	1015.1	1005.3	N	2.2	12.9	12.7			−3	7.7	69	49	160	11	75
June	1013.1	1003.0	SW, NW	3.0	14.0	15.6			0	9.3	70	69	158	8	48
July	1012.1	1002.1	SW, W	2.8	13.0	17.8			5	11.1	75	98	218	18	77
Aug.	1012.6	1002.6	SW	2.8	10.3	16.4			4	10.7	78	92	242	20	71
Sept.	1016.0	1005.8	SW	3.0	6.8	12.6			−3	8.6	81	57	109	8	45
Oct.	1016.0	1005.8	S, SW	3.6	3.4	7.0			−10	6.6	85	48	120	6	34
Nov.	1016.0	1005.7	SE	4.2	1.2	1.2			−18	4.8	88	39	109	8	22
Dec.	1016.4	1005.9	S	4.5	0.8	−2.8			−27	3.6	88	36	83	3	25
Year	1015.0	1004.5	SW	3.6	82.2	6.5		33	−34		80	625	915	408	77

Month	Number of days							Snow cover max. 10-day average depth (cm)	Cloudiness (tenths)			Mean sunshine (h)	Evaporation (mm)	
	wind (speed ≥15 m/s)	frost-free period	fog	precip. (≥0.1 mm)	thunderstorm	snow-cover	clouds (8–10 tenths)		07h	13h	19h		actual	potent.
Jan.	0.7		7	15.9		31	19.4		8.7	8.4	8.1	41	<10	
Feb.	0.7		7	14.7		28	16.9		8.9	8.2	7.3	57	<10	
Mar.	0.6		7	14.2	0.1	15	15.6		7.2	6.9	6.3	137	10	
Apr.	0.4		2	13.1	0.7		10.8		7.2	7.7	7.0	185	>40	
May	0.4		2	13.0	4		8.8		6.7	7.7	7.0	258	>70	
June	0.4		1	13.1	4		8.7		6.4	7.4	6.3	276	>70	
July	0.4		3	14.8	5		9.0		6.9	7.7	6.8	272	>70	
Aug.	0.5		4	15.3	4		9.0		7.2	7.5	6.8	232	>60	
Sept.	0.4		5	12.8	1		9.4		7.7	7.6	6.4	167	>30	
Oct.	0.4		6	14.2	0.2		15.6		8.5	8.1	6.5	98	>10	
Nov.	0.4		9	15.7			19.7		9.0	9.0	8.3	37	<10	
Dec.	0.6		8	15.8		1	21.8		8.8	8.9	8.4	30	<10	
Year	5.9	162	61	172.6	19	87	165	18	7.8	7.9	7.1	1790	400	600

TABLE XXXI

CLIMATIC TABLE FOR KAZAN'

Latitude 55°47'N longitude 49°11'E elevation 64 m

Month	Air press. (mbar) sea level	Air press. (mbar) station	Wind preval. direct.	Wind mean speed (m/s)	Global radiat. (kcal./cm²)	Temperature (°C) mean	Temperature (°C) mean daily range	Temperature (°C) max.	Temperature (°C) min.	Humidity vapor press. (mm)	Humidity relat. (%)	Precipitation (mm) mean	Precipitation (mm) max.	Precipitation (mm) min.	Precipitation (mm) max. in 24 h
Jan.	1021.9		S	4.8		−13.0	7.2	4	−44	1.7	84	16			19
Feb.	1020.8		S	4.6		−12.1	9.7	4	−37	1.7	81	17			16
Mar.	1018.8		S	4.7		−6.2	10.7	12	−31	2.5	80	18			23
Apr.	1018.0		S	4.1		3.9	10.1	31	−19	4.4	74	25			26
May	1015.4		W	4.4		13.0	13.5	34	−6	6.6	62	46			38
June	1012.2		W	3.7		17.6	13.7	38	−1	9.4	62	50			75
July	1009.8		NW	3.4		20.0	13.9	39	5	10.9	67	66			121
Aug.	1013.0		N	3.5		17.6	13.5	38	2	9.9	70	63			74
Sept.	1015.2		S	4.0		11.3	14.5	36	−5	7.4	75	45			101
Oct.	1018.0		S	4.5		3.9	11.4	23	−23	4.9	81	45			45
Nov.	1024.3		S	4.5		−3.0	7.4	16	−37	3.2	84	21			24
Dec.	1022.4		S	4.8		−10.5	6.4	6	−39	2.0	85	23			33
Year	1017.5		S	4.2		3.5		39	−44		75	435			121

Month	Number of days wind (speed ≥15 m/s)	frost-free period	fog	precip. (≥0.1 mm)	thunder-storm	snow-cover	clouds (8–10 tenths)	Snow cover max. 10-day average depth (cm)
Jan.			2	17.2			18.0	
Feb.			2	13.9			13.2	
Mar.			3	13.0			13.4	
Apr.			3	9.8	0.4		10.8	
May			0.3	10.9	2		9.8	
June			0.8	11.8	6		6.8	
July			0.9	12.1	7		8.4	
Aug.			1	11.7	3		8.1	
Sept.			2	13.3	0.8		11.6	
Oct.			4	14.7	0.1		17.9	
Nov.			5	15.6			19.0	
Dec.			4	17.5			19.3	
Year		151	28	162	19	154	156	52

TABLE XXXII

CLIMATIC TABLE FOR KEM'

Latitude 65°00'N longitude 34°48'E elevation 9.9 m

Month	Air press. (mbar) sea level	Air press. (mbar) station	Wind preval. direct.	Wind mean speed (m/s)	Global radiat. (kcal./cm²)	Temperature (°C) mean	Temperature (°C) mean daily range	Temperature (°C) max.	Temperature (°C) min.	Humidity vapor press. (mm)	Humidity relat. (%)	Precipitation (mm) mean	Precipitation (mm) max.	Precipitation (mm) min.	Precipitation (mm) max. in 24 h
Jan.	1010.9	1009.2	W	5.3		−10.7	7.1	6	−43	2.0	87	24	58	8	9
Feb.	1011.6	1009.9	S, SW, W	5.0		−10.9	6.9	5	−40	1.9	87	19	54	8	11
Mar.	1011.4	1009.7	W	5.2		−7.4	8.3	10	−32	2.3	81	18	63	5	14
Apr.	1013.8	1012.2	SW	4.7		−0.8	7.6	20	−26	3.6	77	21	65	5	23
May	1015.1	1013.5	N	4.8		4.8	7.6	26	−12	4.9	74	28	100	3	32
June	1012.2	1010.6	N	4.8		10.5	8.1	30	−4	7.1	75	53	145	3	60
July	1011.0	1009.4	NE	4.4		14.3	7.5	32	1	9.4	78	56	127	9	53
Aug.	1010.7	1009.1	SE	4.3		12.3	7.4	30	−2	8.9	81	60	146	20	51
Sept.	1011.0	1009.4	W	5.0		7.5	6.1	25	−7	6.7	83	53	132	17	62
Oct.	1011.1	1009.5	W	5.5		1.5	4.8	16	−19	4.7	85	39	148	10	58
Nov.	1010.7	1009.0	SW, W	5.7		−4.0	4.7	10	−31	3.3	88	30	77	6	14
Dec.	1011.5	1009.8	SW, W	5.3		−8.1	6.0	7	−34	2.5	87	24	63	10	14
Year	1011.8	1010.1	W	5.0		0.8		32	−43		82	425	714	345	62

Month	Number of days wind (speed ≥15 m/s)	frost-free period	fog	precip. (≥0.1 mm)	thunder-storm	snow-cover	clouds (8–10 tenths)	Snow cover max. 10-day average depth (cm)	Cloudiness (tenths) 07h	Cloudiness (tenths) 13h	Cloudiness (tenths) 19h	Mean sun-shine (h)	Evaporation (mm) actual	Evaporation (mm) potent.
Jan.	1.8		3	17.6		31	16.9		7.7	8.3	7.3	14	<10	
Feb.	1.2		3	15.8		28	14.6		7.9	7.8	7.1	52	<10	
Mar.	2.0		3	15.9		31	12.5		7.5	6.9	6.8	140	<10	
Apr.	1.2		3	13.4	0.03	26	11.8		7.1	7.2	7.3	196	<30	
May	1.3		3	13.2	1		13.5		7.2	7.6	7.1	242	70	
June	1.3		3	14.4	3		12.1		6.9	7.2	7.0	296	>70	
July	0.8		4	13.6	4		11.1		6.6	6.6	6.2	299	>70	
Aug.	1.0		3	15.2	2		12.6		7.0	7.4	6.8	225	>50	
Sept.	1.7		3	18.0	0.3		14.6		8.0	8.2	7.6	124	30	
Oct.	2.2		2	17.7			17.2		8.3	8.4	7.7	64	>10	
Nov.	1.9		2	18.8		16	19.5		8.5	8.8	8.1	21	<10	
Dec.	1.6		3	19.1		31	18.5		8.1	8.4	7.7	6	<10	
Year	18	121	35	193	10	181	175	46	7.6	7.7	7.2	1679	>300	400

TABLE XXXIII

CLIMATIC TABLE FOR KEMEROVO

Latitude 55°23′N longitude 86°04′E elevation 154 m

Month	Air press. (mbar)		Wind		Global radiat. (kcal./cm²)	Temperature (°C)				Humidity		Precipitation (mm)			
	sea level	station	preval. direct.	mean speed (m/s)		mean	mean daily range	max.	min.	vapor press. (mm)	relat. (%)	mean	max.	min.	max. in 24 h
Jan.	1029.3	1008.2			1.9	–19.2	97	5	–55	1.5	82	14			
Feb.	1029.0	1008.1			3.8	–17.0	11.0	8	–48	1.7	81	10			
Mar.	1026.2	1005.8			8.0	–10.6	11.3	13	–42	2.6	80	10			
Apr.	1020.8	1001.3			11.7	0.0	10.0	27	–29	5.0	73	18			
May	1015.0	996.2			14.0	9.2	13.1	36	–18	7.3	62	41			
June	1009.5	991.4			15.8	15.8	13.2	37	–3	12.0	66	49			
July	1007.1	989.1			14.0	18.4	12.4	38	–1	15.4	73	77			
Aug.	1009.6	991.5			11.7	15.5	12.1	35	–3	13.4	77	57			
Sept.	1016.6	997.8			8.8	9.3	11.8	31	–10	9.1	78	48			
Oct.	1021.1	1001.7			4.0	1.1	8.6	26	–39	5.6	80	31			
Nov.	1025.3	1005.0			2.9	–9.8	8.1	12	–46	2.8	82	21			
Dec.	1029.3	1008.0			1.6	–17.0	9.3	6	–50	1.8	83	17			
Year	1019.9	1000.1			98.2	–0.4	–	38	–55	6.5	76	393			

Month	Number of days							Snow cover max. 10-day average depth (cm)	Cloudiness (tenths)			Mean sunshine (h)	Evaporation (mm)	
	wind (speed ≥15 m/s)	frost-free period	fog	precip. (≥0.1 mm)	thunderstorm	snow-cover	clouds (8–10 tenths)		07h	13h	19h		actual	potent.
Jan.			2	15.4			12.9		6.7	7.1	6.1	66	0	
Feb.			1	12.6			11.0		7.2	7.0	5.7	83	0	
Mar.			1	11.5			11.6		7.4	6.8	6.2	135	0.2	
Apr.			1	10.2	0.2		10.8		7.0	7.1	6.5	174	8	
May			0.7	11.4	4		11.2		7.0	7.6	7.1	218	66	
June			1	11.9	6		10.7		6.4	7.6	6.8	245	73	
July			3	13.3	10		9.3		6.4	7.7	6.9	254	70	
Aug.			6	13.8	6		10.7		7.1	7.7	6.6	230	53	
Sept.			4	13.7	1		12.7		7.7	7.8	6.6	148	32	
Oct.			1	15.1	0.1		18.4		8.5	8.6	7.3	81	12	
Nov.			1	17.7			17.5		8.3	8.1	7.2	55	4	
Dec.			1	17.7			15.9		7.5	7.7	6.7	33	0	
Year		108	23	164.3	27	167	153	48	7.3	7.6	6.6	1722	318.2	528.8

TABLE XXXIV

CLIMATIC TABLE FOR KHABAROVSK

Latitude 48°31′N longitude 135°07′E elevation 86 m

Month	Air press. (mbar)		Wind		Global radiat. (kcal./cm²)	Temperature (°C)				Humidity		Precipitation (mm)			
	sea level	station	preval. direct.	mean speed (m/s)		mean	mean daily range	max.	min.	vapor press. (mm)	relat. (%)	mean	max.	min.	max. in 24 h
Jan.	1020.5	1013.5	SW	3.1	4.1	–22.7	8.2	0	–43	0.7	76	10	37	1	11
Feb.	1019.5	1012.7	SW	3.3	6.5	–17.6	10.0	6	–41	0.9	72	7	32	0	7
Mar.	1015.6	1003.1	SW	4.1	10.6	–8.8	11.5	12	–30	1.7	68	12	41	0	16
Apr.	1010.3	1004.1	SW	4.6	12.2	–2.8	11.4	25	–17	3.4	62	32	107	6	32
May	1007.4	1001.4	SW	4.3	15.0	11.2	13.4	31	–4	6.3	63	53	103	15	47
June	1006.0	999.9	SW	3.7	15.5	17.1	12.2	35	2	10.8	72	74	152	12	67
July	1006.1	1000.1	SW	3.5	13.2	21.0	10.8	40	5	14.5	78	11[illegible]	301	9	83
Aug.	1007.9	1001.9	SW	3.4	12.5	19.9	10.2	36	7	14.1	79	118	308	30	85
Sept.	1012.5	1006.1	SW	4.5	9.1	13.9	11.0	29	–4	9.2	77	82	304	18	99
Oct.	1015.9	1008.7	SW	5.4	6.5	–4.8	10.4	25	–15	4.5	66	37	93	8	42
Nov.	1019.2	1011.1	SW	5.5	4.4	–8.0	8.1	15	–29	2.0	69	20	49	1	21
Dec.	1020.4	1012.2	SW	4.5	3.5	–18.6	7.7	7	–38	0.9	72	13	40	1	17
Year	1013.8	1006.7	SW	4.1	113.1	1.2		40	–43		71	569	934	334	99

Month	Number of days							Snow cover max. 10-day average depth (cm)	Cloudiness (tenths)			Mean sunshine (h)	Evaporation (mm)	
	wind (speed ≥15 m/s)	frost-free period	fog	precip. (≥0.1 mm)	thunderstorm	snow-cover	clouds (8–10 tenths)		07h	13h	19h		actual	potent.
Jan.	1.3		1	8		31			4.1	4.6	3.6	165	5	
Feb.	2.0		1	6		28			4.5	4.8	3.9	191	5	
Mar.	2.7		2	7		28	0.1		5.7	5.6	5.0	222	10	
Apr.	3.5		1	9	2		1		6.6	7.3	6.9	195	30	
May	3.9		0.8	12	2		2		7.1	7.8	7.7	220	<60	
June	1.2		2	13	5		3		7.2	7.2	7.2	239	70	
July	1.5		2	12	5		4		7.6	7.7	7.5	245	<60	
Aug.	1.5		3	14	4		3		7.4	7.4	7.2	217	60	
Sept.	1.7		2	13	2		3		6.7	7.0	6.0	196	>50	
Oct.	2.1		1	3	0.2		1		5.4	5.8	4.6	189	20	
Nov.	3.5		1	8		15	0.4		4.6	5.3	3.8	158	<10	
Dec.	2.6		0.8	7		31	0.1		4.4	4.7	3.6	153	5	
Year	27.5	159	18	112	18	156	17.6	23	5.9	6.3	5.6	2390	<400	600

TABLE XXXV

CLIMATIC TABLE FOR KHAR'KOV

Latitude 49°56'N longitude 36°17'E elevation 152 m

Month	Air press. (mbar)		Wind		Global radiat.	Temperature (°C)				Humidity		Precipitation (mm)			
	sea level	station	preval. direct.	mean speed (m/s)	(kcal./cm²)	mean	mean daily range	max.	min.	vapor press. (mm)	relat. (%)	mean	max.	min.	max. in 24 h
Jan.	1021.4	1003.4	SE	5.6		−7.4	7.7	10	−36	2.6	87	36	114	5	35
Feb.	1020.6	1002.6	E	5.9		−7.0	9.7	10	−36	2.6	84	33	75	4	26
Mar.	1018.2	1000.5	E	5.9		−1.6	10.1	19	−32	3.5	82	32	100	3	37
Apr.	1015.9	998.9	SE, E	5.4		7.1	14.3	29	−13	5.3	69	33	105	0	28
May	1015.4	998.9	E	5.0		15.0	15.6	33	−5	7.7	58	50	117	6	44
June	1011.9	995.7	NW	4.3		18.1	14.6	38	−1	10.4	62	60	133	4	68
July	1011.0	994.9	NW	4.1		20.3	13.9	39	5	11.6	65	75	148	7	83
Aug.	1013.3	997.0	NW	3.8		18.9	15.1	38	1	10.3	65	48	137	1	66
Sept.	1017.6	1001.0	NW	3.9		13.5	15.6	37	−4	8.0	68	34	97	0	43
Oct.	1020.4	1003.4	E	4.6		7.2	14.6	30	−18	6.0	75	42	111	0	48
Nov.	1021.8	1004.2	E, SE	5.4		0.4	10.2	21	−28	4.4	84	39	149	2	35
Dec.	1021.0	1003.1	SE	5.8		−5.2	8.4	12	−31	3.0	89	37	122	3	34
Year	1017.4	1000.3	E	5.0		6.6		39	−36		74	519	744	331	83

Month	Number of days							Snow cover max. 10-day average depth (cm)	Cloudiness (tenths)			Mean sun-shine (h)	Evaporation (mm)	
	wind (speed ≥15 m/s)	frost-free period	fog	precip. (≥0.1 mm)	thunder-storm	snow-cover	clouds (8–10 tenths)		07h	13h	19h		actual	potent.
Jan.	0.8		7	15.9		31	21.8		8.4	8.0	7.6	40	<10	
Feb.	1.1		5	14.5		28	17.5		8.4	8.1	7.4	67	<10	
Mar.	0.7		7	12.9	0.1	14	17.3		8.0	7.7	7.4	116	10	
Apr.	0.9		3	10.0	0.8		12.6		6.8	7.5	6.7	182	50	
May	0.4		0.7	9.8	4		8.9		6.0	7.3	6.6	254	>70	
June	0.7		1	9.8	7		8.2		5.1	7.0	6.1	279	>70	
July	0.5		2	10.3	7		6.4		4.5	6.5	5.4	290	>70	
Aug.	0.5		2	8.6	5		6.0		4.7	6.3	5.1	260	>50	
Sept.	0.3		3	7.7	1		7.4		5.7	6.4	5.0	195	>30	
Oct.	0.4		6	10.5	0.3		13.4		7.2	7.4	6.1	122	>20	
Nov.	0.6		8	12.2			20.2		8.4	8.2	8.1	54	<10	
Dec.	0.7		9	14.5		14	22.2		8.6	8.6	8.1	28	<10	
Year	7.6	173	54	137	25	105	182	26	6.8	7.4	6.6	1887	>400	700

TABLE XXXVI

CLIMATIC TABLE FOR KHATANGA

Latitude 71°50'N longitude 102°28'E elevation 24 m

Month	Air press. (mbar)		Wind		Global radiat.	Temperature (°C)				Humidity		Precipitation (mm)			
	sea level	station	preval. direct.	mean speed (m/s)	(kcal./cm²)	mean	mean daily range	max.	min.	vapor press. (mm)	relat. (%)	mean	max.	min.	max. in 24 h
Jan.	1017.9	1013.5	SW	4.2		−34.9		−4	−61	0.6	79	17			
Feb.	1018.6	1014.3	SW	3.8		−31.4		0	−61	0.6	79	13			
Mar.	1017.3	1013.0	NE	4.2		−29.0		4	−57	0.7	78	12			
Apr.	1014.3	1010.2	NE	5.1		−19.2		9	−44	1.5	79	12			
May	1012.1	1008.2	NE	4.7		−7.2		15	−37	3.4	80	16			
June	1009.3	1005.6	NE	5.0		3.9		27	−14	6.8	75	27	78	5	
July	1008.5	1004.9	NE	4.8		11.8		34	−1	10.1	69	38	134	7	
Aug.	1009.5	1005.8	NE	4.6		9.1		29	−2	9.1	79	48	122	8	
Sept.	1011.1	1007.3	SW	4.4		1.7		20	−23	6.3	83	39	56	5	
Oct.	1009.9	1005.9	SW	4.1		−11.5		12	−44	2.9	85	31			
Nov.	1014.7	1010.5	SW	4.2		−28.0		2	−51	0.9	81	24			
Dec.	1017.9	1013.6	SW	4.2		−31.2		−2	−57	0.7	80	19			
Year	1013.4	1009.4	SW	4.4		−13.8		34	−61	3.6	79	296			

Month	Number of days							Snow cover max. 10-day average depth (cm)	Cloudiness (tenths)			Mean sun-shine (h)	Evaporation (mm)	
	wind (speed ≥15 m/s)	frost-free period	fog	precip. (≥0.1 mm)	thunder-storm	snow-cover	clouds (8–10 tenths)		07h	13h	19h		actual	potent.
Jan.	2.0		2	11.7		31	10					0	5	
Feb.	1.8		2	10.9		28	9					24	5	
Mar.	1.5		2	9.2		31	8					154	5	
Apr.	2.3		2	11.2		30	10					257	5	
May	1.3		3	10.9		31	17					247	5	
June	1.9		3	12.7	0.4	4	18					265	70	
July	1.3		2	13.5	2		15					320	<50	
Aug.	1.0		5	18.0	0.3		20					196	<20	
Sept.	1.4		6	14.4		1	17					99	>10	
Oct.	1.8		4	18.4		31	18					46	<10	
Nov.	1.5		2	12.4		30	11					1	5	
Dec.	2.3		3	12.0		31	10					0	5	
Year	20.1	57	36	155.3	3	246	163	37				1609	>100	200

TABLE XXXVII

CLIMATIC TABLE FOR KHOLMSK

Latitude 47°03′N longitude 142°03′E elevation 28.8 m

Month	Air press. (mbar)		Wind		Global radiat.	Temperature (°C)				Humidity		Precipitation (mm)			
	sea level	station	preval. direct.	mean speed (m/s)	(kcal./cm²)	mean	mean daily range	max.	min.	vapor press. (mm)	relat. (%)	mean	max.	min.	max. in 24 h
Jan.			NE	6.8		-9.5		10	-29	1.8	72	40	102	11	21
Feb.			SE	6.3		-8.4		6	-22	1.9	70	26	66	5	27
Mar.			SE	6.1		-4.2		15	-22	2.5	72	32	83	2	33
Apr.			SE	6.0		2.2		20	-14	4.0	74	50	143	10	50
May			SW	5.2		6.5		24	-6	5.6	75	59	156	5	51
June			SW	3.9		10.9		28	-1	8.2	80	63	121	3	60
July			SW	3.5		15.6		30	3	11.5	84	90	207	8	105
Aug.			SE	3.8		17.8		30	5	12.9	82	92	248	22	137
Sept.			SE	5.3		14.1		28	-1	9.7	79	98	177	38	59
Oct.			SE	6.1		7.7		22	-9	6.1	72	88	199	21	73
Nov.			NW	7.1		-0.2		16	-15	3.5	69	76	158	8	58
Dec.			NW	7.8		-6.0		10	-22	2.3	71	63	116	22	36
Year			SE	5.7		3.1		30	-29		75	777	1092	524	137

Month	Number of days							Snow cover max. 10-day average depth (cm)	Cloudiness (tenths)			Mean sunshine (h)	Evaporation (mm)	
	wind (speed ⩾15 m/s)	frost-free period	fog	precip. (⩾0.1 mm)	thunderstorm	snow-cover	clouds (8–10 tenths)		07h	13h	19h		actual	potent.
Jan.	7.7			22.0		31			8.7	8.7	8.5	54	<10	
Feb.	4.9			16.9		28			8.3	8.2	7.3	96	<10	
Mar.	5.8		0.1	15.5		31			7.5	7.5	6.6	155	<10	
Apr.	4.8		0.9	13.1	0.03	6			7.7	7.6	7.1	168	<10	
May	4.2		5	14.4	0.1				7.9	7.6	7.8	188	50	
June	1.4		8	14.9	0.7				8.1	7.8	7.7	165	70	
July	0.9		7	14.8	0.7				8.5	8.4	8.2	148	70	
Aug.	0.9		2	14.4	0.9				8.0	8.0	7.8	165	70	
Sept.	3.1		0.4	14.9	2				7.4	7.3	6.6	178	50	
Oct.	6.2		0.2	15.2	1				7.0	7.2	5.5	156	20	
Nov.	7.2			19.1	0.2	7			8.0	8.3	7.1	86	<10	
Dec.	10.2			23.8		31			8.9	9.1	8.3	45	<10	
Year	57.3	153	24	199	6	147		64	8.0	8.0	7.4	1604	<400	400

TABLE XXXVIII

CLIMATIC TABLE FOR KHOROG

Latitude 37°30′N longitude 71°30′E elevation 2,080 m

Month	Air press. (mbar)		Wind		Global radiat.	Temperature (°C)				Humidity		Precipitation (mm)			
	sea level	station	preval. direct.	mean speed (m/s)	(kcal./cm²)	mean	mean daily range	max.	min.	vapor press. (mm)	relat. (%)	mean	max.	min.	max. in 24 h
Jan.		796.1	NE	1.6		-7.9	10.0	10	-32	2.4	70	31	88	2	31
Feb.		794.9	NE	1.8		-5.8	10.2	14	-32	3.0	69	32	174	0	32
Mar.		794.5	NE	2.6		0.8	8.4	21	-23	4.3	63	39	123	2	29
Apr.		795.2	NE	2.7		9.2	10.1	29	-10	5.7	51	42	198	0	70
May		794.9	W	2.3		14.9	12.1	34	-0.6	6.8	44	24	101	2	28
June		791.7	W	2.6		19.0	13.7	37	2	7.4	39	9	145	0	16
July		789.4	W	2.8		22.8	14.6	38	5	8.6	35	3	25	0	17
Aug.		789.9	W	2.8		22.6	15.0	38	5	7.4	30	0	11	0	11
Sept.		793.2	W	2.4		18.3	15.9	36	0	5.3	28	1	20	0	12
Oct.		796.5	W	2.0		10.9	13.7	30	-14	4.5	38	11	67	0	23
Nov.		799.7	E	1.9		3.4	9.3	23	-18	3.8	51	19	84	0	20
Dec.		797.3	NE	1.8		-3.8	8.4	14	-27	2.9	64	24	91	0	26
Year		794.4	W	2.3		8.7	–	38	-32	5.2	48	235	431	97	70

Month	Number of days							Snow cover max. 10-day average depth (cm)	Cloudiness (tenths)			Mean sunshine (h)	Evaporation (mm)	
	wind (speed ⩾15 m/s)	frost-free period	fog	precip. (⩾0.1 mm)	thunderstorm	snow-cover	clouds (8–10 tenths)		07h	13h	19h		actual	potent.
Jan.	0.1		0.3	8.4			11.9	27	6.9	6.8	5.8	93		
Feb.	0.04		0.3	8.5			12.8	38	7.7	7.3	6.4	101		
Mar.	0.1		0.1	8.9			15.5	33	7.7	7.2	7.0	144		
Apr.	0.4			8.3	0.5		13.8		7.2	7.0	7.1	168		
May	0.5			6.7	1		6.2		5.5	5.5	5.9	218		
June	0.7			3.1	2		2.4		3.0	4.4	4.1	266		
July	0.3			1.3	1		1.2		2.2	3.2	2.5	299		
Aug.	0.3			0.3	0.2		1.3		1.7	2.4	1.8	282		
Sept.	0.2			0.5	0.1		0.9		2.6	3.2	1.5	244		
Oct.	0.2			3.2	0.1		3.4		3.6	4.1	2.7	188		
Nov.	0.1		0.03	4.8			8.6		5.4	5.7	4.6	119		
Dec.	0.1		0.2	7.4			12.1	11	6.6	6.7	5.8	95		
Year	3.04	206	0.9	61.4	5	98	9.2	38	5.0	5.3	4.6	2217		846.5

TABLE XXXIX

CLIMATIC TABLE FOR KIEV

Latitude 50°24′N longitude 30°27′E elevation 179 m

Month	Air press. (mbar) sea level	Air press. (mbar) station	Wind preval. direct.	Wind mean speed (m/s)	Global radiat. (kcal./cm²)	Temperature (°C) mean	Temperature (°C) mean daily range	Temperature (°C) max.	Temperature (°C) min.	Humidity vapor press. (mm)	Humidity relat. (%)	Precipitation (mm) mean	Precipitation (mm) max.	Precipitation (mm) min.	Precipitation (mm) max. in 24 h
Jan.	1020.7	997.3	W	4.3	2.2	−5.9	9.0	9	−33	2.9	87	43	88	3	19
Feb.	1019.4	996.1	NW	4.5	3.9	−5.3	10.2	12	−34	2.9	84	39	113	5	30
Mar.	1017.0	994.1	NW	4.3	7.1	−0.5	9.8	21	−26	3.8	79	35	88	2	28
Apr.	1015.0	992.7	NW	4.3	9.2	7.1	12.4	29	−11	5.4	69	46	142	1	37
May	1015.0	993.5	N	3.8	13.8	14.7	12.0	33	−3	8.0	63	56	144	4	50
June	1012.6	991.3	NW	3.7	15.6	17.4	12.4	34	2	10.2	65	66	239	7	65
July	1011.8	990.6	NW	3.5	15.8	19.3	12.7	39	5	11.5	67	70	195	5	74
Aug.	1013.7	992.3	NW	3.5	12.8	18.2	12.6	39	3	10.7	70	72	223	5	48
Sept.	1017.5	995.9	NW	3.7	8.7	13.6	12.7	33	−4	8.4	73	47	149	2	57
Oct.	1019.4	997.1	W	3.9	5.0	7.7	11.8	27	−19	6.4	80	47	141	1	36
Nov.	1020.2	997.4	SE	4.3	2.1	1.1	9.4	23	−23	4.5	86	53	127	2	41
Dec.	1019.8	996.6	SE, W	4.2	1.6	−3.7	7.5	13	−31	3.4	88	41	103	5	32
Year	1016.8	994.6	SW	4.0	97.8	7.0		39	−34		76	615	925	405	74

Month	Number of days wind (speed ⩾15 m/s)	frost-free period	fog	precip. (⩾0.1 mm)	thunder-storm	snow-cover	clouds (8–10 tenths)	Snow cover max. 10-day average depth (cm)	Cloudiness (tenths) 07h	13h	19h	Mean sun-shine (h)	Evaporation (mm) actual	potent.
Jan.	1.0		11	17.2		31	19.4		8.5	8.3	7.7	42	<10	
Feb.	1.3		8	14.9		28	15.8		8.4	8.2	7.3	64	<10	
Mar.	1.7		9	14.1	0.1	14	15.6		7.7	7.5	6.8	112	>10	
Apr.	1.1		5	12.0	0.9		10.9		6.8	7.2	6.5	162	50	
May	1.2		1	12.6	5		7.9		5.9	7.1	6.4	257	<70	
June	0.9		0.9	12.7	7		7.8		5.1	6.8	6.2	273	>70	
July	0.7		1	13.3	6		7.4		5.2	6.7	5.7	287	>70	
Aug.	0.6		3	12.2	4		6.4		5.3	6.4	5.3	252	>50	
Sept.	0.4		5	10.1	2		7.3		6.1	6.4	5.3	189	>30	
Oct.	0.7		9	11.6	0.4		12.5		7.4	7.3	5.9	123	20	
Nov.	0.5		13	15.0	0.04		18.6		8.8	8.5	8.3	51	<10	
Dec.	0.6		11	17.7	0.02	16	22.0		8.6	8.6	8.3	31	<10	
Year	10.7	179	77	163	25	105	152	29	7.0	7.4	6.6	1843	400	700

TABLE XL

CLIMATIC TABLE FOR KIRENSK

Latitude 57°46′N longitude 108°07′E elevation 256 m

Month	Air press. (mbar) sea level	Air press. (mbar) station	Wind preval. direct.	Wind mean speed (m/s)	Global radiat. (kcal./cm²)	Temperature (°C) mean	Temperature (°C) mean daily range	Temperature (°C) max.	Temperature (°C) min.	Humidity vapor press. (mm)	Humidity relat. (%)	Precipitation (mm) mean	Precipitation (mm) max.	Precipitation (mm) min.	Precipitation (mm) max. in 24 h
Jan.	1031.0		SW	2.1	1.4	−26.9	11.4	3	−58	0.6	78	19	40	4	12
Feb.	1030.2		SW	2.0	3.4	−22.4	14.1	7	−53	0.8	77	11	26	0	8
Mar.	1024.3		SW	2.3	7.9	−11.5	18.9	13	−47	1.4	71	7	25	2	6
Apr.	1016.7		SW	2.7	10.7	−2.1	14.4	24	−35	2.8	65	14	30	3	8
May	1012.2		SW	2.9	12.1	6.8	13.8	32	−15	4.7	60	27	56	3	23
June	1007.0		SW	2.4	14.0	15.0	14.5	36	−4	8.3	68	66	115	10	66
July	1005.3		NE	2.0	14.5	18.8	13.2	37	0	11.1	74	67	142	10	58
Aug.	1008.6		N	2.0	10.2	15.2	12.1	36	−5	9.9	78	56	165	2	61
Sept.	1016.0		SW	2.2	6.8	7.2	11.2	28	−11	6.1	80	36	89	7	29
Oct.	1020.0		SW	2.6	3.5	−2.3	8.0	21	−34	3.2	79	28	53	8	13
Nov.	1021.9		SW	2.6	1.8	−14.8	10.6	8	−48	1.4	80	25	42	11	10
Dec.	1029.8		SW	2.0	0.9	−25.1	10.2	4	−57	0.8	79	25	49	5	8
Year	1018.6		SW	2.3	87.2	−3.6		37	−58		74	381	538	243	66

Month	Number of days wind (speed ⩾15 m/s)	frost-free period	fog	precip. (⩾0.1 mm)	thunder-storm	snow-cover	clouds (8–10 tenths)	Snow cover max. 10-day average depth (cm)	Cloudiness (tenths) 07h	13h	19h	Mean sun-shine (h)	Evaporation (mm) actual	potent.
Jan.	0.5		10	17.5		31	14		7.5	7.8	6.9	42	<10	
Feb.	0.5		6	14.3		28	10		7.4	7.0	5.8	88	<10	
Mar.	2.7		1	12.6		31	10		6.7	7.0	5.8	184	<10	
Apr.	2.5		0.9	10.9	0.03	26	10		7.0	7.7	6.9	200	>10	
May	3.2		2	12.3	1.0		12		7.3	8.1	7.6	211	>30	
June	1.8		6	13.6	5.0		10		7.1	7.5	7.4	258	<60	
July	1.1		9	12.2	6.0		9		7.1	7.0	6.8	276	70	
Aug.	1.0		13	14.0	4.0		13		8.4	7.1	6.5	184	<30	
Sept.	0.2		11	14.3	0.5		14		9.0	7.5	6.8	127	<30	
Oct.	1.1		4	15.8		10	19		9.0	8.6	7.4	77	<10	
Nov.	0.8		3	18.7		30	18		8.2	8.0	6.9	46	<10	
Dec.	0.8		9	17.2		31	14		7.8	8.2	7.3	22	<10	
Year	16.2	99	75	173.4	16	195	153	48	7.7	7.6	6.8	1715	<300	400

TABLE XLI

CLIMATIC TABLE FOR KIROV

Latitude 58°39′N longitude 49°37′E elevation 164 m

Month	Air press. (mbar)		Wind		Global radiat. (kcal./cm²)	Temperature (°C)				Humidity		Precipitation (mm)			
	sea level	station	preval. direct.	mean speed (m/s)		mean	mean daily range	max.	min.	vapor press. (mm)	relat. (%)	mean	max.	min.	max. in 24 h
Jan.	1019.9		SE	5.2		−14.4	8.7	3	−41	1.7	88	33			
Feb.	1019.2		SE	5.3		−13.0	9.0	4	−41	1.7	84	24			
Mar.	1017.2		SW, W	5.4		−7.1	11.2	11	−34	2.4	79	26			
Apr.	1016.8		W	4.9		1.9	9.8	27	−20	3.8	72	28			23
May	1015.1		NW	5.1		9.9	12.7	31	−11	6.2	62	45			31
June	1011.1		W	4.7		15.1	12.3	37	−2	8.8	62	58			44
July	1009.8		NW	4.0		18.0	12.3	35	3	10.7	69	72			96
Aug.	1012.7		NW	3.9		15.1	12.4	36	0	10.0	74	69			43
Sept.	1013.5		NW	4.6		8.8	12.4	29	−8	7.2	81	54			29
Oct.	1016.2		W	5.2		1.5	8.5	21	−23	4.7	86	56			
Nov.	1022.6		W	5.4		−6.1	7.7	11	−40	3.0	88	38			
Dec.	1020.2		SW	5.1		−12.3	6.3	4	−45	1.9	88	35			
Year	1016.2		W	4.9		1.4		37	−45		78	538			

Month	Number of days							Snow cover max. 10-day average depth (cm)	Cloudiness (tenths)			Mean sunshine (h)	Evaporation (mm)	
	wind (speed ≥15 m/s)	frost-free period	fog	precip. (≥0.1 mm)	thunderstorm	snow-cover	clouds (8–10 tenths)		07h	13h	19h		actual	potent.
Jan.	1.3		5	21.1	0.03	31	19.5		8.1	8.3	7.7	31	<10	
Feb.	0.9		3	17.1	0.03	28	14.1		7.9	7.4	6.6	70	<10	
Mar.	0.9		3	15.7		31	13.9		7.6	7.1	6.7	130	<10	
Apr.	0.8		3	11.5	0.2	18	10.1		6.5	6.6	6.5	138	>20	
May	1.6		1	13.5	3		10.4		6.5	7.2	6.7	258	>70	
June	1.5		0.9	14.5	7		8.7		5.9	6.8	6.2	287	>70	
July	0.4		2	13.9	8		9.5		5.8	7.0	6.1	286	>70	
Aug.	0.4		2	15.5	5		8.8		6.3	7.0	6.1	252	>50	
Sept.	0.5		4	16.9	2		13.8		7.9	8.1	7.2	126	>30	
Oct.	0.6		5	18.8			20.8		8.8	8.8	8.1	55	>10	
Nov.	0.8		5	20.9		19	20.8		8.7	8.8	8.2	31	<10	
Dec.	0.8		6	21.2	0.03	31	20.6		8.4	8.6	8.0	23	<10	
Year	10.5	122	40	200.6	25	172	171	69	7.4	7.6	7.0	1747	<400	500

TABLE XLII

CLIMATIC TABLE FOR KISHINEV

Latitude 47°01′N longitude 28°52′E elevation 95 m

Month	Air press. (mbar)		Wind		Global radiat. (kcal./cm²)	Temperature (°C)				Humidity		Precipitation (mm)			
	sea level	station	preval. direct.	mean speed (m/s)		mean	mean daily range	max.	min.	vapor press. (mm)	relat. (%)	mean	max.	min.	max. in 24 h
Jan.	1021.9	1009.5	NW	4.2	2.6	−3.6	6.4	15	−30	3.2	84	26.4	154	0	28
Feb.	1020.3	1008.1	NW	4.2	4.1	−2.5	6.4	16	−32	3.4	81	34.8	83	2	26
Mar.	1017.5	1005.5	NW	4.2	7.8	2.7	8.3	25	−23	4.4	79	16.6	107	0	24
Apr.	1015.0	1003.4	NW	4.3	10.4	9.2	11.3	32	−9	5.8	67	32.4	104	1	34
May	1015.0	1003.7	NW	3.7	14.4	15.9	12.0	36	−2	8.4	62	39.0	155	3	68
June	1013.4	1002.3	NW	3.2	16.3	19.0	11.7	37	4	10.8	66	79.3	221	9	91
July	1012.7	1001.8	NW	3.0	17.7	21.2	12.2	39	8	12.0	63	36.2	307	2	91
Aug.	1014.5	1003.4	NW	2.7	13.8	20.3	12.5	39	7	11.2	64	48.5	153	4	84
Sept.	1018.3	1006.9	NW	3.1	10.5	15.8	12.3	37	−1	9.2	68	43.4	123	0	58
Oct.	1020.0	1008.3	NW	2.8	6.3	10.3	10.4	33	−16	7.4	76	24.6	151	0	40
Nov.	1021.4	1009.3	NW	3.9	2.4	3.8	6.7	28	−22	5.4	84	55.6	146	0	77
Dec.	1021.0	1008.7	NW	3.8	2.0	−1.2	5.9	16	−22	4.0	86	34.0	97	0	41
Year	1017.6	1005.9	NW	3.6	108.3	9.2		39	−32		73	470.8	853	279	91

Month	Number of days							Snow cover max. 10-day average depth (cm)	Cloudiness (tenths)			Mean sunshine (h)	Evaporation (mm)	
	wind (speed ≥15 m/s)	frost-free period	fog	precip. (≥0.1 mm)	thunderstorm	snow-cover	clouds (8–10 tenths)		07h	13h	19h		actual	potent.
Jan.	1.3		10	12	0	31	16.5		7.7	7.7	6.9	70	<10	
Feb.	0.8		8	12	0	28	14.9		8.3	8.2	7.0	79	<10	
Mar.	0.8		7	12	0.1	20	13.1		7.7	7.8	6.9	146	>10	
Apr.	0.3		3	9	1		10.1		6.8	7.4	6.7	201	50	
May	0.2		1	11	5		8.2		6.3	7.8	7.2	258	>70	
June	0.3		1	12	8		7.0		5.3	7.2	6.5	297	>70	
July	0.4		1	10	6		5.5		4.2	5.9	5.1	329	>70	
Aug.	0.6		2	8	3		3.7		4.1	5.7	4.9	307	>50	
Sept.	0.2		3	7	1		4.6		4.9	5.9	4.3	232	>30	
Oct.	0.2		8	8	0.4		9.8		6.8	6.8	4.9	168	>10	
Nov.	0.6		10	10	0.1		16.0		8.6	8.4	7.2	74	<10	
Dec.	0.6		11	12	0	27	19.3		8.4	8.3	7.3	54	<10	
Year	6.3	191	65	123	25	52	129	20	6.6	7.3	6.2	2215	400	800

TABLE XLIII

CLIMATIC TABLE FOR KLYUCHY

Latitude 56°19′N longitude 160°50′E elevation 25 m

Month	Air press. (mbar)		Wind		Global radiat. (kcal./cm²)	Temperature (°C)				Humidity		Precipitation (mm)			
	sea level	station	preval. direct.	mean speed (m/s)		mean	mean daily range	max.	min.	vapor press. (mm)	relat. (%)	mean	max.	min.	max. in 24 h
Jan.	1007.4	1004.1	W	5.1	1.7	−16.9	8.9	6	−49	1.9	82	60	116	12	32
Feb.	1007.5	1004.2	W	4.9	3.6	−15.0	8.7	2	−42	2.0	79	48	151	6	43
Mar.	1010.6	1007.3	W	5.3	8.8	−10.5	9.9	8	−37	2.4	74	40	165	3	48
Apr.	1010.5	1007.3	W	5.3	11.7	−2.7	8.4	14	−26	3.6	71	27	59	2	24
May	1012.5	1009.3	E	4.4	13.8	4.1	8.5	25	−14	5.5	69	29	82	2	23
June	1010.7	1007.3	E	3.9	14.1	10.8	10.8	30	−4	8.7	69	30	80	5	24
July	1009.7	1006.6	E	3.6	13.7	14.7	9.6	34	2	12.7	76	56	120	11	62
Aug.	1010.2	1007.1	W	3.5	10.9	13.8	9.2	29	−1	12.7	80	61	152	8	58
Sept.	1011.5	1008.5	W	4.0	7.9	8.6	9.0	24	−7	8.9	79	45	96	5	42
Oct.	1009.1	1006.1	W	5.6	4.8	1.4	7.5	18	−19	5.0	72	50	225	1	49
Nov.	1007.9	1004.7	W	5.4	2.0	−7.5	7.5	11	−34	3.0	79	52	122	9	34
Dec.	1005.1	1001.8	W	4.7	1.1	−14.6	8.1	5	−44	2.0	83	64	196	13	30
Year	1009.4	1006.2	W	4.6	94.1	−1.2	–	34	−49	5.7	76	562	882	399	62

Month	Number of days							Snow cover max. 10-day average depth (cm)	Cloudiness (tenths)			Mean sunshine (h)	Evaporation (mm)	
	wind (speed ⩾15 m/s)	frost-free period	fog	precip. (⩾0.1 mm)	thunderstorm	snow-cover	clouds (8–10 tenths)		07h	13h	19h		actual	potent.
Jan.	6.6		7	16.0			16.6	75	7.6	7.9	7.0	30	1.2	
Feb.	4.4		5	14.5			14.7	98	7.9	7.6	7.6	71	3.5	
Mar.	5.9		2	12.2			13.3	104	7.4	6.9	6.5	142	4.2	
Apr.	5.3		0.4	9.9			14.3	100	7.5	7.3	7.4	167	12	
May	2.8		0.1	9.5			16.4	26	7.6	7.6	7.7	181	30	
June	1.3		0.1	9.7	0.3		17.3		7.9	7.6	7.8	197	45	
July	1.2		1	12.6	0.5		17.4		8.0	7.6	7.7	194	48	
Aug.	1.8		3	12.6	0.5		15.9		8.1	7.6	7.8	185	37	
Sept.	3.8		5	11.0	0.2		13.5		7.9	7.2	7.1	152	30	
Oct.	7.9		1	9.5	0.07		10.8	4	7.2	7.4	6.1	129	12	
Nov.	7.4		2	11.9			12.9	22	7.3	7.6	6.5	58	8	
Dec.	5.7		7	16.0			15.5	51	7.4	7.9	6.8	34	4	
Year	54.1	109	34	145.4	1.57	191	179	104	7.6	7.5	7.1	1540	234.9	411.5

TABLE XLIV

CLIMATIC TABLE FOR KOLPASHEVO

Latitude 58°18′N longitude 82°54′E elevation 76 m

Month	Air press. (mbar)		Wind		Global radiat. (kcal./cm²)	Temperature (°C)				Humidity		Precipitation (mm)			
	sea level	station	preval. direct.	mean speed (m/s)		mean	mean daily range	max.	min.	vapor press. (mm)	relat. (%)	mean	max.	min.	max. in 24 h
Jan.	1025.5	1014.6	S	3.5	1.3	−20.8	8.6	3	−51	1.0	80	18	36	5	
Feb.	1025.6	1014.9	S	3.8	3.0	−17.8	10.0	7	−51	1.0	78	12	38	3	
Mar.	1022.3	1011.8	S	4.3	8.0	−10.5	11.5	11	−42	1.7	73	15	41	5	
Apr.	1018.4	1008.5	NW	4.0	11.8	−1.3	10.3	24	−31	3.3	66	21	52	2	
May	1013.8	1004.1	NW	4.6	12.3	6.8	10.4	33	−19	5.1	62	51	110	14	
June	1008.9	999.7	NW	3.8	15.6	15.5	10.5	36	−4	8.9	66	58	113	10	
July	1007.1	998.1	S	2.9	14.5	18.3	10.6	36	1	11.3	72	77	183	17	
Aug.	1009.4	1000.2	NW	2.9	11.7	14.9	9.8	32	−2	10.3	79	79	158	14	
Sept.	1014.6	1005.0	S, SW	3.5	6.7	9.0	9.1	29	−9	6.9	79	52	109	7	
Oct.	1017.6	1007.7	SW	4.6	2.4	−0.2	6.4	23	−28	4.0	79	37	74	14	
Nov.	1020.8	1010.3	SW	4.3	1.4	−11.7	7.6	10	−38	2.0	82	31	63	15	
Dec.	1024.3	1013.4	S, SW	4.0	0.8	−16.6	8.5	5	−54	1.1	81	24	55	11	
Year	1017.5	1007.5	S	3.8	89.8	−1.4		36	−54		75	475	624	311	

Month	Number of days							Snow cover max. 10-day average depth (cm)	Cloudiness (tenths)			Mean sunshine (h)	Evaporation (mm)	
	wind (speed ⩾15 m/s)	frost-free period	fog	precip. (⩾0.1 mm)	thunderstorm	snow-cover	clouds (8–10 tenths)		07h	13h	19h		actual	potent.
Jan.	1.1		3			31			7.0	7.6	5.4	48	<10	
Feb.	1.1		2			28			7.6	7.2	6.1	83	<10	
Mar.	3.4		1			31			7.4	7.0	6.7	153	<10	
Apr.	2.9		2		0.2	24			6.9	6.8	6.9	197	>10	
May	4.2		1		2				7.2	7.3	7.1	238	>50	
June	2.7		0.7		6				6.8	7.2	7.0	273	>70	
July	1.4		2		6				6.4	7.0	6.7	276	>70	
Aug.	0.7		4		5				7.1	7.6	7.1	205	>50	
Sept.	1.8		3		1				7.8	8.0	7.3	136	30	
Oct.	2.7		1		0	4			8.9	8.9	7.9	57	<10	
Nov.	1.8		1			30			8.1	8.1	7.7	42	<10	
Dec.	1.9		2			31			7.4	7.8	7.0	23	<10	
Year	25.7	113	23		20	188		61	7.4	7.5	7.0	1731	>300	500

TABLE XLV

CLIMATIC TABLE FOR KRASNODAR

Latitude 45°02′N longitude 39°09′E elevation 33 m

Month	Air press. (mbar)		Wind		Global radiat.	Temperature (°C)				Humidity		Precipitation (mm)			
	sea level	station	preval. direct.	mean speed (m/s)	(kcal./cm²)	mean	mean daily range	max.	min.	vapor press. (mm)	relat. (%)	mean	max.	min.	max. in 24 h
Jan.			E	3.9	3.8	−2.1	7.4	21	−36	5.2	85	52	153	7	43
Feb.			E	4.4	5.3	−1.1	8.3	22	−34	5.4	83	52	131	6	34
Mar.			E	5.3	8.9	4.0	9.5	28	−22	6.0	78	50	143	1	55
Apr.			E	4.8	11.2	10.7	12.1	36	−10	8.8	69	50	182	9	60
May			SW	3.9	16.5	16.5	12.5	37	−3	12.8	69	58	177	4	99
June			SW	3.5	16.9	20.2	12.6	37	3	16.4	68	64	158	10	68
July			SW	3.1	17.1	22.9	13.0	39	8	18.1	66	58	164	0	67
Aug.			NE	3.0	16.0	22.5	13.4	41	4	17.2	66	45	155	2	62
Sept.			NE	3.2	11.9	17.2	13.8	38	−3	13.1	70	38	200	2	58
Oct.			NE	3.4	8.0	11.5	11.9	34	−9	10.0	77	51	181	5	49
Nov.			E	3.9	4.6	5.0	9.5	26	−24	7.8	84	58	201	4	51
Dec.			E	4.2	2.9	0.2	8.1	22	−29	5.9	85	64	175	3	47
Year			E	3.9	123.1	10.6	–	41	−36	10.6	75	640	1020	458	99

Month	Number of days							Snow cover	Cloudiness (tenths)			Mean sun-shine	Evaporation (mm)	
	wind (speed ≥15 m/s)	frost-free period	fog	precip. (≥0.1 mm)	thunder-storm	snow-cover	clouds (8–10 tenths)	max. 10-day average depth (cm)	07h	13h	19h	(h)	actual	potent.
Jan.	1.0		7	14.0	0.1		17.7	6	8.1	8.1	7.3	64	2.5	
Feb.	1.6		6	13.0	0.1		15.0	8	8.2	7.9	7.0	80	4	
Mar.	3.0		4	12.4	0.1		15.3	5	7.9	7.7	6.8	123	20	
Apr.	1.8		3	11.2	0.4		10.3		6.6	7.0	6.4	174	55	
May	0.8		3	11.0	4		8.4		6.2	6.6	6.1	239	100	
June	0.4		2	10.9	8		4.9		4.6	5.8	5.1	289	110	
July	0.2		1	9.1	6		3.0		3.1	4.8	3.8	322	70	
Aug.	0.4		1	7.2	6		2.7		3.6	4.5	3.8	294	52	
Sept.	0.7		3	7.6	2		3.5		4.3	4.7	3.9	235	42	
Oct.	0.6		5	10.4	0.7		8.4		6.5	6.2	4.5	170	23	
Nov.	1.1		6	12.2	0.5		12.2		7.6	7.4	5.9	95	10	
Dec.	1.2		7	13.5	0.2		15.0	4	7.9	8.1	6.8	61	5.9	
Year	12.8	186	48	132	28	42	116	8	6.2	6.6	5.6	2146	494.4	728.9

TABLE XLVI

CLIMATIC TABLE FOR KRASNOVODSK

Latitude 40°02′N longitude 52°59′E elevation −10 m

Month	Air press. (mbar)		Wind		Global radiat.	Temperature (°C)				Humidity		Precipitation (mm)			
	sea level	station	preval. direct.	mean speed (m/s)	(kcal./cm²)	mean	mean daily range	max.	min.	vapor press. (mm)	relat. (%)	mean	max.	min.	max. in 24 h
Jan.	1020.6	1022.3	E	3.4		2.1	6.5	18	−17	4.3	77	11			17
Feb.	1019.1	1021.0	E	3.4		3.0	8.1	22	−13	4.6	75	13			20
Mar.	1018.8	1019.0	NW	3.5		6.0	9.1	30	−10	5.6	71	15			21
Apr.	1016.4	1015.5	NW	4.0		12.6	10.2	34	−4	7.3	54	12			29
May	1013.5	1014.2	NW	4.0		19.3	11.5	38	8	9.7	56	7			28
June	1010.2	1010.5	NW	4.5		24.1	11.4	43	10	12.1	50	2			77
July	1007.3	1008.1	NW	4.6		27.6	10.4	44	12	14.4	48	2			46
Aug.	1009.4	1010.2	NW	4.5		27.5	10.2	43	14	13.2	46	3			17
Sept.	1014.6	1015.4	N	3.9		22.3	10.6	39	7	10.8	48	3			14
Oct.	1019.8	1020.4	N	3.2		15.3	8.9	32	0	8.4	60	6			26
Nov.	1022.6	1022.0	E	3.0		8.2	9.8	27	−13	6.3	65	9			18
Dec.	1022.4	1022.3	E	3.0		4.1	9.2	21	−16	5.1	74	11			15
Year	1016.2	1016.7	E	2.7		14.3		44	−17		61	92			46

Month	Number of days							Snow cover	Cloudiness (tenths)			Mean sun-shine	Evaporation (mm)	
	wind (speed ≥15 m/s)	frost-free period	fog	precip. (≥0.1 mm)	thunder-storm	snow-cover	clouds (8–10 tenths)	max. 10-day average depth (cm)	07h	13h	19h	(h)	actual	potent.
Jan.	2.6		3	6.7	0		11		7.0	7.4	5.7		<10	
Feb.	2.1		2	5.6	0		10		7.0	7.0	5.2		10	
Mar.	3.2		2	7.6	0		11		7.1	6.4	5.5		30	
Apr.	3.9		2	5.5	0.1		8		6.2	5.8	5.4		50	
May	4.0		1	3.2	1		5		5.5	4.6	4.4		30	
June	4.5		1	1.8	1		2		3.5	2.9	2.5		<10	
July	5.2		1	1.2	1		2		3.7	2.7	2.8		<10	
Aug.	4.5		1	1.1	1		1		2.8	2.1	2.1		<10	
Sept.	3.3		1	0.9	0.3		2		2.8	2.2	2.0		<10	
Oct.	2.6		2	3.0	0.5		4		4.6	4.4	2.9		<10	
Nov.	2.2		1	4.5	0.1		8		6.3	6.2	4.5		<10	
Dec.	1.5		2	7.6	0.1		10		6.8	6.8	5.2		<10	
Year	39.6	278	19	48.7	5	5	74	1	5.3	4.9	4.0		<200	1200

TABLE XLVII

CLIMATIC TABLE FOR KRASNOYARSK

Latitude 56°00'N longitude 92°53'E elevation 156 m

Month	Air press. (mbar)		Wind		Global radiat.	Temperature (°C)				Humidity		Precipitation (mm)			
	sea level	station	preval. direct.	mean speed (m/s)	(kcal./cm²)	mean	mean daily range	max.	min.	vapor press. (mm)	relat. (%)	mean	max.	min.	max. in 24 h
Jan.	1028.8	991.9	SW	2.5		−17.4	7.9	6	−49	1.2	72	12	28	2	13
Feb.	1029.0	992.3	SW	1.8		−16.0	8.9	8	−43	1.4	72	9	27	0	9
Mar.	1025.4	989.8	SW	2.4		−8.0	9.4	20	−40	2.2	66	10	25	0	12
Apr.	1019.8	985.5	SW	2.7		−1.6	9.5	28	−28	3.7	58	22	68	1	21
May	1014.5	981.3	SW	2.7		9.5	11.4	34	−17	6.2	54	38	84	4	45
June	1009.0	976.9	SW	2.1		16.7	11.6	38	−4	10.0	62	58	106	15	45
July	1007.0	975.3	SW	1.7		19.9	10.9	41	4	13.2	71	83	171	17	67
Aug.	1009.7	977.5	SW	1.7		16.6	10.2	37	−1	11.8	76	65	157	14	47
Sept.	1017.0	983.8	SW	2.2		9.9	9.3	32	−9	7.6	75	47	88	15	32
Oct.	1021.2	987.0	SW	2.5		1.6	7.6	26	−32	4.3	68	34	78	3	22
Nov.	1024.8	989.1	SW	2.8		−8.3	7.1	13	−43	2.4	72	25	91	4	32
Dec.	1029.0	992.2	SW	2.3		−15.9	7.6	10	−45	1.4	72	17	59	3	8
Year	1019.6	985.3	SW	2.3		−0.8		41	−49		68	419	585	260	67

Month	Number of days							Snow cover	Cloudiness (tenths)			Mean sun-shine (h)	Evaporation (mm)	
	wind (speed ≥15 m/s)	frost-free period	fog	precip. (≥0.1 mm)	thunder-storm	snow-cover	clouds (8–10 tenths)	max. 10-day average depth (cm)	07h	13h	19h		actual	potent.
Jan.	3.7		2	10.8		31			6.9	7.2	6.0	54	5	
Feb.	1.7		1	7.9		28			7.2	7.0	5.6	84	<10	
Mar.	2.1		0.7	8.0		31			7.4	7.3	6.3	163	<10	
Apr.	1.6		0.5	7.9	0.1	4			7.0	7.5	6.8	191	>10	
May	1.8		0.5	12.4	1.0				6.9	7.7	7.1	221	>50	
June	0.8		0.7	13.6	3.0				6.6	7.4	6.8	280	>60	
July	0.1		2	13.2	6.0				6.4	7.3	6.8	267	70	
Aug.	0.2		3	14.9	4.0				7.0	7.7	6.8	212	50	
Sept.	0.9		3	13.1	0.5				7.6	7.6	6.8	160	30	
Oct.	2.1		1	10.8					8.5	8.3	7.1	97	<10	
Nov.	3.9		0.8	13.1		26			8.3	8.2	7.2	47	<10	
Dec.	3.6		2	11.5		31			7.4	7.9	6.7	30	5	
Year	22.5	119	18	137.2	15	150		23	7.3	7.6	6.7	1806	300	500

TABLE XLVIII

CLIMATIC TABLE FOR KUSHKA

Latitude 35°17'N longitude 62°21'E elevation 630 m

Month	Air press. (mbar)		Wind		Global radiat.	Temperature (°C)				Humidity		Precipitation (mm)			
	sea level	station	preval. direct.	mean speed (m/s)	(kcal./cm²)	mean	mean daily range	max.	min.	vapor press. (mm)	relat. (%)	mean	max.	min.	max. in 24 h
Jan.	1025.1	948.7	SW	1.9	7.8	2.3	11.6	27	−33	5.1	75	40			
Feb.	1022.4	946.9	SW	1.9	10.0	4.8	12.2	31	−26	6.3	74	42			
Mar.	1018.9	945.1	NE	2.0	13.8	8.7	11.9	32	−26	8.0	74	60			
Apr.	1014.9	942.9	NE	1.8	17.0	14.3	14.0	37	−6	10.8	63	30			
May	1011.8	941.5	N, NE	2.2	21.0	20.5	17.5	44	1	10.7	47	12			
June	1007.1	936.8	NE	3.1	23.9	25.3	19.0	45	4	10.8	33	2			
July	1003.6	935.3	N	3.5	23.0	27.6	18.8	46	10	10.9	30	0			
Aug.	1006.7	937.7	N	3.3	22.0	25.3	19.4	43	7	9.3	29	0			
Sept.	1014.1	943.3	N	2.4	18.0	19.3	20.2	42	−4	7.5	35	1			
Oct.	1020.6	947.8	NE, SW	1.8	14.0	13.4	19.0	37	−9	6.1	46	9			
Nov.	1023.6	949.1	NE, SW	1.6	10.2	7.7	16.2	35	−19	5.7	57	25			
Dec.	1024.4	948.9	SW	1.6	8.0	3.9	12.2	32	−24	5.6	70	25			
Year	1016.1	943.8	NE	2.3	188.7	14.4	–	46	−33	8.1	53	246			

Month	Number of days							Snow cover	Cloudiness (tenths)			Mean sun-shine (h)	Evaporation (mm)	
	wind (speed ≥15 m/s)	frost-free period	fog	precip. (≥0.1 mm)	thunder-storm	snow-cover	clouds (8–10 tenths)	max. 10-day average depth (cm)	07h	13h	19h		actual	potent.
Jan.	1.1		3	8.9	0.4		9.8		6.5	6.1	5.3	128	10	
Feb.	1.5		2	9.3	1		10.6		6.8	6.6	5.8	116	15	
Mar.	1.3		2	10.2	3		12.8		7.3	6.9	6.6	145	25	
Apr.	1.1		0.6	7.1	3		9.4		5.9	6.0	5.9	228	37	
May	1.5		0.1	3.3	2		3.7		3.2	3.6	3.9	320	45	
June	2.2			0.2	0.03		0.4		0.8	1.1	1.0	377	7	
July	2.2			0.0			0.1		0.5	0.5	0.5	386	4	
Aug.	2.6			0.0			0.1		0.5	0.4	0.4	368	0	
Sept.	1.2			0.0			0.1		0.4	0.4	0.3	319	0	
Oct.	0.4		0.1	1.5	0.1		4.5		2.2	1.9	1.5	266	0	
Nov.	0.7		0.9	4.3	0.2		4.7		4.6	4.5	3.6	190	0.3	
Dec.	0.9		3	7.2	0.2		9.4		6.2	6.0	4.9	141	0.5	
Year	16.7	207	12	52	10		63		3.7	3.7	3.3	2986	143.8	1083.7

TABLE IXL

CLIMATIC TABLE FOR KUSTANAY

Latitude 53°13′N longitude 63°37′E elevation 171 m

Month	Air press. (mbar)		Wind		Global radiat.	Temperature (°C)				Humidity		Precipitation (mm)			
	sea level	station	preval. direct.	mean speed (m/s)	(kcal./cm^2)	mean	mean daily range	max.	min.	vapor press. (mm)	relat. (%)	mean	max.	min.	max. in 24 h
Jan.	1026.0		S	4.8	2.5	–17.8	9.7	6	–51	1.4	81	10			
Feb.	1025.1		S	4.9	4.3	–17.0	10.3	2	–48	1.3	81	9			
Mar.	1022.2		SW	4.8	9.2	–10.7	11.0	13	–37	2.0	82	9			
Apr.	1019.4		SW	4.6	11.4	1.8	11.5	31	–28	4.4	72	18	76	0	24
May	1015.1		N, NW	5.4	16.0	12.9	14.5	37	–9	6.6	56	26	73	2	42
June	1010.6		N	4.2	16.6	18.4	14.3	40	–4	9.7	57	35	111	6	54
July	1007.0		N	3.5	16.5	20.4	13.4	42	3	10.5	61	46	143	4	69
Aug.	1011.8		SW	3.5	12.5	18.1	13.7	39	0	9.5	62	34	104	4	55
Sept.	1015.2		SW	4.2	7.9	11.9	13.2	34	–9	6.7	66	25	117	0	36
Oct.	1019.5		SW	4.7	4.2	3.0	10.2	28	–22	4.3	74	28	69	1	28
Nov.	1026.3		SW	4.6	2.2	–6.4	8.9	18	–38	2.8	81	15			
Dec.	1026.4		SW	4.4	1.7	–14.9	9.4	3	–44	1.5	81	13			
Year	1018.7		SW	4.5	105.0	1.6		42	–51		71	268			

Month	Number of days							Snow cover	Cloudiness (tenths)			Mean sun-	Evaporation (mm)	
	wind (speed ≥15 m/s)	frost-free period	fog	precip. (≥0.1 mm)	thunder-storm	snow-cover	clouds (8–10 tenths)	max. 10-day average depth (cm)	07h	13h	19h	shine (h)	actual	potent.
Jan.	2.6		4			31	10.3		6.6	6.9	5.5	76	5	
Feb.	2.3		4			28	9.4		6.6	6.3	5.0	107	5	
Mar.	3.4		5		0.05	31	8.9		6.7	6.3	5.8	162	<10	
Apr.	2.6		4	7.1	0.1	8	7.7		6.2	6.8	6.2	220	>10	
May	3.6		0.3	9.9	3		7.2		5.6	6.9	6.0	282	>40	
June	2.8		0.7	9.9	6		5.7		5.2	6.9	6.2	307	70	
July	1.8		0.6	10.8	7		7.3		5.7	7.2	6.2	293	70	
Aug.	1.7		2	9.7	4		6.5		5.5	6.8	5.9	261	30	
Sept.	2.0		2	8.7	1		7.9		6.3	7.1	6.1	186	20	
Oct.	2.6		3	9.9			11.6		7.6	8.0	6.6	110	<10	
Nov.	2.2		4			16	12.0		7.2	7.3	6.2	75	5	
Dec.	2.6		5			31	11.1		7.2	7.3	6.1	58	5	
Year	30.2	117	35		21	150	106	24	6.4	7.0	6.0	2137	<300	600

TABLE L

CLIMATIC TABLE FOR KUTAISI

Latitude 42°16′N longitude 42°38′E elevation 116.1 m

Month	Air press. (mbar)		Wind		Global radiat.	Temperature (°C)				Humidity		Precipitation (mm)			
	sea level	station	preval. direct.	mean speed (m/s)	(kcal./cm^2)	mean	mean daily range	max.	min.	vapor press. (mm)	relat. (%)	mean	max.	min.	max. in 24 h
Jan.	1020.4	1001.3	E	5.6		4.7	6.5	22	–17	4.4	67	106	310	20	
Feb.	1019.0	999.9	E	5.6		5.5	7.1	26	–14	4.3	66	129	324	23	
Mar.	1017.0	998.3	E	5.9		9.1	8.5	32	–10	5.3	64	100	267	20	
Apr.	1014.9	996.5	E	5.7		12.9	10.5	34	–3	6.9	66	112	245	10	
May	1013.9	995.8	W	4.6		17.7	10.9	37	2	10.0	68	85	168	8	
June	1012.5	994.6	W	3.7		20.7	10.5	40	7	12.7	71	105	271	33	
July	1010.9	993.0	W	3.0		22.9	9.2	40	10	15.9	71	106	463	16	
Aug.	1011.4	993.8	W	3.4		23.4	9.4	40	10	15.5	74	86	205	2	
Sept.	1015.3	997.4	E	3.6		20.7	9.7	40	3	11.2	70	116	326	2	
Oct.	1018.5	1000.2	E	4.8		16.9	9.4	35	–3	9.5	68	108	305	25	
Nov.	1020.1	1001.5	E	7.2		11.5	7.7	29	–11	6.8	68	141	397	1	
Dec.	1020.2	1001.4	E	6.7		7.4	6.8	25	–14	5.2	68	139	321	11	
Year	1016.2	997.8	E	5.0		14.4		40	–17		69	1333	1971	576	

Month	Number of days							Snow cover	Cloudiness (tenths)			Mean sun-	Evaporation (mm)	
	wind (speed ≥15 m/s)	frost-free period	fog	precip. (≥0.1 mm)	thunder-storm	snow-cover	clouds (8–10 tenths)	max. 10-day average depth (cm)	07h	13h	19h	shine (h)	actual	potent.
Jan.	7.5		0.7	11.7	0.1		15.1		6.9	7.4	6.7			
Feb.	6.9		0.8	13.8	0.1		14.6		7.5	7.6	6.9			
Mar.	8.5		1	13.8	0		14.8		7.6	7.5	7.2			
Apr.	8.1		1	13.3	1		12.3		7.1	7.0	7.0			
May	7.1		3	12.1	5		12.0		6.7	6.5	7.0			
June	4.2		2	11.9	7		11.2		5.8	5.6	6.6			
July	2.7		0.8	13.6	7		12.4		6.3	5.7	7.1			
Aug.	4.3		1	11.6	5		12.1		5.5	5.1	6.4			
Sept.	5.4		0.6	10.8	3		9.6		5.1	4.8	5.4			
Oct.	7.2		0.5	10.3	2		10.2		5.8	5.8	5.1			
Nov.	9.6		0.4	11.8	0.2		12.2		6.2	6.3	5.3			
Dec.	9.6		0.9	14.5	0.4		12.8		6.7	6.9	6.0			
Year	81.1	288	13	149.2	31	25	149	19	6.4	6.4	6.4			

TABLE LI

CLIMATIC TABLE FOR KUYBYSHEV

Latitude 53°15′N longitude 50°27′E elevation 44 m

Month	Air press. (mbar)		Wind		Global radiat.	Temperature (°C)				Humidity		Precipitation (mm)			
	sea level	station	preval. direct.	mean speed (m/s)	(kcal./cm²)	mean	mean daily range	max.	min.	vapor press. (mm)	relat. (%)	mean	max.	min.	max. in 24 h
Jan.		1006.7	E	4.7	2.1	−13.8	6.6	4	−43	2.2	85	33	85	6	25
Feb.		1003.6	E	5.0	4.0	−13.0	7.4	4	−37	2.2	82	24	84	1	32
Mar.		1001.2	SW	5.3	7.6	−6.8	7.6	13	−29	3.5	82	30	92	3	30
Apr.		999.9	SW	4.6	11.2	4.6	9.0	31	−25	6.2	69	32	67	1	41
May		996.9	W	4.5	15.2	14.0	11.0	34	−8	8.5	54	43	100	6	30
June		996.2	W	3.8	16.2	18.7	11.3	38	0	12.1	56	40	168	3	59
July		993.3	N, W, NW	3.5	15.6	20.7	10.7	39	5	14.6	62	50	159	1	55
Aug.		997.7	E	3.5	12.9	19.0	11.0	38	2	13.3	62	44	96	2	54
Sept.		1001.9	W	3.8	8.2	12.4	10.1	37	−7	9.6	68	41	82	7	72
Oct.		1003.9	W	4.5	4.4	4.2	7.1	26	−21	6.3	76	46	108	7	27
Nov.		1007.1	SW	4.5	2.3	−4.1	5.8	15	−37	4.2	84	33	90	1	34
Dec.		1008.4	SW	4.8	1.5	−10.7	6.1	6	−38	2.9	86	33	80	4	26
Year		1002.2	E, SW, W	4.4	101.2	3.8	–	39	−43	7.1	72	449	649	293	72

Month	Number of days							Snow cover	Cloudiness (tenths)			Mean sunshine (h)	Evaporation (mm)	
	wind (speed ≥15 m/s)	frost-free period	fog	precip. (≥0.1 mm)	thunderstorm	snow-cover	clouds (8–10 tenths)	max. 10-day average depth (cm)	07h	13h	19h		actual	potent.
Jan.	1.1		6	17.0			16.2	26	7.7	7.7	6.9	59	0.2	
Feb.	1.1		4	12.7			12.3	31	7.2	7.2	6.2	92	0.5	
Mar.	1.7		6	12.8			13.7	31	7.7	7.1	6.7	140	0.8	
Apr.	1.2		4	9.2	0.4		11.0	11	6.8	7.2	6.5	209	25	
May	1.3		0.7	9.6	4		8.5		6.1	7.0	6.6	285	63	
June	0.7		0.6	9.8	7		5.7		5.1	6.2	5.8	329	60	
July	0.6		0.9	10.3	9		6.2		4.9	6.4	5.7	317	65	
Aug.	0.3		1	9.4	5		6.1		5.1	6.0	5.5	278	45	
Sept.	0.4		2	10.4	2		8.8		6.5	6.9	6.1	188	28	
Oct.	1.0		5	12.3	0.03		15.1		8.0	7.9	6.7	109	12	
Nov.	0.6		9	13.4			16.7	5	7.7	7.9	7.0	69	4	
Dec.	1.3		8	16.8			18.3	13	8.0	8.1	7.4	38	0.5	
Year	11.3	157	47.2	144	27.4	142	139	31	6.7	7.1	6.4	2113	304	684

TABLE LII

CLIMATIC TABLE FOR KZYL-ORDA

Latitude 44°46′N longitude 65°32′E elevation 129 m

Month	Air press. (mbar)		Wind		Global radiat.	Temperature (°C)				Humidity		Precipitation (mm)			
	sea level	station	preval. direct.	mean speed (m/s)	(kcal./cm²)	mean	mean daily range	max.	min.	vapor press. (mm)	relat. (%)	mean	max.	min.	max. in 24 h
Jan.		1010.9	NE	5.5	4.9	−9.6	8.7	15	−38	3.0	80	13	36	1	16
Feb.		1008.2	NE	5.7	7.2	−7.5	9.5	24	−33	3.4	78	15	55	0	16
Mar.		1005.8	NE	5.7	10.9	0.5	11.1	30	−30	4.8	70	14	43	0	24
Apr.		1002.2	NE	6.1	14.4	11.2	13.8	36	−11	6.5	49	14	44	0	25
May		999.4	NE	5.5	19.4	18.8	15.3	41	−2	8.5	39	11	63	0	41
June		993.6	NE	4.8	20.5	23.6	15.7	44	3	10.9	38	5	34	0	27
July		991.8	NE	4.1	20.0	24.6	15.9	46	8	12.7	38	4	45	0	35
Aug.		993.9	NE	4.1	17.0	22.5	16.3	44	5	11.0	39	3	28	0	22
Sept.		1001.1	NE	4.1	14.0	15.8	16.6	42	−6	7.9	44	4	30	0	23
Oct.		1006.4	NE	4.2	9.5	7.8	14.5	34	−12	5.7	54	7	31	0	20
Nov.		1011.6	NE	4.5	5.2	−0.6	10.7	26	−28	4.2	69	10	47	0	30
Dec.		1012.3	NE	5.0	2.8	−7.0	8.4	17	−33	3.3	78	14	45	1	20
Year		1003.1	NE	4.9	146.8	8.3	–	46	−38	6.8	56	114	187	46	41

Month	Number of days							Snow cover	Cloudiness (tenths)			Mean sunshine (h)	Evaporation (mm)	
	wind (speed ≥15 m/s)	frost-free period	fog	precip. (≥0.1 mm)	thunderstorm	snow-cover	clouds (8–10 tenths)	max. 10-day average depth (cm)	07h	13h	19h		actual	potent.
Jan.	2.9		5	7.8			11.7	6	6.5	6.6	5.4	113	3	
Feb.	3.1		4	6.7			9.1	6	6.4	6.0	5.1	143	5	
Mar.	3.4		2	6.5	0.3		7.3	2	6.4	6.2	5.5	204	13	
Apr.	6.4		0.5	4.6	0.7		5.9		5.3	6.0	5.6	253	13	
May	5.3		0.07	3.5	2		2.5		3.8	4.8	4.6	348	10	
June	2.7		0.03	1.9	2		0.8		2.8	3.8	3.6	374	5	
July	2.0			1.3	2		0.2		1.9	2.9	2.6	424	4	
Aug.	3.4		0.03	1.1	0.8		0.4		1.6	2.1	1.8	375	2	
Sept.	2.0		0.07	1.5	0.4		0.4		2.2	2.5	1.8	326	3	
Oct.	3.3		0.6	3.3	0.07		3.5		4.3	4.4	3.1	240	4	
Nov.	2.2		2	5.0	0.03		6.2		5.6	5.6	4.3	150	9	
Dec.	1.9		6	7.0			10.0	3	6.3	6.1	5.2	112	13	
Year	38.6	173	21	50	8	62	58	6	4.4	4.8	4.1	3062	84	924.9

TABLE LIII

CLIMATIC TABLE FOR LENINGRAD

Latitude 59°58′N longitude 30°18′E elevation 4 m

Month	Air press. (mbar)		Wind		Global radiat.	Temperature (°C)				Humidity		Precipitation (mm)			
	sea level	station	preval. direct.	mean speed (m/s)	(kcal./cm²)	mean	mean daily range	max.	min.	vapor press. (mm)	relat. (%)	mean	max.	min.	max. in 24 h
Jan.	1014.2	1013.7	S	3.4	0.4	−7.5	7.0	6	−36	2.5	86	36	92	5	23
Feb.	1014.5	1013.9	S	3.1	1.5	−7.9	8.6	6	−35	2.4	84	32	67	8	13
Mar.	1013.3	1021.7	W	3.0	4.7	−4.1	10.1	13	−28	2.9	79	25	72	1	19
Apr.	1014.2	1013.7	W	2.9	7.7	2.9	10.9	24	−17	5.2	73	34	72	8	24
May	1015.1	1014.6	W	2.8	11.6	9.6	11.6	31	−6	6.0	66	41	115	3	56
June	1011.7	1009.7	W	2.9	13.3	14.5	11.6	32	0	8.2	68	54	146	11	42
July	1010.5	1008.5	W	2.6	12.7	17.7	11.4	33	6	10.4	71	69	154	5	56
Aug.	1010.6	1008.5	W	2.4	8.7	15.7	11.2	32	3	10.0	76	77	203	1	76
Sept.	1013.1	1011.0	W	2.7	5.4	10.7	2.9	28	−3	7.7	81	58	178	11	34
Oct.	1013.8	1011.7	W	3.2	2.0	4.7	8.4	21	−13	5.6	74	52	110	11	28
Nov.	1013.7	1011.5	S	3.3	0.5	−0.6	6.0	12	−17	4.0	87	45	106	9	28
Dec.	1013.8	1011.7	SW	3.4	0.2	−5.3	5.6	9	−33	3.0	88	36	93	11	17
Year	1013.2	1011.1	W	3.0	68.7	4.2		33	−36		79	559	825	417	76

Month	Number of days							Snow cover	Cloudiness (tenths)			Mean sun-	Evaporation (mm)	
	wind (speed ≥15 m/s)	frost-free period	fog	precip. (≥0.1 mm)	thunder-storm	snow-cover	clouds (8–10 tenths)	max. 10-day average depth (cm)	07h	13h	19h	shine (h)	actual	potent.
Jan.	0.2		5	21.0		31	20.1		8.3	8.6	8.2	17	<10	
Feb.	0.0		6	17.7		28	17.2		8.0	8.1	7.4	38	<10	
Mar.	0.2		6	13.9		31	14.0		7.3	6.8	6.4	111	<10	
Apr.	0.0		5	12.7	0.1		11.7		6.8	7.1	6.8	166	30	
May	0.5		2	12.8	2		9.2		6.2	6.8	6.2	253	70	
June	0.1		1	13.8	3		8.8		6.1	6.7	6.1	263	>70	
July	0.1		1	13.9	5		8.5		6.2	6.7	6.1	277	>70	
Aug.	0.1		3	15.5	4		9.8		6.7	7.1	6.5	212	50	
Sept.	0.2		6	16.4	1		11.4		7.4	7.5	6.7	130	30	
Oct.	0.4		8	16.8			17.4		8.5	8.4	7.5	66	>10	
Nov.	0.3		7	18.6			21.5		8.7	8.8	8.4	21	<10	
Dec.	0.3		7	20.6		25	22.8		8.5	8.9	8.4	9	<10	
Year	2.4	159	57	194	15	132	172	32	7.4	7.6	7.1	1563	<400	500

TABLE LIV

CLIMATIC TABLE FOR LOUKHY

Latitude 66°05′N longitude 32°59′E elevation 94 m

Month	Air press. (mbar)		Wind		Global radiat.	Temperature (°C)				Humidity		Precipitation (mm)			
	sea level	station	preval. direct.	mean speed (m/s)	(kcal./cm²)	mean	mean daily range	max.	min.	vapor press. (mm)	relat. (%)	mean	max.	min.	max. in 24 h
Jan.			SW	3.0	0.8	−12.0		7	−44	2.5	86	27	61	10	13
Feb.			SW	3.0	1.6	−12.2		6	−46	2.4	85	23	45	7	8
Mar.			SW	3.2	4.8	−8.6		12	−42	2.7	78	20	47	2	10
Apr.			SW	3.1	7.8	−1.9		21	−36	4.2	72	25	60	2	15
May			SW	3.3	12.2	4.4		29	−14	5.8	66	34	66	8	29
June			SW	3.4	14.0	10.8		51	−7	9.2	66	50	116	10	29
July			NE	3.0	12.8	14.4		31	−3	11.8	70	64	170	15	47
Aug.			SW	2.8	8.6	12.3		29	−6	11.8	78	68	166	10	60
Sept.			SW	3.1	4.9	7.0		24	−11	8.8	83	51	120	17	41
Oct.			SW	3.3	2.3	0.6		16	−24	6.0	86	43	120	11	44
Nov.			SW	3.4	0.9	−5.0		9	−36	4.4	88	35	65	13	16
Dec.			SW	3.2	0.0	−8.9		8	−42	3.2	87	27	64	12	13
Year			SW	3.2	70.7	0.1		31	−46	6.1	79	467	650	324	60

Month	Number of days							Snow cover	Cloudiness (tenths)			Mean sun-	Evaporation (mm)	
	wind (speed ≥15 m/s)	frost-free period	fog	precip. (≥0.1 mm)	thunder-storm	snow-cover	clouds (8–10 tenths)	max. 10-day average depth (cm)	07h	13h	19h	shine (h)	actual	potent.
Jan.	0.3		3	18.5			15.5	33	7.3	8.0	7.0	6	0.2	
Feb.	0.4		3	16.2			13.4	44	7.8	7.7	6.7	47	0.8	
Mar.	0.7		3	14.4			11.3	47	7.3	6.6	6.3	142	0.9	
Apr.	0.3		2	13.0	0.03		12.3	45	7.3	7.5	7.4	197	12	
May	0.3		2	12.8	0.4		14.6	8	7.2	7.7	7.2	232	45	
June	0.1		1	13.9	3		12.7		7.0	7.6	7.0	269	60	
July	0.1		2	13.1	4		10.6		6.5	6.8	6.1	286	80	
Aug.	0.2		4	15.7	3		14.2		7.4	7.7	7.0	208	73	
Sept.	0.3		4	16.0	0.2		14.4		8.0	8.2	7.4	114	48	
Oct.	0.4		4	17.3			17.0	2	8.1	8.4	7.4	56	9	
Nov.	0.1		5	18.8			19.0	12	8.3	8.5	8.0	9	0.8	
Dec.	0.4		3	19.4			18.6	21	7.8	8.3	7.7	0	0.5	
Year	3.6	81	36	189.1	11	186	174	41	7.5	7.7	7.0	1566	330.2	374.0

TABLE LV

CLIMATIC TABLE FOR L'VOV

Latitude 49°49'N longitude 23°57'E elevation 325 m

Month	Air press. (mbar) sea level	station	Wind preval. direct.	mean speed (m/s)	Global radiat. (kcal./cm²)	Temperature (°C) mean	mean daily range	max.	min.	Humidity vapor press. (mm)	relat. (%)	Precipitation (mm) mean	max.	min.	max. in 24 h
Jan.	1020.2	982.6	W	4.8		−3.8	9.3	13	−33	3.3	86	33	60	6	21
Feb.	1018.6	981.1	W	5.0		−2.8	9.5	13	−32	3.3	84	33	67	4	35
Mar.	1015.6	978.9	W	4.8		1.8	10.8	23	−29	4.2	79	30	83	6	44
Apr.	1013.4	977.5	W	4.0		7.6	12.9	30	−12	5.7	76	47	126	11	39
May	1014.5	979.4	W	3.5		13.8	13.5	33	−2	8.2	72	66	172	9	52
June	1013.2	978.6	W, NW	3.3		16.6	13.8	33	1	10.4	74	77	217	16	72
July	1013.0	978.6	W	3.0		18.3	14.1	35	4	11.9	76	96	301	25	80
Aug.	1014.4	979.8	W	3.0		17.5	14.7	37	2	11.4	78	91	215	30	65
Sept.	1017.4	982.2	W	3.1		13.6	13.0	32	−2	9.1	79	47	234	8	71
Oct.	1017.7	981.9	W	3.8		8.8	12.3	27	−18	6.9	83	52	121	8	53
Nov.	1018.8	982.1	SE	4.6		2.7	9.0	24	−24	4.9	88	46	161	10	30
Dec.	1018.3	981.0	SE	4.7		−1.4	9.0	15	−24	3.8	89	37	70	8	19
Year	1016.3	980.3	W	4.0		7.7		37	−33		80	655	1320	437	

Month	Number of days: wind (speed ⩾15 m/s)	frost-free period	fog	precip. (⩾0.1 mm)	thunder-storm	snow-cover	clouds (8–10 tenths)	Snow cover max. 10-day average depth (cm)	Cloudiness (tenths) 07h	13h	19h	Mean sun-shine (h)	Evaporation (mm) actual	potent.
Jan.	5.0			17.4		31	17.4		7.7	7.3	7.1	58	<10	
Feb.	4.6			17.6		28	15.7		8.2	7.6	7.2	67	<10	
Mar.	4.6			15.0	0.06	1	14.3		7.7	7.0	6.6	116	>10	
Apr.	2.7			13.1	1		11.4		6.1	6.6	6.0	167	>50	
May	2.8			14.2	5		8.6		6.2	7.1	6.5	238	>70	
June	2.1			13.0	6		8.6		5.7	6.8	6.0	239	>70	
July	1.7			14.3	7		7.0		5.7	6.6	5.6	253	>70	
Aug.	1.7			12.5	5		7.4		5.8	6.3	5.4	227	>50	
Sept.	1.4			11.2	2		8.1		5.8	6.3	5.2	191	>30	
Oct.	2.9			11.2	0.1		12.4		7.1	6.5	5.1	139	>20	
Nov.	4.9			15.6			16.4		8.3	8.1	7.5	52	<10	
Dec.	4.1			17.1		8	19.6		8.1	7.9	7.5	45	<10	
Year	38.5	158		172	26	85	147		6.9	7.0	6.3	1792	>400	600

TABLE LVI

CLIMATIC TABLE FOR MALYE-KARMAEULY (NOVAYA ZEMLYA)

Latitude 72°23'N longitude 52°44'E elevation 16 m

Month	Air press. (mbar) sea level	station	Wind preval. direct.	mean speed (m/s)	Global radiat. (kcal./cm²)	Temperature (°C) mean	mean daily range	max.	min.	Humidity vapor press. (mm)	relat. (%)	Precipitation (mm) mean	max.	min.	max. in 24 h
Jan.		1005.7	SE	10.3	0.1	−15.0		1	−41	2.4	80	26			
Feb.		1006.9	SE	10.4	0.9	−14.5		1	−40	2.0	78	18			
Mar.		1006.2	SE	9.2	3.3	−15.4		1	−44	2.0	78	19			
Apr.		1001.6	SE	8.1	8.0	−10.8		6	−32	3.1	80	18			
May		1007.9	N	7.2	10.9	−4.5		13	−24	3.9	80	20			
June		1012.0	N	6.9	12.6	1.4		20	−17	5.9	82	24	59	2	28
July		1009.6	N	6.3	12.2	6.4		24	−10	8.2	82	30	82	2	27
Aug.		1011.3	N	6.9	7.9	6.3		20	−1	8.2	84	36	95	7	27
Sept.		1008.4	SE	7.1	3.9	2.7		17	−13	6.6	83	41	88	8	28
Oct.		1002.6	SE	7.9	1.5	−2.7		10	−18	4.7	82	35			
Nov.		1006.9	SE	9.0	0.7	−9.0		3	−34	3.6	82	26			
Dec.		1008.0	SE	10.4	0.5	−13.0		2	−36	3.0	82	24			
Year		1007.3	SE	8.3	62.5	−5.7		24	−44	4.5	81	317			28

Month	Number of days: wind (speed ⩾15 m/s)	frost-free period	fog	precip. (⩾0.1 mm)	thunder-storm	snow-cover	clouds (8–10 tenths)	Snow cover max. 10-day average depth (cm)	Cloudiness (tenths) mean	Mean sun-shine (h)	Evaporation (mm) actual	potent.
Jan.	15.7		0.5	19.7			14.0		7.1	0	0	
Feb.	13.8		1	14.0			11.6		6.4	21	0	
Mar.	13.9		0.7	18.3			12.5		6.7	95	0	
Apr.	11.0		3	15.3			17.4		7.6	155	0	
May	8.2		5	14.2			21.4		8.3	175	0	
June	7.0		7	12.5	0.3		19.1		8.1	178	15	
July	5.5		10	12.7	0.5		18.5		7.6	194	18	
Aug.	5.4		8	16.9	0.5		22.4		8.5	146	15	
Sept.	7.4		5	17.8	0.04		21.2		8.6	76	9	
Oct.	10.2		3	17.0			19.2		8.2	44	8	
Nov.	12.1		2	14.8			17.2		7.8	2	0	
Dec.	14.0		1	17.9			16.7		7.3	0	0	
Year	124.2	68	46.2	191	1.3	238	211.2	31	7.7	1086	65	

TABLE LVII

CLIMATIC TABLE FOR MINSK

Latitude 53°52′N longitude 27°32′E elevation 234 m

Month	Air press. (mbar)		Wind		Global radiat.	Temperature (°C)				Humidity		Precipitation (mm)			
	sea level	station	preval. direct.	mean speed (m/s)	(kcal./cm²)	mean	mean daily range	max.	min.	vapor press. (mm)	relat. (%)	mean	max.	min.	max. in 24 h
Jan.	1017.9		SW	4.9	1.6	-6.6	9.1	6	-39	2.8	88	34	101	6	21
Feb.	1015.8		SE	5.0	2.9	-6.2	10.8	8	-35	2.8	85	30	70	15	24
Mar.	1017.1		W	4.7	7.2	-2.1	11.3	19	-39	3.5	80	26	87	6	18
Apr.	1014.9		SE	4.4	9.0	5.1	12.5	26	-18	5.0	74	42	109	9	33
May	1015.5		NW	4.1	13.8	12.5	13.7	31	-6	7.8	67	59	120	12	62
June	1013.8		NW	3.9	14.9	15.6	13.8	33	0	9.8	69	72	221	18	61
July	1012.2		NW	3.6	14.1	17.6	13.8	35	5	11.5	73	83	206	13	57
Aug.	1013.4		W, NW	3.5	10.9	15.9	14.2	34	2	10.7	76	81	158	19	61
Sept.	1015.9		W	3.7	6.9	11.4	14.5	29	-5	8.4	80	62	129	5	40
Oct.	1017.5		W	4.3	3.7	5.7	11.6	25	-20	6.2	85	47	108	3	30
Nov.	1019.2		SE	4.8	1.5	-0.2	8.4	16	-27	4.4	90	40	140	5	43
Dec.	1014.6		SE	5.0	0.9	-4.7	7.4	10	-31	3.2	89	30	94	7	26
Year	1015.7		W	4.3	87.4	5.3		35	-39		80	606	896	369	62

Month	Number of days							Snow cover	Cloudiness (tenths)			Mean sun-shine	Evaporation (mm)	
	wind (speed ⩾15 m/s)	frost-free period	fog	precip. (⩾0.1 mm)	thunder-storm	snow-cover	clouds (8–10 tenths)	max. 10-day average depth (cm)	07h	13h	19h	(h)	actual	potent.
Jan.	0.8		10	18.5	0.04	31	21.3		8.5	8.4	7.9	40	<10	
Feb.	0.6		7	16.6		28	18.4		8.7	8.0	7.6	60	<10	
Mar.	0.8		8	15.2	0.1	24	16.8		7.6	7.0	6.6	146	>10	
Apr.	0.7		4	14.5	1		11.6		6.9	7.3	6.7	193	>50	
May	0.4		2	13.6	4		10.4		6.0	7.1	6.3	261	>70	
June	0.6		0.9	14.6	7		9.5		5.7	7.2	6.1	286	>70	
July	0.3		2	15.6	7		8.5		5.8	7.4	6.5	272	>70	
Aug.	0.3		2	15.2	6		9.5		6.6	7.3	6.2	238	>50	
Sept.	0.4		6	14.0	2		10.3		7.3	7.3	6.0	170	>30	
Oct.	0.4		10	15.9	0.2		16.8		8.4	8.2	6.9	92	>20	
Nov.	0.4		10	19.0			21.4		9.1	9.0	8.4	32	<10	
Dec.	0.5		10	19.7	0.04	19	23.8		8.9	8.8	8.4	25	<10	
Year	6.2	152	72	192	27	111	178	34	7.5	7.8	7.0	1815	>400	600

TABLE LVIII

CLIMATIC TABLE FOR MINUSINSK

Latitude 53°42′N longitude 91°42′E elevation 251 m

Month	Air press. (mbar)		Wind		Global radiat.	Temperature (°C)				Humidity		Precipitation (mm)			
	sea level	station	preval. direct.	mean speed (m/s)	(kcal./cm²)	mean	mean daily range	max.	min.	vapor press. (mm)	relat. (%)	mean	max.	min.	max. in 24 h
Jan.	1032.4	998.1	SW	2.0		-20.3	12.4	6	-52	1.0	77	9	18	0	8
Feb.	1032.4	998.1	SW	1.5		-19.9	13.9	8	-50	1.1	78	9	23	0	7
Mar.	1027.9	994.9	SW	2.2		-10.9	14.1	21	-47	1.9	74	10	22	0	15
Apr.	1021.0	989.8	SW	3.2		2.2	13.3	27	-32	3.7	63	15	34	1	12
May	1015.4	985.3	SW	3.3		10.3	14.7	34	-18	5.8	56	32	72	7	25
June	1009.5	980.3	SW	2.3		17.1	14.5	37	-4	9.6	62	54	99	6	52
July	1006.6	977.8	SW	1.9		19.7	13.5	39	2	12.3	67	66	137	18	60
Aug.	1009.9	980.7	SW	2.0		16.6	13.7	38	-3	10.6	70	43	118	10	49
Sept.	1017.9	987.7	SW	2.2		9.5	13.4	32	-12	6.4	75	33	85	7	29
Oct.	1023.6	992.2	SW	2.8		1.0	11.6	26	-31	4.3	73	16	59	1	15
Nov.	1028.1	995.4	SW	3.3		-9.0	10.5	15	-48	2.3	76	17	43	1	12
Dec.	1032.1	998.1	SW	2.2		-17.8	11.3	7	-53	1.2	77	13	24	0	28
Year	1021.4	989.8	SW	2.4		-0.1		39	-53		71	316	493	155	60

Month	Number of days							Snow cover	Cloudiness (tenths)			Mean sun-shine	Evaporation (mm)	
	wind (speed ⩾15 m/s)	frost-free period	fog	precip. (⩾0.1 mm)	thunder-storm	snow-cover	clouds (8–10 tenths)	max. 10-day average depth (cm)	07h	13h	19h	(h)	actual	potent.
Jan.	0.6		6			31	10		6.2	6.8	5.3	57	5	
Feb.	0.5		5			29	10		7.0	6.5	5.1	92	5	
Mar.	0.8		4			24	11		7.2	6.5	5.7	146	5	
Apr.	2.3		1	6.7	0.2		11		7.0	7.0	7.0	164	>10	
May	2.5		0.7	11.6	2.0		12		6.8	7.2	6.9	199	>40	
June	1.3		1	12.2	4.0		9		6.5	6.7	6.8	232	>70	
July	0.8		2	12.5	6.0		10		6.3	6.5	6.9	251	70	
Aug.	0.4		3	14.4	4.0		8		6.4	6.3	6.4	225	50	
Sept.	0.7		7	12.7	0.4		12		6.8	6.7	6.5	162	30	
Oct.	1.5		5	10.0	0.06		14		7.8	7.5	6.1	100	5	
Nov.	1.2		4			15	17		8.1	7.8	6.4	51	5	
Dec.	1.4		7			31	15		7.1	7.5	6.2	37	5	
Year	14	116	45		17.0	144	139	24	6.9	6.9	6.3	1716	300	400

TABLE LIX

CLIMATIC TABLE FOR MOSCOW

Latitude 55°45′N longitude 37°34′E elevation 156 m

Month	Air press. (mbar)		Wind		Global radiat.	Temperature (°C)				Humidity		Precipitation (mm)			
	sea level	station	preval. direct.	mean speed (m/s)	(kcal./cm²)	mean	mean daily range	max.	min.	vapor press. (mm)	relat. (%)	mean	max.	min.	max. in 24 h
Jan.	1019.2	997.9	W	5.0	1.4	−10.3	7.8	4	−42	2.2	85	31	67	8	20
Feb.	1019.0	997.7	SE	4.9	2.9	−9.7	12.0	6	−40	2.1	82	28	75	7	21
Mar.	1017.1	996.2	W	5.2	7.3	−5.0	11.0	15	−32	2.8	77	33	98	6	21
Apr.	1016.4	996.2	SE	4.7	8.8	3.7	10.1	28	−19	4.7	71	35	100	3	25
May	1015.8	996.3	N	4.5	13.0	11.7	13.3	32	−7	6.8	64	52	103	2	33
June	1011.5	992.3	NW	3.9	15.6	15.4	13.3	35	−2	9.4	66	67	174	5	51
July	1010.5	991.5	NW	3.5	15.3	17.8	11.8	37	4	11.2	69	74	169	25	79
Aug.	1011.8	992.7	NW	3.5	11.8	15.8	13.0	37	1	10.2	74	74	164	1	63
Sept.	1015.8	996.2	W	4.3	6.7	10.4	13.1	32	−5	7.7	79	58	131	7	52
Oct.	1018.0	997.9	W	4.7	3.0	4.1	11.8	24	−20	5.3	82	51	143	6	42
Nov.	1018.6	997.9	SW	4.9	1.6	−2.3	7.3	13	−33	3.7	85	36	114	11	22
Dec.	1019.2	997.9	S	4.7	1.0	−8.0	7.0	8	−39	2.5	86	36	82	7	26
Year	1016.0	995.9		4.5	88.4	3.6		37	−42		77	575	819	354	79

Month	Number of days							Snow cover	Cloudiness (tenths)			Mean sun-	Evaporation (mm)	
	wind (speed ≥15 m/s)	frost-free period	fog	precip. (≥0.1 mm)	thunder-storm	snow-cover	clouds (8–10 tenths)	max. 10-day average depth (cm)	07h	13h	19h	shine (h)	actual	potent.
Jan.	1.7		3	17		31	19.5		8.4	8.5	7.8	30	5	
Feb.	1.3		3	15		28	16.7		8.6	8.0	7.5	58	5	
Mar.	1.8		3	14	0.04	31	14.3		7.6	7.5	6.8	113	10	
Apr.	1.3		2	13	0.5	6	11.5		6.8	7.4	6.8	161	30	
May	1.5		0.7	12	4		10.0		6.2	7.2	6.5	242	70	
June	1.4		0.6	15	6		8.0		5.5	7.3	6.1	256	70	
July	1.1		1	16	8		7.5		5.9	7.3	6.4	258	70	
Aug.	0.7		3	16	4		8.8		6.4	7.4	6.5	218	40	
Sept.	0.9		3	17	1		11.7		7.4	7.6	6.4	136	20	
Oct.	1.3		4	16	0.1		18.3		8.6	8.4	7.4	73	10	
Nov.	1.5		4	17		4	20.9		8.5	8.6	8.2	32	5	
Dec.	1.6		3	19		31	23.4		8.7	8.9	8.5	20	5	
Year	16.1	141	31	187	23	146	171	52	7.4	7.8	7.1	1597	300	600

TABLE LX

CLIMATIC TABLE FOR MURMANSK

Latitude 68°58′N longitude 33°03′E elevation 46 m

Month	Air press. (mbar)		Wind		Global radiat.	Temperature (°C)				Humidity		Precipitation (mm)			
	sea level	station	preval. direct.	mean speed (m/s)	(kcal./cm²)	mean	mean daily range	max.	min.	vapor press. (mm)	relat. (%)	mean	max.	min.	max. in 24 h
Jan.	1007.2	1004.3	S	5.7		−9.9	7.5	7	−37	2.2	83	19			
Feb.	1008.0	1005.1	S	5.3		−9.9	6.0	6	−38	2.0	82	16			
Mar.	1009.6	1006.7	S	4.7		−7.0	8.5	9	−36	2.4	78	18			
Apr.	1013.0	1010.2	S	4.2		−1.2	9.8	17	−27	3.3	73	19			
May	1014.8	1012.1	N	4.4		3.5	10.8	27	−12	4.5	69	25	67	7	
June	1012.5	1009.8	N	4.1		8.9	12.3	31	−4	6.2	68	40	134	4	
July	1011.7	1009.0	N	3.5		12.8	12.8	33	0	8.2	73	54	136	10	
Aug.	1011.6	1008.9	S	3.4		10.9	12.9	30	−1	8.0	77	60	139	16	
Sept.	1009.3	1006.6	S	4.0		6.4	8.3	24	−10	6.2	81	44	108	18	
Oct.	1009.1	1006.3	S	4.2		0.3	6.0	14	−21	4.3	83	30	90	8	
Nov.	1008.1	1005.3	S	4.7		−5.1	6.6	9	−32	3.1	84	28			
Dec.	1009.1	1006.3	S	5.0		−8.6	7.0	7	−37	2.4	84	33			
Year	1010.3	1007.6	S	4.4		0.1		33	−38		78	376			

Month	Number of days							Snow cover	Cloudiness (tenths)			Mean sun-	Evaporation (mm)	
	wind (speed ≥15 m/s)	frost-free period	fog	precip. (≥0.1 mm)	thunder-storm	snow-cover	clouds (8–10 tenths)	max. 10-day average depth (cm)	07h	13h	19h	shine (h)	actual	potent.
Jan.	8.7		0.9			31	14.2		7.6	7.8	7.1	1	<10	
Feb.	6.4		1			28	13.0		7.5	7.5	6.9	32	<10	
Mar.	7.2		3			31	12.8		7.7	6.9	7.0	121	<10	
Apr.	3.5		1	14.6	0.02	30	15.0		7.9	7.4	7.6	203	<10	
May	2.6		1	15.2	0.05	6	19.6		8.3	8.0	8.1	197	30	
June	2.3		2	15.1	1		'6.6		7.8	7.8	7.3	246	>70	
July	1.4		3	14.1	2		15.7		7.6	7.3	6.7	236	70	
Aug.	1.3		6	16.6	1		18.4		8.0	8.2	7.6	146	50	
Sept.	2.5		6	17.4	0.1		19.5		8.6	8.5	8.1	73	20	
Oct.	4.9		4	17.7			18.6		8.5	8.4	7.7	43	<10	
Nov.	6.0		3			20	18.2		8.3	8.2	7.9	3	<10	
Dec.	7.4		2			31	16.1		7.6	8.2	7.5	0	<10	
Year	54.2	104	33		4	195	ı98	64	7.9	7.9	7.5	1297	300	300

TABLE LXI

CLIMATIC TABLE FOR MYS CHELYUSKIN

Latitude 77°43′N longitude 104°17′E elevation 6 m

Month	Air press. (mbar)		Wind		Global radiat.	Temperature (°C)				Humidity		Precipitation (mm)			
	sea level	station	preval. direct.	mean speed (m/s)	(kcal./cm²)	mean	mean daily range	max.	min.	vapor press. (mm)	relat. (%)	mean	max.	min.	max. in 24 h
Jan.	1013.3	1012.3	SW	7.2	0.0	−31.1		0	−49	0.6	84	12			3
Feb.	1015.7	1014.7	SW	6.4	0.1	−29.7		0	−46	0.6	86	14			1
Mar.	1016.8	1015.8	E	6.6	2.6	−28.5		−3	−42	0.4	85	18			4
Apr.	1014.3	1013.4	E, SW	5.7	9.3	−22.6		1	−42	0.9	85	21			4
May	1013.0	1012.1	E	6.1	16.0	−11.2		8	−26	2.1	88	24			4
June	1011.6	1010.7	E	6.1	16.7	−2.8		12	−13	3.9	91	25	61	2	24
July	1010.6	1009.7	E	6.0	12.5	0.8		24	−6	4.7	93	27	58	2	24
Aug.	1009.9	1009.0	W	6.2	6.5	0.6		20	−7	4.6	94	28	76	7	33
Sept.	1009.9	1009.0	SW	6.7	2.5	−2.9		13	−16	3.8	91	24	55	3	29
Oct.	1008.0	1007.1	SW	6.6	0.4	−10.3		4	−30	2.0	87	21			10
Nov.	1012.8	1011.9	SW	6.9	0.0	−22.6		0	−39	0.9	86	16			3
Dec.	1016.2	1015.1	SW	7.6	0.0	−27.4		−1	−44	0.6	84	14			2
Year	1012.7	1011.7	E	6.5	66.6	−15.6		24	−49		88	294			33

Month	Number of days							Snow cover max. 10-day average depth (cm)	Cloudiness (tenths)			Mean sun-shine (h)	Evaporation (mm)	
	wind (speed ≥15 m/s)	frost-free period	fog	precip. (≥0.1 mm)	thunder-storm	snow-cover	clouds (8–10 tenths)		07h	13h	19h		actual	potent.
Jan.	7.1		2	13.0		31	10						<5	
Feb.	6.1		3	10.9		28	7					2	<5	
Mar.	5.7		5	8.5		31	8					95	<10	
Apr.	4.2		6	8.2		30	12					223	<10	
May	3.4		6	8.3		31	20					197	<10	
June	3.6		12	7.9		1	20					172	<10	
July	3.5		21	11.4			22					187	<70	
Aug.	2.8		22	16.1			25					91	<30	
Sept.	4.5		12	17.4		14	25					28	<10	
Oct.	6.1		4	16.6		31	22					7	<10	
Nov.	4.8		3	14.6		30	13						<10	
Dec.	6.6		1	13.7		31	9						<5	
Year	58.4	0	97	146.6	0	287	193	33				1002	<200	

TABLE LXII

CLIMATIC TABLE FOR MYS SHMIDTA

Latitude 68°55′N longitude 179°29′W elevation 7 m

Month	Air press. (mbar)		Wind		Global radiat.	Temperature (°C)				Humidity		Precipitation (mm)			
	sea level	station	preval. direct.	mean speed (m/s)	(kcal./cm²)	mean	mean daily range	max.	min.	vapor press. (mm)	relat. (%)	mean	max.	min.	max. in 24 h
Jan.	1021.5		W	6.7	0.0	−26.4		5	−47	0.7	85	23			5
Feb.	1022.5		W	6.1	1.1	−27.3		5	−46	0.6	86	21			6
Mar.	1021.0		W, SW	6.5	5.2	−25.7		3	−45	0.6	86	17			6
Apr.	1018.1		SE	5.1	10.5	−17.6		5	−40	1.0	88	17			7
May	1016.8		SE	4.8	15.2	−7.8		11	−30	2.6	90	19			6
June	1011.8		SE	4.8	15.6	1.3		23	−12	4.6	89	25	36	2	16
July	1009.9		SE	4.8	12.3	3.6		26	−5	5.6	90	33	78	2	26
Aug.	1011.0		NW	5.0	7.7	2.8		26	−7	5.3	92	42	125	14	39
Sept.	1014.0		NW	5.5	4.3	−0.3		17	−14	4.2	90	30	161	3	21
Oct.	1011.5		NW	7.0	1.7	−7.7		12	−30	2.6	86	25			14
Nov.	1016.5		W	7.3	0.2	−16.5		6	−39	1.5	87	20			19
Dec.	1018.5		W	6.5	0.0	−23.4		4	−43	0.8	85	24			6
Year	1016.0		NW	5.8	73.8	−12.1		26	−47		88	296			39

Month	Number of days							Snow cover max. 10-day average depth (cm)	Cloudiness (tenths)			Mean sun-shine (h)	Evaporation (mm)	
	wind (speed ≥15 m/s)	frost-free period	fog	precip. (≥0.1 mm)	thunder-storm	snow-cover	clouds (8–10 tenths)		07h	13h	19h		actual	potent.
Jan.	6.5		3	8.8		31	10					1	<5	
Feb.	5.1		3	9.3		28	11					45	<5	
Mar.	5.1		3	10.6		31	11					146	<5	
Apr.	3.0		5	8.6		30	13					200	<10	
May	1.3		12	9.4		31	19					169	<10	
June	0.8		14	8.6	0.05	20	18					225	<40	
July	1.7		18	12.0	0.2		20					197	<60	
Aug.	1.9		18	15.6	0.05		22					103	<20	
Sept.	2.9		9	15.2		3	22					74	<10	
Oct.	7.2		4	15.1		31	21					45	<10	
Nov.	6.5		3	13.1		30	17					4	<5	
Dec.	6.0		4	11.2		31	14						<5	
Year	48	22	96	137.5	0.3	254	198	50				1209	<200	

TABLE LXIII

CLIMATIC TABLE FOR NAR'YAN-MAR

Latitude 67°39'N longitude 53°01'E elevation 6.5 m

Month	Air press. (mbar) sea level	Air press. (mbar) station	Wind preval. direct.	Wind mean speed (m/s)	Global radiat. (kcal./cm²)	Temperature (°C) mean	Temperature (°C) mean daily range	Temperature (°C) max.	Temperature (°C) min.	Humidity vapor press. (mm)	Humidity relat. (%)	Precipitation (mm) mean	Precipitation (mm) max.	Precipitation (mm) min.	Precipitation (mm) max. in 24 h
Jan.	1010.8	1009.4	S	5.4		–17.3	8.7	5	–51	1.5	85	20			
Feb.	1011.3	1009.9	S	5.0		–16.7	8.7	3	–47	1.6	84	14			
Mar.	1011.9	1010.5	SW	5.2		–14.2	9.6	5	–47	1.6	83	17			
Apr.	1013.8	1012.5	SW	5.0		–7.2	8.2	15	–37	2.6	82	21			
May	1013.8	1012.5	NE	5.2		–0.6	7.0	26	–24	3.7	78	24			
June	1011.4	1010.1	NE	5.5		7.1	9.0	33	–7	5.9	74	38			
July	1010.5	1009.3	NE	4.8		12.0	9.0	33	–2	8.2	76	41			
Aug.	1010.4	1009.1	NE	4.4		10.2	8.1	33	–3	8.0	82	59			
Sept.	1009.0	1007.7	S	4.8		5.5	5.9	23	–11	6.2	76	52			
Oct.	1009.2	1007.9	SW	4.8		–1.7	4.7	14	–26	4.0	88	39			
Nov.	1008.8	1007.4	SW	4.8		–9.2	6.6	6	–45	2.5	88	41			
Dec.	1011.7	1010.3	SW	5.1		–14.5	7.8	7	–48	1.8	86	22			
Year	1011.0	1009.7	SW	5.0		–3.9		33	–51		83	378			

Month	Number of days: wind (speed ⩾15 m/s)	frost-free period	fog	precip. (⩾0.1 mm)	thunder-storm	snow-cover	clouds (8–10 tenths)	Snow cover max. 10-day average depth (cm)	Cloudiness (tenths) 07h	13h	19h	Mean sunshine (h)	Evaporation (mm) actual	potent.
Jan.	4.1		5	18.6		31	16.4		7.4	8.3	7.2	3	<5	
Feb.	2.9		4	16.2		28	12.9		7.5	7.2	6.5	43	<5	
Mar.	3.6		5	16.7		31	11.6		7.5	6.8	6.8	146	<10	
Apr.	2.7		4	14.8		30	15.2		7.8	7.3	7.4	185	<10	
May	2.9		4	15.7	0.5	12	20.9		8.4	8.1	8.1	189	20	
June	3.4		3	13.9	2		16.8		8.2	7.8	7.4	230	70	
July	1.6		3	11.9	3		15.7		7.6	7.4	7.1	269	70	
Aug.	1.4		4	13.5	2		19.1		8.2	8.1	7.7	160	50	
Sept.	2.2		3	19.3	0.4		20.9		8.9	8.8	8.5	74	20	
Oct.	2.2		5	20.5		4	21.2		9.0	8.8	8.2	39	<10	
Nov.	3.2		5	19.5		30	18.9		8.2	8.5	7.9	10	<5	
Dec.	3.1		5	18.5		31	18.2		7.6	8.5	7.5	0	<5	
Year	33.3	92	50	201	8	221	208	64	8.0	8.0	7.5	1348	<300	300

TABLE LXIV

CLIMATIC TABLE FOR NIKOLAYEVSK-NA-AMURE

Latitude 53°09'N longitude 140°42'E elevation 46 m

Month	Air press. (mbar) sea level	Air press. (mbar) station	Wind preval. direct.	Wind mean speed (m/s)	Global radiat. (kcal./cm²)	Temperature (°C) mean	Temperature (°C) mean daily range	Temperature (°C) max.	Temperature (°C) min.	Humidity vapor press. (mm)	Humidity relat. (%)	Precipitation (mm) mean	Precipitation (mm) max.	Precipitation (mm) min.	Precipitation (mm) max. in 24 h
Jan.	1016.4	1014.2	W	4.0		–25.8	9.7	0	–47	0.6	77	19	94	0	38
Feb.	1017.2	1015.0	W	3.7		–20.3	11.4	2	–46	0.8	77	20	82	0	24
Mar.	1014.3	1012.2	W	3.8		–11.1	11.5	10	–38	1.4	74	18	76	0	24
Apr.	1010.2	1008.2	E	3.7		0.7	9.5	20	–29	3.0	75	32	85	5	29
May	1009.8	1007.8	E	3.8		8.8	11.3	29	–12	4.7	78	43	100	1	24
June	1008.1	1006.2	E	3.8		14.8	11.0	32	–4	7.9	78	41	105	0	55
July	1006.9	1005.0	E	3.6		19.6	9.7	34	1	11.1	80	55	208	1	60
Aug.	1008.5	1006.6	E	3.3		18.0	8.9	35	1	11.2	82	72	170	6	74
Sept.	1010.8	1008.9	E	3.3		12.3	8.7	30	–6	8.0	82	73	200	2	54
Oct.	1011.9	1009.9	W	3.5		3.4	8.5	22	–25	4.1	76	55	212	10	82
Nov.	1012.6	1010.5	W	4.1		–10.0	8.4	11	–34	1.8	76	43	152	3	40
Dec.	1013.3	1010.5	W	4.2		–22.4	8.8	5	–44	0.9	79	31	112	0	60
Year	1011.7	1009.7	W	3.7		–1.1		35	–47		78	503	703	271	82

Month	Number of days: wind (speed ⩾15 m/s)	frost-free period	fog	precip. (⩾0.1 mm)	thunder-storm	snow-cover	clouds (8–10 tenths)	Snow cover max. 10-day average depth (cm)	Cloudiness (tenths) 07h	13h	19h	Mean sunshine (h)	Evaporation (mm) actual	potent.
Jan.	1.8		0.4	12		31	7.9		4.5	5.1	4.2	122	5	
Feb.	3.2		0.4	10		28	7.5		5.4	5.7	4.6	166	5	
Mar.	2.0		1	12		31	9.7		5.7	5.9	5.5	217	5	
Apr.	3.7		3	12	0.3	30	12.6		6.9	7.0	7.1	182	10	
May	4.5		5	11	0.4	4	15.3		7.8	7.4	7.3	206	<40	
June	3.5		2	10	2		13.2		7.6	7.0	6.8	216	50	
July	1.9		2	11	2		13.8		7.5	6.9	6.6	214	<60	
Aug.	2.2		3	13	2		13.4		7.7	7.1	6.7	224	65	
Sept.	2.4		4	15	1		13.5		7.6	7.4	6.7	166	45	
Oct.	3.7		3	14			10.3		6.9	7.0	5.6	164	15	
Nov.	4.4		0.9	12		29	9.4		6.4	6.4	5.5	121	5	
Dec.	5.1		0.3	12		31	9.1		5.3	6.1	4.7	106	5	
Year	38.4	119	25	144	7	187	136	89 c...	6.6	6.6	5.9	2106	300	400

TABLE LXV

CLIMATIC TABLE FOR NOVGOROD

Latitude 58°21′N longitude 3 °15′E elevation 25 m

Month	Air press. (mbar)		Wind		Global radiat.	Temperature (°C)				Humidity		Precipitation (mm)			
	sea level	station	preval. direct.	mean speed (m/s)	(kcal./cm²)	mean	mean daily range	max.	min.	vapor press. (mm)	relat. (%)	mean	max.	min.	max. in 24 h
Jan.	1015.8	1011.4	S	5.8		−8.4		6	−49	2.5	86	28	90	5	22
Feb.	1015.6	1011.3	S	5.1		−8.4		6	−41	2.4	85	26	103	3	15
Mar.	1014.2	1009.9	S	4.7		−4.5		13	−34	2.8	81	24	54	4	14
Apr.	1014.7	1010.6	S	4.3		3.4		26	−24	4.6	76	37	86	1	20
May	1015.2	1011.3	NE	4.6		10.6		31	−9	6.6	67	46	151	4	58
June	1011.9	1007.9	W, NW	4.5		14.6		32	−4	8.9	71	67	156	22	50
July	1010.6	1006.6	W	3.9		17.0		34	−1	10.9	76	74	172	6	65
Aug.	1010.9	1006.9	SW	3.9		14.7		33	−4	10.1	81	86	174	10	46
Sept.	1014.1	1010.1	SW	4.3		9.8		29	−13	7.7	85	66	163	10	55
Oct.	1015.1	1011.0	SW	5.0		4.0		21	−25	5.6	87	50	114	12	31
Nov.	1015.4	1011.1	S	5.7		−1.4		14	−27	4.0	89	44	89	9	22
Dec.	1015.4	1011.0	S	5.6		−6.3		10	−39	3.0	88	33	78	8	17
Year	1014.1	1009.9	S	4.8		3.8		34	−45		81	581	778	380	65

Month	Number of days							Snow cover max. 10-day average depth (cm)	Cloudiness (tenths)			Mean sunshine (h)	Evaporation (mm)	
	wind (speed ≥15 m/s)	frost-free period	fog	precip. (≥0.1 mm)	thunderstorm	snow-cover	clouds (8–10 tenths)		07h	13h	19h		actual	potent.
Jan.	2.6		2	17.2		31	19.5					28	5	
Feb.	1.7		2	14.7		28	15.2					56	5	
Mar.	1.4		3	13.1	0.1	31	11.3					145	5	
Apr.	0.3		3	12.1	1	4	10.8					183	<30	
May	1.8		1	11.5	2		8.9					252	>70	
June	0.9		1	13.9	3		8.3					283	>70	
July	0.9		2	13.9	5		7.8					276	>70	
Aug.	0.8		6	15.3	3		9.2					222	>40	
Sept.	0.8		6	15.8	1		11.1					145	>20	
Oct.	1.3		6	16.4	0.1		16.9					61	>10	
Nov.	0.6		6	17.7			21.0					23	<10	
Dec.	1.6		4	18.2		25	21.8					17	5	
Year	14.7	99	42	180	15	137	162	42				1691	>300	500

TABLE LXVI

CLIMATIC TABLE FOR NOVOROSSIYSK

Latitude 44°42′N longitude 37°48′E elevation 37 m

Month	Air press. (mbar)		Wind		Global radiat.	Temperature (°C)				Humidity		Precipitation (mm)			
	sea level	station	preval. direct.	mean speed (m/s)	(kcal./cm²)	mean	mean daily range	max.	min.	vapor press. (mm)	relat. (%)	mean	max.	min.	max. in 24 h
Jan.				6.4		2.5				4.8	77	64			
Feb.				6.1		2.6				4.3	75	60			
Mar.				5.6		6.1				6.3	74	48			
Apr.				4.9		10.6				7.1	73	49			
May				3.9		15.9				10.0	75	41			
June				3.4		20.2				12.8	72	54			
July				3.5		23.6				14.5	69	70			
Aug.				4.2		23.7				11.0	66	40			
Sept.				4.8		19.2				11.1	70	58			
Oct.				5.1		14.7				9.0	71	48			
Nov.				6.2		8.6				6.8	73	70			
Dec.				6.1		4.8				5.3	76	86			
Year				5.0		12.7		37	−24		72	688			

Month	Number of days							Snow cover max. 10-day average depth (cm)
	wind (speed ≥15 m/s)	frost-free period	fog	precip. (≥0.1 mm)	thunderstorm	snow-cover	clouds (8–10 tenths)	
Jan.			2	14.0	0.1			
Feb.			3	13.5	0.1			
Mar.			3	11.9	0.1			
Apr.			5	12.2	0.4			
May			5	10.1	2			
June			2	9.1	5			
July			1	8.0	6			
Aug.			2	6.3	4			
Sept.			3	7.2	3			
Oct.			4	9.0	2			
Nov.			3	10.4	0.6			
Dec.			3	13.7	0.2			
Year		231	36	125.4	24	14		2

TABLE LXVII

CLIMATIC TABLE FOR NOVOSIBIRSK

Latitude 55°02'N longitude 82°54'E elevation 162 m

Month	Air press. (mbar)		Wind		Global radiat.	Temperature (°C)				Humidity		Precipitation (mm)			
	sea level	station	preval. direct.	mean speed (m/s)	(kcal./cm²)	mean	mean daily range	max.	min.	vapor press. (mm)	relat. (%)	mean	max.	min.	max. in 24 h
Jan.	1028.0	1013.4	SW	4.1	1.9	−19.0	9.3	6	−50	1.4	80	16	52	14	14
Feb.	1028.2	1013.7	SW	3.2	4.2	−17.2	10.0	6	−47	1.6	78	12	60	3	14
Mar.	1025.4	1011.3	SW	4.6	8.4	−10.7	10.5	10	−41	2.5	78	13	34	4	14
Apr.	1020.8	1007.3	SW	4.2	11.7	−0.2	9.9	28	−33	5.0	70	22	44	7	24
May	1014.7	1001.8	SW	4.2	14.0	10.0	12.8	36	−17	7.5	59	34	79	15	32
June	1009.5	996.9	SW	3.4	15.7	16.3	12.6	38	−2	12.4	66	60	100	18	95
July	1006.7	994.3	N	2.9	13.9	18.7	11.4	38	2	15.7	72	74	140	11	55
Aug.	1009.5	996.9	SW	3.1	11.9	16.0	11.0	35	−2	13.6	76	60	128	8	45
Sept.	1016.0	1003.1	SW	3.4	8.4	9.9	11.0	33	−9	9.3	76	45	66	13	44
Oct.	1020.5	1007.1	SW	4.2	4.0	1.5	8.5	27	−29	5.6	77	35	81	12	22
Nov.	1024.5	1010.5	SW	4.4	2.3	−9.7	7.7	11	−46	2.8	82	30	90	24	31
Dec.	1028.0	1013.5	SW	4.3	1.7	−16.9	8.7	7	−48	1.7	82	24	60	19	15
Year	1019.3	1005.8	SW	3.8	98.1	−0.2	–	38	−50	6.6	75	425	512	339	95

Month	Number of days							Snow cover max. 10-day average depth (cm)	Cloudiness (tenths)			Mean sun-shine (h)	Evaporation (mm)	
	wind (speed >15 m/s)	frost-free period	fog	precip. (>0.1 mm)	thunder-storm	snow-cover	clouds (8–10 tenths)		07h	13h	19h		actual	potent.
Jan.			3	17.2			13.0	27	7.2	7.3	6.2	67	0	
Feb.			3	13.5			10.9	32	7.3	7.0	5.8	107	0	
Mar.			2	16.9			12.0	34	7.4	7.0	6.4	166	0.7	
Apr.			2	19.4	0.3		11.1	18	6.9	7.0	6.6	213	13	
May			0.4	13.2	4		12.0		6.9	7.4	6.8	264	69	
June			0.9	14.4	8		9.2		6.3	7.5	6.5	302	75	
July			2	15.5	10		8.7		6.2	7.2	6.6	304	88	
Aug.			4	16.2	6		11.4		7.0	7.6	6.6	245	50	
Sept.			4	15,5	1		12.0		7.3	7.4	6.2	170	30	
Oct.			1	16.1	0.1		17.6	1	8.4	8.4	7.2	100	10	
Nov.			2	20.2			17.5	12	8.2	8.0	7.2	58	5	
Dec.			3	19.3			16.2	21	7.7	8.0	6.8	45	0	
Year		120	27	188	29	168	152	34	7.2	7.5	6.6	2041	320.7	556.7

TABLE LXVIII

CLIMATIC TABLE FOR ODESSA

Latitude 46°29'N longitude 30°38'E elevation 64 m

Month	Air press. (mbar)		Wind		Global radiat.	Temperature (°C)				Humidity		Precipitation (mm)			
	sea level	station	preval. direct.	mean speed (m/s)	(kcal./cm²)	mean	mean daily range	max.	min.	vapor press. (mm)	relat. (%)	mean	max.	min.	max. in 24 h
Jan.	1020.7	1015.2	N	6.2	2.8	−2.8	7.2	15	−28	3.6	86	28	90	1	26
Feb.	1019.2	1013.8	NW	6.1	4.6	−2.3	8.3	19	−30	3.6	83	26	102	1	17
Mar.	1017.1	1011.7	N, NW	6.2	8.7	2.0	7.9	25	−18	4.5	80	20	58	0	25
Apr.	1014.1	1009.4	S	5.2	11.7	8.0	8.2	31	−8	6.2	75	27	58	0	23
May	1014.7	1009.5	S	4.7	16.2	15.0	7.8	34	−2	9.4	72	34	103	1	57
June	1012.9	1007.8	N	4.4	17.3	19.2	8.6	37	4	11.7	69	45	113	2	66
July	1012.2	1007.1	N, NW	4.2	18.3	22.1	9.1	38	6	12.6	63	34	125	1	40
Aug.	1014.1	1009.0	NW	4.2	15.2	21.4	8.9	38	6	12.0	66	37	139	0	46
Sept.	1017.8	1012.6	N	4.6	11.6	16.7	8.6	34	−3	10.0	70	29	139	0	49
Oct.	1019.2	1014.1	N, E, NW	5.6	7.0	11.5	8.0	32	−16	8.1	77	35	194	1	54
Nov.	1020.6	1015.1	E	6.3	2.7	4.9	6.9	27	−16	5.6	84	43	150	0	32
Dec.	1020.0	1014.6	NE	6.5	2.0	0.0	7.9	16	−22	4.3	86	31	98	0	29
Year	1016.9	1011.7	W	5.4	118.3	9.6		38	−30		76	389	599	192	66

Month	Number of days							Snow cover max. 10-day average depth (cm)	Cloudiness (tenths)			Mean sun-shine (h)	Evaporation (mm)	
	wind (speed >15 m/s)	frost-free period	fog	precip. (>0.1 mm)	thunder-storm	snow-cover	clouds (8–10 tenths)		07h	13h	19h		actual	potent.
Jan.	5.0		11	11.2	0		18.7		8.4	8.2	7.1	70	5	
Feb.	3.9		8	9.5	0		15.7		8.3	8.2	7.1	80	10	
Mar.	3.8		9	9.6	0.2		13.5		7.9	7.8	6.8	143	15	
Apr.	2.6		6	9.2	0.7		8.8		6.5	6.8	6.4	208	40	
May	1.7		3	8.6	4		6.7		5.9	6.3	6.4	277	60	
June	1.3		1	9.2	7		5.3		5.0	5.4	5.7	305	60	
July	0.9		0.6	7.3	6		3.3		3.4	4.5	4.6	349	60	
Aug.	1.0		1	6.1	4		2.1		3.7	4.5	4.2	322	20	
Sept.	1.3		2	5.7	2		4.2		4.4	4.8	4.2	250	20	
Oct.	3.2		6	7.7	0.6		10.3		6.9	6.7	5.3	175	>10	
Nov.	4.4		8	9.6	0.3		16.6		8.4	8.4	7.4	69	5	
Dec.	4.4		10	10.9	0		19.7		8.7	8.6	7.7	60	5	
Year	33.5	215	66	10.5	25	33	125	5	6.5	6.7	6.1	2308	300	800

TABLE LXIX

CLIMATIC TABLE FOR OKHOTSK

Latitude 59°22'N longitude 143°12'E elevation 6 m

Month	Air press. (mbar)		Wind		Global radiat.	Temperature (°C)				Humidity		Precipitation (mm)			
	sea level	station	preval. direct.	mean speed (m/s)	(kcal./cm²)	mean	mean daily range	max.	min.	vapor press. (mm)	relat. (%)	mean	max.	min.	max. in 24 h
Jan.	1012.9		N	4.4	1.3	−24.5	6.5	3	−40	0.5	65	11	61	0	15
Feb.	1015.1		N	3.8	3.2	−20.5	8.2	0	−45	0.6	63	6	25	0	16
Mar.	1014.5		N	3.7	8.4	−14.2	11.4	4	−37	1.2	67	14	60	0	16
Apr.	1011.1		N	3.6	11.7	−5.7	9.2	9	−34	2.4	74	17	50	0	18
May	1012.1		SE	3.4	14.2	0.7	6.2	26	−16	4.1	83	38	100	2	70
June	1010.5		SE	3.5	12.6	5.6	5.2	31	−2	6.2	89	44	117	4	60
July	1008.9		SE	3.4	10.7	11.9	5.4	31	2	9.5	89	65	284	11	76
Aug.	1009.5		SE	3.4	9.7	12.9	6.6	32	1	9.5	87	55	148	5	52
Sept.	1012.1		N	3.5	7.1	8.3	7.6	24	−9	6.8	81	54	270	2	80
Oct.	1011.3		N	4.5	4.5	−2.2	7.4	16	−21	2.7	66	39	100	0	34
Nov.	1011.1		N	4.9	1.7	−14.7	5.4	5	−37	1.2	64	25	48	0	24
Dec.	1011.0		N	4.7	0.8	−21.2	5.7	1	−40	0.7	63	10	96	0	36
Year	1011.7		N	3.9	85.9	−5.3		32	−45		74	378	671	165	80

Month	Number of days							Snow cover max. 10-day average depth (cm)	Cloudiness (tenths)			Mean sunshine (h)	Evaporation (mm)	
	wind (speed ⩾15 m/s)	frost-free period	fog	precip. (⩾0.1 mm)	thunderstorm	snow-cover	clouds (8–10 tenths)		07h	13h	19h		actual	potent.
Jan.	0.8		0.08			31	1					102	5	
Feb.	0.4		0.04			28	1					156	5	
Mar.	1.4		0.6			31	2					228	5	
Apr.	1.8		3	8	0.0	30	4					229	10	
May	1.0		7	9	0.0	7	8					214	15	
June	1.2		11	10	0.2		12					148	<50	
July	0.9		12	13	1		11					170	<50	
Aug.	0.8		10	12	0.7		9					201	40	
Sept.	1.2		3	10	0.2		6					173	20	
Oct.	2.5		0.4	6	0.08	6	4					187	5	
Nov.	1.7		0.2			30	4					118	5	
Dec.	1.3		0.2			31	1					84	5	
Year	15	107	48		3	199	63	45				2010	200	300

TABLE LXX

CLIMATIC TABLE FOR OLENEK

Latitude 68°30'N longitude 112°36'E elevation 130 m

Month	Air press. (mbar)		Wind		Global radiat.	Temperature (°C)				Humidity		Precipitation (mm)			
	sea level	station	preval. direct.	mean speed (m/s)	(kcal./cm²)	mean	mean daily range	max.	min.	vapor press. (mm)	relat. (%)	mean	max.	min.	max. in 24 h
Jan.	1022.3		N	0.8	0.0	−40.9	8.9	−2	−65	0.2	73	10	22	3	6
Feb.	1024.8		N	0.8	1.1	−35.8	10.6	−1	−64	0.3	74	7	12	1	4
Mar.	1018.8		N	1.4	5.3	−26.7	16.4	5	−57	0.4	70	7	18	2	8
Apr.	1014.9		N	2.4	11.1	−13.0	16.3	12	−44	1.4	64	10	32	1	8
May	1012.5		N	3.0	15.0	−0.9	11.3	27	−29	2.9	61	20	49	4	20
June	1007.8		N	3.2	16.1	10.8	11.9	34	−15	5.7	60	38	88	8	34
July	1008.1		N, NE	2.6	16.4	14.1	13.2	36	−4	8.0	64	54	110	2	60
Aug.	1010.3		N, NE	2.3	9.2	9.7	11.9	33	−12	6.8	73	55	114	10	42
Sept.	1013.5		N	2.0	4.2	2.5	10.0	25	−24	4.4	77	26	63	5	19
Oct.	1012.9		N	1.8	2.0	−11.2	8.9	13	−44	1.9	80	22	43	6	14
Nov.	1020.0		N	1.0	0.3	−31.0	9.4	1	−57	0.4	77	13	23	3	4
Dec.	1020.8		N	1.0	0.0	−37.4	9.8	−2	−62	0.2	75	13	27	3	5
Year	1015.6		N	1.9	80.7	−13.3		36	−65		71	275	362	145	60

Month	Number of days							Snow cover max. 10-day average depth (cm)	Cloudiness (tenths)			Mean sunshine (h)	Evaporation (mm)	
	wind (speed ⩾15 m/s)	frost-free period	fog	precip. (⩾0.1 mm)	thunderstorm	snow-cover	clouds (8–10 tenths)		07h	13h	19h		actual	potent.
Jan.	0.0		9	11.8		31	7		4.9	6.6	5.0	0	5	
Feb.	0.1		7	10.4		28	4		5.7	6.1	4.4	48	5	
Mar.	0.1		2	10.9		31	7		6.7	6.1	5.9	187	5	
Apr.	0.5		0.2	9.8		30	10		6.2	6.3	6.5	268	5	
May	0.2		0.1	9.3		20	15		6.9	7.2	7.0	276	>10	
June	0.9		0.1	13.0	2.0		13		7.4	7.8	7.3	302	>70	
July	0.2		0.7	14.3	3.0		14		6.9	7.3	7.0	323	<60	
Aug.	0.4		4	16.2	2.0		17		7.8	7.9	7.6	194	<20	
Sept.	0.1		4	13.2	0.3		17		8.6	8.3	7.9	97	>10	
Oct.	0.0		0.5	17.5		30	17		8.4	8.3	7.3	50	<10	
Nov.	0.0		3	13.1		30	8		5.9	7.0	5.5	10	<10	
Dec.	0.1		4	14.2		31	8		5.4	7.1	5.4	0	5	
Year	2.6	47	35	153.7	7.3	233	137	41	6.7	7.2	6.4	1755	>200	300

TABLE LXXI

CLIMATIC TABLE FOR OMSK

Latitude 54°56′N longitude 73°24′E elevation 105 m

Month	Air press. (mbar) sea level	station	Wind preval. direct.	mean speed (m/s)	Global radiat. (kcal./cm²)	Temperature (°C) mean	mean daily range	max.	min.	Humidity vapor press. (mm)	relat. (%)	Precipitation (mm) mean	max.	min.	max. in 24 h
Jan.	1026.8		SW	4.7	2.1	−18.9	8.9	4	−47	0.9	81	8	35	4	12
Feb.	1026.3		SW	5.0	3.9	−17.6	9.6	5	−49	1.0	80	6	23	0	13
Mar.	1022.2		SW	5.2	8.8	−11.1	10.1	14	−43	2.0	80	9	28	0	6
Apr.	1019.2		SW	5.0	12.5	0.5	10.0	29	−27	4.0	71	18	58	0	24
May	1014.1		SW	5.0	15.0	11.2	13.4	35	−13	6.4	54	30	90	3	21
June	1009.8		N	4.5	16.2	17.1	13.4	40	−2	9.8	59	53	134	2	61
July	1006.1		NW	3.7	14.7	19.5	12.1	40	2	11.6	68	72	199	20	29
Aug.	1010.6		NW	3.6	12.0	16.5	12.1	37	−2	10.1	71	46	121	5	40
Sept.	1014.2		SW	4.0	7.9	10.8	11.4	32	−8	6.9	70	33	108	5	25
Oct.	1018.2		SW	4.7	3.9	1.8	8.2	24	−28	4.3	75	23	56	1	29
Nov.	1025.2		SW	4.8	2.1	−8.5	7.7	15	−41	2.3	82	15	52	4	11
Dec.	1026.6		SW	4.9	1.4	−16.3	8.5	3	−43	1.3	81	12	29	3	10
Year	1018.2		SW	4.6	100.5	0.4		40	−49		73	325	524	211	40

Month	Number of days wind (speed ≥15 m/s)	frost-free period	fog	precip. (≥0.1 mm)	thunder-storm	snow-cover	clouds (8–10 tenths)	Snow cover max. 10-day average depth (cm)	Cloudiness (tenths) 07h	13h	19h	Mean sun-shine (h)	Evaporation (mm) actual	potent.
Jan.	1.4		3	11.4		31	11.9		7.0	7.3	5.6	82	5	
Feb.	1.4		5	8.1		28	8.8		6.9	6.5	4.7	122	5	
Mar.	1.9		5	9.1		31	10.2		6.9	6.1	5.7	192	5	
Apr.	1.9		4	7.4	0.2	8	9.3		6.5	6.9	6.4	249	>10	
May	3.2		0.9	10.0	2		9.1		6.1	7.1	6.4	290	60	
June	2.3		0.8	11.2	6		6.6		5.8	7.1	6.4	318	>70	
July	0.7		2	12.5	7		9.1		6.1	7.5	6.6	299	70	
Aug.	0.7		2	11.6	4		8.9		6.5	7.4	6.3	252	40	
Sept.	1.0		4	9.1	1		9.8		6.9	7.5	6.5	191	30	
Oct.	1.6		3	11.7	0.1		15.4		8.1	8.4	6.8	97	5	
Nov.	1.2		3	13.8		22	19.0		8.0	7.9	6.8	71	5	
Dec.	1.5		4	13.4		31	15.8		7.4	7.6	6.3	60	5	
Year	18.8	115	37	129.3	20	157	134	31	6.9	7.3	6.2	2923	300	600

TABLE LXXII

CLIMATIC TABLE FOR ONOR

Latitude 50°14′N longitude 142°35′E elevation 180 m

Month	Air press. (mbar) sea level	station	Wind preval. direct.	mean speed (m/s)	Global radiat. (kcal./cm²)	Temperature (°C) mean	mean daily range	max.	min.	Humidity vapor press. (mm)	relat. (%)	Precipitation (mm) mean	max.	min.	max. in 24 h
Jan.		988.9	N	2.8	3.1	−20.5		4	−46	1.0	75	13	49	2	
Feb.		990.6	N	2.7	5.1	−17.0		3	−38	1.3	73	11	50	2	
Mar.		990.2	NW	2.9	8.6	−10.1		8	−33	2.1	70	22	78	5	
Apr.		987.0	SE	3.3	10.0	−1.0		22	−27	4.0	70	40	82	12	
May		986.5	SE	3.5	10.9	5.0		31	−10	6.2	72	48	83	14	
June		988.5	SE	3.0	11.3	10.7		34	−4	9.6	76	42	101	14	
July		987.4	SE	2.7	10.6	15.2		36	−1	13.9	81	65	174	12	
Aug.		988.6	SE	2.5	9.5	15.6		37	0	14.7	83	73	208	11	
Sept.		988.8	SE	2.5	8.4	10.8		30	−6	10.4	82	97	239	43	
Oct.		991.4	NW	2.6	5.7	2.5		21	−17	5.8	76	79	108	21	
Nov.		989.8	NW	2.6	3.8	−7.6		14	−30	2.8	77	49	132	10	
Dec.		988.6	NW	2.7	1.7	−16.1		5	−41	1.6	79	31	71	3	
Year		988.8	SE	2.8	88.7	−1.0		37	−46	6.1	76	570	692	448	

Month	Number of days wind (speed ≥15 m/s)	frost-free period	fog	precip. (≥0.1 mm)	thunder-storm	snow-cover	clouds (8–10 tenths)	Snow cover max. 10-day average depth (cm)	Cloudiness (tenths) 07h	13h	19h	Mean sun-shine (h)	Evaporation (mm) actual	potent.
Jan.	0.6		0.04	10.0			6.4		4.9	5.8	4.5		0.2	
Feb.	0.9		0.1	9.8			6.9		5.8	6.3	4.7		0.6	
Mar.	0.7		0.8	11.5			10.2		6.3	6.7	5.6		6	
Apr.	0.6		3	12.1			12.8		7.1	7.2	6.9		12	
May	0.7		7	13.8	0.4		15.4		8.1	7.8	7.4		45	
June	0.0		13	12.6	1.1		16.1		8.4	7.2	7.3		65	
July	0.0		16	13.4	1.3		17.2		8.7	7.2	7.2		68	
Aug.	0.1		12	14.3	1.2		16.8		8.5	7.5	7.5		67	
Sept.	0.3		6	15.9	1.2		14.3		7.9	7.8	6.9		55	
Oct.	0.7		3	15.6	0.2		10.7		7.0	7.1	5.7		25	
Nov.	0.7		1	17.7			10.7		7.5	7.6	6.2		1	
Dec.	0.8		0.2	15.5			8.8		6.2	6.8	5.4		0	
Year	6.1	108	62.1	162	5.4	184	146.3	41	7.2	7.1	6.3		344.8	444.8

TABLE LXXIII

CLIMATIC TABLE FOR ORENBURG

Latitude 51°45′N longitude 55°06′E elevation 109 m

Month	Air press. (mbar)		Wind		Global radiat. (kcal./cm²)	Temperature (°C)				Humidity		Precipitation (mm)			
	sea level	station	preval. direct.	mean speed (m/s)		mean	mean daily range	max.	min.	vapor press. (mm)	relat. (%)	mean	max.	min.	max. in 24 h
Jan.	1026.2		E	4.6		−15.0	8.1	5	−44	1.5	80	21	39	1	22
Feb.	1025.2		E	4.5		−14.3	10.2	6	−44	1.5	80	20	50	1	24
Mar.	1022.4		E	4.7		−7.7	11.3	17	−40	2.5	82	26	65	5	20
Apr.	1019.6		E	4.4		4.3	14.1	31	−28	4.5	70	22	104	0	36
May	1015.6		W	4.1		14.7	15.7	37	−8	7.2	55	38	128	1	43
June	1011.7		N, E	3.1		19.7	16.0	39	−3	9.8	53	37	168	0	60
July	1008.6		N, NW	3.8		22.0	15.1	41	2	10.8	55	43	129	8	53
Aug.	1012.3		E	3.4		19.8	16.5	40	0	9.6	55	30	138	0	50
Sept.	1016.8		W	3.7		13.1	16.5	36	−8	6.9	60	30	92	0	31
Oct.	1021.2		W	4.1		4.7	14.8	27	−22	4.9	72	41	80	5	35
Nov.	1027.6		E	4.2		−4.1	10.2	19	−38	3.2	81	25	85	2	25
Dec.	1026.8		E	4.4		−11.8	7.5	6	−41	1.9	83	25	74	0	29
Year	1019.5		E	4.1		3.8		41	−44		69	358	749	200	60

Month	Number of days							Snow cover max. 10-day average depth (cm)	Cloudiness (tenths)			Mean sunshine (h)	Evaporation (mm)	
	wind (speed ≥15 m/s)	frost-free period	fog	precip. (≥0.1 mm)	thunderstorm	snow-cover	clouds (8–10 tenths)		07h	13h	19h		actual	potent.
Jan.	2.4		5	14		31	14.1		7.3	7.1	6.3	71	5	
Feb.	2.1		5	11		28	11.4		7.2	6.7	5.6	98	5	
Mar.	2.6		7	12		31	13.2		7.6	6.9	6.4	129	<10	
Apr.	1.8		4	8	0.2	8	10.4		6.8	7.0	6.3	209	20	
May	1.7		1	9	3		8.1		5.9	6.8	6.5	288	<40	
June	0.8		1	9	7		5.0		4.8	6.2	5.7	315	>40	
July	0.8		0.9	10	7		5.4		4.5	6.4	5.7	320	<50	
Aug.	0.7		2	9	4		5.1		4.7	5.7	5.0	293	<30	
Sept.	1.4		4	9	1		6.8		6.1	6.4	5.3	200	20	
Oct.	1.1		5	10	0.1		13.6		7.5	7.7	6.2	120	<10	
Nov.	1.3		6	13		9	15.0		7.4	7.5	6.6	69	5	
Dec.	2.3		6	15		31	16.6		7.6	7.7	6.8	53	5	
Year	19	147	47	129	22	146	125	52	6.4	6.8	6.0	2165	<300	700

TABLE LXXIV

CLIMATIC TABLE FOR OSTROV BERINGA (Komandorskiye Islands)

Latitude longitude elevation

Month	Air press. (mbar)		Wind		Global radiat. (kcal./cm²)	Temperature (°C)				Humidity		Precipitation (mm)			
	sea level	station	preval. direct.	mean speed (m/s)		mean	mean daily range	max.	min.	vapor press. (mm)	relat. (%)	mean	max.	min.	max. in 24 h
Jan.	999.8					−3.5					83	38			
Feb.	1001.4					−3.9					83	23			
Mar.	1004.9					−3.2					84	30			
Apr.	1006.5					−1.0					85	24			
May	1010.1					2.0					87	31			
June	1011.3					5.1					90	28			
July	1010.1					8.7					93	48			
Aug.	1009.4					10.6					92	59			
Sept.	1011.1					9.1					87	53			
Oct.	1007.0					4.6					81	72			
Nov.	1003.8					0.3					82	64			
Dec.	998.9					−2.4					83	46			
Year	1005.8					2.2					86	516			

TABLE LXXV

CLIMATIC TABLE FOR OSTROV CHETYREKHSTOLBOVOY

Latitude 70°38'N longitude 162°24'E elevation 30 m

Month	Air press. (mbar)		Wind		Global radiat.	Temperature (°C)				Humidity		Precipitation (mm)			
	sea level	station	preval. direct.	mean speed (m/s)	(kcal./cm²)	mean	mean daily range	max.	min.	vapor press. (mm)	relat. (%)	mean	max.	min.	max. in 24 h
Jan.	1020.9		SW	6.9	0.0	−29.5		−4	−49	0.6	89	9			1
Feb.	1023.4		NE	6.6	0.8	−29.9		−6	−51	0.6	88	10			4
Mar.	1021.1		NE, E	6.1	4.8	−27.0		−1	−50	0.7	89	8			3
Apr.	1017.0		E	6.0	10.9	−20.6		2	−37	1.5	90	8			1
May	1015.7		E	6.0	18.0	−9.1		10	−31	3.5	90	10			8
June	1010.3		E	6.4	15.6	−0.2		21	−13	5.7	89	13			15
July	1009.7		E	6.1	13.8	2.5		23	−6	6.4	90	21			20
Aug.	1010.9		E	6.0	7.9	2.0		19	−10	6.4	92	19			12
Sept.	1014.3		E	5.9	3.6	−1.5		10	−15	5.3	88	16			8
Oct.	1013.2		W	6.7	1.4	−10.5		8	−28	3.1	86	15			9
Nov.	1018.6		NE, SW	6.7	0.1	−21.9		−2	−39	1.3	87	14			2
Dec.	1019.7		NE	6.8	0.0	−26.4		1	−47	0.8	87	12			3
Year	1016.2		E	6.4	74.9	−14.3		23	−51	3.0	89	155			20

Month	Number of days							Snow cover	Cloudiness (tenths)			Mean sunshine	Evaporation (mm)	
	wind (speed ≥15 m/s)	frost-free period	fog	precip. (≥0.1 mm)	thunderstorm	snow-cover	clouds (8–10 tenths)	max. 10-day average depth (cm)	07h	13h	19h	(h)	actual	potent.
Jan.	6.0		1	6.4		31	7					0	<5	
Feb.	4.2		2	6.3		28	7					41	<5	
Mar.	4.0		2	5.9		31	6					180	<10	
Apr.	2.5		3	4.8		30	8					258	<10	
May	1.5		10	5.0		31	17					203	<10	
June	1.8		15	7.2	0.2		18					211	<40	
July	1.3		20	10.3	0.5		19					215	<60	
Aug.	1.2		19	11.8	0.2		22					111	<30	
Sept.	1.9		9	12.4	0.2	4	22					60	<10	
Oct.	3.9		3	12.6		31	20					26	<10	
Nov.	3.7		2	9.5		30	12					2	<5	
Dec.	5.0		1	7.1		31	9					0	<5	
Year	37	0	87	99.3	1	245	167					1308	<200	

TABLE LXXVI

CLIMATIC TABLE FOR OSTROV DIKSON

Latitude 73°30'N longitude 80°14'E elevation 22 m

Month	Air press. (mbar)		Wind		Global radiat.	Temperature (°C)				Humidity		Precipitation (mm)			
	sea level	station	preval. direct.	mean speed (m/s)	(kcal./cm²)	mean	mean daily range	max.	min.	vapor press. (mm)	relat. (%)	mean	max.	min.	max. in 24 h
Jan.	1012.2	1009.3	S	8.4	0.0	−27.5		0	−51	0.8	86	20			3
Feb.	1013.1	1010.3	S	8.1	0.5	−25.9		0	−52	0.8	86	13			4
Mar.	1013.4	1010.6	S	7.4	4.0	−25.0		0	−50	0.7	84	17			4
Apr.	1012.4	1009.6	S	7.3	9.9	−18.1		7	−40	1.2	86	9			6
May	1011.4	1008.8	NE	7.1	14.8	−8.2		11	−33	2.3	87	11			9
June	1010.3	1007.7	NE	6.8	13.9	−1.0		22	−15	4.3	90	23	67	1	13
July	1009.7	1007.2	NE	6.6	12.2	3.6		24	−3	5.8	90	32	84	4	25
Aug.	1010.0	1007.5	NE	6.7	7.4	4.8		27	−3	6.0	89	46	82	8	31
Sept.	1008.6	1006.0	S	7.3	3.2	0.8		14	−19	4.8	89	42	83	7	24
Oct.	1006.5	1003.9	S	8.1	1.0	−7.2		8	−36	2.7	87	21			11
Nov.	1009.4	1006.6	S	8.1	0.0	−19.2		2	−40	1.6	88	14			26
Dec.	1012.9	1010.1	S	8.0	0.0	−24.3		0	−48	0.9	86	18			7
Year	1010.8	1008.1	S	7.5	66.9	−12.3		27	−52		87	266			31

Month	Number of days							Snow cover	Cloudiness (tenths)			Mean sunshine	Evaporation (mm)	
	wind (speed ≥15 m/s)	frost-free period	fog	precip. (≥0.1 mm)	thunderstorm	snow-cover	clouds (8–10 tenths)	max. 10-day average depth (cm)	07h	13h	19h	(h)	actual	potent.
Jan.	11.5		4	13.3		31	13					0	<5	
Feb.	8.7		5	12.0		28	12					16	<5	
Mar.	9.1		6	10.9		31	11					129	<5	
Apr.	7.2		5	10.5		30	14					211	<5	
May	6.2		6	11.4		31	22					152	<10	
June	3.8		13	12.0	0.1	11	23					149	<30	
July	2.4		19	12.9	0.7		20					196	<70	
Aug.	3.6		13	16.8	0.4		24					122	<30	
Sept.	4.8		10	19.4			23					59	<10	
Oct.	8.4		5	18.3		29	22					22	<10	
Nov.	8.5		3	15.2		30	16					0	<5	
Dec.	10.2		3	13.1		31	12					0	<5	
Year	84.4	44	92	165.8	1	253	212	32				1056	<200	

TABLE LXXVII

CLIMATIC TABLE FOR OSTROV KOTEL'NYY

Latitude 76°00'N longitude 137°54'E elevation 11 m

Month	Air press. (mbar)		Wind		Global radiat.	Temperature (°C)				Humidity		Precipitation (mm)			
	sea level	station	preval. direct.	mean speed (m/s)	(kcal./cm²)	mean	mean daily range	max.	min.	vapor press. (mm)	relat. (%)	mean	max.	min.	max. in 24 h
Jan.	1019.7		SW	5.7	0.0	−29.5		−9	−47	0.5	79	8			5
Feb.	1022.4		SW	5.7	0.2	−29.9		−11	−49	0.5	79	8			3
Mar.	1019.8		SW	5.7	3.1	−27.0		−10	−47	0.6	78	6			3
Apr.	1015.6		SE	5.6	9.8	−20.6		−2	−42	1.2	79	9			3
May	1016.0		SE	6.2	16.2	−9.1		8	−30	2.9	80	10			7
June	1011.7		E	6.2	15.6	−0.2		21	−12	5.5	75	11	25	1	12
July	1009.8		N, W	6.2	11.1	2.5		22	−4	6.9	69	20	67	3	17
Aug.	1009.9		W	6.7	6.9	2.0		19	−6	6.6	79	19	57	6	26
Sept.	1011.5		SE	6.8	3.0	−1.5		11	−20	5.1	83	11	36	2	9
Oct.	1010.7		SE	6.4	0.6	−10.5		8	−29	2.8	85	11			5
Nov.	1018.8		SE	5.8	0.0	−21.9		−4	−40	1.2	81	10			2
Dec.	1019.9		SE	5.4	0.0	−26.4		−3	−42	0.8	80	8			2
Year	1015.5		SE	6.0	66.5	−14.3		22	−49	2.9	79	131			26

Month	Number of days							Snow cover	Cloudiness (tenths)			Mean sun-shine	Evaporation (mm)	
	wind (speed ≥15 m/s)	frost-free period	fog	precip. (≥0.1 mm)	thunder-storm	snow-cover	clouds (8–10 tenths)	max. 10-day average depth (cm)	07h	13h	19h	(h)	actual	potent.
Jan.	2.9		2	8.1		31	6						<5	
Feb.	2.6		4	7.9		28	6					8	<5	
Mar.	3.6		6	7.9		31	6					125	<5	
Apr.	3.7		5	7.8		30	11					237	<5	
May	3.4		6	7.9		31	21					190	<10	
June	3.1		11	8.9	0.06	16	23					172	<40	
July	2.7		16	12.3	0.1		24					146	<60	
Aug.	3.5		13	14.1	0.06		25					98	<30	
Sept.	3.4		7	14.9		14	25					41	<10	
Oct.	3.4		3	14.9		31	21					10	<10	
Nov.	2.6		2	10.2		30	11					0	<5	
Dec.	3.4		1	9.1		31	6						<5	
Year	38.3	10	76	124.0	0.2	269	185	20				1027	<200	

TABLE LXXVIII

CLIMATIC TABLE FOR OYMYAKON

Latitude 63°16'N longitude 143°09'E elevation 740 m

Month	Air press. (mbar)		Wind		Global radiat.	Temperature (°C)				Humidity		Precipitation (mm)			
	sea level	station	preval. direct.	mean speed (m/s)	(kcal./cm²)	mean	mean daily range	max.	min.	vapor press. (mm)	relat. (%)	mean	max.	min.	max. in 24 h
Jan.	1023.5		SE	0.3	0.6	−50.1	6.3	−16	−70	0.1	74	7	24	2	4
Feb.	1025.1		SE	0.3	2.5	−44.3	10.3	−11	−71	0.1	74	6	19	0	5
Mar.	1026.3		SE	0.8	8.1	−32.0	19.6	4	−64	0.3	72	5	14	0	4
Apr.	1014.9		SE	1.5	13.3	−14.8	25.4	11	−57	1.1	67	4	19	1	16
May	1008.9		E	2.4	16.0	1.7	20.1	28	−30	3.1	59	10	33	0	14
June	1004.3		E	2.4	15.3	11.4	23.2	31	−10	5.7	60	32	67	5	25
July	1005.0		E	1.9	14.5	14.5	23.6	33	−5	8.1	67	40	90	17	29
Aug.	1008.3		NW	1.5	13.0	10.4	24.4	31	−11	6.8	72	37	78	5	29
Sept.	1013.7		NW	1.6	6.7	2.4	21.5	24	−25	4.0	74	20	65	3	16
Oct.	1018.3		NW	1.5	3.6	−14.8	14.5	11	−48	1.6	78	12	29	2	16
Nov.	1019.6		NW	0.6	1.1	−26.1	7.1	3	−62	0.3	77	11	32	1	11
Dec.	1022.8		SE	0.3	0.3	−47.1	4.1	−8	−68	0.1	76	9	38	1	15
Year	1015.4		NW	1.3	95.0	−16.5		33	−71		71	193	316	136	29

Month	Number of days							Snow cover	Cloudiness (tenths)			Mean sun-shine	Evaporation (mm)	
	wind (speed ≥15 m/s)	frost-free period	fog	precip. (≥0.1 mm)	thunder-storm	snow-cover	clouds (8–10 tenths)	max. 10-day average depth (cm)	07h	13h	19h	(h)	actual	potent.
Jan.	0.0		6	11.6		31	7.1		5.3	5.8	4.9	29		
Feb.	0.0		2	10.0		28	7.1		5.8	5.6	5.3	115		
Mar.	0.0		0.6	7.7		31	6.9		5.4	4.9	4.6	224		
Apr.	0.5		0.1	4.5		30	8.5		6.1	5.9	6.0	268		
May	0.5		0.0	8.4	0.4	9	13.7		6.9	7.7	6.4	261		
June	0.8		1	11.6	4		14.5		6.7	8.1	7.6	268		
July	0.6		1	12.2	4		13.1		6.8	7.5	7.2	273		
Aug.	0.2		4	11.9	2		12.7		6.6	6.8	6.4	214		
Sept.	0.2		2	9.7			14.2		7.5	7.7	7.2	140		
Oct.	0.3		0.9	11.2		28	13.5		7.5	6.8	6.1	75		
Nov.	0.0		6	14.5		30	9.8		6.2	6.2	5.4	42		
Dec.	0.0		7	12.0		31	6.6		5.8	6.4	5.4	2		
Year	3.1	<30	31	125	10	229	128	29	6.4	6.6	6.0	1911		

TABEL LXXIX

CLIMATIC TABLE FOR PENZA

Latitude 53°11'N longitude 45°01'E elevation 235 m

Month	Air press. (mbar)		Wind		Global radiat.	Temperature (°C)				Humidity		Precipitation (mm)			
	sea level	station	preval. direct.	mean speed (m/s)	(kcal./cm²)	mean	mean daily range	max.	min.	vapor press. (mm)	relat. (%)	mean	max.	min.	max. in 24 h
Jan.	1022.0		S	4.9		−12.0	7.9	4	−43	1.9	86	40	103	7	25
Feb.	1022.2		S	4.8		−11.5	9.7	5	−39	1.9	83	34	132	5	28
Mar.	1019.8		S	5.0		−5.8	10.8	14	−26	2.7	80	36	98	2	30
Apr.	1017.9		S	4.6		4.3	11.8	30	−17	4.7	70	32	71	2	43
May	1016.2		NW	4.5		13.5	13.6	33	−8	7.0	60	52	138	4	71
June	1011.4		NW	4.0		17.5	13.2	38	−1	9.9	61	55	121	2	69
July	1010.2		NW	3.7		18.8	13.6	38	5	11.4	66	64	188	11	100
Aug.	1012.3		NW	3.7		17.9	13.6	37	2	10.2	68	56	157	7	49
Sept.	1016.8		NW	4.5		11.7	13.6	35	−4	7.6	70	50	124	7	50
Oct.	1020.6		NW	5.0		4.5	12.1	25	−18	5.2	79	48	129	2	41
Nov.	1021.5		S	4.7		−3.3	8.7	15	−36	3.5	84	46	104	5	38
Dec.	1022.4		S	4.9		−9.6	6.8	7	−35	2.2	87	46	129	10	29
Year	1017.8		NW	4.5		3.9		38	−43		74	559	781	344	100

Month	Number of days							Snow cover	Cloudiness (tenths)			Mean sun-shine	Evaporation (mm)	
	wind (speed ⩾15 m/s)	frost-free period	fog	precip. (⩾0.1 mm)	thunder-storm	snow-cover	clouds (8–10 tenths)	max. 10-day average depth (cm)	07h	13h	19h	(h)	actual	potent.
Jan.	2.0		5	17.0		31	17.1		7.8	7.8	7.1	45	5	
Feb.	1.4		4	13.8		28	14.3		7.6	7.1	6.4	79	5	
Mar.	2.3		8	13.5	0.2	31	13.5		7.6	7.0	6.7	123	<10	
Apr.	1.6		6	10.2	0.5	6	11.1		6.6	7.1	6.9	184	20	
May	2.1		2	11.2	3		9.4		6.1	7.5	6.7	214	>70	
June	1.1		1	11.8	6		6.2		5.2	6.8	6.1	274	>70	
July	0.5		1	12.7	7		6.3		5.3	7.2	6.0	288	60	
Aug.	0.3		3	11.7	4		6.3		5.5	6.8	5.7	238	>50	
Sept.	0.8		4	12.3	1		9.6		6.6	7.1	6.1	152	>30	
Oct.	2.0		7	13.2	0.1		15.9		7.9	8.0	6.7	95	>10	
Nov.	0.9		9	15.9		7	18.4		8.0	8.0	7.4	49	5	
Dec.	2.0		6	17.5		31	19.8		8.0	8.1	7.8	36	5	
Year	17	151	56	161		145	148	60 c	6.8	7.4	6.6	1807	>300	600

TABLE LXXX

CLIMATIC TABLE FOR PERM'

Latitude 57°57'N longitude 56°13'E elevation 169.7 m

Month	Air press. (mbar)		Wind		Global radiat.	Temperature (°C)				Humidity		Precipitation (mm)			
	sea level	station	preval. direct.	mean speed (m/s)	(kcal./cm²)	mean	mean daily range	max.	min.	vapor press. (mm)	relat. (%)	mean	max.	min.	max. in 24 h
Jan.	1020.2		S	3.4		−15.4	6.7	4	−45	1.4	82	38	77	15	15
Feb.	1020.9		SE, SW	3.3		−13.4	7.9	6	−41	1.5	78	27	73	4	15
Mar.	1018.7		SW	3.4		−7.2	8.8	14	−35	2.2	75	31	108	0	28
Apr.	1018.0		SW	3.1		2.2	8.9	27	−24	3.9	68	35	99	1	22
May	1014.8		SE, SE	3.6		10.0	11.0	35	−13	6.1	60	47	103	16	48
June	1010.1		W	3.5		15.6	11.6	36	−3	9.1	62	64	137	15	72
July	1008.9		SE	2.7		18.0	11.1	37	2	10.9	68	68	179	19	55
Aug.	1010.5		N	2.8		15.3	10.6	37	−1	9.9	72	62	183	18	48
Sept.	1014.2		S, SW	3.1		9.2	8.3	30	−8	7.2	78	59	120	17	46
Oct.	1017.2		SW	3.6		1.6	5.2	22	−21	4.5	83	55	113	13	26
Nov.	1018.7		SW	3.5		−6.7	5.3	12	−38	2.8	83	43	151	20	24
Dec.	1021.2		SW	3.3		−13.2	6.5	3	−44	1.7	83	41	128	11	25
Year	1016.1		SW	3.3		1.3		37	−45		74	570	912	451	72

Month	Number of days							Snow cover	Cloudiness (tenths)			Mean sun-shine	Evaporation (mm)	
	wind (speed ⩾15 m/s)	frost-free period	fog	precip. (⩾0.1 mm)	thunder-storm	snow-cover	clouds (8–10 tenths)	max. 10-day average depth (cm)	07h	13h	19h	(h)	actual	potent.
Jan.	2.5		6	20.8		31	19.2					24	5	
Feb.	2.0		6	16.8		28	13.6					75	5	
Mar.	3.2		3	14.5		31	13.4					146	5	
Apr.	1.9		2	10.8	0.2	18	11.1					210	>20	
May	2.7		2	13.3	3		11.2					268	<70	
June	1.9		0.8	14.5	6		8.4					303	>70	
July	1.2		2	15.2	7		10.5					309	70	
Aug.	1.1		3	15.7	4		10.3					236	<40	
Sept.	1.5		3	16.1	0.6		16.1					130	30	
Oct.	2.4		3	18.7	0.1		22.2					48	<10	
Nov.	2.4		3	20.6		27	19.8					33	5	
Dec.	1.8		3	20.7		31	20.7					17	5	
Year	24.6	118	37	198	21	176	176	76 cm				1799	>300	500

TABLE LXXXI

CLIMATIC TABLE FOR PETROPAVLOVSK-KAMCHATSKIY

Latitude 52°58′N longitude 158°45′E elevation 32 m

Month	Air press. (mbar)		Wind		Global radiat.	Temperature (°C)				Humidity		Precipitation (mm)			
	sea level	station	preval. direct.	mean speed (m/s)	(kcal./cm²)	mean	mean daily range	max.	min.	vapor press. (mm)	relat. (%)	mean	max.	min.	max. in 24 h
Jan.	1002.3	990.3	N	2.3	2.3	−8.3	8.4	6	−34	1.7	64	111	432	25	26
Feb.	1003.0	991.0	N	2.8	4.2	−8.7	10.6	6	−28	1.8	65	88	229	7	25
Mar.	1007.0	994.9	N	3.6	8.8	−5.3	12.1	10	−24	2.4	67	174	657	4	39
Apr.	1009.3	997.4	NW	2.9	12.1	−0.6	13.1	15	−16	3.4	74	107	186	14	29
May	1011.5	999.8	S	2.5	12.8	3.9	16.1	22	−10	4.6	76	76	156	3	26
June	1011.4	999.9	S	2.5	13.8	8.7	16.2	27	−1	6.8	79	58	138	5	20
July	1010.2	998.9	S	2.2	11.4	12.8	15.5	31	2	9.2	81	73	182	20	26
Aug.	1010.5	999.3	N, S	2.1	10.6	13.5	15.6	27	3	9.5	82	106	178	26	34
Sept.	1011.5	1000.2	N	2.1	8.6	10.4	17.1	24	−1	7.5	78	102	239	18	41
Oct.	1009.5	997.9	N	3.1	4.8	4.9	15.5	19	−8	4.7	68	143	378	27	57
Nov.	1006.1	994.2	N	3.0	2.4	−1.6	9.3	13	−16	3.1	68	182	733	23	59
Dec.	1001.8	989.8	N	3.5	1.8	−6.1	9.0	6	−26	2.1	67	115	554	6	36
Year	1007.9	996.1		2.8	93.6	2.0		31	−34		72	1335	2170	875	59

Month	Number of days							Snow cover max. 10-day average depth (cm)	Cloudiness (tenths)			Mean sunshine (h)	Evaporation (mm)	
	wind (speed ≥15 m/s)	frost-free period	fog	precip. (≥0.1 mm)	thunderstorm	snow-cover	clouds (8–10 tenths)		07h	13h	19h		actual	potent.
Jan.	6.3		0	12		31	4		6.3	6.6	5.5	83	5	
Feb.	5.8		0.7	11		28	5		7.0	6.9	5.8	103	5	
Mar.	6.9		1	13		31	5		6.9	6.8	6.4	154	5	
Apr.	5.3		0.8	11		30	5		6.8	6.7	7.0	182	10	
May	2.5		4	11		16	8		7.4	7.3	7.4	191	<20	
June	1.2		10	10	0.09		9		8.4	8.1	8.3	177	<40	
July	2.1		11	13	0.3		9		8.1	7.9	8.0	179	<40	
Aug.	2.0		10	14	0.1		9		7.9	7.4	7.7	175	40	
Sept.	3.4		4	12	0.04		8		7.2	7.0	6.9	158	20	
Oct.	6.6		2	11			4		6.0	6.3	5.1	149	5	
Nov.	6.7		2	11		21	4		6.0	6.3	5.0	103	5	
Dec.	5.7		0.5	12		31	4		5.7	6.3	5.2	80	5	
Year	54.5	149	46	141	0.5	193	74	99 c..	7.0	7.0	6.5	1734	200	300

TABLE LXXXII

CLIMATIC TABLE FOR PETROŽAVODSK

Latitude 61°49′N longitude 34°16′E elevation 40 m

Month	Air press. (mbar)		Wind		Global radiat.	Temperature (°C)				Humidity		Precipitation (mm)			
	sea level	station	preval. direct.	mean speed (m/s)	(kcal./cm²)	mean	mean daily range	max.	min.	vapor press. (mm)	relat. (%)	mean	max.	min.	max. in 24 h
Jan.	1013.7	1008.3	SW	4.4	0.6	−9.7		5	−40	2.2	86	35.1			17
Feb.	1014.1	1008.7	SW	4.1	1.7	−9.9		6	−38	2.2	84	25.6			12
Mar.	1013.1	1007.8	SW	4.4	5.8	−5.8		12	−33	2.7	76	17.1			21
Apr.	1014.6	1009.4	SW	3.9	8.7	−1.3		24	−25	3.9	70	26.8			21
May	1015.2	1010.2	E	3.8	12.0	7.7		30	−10	5.3	65	50.8			64
June	1011.8	1006.9	SW	3.8	14.2	13.3		33	−2	7.7	68	53.5			53
July	1010.5	1005.5	SW	3.5	13.4	16.6		35	2	9.8	73	76.6			73
Aug.	1010.7	1005.8	SW	3.1	9.4	14.3		33	1	9.6	78	72.7			71
Sept.	1012.5	1007.4	SW	3.7	4.9	9.1		27	−7	7.1	83	73.7			46
Oct.	1013.3	1008.1	SW	4.2	2.0	3.2		18	−16	5.0	85	58.0			27
Nov.	1013.4	1008.1	SW	4.3	0.7	−2.2		11	−27	3.7	88	39.0			22
Dec.	1013.9	1008.6	SW	4.2	0.2	−7.0		8	−37	2.8	88	30.3			15
Year	1013.1	1007.9	SW	3.9	73.6	2.6		35	−40		79	559.2			73

Month	Number of days							Snow cover max. 10-day average depth (cm)	Cloudiness (tenths)			Mean sunshine (h)	Evaporation (mm)	
	wind (speed ≥15 m/s)	frost-free period	fog	precip. (≥0.1 mm)	thunderstorm	snow-cover	clouds (8–10 tenths)		07h	13h	19h		actual	potent.
Jan.			2				19.4					19	<10	
Feb.			3				16.9					52	<10	
Mar.			3				12.0					162	<10	
Apr.			4	11.3	0.2		12.4					200	30	
May			3	12.1	2		12.0					249	70	
June			1	12.5	4		10.9					300	>70	
July			2	12.9	5		11.6					294	70	
Aug.			2	15.8	4		12.7					219	50	
Sept.			4	15.8	1		14.9					128	30	
Oct.			3	15.0			19.8					58	10	
Nov.			3				22.8					26	<10	
Dec.			3		0.06		22.0					12	<10	
Year		126	33		16	149	187	39 cr				1719	<400	400

TABLE LXXXIII

CLIMATIC TABLE FOR PODKAMENNAYA TUNGUSKA

Latitude 61°36'N longitude 90°00'E elevation 60.2 m

Month	Air press. (mbar)		Wind		Global radiat.	Temperature (°C)				Humidity		Precipitation (mm)			
	sea level	station	preval. direct.	mean speed (m/s)	(kcal./cm²)	mean	mean daily range	max.	min.	vapor press. (mm)	relat. (%)	mean	max.	min.	max. in 24 h
Jan.	1025.1	1017.9	SW	2.8		−24.2	8.7	2	−58		78	28			
Feb.	1025.3	1018.2	E	2.9		−21.1	10.6	4	−54		76	17			
Mar.	1021.8	1014.9	SW	3.4		−13.9	13.5	10	−51		72	18			
Apr.	1016.5	1009.8	SW	3.3		−2.4	12.4	21	−36		62	27			
May	1012.3	1005.9	NW	3.3		4.9	11.2	33	−25		63	48			
June	1008.3	1002.1	NW	3.0		13.6	12.2	36	−6		67	66			
July	1007.8	1001.7	E	2.4		17.4	12.1	37	−1		72	70			
Aug.	1008.9	1002.7	E	2.5		13.8	10.5	34	−3		79	73			
Sept.	1014.1	1007.8	SW	2.9		7.5	8.7	26	−13		80	56			
Oct.	1016.1	1009.4	SW	3.8		−1.8	5.8	22	−33		82	52			
Nov.	1020.4	1013.4	SW	3.3		−16.0	8.0	7	−50		82	40			
Dec.	1024.4	1017.2	E	3.0		−23.6	8.7	4	−58		80	30			
Year	1016.8	1010.1	SW	3.0		−3.8		37	−58		74	527			

Month	Number of days							Snow cover max. 10-day average depth (cm)	Cloudiness (tenths)			Mean sunshine (h)	Evaporation (mm)	
	wind (speed ≥15 m/s)	frost-free period	fog	precip. (≥0.1 mm)	thunder-storm	snow-cover	clouds (8–10 tenths)		07h	13h	19h		actual	potent.
Jan.	0.5					31			7.1	7.7	6.7	26	5	
Feb.	0.8					28			7.0	7.0	6.0	76	5	
Mar.	1.0					31			7.2	7.1	6.7	144	5	
Apr.	1.3					30			6.9	7.2	7.1	194	10	
May	1.7				0.8	5			7.3	7.7	7.4	230	50	
June	1.2				5				7.0	7.6	7.2	266	>70	
July	0.3				5				6.2	7.0	6.5	303	>70	
Aug.	0.7				4				7.5	7.7	6.9	211	50	
Sept.	0.7				0.7				8.0	8.0	7.4	108	30	
Oct.	1.1					14			9.1	8.9	8.4	42	<10	
Nov.	0.8					30			7.9	8.0	7.4	35	5	
Dec.	0.6					31			7.2	7.7	6.8	11	5	
Year	10.7	92			16	203		80	7.4	7.6	7.0	1646	300	400

TABLE LXXXIV

CLIMATIC TABLE FOR PYATIGORSK

Latitude 44°03'N longitude 43°02'E elevation 573 m

Month	Air press. (mbar)		Wind		Global radiat.	Temperature (°C)				Humidity		Precipitation (mm)			
	sea level	station	preval. direct.	mean speed (m/s)	(kcal./cm²)	mean	mean daily range	max.	min.	vapor press. (mm)	relat. (%)	mean	max.	min.	max. in 24 h
Jan.	1021.4		E	3.2		−4.3		18	−31	3.0	85	13			
Feb.	1020.3		E	3.6		−3.2		18	−28	3.3	84	15			
Mar.	1020.7		E	4.0		−1.6		34	−24	4.3	82	25			
Apr.	1017.1		E	4.1		8.1		32	−13	6.1	70	37			
May	1014.3		E	3.3		14.7		33	−3	9.0	69	65			
June	1012.5		E	2.8		18.7		40	2	11.6	68	77			
July	1010.1		NW	2.9		21.6		41	7	13.3	65	70			
Aug.	1011.4		E	2.9		20.8		40	5	12.9	65	50			
Sept.	1015.9		E	3.0		15.5		40	−3	10.1	73	51			
Oct.	1019.8		E	3.3		9.6		36	−13	7.4	79	39			
Nov.	1024.8		E	3.5		2.7		27	−28	5.0	86	25			
Dec.	1023.1		E	3.5		−2.0		20	−33	3.7	85	15			
Year	1017.6		E	3.3		8.6		41	−33		76	482			

Month	Number of days							Snow cover max. 10-day average depth (cm)	Cloudiness (tenths)			Mean sunshine (h)	Evaporation (mm)	
	wind (speed ≥15 m/s)	frost-free period	fog	precip. (≥0.1 mm)	thunder-storm	snow-cover	clouds (8–10 tenths)		07h	13h	19h		actual	potent.
Jan.	2.8		16	13.9		31	18.1		8.2	7.9	7.6	77	5	
Feb.	2.6		14	13.9		28	17.1		8.5	7.8	7.9	74	5	
Mar.	3.9		13	13.2	0.4	1	17.8		8.3	8.0	7.8	106	20	
Apr.	3.5		7	11.1	1		14.2		7.3	7.4	7.5	136	40	
May	2.6		3	13.1	5		11.4		6.5	7.3	7.6	185	>50	
June	2.2		0.9	13.2	6		8.3		5.2	6.2	6.9	230	90	
July	1.9		1	11.3	6		6.7		4.9	5.5	6.0	255	>100	
Aug.	1.7		2	10.0	3		5.9		4.6	5.0	5.8	244	>40	
Sept.	1.7		6	9.3	1		8.5		5.6	5.6	5.5	171	30	
Oct.	2.0		10	12.3	0.4		12.9		7.1	6.7	6.0	133	20	
Nov.	2.0		12	14.4	0.03		17.1		8.2	7.8	7.5	83	5	
Dec.	2.2		15	14.4		13	18.4		8.0	8.2	7.7	62	5	
Year	29.1	175	100	150	23	70	156	11	6.9	7.9	7.0	1756	400	600

TABLE LXXXV

CLIMATIC TABLE FOR RIGA

Latitude 56°58'N longitude 24°04'E elevation 3 m

Month	Air press. (mbar)		Wind		Global radiat. (kcal./cm²)	Temperature (°C)				Humidity		Precipitation (mm)			
	sea level	station	preval. direct.	mean speed (m/s)		mean	mean daily range	max.	min.	vapor press. (mm)	relat. (%)	mean	max.	min.	max. in 24 h
Jan.			S	4.2	0.9	−5.0	5.6	7	−32	4.1	85	32	79	12	26
Feb.			S	3.9	2.2	−4.8	5.9	11	−35	3.8	83	28	86	4	21
Mar.			S, SE	3.6	5.8	−2.0	7.7	17	−30	4.2	77	24	91	2	24
Apr.			S	3.8	9.0	4.6	8.8	25	−13	6.6	74	33	127	11	25
May			N	3.6	13.7	10.7	10.5	30	−7	9.2	71	42	107	0	34
June			N	3.5	14.7	14.3	10.2	32	−2	12.5	73	60	173	13	44
July			N	3.1	14.3	17.1	9.8	34	3	14.7	76	78	249	14	61
Aug.			SW	3.0	10.7	15.7	9.5	33	2	14.6	80	71	167	17	50
Sept.			S	3.0	6.5	11.7	9.0	28	−5	11.7	83	60	144	0	41
Oct.			S	3.6	3.2	6.2	6.4	22	−11	8.6	83	53	131	12	36
Nov.			S	3.9	1.1	1.5	4.4	14	−22	6.2	86	47	142	13	33
Dec.			S	4.0	0.6	−2.6	4.5	12	−26	5.0	87	38	100	11	20
Year			S	3.6	82.7	5.6	–	34	−35	8.4	80	566	891	436	61

Month	Number of days							Snow cover max. 10-day average depth (cm)	Cloudiness (tenths)			Mean sun-shine (h)	Evaporation (mm)	
	wind (speed ⩾15 m/s)	frost-free period	fog	precip. (⩾0.1 mm)	thunder-storm	snow-cover	clouds (8–10 tenths)		07h	13h	19h		actual	potent.
Jan.	1.0		4	19.0	0.05		20.6	9	8.5	8.7	8.0	36	0.7	
Feb.	0.7		4	15.4	–		17.0	11	8.6	8.3	7.3	61	1.0	
Mar.	0.9		4	11.9	0.05		12.5	10	7.5	6.9	6.6	140	8.0	
Apr.	0.7		4	13.2	1		12.9		7.4	7.6	7.2	197	41	
May	0.3		2	12.2	3		10.2		6.9	6.9	6.5	268	55	
June	0.5		2	12.7	5		11.0		7.2	7.4	6.8	282	65	
July	0.1		2	14.3	6		12.0		6.9	7.6	6.9	276	85	
Aug.	0.0		3	15.2	4		10.8		7.5	7.7	7.1	235	78	
Sept.	0.3		5	15.5	2		10.9		7.7	7.7	6.9	166	54	
Oct.	0.6		5	15.9	0.2		16.9		8.5	8.5	7.4	91	40	
Nov.	0.6		5	16.8	0.2		22.2	1	9.1	9.1	8.3	35	15	
Dec.	0.6		4	18.2	0.1		21.5	5	8.5	9.0	8.3	25	2.0	
Year	6.3	133	44	180.3	22	93	178.5	11	7.9	8.0	7.3	1812	444.7	552.1

TABLE LXXXVI

CLIMATIC TABLE FOR ROSTOV-NA-DONU

Latitude 47°15'N longitude 39°49'E elevation 77 m

Month	Air press. (mbar)		Wind		Global radiat. (kcal./cm²)	Temperature (°C)				Humidity		Precipitation (mm)			
	sea level	station	preval. direct.	mean speed (m/s)		mean	mean daily range	max.	min.	vapor press. (mm)	relat. (%)	mean	max.	min.	max. in 24 h
Jan.	1020.3		E	6.5		−6.3	7.7	15	−33	2.9	87	38	109	4	26
Feb.	1018.3		E	7.0		−5.5	9.3	19	−30	3.1	85	41	125	1	25
Mar.	1018.3		E	6.8		0.2	11.2	28	−28	4.2	80	32	94	1	26
Apr.	1016.3		E	6.4		8.6	14.2	32	−6	6.0	67	39	111	4	39
May	1014.5		E	5.6		15.9	13.8	34	−2	8.9	59	36	157	1	78
June	1012.2		E	4.6		19.6	13.8	38	0	11.8	61	58	176	4	100
July	1010.2		W	4.3		22.7	14.0	40	8	12.8	58	49	186	1	78
Aug.	1011.9		NE, E	4.2		21.8	14.3	40	5	11.3	58	37	112	1	57
Sept.	1016.7		E	4.4		15.8	14.3	36	−4	8.8	62	32	123	0	49
Oct.	1019.9		E	5.4		9.3	12.7	33	−10	6.9	75	44	105	0	49
Nov.	1022.4		E	7.0		2.0	9.8	25	−23	5.0	82	40	138	2	35
Dec.	1021.9		E	7.0		−3.5	8.2	15	−28	3.6	87	37	134	1	31
Year	1017.0		E	5.8		8.4		40	−33		72	483	758	29.9	100

Month	Number of days							Snow cover max. 10-day average depth (cm)	Cloudiness (tenths)			Mean sun-shine (h)	Evaporation (mm)	
	wind (speed ⩾15 m/s)	frost-free period	fog	precip. (⩾0.1 mm)	thunder-storm	snow-cover	clouds (8–10 tenths)		07h	13h	19h		actual	potent.
Jan.	2.9		9	13.2	0.0	31	20.5		8.4	8.4	7.8	47	<10	
Feb.	3.4		8	12.5	0.0	28	16.0		8.4	8.1	7.5	68	<10	
Mar.	3.6		9	11.5	0.1		14.7		8.0	7.8	7.0	132	20	
Apr.	3.3		4	9.6	1		8.8		6.7	6.8	6.4	189	50	
May	1.9		1	8.9	4		6.0		5.9	6.2	6.1	270	>70	
June	1.1		2	9.6	8		3.5		4.9	5.8	5.4	297	70	
July	1.2		2	8.2	7		2.4		3.5	4.9	4.2	330	<60	
Aug.	1.3		2	6.6	4		1.8		3.6	4.5	4.0	304	<40	
Sept.	1.1		2	6.3	2		3.6		4.3	4.8	4.0	245	<30	
Oct.	2.4		8	8.7	0.7		8.7		6.7	6.5	5.1	152	>10	
Nov.	2.5		11	10.4	0.07		16.4		8.3	7.9	6.9	79	<10	
Dec.	3.2		11	12.9	0.0	4	20.2		8.8	8.6	7.9	36	<10	
Year	27.9	181	69	118	27	70	123	18	6.5	6.7	6.0	214	<400	800

TABLE LXXXVII

CLIMATIC TABLE FOR RUDNAYA PRISTAN'

Latitude 44°22'N longitude 135°51'E elevation 7 m

Month	Air press. (mbar) sea level	Air press. (mbar) station	Wind preval. direct.	Wind mean speed (m/s)	Global radiat. (kcal./cm²)	Temperature (°C) mean	Temperature (°C) mean daily range	Temperature (°C) max.	Temperature (°C) min.	Humidity vapor press. (mm)	Humidity relat. (%)	Precipitation (mm) mean	Precipitation (mm) max.	Precipitation (mm) min.	Precipitation (mm) max. in 24 h
Jan.		1016.0	W	6.8		−12.9		6	−34	1.2	46	12			
Feb.		1015.2	W	5.3		−9.3		12	−30	1.6	48	12			
Mar.		1014.6	W	3.9		−3.6		17	−28	2.9	58	27			
Apr.		1812.4	W	3.0		2.9		24	−13	5.2	70	51			
May		1010.4	E	2.7		6.9		30	−4	7.7	77	71			
June		1009.0	E	2.5		10.9		33	0	11.4	86	85			
July		1007.6	E	2.1		15.8		38	3	16.2	89	117			
Aug.		1008.7	E	2.3		18.4		36	3	18.5	87	110			
Sept.		1010.4	W	3.0		14.1		31	−3	12.8	77	125			
Oct.		1014.2	W	3.4		6.9		26	−12	6.7	64	70			
Nov.		1015.1	W	4.2		−2.4		18	−22	3.1	52	42			
Dec.		1016.5	W	5.7		−10.1		10	−30	1.5	47	20			
Year		1012.5	W	3.7		3.1		38	−34	7.4	67	742			

Month	Number of days: wind (speed ⩾15 m/s)	frost-free period	fog	precip. (⩾0.1 mm)	thunder-storm	snow-cover	clouds (8–10 tenths)	Snow cover max. 10-day average depth (cm)	Cloudiness (tenths) 07h	Cloudiness 13h	Cloudiness 19h	Mean sun-shine (h)	Evaporation (mm) actual	Evaporation (mm) potent.
Jan.	7.8						2.4	3	3.3	3.6	2.5	194	1.2	
Feb.	4.4		0.1				2.8	3	3.4	3.8	2.6	196	3.2	
Mar.	2.4		1		0.03		5.6	3	4.8	5.4	4.2	223	8	
Apr.	1.5		6		0.03		8.0		6.0	6.5	5.8	191	15	
May	0.9		8		1		12.5		6.9	7.2	7.3	186	20	
June	0.2		12		2		16.2		7.6	7.6	7.9	132	42	
July	0.2		13		1		18.9		8.1	8.0	8.1	123	52	
Aug.	0.1		9		1		16.2		7.6	7.5	7.6	141	60	
Sept.	1.1		2		1		7.6		5.4	5.8	5.5	183	35	
Oct.	1.6		0.8		0.4		4.0		4.1	4.4	3.4	216	20	
Nov.	3.3		0.5		0.2		2.3	1	3.4	3.8	2.5	184	10	
Dec.	6.4		0.03				2.5	3	3.1	3.4	2.4	186	3	
Year	29.9	151	52.4		7	90	99	4	5.3	5.6	4.8	2155	269.4	520.0

TABLE LXXXVIII

CLIMATIC TABLE FOR SALEKHARD

Latitude 66°32'N longitude 66°32'E elevation 35 m

Month	Air press. (mbar) sea level	Air press. (mbar) station	Wind preval. direct.	Wind mean speed (m/s)	Global radiat. (kcal./cm²)	Temperature (°C) mean	Temperature (°C) mean daily range	Temperature (°C) max.	Temperature (°C) min.	Humidity vapor press. (mm)	Humidity relat. (%)	Precipitation (mm) mean	Precipitation (mm) max.	Precipitation (mm) min.	Precipitation (mm) max. in 24 h
Jan.	1014.1	1012.2	S, SW	3.9	0.2	−24.4	7.0	2	−50	0.8	84	24	45	4	6
Feb.	1013.8	1011.9	NE	4.0	1.3	−21.9	8.2	2	−54	0.8	84	20	46	3	6
Mar.	1013.8	1011.9	NE	4.4	5.6	−17.9	10.1	5	−47	1.1	81	24	39	5	16
Apr.	1013.9	1012.2	NE	5.1	10.7	−10.2	11.9	11	−33	2.1	80	32	71	5	12
May	1011.9	1010.2	NE	5.7	13.4	−2.1	11.6	24	−26	3.4	77	39	106	10	32
June	1009.7	1007.9	NE	6.0	15.0	7.1	11.5	30	−11	6.1	71	51	114	2	44
July	1008.6	1006.9	NE	5.0	16.0	13.8	11.6	31	0	8.8	70	57	135	9	48
Aug.	1009.4	1007.7	NE	5.2	9.4	11.2	11.2	30	−2	8.2	77	57	165	12	42
Sept.	1009.5	1007.8	NE	4.5	3.8	5.2	9.8	22	−10	5.8	83	54	130	11	25
Oct.	1009.3	1007.5	SW	5.5	1.8	−1.1	6.6	15	−28	3.2	85	46	71	14	22
Nov.	1010.9	1009.0	SW	4.2	0.4	−15.8	6.7	5	−40	1.4	85	31	71	1	11
Dec.	1014.7	1012.9	SW	4.2	0.3	−21.5	8.1	2	−52	0.8	84	29	53	7	12
Year	1011.6	1009.8	NE	4.8	77.9	−6.7		31	−54		80	464	696	265	48

Month	Number of days: wind (speed ⩾15 m/s)	frost-free period	fog	precip. (⩾0.1 mm)	thunder-storm	snow-cover	clouds (8–10 tenths)	Snow cover max. 10-day average depth (cm)	Cloudiness (tenths) 07h	Cloudiness 13h	Cloudiness 19h	Mean sun-shine (h)	Evaporation (mm) actual	Evaporation (mm) potent.
Jan.	3.0		6	19.3		31	12.7		6.6	7.6	6.2	5	5	
Feb.	2.6		4	17.1		28	10.0		7.2	7.2	6.0	46	5	
Mar.	4.2		4	16.2		31	10.1		7.0	6.3	6.4	145	5	
Apr.	3.6		3	11.2		30	11.1		7.3	7.0	7.1	191	5	
May	4.3		2	13.2	0.1	20	14.1		7.9	7.6	7.3	223	<10	
June	4.2		1	13.3	1		12.6		7.3	7.3	7.0	245	70	
July	2.5		0.9	11.1	3		10.0		7.0	7.0	6.8	299	>60	
Aug.	2.7		3	13.9	2		14.3		7.4	7.5	7.2	190	50	
Sept.	2.9		4	12.2	0.2		16.0		8.2	8.4	7.9	95	20	
Oct.	4.2		3	16.2		15	17.4		8.5	8.4	7.6	53	<10	
Nov.	3.1		6	18.4		30	15.2		7.6	7.7	6.7	20	5	
Dec.	3.1		6	20.3		31	13.3		6.8	7.9	6.4	0	5	
Year	40.4	94	43	182.4	6	233	157	59	7.4	7.5	6.9	1512	200	300

TABLE LXXXIX

CLIMATIC TABLE FOR SEMIPALATINSK

Latitude 50°24′N longitude 80°13′E elevation 202 m

Month	Air press. (mbar)		Wind		Global radiat.	Temperature (°C)				Humidity		Precipitation (mm)			
	sea level	station	preval. direct.	mean speed (m/s)	(kcal./cm²)	mean	mean daily range	max.	min.	vapor press. (mm)	relat. (%)	mean	max.	min.	max. in 24 h
Jan.	1032.8		E	4.1	3.7	−16.2	9.7	6	−47	1.2	77	14	56	3	16
Feb.	1030.2		E	3.8	5.8	−15.6	10.8	7	−45	1.2	77	15	36	0	13
Mar.	1027.0		E	4.2	10.5	−8.9	11.0	24	−41	2.1	78	17	63	1	14
Apr.	1021.8		E, W	4.0	10.5	3.8	13.0	31	−24	4.0	64	19	62	0	22
May	1015.2		W	3.7	14.6	13.9	15.0	37	−10	6.3	51	22	82	3	26
June	1009.1		W	3.4	17.3	20.0	14.7	41	−1	9.1	51	30	91	3	40
July	1005.4		W	3.1	16.2	22.1	14.1	42	5	10.4	55	32	120	1	38
Aug.	1009.5		W	2.9	13.2	19.9	14.6	42	1	9.4	57	23	69	0	26
Sept.	1016.0		E	3.2	11.5	13.2	15.1	37	−8	6.4	59	21	54	0	27
Oct.	1023.8		E	3.6	5.7	4.8	11.6	29	−18	4.1	66	22	71	3	21
Nov.	1030.4		E	4.4	2.9	−6.1	8.9	17	−49	2.5	76	27	83	3	30
Dec.	1032.6		E	4.0	2.7	−13.5	9.2	8	−46	1.6	77	22	58	6	15
Year	1021.1		E	3.7	114.6	3.1		42	−49		66	264	418	142	40

Month	Number of days							Snow cover max. 10-day average depth (cm)	Cloudiness (tenths)			Mean sun-shine (h)	Evaporation (mm)	
	wind (speed ⩾15 m/s)	frost-free period	fog	precip. (⩾0.1 mm)	thunder-storm	snow-cover	clouds (8–10 tenths)		07h	13h	19h		actual	potent.
Jan.	1.8		5	12.0	0		11		6.3	6.1	5.0	113	<10	
Feb.	1.3		5	9.4	0		7		6.3	5.8	4.6	137	<10	
Mar.	2.0		4	10.4	0.04		8		6.8	6.2	5.3	197	<10	
Apr.	1.7		1	5.7	0.3		6		6.3	6.4	5.8	238	>10	
May	2.4		0.3	8.0	2		6		5.4	6.2	5.9	302	70	
June	2.0		0.1	7.8	5		5		5.1	6.2	5.8	314	50	
July	2.1		0.4	8.3	6		5		4.9	6.2	5.7	321	<50	
Aug.	1.4		0.9	7.9	5		4		4.8	5.9	5.3	305	<20	
Sept.	1.4		1	6.5	0.7		5		5.3	5.8	4.8	251	<20	
Oct.	2.1		2	8.5	0.1		11		7.2	7.3	5.6	153	<10	
Nov.	1.3		2	12.9	0.04		14		7.4	7.2	6.1	103	<10	
Dec.	1.7		3	13.3	0		11		7.0	7.1	5.8	89	<10	
Year	21.2	140	25	110.7	19	147	93	30	6.1	6.4	5.5	2523	<300	600

TABLE XC

CLIMATIC TABLE FOR SEROV

Latitude 59°36′N longitude 60°32′E elevation 132 m

Month	Air press. (mbar)		Wind		Global radiat.	Temperature (°C)				Humidity		Precipitation (mm)			
	sea level	station	preval. direct.	mean speed (m/s)	(kcal./cm²)	mean	mean daily range	max.	min.	vapor press. (mm)	relat. (%)	mean	max.	min.	max. in 24 h
Jan.		1008.6	W	2.7	1.6	−16.9		2	−45	1.7	80	18			
Feb.		1005.5	W	3.0	2.9	−15.1		6	−45	1.6	76	15			
Mar.		1002.3	W	3.4	6.8	−7.9		18	−40	2.5	70	19			
Apr.		1001.5	W	3.5	11.0	2.0		26	−27	4.7	62	20			
May		1000.4	W	3.9	13.2	8.0		34	−12	6.6	59	41			
June		996.9	W	3.6	15.4	14.5		36	−4	10.6	61	64			
July		996.4	W	3.1	13.7	16.7		35	−1	13.3	69	69			
Aug.		996.7	W	2.8	10.8	14.4		34	−3	12.5	74	65			
Sept.		996.2	W	3.1	6.8	8.1		30	−10	8.7	77	46			
Oct.		1003.4	W	3.4	3.4	0.0		22	−25	5.2	78	35			
Nov.		1005.2	W	3.3	1.7	−8.2		12	−45	2.9	77	30			
Dec.		1011.2	W	2.7	1.2	−15.3		4	−51	1.9	80	24			
Year		1002.0	W	3.2	88.5	0.0		36	−51	6.0	72	446			

Month	Number of days							Snow cover max. 10-day average depth (cm)	Cloudiness (tenths)			Mean sun-shine (h)	Evaporation (mm)	
	wind (speed ⩾15 m/s)	frost-free period	fog	precip. (⩾0.1 mm)	thunder-storm	snow-cover	clouds (8–10 tenths)		07h	13h	19h		actual	potent.
Jan.	0.2		2				10.7	36	6.4	7.3	5.7		0	
Feb.	0.4		2				10.4	42	7.1	6.8	5.7		0	
Mar.	0.6		1				11.1	41	7.0	7.1	6.6		0.4	
Apr.	1.2		1		0.2		10.3	17	7.2	7.6	7.0		10	
May	1.3		1		2		12.3		7.0	7.7	7.1		58	
June	1.0		1		7		9.6		6.1	7.5	6.7		65	
July	0.4		1		8		11.8		6.3	7.7	6.9		70	
Aug.	0.1		1		5		9.8		6.3	7.4	6.4		55	
Sept.	0.4		2		1		12.7		7.6	8.3	7.2		33	
Oct.	0.5		1				14.6		8.2	8.6	7.1		10	
Nov.	0.2		1				12.3	13	7.2	7.5	6.4		4	
Dec.	0.2		1				11.7	26	6.5	7.3	6.3		0	
Year	6.5	93	15		23.2		137.3	42	6.9	7.6	6.6		305.4	478.5

TABLE XCI

CLIMATIC TABLE FOR SEYMCHAN

Latitude 62°55′N longitude 152°25′E elevation 211.2 m

Month	Air press. (mbar)		Wind		Global radiat. (kcal./cm²)	Temperature (°C)				Humidity		Precipitation (mm)			
	sea level	station	preval. direct.	mean speed (m/s)		mean	mean daily range	max.	min.	vapor press. (mm)	relat. (%)	mean	max.	min.	max. in 24 h
Jan.	1022.3		C	1.1		−39.5	8.4	−14	−62	−	73	28	61	10	9
Feb.	1024.8		C	1.2		−34.5	11.2	−3	−61	−	75	21	41	5	10
Mar.	1021.8		C	1.5		−27.1	17.1	3	−60	−	69	11	25	0	8
Apr.	1015.1		C	2.3		−11.9	15.7	8	−45	−	66	8	34	0	11
May	1012.1		C	2.5		2.1	12.4	27	−26	3.4	59	11	27	2	14
June	1007.5		C	2.4		12.7	14.4	35	−6	6.4	59	34	74	0	24
July	1007.1		C	2.3		15.5	14.3	37	−4	8.9	67	41	80	13	37
Aug.	1009.5		C	2.0		11.8	14.8	35	−9	7.7	71	39	95	16	25
Sept.	1013.4		C	1.9		3.9	13.1	25	−22	4.6	73	25	83	2	30
Oct.	1015.4		C	1.5		−11.8	11.2	14	−40	−	80	20	40	2	19
Nov.	1019.6		C	1.3		−27.5	9.5	6	−55	−	80	31	55	12	18
Dec.	1021.8		C	1.1		−36.3	8.0	−5	−62	−	78	27	50	5	10
Year	1015.9		C	1.8		−11.9		37	−62	−	71	296	409	66	37

Month	Number of days							Snow cover	Cloudiness (tenths)			Mean sunshine (h)	Evaporation (mm)	
	wind (speed ≥15 m/s)	frost-free period	fog	precip. (≥0.1 mm)	thunderstorm	snow-cover	clouds (8–10 tenths)	max. 10-day average depth (cm)	07h	13h	19h		actual	potent.
Jan.	0.3		17	18.4		31	15.4					20	5	
Feb.	0.4		10	16.1		28	14.1					74	5	
Mar.	0.4		5	10.9		31	9.2					214	5	
Apr.	0.4		1	7.7		30	10.9					267	5	
May	0.2		0.2	6.6	0.1	11	13.6					301	>20	
June	0.4		1	9.7	2		11.8					292	>70	
July	0.4		1	11.4	3		14.3					277	70	
Aug.	0.1		2	10.4	1		12.9					229	50	
Sept.	0.4		2	8.9			12.9					162	20	
Oct.	0.3		3	12.5		23	12.4					106	5	
Nov.	0.6		6	17.8		30	15.7					45	5	
Dec.	0.4		18	18.1		31	14.7					15	5	
Year	4.3	51	66	149	6	221	158	53				2002	>200	400

TABLE XCII

CLIMATIC TABLE FOR SIMFEROPOL'

Latitude 45°01′N longitude 33°59′E elevation 205 m

Month	Air press. (mbar)		Wind		Global radiat. (kcal./cm²)	Temperature (°C)				Humidity		Precipitation (mm)			
	sea level	station	preval. direct.	mean speed (m/s)		mean	mean daily range	max.	min.	vapor press. (mm)	relat. (%)	mean	max.	min.	max. in 24 h
Jan.	1019.6	986.2	NE	2.9		−0.7	8.2	20	−26	3.8	85	44	102	2	29
Feb.	1018.6	985.1	SW	3.3		−0.6	8.7	23	−29	3.7	83	40	103	0	31
Mar.	1016.2	983.4	NE	3.7		3.6	10.0	29	−20	4.4	77	35	84	0	20
Apr.	1014.2	981.9	NE, SE	3.5		8.5	12.5	32	−10	5.5	67	36	91	0	43
May	1014.3	982.9	SE	2.8		14.3	13.4	36	−3	8.4	67	49	111	2	39
June	1012.3	981.4	E, SE, SW	2.5		18.0	13.1	37	3	10.6	67	75	221	8	101
July	1011.4	980.7	SE	2.2		20.6	14.0	40	6	11.8	63	58	312	4	122
Aug.	1013.1	982.5	SE	2.2		20.0	14.2	40	6	11.2	63	34	101	2	40
Sept.	1016.8	985.4	SE	2.3		15.3	14.1	38	−3	9.0	68	31	114	2	58
Oct.	1018.7	986.7	SE	2.9		11.1	11.9	34	−10	7.4	75	41	114	1	42
Nov.	1020.2	987.5	NE, SE	3.3		5.3	9.6	30	−18	5.5	83	44	139	2	52
Dec.	1019.5	986.3	SE	3.2		1.7	8.2	23	−22	4.4	85	41	162	3	43
Year	1016.2	984.2	SE	2.8		9.8	−	40	−29		74	528	768	318	122

Month	Number of days							Snow cover	Cloudiness (tenths)			Mean sunshine (h)	Evaporation (mm)	
	wind (speed ≥15 m/s)	frost-free period	fog	precip. (≥0.1 mm)	thunderstorm	snow-cover	clouds (8–10 tenths)	max. 10-day average depth (cm)	07h	13h	19h		actual	potent.
Jan.	4.8			13.4		31	14.5		7.7	7.7	6.9	83	5	
Feb.	3.8			13.0		28	12.3		8.2	8.1	7.2	90	5	
Mar.	6.4			10.9		22	10.5		7.6	7.4	6.6	154	<15	
Apr.	5.3			9.3			6.8		6.3	6.5	5.7	214	50	
May	2.1			9.2			3.6		5.5	6.1	5.3	282	70	
June	1.4			9.6			2.4		4.3	5.9	4.4	315	70	
July	1.4			7.6			1.0		2.7	4.5	3.0	357	70	
Aug.	1.4			6.2			1.0		2.7	4.5	3.1	334	<40	
Sept.	1.6			6.6			1.9		3.3	4.4	3.0	255	30	
Oct.	2.5			8.8			5.0		5.7	5.9	4.4	190	20	
Nov.	2.8			11.3		28	9.3		7.0	7.1	6.1	117	10	
Dec.	4.7			12.8		31	14.5		7.8	7.8	6.9	78	10	
Year	38.2	194		119.0		39	83	12	5.7	6.3	5.2	2469	<400	900

TABLE XCIII

CLIMATIC TABLE FOR SIMUSHIR (Kuril Islands)

Latitude 47°15'N longitude 150°47'E

Month	Air press. (mbar)		Wind		Global radiat.	Temperature (°C)				Humidity		Precipitation (mm)			
	sea level	station	preval. direct.	mean speed (m/s)	(kcal./cm²)	mean	mean daily range	max.	min.	vapor press. (mm)	relat. (%)	mean	max.	min.	max. in 24 h
Jan.	1006.7										80	102			
Feb.	1009.1										80	63			
Mar.	1010.2										82	90			
Apr.	1010.7										83	117			
May	1010.7										85	119			
June	1009.9										92	70			
July	1008.6										94	112			
Aug.	1011.4										96	162			
Sept.	1012.2										89	198			
Oct.	1015.4										81	175			
Nov.	1012.7										79	123			
Dec.	1008.6										80	130			
Year	1010.5										85	1461			

TABLE XCIV

CLIMATIC TABLE FOR SOCHI

Latitude 43°35'N longitude 39°43'E elevation 31 m

Month	Air press. (mbar)		Wind		Global radiat.	Temperature (°C)				Humidity		Precipitation (mm)			
	sea level	station	preval. direct.	mean speed (m/s)	(kcal./cm²)	mean	mean daily range	max.	min.	vapor press. (mm)	relat. (%)	mean	max.	min.	max. in 24 h
Jan.	1019.8		NE	2.5	3.3	5.7	6.3	21	–14	6.9	70	145	354	2	145
Feb.	1018.3		NE	2.8	4.8	5.7	6.7	24	–14	7.0	71	126	310	10	126
Mar.	1016.7		SE	2.5	6.1	8.4	7.1	30	–11	7.5	70	99	274	15	99
Apr.	1014.9		SE	2.3	10.0	11.5	7.6	31	–2	10.0	74	92	273	13	92
May	1014.5		SE	2.0	14.1	16.0	7.6	34	4	14.3	78	71	205	3	71
June	1012.9		NE	2.0	16.5	19.9	7.7	35	9	18.7	78	78	255	8	78
July	1010.7		NE	1.9	17.0	22.7	7.7	35	11	22.1	78	94	269	2	94
Aug.	1011.7		NE	2.0	15.1	23.0	7.9	38	10	21.5	76	84	581	5	84
Sept.	1015.4		NE	2.2	10.5	19.7	8.2	36	3	17.3	74	127	380	3	127
Oct.	1018.3		NE	2.3	7.0	16.1	8.3	34	–5	12.8	72	127	350	0	127
Nov.	1019.5		NE	2.4	3.8	11.3	7.4	29	–5	10.0	70	143	407	14	143
Dec.	1019.2		NE	2.8	2.7	7.9	6.8	23	–9	7.8	69	170	425	8	170
Year	1015.9		NE	2.3	110.9	14.0		38	–14	13.0	73	1356	2762	1100	1356

Month	Number of days							Snow cover	Cloudiness (tenths)			Mean sun-shine	Evaporation (mm)	
	wind (speed ⩾15 m/s)	frost-free period	fog	precip. (⩾0.1 mm)	thunder-storm	snow-cover	clouds (8–10 tenths)	max. 10-day average depth (cm)	07h	13h	19h	(h)	actual	potent.
Jan.	3.3		0.3	15.3	0.8		14.9		7.9	8.1	7.0	84	10	
Feb.	3.1		0.4	14.4	0.4		14.8		8.1	8.0	7.2	98	30	
Mar.	2.7		2	14.5	0.4		14.6		8.1	7.9	7.4	128	30	
Apr.	1.6		4	14.2	0.8		13.5		7.9	7.3	7.4	159	50	
May	0.7		4	12.9	2		10.3		7.2	6.7	6.8	223	>60	
June	0.4		0.6	10.7	5		6.5		5.3	5.0	5.2	283	70	
July	0.4		0.1	9.9	6		4.2		4.4	3.8	3.7	313	70	
Aug.	0.6		0.1	8.3	6		2.8		4.6	3.6	3.8	305	50	
Sept.	0.9		0.1	9.4	5		4.6		5.0	4.3	4.3	252	30	
Oct.	0.9		0.0	11.0	3		7.8		6.4	6.2	5.0	194	20	
Nov.	1.7		0.5	12.5	1		10.5		6.8	6.7	5.8	121	20	
Dec.	3.1		0.3	14.9	1		14.1		7.3	7.6	6.4	94	10	
Year	19.4	289	13	148.0	31	8	119	6	6.6	6.3	5.8	2253	400	1000

TABLE XCV

CLIMATIC TABLE FOR SUKHUMI

Latitude 43°10'N longitude 41°05'E elevation 37 m

Month	Air press. (mbar)		Wind		Global radiat. (kcal./cm²)	Temperature (°C)				Humidity		Precipitation (mm)			
	sea level	station	preval. direct.	mean speed (m/s)		mean	mean daily range	max.	min.	vapor press. (mm)	relat. (%)	mean	max.	min.	max. in 24 h
Jan.	1020.0	1015.4	NE	1.7		5.7	7.3	24	−14	6.7	70	114	378	8	
Feb.	1018.7	1014.1	NE	1.8		6.3	7.6	28	−13	6.8	71	118	281	14	
Mar.	1017.1	1012.5	NE	1.7		9.4	8.1	31	−10	8.0	69	112	295	29	
Apr.	1015.1	1010.6	NE	1.6		12.7	9.0	34	−2	10.4	72	122	234	16	
May	1014.5	1010.1	W	1.5		17.1	8.6	37	4	14.5	75	97	244	2	
June	1012.9	1008.5	W	1.5		20.8	8.5	39	9	18.5	75	97	273	9	
July	1010.9	1006.6	W	1.5		23.5	8.2	40	12	21.6	79	112	307	14	
Aug.	1011.8	1007.5	W	1.4		23.9	8.8	41	11	21.9	75	114	492	7	
Sept.	1015.4	1011.1	N	1.6		20.6	9.4	39	3	17.5	73	134	326	20	
Oct.	1018.6	1014.1	N	1.8		16.7	9.4	37	−5	13.2	71	107	392	3	
Nov.	1019.6	1015.1	NE	1.8		12.0	8.7	32	−5	9.7	69	128	294	5	
Dec.	1019.7	1015.1	NE	1.6		8.2	8.0	24	−8	7.6	69	135	299	11	
Year	1016.2	1011.7	NE	1.6		14.7		41	−14	13.0	73	1390	1986	934	

Month	Number of days							Snow cover max. 10-day average depth (cm)	Cloudiness (tenths)			Mean sunshine (h)	Evaporation (mm)	
	wind (speed ≥15 m/s)	frost-free period	fog	precip. (≥0.1 mm)	thunder-storm	snow-cover	clouds (8–10 tenths)		07h	13h	19h		actual	potent.
Jan.	0.6		0.6	13.6	0.4		15.3		7.7	7.9	7.1	94	20	
Feb.	0.6		0.7	13.9	0.2		14.8		8.2	8.1	7.2	99	30	
Mar.	0.9		2	14.5	0.2		14.6		8.1	8.0	7.6	130	50	
Apr.	0.7		3	14.8	0.8		14.1		7.8	7.5	7.6	155	>50	
May	1.1		3	13.3	3		12.3		7.3	6.9	7.5	202	>80	
June	0.6		0.5	11.9	5		7.9		5.8	5.4	6.0	245	>80	
July	0.3		0.1	10.8	6		7.9		5.7	5.0	5.7	259	70	
Aug.	0.2		0.0	10.0	6		6.6		5.1	4.5	5.2	264	70	
Sept.	0.5		0.1	11.0	6		6.5		5.1	4.9	4.8	228	50	
Oct.	0.5		0.2	10.9	3		8.4		6.3	6.2	5.1	181	30	
Nov.	0.4		0.2	12.1	1		11.8		7.0	7.0	5.9	131	30	
Dec.	0.5		0.2	13.5	0.7		14.0		7.5	7.7	6.6	101	30	
Year	6.9	303	11	150.3	32	11	134	9	6.8	6.6	6.4	2089	600	1000

TABLE XCVI

CLIMATIC TABLE FOR SURGUT

Latitude 61°15'N longitude 73°30'E elevation 40 m

Month	Air press. (mbar)		Wind		Global radiat. (kcal./cm²)	Temperature (°C)				Humidity		Precipitation (mm)			
	sea level	station	preval. direct.	mean speed (m/s)		mean	mean daily range	max.	min.	vapor press. (mm)	relat. (%)	mean	max.	min.	max. in 24 h
Jan.	1018.6	1013.5	SW	4.7		−22.2	9.7	3	−52	0.9	79	24	62	6	11
Feb.	1019.1	1014.1	W	5.2		−19.3	10.5	6	−55	1.0	78	19	45	2	6
Mar.	1017.5	1012.6	W	5.4		−12.8	12.9	10	−49	1.6	74	23	64	4	12
Apr.	1015.9	1011.1	W	5.2		−4.4	10.6	22	−37	2.8	69	30	73	2	19
May	1012.5	1007.8	N, W	5.4		3.6	10.0	32	−22	4.6	67	48	121	5	37
June	1008.3	1003.9	N	5.0		12.6	10.1	34	−7	8.0	66	55	139	8	24
July	1006.9	1002.5	N	4.3		16.8	9.7	34	−1	10.4	70	68	189	21	68
Aug.	1008.2	1003.8	N	4.5		13.9	9.2	30	−4	9.6	78	57	186	0	87
Sept.	1011.4	1006.9	SW, W	5.2		7.4	7.9	27	−10	6.7	81	60	128	8	34
Oct.	1012.7	1007.9	SW	5.8		−1.7	5.8	21	−30	3.8	83	49	92	16	21
Nov.	1015.5	1010.6	W	5.8		−13.3	8.4	8	−47	1.9	82	30	68	8	15
Dec.	1019.4	1014.3	SW, W	5.2		−20.2	9.5	2	−55	1.0	81	29	62	10	12
Year	1013.8	1009.0	W	5.2		−3.3		34	−55		76	492	665	298	87

Month	Number of days							Snow cover max. 10-day average depth (cm)	Cloudiness (tenths)			Mean sunshine (h)	Evaporation (mm)	
	wind (speed ≥15 m/s)	frost-free period	fog	precip. (≥0.1 mm)	thunder-storm	snow-cover	clouds (8–10 tenths)		07h	13h	19h		actual	potent.
Jan.	1.1			22.8		31	13.5		6.8	7.6	6.1	30	5	
Feb.	1.3			16.4		28	9.8		7.3	6.5	5.4	80	5	
Mar.	2.4			16.2		31	9.1		7.4	6.5	6.6	151	5	
Apr.	2.2			14.0	0.02	30	10.6		6.8	6.7	6.8	193	10	
May	2.6			15.9	1	4	13.4		7.3	7.4	7.3	217	50	
June	1.5			14.6	3		11.2		6.9	7.5	7.1	253	>70	
July	0.7			14.5	5		10.9		6.3	7.2	6.8	275	>70	
Aug.	0.9			17.4	4		12.2		7.1	7.7	7.2	204	>50	
Sept.	1.0			16.2	0.6		14.4		8.2	8.3	7.6	120	30	
Oct.	1.4			20.4	0	8	17.6		8.8	8.7	8.0	59	5	
Nov.	1.7			21.9		30	10.9		8.2	8.0	6.9	36	5	
Dec.	1.4			23.0		31	14.3		7.1	7.7	6.6	14	5	
Year	18.2	102		213.3	14	205	154	75	7.4	7.5	6.9	1632	300	400

TABLE XCVII

CLIMATIC TABLE FOR SVERDLOVSK

Latitude 56°44′N longitude 61°04′E elevation 282 m

Month	Air press. (mbar) sea level	Air press. (mbar) station	Wind preval. direct.	Wind mean speed (m/s)	Global radiat. (kcal./cm²)	Temperature (°C) mean	Temperature (°C) mean daily range	Temperature (°C) max.	Temperature (°C) min.	Humidity vapor press. (mm)	Humidity relat. (%)	Precipitation (mm) mean	Precipitation (mm) max.	Precipitation (mm) min.	Precipitation (mm) max. in 24 h
Jan.	1023.8		W	4.5	1.5	−15.6	8.7	4	−42	1.3	80	15	56	1	8.7
Feb.	1023.4		W	4.4	3.5	−13.6	12.3	8	−42	1.5	77	17	39	1	12.3
Mar.	1019.6		W	4.8	8.4	−7.4	14.8	17	−39	2.1	71	17	62	0	14.8
Apr.	1018.6		W	4.7	11.2	2.1	14.1	27	−21	3.6	65	20	73	0	14.1
May	1014.7		W	4.3	14.0	9.9	14.5	33	−14	5.6	58	40	115	2	14.5
June	1014.5		W	4.0	15.8	15.2	15.3	35	−2	8.7	61	59	160	6	15.3
July	1008.5		W	3.8	14.7	17.3	14.3	38	2	10.4	68	80	221	21	14.3
Aug.	1012.2		W	3.9	11.3	14.8	14.2	37	−1	9.3	72	82	212	10	14.2
Sept.	1015.0		W	4.4	6.7	9.0	14.4	30	−9	6.6	75	49	119	5	14.4
Oct.	1018.4		W	5.1	3.3	1.2	12.6	25	−22	4.1	77	29	106	5	12.6
Nov.	1024.0		W	5.1	1.7	−7.1	8.9	14	−39	2.6	79	25	56	5	8.9
Dec.	1022.7		W	4.4	1.0	−13.6	7.8	6	−43	1.5	82	27	95	2	7.8
Year	1017.9		W	4.4	93.1	1.0		38	−43		72	462	743	314	

Month	Number of days wind (speed ≥15 m/s)	Number of days frost-free period	Number of days fog	Number of days precip. (≥0.1 mm)	Number of days thunder-storm	Number of days snow-cover	Number of days clouds (8–10 tenths)	Snow cover max. 10-day average depth (cm)	Cloudiness (tenths) 07h	Cloudiness (tenths) 13h	Cloudiness (tenths) 19h	Mean sun-shine (h)	Evaporation (mm) actual	Evaporation (mm) potent.
Jan.	0.9		3	13.8		31	14.8		7.5	7.9	6.9	52	5	
Feb.	1.0		4	11.1		28	11.6		7.8	7.3	5.8	90	5	
Mar.	1.4		3	10.4		31	11.8		7.5	7.2	6.5	144	<10	
Apr.	1.2		2	8.6	0.2	8	9.9		7.1	7.5	7.0	202	20	
May	2.3		2	13.2	3		10.1		6.5	7.7	7.0	245	<60	
June	1.7		3	14.3	8		9.2		6.1	7.5	6.7	269	>70	
July	0.9		3	15.0	9		10.0		6.4	7.7	7.0	267	<60	
Aug.	0.7		5	15.0	5		11.0		6.6	7.5	6.6	220	<40	
Sept.	1.0		4	14.3	0.7		12.5		8.0	8.4	7.3	133	<20	
Oct.	1.3		3	14.6	0.04		18.5		8.7	8.7	7.9	73	<10	
Nov.	1.3		2	15.2		24	19.3		7.9	8.0	7.0	46	<10	
Dec.	1.0		4	15.8		31	16.9		7.9	8.1	7.4	40	5	
Year	14.7	110	38	161.3	26	166	156	44	7.3	7.8	6.9	1781	300	500

TABLE XCVIII

CLIMATIC TABLE FOR SYKTYVKAR

Latitude 61°40′N longitude 50°51′E elevation 96.2 m

Month	Air press. (mbar) sea level	Air press. (mbar) station	Wind preval. direct.	Wind mean speed (m/s)	Global radiat. (kcal./cm²)	Temperature (°C) mean	Temperature (°C) mean daily range	Temperature (°C) max.	Temperature (°C) min.	Humidity vapor press. (mm)	Humidity relat. (%)	Precipitation (mm) mean	Precipitation (mm) max.	Precipitation (mm) min.	Precipitation (mm) max. in 24 h
Jan.	1016.4	999.1	SW	4.5		−15.2	7.3	3	−51	1.4	85	24	55	6	21
Feb.	1016.8	999.7	S	4.1		−13.8	11.2	3	−45	1.5	82	19	49	4	15
Mar.	1015.4	998.6	SW	4.3		−7.8	15.8	13	−39	2.1	76	23	70	0	15
Apr.	1015.6	999.4	SW	3.8		1.1	13.4	26	−27	3.4	69	30	80	1	16
May	1014.1	998.3	SW	4.3		7.7	14.8	30	−15	5.4	64	48	118	6	23
June	1010.2	994.7	SW	3.9		13.8	15.2	35	−5	7.9	64	55	132	8	28
July	1009.4	994.1	N	3.3		16.6	14.4	35	−1	10.0	69	68	166	13	37
Aug.	1010.7	995.3	SW	3.2		13.8	13.6	35	−2	9.2	77	57	156	3	44
Sept.	1012.3	996.6	SW	3.8		7.8	13.4	29	−9	6.7	84	60	124	3	31
Oct.	1014.2	997.8	SW	4.3		0.3	10.7	20	−30	4.4	86	49	111	14	21
Nov.	1015.5	998.7	SW	4.7		−7.2	7.1	10	−44	2.8	86	30	96	8	13
Dec.	1017.0	999.8	SW	4.5		−13.6	6.6	3	−45	1.8	86	29	50	7	11
Year	1014.0	997.7	SW	4.1		0.3		35	−51		78	492	791	340	44

Month	Number of days wind (speed ≥15 m/s)	Number of days frost-free period	Number of days fog	Number of days precip. (≥0.1 mm)	Number of days thunder-storm	Number of days snow-cover	Number of days clouds (8–10 tenths)	Snow cover max. 10-day average depth (cm)	Cloudiness (tenths) 07h	Cloudiness (tenths) 13h	Cloudiness (tenths) 19h	Mean sun-shine (h)	Evaporation (mm) actual	Evaporation (mm) potent.
Jan.	1.9		3	19.8		31	19.8		8.1	8.4	7.6	18	<10	
Feb.	1.8		2	16.3		28	16.3		8.1	8.0	7.0	52	<10	
Mar.	2.1		1	16.5		31	16.5		7.9	7.6	7.2	100	<10	
Apr.	2.2		2	13.0	0.09	22	13.0		7.4	7.4	7.1	168	20	
May	2.5		1	14.0	2		14.0		7.3	7.9	7.2	212	50	
June	2.2		0.6	14.1	4		14.1		6.6	7.3	7.0	272	>70	
July	1.4		1	13.8	6		13.8		6.3	7.2	6.8	292	>70	
Aug.	0.8		2	13.4	3		13.4		7.0	7.6	6.7	220	>50	
Sept.	1.7		5	18.2	0.5		18.2		8.6	8.8	8.0	95	<30	
Oct.	1.4		5	21.0	0.02		21.0		9.2	9.2	8.4	39	<10	
Nov.	1.4		4	19.3		28	19.3		8.9	8.9	8.4	17	<10	
Dec.	1.2		3	21.3		31	21.3		8.4	8.7	8.0	11	<10	
Year	20.6	103	30	201	16	186	201	56	7.8	8.1	7.4	1496	>300	400

TABLE XCIX

CLIMATIC TABLE FOR TALLIN

Latitude 59°25′N longitude 24°48′E elevation 44 m

Month	Air press. (mbar) sea level	Air press. (mbar) station	Wind preval. direct.	Wind mean speed (m/s)	Global radiat. (kcal./cm²)	Temperature (°C) mean	Temperature (°C) mean daily range	Temperature (°C) max.	Temperature (°C) min.	Humidity vapor press. (mm)	Humidity relat. (%)	Precipitation (mm) mean	Precipitation (mm) max.	Precipitation (mm) min.	Precipitation (mm) max. in 24 h
Jan.	1013.3	1012.5	S	6.3		−5.0	6.7	7	−32	3.1	85	33	86	5	28
Feb.	1013.5	1012.7	SW	5.4		−5.8	8.4	8	−32	2.8	83	26	107	5	25
Mar.	1012.5	1011.7	SW	5.3		−3.0	8.7	15	−24	3.4	80	24	81	2	18
Apr.	1013.4	1012.7	S, SW	5.4		2.6	8.6	24	−17	4.5	78	32	75	2	24
May	1014.8	1014.1	W	5.1		8.4	9.7	30	−5	6.7	74	41	91	8	44
June	1012.1	1011.4	W	5.0		13.1	9.9	31	1	8.9	77	49	135	12	82
July	1010.5	1009.8	W	4.8		16.4	8.4	33	5	11.1	79	71	139	19	41
Aug.	1010.5	1009.8	SW, W	4.7		15.0	10.9	33	4	10.7	80	68	173	6	57
Sept.	1013.2	1012.5	SW	5.0		11.0	10.9	28	−2	8.3	83	75	177	1	47
Oct.	1012.9	1012.2	SW	5.9		5.6	8.9	21	−11	6.0	84	65	155	3	40
Nov.	1012.9	1012.2	SW	6.3		0.8	6.6	12	−17	4.5	87	45	153	10	24
Dec.	1012.7	1011.9	SE	6.4		−3.0	5.6	10	−27	3.6	87	39	104	7	24
Year	1012.7	1012.0	SW	5.5		4.7		33	−32		81	568	813	363	82

Month	Number of days wind (speed ≥15 m/s)	frost-free period	fog	precip. (≥0.1 mm)	thunder-storm	snow-cover	clouds (8–10 tenths)	Snow cover max. 10-day average depth (cm)	Cloudiness (tenths) 07h	13h	19h	Mean sunshine (h)	Evaporation (mm) actual	potent.
Jan.	3.2		3	16.9		31	20.3		8.6	8.5	8.1	23	<10	
Feb.	2.1		4	14.5		28	14.8		8.1	7.7	7.2	51	<10	
Mar.	1.6		4	12.1		29	10.4		6.8	6.2	5.7	148	<10	
Apr.	1.5		5	11.9	0.4		10.9		7.0	6.6	6.5	192	>30	
May	0.8		4	11.5	2		9.2		6.4	6.1	5.8	262	>75	
June	0.4		2	11.7	3		8.4		6.1	6.4	6.0	287	>80	
July	0.3		2	13.2	5		10.1		6.2	6.4	6.3	281	70	
Aug.	1.4		3	14.5	3		8.4		6.6	6.6	6.2	234	>55	
Sept.	1.3		2	14.6	1		10.8		7.5	7.2	6.8	156	>30	
Oct.	2.2		3	17.4	0.1		15.2		8.2	8.1	7.2	75	>10	
Nov.	2.0		3	16.8	0.05		20.5		8.8	8.7	8.3	28	<10	
Dec.	2.7		3	17.2	0.1	6	21.6		8.6	8.8	8.4	16	<10	
Year	19.5	164	38	172.3	15	103	161	27	7.4	7.3	6.9	1753	400	500

TABLE C

CLIMATIC TABLE FOR TAMBOV

Latitude 52°44′N longitude 41°28′E elevation 139 m

Month	Air press. (mbar) sea level	Air press. (mbar) station	Wind preval. direct.	Wind mean speed (m/s)	Global radiat. (kcal./cm²)	Temperature (°C) mean	Temperature (°C) mean daily range	Temperature (°C) max.	Temperature (°C) min.	Humidity vapor press. (mm)	Humidity relat. (%)	Precipitation (mm) mean	Precipitation (mm) max.	Precipitation (mm) min.	Precipitation (mm) max. in 24 h
Jan.			SE	4.2	2.2	−10.8	6.9	4	−39	2.7	85	34	88	9	
Feb.			SE	4.4	4.1	−10.2	7.6	5	−37	2.9	84	29	105	1	
Mar.			SE, S	4.4	7.7	−5.1	7.8	18	−30	3.8	83	30	62	8	
Apr.			SE	3.8	11.4	5.1	9.4	30	−18	6.7	72	34	67	3	
May			N, SE	3.8	14.7	13.9	12.0	34	−8	9.3	60	50	125	5	
June			N	3.4	16.2	18.0	12.2	38	−1	12.7	60	53	121	6	
July			NW	3.1	17.7	20.2	11.9	40	4	15.1	65	64	172	3	
Aug.			N	2.9	13.7	18.5	11.8	38	1	14.3	68	53	155	1	
Sept.			W	3.2	9.8	12.2	11.1	37	−7	10.3	72	45	114	1	
Oct.			W	3.7	5.6	5.3	7.8	27	−19	7.0	80	44	106	0	
Nov.			SE	4.1	3.0	−2.0	5.8	16	−34	4.8	84	39	104	9	
Dec.			SE	4.2	1.8	−7.7	5.8	8	−37	3.6	86	38	117	2	
Year			SE	3.8	107.9	4.8	–	40	−39	7.8	75	513	735	337	

Month	Number of days wind (speed ≥15 m/s)	frost-free period	fog	precip. (≥0.1 mm)	thunder-storm	snow-cover	clouds (8–10 tenths)	Snow cover max. 10-day average depth (cm)	Cloudiness (tenths) 07h	13h	19h	Mean sunshine (h)	Evaporation (mm) actual	potent.
Jan.	0.6		7	17.4			17.5	20	8.2	8.1	7.1	36	0.3	
Feb.	0.5		6	14.7			14.6	25	7.9	7.5	6.9	76	3.5	
Mar.	0.9		7	13.6	0.04		13.8	24	7.7	6.9	6.7	121	7.0	
Apr.	0.6		4	10.9	0.4		10.5		6.8	7.4	6.6	180	30	
May	1.5		0.8	11.6	3		7.8		5.9	7.1	6.2	265	53	
June	0.6		0.3	12.0	6		8.0		4.9	6.5	5.7	284	62	
July	0.6		0.7	12.4	8		6.0		4.9	6.9	5.9	280	75	
Aug.	0.5		2	11.4	6		5.9		5.1	6.6	5.4	239	65	
Sept.	0.6		0.7	11.5	2		7.6		5.9	6.9	5.4	178	48	
Oct.	0.9		4	12.6	0.08		14.6		7.8	7.9	6.6	93	21	
Nov.	0.5		7	14.9			17.2	6	8.0	8.0	7.2	50	10	
Dec.	0.9		9	18.0			20.9	12	8.4	8.5	7.8	26	0.8	
Year	8.7	152	48.5	161	26	135	142.4	25	6.8	7.4	6.4	1828	375.6	675.9

TABLE CI

CLIMATIC TABLE FOR TASHKENT

Latitude 41°16′N longitude 69°16′E elevation 479 m

Month	Air press. (mbar)		Wind		Global radiat.	Temperature (°C)				Humidity		Precipitation (mm)			
	sea level	station	preval. direct.	mean speed (m/s)	(kcal./cm²)	mean	mean daily range	max.	min.	vapor press. (mm)	relat. (%)	mean	max.	min.	max. in 24 h
Jan.	1026.1	966.6	NE	1.3	4.8	−1.1	12.8	22	−28	3.2	74	49	107	8	30
Feb.	1023.9	965.0	NE	1.6	6.5	1.5	12.9	26	−26	3.7	69	51	110	0	32
Mar.	1020.4	963.1	NE	1.7	8.9	7.8	14.2	33	−20	4.9	67	81	165	6	40
Apr.	1016.0	960.3	NE	1.6	13.1	14.7	15.3	35	−6	7.4	60	58	154	0	38
May	1012.5	958.1	NE	1.5	18.0	20.2	16.3	42	0	9.2	55	32	89	0	50
June	1006.9	953.7	NE	1.4	19.3	25.3	17.1	44	4	9.9	44	12	67	0	33
July	1003.6	950.0	C, NE	1.3	19.4	27.4	17.4	44	8	10.7	40	4	36	0	25
Aug.	1006.4	953.3	NE	1.3	17.2	25.5	18.4	43	7	9.7	44	3	22	0	10
Sept.	1014.2	959.5	NE	1.4	13.5	19.7	18.7	40	0	7.0	46	3	34	0	19
Oct.	1021.7	965.3	NE	1.2	9.4	12.7	17.5	38	−11	5.6	56	23	117	0	30
Nov.	1025.2	967.4	NE	1.2	5.5	6.7	16.1	31	−22	4.3	67	44	148	1	47
Dec.	1026.2	067.3	NE	1.2	3.7	1.8	13.2	24	−30	3.6	75	57	164	6	37
Year	1016.9	960.0	NE	1.4	139.3	13.5		44	−30		58	417	643	141	50

Month	Number of days							Snow cover	Cloudiness (tenths)			Mean sun-shine (h)	Evaporation (mm)	
	wind (speed ⩾15 m/s)	frost-free period	fog	precip. (⩾0.1 mm)	thunder-storm	snow-cover	clouds (8–10 tenths)	max. 10-day average depth (cm)	07h	13h	19h		actual	potent.
Jan.	0.03		8	10.5	0		14		7.2	7.1	6.7	116	<10	
Feb.	0.1		5	9.8	0.1		12		7.4	7.4	7.0	125	10	
Mar.	0.2		3	11.4	0.8		13		7.4	7.4	7.2	165	20	
Apr.	0.2		0.8	9.9	3		11		6.4	6.8	6.8	229	>30	
May	0.3		0.3	6.8	5		6		4.8	5.1	5.6	312	30	
June	0.2		0.1	3.6	6		2		2.8	2.9	3.5	359	<10	
July	0.05		0.04	1.4	3		0.9		1.9	1.7	1.9	390	<10	
Aug.	0.03		0.2	0.6	0.9		0.5		1.5	1.1	1.4	371	<10	
Sept.	0.03		0.2	1.1	0.2		1		2.0	1.8	1.8	304	<10	
Oct.	0.1		1	4.9	0.1		5		4.3	4.3	3.6	233	<10	
Nov.	0.03		4	7.9	0		10		6.2	6.0	5.3	156	<10	
Dec.	0.03		10	10.1	0		14		6.9	6.8	6.4	110	<10	
Year	1.3	204	33	78.0	15	43	89	11	4.9	4.9	4.8	2820	>100	1000

TABLE CII

CLIMATIC TABLE FOR TBILISI

Latitude 41°41′N longitude 44°57′E elevation 404 and 490 m

Month	Air press. (mbar)		Wind		Global radiat.	Temperature (°C)				Humidity		Precipitation (mm)			
	sea level	station	preval. direct.	mean speed (m/s)	(kcal./cm²)	mean	mean daily range	max.	min.	vapor press. (mm)	relat. (%)	mean	max.	min.	max. in 24 h
Jan.	1023.1	973.1	NW	2.9	4.4	0.5	11.6	18	−23	3.5	74	16	68	0	29
Feb.	1021.4	971.7	NW	3.4	6.2	2.3	12.3	22	−14	3.7	69	22	87	0	44
Mar.	1018.9	970.2	SE	3.6	8.8	6.8	14.9	29	−13	4.5	65	29	88	1	30
Apr.	1015.8	968.2	SE	3.7	12.2	11.8	15.1	32	−4	6.2	61	56	130	5	99
May	1014.6	967.9	NW	3.1	16.1	17.1	14.2	35	1	9.1	63	92	198	5	65
June	1011.8	965.8	N	3.4	17.0	21.0	13.9	38	7	10.6	60	72	220	3	68
July	1009.6	964.3	N	3.5	17.2	24.2	14.3	40	9	12.1	57	50	175	1	80
Aug.	1011.3	065.9	N	3.2	16.2	24.1	14.8	40	9	11.8	57	34	203	0	65
Sept.	1016.3	060.8	NW	3.1	11.5	19.4	14.6	38	1	10.1	64	43	179	1	130
Oct.	1020.6	975.1	SE, NW	2.7	8.5	13.8	14.4	33	−5	8.0	72	42	139	4	67
Nov.	1022.9	974.1	SE	2.5	4.7	7.4	14.0	27	−7	5.7	77	39	126	1	76
Dec.	1023.1	973.5	NW	2.5	3.6	2.8	11.5	22	−19	4.2	75	23	83	0	46
Year	1017.4	969.8	NW	3.1	126.4	12.6		40	−23		66	513	767	241	130

Month	Number of days							Snow cover	Cloudiness (tenths)			Mean sun-shine (h)	Evaporation (mm)	
	wind (speed ⩾15 m/s)	frost-free period	fog	precip. (⩾0.1 mm)	thunder-storm	snow-cover	clouds (8–10 tenths)	max. 10-day average depth (cm)	07h	13h	19h		actual	potent.
Jan.	2.0		2	6.1	0.0		12.0		7.0	7.3	6.1	104		
Feb.	2.2		1	7.4	0.0		12.2		7.7	7.6	5.9	110		
Mar.	2.9		0.8	8.4	0.3		12.3		7.7	7.5	6.6	149		
Apr.	2.5		0.4	12.1	2		11.6		7.2	7.1	7.0	170		
May	1.4		0.1	15.8	9		10.3		6.7	6.4	7.6	211		
June	1.1		0.1	11.4	9		6.3		5.4	5.0	6.7	253		
July	1.0		0.1	9.9	7		6.3		5.6	4.7	5.0	272		
Aug.	1.1		0.1	7.8	5		5.2		5.4	4.1	5.0	264		
Sept.	1.0		0.2	9.2	4		7.1		5.9	4.8	4.9	206		
Oct.	1.0		0.4	8.6	2		9.4		6.5	5.9	5.0	170		
Nov.	1.2		2	9.4	0.1		12.0		7.6	7.2	5.8	110		
Dec.	1.3		3	7.2	0.0		12.6		7.2	7.1	5.8	93		
Year	18.7	235	10	113.3	38	16	117	6	6.7	6.2	6.0	2112		

TABLE CIII

CLIMATIC TABLE FOR TERMEZ

Latitude 37°17′N longitude 67°19′E elevation 302 m

Month	Air press. (mbar) sea level	Air press. (mbar) station	Wind preval. direct.	Wind mean speed (m/s)	Global radiat. (kcal./cm²)	Temperature (°C) mean	Temperature (°C) mean daily range	Temperature (°C) max.	Temperature (°C) min.	Humidity vapor press. (mm)	Humidity relat. (%)	Precipitation (mm) mean	Precipitation (mm) max.	Precipitation (mm) min.	Precipitation (mm) max. in 24 h
Jan.	1025.3	987.9	SW	2.1	5.6	2.8	14.4	24	−24	4.2	79	21	48	0	17
Feb.	1022.9	985.9	NE	3.0	7.6	5.7	15.4	30	−20	5.0	74	23	56	0	30
Mar.	1019.4	983.4	NE	3.8	10.6	11.5	15.9	32	−14	6.4	66	30	59	1	29
Apr.	1014.7	979.7	NE, SW	3.4	14.3	18.5	17.2	41	−3	9.0	60	19	56	0	24
May	1010.8	976.7	SW	2.9	18.2	24.5	18.4	46	4	9.8	47	10	71	0	28
June	1004.7	971.3	SW	2.7	20.0	29.3	19.4	50	7	10.6	40	1	10	0	10
July	1000.8	967.8	SW	2.6	19.6	31.4	19.2	50	11	10.4	34	0	2	0	2
Aug.	1004.1	970.7	SW	2.5	18.4	29.6	19.8	50	10	9.1	33	0	0	0	0
Sept.	1012.1	977.8	SW	2.0	13.1	23.3	20.0	41	2	7.3	39	0	1	0	0
Oct.	1019.7	984.3	SW	2.0	11.1	16.9	20.2	40	−9	6.2	50	3	12	0	25
Nov.	1023.8	987.5	NE	2.1	6.9	10.1	17.5	36	−17	5.0	62	0	74	0	16
Dec.	1025.4	988.2	NE	2.0	4.9	4.8	15.5	27	−25	4.6	75	17	64	0	22
Year	1015.3	980.1	SW	2.6	150.3	17.4		50	−25		55	133	257	62	30

Month	Number of days: wind (speed $\geq$15 m/s)	frost-free period	fog	precip. ($\geq$0.1 mm)	thunder-storm	snow-cover	clouds (8–10 tenths)	Snow cover max. 10-day average depth (cm)	Cloudiness (tenths) 07h	Cloudiness (tenths) 13h	Cloudiness (tenths) 19h	Mean sunshine (h)	Evaporation (mm) actual	Evaporation (mm) potent.
Jan.	1.5		3	7.6	0.1		10.9					140	5	
Feb.	1.9		1	7.1	0.4		10.5					148	10	
Mar.	2.5		0.5	7.7	0.5		8.7					176	25	
Apr.	2.0		0.2	5.4	2		6.3					235	45	
May	2.1		0.1	2.8	2		2.4					326	25	
June	1.1			0.5	0.7		0.7					373	<10	
July	0.9			0.0	0.2		0.2					388	<10	
Aug.	0.9			0.0	0.1		0.1					362	<10	
Sept.	0.5			0.2			0.2					312	<10	
Oct.	1.4		0.2	1.0	0.1		2.3					361	5	
Nov.	1.1		1	2.9	0.03		5.3					184	5	
Dec.	0.8		3	6.7	0.1		9.2					138	5	
Year	16.7	246	9	41.9	6	15	57	1				3043	>100	1000

TABLE CIV

CLIMATIC TABLE FOR TOBOL'SK

Latitude 58°12′N longitude 69°14′E elevation 64 m

Month	Air press. (mbar) sea level	Air press. (mbar) station	Wind preval. direct.	Wind mean speed (m/s)	Global radiat. (kcal./cm²)	Temperature (°C) mean	Temperature (°C) mean daily range	Temperature (°C) max.	Temperature (°C) min.	Humidity vapor press. (mm)	Humidity relat. (%)	Precipitation (mm) mean	Precipitation (mm) max.	Precipitation (mm) min.	Precipitation (mm) max. in 24 h
Jan.	1020.3	1007.1	SE	3.8		−18.3	7.5	44	1	1.1	82	10	44	1	9
Feb.	1021.1	1008.1	SE	4.1		−15.9	8.8	40	0	1.2	79	16	40	0	10
Mar.	1018.8	1006.3	SE	4.4		−9.1	9.9	50	2	2.0	75	21	50	2	11
Apr.	1017.2	1005.3	SE	4.1		1.0	9.4	90	0	3.5	72	26	90	0	40
May	1013.4	1001.8	NW	4.5		9.1	10.7	112	15	5.5	64	42	112	15	50
June	1009.1	997.8	NW	4.1		16.0	10.9	200	6	8.9	68	60	200	6	65
July	1007.1	995.9	N	3.6		18.2	9.7	219	10	10.9	74	73	219	10	53
Aug.	1009.1	997.8	NW	3.4		15.4	9.7	198	7	9.8	78	63	198	7	48
Sept.	1012.9	1001.3	W	3.2		9.5	8.7	126	10	7.0	81	54	126	10	38
Oct.	1015.5	1003.4	W	3.9		0.7	6.3	88	6	14.0	81	39	88	6	24
Nov.	1018.0	1005.5	W	4.0		−9.3	6.8	65	7	2.4	83	36	65	7	14
Dec.	1031.1	1008.1	SE	3.6		−16.4	7.6	48	6	1.4	83	26	48	6	17
Year	1015.3	1003.1	SE	3.9		0.1		706	288		77	475	706	288	65

Month	Number of days: wind (speed $\geq$15 m/s)	frost-free period	fog	precip. ($\geq$0.1 mm)	thunder-storm	snow-cover	clouds (8–10 tenths)	Snow cover max. 10-day average depth (cm)	Cloudiness (tenths) 07h	Cloudiness (tenths) 13h	Cloudiness (tenths) 19h	Mean sunshine (h)	Evaporation (mm) actual	Evaporation (mm) potent.
Jan.	2.0		1	14.7		31	10.6		6.9	7.5	5.7	38	<10	
Feb.	2.4		0.7	11.6		28	8.1		6.8	6.5	4.8	88	<10	
Mar.	3.1		0.9	11.5		21	7.4		6.7	6.5	6.1	150	<10	
Apr.	2.6		0.9	9.6	0.2	18	7.7		6.9	6.9	7.0	184	10	
May	3.4		0.3	13.1	2		8.5		6.6	7.2	7.0	239	>50	
June	1.9		0.6	12.1	6		7.0		6.0	7.4	6.4	298	>70	
July	1.3		0.6	14.5	7		6.9		6.5	7.5	6.9	267	>70	
Aug.	1.0		2	15.0	4		8.2		6.6	7.2	6.7	217	>50	
Sept.	1.7		3	1.47	0.9		9.6		8.0	8.3	7.3	138	30	
Oct.	2.1		3	16.1			15.1		8.3	8.4	7.3	65	<10	
Nov.	1.1		2	16.6		29	16.5		7.5	8.0	6.8	38	<10	
Dec.	2.0		2	16.3		31	13.4		7.2	7.5	6.3	26	<10	
Year	24.6	120	17	165.8	20	175	11.9	73	7.0	7.4	6.5	1748	>300	500

TABLE CV

CLIMATIC TABLE FOR TURA

Latitude 64°10′N longitude 100°04′E elevation 130 m

Month	Air press. (mbar) sea level	Air press. (mbar) station	Wind preval. direct.	Wind mean speed (m/s)	Global radiat. (kcal./cm²)	Temperature (°C) mean	Temperature (°C) mean daily range	Temperature (°C) max.	Temperature (°C) min.	Humidity vapor press. (mm)	Humidity relat. (%)	Precipitation (mm) mean	Precipitation (mm) max.	Precipitation (mm) min.	Precipitation (mm) max. in 24 h
Jan.	1026.2	1002.6	E	1.5	0.3	−36.8	8.0	−2	−66		77	10	21	5	
Feb.	1026.5	1003.7	NW	1.4	1.9	−29.0	12.3	3	−67		78	10	20	2	
Mar.	1023.0	1001.1	NW	1.8	5.8	−19.8	20.4	11	−56		72	9	14	3	
Apr.	1016.2	995.4	NW	2.5	11.3	−8.2	22.0	16	−41		64	12	26	4	
May	1011.5	991.7	NW	2.7	13.4	3.8	17.7	29	−30	3.5	62	22	42	4	
June	1007.1	987.9	NW	2.4	14.9	12.2	19.4	34	−8	7.0	64	48	117	11	
July	1006.7	987.8	NW	2.0	13.0	15.8	20.0	35	−2	9.9	70	62	132	15	
Aug.	1008.7	989.5	W	2.0	9.6	12.2	19.0	32	−7	8.3	75	48	160	13	
Sept.	1012.8	993.0	W	2.0	4.7	5.1	18.0	25	−21	5.6	77	34	83	8	
Oct.	1016.4	995.7	W	2.3	2.4	−7.0	12.6	17	−47		80	24	45	15	
Nov.	1021.8	999.5	E, NW	1.8	0.7	−24.5	7.6	4	−59		79	20	44	9	
Dec.	1026.8	1003.4	E, NW	1.7	0.1	−34.4	7.4	−1	−65		78	16	30	4	
Year	1017.0	995.9	NW	2.0	78.1	−9.2		35	−67		73	317	424	270	

Month	Number of days wind (speed ≥15 m/s)	Number of days frost-free period	Number of days fog	Number of days precip. (≥0.1 mm)	Number of days thunder-storm	Number of days snow-cover	Number of days clouds (8–10 tenths)	Snow cover max. 10-day average depth (cm)	Cloudiness (tenths) 07h	Cloudiness (tenths) 13h	Cloudiness (tenths) 19h	Mean sun-shine (h)	Evaporation (mm) actual	Evaporation (mm) potent.
Jan.	0.2		9	12.9		31	8		6.2	7.4	6.2	3	<10	
Feb.	0.1		4	11.7		28	7		6.7	6.4	5.2	65	<10	
Mar.	0.7		2	11.4		31	8		7.0	6.6	6.3	168	<10	
Apr.	0.7		0	9.6		30	11		6.8	7.2	7.0	229	<10	
May	0.6		0.6	10.1		6	15		7.4	8.0	7.8	232	>20	
June	0.7		1	13.2			13		7.2	8.4	7.8	249	>70	
July	0.5		6	14.1			15		7.1	7.8	7.2	253	>70	
Aug.	0.4		10	14.2			15		8.7	7.9	7.3	177	<30	
Sept.	0.2		9	12.7			18		9.0	8.2	7.7	98	>10	
Oct.	0.3		0.5	16.0		19	18		8.6	8.6	7.8	56	<10	
Nov.	0.4		3	15.3		30	11		7.3	7.7	6.3	22	<10	
Dec.	0.4		8	15.2		31	9		6.6	7.7	6.6	0.06	<10	
Year	5.2	70	53	156.4		210	148	44	7.4	7.7	6.9	1552	>200	400

TABLE CVI

CLIMATIC TABLE FOR TURGAY

Latitude 49°38′N longitude 63°30′E elevation 123 m

Month	Air press. (mbar) sea level	Air press. (mbar) station	Wind preval. direct.	Wind mean speed (m/s)	Global radiat. (kcal./cm²)	Temperature (°C) mean	Temperature (°C) mean daily range	Temperature (°C) max.	Temperature (°C) min.	Humidity vapor press. (mm)	Humidity relat. (%)	Precipitation (mm) mean	Precipitation (mm) max.	Precipitation (mm) min.	Precipitation (mm) max. in 24 h
Jan.	1029.0		N, NE	5.2		−17.2	9.3	7	−44	1.4	81	12	76	1	23
Feb.	1027.6		N	5.8		−16.0	9.1	5	−41	1.2	82	10	62	0	15
Mar.	1025.0		NE	5.8		−8.6	10.0	24	−35	2.0	82	9	41	0	22
Apr.	1021.2		NE	5.6		5.1	11.7	31	−24	4.3	65	17	63	0	31
May	1016.3		NE	5.2		15.9	13.7	39	−6	6.6	50	16	52	0	23
June	1011.3		N, NE	4.7		21.8	14.6	42	0	8.0	43	18	65	0	35
July	1007.9		N	4.5		24.2	13.9	43	6	8.7	46	27	116	0	93
Aug.	1011.9		N	4.5		21.6	14.2	42	3	7.6	48	14	69	0	59
Sept.	1018.0		W	4.8		14.9	13.8	36	−6	5.5	48	9	36	0	25
Oct.	1023.1		W	4.8		5.6	10.9	30	−20	4.1	64	19	72	0	19
Nov.	1029.2		SW	4.7		−4.4	8.5	19	−33	2.7	78	12	49	0	27
Dec.	1029.5		SW	4.8		−13.0	8.9	7	−40	1.7	84	14	61	0	9
Year	1020.8		N, NE	5.0		4.2		43	−44		64	177	318	78	93

Month	Number of days wind (speed ≥15 m/s)	Number of days frost-free period	Number of days fog	Number of days precip. (≥0.1 mm)	Number of days thunder-storm	Number of days snow-cover	Number of days clouds (8–10 tenths)	Snow cover max. 10-day average depth (cm)	Cloudiness (tenths) 07h	Cloudiness (tenths) 13h	Cloudiness (tenths) 19h	Mean sun-shine (h)	Evaporation (mm) actual	Evaporation (mm) potent.
Jan.	1.3		4	9.5	0	31	14		6.4	6.5	4.7	93	<10	
Feb.	1.7		3	7.7	0	28	10		6.0	5.8	4.5	129	<10	
Mar.	3.1		4	6.9	0	30	11		6.5	6.0	5.2	166	<10	
Apr.	1.6		3	5.4	0.3	3	7		5.4	5.9	5.7	234	20	
May	2.3		0.3	6.2	3		4		4.7	5.6	5.4	314	>50	
June	2.0		0.1	5.4	4		3		4.0	5.8	5.4	326	<45	
July	1.7		0.1	5.5	5		4		3.8	5.6	4.9	336	20	
Aug.	2.0		0.2	4.8	2		4		3.6	4.8	4.4	312	20	
Sept.	2.2		0.2	4.1	0.6		5		4.2	5.1	4.3	242	<15	
Oct.	1.4		1	6.1	0		10		6.4	6.6	5.2	157	<10	
Nov.	0.9		4	6.8	0		12		6.6	6.5	5.5	104	<10	
Dec.	1.6		6	9.5	0	29	14		6.5	6.4	5.4	78	<10	
Year	21.8	158	26	78	15	125	98	17	5.3	5.9	5.0	2491	200	700

TABLE CVII

CLIMATIC TABLE FOR TURUKHANSK

Latitude 65°47'N longitude 87°57'E elevation 45 m

Month	Air press. (mbar)		Wind		Global radiat.	Temperature (°C)				Humidity		Precipitation (mm)			
	sea level	station	preval. direct.	mean speed (m/s)	(kcal./cm²)	mean	mean daily range	max.	min.	vapor press. (mm)	relat. (%)	mean	max.	min.	max. in 24 h
Jan.	1019.9	1015.2	S	4.0	0.2	−28.4	7.9	−2	−60	0.6	80	23	55	0	15
Feb.	1020.9	1016.3	E	3.9	1.6	−23.8	8.3	1	−61	0.9	79	16	41	0	6
Mar.	1017.7	1013.3	SE	4.1	5.3	−17.2	11.2	10	−53	1.2	76	20	48	1	10
Apr.	1013.6	1009.4	NW	4.0	11.5	−9.6	11.6	14	−41	2.0	69	24	78	3	16
May	1011.0	1006.9	NW	3.8	13.9	−0.8	8.8	28	−29	3.2	67	32	90	2	23
June	1006.8	1002.9	NW	3.9	15.0	8.8	9.6	32	−8	6.4	65	55	116	14	37
July	1006.5	1002.7	NW	3.4	14.8	15.4	9.9	34	0	9.6	70	67	183	6	41
Aug.	1007.6	1003.7	NW	3.6	10.1	12.6	8.6	31	−6	8.8	77	74	163	9	41
Sept.	1011.0	1007.0	S	4.0	3.4	5.1	6.8	24	−17	5.5	80	67	146	14	33
Oct.	1012.0	1007.8	S	4.8	2.2	−6.0	5.2	14	−43	2.9	83	55	108	23	15
Nov.	1016.3	1011.8	S	3.9	0.5	−19.9	7.7	6	−57	1.2	82	32	72	0	11
Dec.	1020.0	1015.4	S	4.0	0.1	−27.4	8.0	1	−60	0.6	80	31	64	12	8
Year	1013.6	1009.4	S	4.0	78.6	−7.6		34	−61		76	496	692	293	41

Month	Number of days							Snow cover max. 10-day average depth (cm)	Cloudiness (tenths)			Mean sunshine (h)	Evaporation (mm)	
	wind (speed ⩾15 m/s)	frost-free period	fog	precip. (⩾0.1 mm)	thunderstorm	snow-cover	clouds (8–10 tenths)		07h	13h	19h		actual	potent.
Jan.	1.4			18.5		31	15		7.3	8.1	7.1	4	<10	
Feb.	1.2			16.1		28	13		7.0	7.2	6.3	59	<10	
Mar.	1.6			16.8		31	12		7.5	7.2	6.9	137	<10	
Apr.	1.9			13.4		30	12		7.2	7.1	7.3	222	<10	
May	0.8			14.3		22	16		7.6	7.6	7.4	252	15	
June	0.8			14.4	3.0		13		7.8	7.7	7.7	253	50	
July	0.9			14.2	4.0		12		6.7	7.2	6.8	322	>70	
Aug.	0.5			16.2	3.0		14		7.8	7.7	7.4	203	50	
Sept.	1.4			19.0	0.2		16		8.5	8.3	8.1	107	20	
Oct.	1.7			22.1		23	23		9.1	8.8	8.4	42	<10	
Nov.	0.7			19.8		30	18		7.6	7.9	7.0	17	<10	
Dec.	0.9			17.5		31	15		6.9	7.9	6.9	0.0	<10	
Year	13.8	81		202.3	10.2	229	179	85	7.6	7.7	7.3	1618	<300	300

TABLE CVIII

CLIMATIC TABLE FOR TYAN SHANS

Latitude 41°55'N longitude 78°14'E elevation 3,614 m

Month	Air press. (mbar)		Wind		Global radiat.	Temperature (°C)				Humidity		Precipitation (mm)			
	sea level	station	preval. direct.	mean speed (m/s)	(kcal./cm²)	mean	mean daily range	max.	min.	vapor press. (mm)	relat. (%)	mean	max.	min.	max. in 24 h
Jan.	1037.2	647.1	N	1.4	6.9	−21.9		−4	−48	0.9	72	5	13	0.1	5
Feb.	1031.7	647.0	N	1.9	9.1	−19.0		−1	−46	1.0	70	5	16	1	6
Mar.	1027.8	650.1	N	2.1	13.8	−13.4		8	−41	1.6	69	13	30	3	7
Apr.	1020.0	653.3	N	2.3	17.5	−7.5		10	−37	2.6	69	23	58	5	16
May	1014.7	655.8	N	2.6	20.0	−1.6		15	−26	3.9	70	42	76	5	24
June	1010.0	656.3	N	2.4	19.5	1.9		16	−16	4.8	70	55	98	19	43
July	1006.9	656.5	N	2.3	19.1	4.5		19	−15	5.4	68	60	106	24	28
Aug.	1009.2	657.5	N	2.2	17.1	3.7		19	−14	5.2	68	52	98	13	32
Sept.	1016.1	657.7	SW	2.3	13.6	2.3		16	−18	3.9	67	26	53	6	22
Oct.	1027.3	656.9	SW	2.3	10.9	−6.7		13	−32	2.5	66	11	35	0.3	15
Nov.	1031.9	653.1	SW	2.0	7.8	−15.1		6	−42	1.4	70	8	20	0.2	9
Dec.	1035.4	649.8	N	1.5	6.1	−19.3		0	−43	1.0	72	6	16	0.2	8
Year	1022.4	653.4	N	2.1	161.4	−7.7		19	−48	2.8	69	306	382	214	43

Month	Number of days							Snow cover max. 10-day average depth (cm)	Cloudiness (tenths)			Mean sunshine (h)	Evaporation (mm)	
	wind (speed ⩾15 m/s)	frost-free period	fog	precip. (⩾0.1 mm)	thunderstorm	snow-cover	clouds (8–10 tenths)		07h	13h	19h		actual	potent.
Jan.	0.1		0.4	8.4			9.0	20	6.3	6.7	5.0			
Feb.	0.4		0.3	8.3			10.1	24	7.0	7.2	5.6			
Mar.	0.5		0.2	12.0			13.7	30	7.3	7.6	7.0			
Apr.	0.5		0.6	13.2	0.2		13.4	28	6.9	7.5	7.4			
May	0.9		0.8	17.6	2		13.5	8	6.7	7.7	7.9			
June	0.9		1.0	18.7	5		10.3		6.0	7.8	7.8			
July	1.1		1.0	18.7	6		8.5		5.4	7.5	7.1			
Aug.	0.9		1.0	17.4	5		6.3		4.7	6.8	6.0			
Sept.	0.8		1.0	12.8	2		5.1		4.1	5.9	4.9			
Oct.	0.6		0.5	9.6	0.1		6.2		5.0	6.2	4.4			
Nov.	0.4		0.2	8.3			7.5	12	5.8	6.2	4.5			
Dec.	0.2		0.4	8.9			9.0	18	6.2	6.3	4.9			
Year	7.3		7.4	154.9	20	212	113	30	6.0	7.0	6.0		210.0	

TABLE CIX

CLIMATIC TABLE FOR UELEN

Latitude 66°10′N longitude 169°50′W elevation 7 m

Month	Air press. (mbar)		Wind		Global radiat.	Temperature (°C)				Humidity		Precipitation (mm)			
	sea level	station	preval. direct.	mean speed (m/s)	(kcal./cm²)	mean	mean daily range	max.	min.	vapor press. (mm)	relat. (%)	mean	max.	min.	max. in 24 h
Jan.	1016.7		N	5.6	0.2	−21.7		6	−45	1.0	84	26			17
Feb.	1017.9		N	5.2	1.1	−21.5		4	−44	0.9	83	27			16
Mar.	1018.0		N	5.0	5.0	−20.9		3	−42	1.0	83	24			13
Apr.	1015.1		N	5.3	8.7	−13.7		5	−38	1.7	86	23			12
May	1015.4		N	4.6	11.9	−4.7		7	−28	3.2	90	25			13
June	1014.2		S	4.8	14.1	1.6		18	−8	4.9	91	27	46	0	16
July	1011.8		N	6.6	11.0	5.4		21	−2	6.3	91	36	62	4	17
Aug.	1010.7		N	5.9	7.6	5.0		19	−3	6.2	92	57	150	18	49
Sept.	1011.4		N	6.6	3.7	2.7		14	−7	5.2	90	45	128	11	36
Oct.	1007.3		N	7.3	1.5	−1.9		9	−25	3.8	87	36			53
Nov.	1011.1		N	7.8	0.3	−10.1		6	−38	2.4	86	35			24
Dec.	1013.0		N	6.0	0.1	−18.1		2	−40	1.1	85	33			26
Year	1013.5		N	5.9	65.2	−8.2		21	−45		87	394			53

Month	Number of days							Snow cover	Cloudiness (tenths)			Mean sun-shine	Evaporation (mm)	
	wind (speed ≥15 m/s)	frost-free period	fog	precip. (≥0.1 mm)	thunder-storm	snow-cover	clouds (8–10 tenths)	max. 10-day average depth (cm)	07h	13h	19h	(h)	actual	potent.
Jan.	6.2		4	8.4		31	10					9	<5	
Feb.	4.1		3	8.8		28	11					53	<5	
Mar.	3.7		4	8.7		31	12					137	<5	
Apr.	4.1		5	9.1		30	16					146	<10	
May	2.2		9	10.3		31	22					112	<10	
June	3.7		14	6.9	0.1	8	16					210	50	
July	8.3		17	12.1	0.04		20					173	<60	
Aug.	6.2		14	15.1			22					102	20	
Sept.	6.1		9	16.1			22					58	10	
Oct.	8.7		4	15.6		14	21					30	<10	
Nov.	9.2		2	12.7		30	19					7	<5	
Dec.	5.6		2	10.9		31	14					1	<5	
Year	68.1	76	87	134.7	0.1	237	205	54				1038	<200	

TABLE CX

CLIMATIC TABLE FOR UFA

Latitude 54°45′N longitude 56°00′E elevation 197 m

Month	Air press. (mbar)		Wind		Global radiat.	Temperature (°C)				Humidity		Precipitation (mm)			
	sea level	station	preval. direct.	mean speed (m/s)	(kcal./cm²)	mean	mean daily range	max.	min.	vapor press. (mm)	relat. (%)	mean	max.	min.	max. in 24 h
Jan.		1005.5	S	3.9	1.5	−14.6	6.8	4	−43	2.1	82	22			
Feb.		1004.6	S	3.8	3.1	−13.7	8.0	9	−39	2.1	80	16			
Mar.		1002.4	SW	3.8	6.3	−7.4	8.4	14	−34	3.0	78	19			
Apr.		998.4	SW	3.0	10.9	3.2	9.1	31	−30	5.9	71	27			
May		996.7	SW	3.4	12.8	12.5	11.3	36	−10	9.0	59	33			
June		991.5	SW	3.0	15.4	17.7	11.8	38	−1	12.9	62	48			
July		988.0	N	2.3	13.6	19.0	10.8	39	4	15.1	71	57			
Aug.		991.0	SW	2.4	10.5	17.0	11.4	36	0	13.9	72	44			
Sept.		994.9	SW	2.9	6.8	10.9	10.1	32	−5	9.9	75	44			
Oct.		1000.2	SW	3.6	3.4	2.7	6.7	25	−22	6.3	81	47			
Nov.		1001.3	SW	3.6	1.6	−5.6	6.0	14	−38	3.5	83	32			
Dec.		1002.7	S	4.4	1.0	−11.9	6.7	5	−44	2.6	85	30			
Year		998.1	SW	3.3	86.9	2.5	–	39	−44	7.2	75	419			

Month	Number of days							Snow cover	Cloudiness (tenths)			Mean sun-shine	Evaporation (mm)	
	wind (speed ≥15 m/s)	frost-free period	fog	precip. (≥0.1 mm)	thunder-storm	snow-cover	clouds (8–10 tenths)	max. 10-day average depth (cm)	07h	13h	19h	(h)	actual	potent.
Jan.	3.4		4	20.0			16.3	58	7.4	7.8	6.7		0	
Feb.	2.9		2	15.3			11.6	71	7.1	7.0	6.0		0	
Mar.	3.8		4	15.2			13.5	74	7.4	7.0	6.6		0.5	
Apr.	2.3		3	11.2	0.5		11.0	57	6.7	7.0	6.6		15	
May	3.1		1	12.2	4		9.0		6.4	7.1	6.5		61	
June	3.0		1	13.3	7		6.7		5.4	6.7	6.0		65	
July	1.5		2	14.1	9		8.2		5.5	6.9	6.3		66	
Aug.	1.4		2	14.2	6		6.8		5.5	6.4	5.7		37	
Sept.	1.8		3	14.8	1		12.1		7.1	7.6	6.7		26	
Oct.	2.9		4	16.8	0.1		18.0		8.3	8.6	7.5		13	
Nov.	1.9		6	19.7			17.6	15	8.1	8.1	7.2		3	
Dec.	3.1		5	21.6			17.2	37	7.7	7.8	7.0		0	
Year	31.1	137	37	188.4	28	164	148	74	6.9	7.3	6.6		286.5	611

TABLE CXI

CLIMATIC TABLE FOR ULANUDE

Latitude 51°48'N longitude 107°26'E elevation 510 m

Month	Air press. (mbar)		Wind		Global radiat.	Temperature (°C)				Humidity		Precipitation (mm)			
	sea level	station	preval. direct.	mean speed (m/s)	(kcal./cm²)	mean	mean daily range	max.	min.	vapor press. (mm)	relat. (%)	mean	max.	min.	max. in 24 h
Jan.	1037.4	966.6	W	1.9	3.2	−25.4	9.7	0	−51	0.8	75	6	13	0	7
Feb.	1035.2	965.8	W	1.9	6.3	−20.9	12.7	1	−45	1.0	72	2	8	0	3
Mar.	1028.9	962.5	E	2.7	9.8	−10.6	13.8	15	−40	2.0	64	3	15	0	12
Apr.	1019.8	957.1	NW	3.7	12.3	1.2	12.9	28	−28	3.5	54	6	19	0	12
May	1013.5	952.9	NW	3.8	14.5	8.8	15.2	32	−15	5.5	49	13	47	0	28
June	1008.5	949.8	NW	3.4	14.9	16.2	14.9	39	−4	10.2	57	32	113	1	42
July	1005.5	947.8	NW	2.8	14.0	19.4	12.9	39	1	14.2	65	69	185	9	92
Aug.	1009.9	951.1	NW	2.6	11.9	16.5	12.3	40	−4	12.9	69	62	153	2	55
Sept.	1018.2	957.3	NW	2.5	9.8	8.8	12.4	30	−11	7.7	69	27	80	3	41
Oct.	1024.6	961.3	E	2.5	6.2	−0.1	11.3	23	−28	4.2	68	8	30	1	12
Nov.	1029.4	963.0	E	2.6	4.0	−12.7	9.7	10	−38	2.0	75	9	43	0	15
Dec.	1033.8	964.7	SW	2.1	2.0	−21.9	9.0	5	−49	1.1	78	9	34	0	9
Year	1022.0	958.3	NW	2.7	108.9	−1.7	–	40	−51	5.4	66	246	413	153	92

Month	Number of days							Snow cover max. 10-day average depth (cm)	Cloudiness (tenths)			Mean sunshine (h)	Evaporation (mm)	
	wind (speed ⩾15 m/s)	frost-free period	fog	precip. (⩾0.1 mm)	thunderstorm	snow-cover	clouds (8–10 tenths)		07h	13h	19h		actual	potent.
Jan.	1.0		3	7.6			8.7	16					0	
Feb.	1.0		2	4.3			7.1	17					0.3	
Mar.	2.7		0.1	3.2			7.6	14					0.6	
Apr.	6.1		0.2	4.0			9.8	1					8	
May	6.1		0.3	5.9	0.5		10.0						12	
June	3.9		0.8	9.4	3		10.8						30	
July	1.8		0.6	11.6	6		12.1						52	
Aug.	1.4		4	11.3	4		11.9						49	
Sept.	2.0		5	7.9	0.5		9.4						20	
Oct.	2.1		1	5.6			10.0						11	
Nov.	2.0		0.6	8.6			11.1	7					4	
Dec.	1.2		1	10.1			11.9	12					0	
Year	31.3	102	19	90	14	148	120	17					186.9	536.1

TABLE CXII

CLIMATIC TABLE FOR UST KAMCHATSK

Latitude 56°14'N longitude 162°28'E elevation 6.4 m

Month	Air press. (mbar)		Wind		Global radiat.	Temperature (°C)				Humidity		Precipitation (mm)			
	sea level	station	preval. direct.	mean speed (m/s)	(kcal./cm²)	mean	mean daily range	max.	min.	vapor press. (mm)	relat. (%)	mean	max.	min.	max. in 24 h
Jan.	1005.7	1004.9	NW	5.9		−12.6	8.6	5	−42	1.7	81	82	196	11	14
Feb.	1006.9	1006.1	N	5.6		−12.8	8.2	5	−40	1.8	82	74	221	3	14
Mar.	1009.8	1009.0	N	5.2		−9.9	9.3	7	−37	2.1	80	51	130	9	10
Apr.	1010.1	1009.3	N	4.4		−3.8	7.3	12	−29	3.1	82	39	98	6	9
May	1012.5	1011.7	S	3.9		1.6	5.3	20	−16	4.5	86	34	96	3	10
June	1011.7	1010.9	S	4.5		6.8	5.8	29	−2	6.6	87	28	79	2	9
July	1010.1	1009.3	S	4.1		11.2	5.9	30	3	8.8	88	52	107	10	15
Aug.	1010.3	1009.5	S	3.8		12.2	6.3	26	1	9.2	86	58	125	10	19
Sept.	1011.5	1010.7	S	4.0		9.1	7.1	23	−3	7.5	84	49	108	4	13
Oct.	1009.7	1007.9	NW	4.2		2.6	6.8	16	−13	4.6	77	59	178	10	16
Nov.	1006.6	1005.8	NW	4.9		−4.4	6.4	12	−25	2.8	78	66	160	3	15
Dec.	1003.8	1003.0	NW	5.4		−10.6	8.2	7	−38	2.0	81	78	152	9	12
Year	1009.0	1008.2	N	4.7		−0.9		30	−42		83	670	892	353	19

Month	Number of days							Snow cover max. 10-day average depth (cm)	Cloudiness (tenths)			Mean sunshine (h)	Evaporation (mm)	
	wind (speed ⩾15 m/s)	frost-free period	fog	precip. (⩾0.1 mm)	thunderstorm	snow-cover	clouds (8–10 tenths)		07h	13h	19h		actual	potent.
Jan.	4.6		0.9	18.5		31	16.2		7.4	8.0	7.1	46	5	
Feb.	3.3		0.8	16.8		28	14.4		7.8	7.6	6.8	72	5	
Mar.	4.1		1	15.0		31	15.2		7.7	7.1	7.0	130	5	
Apr.	2.4		2	12.7		30	14.6		7.8	7.4	7.6	144	10	
May	0.6		7	11.6		19	19.2		8.1	7.9	8.0	158	30	
June	0.5		9	10.3			21.2		8.8	8.4	8.5	147	50	
July	0.6		8	13.3	0.8		21.1		8.8	8.5	8.5	130	50	
Aug.	0.8		6	13.3	0.3		20.4		8.3	8.1	8.2	144	50	
Sept.	1.8		4	12.5	0.2		14.6		7.9	7.6	7.4	142	30	
Oct.	2.1		1	11.3			10.0		7.1	7.2	6.0	124	5	
Nov.	3.1		1	13.2		17	13.0		7.1	7.4	6.2	79	5	
Dec.	3.9		1	17.4		31	14.5		7.1	7.6	6.6	47	5	
Year	27.8	93	42	165.9	1	188	194	107	7.8	8.3	7.3	1363	200	200

TABLE CXIII

CLIMATIC TABLE FOR UZHGOROD

Latitude 48°38′N longitude 22°16′E elevation 118 m

Month	Air press. (mbar)		Wind		Global radiat.	Temperature (°C)				Humidity		Precipitation (mm)			
	sea level	station	preval. direct.	mean speed (m/s)	(kcal./cm²)	mean	mean daily range	max.	min.	vapor press. (mm)	relat. (%)	mean	max.	min.	max. in 24 h
Jan.	1021.3	1004.9	SE	2.4		−2.9	6.6	13	−28	3.5	81	55.0	116	5	32
Feb.	1018.9	1002.7	SE	2.5		−1.4	6.7	17	−28	3.7	79	49.4	95	6	24
Mar.	1015.2	999.4	SE	2.9		4.3	9.2	27	−24	4.7	69	31.3	110	5	27
Apr.	1012.8	997.3	SE	2.8		10.0	10.6	32	−12	6.2	63	32.9	100	9	42
May	1014.2	999.0	NE	2.6		15.4	11.6	33	−2	9.4	64	60.2	136	3	47
June	1013.5	998.5	NW	2.5		17.9	11.1	37	3	11.6	67	98.4	233	30	75
July	1013.4	998.5	NW	2.2		19.9	11.6	39	6	12.3	66	68.9	187	16	50
Aug.	1014.8	999.8	NW	2.2		19.0	11.6	40	4	12.1	68	91.2*	224	12	67
Sept.	1017.4	1002.2	SE	1.9		15.0	11.7	34	−1	10.1	71	49.0	157	9	44
Oct.	1017.7	1002.2	SE	2.0		10.1	10.5	31	−18	8.0	75	43.7	169	0	52
Nov.	1019.3	1003.4	SE	2.2		4.3	6.4	29	−22	5.6	81	49.1	133	3	43
Dec.	1019.1	1003.0	SE	2.2		−0.2	5.4	17	−25	4.1	84	76.2	126	6	30
Year	1016.5	1000.9	SE	2.4		9.3		40	−28		72	705.3	1068	416	75

Month	Number of days							Snow cover max. 10-day average depth (cm)	Cloudiness (tenths)			Mean sunshine (h)	Evaporation (mm)	
	wind (speed ≥15 m/s)	frost-free period	fog	precip. (≥0.1 mm)	thunder-storm	snow-cover	clouds (8–10 tenths)		07h	13h	19h		actual	potent.
Jan.	0.2		8	14.5	0.1	31	15.0		7.8	7.7	7.2	52	<10	
Feb.	0.3		5	12.9	0.1	17	11.9		8.0	7.6	6.8	67	<10	
Mar.	1.2		2	11.2	0.8		8.7		7.0	6.8	6.1	141	10	
Apr.	0.5		1	11.2	2		6.2		6.4	6.5	6.0	215	50	
May	0.3		0.7	12.3	6		4.7		6.3	6.8	6.7	256	>70	
June	0.3		0.4	13.6	7		5.2		6.0	6.5	6.5	249	>70	
July	0.2		0.6	12.8	7		4.0		5.3	5.9	5.6	285	>70	
Aug.	0.7		0.7	11.8	4		2.7		5.4	5.7	5.7	261	>50	
Sept.	0.0		0.9	10.9	2		4.9		5.4	5.7	4.9	222	>30	
Oct.	0.1		3	10.9	0.7		8.6		6.3	6.0	4.3	166	>20	
Nov.	0.5		7	13.4	0.1		13.6		8.3	8.1	7.1	61	<10	
Dec.	0.2		8	15.1	0.0	7	16.3		8.4	8.4	7.7	35	<10	
Year	4.5	172	37	150.6	30	52	102		6.7	6.8	6.2	2040	>400	700

*NOTE: The August precipitation is as shown in CLINO tables book for Europe but not as shown graphically in Klimat SSSR, Eur., p. 345.

TABLE CXIV

CLIMATIC TABLE FOR VERKHOYANSK

Latitude 67°33′N longitude 133°23′E elevation 137 m

Month	Air press. (mbar)		Wind		Global radiat.	Temperature (°C)				Humidity		Precipitation (mm)			
	sea level	station	preval. direct.	mean speed (m/s)	(kcal./cm²)	mean	mean daily range	max.	min.	vapor press. (mm)	relat. (%)	mean	max.	min.	max. in 24 h
Jan.	1026.2		SW	0.7	0.1	−48.9	4.3	−12	−66	0.1	75	7	16	0	5
Feb.	1027.5		SW	0.8	0.7	−43.7	7.9	0	−68	0.1	75	5	10	0	9
Mar.	1021.6		SW	1.0	5.6	−29.9	20.7	5	−60	0.4	71	5	10	0	3
Apr.	1015.0		NE	1.8	11.7	−13.0	24.8	14	−54	1.2	65	4	19	0	16
May	1011.1		NE	2.9	14.6	2.0	17.8	31	−29	3.2	58	5	23	0	12
June	1007.1		NE	3.2	16.0	12.2	16.8	34	−7	6.2	58	25	87	3	28
July	1007.0		NE	2.6	14.2	15.3	17.9	35	−3	8.2	62	33	74	0	33
Aug.	1009.8		SW	2.2	10.3	11.0	20.2	33	−10	6.9	70	30	65	5	46
Sept.	1014.1		SW	1.8	5.2	2.6	18.0	25	−22	4.2	74	13	44	0	28
Oct.	1015.6		SW	1.4	2.6	−14.1	12.1	14	−45	1.5	80	11	40	2	20
Nov.	1022.3		SW	0.8	0.3	−36.1	5.6	1	−57	0.3	80	10	25	0	10
Dec.	1025.1		SW	0.7	0.0	−45.6	3.9	−7	−64	0.1	78	7	17	0	3
Year	1016.9		SW	1.7	82.3	−15.6		35	−68		70	155	237	51	46

Month	Number of days							Snow cover max. 10-day average depth (cm)	Cloudiness (tenths)			Mean sunshine (h)	Evaporation (mm)	
	wind (speed ≥15 m/s)	frost-free period	fog	precip. (≥0.1 mm)	thunder-storm	snow-cover	clouds (8–10 tenths)		07h	13h	19h		actual	potent.
Jan.	0.0		10	9.1		31	6		4.1	5.3	4.1	1	<10	
Feb.	0.0		10	7.6		28	4		4.5	4.6	3.7	79	<10	
Mar.	0.1		7	6.8		31	7		5.2	5.0	4.7	215	<10	
Apr.	0.3		0.6	5.0	0.02	30	9		5.2	5.2	5.4	298	<10	
May	0.5		0.7	5.8	0.3	12	13		6.4	7.1	6.7	300	>20	
June	0.9		0.3	9.6	3.0		13		6.8	7.3	6.9	309	70	
July	0.5		0.8	9.4	2.0		13		6.8	6.9	6.6	300	70	
Aug.	0.4		4	9.4	1.0		13		7.2	7.0	6.5	232	<30	
Sept.	0.2		3	8.2	0.2		14		7.9	7.5	7.2	126	>10	
Oct.	0.2		2	10.4		27	14		7.4	7.1	6.5	73	<10	
Nov.	0.0		6	11.9		30	8		5.1	5.9	4.6	20	<10	
Dec.	0.0		8	10.5		31	7		4.2	5.5	4.4	0	<10	
Year	3.1	69	52	103.7	6.0	223	121	25	5.9	6.2	5.6	1953	>200	400

TABLE CXV

CLIMATIC TABLE FOR VILYUYSK

Latitude 63°46′N longitude 121°37′E elevation 107 m

Month	Air press. (mbar)		Wind		Global radiat.	Temperature (°C)				Humidity		Precipitation (mm)			
	sea level	station	preval. direct.	mean speed (m/s)	(kcal./cm²)	mean	mean daily range	max.	min.	vapor press. (mm)	relat. (%)	mean	max.	min.	max. in 24 h
Jan.	1024.4		SW	1.8		−38.2	6.0	−5	−61	0.2	76	9	24	2	6
Feb.	1026.0		SW	1.7		−31.2	10.5	0	−60	0.4	76	7	28	0	6
Mar.	1020.2		SW	2.1		−19.7	17.2	8	−50	0.8	71	6	17	1	6
Apr.	1013.3		SW	2.6		−7.6	17.1	19	−40	1.9	62	11	36	0	8
May	1010.1		N	2.8		4.2	15.0	32	−23	3.8	56	16	67	0	27
June	.006.5		N	2.5		14.3	15.1	36	−4	7.4	57	23	82	1	60
July	1006.9		N	2.3		18.0	14.5	37	0	9.9	63	43	106	1	49
Aug.	1009.1		N	2.1		13.9	15.4	35	−6	8.6	69	36	135	2	71
Sept.	1013.5		SW	2.2		5.4	14.8	28	−15	5.2	73	24	70	0	20
Oct.	1015.5		SW	2.4		−7.7	11.8	18	−38	2.3	79	18	36	4	11
Nov.	1020.0		SW	2.1		−25.9	7.7	3	−53	0.6	80	13	28	2	6
Dec.	1022.7		SW	1.8		−35.8	5.2	−3	−58	0.2	77	10	31	4	11
Year	1015.6		SW	2.2		−9.2		37	−61		70	226	353	127	71

Month	Number of days							Snow cover	Cloudiness (tenths)			Mean sun-shine	Evaporation (mm)	
	wind (speed ≥15 m/s)	frost-free period	fog	precip. (≥0.1 mm)	thunder-storm	snow-cover	clouds (8–10 tenths)	max. 10-day average depth (cm)	07h	13h	19h	(h)	actual	potent.
Jan.	0.2		5	12.9		31	7		5.5	6.5	5.0	36	<10	
Feb.	0.2		3	10.5		28	6		5.4	5.5	4.5	90	<10	
Mar.	0.6		1	8.8		31	6		5.8	5.3	5.1	224	<10	
Apr.	1.5		0.4	8.7		20	10		5.5	6.0	6.2	266	<10	
May	1.2		0.3	7.6	0.3	6	12		6.5	7.3	6.9	281	20	
June	1.5		0.4	9.4	3.0		12		6.5	7.4	7.1	321	70	
July	1.4		0.4	9.8	4.0		11		6.3	7.3	6.6	299	70	
Aug.	1.3		1	9.2	1.0		12		6.3	6.9	6.5	280	<30	
Sept.	0.8		2	10.7	0.2		13		7.3	7.5	7.4	142	>10	
Oct.	1.0		1	14.7		21	17		8.3	8.1	7.4	74	<10	
Nov.	0.3		2	14.7		30	10		6.2	6.8	5.6	63	<10	
Dec.	0.2		6	13.0		31	8		5.5	6.4	4.9	3	<10	
Year	10.2	96	22	130.0	8.0	216	124	43	6.3	6.8	6.1	2079	>200	400

TABLE CXVI

CLIMATIC TABLE FOR VLADIVOSTOK

Latitude 43°07′N longitude 131°54′E elevation 138 m

Month	Air press. (mbar)		Wind		Global radiat.	Temperature (°C)				Humidity		Precipitation (mm)			
	sea level	station	preval. direct.	mean speed (m/s)	(kcal./cm²)	mean	mean daily range	max.	min.	vapor press. (mm)	relat. (%)	mean	max.	min.	max. in 24 h
Jan.	1022.6	1005.5	N	8.1	5.4	−14.7	8.4	5	−31	1.1	64	10	53	0	48
Feb.	1021.2	1004.3	N	7.5	7.6	−10.9	9.0	10	−29	1.6	64	13	57	0	36
Mar.	1017.9	1001.6	N	6.9	10.6	−3.9	9.5	15	−22	2.4	67	20	78	0	35
Apr.	1013.5	997.7	SE	7.0	11.8	4.1	10.2	21	−9	4.1	69	44	122	4	69
May	1010.1	994.6	SE	6.6	13.2	8.9	10.3	30	−1	6.5	77	69	191	6	71
June	1008.1	992.7	SE	6.7	11.6	13.0	11.5	32	4	9.9	88	88	271	11	138
July	1007.9	992.7	SE	6.4	10.4	17.5	9.0	36	8	13.7	92	101	234	16	108
Aug.	1009.0	994.2	SE	6.3	10.9	20.0	9.7	32	10	15.0	88	145	423	10	153
Sept.	1013.8	998.6	SE	6.5	10.4	15.8	9.5	29	2	10.7	78	126	361	27	178
Oct.	1017.8	1002.2	N	7.3	8.2	8.7	9.0	23	−8	5.9	68	57	159	5	76
Nov.	1020.4	1004.2	N	7.8	5.7	−1.1	7.3	18	−20	3.0	62	31	211	0	127
Dec.	1021.6	1004.7	N	8.0	4.7	−10.5	7.3	9	−28	1.5	63	17	83	0	38
Year	1015.3	999.4	N	7.1	110.5	3.9		36	−31		73	721	1076	371	178

Month	Number of days							Snow cover	Cloudiness (tenths)			Mean sun-shine	Evaporation (mm)	
	wind (speed ≥15 m/s)	frost-free period	fog	precip. (≥0.1 mm)	thunder-storm	snow-cover	clouds (8–10 tenths)	max. 10-day average depth (cm)	07h	13h	19h	(h)	actual	potent.
Jan.	8.1		1	4.7		31	0.8		3.2	3.8	2.4	192	<10	
Feb.	6.9		1	4.4		19	0.8		3.7	4.1	2.7	194	<10	
Mar.	6.2		4	6.9			2.3		5.3	5.7	4.7	206	>10	
Apr.	6.8		8	8.8			4.6		6.4	6.9	6.3	186	>30	
May	4.7		13	13.3	0.8		7.5		7.4	7.2	7.5	178	>70	
June	3.1		16	16.3	2		13.6		8.5	7.8	8.4	136	70	
July	2.5		20	17.0	2		16.0		9.1	8.2	8.6	125	<70	
Aug.	3.8		9	13.8	2		11.6		8.5	7.7	7.8	163	70	
Sept.	5.0		3	10.6	1		4.1		6.5	6.2	5.7	204	70	
Oct.	7.7		2	8.0	1		2.7		5.1	4.9	4.0	205	50	
Nov.	8.6		3	6.2	0.3		1.8		4.5	4.8	3.5	169	10	
Dec.	8.4		1	5.2		13	1.1		3.5	4.1	2.8	173	<10	
Year	71.8	187	81	115	9	80	66.9	18	6.0	6.0	5.4	2131	>400	600

TABLE CXVII

CLIMATIC TABLE FOR VOLGOGRAD

Latitude 48°42'N longitude 44°31'E elevation 42 m

Month	Air press. (mbar)		Wind		Global radiat. (kcal./cm²)	Temperature (°C)				Humidity		Precipitation (mm)			
	sea level	station	preval. direct.	mean speed (m/s)		mean	mean daily range	max.	min.	vapor press. (mm)	relat. (%)	mean	max.	min.	max. in 24 h
Jan.	1023.6		NE	6.1	2.6	−9.6	7.3	11	−35	2.1	85	23	60	1	
Feb.	1022.7		NE	7.1	4.2	−8.9	9.2	10	−31	2.1	85	20	63	1	
Mar.	1020.4		NE	5.3	8.7	−2.6	10.2	23	−26	3.3	84	18	67	4	
Apr.	1017.8		NE	5.2	11.8	8.2	12.4	31	−14	5.3	65	19	63	0	
May	1015.8		E	4.8	16.3	17.0	13.3	35	−4	7.6	56	27	79	0	
June	1011.4		NE	4.9	16.9	21.4	13.0	40	4	10.2	49	40	137	0	
July	1010.1		N, NW	4.2	16.9	24.2	13.2	42	9	11.4	47	33	103	0	
Aug.	1012.6		N	4.6	14.7	22.7	13.5	43	6	10.2	51	23	116	0	
Sept.	1017.8		N, SE, S	4.3	10.3	15.9	13.6	36	−2	8.0	57	27	74	1	
Oct.	1021.9		W	4.6	6.1	8.2	12.9	32	−14	5.9	71	23	66	0	
Nov.	1023.4		NE	5.0	3.2	0.2	8.9	22	−25	4.3	82	34	76	1	
Dec.	1023.5		NE	5.8	1.7	−6.3	7.4	12	−31	2.8	86	31	117	3	
Year	1018.4		NE	5.1	113.4	7.5		43	−35		68	318	571	156	

Month	Number of days							Snow cover max. 10-day average depth (cm)	Cloudiness (tenths)			Mean sunshine (h)	Evaporation (mm)	
	wind (speed ≥15 m/s)	frost-free period	fog	precip. (≥0.1 mm)	thunderstorm	snow-cover	clouds (8–10 tenths)		07h	13h	19h		actual	potent.
Jan.	5.8		12	19		31	16.9					54	5	
Feb.	7.0		10	15		28	14.7					82	5	
Mar.	6.7		9	14		20	14.5					135	5	
Apr.	6.2		2	9	0.4		8.2					209	30	
May	4.6		0.5	7	3		6.2					279	70	
June	4.4		0.1	10	7		5.2					313	70	
July	4.6		0.0	9	6		3.8					332	<50	
Aug.	3.7		0.2	8	4		2.9					311	30	
Sept.	3.1		0.9	8	2		4.3					242	>20	
Oct.	3.8		5	10	0.06		7.9					139	5	
Nov.	4.2		10	13			15.9					83	5	
Dec.	5.8		10	18		17	19.9					46	5	
Year	59.9	181	60	140	22	101	121	15				2225	300	800

TABLE CXVIII

CLIMATIC TABLE FOR VOLOGDA

Latitude 59°17'N longitude 39°52'E elevation 117.9 m

Month	Air press. (mbar)		Wind		Global radiat. (kcal./cm²)	Temperature (°C)				Humidity		Precipitation (mm)			
	sea level	station	preval. direct.	mean speed (m/s)		mean	mean daily range	max.	min.	vapor press. (mm)	relat. (%)	mean	max.	min.	max. in 24 h
Jan.	1017.0	1001.3	SW	5.3	0.9	−11.7	9.0	5	−48	1.9	86	2	95	8	
Feb.	1017.1	1001.4	SE	5.2	2.6	−11.0	13.1	5	−43	1.9	83	2	44	3	
Mar.	1015.4	1000.1	SW	5.2	7.2	−6.2	13.0	13	−36	2.5	78	4	82	5	
Apr.	1015.6	1000.9	SW	4.6	9.5	2.4	12.1	28	−24	4.1	73	10	79	2	
May	1015.4	1001.0	NW	4.7	12.3	9.8	14.0	31	−11	6.5	67	10	138	2	
June	1011.3	997.1	NW	4.3	13.8	14.5	15.3	32	−4	8.9	70	48	150	18	
July	1010.1	996.2	NE	3.6	13.9	17.1	15.2	35	1	10.0	75	118	181	11	
Aug.	1011.1	997.0	SW	3.5	10.3	14.6	16.5	35	−2	10.0	79	104	175	11	
Sept.	1013.4	999.0	SW	4.1	5.5	9.0	13.9	29	−6	7.4	84	38	146	12	
Oct.	1015.5	1000.7	SW	4.8	2.3	2.7	12.0	23	−25	5.2	87	9	113	7	
Nov.	1016.0	1000.9	SW	5.2	1.1	−3.5	7.9	11	−32	3.5	88	6	75	5	
Dec.	1016.6	1001.1	SW	5.3	0.6	−9.2	6.2	6	−40	2.2	88	4	77	3	
Year	1014.5	999.7	SW	4.7	81.0	2.4		35	−48		80	374	852	382	

Month	Number of days							Snow cover max. 10-day average depth (cm)	Cloudiness (tenths)			Mean sunshine (h)	Evaporation (mm)	
	wind (speed ≥15 m/s)	frost-free period	fog	precip. (≥0.1 mm)	thunderstorm	snow-cover	clouds (8–10 tenths)		07h	13h	19h		actual	potent.
Jan.	1.3		4			31	20.8		8.3	8.3	7.6	27	<10	
Feb.	1.3		4			28	15.1		8.1	8.0	6.9	59	<10	
Mar.	1.4		3		0.02	31	14.9		7.8	7.1	7.0	134	<10	
Apr.	1.0		4		0.1	17	11.7		7.0	7.4	6.9	187	>20	
May	1.1		2		3		10.5		6.9	7.6	6.8	264	>70	
June	0.6		1		5		10.9		6.3	7.6	6.8	271	>70	
July	0.1		2		8		11.0		6.3	7.7	6.6	292	>70	
Aug.	0.2		4		4		11.4		6.9	7.9	6.5	241	>50	
Sept.	0.7		5		0.7		14.4		8.1	8.4	7.1	124	>30	
Oct.	1.1		6		0.06		20.2		9.0	8.9	7.8	55	>10	
Nov.	0.9		6			12	20.7		8.8	8.8	8.2	26	<10	
Dec.	1.5		5			31	23.0		8.7	8.9	8.4	13	<10	
Year	11.2	118	46		21	166	185	59	7.7	8.0	7.2	1693	>300	500

TABLE CXIX

CLIMATIC TABLE FOR VYAKHTU

Latitude 51°36′N longitude 141°54′E elevation 14 m

Month	Air press. (mbar)		Wind		Global radiat.	Temperature (°C)				Humidity		Precipitation (mm)			
	sea level	station	preval. direct.	mean speed (m/s)	(kcal./cm²)	mean	mean daily range	max.	min.	vapor press. (mm)	relat. (%)	mean	max.	min.	max. in 24 h
Jan.		1011.8	N	4.3	2.9	−20.3		4	−48	1.1	75	25			
Feb.		1012.8	N	4.8	5.1	−17.3		2	−41	1.4	75	22			
Mar.		1012.7	N	4.9	8.6	−10.6		6	−35	2.2	75	30			
Apr.		1009.0	S	5.3	10.0	−2.1		15	−26	4.4	80	39			
May		1007.8	S	5.5	11.5	3.2		24	−10	6.5	82	50			
June		1007.9	S	5.4	11.8	9.0		26	−3	9.8	85	51			
July		1007.3	S	5.4	10.9	13.6		30	1	13.8	87	68			
Aug.		1007.7	S	5.1	9.4	15.4		33	3	15.4	87	78			
Sept.		1005.1	S	5.7	8.1	11.6		26	−5	11.7	83	82			
Oct.		1011.5	S	5.6	5.4	3.5		16	−21	6.6	77	63			
Nov.		1010.8	W	6.0	3.1	−6.8		11	−33	3.0	73	55			
Dec.		1011.0	NW	5.6	2.0	−15.6		6	−43	1.7	76	44			
Year		1009.6	S	5.3	88.8	−1.4		33	−48	6.5	80	607			

Month	Number of days							Snow cover max. 10-day average depth (cm)	Cloudiness (tenths)			Mean sunshine (h)	Evaporation (mm)	
	wind (speed ⩾15 m/s)	frost-free period	fog	precip. (⩾0.1 mm)	thunderstorm	snow-cover	clouds (8–10 tenths)		07h	13h	19h		actual	potent.
Jan.	3.1		0.2				7.6	18	5.2	5.7	4.2		0	
Feb.	4.9		0.2				7.2	22	5.7	5.6	4.9		0.4	
Mar.	5.7		0.8				8.9	23	6.1	5.6	5.1		5	
Apr.	6.2		4				13.6	20	7.4	6.9	7.0		10	
May	6.7		8		0.5		14.8	1	7.8	7.6	7.5		35	
June	4.7		8		1.3		16.7		8.1	7.5	7.6		48	
July	4.5		10		1.6		17.9		8.3	7.9	7.6		53	
Aug.	4.8		7		1.9		14.6		8.2	7.5	7.3		65	
Sept.	6.5		2		1.4		13.2		7.5	7.5	7.0		50	
Oct.	7.6		2		0.2		10.1	2	6.5	6.8	5.6		22	
Nov.	8.2		0.3				8.3	6	6.1	6.5	5.3		0.8	
Dec.	8.3		0.2				9.2	16	6.0	6.7	5.2		0	
Year	71.2	123	43		7	173	142.1	23	6.9	6.8	6.2		289.2	422.0

TABLE CXX

CLIMATIC TABLE FOR YAKUTSK

Latitude 62°05′N longitude 129°45′E elevation 100 m

Month	Air press. (mbar)		Wind		Global radiat.	Temperature (°C)				Humidity		Precipitation (mm)			
	sea level	station	preval. direct.	mean speed (m/s)	(kcal./cm²)	mean	mean daily range	max.	min.	vapor press. (mm)	relat. (%)	mean	max.	min.	max. in 24 h
Jan.	1026.0		N	1.4	0.7	−43.2	8.0	−8	−63	0.1	74	7	21	0	7
Feb.	1026.3		N	1.5	2.5	−35.8	12.0	−7	−64	0.2	74	6	21	1	10
Mar.	1020.6		N	1.9	7.7	−22.0	16.5	6	−55	0.6	70	5	13	0	7
Apr.	1013.0		NW	2.7	12.8	−7.4	14.3	21	−41	1.9	61	7	24	0	13
May	1008.9		NW	3.5	14.6	5.6	12.6	33	−21	3.9	53	16	44	1	17
June	1005.5		NW	3.1	16.8	15.4	14.3	35	−4	7.4	54	31	117	2	47
July	1005.1		NW	2.9	14.9	18.8	14.1	38	−1	9.9	60	43	131	4	49
Aug.	1008.3		NW	2.8	11.8	14.8	13.8	35	−9	8.7	67	38	118	4	37
Sept.	1013.3		NW	2.6	6.9	6.2	11.8	27	−12	5.1	70	22	71	3	22
Oct.	1016.5		NW	2.6	3.4	−7.8	9.3	19	−41	2.2	78	16	41	0	11
Nov.	1021.6		N	1.9	1.2	−27.7	9.9	2	−55	0.5	78	13	31	1	6
Dec.	1024.4		N	1.5	0.4	−39.6	8.3	−2	−60	0.1	75	9	18	0	8
Year	1015.8		N, NW	2.4	93.7	−10.2		38	−64		68	213	333	104	49

Month	Number of days							Snow cover max. 10-day average depth (cm)	Cloudiness (tenths)			Mean sunshine (h)	Evaporation (mm)	
	wind (speed ⩾15 m/s)	frost-free period	fog	precip. (⩾0.1 mm)	thunderstorm	snow-cover	clouds (8–10 tenths)		07h	13h	19h		actual	potent.
Jan.	0.1		17	13.3		31	9		6.7	7.4	6.2	40	<10	
Feb.	0.1		10	10.6		28	7		6.2	6.2	5.2	126	<10	
Mar.	0.7		2	6.9		31	9		6.3	5.8	5.5	247	<10	
Apr.	1.1		0.5	5.9		30	10		6.3	6.2	6.4	286	10	
May	1.8		0.3	7.5	0.6		15		7.2	8.0	7.4	293	<30	
June	2.1		0.9	8.9	3.0		12		6.5	7.4	7.0	344	70	
July	1.2		0.8	9.7	5.0		11		6.4	6.9	6.8	338	70	
Aug.	1.1		2	10.2	3.0		11		6.6	7.1	6.8	268	<30	
Sept.	0.8		2	9.2	0.3		12		7.3	7.5	6.9	182	20	
Oct.	0.8		2	13.0		19	16		8.5	8.2	7.4	93	<10	
Nov.	0.2		5	14.2		30	12		7.5	7.6	6.2	61	<10	
Dec.	0.1		16	15.1		31	9		6.6	7.8	6.3	16	<10	
Year	10.1	95	59	124.5	12.0	205	133	30	6.8	7.2	6.5	2294	>200	400

TABLE CXXI

CLIMATIC TABLE FOR YALTA

Latitude 44°29′N longitude 34°10′E elevation 68.9 m

Month	Air press. (mbar)		Wind		Global radiat.	Temperature (°C)				Humidity		Precipitation (mm)			
	sea level	station	preval. direct.	mean speed (m/s)	(kcal./cm²)	mean	mean daily range	max.	min.	vapor press. (mm)	relat. (%)	mean	max.	min.	max. in 24 h
Jan.	1019.2	1018.7	N	3.3		3.8	6.0	21	−15	4.7	76	75	194	14	100
Feb.	1017.8	1017.2	N	3.2		3.5	6.1	20	−15	4.4	75	57	148	1	87
Mar.	1016.3	1015.8	N	3.6		6.1	6.7	29	−11	5.2	72	46	167	0	59
Apr.	1014.6	1014.1	N	2.7		10.4	7.3	28	−4	6.4	72	29	76	0	32
May	1014.3	1013.8	N, E	2.3		16.0	7.6	31	1	9.3	72	27	72	2	48
June	1012.5	1011.9	N	2.4		20.4	8.0	35	7	11.8	67	42	182	1	85
July	1011.1	1010.6	N	2.5		23.9	8.4	39	11	13.3	61	42	274	1	135
Aug.	1012.6	1012.1	N	2.8		23.7	8.5	37	11	12.5	61	29	137	3	94
Sept.	1016.4	1015.9	N	3.2		19.0	8.5	34	4	10.3	63	39	125	1	79
Oct.	1018.4	1017.9	NW	3.2		14.2	7.7	31	−4	8.8	72	48	160	1	52
Nov.	1019.4	1018.8	N	3.1		9.1	6.7	25	−8	6.6	77	59	292	0	95
Dec.	1018.4	1017.9	N	3.2		6.0	6.2	22	−12	5.4	76	76	285	6	154
Year	1015.9	1015.4	N	3.0		13.0		39	−15		70	560	1027	283	154

Month	Number of days							Snow cover max. 10-day average depth (cm)	Cloudiness (tenths)			Mean sun-shine (h)	Evaporation (mm)	
	wind (speed ≥15 m/s)	frost-free period	fog	precip. (≥0.1 mm)	thunder-storm	snow-cover	clouds (8–10 tenths)		07h	13h	19h		actual	potent.
Jan.	1.9			15.1			15.9		7.8	8.2	7.4	76	10	
Feb.	3.2			13.4			14.9		8.3	8.3	7.5	83	10	
Mar.	3.3			11.5			10.6		7.5	7.6	7.1	140	30	
Apr.	1.3			8.5			7.8		6.7	7.0	6.4	187	50	
May	0.7			8.5			6.1		6.1	6.4	6.0	242	60	
June	0.8			8.5			2.6		4.0	5.3	4.1	288	70	
July	0.7			7.2			0.9		2.5	3.9	2.9	328	70	
Aug.	1.7			5.6			1.0		2.6	3.9	2.8	310	30	
Sept.	1.7			5.9			2.0		3.5	4.4	3.7	241	20	
Oct.	1.6			8.0			6.4		5.8	6.2	4.9	179	20	
Nov.	1.7			11.0			10.9		7.2	7.3	6.4	103	10	
Dec.	2.3			13.9			16.8		7.6	8.1	7.0	73	10	
Year	20.9	247		118		11	95	6	5.8	6.4	5.5	2250	<400	900

TABLE CXXII

CLIMATIC TABLE FOR YENISEYSK

Latitude 58°27′N longitude 92°10′E elevation 78 m

Month	Air press. (mbar)		Wind		Global radiat.	Temperature (°C)				Humidity		Precipitation (mm)			
	sea level	station	preval. direct.	mean speed (m/s)	(kcal./cm²)	mean	mean daily range	max.	min.	vapor press. (mm)	relat. (%)	mean	max.	min.	max. in 24 h
Jan.	1027.2	1016.3	SE	2.2	1.2	−22.0	12.1	4	−59	0.9	80	24	54	5	11
Feb.	1027.5	1016.7	SE	1.8	3.3	−19.3	18.5	7	−53	1.5	77	17	44	1	12
Mar.	1023.3	1021.9	SW	2.7	7.5	−10.9	23.1	13	−47	1.7	71	17	36	2	9
Apr.	1018.1	1008.1	SW	2.8	11.5	−1.8	17.8	23	−35	2.9	63	19	52	5	19
May	1012.9	1003.3	W	3.1	12.6	6.5	17.1	33	−17	4.8	61	44	102	9	18
June	1008.2	999.0	SW	2.3	15.8	14.4	16.8	36	−4	8.8	67	57	152	10	40
July	1006.7	997.5	SW	1.7	15.2	17.8	15.9	37	1	11.9	74	60	115	18	41
Aug.	1009.0	999.8	SW	1.9	10.6	14.7	16.4	34	−3	10.5	79	60	174	15	50
Sept.	1015.6	1006.1	SW	2.2	6.6	7.9	16.7	29	−13	6.6	80	53	95	18	37
Oct.	1019.0	1009.0	SW	2.8	3.1	−0.9	14.7	22	−34	3.8	79	42	90	11	12
Nov.	1022.5	1012.1	SW	2.8	1.4	−12.1	11.4	9	−39	1.8	81	41	76	14	20
Dec.	1026.9	1016.0	SE	2.3	0.8	−20.9	10.5	6	−53	1.0	80	36	63	12	9
Year	1018.1	1008.1	SW	2.4	89.6	−2.2		37	−59		74	470	633	308	50

Month	Number of days							Snow cover max. 10-day average depth (cm)	Cloudiness (tenths)			Mean sun-shine (h)	Evaporation (mm)	
	wind (speed ≥15 m/s)	frost-free period	fog	precip. (≥0.1 mm)	thunder-storm	snow-cover	clouds (8–10 tenths)		07h	13h	19h		actual	potent.
Jan.	0.6		5	20.8		31			7.5	7.9	6.7	27	5	
Feb.	0.9		4	14.1		28			7.6	7.1	5.9	87	5	
Mar.	1.5		2	14.0		31			7.4	7.2	6.5	145	5	
Apr.	2.0		0.6	11.9		24			7.0	7.2	6.9	208	>10	
May	2.4		1	14.8	2.0				7.2	7.9	7.5	224	>50	
June	1.2		1	14.6	5.0				6.9	7.5	7.2	262	>60	
July	0.6		4	14.6	8.0				6.5	7.1	6.4	278	>70	
Aug.	0.4		6	17.7	5.0				7.4	7.6	7.0	194	50	
Sept.	0.7		5	18.8	1.0				8.2	8.0	7.2	118	30	
Oct.	1.5		2	20.1		6			9.0	8.9	7.9	61	5	
Nov.	1.3		2	21.5		30			8.4	8.3	7.5	30	5	
Dec.	0.9		4	21.2		31			7.7	8.2	7.3	16	5	
Year	14	103	37	204.1	21.0	187		76	7.6	7.7	7.0	1650	300	400

TABLE CXXIII

CLIMATIC TABLE FOR YEREVAN

Latitude 40°08′N longitude 44°28′E elevation 907 m

Month	Air press. (mbar) sea level	Air press. (mbar) station	Wind preval. direct.	Wind mean speed (m/s)	Global radiat. (kcal./cm^2)	Temperature (°C) mean	Temperature (°C) mean daily range	Temperature (°C) max.	Temperature (°C) min.	Humidity vapor press. (mm)	Humidity relat. (%)	Precipitation (mm) mean	Precipitation (mm) max.	Precipitation (mm) min.	Precipitation (mm) max. in 24 h
Jan.	1028.0	916.3	SW	0.8	4.7	-4.0	8.3	15	-31	4.1	78	23	74	2	21
Feb.	1025.2	914.6	SW	1.3	6.8	-1.3	9.1	20	-23	4.5	70	24	84	0	21
Mar.	1019.7	912.6	SE	1.7	10.4	5.4	10.9	24	-17	5.4	64	29	103	3	34
Apr.	1014.3	910.6	SE	1.9	13.8	11.8	13.0	30	-7	7.6	55	42	104	5	29
May	1012.6	910.9	S	1.8	18.5	17.0	13.6	34	-1	10.6	56	50	181	13	42
June	1008.1	908.5	NE	2.3	20.3	21.1	15.0	39	5	12.2	50	26	76	0	31
July	1004.3	906.3	NE	2.9	20.0	25.1	15.1	41	7	14.2	45	13	47	0	29
Aug.	1006.3	908.1	NE	2.4	18.6	24.9	15.6	40	7	13.5	44	9	68	0	26
Sept.	1011.9	911.4	NE	1.6	14.5	20.1	15.6	39	-2	11.0	49	12	86	0	39
Oct.	1018.8	915.4	SW	1.0	11.0	13.6	14.0	34	-7	8.5	60	26	133	0	35
Nov.	1023.0	916.5	E	0.8	5.8	6.2	10.6	26	-14	7.0	72	28	79	0	36
Dec.	1026.5	917.1	SW	0.6	3.8	-0.9	8.1	16	-24	4.9	78	22	62	0	28
Year	1016.6	912.3	NE	1.6	148.2	11.6	–	41	-31	8.6	60	304	465	128	42

Month	Number of days wind (speed ≥15 m/s)	frost-free period	fog	precip. (≥0.1 mm)	thunder-storm	snow-cover	clouds (8–10 tenths)	Snow cover max. 10-day average depth (cm)	Cloudiness (tenths) 07h	13h	19h	Mean sun-shine (h)	Evaporation (mm) actual	potent.
Jan.	0.8		13	8.8			14.6	9	7.3	6.8	6.0	89		
Feb.	1.2		7	8.5	0.3		11.2	10	6.9	6.8	5.7	118		
Mar.	2.5		2	8.2	0.6		11.3	4	6.9	7.2	6.3	169		
Apr.	3.2		0.2	10.9	4		9.2		6.1	6.6	6.6	212		
May	3.4		0.1	12.8	10		6.6		5.2	5.6	6.9	283		
June	5.8		0.2	8.0	9		2.2		3.0	3.4	5.4	334		
July	8.1			4.9	5		1.0		2.5	2.2	3.8	359		
Aug.	6.3			2.7	4		0.4		2.3	1.9	3.1	352		
Sept.	3.4		0.1	3.3	2		1.0		2.4	2.2	2.9	300		
Oct.	0.9		1	6.6	2		3.3		4.2	4.1	3.6	246		
Nov.	0.5		4	6.8	0.5		7.9		5.7	5.7	4.7	144		
Dec.	0.4		10	7.7			12.8	3	6.7	6.7	5.7	90		
Year	36.5	213	38	89	37	49	82	10	4.9	4.9	5.1	2696	267.0	946.5

TABLE CXXIV

CLIMATIC TABLE FOR YUZHNO-SAKHALINSK

Latitude 46°58′N longitude 142°43′E elevation 23 m

Month	Air press. (mbar) sea level	Air press. (mbar) station	Wind preval. direct.	Wind mean speed (m/s)	Global radiat. (kcal./cm^2)	Temperature (°C) mean	Temperature (°C) mean daily range	Temperature (°C) max.	Temperature (°C) min.	Humidity vapor press. (mm)	Humidity relat. (%)	Precipitation (mm) mean	Precipitation (mm) max.	Precipitation (mm) min.	Precipitation (mm) max. in 24 h
Jan.			N	3.8	4.0	-12.9	11.0	10	-39	1.9	80	33			60
Feb.			N	4.1	6.6	-12.0	11.9	5	-36	2.0	80	29			32
Mar.			N	3.8	10.9	-6.6	11.2	14	-31	3.0	80	41			36
Apr.			N	4.0	11.5	-0.1	8.7	21	-25	5.2	73	47			33
May			N	4.5	14.6	6.0	11.4	28	-8	7.4	77	63			41
June			N, S	4.0	12.0	10.9	10.4	30	-3	10.8	82	61			47
July			N	3.5	11.6	15.6	9.4	34	0	15.2	86	91			94
Aug.			N	3.1	10.0	17.1	9.3	34	2	16.9	86	92			88
Sept.			N	2.9	8.3	12.7	10.5	28	-6	12.6	84	106			107
Oct.			N	2.9	6.3	5.7	11.3	24	-12	7.6	80	76			43
Nov.			N	3.2	4.0	-1.9	9.6	16	-26	7.2	78	61			50
Dec.			N	3.2	3.2	-9.0	10.1	8	-34	2.7	80	53			32
Year			N	3.6	103.0	2.1		34	-39	7.5	81	753			107

Month	Number of days wind (speed ≥15 m/s)	frost-free period	fog	precip. (≥0.1 mm)	thunder-storm	snow-cover	clouds (8–10 tenths)	Snow cover max. 10-day average depth (cm)	Cloudiness (tenths) 07h	13h	19h	Mean sun-shine (h)	Evaporation (mm) actual	potent.
Jan.	3.3					31			6.7	6.6	5.5	113	<10	
Feb.	3.5					28			7.0	6.8	4.8	140	<10	
Mar.	2.6					31			7.2	7.2	6.0	175	<10	
Apr.	1.9					11			7.5	7.7	7.2	176	<10	
May	2.2								7.9	7.8	7.4	196	50	
June	0.8								8.7	7.7	7.6	176	70	
July	0.4								9.1	8.1	8.0	155	70	
Aug.	0.2								8.6	8.2	8.1	152	70	
Sept.	1.3								7.6	7.8	6.5	168	50	
Oct.	1.2								6.7	7.1	5.1	164	30	
Nov.	1.5					8			7.0	7.2	5.6	119	10	
Dec.	1.5					31			7.6	7.4	6.6	94	<10	
Year	20.4					153			7.6	7.5	6.5	1828	400	500

TABLE CXXV

CLIMATIC TABLE FOR ZAPOROZHYE

Latitude 47°48′N longitude 35°15′E elevation 86 m

Month	Air press. (mbar)		Wind		Global radiat.	Temperature (°C)				Humidity		Precipitation (mm)			
	sea level	station	preval. direct.	mean speed (m/s)	(kcal./cm²)	mean	mean daily range	max.	min.	vapor press. (mm)	relat. (%)	mean	max.	min.	max. in 24 h
Jan.	1021.4	1012.5	NE	4.2	2.9	−4.9		14	−32	4.3	86	31	94	3	20
Feb.	1020.0	1011.1	NE	4.4	5.0	−4.2		16	−34	4.4	84	27	90	1	24
Mar.	1017.6	1009.0	NE	4.6	7.8	1.0		24	−25	5.4	80	26	84	3	27
Apr.	1015.4	1006.9	NE, E	4.2	12.7	9.0		31	−9	7.6	66	35	81	0	51
May	1015.0	1006.6	NE, E	4.0	16.6	16.4		35	−2	10.8	60	39	116	0	45
June	1012.5	1004.2	N, NE	3.3	17.4	20.1		38	3	14.5	61	57	166	1	104
July	1011.4	1003.1	N	3.2	17.5	22.8		39	8	15.6	58	50	142	4	84
Aug.	1013.4	1005.1	N	3.1	15.4	21.6		41	6	14.8	58	45	150	0	61
Sept.	1017.9	1009.5	N	3.0	11.8	16.0		37	−3	11.5	64	30	90	1	62
Oct.	1020.4	1011.9	NE	3.4	7.6	9.3		33	−18	8.7	75	30	79	2	47
Nov.	1021.4	1012.7	E	4.4	3.7	2.8		24	−22	6.8	84	36	91	4	38
Dec.	1020.8	1012.1	E	4.3	2.1	−2.3		15	−26	5.0	87	37	102	3	30
Year	1017.3	1008.7	NE	3.8	120.5	9.0		41	−34	9.1	72	443	660	282	104

Month	Number of days							Snow cover	Cloudiness (tenths)			Mean sunshine	Evaporation (mm)	
	wind (speed ≥15 m/s)	frost-free period	fog	precip. (≥0.1 mm)	thunderstorm	snow-cover	clouds (8–10 tenths)	max. 10-day average depth (cm)	07h	13h	19h	(h)	actual	potent.
Jan.	0.8		9	13.0			18.7	7	8.1	8.1	7.4		0.4	
Feb.	1.6		7	12.3	0.03		16.0	8	8.3	6.4	7.4		0.8	
Mar.	1.6		5	10.0	0.1		14.8	5	7.6	7.7	6.7		10	
Apr.	1.5		2	9.2	0.8		11.0		6.4	7.1	6.5		50	
May	1.5		0.7	9.5	4		7.5		5.7	6.8	6.2		82	
June	0.8		0.4	9.9	7		4.8		4.8	6.2	5.8		75	
July	0.9		0.2	8.0	6		2.7		3.4	5.5	4.6		70	
Aug.	0.7		0.2	7.3	4		2.8		3.6	5.1	4.1		50	
Sept.	0.4		0.9	6.8	2		4.0		4.3	5.3	4.3		31	
Oct.	0.4		3	8.6	0.3		9.9		6.8	6.6	5.0		22	
Nov.	1.7		6	10.8			16.3		8.1	8.1	7.0		8	
Dec.	1.0		9	13.4			20.5	4	8.6	8.6	8.0		1.1	
Year	12.9	187	43.4	119	24	58	129	8	6.3	7.0	6.1		400.3	813.9

TABLE CXXVI

CLIMATIC TABLE FOR ZYRYANKA

Latitude 65°44′N longitude 150°54′E elevation 43 m

Month	Air press. (mbar)		Wind		Global radiat.	Temperature (°C)				Humidity		Precipitation (mm)			
	sea level	station	preval. direct.	mean speed (m/s)	(kcal./cm²)	mean	mean daily range	max.	min.	vapor press. (mm)	relat. (%)	mean	max.	min.	max. in 24 h
Jan.	1023.2		NW	1.9		−38.8	7.3	−10	−60	0.2	77	14	36	2	7
Feb.	1025.9		NW	1.8		−34.7	9.1	−4	−61	0.3	77	13	35	5	7
Mar.	1021.1		NW	2.2		−26.2	14.5	5	−55	0.4	72	8	21	0	6
Apr.	1015.2		N, NW	3.0		−12.5	14.2	13	−51	1.4	69	8	24	1	10
May	1012.5		N	3.4		1.3	11.4	29	−28	3.4	61	8	27	0	13
June	1007.4		N	3.4		12.8	11.1	34	−6	6.7	62	28	80	4	33
July	1007.4		N	3.4		15.3	10.6	36	1	9.0	70	52	167	12	38
Aug.	1009.7		N, NW	3.2		11.2	10.4	33	−4	7.9	74	47	88	13	39
Sept.	1013.8		N, NW	2.8		3.8	9.3	28	−13	9.0	76	21	65	4	21
Oct.	1014.7		NW	2.6		−11.7	7.1	15	−37	1.9	83	19	48	4	18
Nov.	1020.0		NW	2.1		−28.1	7.1	5	−50	0.4	81	19	49	3	9
Dec.	1022.4		NW	1.9		−36.2	6.9	−8	−59	0.2	78	17	48	2	10
Year	1016.1		NW	2.6		−12.0		36	−61		73	254	371	194	39

Month	Number of days							Snow cover	Cloudiness (tenths)			Mean sunshine	Evaporation (mm)	
	wind (speed ≥15 m/s)	frost-free period	fog	precip. (≥0.1 mm)	thunderstorm	snow-cover	clouds (8–10 tenths)	max. 10-day average depth (cm)	07h	13h	19h	(h)	actual	potent.
Jan.	0.7		15	17.6		31	12.8		6.1	7.3	6.3	9	5	
Feb.	0.5		8	16.1		28	12.2		6.5	6.8	6.0	69	5	
Mar.	0.6		5	11.4		31	9.3		6.2	5.8	5.5	214	5	
Apr.	0.8		0.5	8.3		30	10.0		6.2	5.7	5.9	296	5	
May	0.4		0.0	5.4	0.1	13	12.6		6.6	6.5	6.3	349	>15	
June	1.4		0.1	9.1	2		12.7		6.7	6.8	6.7	344	>60	
July	1.8		0.7	12.1	2		14.8		7.3	7.3	7.0	294	>60	
Aug.	0.9		2	11.6	1		14.9		7.4	7.2	6.9	241	40	
Sept.	0.6		3	10.0			13.6		7.7	7.3	7.0	166	10	
Oct.	0.7		3	14.8		30	15.6		8.0	7.6	7.1	83	5	
Nov.	0.4		7	17.3		30	14.1		6.9	7.4	6.5	29	5	
Dec.	0.6		16	17.0		31	12.6		6.0	7.4	6.1	2	5	
Year	9.4	92	60	151	5	234	155	55	6.8	6.9	6.4	2096	>200	400

Reference Index

Geographical Index

Subject Index